TOPOGRAPHIC LASER RANGING AND SCANNING

Principles and Processing

TOPOGRAPHIC LASER RANGING AND SCANNING

Principles and Processing

Edited by

Jie Shan and Charles K. Toth

CRC Press is an imprint of the
Taylor & Francis Group, an informa business

CRC Press
Taylor & Francis Group
6000 Broken Sound Parkway NW, Suite 300
Boca Raton, FL 33487-2742

International Standard Book Number-13: 978-1-4200-5142-1 (Hardcover)

Library of Congress Cataloging-in-Publication Data

Topographic laser ranging and scanning : principles and processing / edited by Jie Shan and Charles K. Toth.
p. cm.
Includes bibliographical references and index.
ISBN-13: 978-1-4200-5142-1 (alk. paper)
ISBN-10: 1-4200-5142-3 (alk. paper)
1. Lasers in surveying. 2. Aerial photogrammetry. I. Shan, Jie. II. Toth, Charles K. III. Title.

TA579.T654 2009
526.9'8--dc22 2008017336

Visit the Taylor & Francis Web site at
http://www.taylorandfrancis.com

and the CRC Press Web site at
http://www.crcpress.com

Contents

Preface

LiDAR is probably the most significant technology introduced in mainstream topographic mapping in the last decade. The main advantage of the technique is that it provides a direct method for 3D data collection. Furthermore, it is highly accurate because of the millimeter- and centimeter-level laser ranging accuracy and precise sensor platform orientation supported by an integrated position and orientation system (POS). Unlike the traditional photogrammetric methods, LiDAR directly collects an accurately georeferenced set of dense point clouds, which can be almost directly used in basic applications. However, the full exploitation of LiDAR's potentials and capabilities challenges for new data processing methods that are fundamentally different from the ones used in traditional photogrammetry. Over the last decade, there have been many significant developments in this field, mainly resulting from multidisciplinary research, including computer vision, computer graphics, electrical engineering, and photogrammetry. Consequently, the conventional image-based photogrammetry and vision is gradually adapting to a new subject, which is primarily concerned with point clouds data collection, calibration, registration, and information extraction.

Although there have been many studies and applications since the introduction of LiDAR, a comprehensive compilation is still missing that (1) describes the basic principles and fundamentals in laser ranging and scanning; (2) reflects the state-of-the-art of laser scanning technologies, systems, and data collection methods; and (3) presents the data processing methods and recent developments reported in different subject areas, which is probably the most challenging task. It is not an overstatement that such a collection has been long overdue. The primary objective of this book is to meet this need and present a manual to the LiDAR research, practice, and education communities.

The collection in this book consists of four major parts, each addressing different topic areas in LiDAR technology and data processing. Part I, Chapters 1 through 3 primarily present a brief introduction and comprehensive summary of LiDAR systems. After a concise introduction to the laser fundamentals, Chapter 1 describes the two laser ranging methods (timed pulse method and phase comparison method) and the principles of laser profiling and scanning. The issues and properties of laser ranging techniques, such as safety concerns, returned power, beam divergence, and reflectivity, are also discussed. Chapters 2 and 3 present a rather comprehensive description of the air- and spaceborne laser devices (profilers and scanners) and terrestrial scanners. Each of these two chapters covers a brief historical review, the taxonomy, and a thorough discussion on the components, working principles, and main specifications of the LiDAR equipment.

Part II addresses the operational principles of different units and ranging methods in a LiDAR system. Chapter 4 discusses the working principles of LiDAR systems and supporting devices and their synchronization. The theory and practices for georegistration and calibration of the boresight are also presented, along with testing examples. Finally, a discussion is offered on the working steps required to produce a digital terrain model (DTM), including flight planning, equipment installation, and data preprocessing. In order to most effectively use the products generated from LiDAR systems, Chapter 5 discusses the operational principles of LiDAR systems and the resulting effects on the acquired data. Laser ranging methods, including waveform recording and single-photon detection approaches, are examined. The relationships between ranging methods, the instrument parameters, and the character of the resulting elevation data are considered, with emphasis on the measurement of ground topography beneath vegetation cover and characterization of forest canopy structure. Chapter 6 starts with an introduction to direct georeferencing technology. The discussion is focused on the analytics of the inertial navigation system (INS), which is important for achieving quality airborne LiDAR data. The principles for the integration of INS, global positioning system (GPS), and LiDAR are also presented. Chapter 7 is about small footprint pulsed laser systems with emphasis on full waveform analysis. Addressed with details in this chapter are approaches for designing a laser system, modeling the spatial and temporal properties of the emitted laser pulses, detecting return pulses, and deriving attributes from the waveform.

The theme of Part III relates to the subsequent geometric processing of LiDAR data, in particular with respect to quality, accuracy, and standards. Chapter 8 starts with a review of the underlying theory for surface representation. The factors that affect strip adjustment to form one integrated consistent data set are studied. The discussion covers both data-driven and sensor calibration-based methods. Chapter 9 identifies the unique properties for quality control and calibration of a LiDAR system. Its error sources and their impact on the resulting surface are identified and discussed. Such a study yields several recommended procedures for quality assurance and control of the LiDAR systems and their derived data. Chapter 10 addresses the data format, organization, storage, and standards, and their effect on production from a practical point of view.

Part IV is about information extraction from the LiDAR data. As the starting chapter of this topic, Chapter 11 focuses on the extraction of the terrain surface from either the original point cloud or gridded data. Emphasis is placed on the principles of different filters for ground separation and their performance. The role of structural lines in terrain model generation is discussed to ensure its fidelity and quality. Chapter 12 presents a broad overview of laser-based forest inventory. This includes user requirements, laser-scattering process in forests, canopy height retrieval, and the corresponding accuracy. Besides tree or stand parameter retrieval, the change detection possibilities using multitemporal laser surveys are also discussed. Chapter 13 introduces algorithms for a multiprimitive and multisensor triangulation environment to assure the precise alignment of the involved LiDAR and imagery data. Such alignment leads to the straightforward production of orthophotos, for which several methodologies using LiDAR and photogrammetric data of urban areas are outlined. Chapters 14 through 18 focus on feature extraction in urban environments. Such a dynamic topic attracts many studies from different angles in different disciplines. These methods include random sample consensus (RANSAC) and Hough transform for model detection and estimation (Chapter 14), clustering approach for segmentation and reconstruction (Chapters 15, 16, and 18), combined use of multiple data sets (Chapters 16 and 17), region growing (Chapters 16 and 18), graph partition (Chapters 14 and 17), and rule-based approaches (Chapter 16). Although each of the above chapters has independent quality evaluation, Chapter 19 presents a comprehensive summary of building extraction qualities based on an organized joint effort of different participants. In summary, through the above independent yet related structure of the book contents, we present the audience a broad discussion on different aspects of laser topographic mapping—principles, systems, operation, geometric processing, and information extraction.

The book is targeted to a variety of audience and is in particular expected to be used as a reference book for senior undergraduate and graduate students, majoring or working in diverse disciplines, such as geomatics, geodesy, natural resources, urban planning, computer vision, and computer graphics. It can also be used by researchers who are interested in developing new methods and need in-depth knowledge of laser scanning and data processing. Other professionals may gain the same from the broad topics addressed in this book.

Despite the remarkable achievements in LiDAR technology and data processing, it is still a relatively young subject and would likely change its course rather quickly in the near future. We expect this book can serve as a durable introduction to the fundamentals, principles, basic problems, and issues in LiDAR technology for topographic applications. It also provides a rather comprehensive and sophisticated coverage on various data processing techniques, which we believe will sustain and ultimately lead to new developments while the technology is advancing.

Finally, we would like to thank all of the contributors and technical staff members at the publisher, without whom this book would not have been possible. It is greatly appreciated that the leaders in their area of expertise agreed to share their recent work and carefully prepared the manuscripts. The publisher's staff members helped the editors and authors throughout the entire publication period. It is our great pleasure to be working with all of them to present this book to the broad LiDAR community and beyond.

Jie Shan
Purdue University

Charles K. Toth
The Ohio State University

Editors

Jie Shan received his PhD in photogrammetry and remote sensing from Wuhan University, China. Since then he has worked as faculty at universities in China, Sweden, and the United States, and has been a research fellow in Germany. Currently, he is an associate professor in geomatics engineering at the School of Civil Engineering of Purdue University, West Lafayette, Indiana, where he teaches subjects in photogrammetry and remote sensing, and geospatial information science and technology. His research interests are digital mapping, geospatial data mining, and urban modeling. He is a recipient of multiple academic awards and author/coauthor of over 100 scientific papers.

Charles K. Toth has PhDs in electrical engineering and geoinformation science from the University of Budapest, Budapest, Hungary, and is currently a senior research scientist with the Ohio State University Center for Mapping. His research expertise covers broad areas of 2D/3D signal processing, spatial information systems, airborne imaging, surface extraction and modeling, integrating and calibrating of multisensor systems, multisensor geospatial data acquisition systems, and mobile mapping technology. His references are represented by over 150 scientific/technical papers.

Contributors

Frédéric Bretar Institut Géographique National, Laboratoire MATIS, Saint Mandé, France

Shu-Ching Chen School of Computing and Information Sciences, Florida International University, Miami, Florida

Simon Clode School of Information Technology and Electrical Engineering, The University of Queensland, Brisbane, Australia

Naser El-Sheimy Schulich School of Engineering, The University of Calgary, Calgary, Alberta, Canada

Lewis Graham GeoCue Corporation, Madison, Alabama

Eberhard Gülch Department of Geomatics, Computer sceince and Mathematics, Stuttgart University of Applied Sciences, Stuttgart, Germany

Ayman Habib Schulich School of Engineering, The University of Calgary, Calgary, Alberta, Canada

David Harding Goddard Space Flight Center, National Aeronautics and Space Administration, Greenbelt, Maryland

Markus Holopainen Finnish Geodetic Institute, Masala, Finland

Xianfeng Huang State Key Laboratory of Information Engineering in Surveying, Mapping and Remote Sensing (LIESMARS), Wuhan University, Wuhan, China

Hannu Hyyppä Helsinki University of Technology, Otakaari, Finland

Juha Hyyppä Finnish Geodetic Institute, Masala, Finland

Boris Jutzi FGAN Research Institute for Optronics and Pattern Recognition, Ettlingen, Germany

Harri Kaartinen Finnish Geodetic Institute, Masala, Finland

Antero Kukko Finnish Geodetic Institute, Masala, Finland

Gottfried Mandlburger Institute of Photogrammetry and Remote Sensing, Vienna University of Technology, Wien, Austria

Gordon Petrie Department of Geographical and Earth Sciences, University of Glasgow, Glasgow, Scotland, UK

Norbert Pfeifer Institute of Photogrammetry and Remote Sensing, Vienna University of Technology, Wien, Austria

Franz Rottensteiner Cooperative Research Centre for Spatial Information, University of Melbourne, Melbourne, Australia

Aparajithan Sampath School of Civil Engineering, Purdue University, West Lafayette, Indiana, USA

Jie Shan School of Civil Engineering, Purdue University, West Lafayette, Indiana

Gunho Sohn Department of Earth and Space Science and Engineering, York University, Toronto, Ontario, Canada

Uwe Stilla Photogrammetry and Remote Sensing, Technical University of Munich, Munich, Germany

Vincent Tao Department of Earth and Space Science and Engineering, York University, Toronto, Ontario, Canada

Charles K. Toth Center for Mapping, The Ohio State University, Columbus, Ohio

Aloysius Wehr Institute of Navigation, University Stuttgart, Stuttgart, Germany

Jianhua Yan School of Computing and information Sciences, Florida International University, Miami, Florida

Xiaowei Yu Finnish Geodetic Institute, Masala, Finland

Keqi Zhang Department of Environmental Studies and International Hurricane Research Center, Florida International University, Miami, Florida

Abbreviations

A/D	analogue-to-digital
AFOV	angular field of view
AGL	average ground level
AIC	aerial industrial camera
ALPS	airborne laser polarization sensor
ALPS	airborne laser profiling system
ALS	airborne laser scanning
ALTMS	airborne LiDAR terrain mapping system
AM	amplitude modulation
AOL	airborne oceanographic LiDAR
APD	avalanche photodiode detectors
ASCII	American Standard Code for Information Exchange
ATLAS	airborne topographic laser altimeter system
BSP-Tree	binary space partitioning tree
CATS	coastal area tactical-mapping system
CCD	charged- coupled device
CCNS	computer-controlled navigation system
CFRP	carbon fiber reinforced plastic
CHM	canopy height model
CLF	compass line filter
CMOS	complementary metal-oxide semiconductor
CORS	continuously operating reference stations
COTS	commercial-off-the-shelf
CSG	constructive solid geometry
CT	contour tree
CW	continuous wave
DCM	digital canopy model
DCM	direction cosine matrix
DEM	digital elevation model
DGPS	differential global positioning system
DSM	digital surface model
DSS	digital sensor system
DTM	digital terrain model
EAARL	experimental advanced airborne research LiDAR
EDM	electronic distance measuring
EGI	extended Gaussian image
EKF	extended Kalman filter
EOP	exterior orientation parameters
EPFL	Federal Institute of Technology of Lausanne
ESSP	earth system science pathfinder
FLI-MAP	fast laser imaging mobile airborne platform
FM	frequency modulation
FOG	fiber-optic gyros
FOV	field-of-view

FWHM	full-width at half the maximum
GCP	ground control point
GIS	geographic information system
GLAS	geoscience laser altimeter system
GML	geography mark-up language
GNSS	global navigation satellite system
GPS	global positioning system
GSD	ground sampling distance
GSD	ground spacing distance
HDTV	high definition television
ICESat	ice, cloud and land elevation satellite
ICP	iterative closest point
IDW	inverse distance weighted
IFOV	instantaneous field of view
IIP	instrument incubator program
IOP	interior orientation parameters
IPAS	inertial position and attitude system
IQR	interquartile range
ISODATA	iterative self-organizing data analysis technique
ITC	individual tree crown
KF	Kalman filter
LADS	laser airborne depth sounder
LCD	liquid crystal display
LiDAR	light detection and ranging
LITE	LiDAR in-space technology experiment
LKF	linearized Kalman filter
LoD	levels of detail
LOLA	lunar orbiter laser altimeter
LRU	laser ranging unit
LSA	least squares adjustment
LSM	least squares matching
LVIS	laser vegetation imaging sensor
MAPL	multiwavelength airborne polarimetric LiDAR
MBLA	multibeam laser altimeter
MBR	minimum bounding rectangle
MCP	microchannel plate
MDL	minimum description length
MLA	MESSENGER laser altimeter
MMLA	multi-kilohertz micro-laser altimeter
MOLA	mars orbiter laser altimeter
MPP	multipurpose platform
NDVI	normalized difference vegetation index
OEM	original equipment manufacturer
OMSD	opto-mechanical scanning device
PALS	portable airborne laser system
PCA	principal component analysis
PCD	phase coded disk
PDA	personal digital assistant
PMT	photomultiplier tube

POS	position and orientation system
PPP	precise point positioning
PPS	pulse per second
PRF	pulse repetition frequency
QA	quality assurance
QC	quality control
RAG	region adjacency graph
RANSAC	random sample consensus
RASCAL	raster scanning airborne LiDAR
RMS	root mean square
RMSE	root mean square error
RTF	recursive terrain fragmentation
SHOALS	scanning hydrographic operational airborne LiDAR survey
SIMP	Swath mapping multi-polarization photon-counting
SLA	shuttle laser altimeter
SLICER	scanning LiDAR imager of canopies by echo recovery
SNR	signal-to-noise ratio
SPAD	single photon avalanche diode
SPCM	single photon counting modules
SRS	stellar reference system
TIN	triangulated irregular network
TOF	time-of-flight
UKF	unscented Kalman filter
UTM	universal transverse mercator
VCL	vegetation canopy LiDAR
VGA	variable gain-state amplifier
VRML	virtual reality modeling language
XML	extensible markup language

1

Introduction to Laser Ranging, Profiling, and Scanning

Gordon Petrie and Charles K. Toth

CONTENTS

1.1 Introduction

Topographic laser profiling and scanning systems have been the subject of phenomenal developments in recent years and, without doubt, they have become the most important geospatial data acquisition technology that has been introduced since the last millennium.

Installed on airborne and land-based platforms, these systems can collect explicit 3D data in large volumes at an unprecedented accuracy. Furthermore, the complexity of the required processing of the measured laser data is relatively modest, which has further fueled the rapid proliferation of this technology to a variety of applications. Although the invention of laser goes back to the early 1960s, the lack of various supporting technologies prevented the exploitation of this device in the mapping field for several decades. The introduction of direct geo-referencing technology in the mid-1990s and the general advancements in computer technology have been the key enabling technologies for developing laser profiling and scanning systems for use in the topographic mapping field that are commercially viable. This chapter provides a brief historical background on the development path and the fundamentals of laser technology before introducing the basic principles and techniques of topographic laser profiling and scanning.

The present widespread use of laser profiling and scanning systems for topographic applications only really began in the mid-1990s after a long prior period of research and the development of the appropriate technology and instrumentation. In this respect, it should be mentioned that NASA played a large part in pioneering and developing the requisite technology through its activities in Arctic topographic mapping from the 1960s onwards. However, prior to these developments, lasers had been used extensively for numerous other applications within the field of surveying—especially within engineering surveying. Indeed, quite soon after the invention of the laser in its various original forms—the solid-state laser (in 1960), the gas laser (in 1961), and the semiconductor laser (in 1962)—the device was adopted by professional land surveyors and civil engineers. Various types of laser-based surveying instruments were devised and soon began to be used in their field survey operations. These early applications of lasers included the alignment operations and deformation measurements being made in tunnels and shafts and on bridges. Another early use was the incorporation of lasers in both manually operated and automatic (self-leveling) laser levels, again principally for engineering survey applications.

1.2 Terrestrial Applications

Turning next to laser ranging, profiling, and scanning, which are the main subjects of the present volume, lasers started to be used by surveyors for distance or range measurements in the mid- to late-1960s. These measurements were made using instruments that were based either on phase comparison methods or on pulse echo techniques. The latter included the powerful solid-state laser rangers that were used for military ranging applications such as gunnery and tracking. On the field surveying side, from the 1970s onwards, lasers started to replace the tungsten or mercury vapor lamps that had been used in early types of EDM (electronic distance measuring) instruments. Initially these laser-based EDM instruments were mainly used as stand-alone devices measuring the distances required for control surveys or geodetic networks using trilateration or traversing methods. The angles required for these operations were measured separately using theodolites. Later, these two types of instruments were merged with the laser-based ranging technology being incorporated into total stations, which were also capable of making precise angular measurements using opto-electronic encoders. These total stations allowed topographic surveys to be undertaken by surveyors with field survey assistants setting up pole-mounted reflectors at the successive positions required for the construction of the topographic map or terrain model. This type of operation is often referred to as electronic tacheometry. With the development of very small and powerful (yet eye-safe) lasers, reflectorless distance measurements then became possible. These allowed manually operated ground-based profiling devices based on laser rangers to be developed, initially for use in quarries and open-cast pits and in tunnels (Petrie, 1990). Given all this prior development of lasers for numerous different field surveying applications, it was a natural and logical development for scanning mechanisms to be added to these laser rangers and profilers. This has culminated in the development of the present series of terrestrial laser scanners that are now being used widely for topographic mapping applications, either from stationary positions when mounted on tripods or in a mobile mode when mounted on vehicular platforms.

1.3 Airborne and Spaceborne Applications

Turning next to airborne platforms, laser altimeters that could be used to measure continuous profiles of the terrain from aircraft had been devised and flown as early as 1965. One early example was based on the use of a gas laser (Miller, 1965), while another was based on a semiconductor laser (Shepherd, 1965). As with ground laser ranging devices, at first, both the phase comparison and pulse echo methods of measuring the distance from the airborne platform to the ground were used for the purpose. A steady development of airborne laser profilers then took place throughout the 1970s and 1980s. However, the limitation in this technique is that it can only acquire elevation data along a single line crossing the terrain during an individual flight. Thus, if a large area of terrain had to be covered, as required for topographic mapping, a very large number of flight lines had to be flown to give complete coverage of the ground. So, in practice, airborne laser profiling was only applicable to those surveys being conducted along individual lines or quite narrow corridors—for example, when carrying out geophysical surveys—where the method could be used in conjunction with other instruments such as gravimeters or magnetometers that had similar linear measuring characteristics. However, once suitable scanning mechanisms had been devised during the early 1990s, airborne laser scanning developed very rapidly and has come into widespread use. The introduction of direct geo-referencing as an enabling technology was essential for the adoption of airborne laser scanners, since, unlike in photogrammetry, there is no viable method to reconstruct the sensor trajectory solely from the laser range measurements and thus to determine the ground coordinates of the laser points. The GPS constellation had been completed by the early/mid-1990s and, with the commercial availability of medium/high performance IMUs by the mid-1990s, integrated GPS/IMU geo-referencing systems were able to deliver airborne platform position and attitude data at an accuracy of 4–7 cm and 20–60 arc-seconds, respectively.

On the spaceborne side, given the huge distances of hundreds of kilometers over which the range has to be measured from an Earth-orbiting satellite—100 times greater than those being measured from an airborne platform—very powerful lasers have to be used. Besides which, the time-of-flight (TOF) of the laser pulse is much greater, so the rate at which measurements can be made is reduced, unless a multipulse technique is used. Furthermore, the platform speed is very high at 29,000 km/h, which again is 100 times greater than that of a survey aircraft. So far, these demanding operational characteristics have limited spaceborne missions using laser ranging instruments to the acquisition of topographic data using profiling rather than scanning techniques.

1.4 Basic Principles of Laser Ranging, Profiling, and Scanning

A very brief introduction to the basic principles of laser ranging, profiling, and scanning will be provided here before going on to the remainder of this chapter, which is concerned with laser fundamentals and laser ranging.

1.4.1 Laser Ranging

All laser ranging, profiling, and scanning operations are based on the use of some type of laser-based ranging instrument—usually described as a laser ranger or laser rangefinder—that can measure distance to a high degree of accuracy. As will be discussed in more detail later in this chapter, this measurement of distance or range, which is always based on the precise measurement of time, can be carried out using one of the two main methods.

1. The first of these involves the accurate measurement of the TOF of a very short but intense pulse of laser radiation to travel from the laser ranger to the object being measured and to return to the instrument after having been reflected from the object—hence the use of the term "pulse echo" mentioned above. Thus, the laser ranging instrument measures the precise time interval that has elapsed between the pulse being emitted by the laser ranger located at point A and its return after reflection from a ground object located at point B (Figure 1.1).

$$R = v \cdot t/2 \tag{1.1}$$

where
R is the slant distance or range
v is the speed of electromagnetic radiation, which is a known value
t is the measured time interval

From this, the following simple relationship can be derived:

$$\Delta R = \Delta v \cdot t/2 + v \cdot \Delta t/2 \tag{1.2}$$

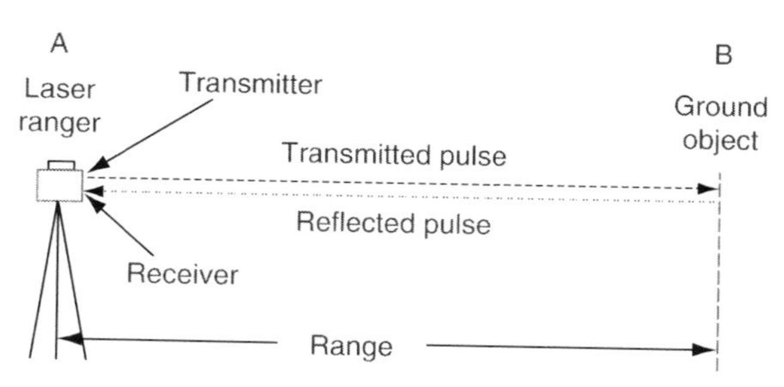

FIGURE 1.1
Basic operation of a laser rangefinder that is using the timed pulse or TOF method.

where

ΔR is the range precision
Δv is the velocity precision
Δt is the corresponding precision value of the time measurement

Since the speed of light is very accurately known, in practice, the range precision or resolution is determined by the precision of the time measurement.

2. In the second (alternative) method, the laser transmits a continuous beam of laser radiation instead of a pulse. In this case, the range value is derived by comparing the transmitted and received versions of the sinusoidal wave pattern of this emitted beam and measuring the phase difference between them. Since the wavelength (λ) of the carrier signal of the emitted beam of laser radiation is quite short—typically around 1 μm, and there is no need for such a measuring accuracy in topographic mapping applications, a modulation signal in the form of a measuring wave pattern is superimposed on the carrier signal and its phase difference can be measured more precisely. Thus, the amplitude (or intensity) of the laser radiation will be modulated by a sinusoidal signal, which has a period T_m and wavelength λ_m. The measurement of the slant distance R is then carried out through the accurate measurement of the phase difference (or the phase angle, φ) between the emitted signal at point A and the signal received at the instrument after its reflection either from the ground itself or from an object that is present on the ground at point B. This phase measurement is usually carried out using a digital pulse counting technique. This gives the fractional part of the total distance ($\Delta\lambda$) (Figure 1.2). By changing the modulation pattern, the integer number

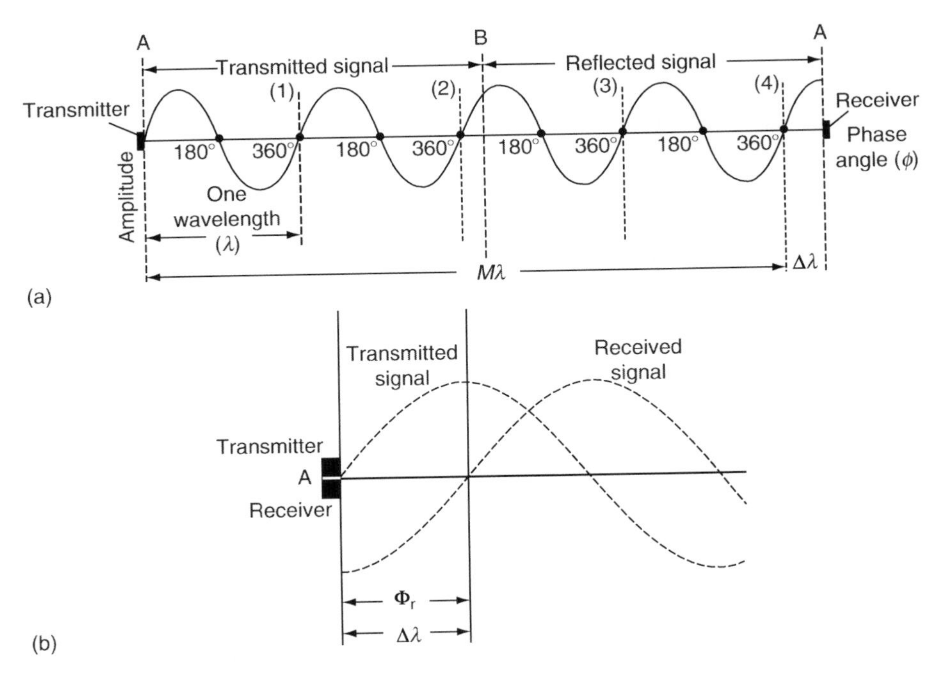

FIGURE 1.2
(a) Phase comparison is carried out between the transmitted and reflected signals from a CW laser and (b) phase comparison between the two signals takes place at the laser rangefinder located at A.

of wavelengths (M) can be determined and added to the fractional values to give the final slant range (R).

$$R = (M\lambda + \Delta\lambda)/2 \tag{1.3}$$

where
M is the integer number of wavelengths
λ is the known value of the wavelength
$\Delta\lambda$ is the fractional part of the wavelength $= (\varphi/2\pi)\cdot\lambda$, where φ is the phase angle

1.4.2 Laser Profiling

The use of a reflectorless laser ranger to measure the distances to a series of closely spaced points located adjacent to one another along a line on the terrain results in a two-dimensional (vertical) profile or vertical cross section of the ground showing the elevations of the ground along that line.

1. In the case of a terrestrial or ground-based laser ranger, the measurement of the terrain profile is executed in a series of steps with the successive measured distances (slant ranges) and vertical angles (V) to each sampled point being recorded and stored digitally (Figure 1.3a). The profile of the terrain along the measured line can then be derived from this measured data by computation using the following quite simple relationships (Figure 1.3b):

$$D = R\cos V \tag{1.4}$$

where
D is the horizontal distance
R is the measured slant range
V is the measured vertical angle

$$\Delta H = R\sin V \tag{1.5}$$

where ΔH is the difference in height between the laser ranger and the ground point being measured.

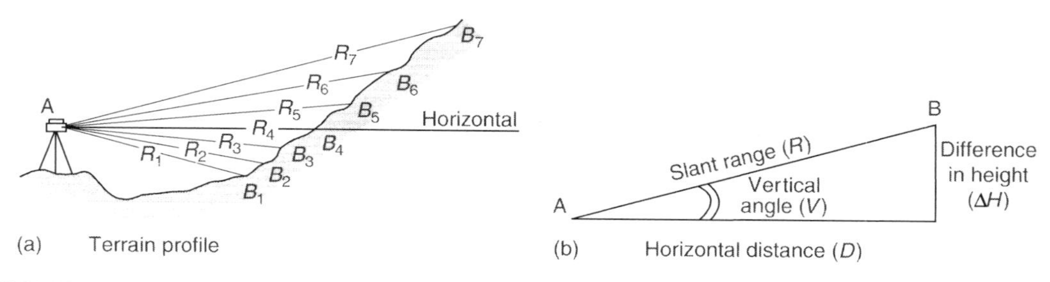

FIGURE 1.3
(a) Measurement of slant ranges (R) and vertical angles by a rangefinder located at A to a series of successive points located along a line on the ground to form a profile. (b) The measured slant ranges (R) and vertical angles (V) are used to compute the horizontal distances and differences in height between the rangefinder at A and each of the ground objects at B.

2. In the case of a simple laser profiler that has been mounted on an airborne or spaceborne platform, the laser ranger, which, in this context, is often called a laser altimeter, is pointed vertically toward the ground to allow a rapid series of measurements of the distances to the ground from the successive positions of the moving platform. The measurements of the vertical distances from the platform to a series of adjacent points along the ground track are made possible through the forward motion of the airborne or spaceborne platform. If the positions and altitudes of the platform at these successive positions in the air or in space are known or can be determined, for example using a GPS/IMU system (or a star-tracker in the spaceborne case), then the corresponding ranges measured at these points will allow their ground elevation values to be determined. Consequently, these allow the terrain profile along the flight line to be constructed (Figure 1.4).

1.4.3 Laser Scanning

With the addition of a scanning mechanism, for example, utilizing a rotating mirror or prism, the laser ranging instrument is upgraded from being a profiler to becoming a scanner that can measure and map the topographic features of an area in detail instead of simply determining elevation values along a line in the terrain—as is done with a laser profiler.

1. With regard to a terrestrial or ground-based laser scanner instrument, the position of the platform is fixed; therefore, motion in two directions is needed to scan an area of the terrain. Thus, besides the vertical motion given by the rotating mirror or prism, the addition of a controlled (and measured) motion in the azimuth direction—usually implemented through the use of a motor drive—allows the measurement of a series of profiles around the vertical axis of the laser ranger. This provides the position and elevation data that will allow a 3D model of the

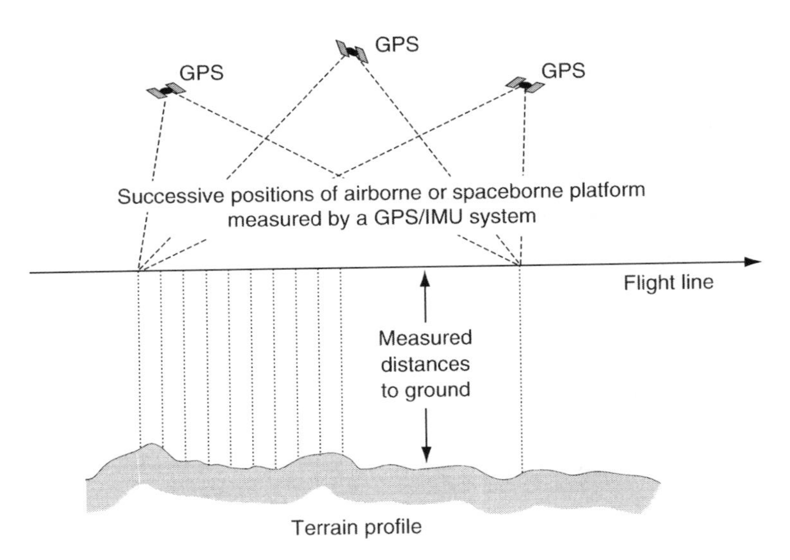

FIGURE 1.4
Profile being measured along a line on the terrain from an airborne or spaceborne platform using a laser altimeter.

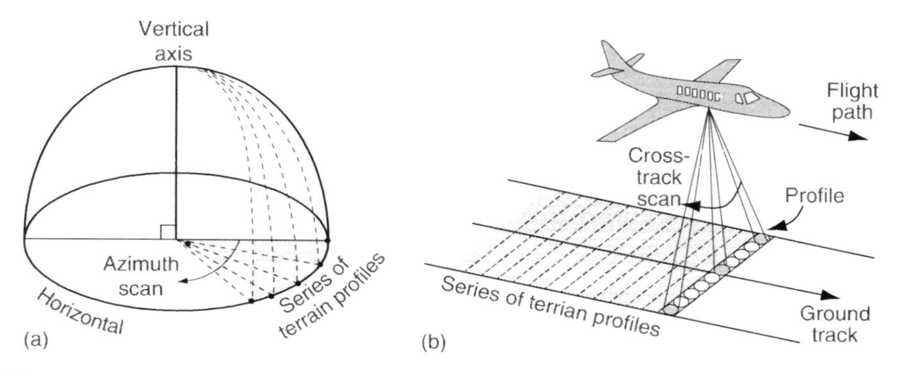

FIGURE 1.5
(a) Coverage of an area of terrain from a fixed ground station is achieved through the measurement of a series of profiles adjacent to one another, the position of each new profile being defined by an angular step in the azimuth direction. (b) The coverage of a swath of terrain being carried out from an airborne platform is achieved through a series of profiles measured in the cross-track direction. Each individual terrain profile is being measured by the laser rangefinder using an optical-mechanical scanning mechanism to point to each successive position that is required along that profile.

terrain and the objects present on it to be formed for the area around the position occupied by the laser scanner (Figure 1.5a).

2. In the case of airborne and spaceborne instruments, the area scanning is achieved by a series of profile measurements in the direction perpendicular to the flight line while the forward motion of the platform provides the second dimension. The addition of a scanning mechanism employing a reflective mirror or prism, whose angular rotation values can be continuously and precisely measured using an angular encoder, enables the additional profiles of the terrain to be measured by the laser ranger in the lateral or cross-track direction, so supplementing the longitudinal profile being measured in the along-track direction of the flight line. Through a series of these profiles measured in the cross-track direction, the positions and elevations of a mesh of points, also called a LiDAR point cloud,is generated covering a swath or area of the terrain instead of the set of elevation values along the ground track of the flight line that is produced by the airborne or spaceborne profiler (Figure 1.5b).

1.5 Laser Fundamentals

The word laser is an acronym for light amplification by stimulated emission of radiation. Essentially a laser is an optical device that, when activated by an external energy source, produces and emits a beam or pulse of monochromatic radiation in which all the waves are coherent and in phase. The emitted radiation is also highly collimated and directional in the sense that it is emitted as a narrow beam or pulse in a specific direction.

In general terms, lasers are usually classified according to the type of material that is being used as the radiation source. The most common types are gas lasers, solid-state lasers, and semiconductor lasers. Other types that are used less frequently are liquid lasers, dye lasers, excimer lasers, etc. For the laser ranging, profiling, and scanning that is being carried out for topographic mapping purposes, where very high energy levels are required to perform distance measurements often over long ranges, only certain types of solid-state and semiconductor lasers have the very specific characteristics—high intensity combined with a high degree of collimation—that are necessary to carry out these operations.

1.5.1 Laser Components

All lasers comprise three main elements:

1. The first of these elements comprises the active material of the laser that contains atoms whose electrons may be excited and raised to a much higher (metastable) energy level by an energy source. Examples of the materials that are being used extensively in the laser-based instruments that have been developed for the ranging, profiling, and scanning operations being carried out for topographic applications include a solid-state crystalline material such as neodymium-doped yttrium aluminum garnet (Nd:YAG) and a semiconductor material such as gallium arsenide (GaAs).
2. The second element that is present in every laser is an energy source. This provides the energy to start up and continue the lasing action. The continuous provision of energy to the laser is usually described as "pumping" the laser. Examples of suitable energy sources include optical sources such as a high-intensity discharge lamp or a laser diode, both of which are used with solid-state lasers. Alternatively, an electrical power unit producing a current that passes directly through the active material may be used as the energy source in the case of semiconductor lasers.
3. The third element is the provision of two mirrors: one that is fully reflective, reflecting 100% of the incident laser radiation; and the other that is semi-reflective (i.e., partly transmissive). Again these are integral components or features of every laser.

1.5.2 Solid-State Materials

The most frequently used Nd:YAG solid-state crystalline material that has already been mentioned above, usually takes the form of a cylindrical rod. The growing and manufacture of the YAG laser rod is a slow and quite complex procedure, so the cost of the rod is comparatively high. The basic YAG material is transparent and colorless. When doped with a very small amount (1%) of Nd, the crystal takes on a light blue color. The Nd atoms act as the actual lasant. An alternative matrix material to the YAG crystalline material is glass. Indeed neodymium-doped glass (Nd:glass) has been used in a number of military rangefinders. Glass is a relatively inexpensive material and can easily be worked to give the desired shape (rod or disk) and dimensions. However, the problem with using glass as the matrix is that it has poor thermal conductivity and a low capability to dissipate heat. Thus, while the Nd:glass laser can produce high energy pulses, the repetition rate has to be kept low in order to keep the heat level low. So it requires cooling if operated at a high power level and/or at a high pulse repetition rate, which will be the case with the airborne laser profiling and scanning carried out for topographic applications, though not in military rangefinders. By contrast, when the Nd:YAG laser is operated in a pulsed mode, very high pulsing rates can be achieved while still providing a very high average output power of 1 kW or more—since the thermal conductivity of the YAG crystal has a much higher rate than that of glass. The end faces of the Nd:YAG rod are made accurately parallel to one another with one of them silvered to make it fully reflective, while the other is semi-silvered and partly reflective. However, more commonly, separate mirrors are utilized to provide the required reflections.

The Nd:YAG laser has been adopted quite widely as the basis of the laser rangefinders used in airborne laser scanners. A less commonly used solid-state laser is that employing yttrium lithium fluoride (YLF) as the matrix material. Again this is doped using a small amount of Nd to form the Nd:YLF laser. This is used for example in NASA's RASCAL airborne laser scanner. Yet another material that has been used in NASA-built and -developed airborne scanners is neodymium-doped yttrium vanadate (Nd:YVO_4). Compared with the Nd:YAG laser, the so-called "vanadate" laser has the advantage that it allows very high pulse repetition rates; on the other hand, it does not allow as high a pulse energy with Q-switching as the Nd:YAG laser. The Terrapoint ALTMS airborne scanners, originally developed with the assistance of NASA, all utilize these "vanadate" lasers.

1.5.3 Laser Action—Solid-State Materials

While the laser action can be activated and maintained using a xenon lamp or krypton discharge tube (Figure 1.6), for most laser rangers, the Nd-doped lasers described above are pumped continuously using a semiconductor diode or diode array (Figure 1.7). The minimum pumping power that is required to begin the lasing action is called the lasing threshold. When the pumping action begins, it energizes a large number of the Nd atoms into an excited state, in which their electrons are raised from their normal ground state to a much higher but unstable energy level. As the electrons cascade down from this high-energy state back toward their normal ground state, they pass through an intermediate energy level called the metastable state. By continuing to pump energy into the active material, most of

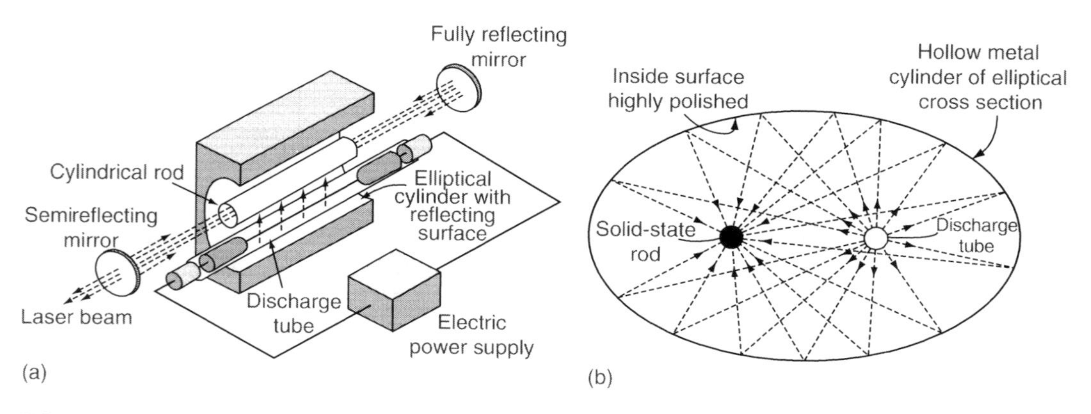

FIGURE 1.6
(a) Construction of a solid-state laser with the Nd-doped cylindrical rod placed at one focal point of a hollow metal cylinder with an elliptical cross section, whose inside surface is highly polished; a high-energy discharge tube is placed at the other focal point to provide the energy needed to start up and maintain the lasing action. (b) The discharge tube is placed at one of the focal points of the elliptical cylinder, in which case, all the emitted light will be reflected by the polished surface to pass into the solid-state rod placed at the other focal point.

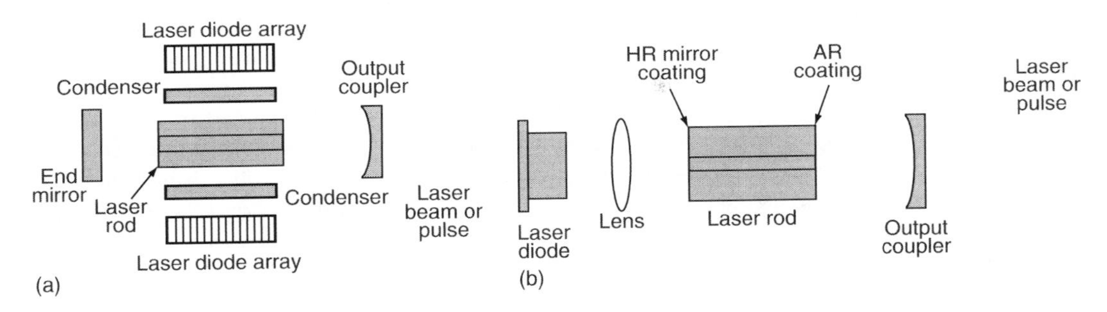

FIGURE 1.7
Different designs of Nd-doped lasers pumped by laser diodes (a) in the form of banks of diodes located above and below the laser rod and (b) using a single diode aligned with the central axis of the laser rod.

the electrons remain in this metastable state and a so-called "population inversion" is achieved. While they are in this metastable state, the electrons may spontaneously emit energy in the form of photons—an action called "spontaneous emission" (Figure 1.8). Other adjacent atoms are then stimulated to emit photons with the same frequency and phase—described as "stimulated emission." Thus, amplification of the radiation is produced by stimulating coherent emissions from the other excited atoms.

For those photons that travel along the axis of the rod or cylinder, when they reach the fully reflective end or mirror, they are reflected back along their paths to the other

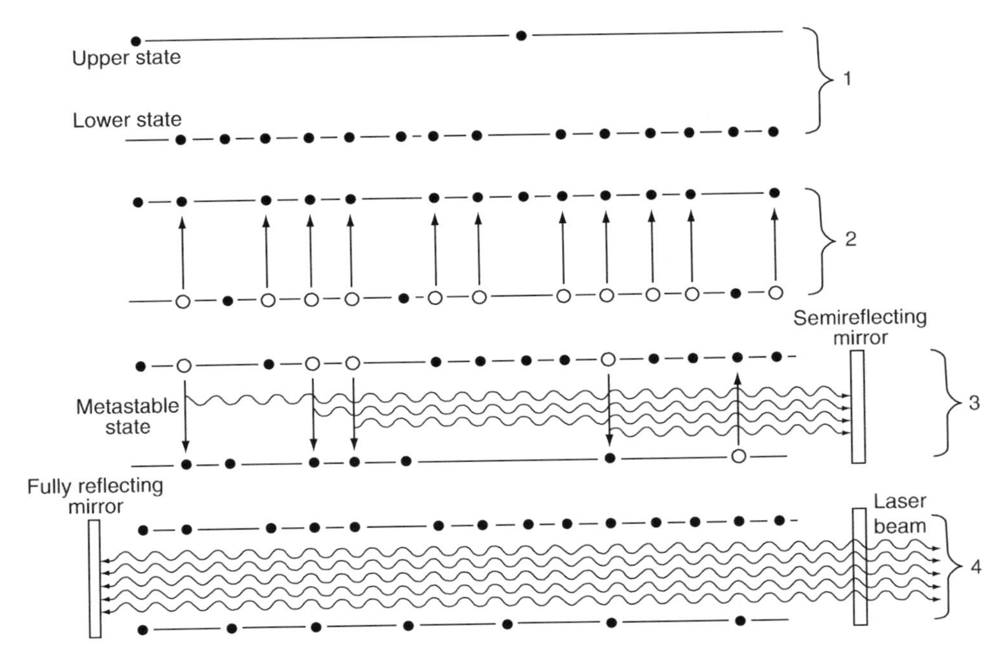

FIGURE 1.8
Simplified representation of the lasing action: (1) Showing the situation prior to the application of the energy source; (2) situation with the application of energy from the external source; (3) "spontaneous emission" takes place with photons being emitted parallel to the main axis of the cylindrically shaped laser rod; and (4) after further "stimulated emission" and reflections to and from the mirrors, the laser beam or pulse is emitted through the semireflective (or partially transmitting) mirror.

end where they are reflected back again. In the course of their repeated forward and backward paths along the cylinder, they are joined by other photons emitted from other excited atoms whose electrons have reached the metastable state (Figure 1.8). The new photons are precisely in phase with the original or primary photons. The overall effect may be envisaged or depicted as comprising an in-phase tidal wave of photons sweeping up and down the axis of the cylinder or rod, increasing in amplitude and intensity as more of the Nd atoms are stimulated into emitting photons by the passing wave. The beam of light (or infrared radiation) builds up very rapidly in intensity until it emerges from the semireflective (partly transmissive) mirror either as a continuous beam or as a short, intense, and highly directional pulse of infrared laser radiation. In the case of the Nd:YAG and Nd:YVO_4 lasers, the radiation is emitted at the wavelength of 1064 nm; other important emission wavelengths for the Nd:YVO_4 (vanadate) lasers are 914 and 1342 nm. With the Nd:YLF laser, the laser radiation is emitted at the wavelength of 1046 nm.

Through the use of frequency doubling techniques, which halve the emitted wavelengths, these three lasers can also be adapted to emit their radiation at wavelengths of 532 nm (Nd:YAG and N:YVO_4) and 523 nm (Nd:YLF), respectively, in the green part of the electromagnetic spectrum. This is achieved by passing the pulse through a block of a special nonlinear crystalline material such as potassium dihydrogen phosphate (KDP) or potassium titanyl phosphate (KTP) which is transparent and has the specific crystalline structure and bi-refringent properties to achieve the half-wavelength that is required. The shorter wavelength laser is used primarily in bathymetric mapping, since it allows for better penetration into water.

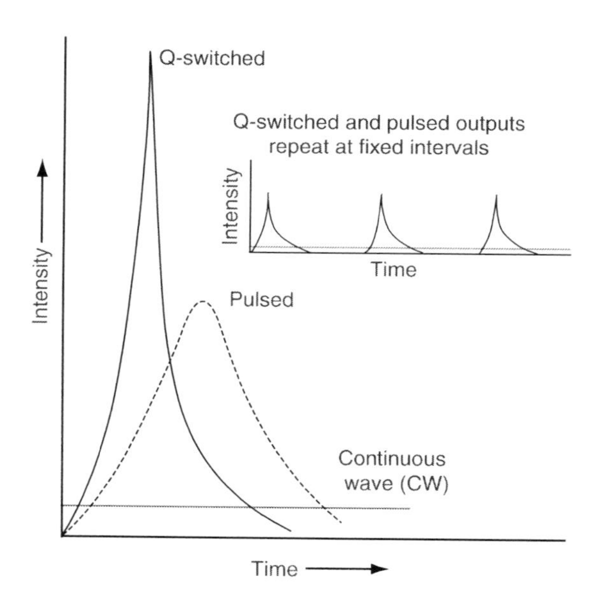

FIGURE 1.9
Laser output modes—CW, pulsed, and Q-switched. CW lasers have low intensity outputs and consequently only require a comparatively small energy input for their operation; pulsed lasers emit pulses with much higher energy levels; while, finally, still higher energies are emitted with Q-switched lasers. (From Price, W.F. and Uren, J., *Laser Surveying*, Van Nostrand Reinhold, 1990. With permission.)

When operating in its pulsed mode, the power of the solid-state laser can be built up still further through the use of a technique called Q-switching. This employs a special type of shutter that delays the release of the energy stored in the pulse until it reaches a very high power level (Figure 1.9). The shutter is then opened to allow the laser emission to take place in the form of a very intense pulse of coherent radiation. The Q-switch can take the form of a Kerr cell with switchable polarizing filters. Alternatively, it may take the form of a light-absorbing dye, which is bleached when the energy reaches a suitably high level, so releasing the pulse. Q-switching is a necessity when a laser ranger or scanner has to operate over distances of hundreds of kilometers; for example, when it is mounted on a spaceborne platform. Typically Q-switched lasers can produce pulses of intense energy over a time range of 10–250 ns.

1.5.4 Semiconductor Materials

Those lasers that are based on semiconductor materials all exhibit the electrical properties that are characteristic of electrical diodes. Thus, they are often referred to as diode lasers as an alternative to the description semiconductor laser. As noted above, gallium arsenide (GaAs) is the main semiconductor material that is used as the basis for the lasers used in rangers and scanners. Under suitable conditions, they will emit an intense beam or pulse of infrared radiation. The basic structure of a simple semiconductor laser is shown in Figure 1.10. It comprises various layers of the semiconductor material with metal conductors attached to the top and the bottom of the diode to allow electrical power to pass into the material. The upper semiconductor layer contains impurities that cause a deficiency in their electrons, leaving positive "holes." This material is therefore called a p (=positive) type semiconductor. The lower layer has impurities that result in an excess of electrons with negative charges. So this material is said to be an n (= negative) type semiconductor. The active element of this type of laser is an additional layer called the p–n junction that is sandwiched between the p-type and the n-type semiconductor layers (Figure 1.10).

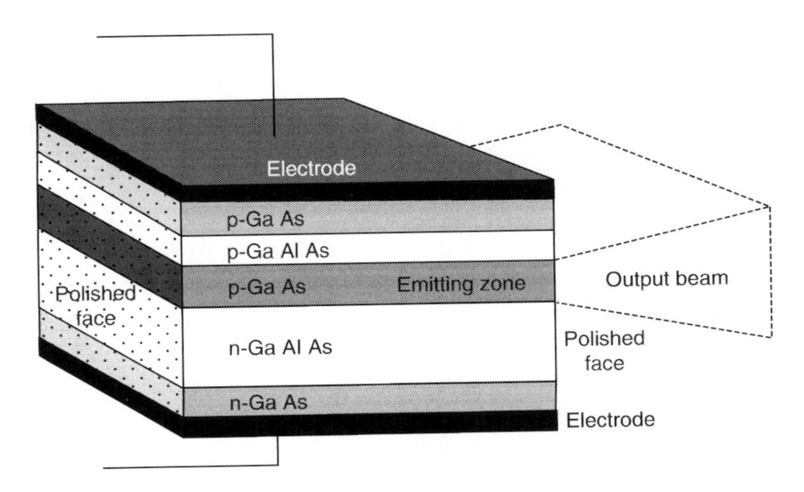

FIGURE 1.10
Basic construction of a semiconductor laser in which the active element is a p–n junction sandwiched between a p-type and an n-type semiconductor layer.

1.5.5 Laser Action—Semiconductor Materials

When stimulated by an electric current, electrons enter the p–n junction from the n-side and holes from the p-side. The two mix and again a population inversion takes place. In a simplified description, the electrons and holes move in the desired direction through the application of an electric voltage across the junction. Therefore, this type of laser is also referred to as an injection laser since the electrons and holes are injected into the junction by the applied voltage. The actual voltage is produced through the connection of a suitable electric power source to the two electrodes affixed to the top and bottom surfaces of the material. Thus, in this case, the pumping action is purely electrical in character. As described above, atoms within the active material of the p–n junction are excited to higher energy states and in falling back to a lower state, they exhibit electroluminescence and emit photons. The photons traveling along the junction region stimulate further photon emission and so produce still more radiation. The actual reflective lasing action can be produced by simply polishing one end face of the material so that it acts as a semi-reflective mirror. For the other (fully reflective) end, a metal film of gold—which is a good reflector at infrared wavelengths—is coated on to the end of the device, ensuring 100% reflection. The wavelength of the radiation being emitted from a GaAs semiconductor laser is normally around 835–840 nm in the near-infrared part of the electromagnetic spectrum. A laser based on a similar material—gallium aluminum arsenide (GaAlAs)—produces its radiation over the much greater range of 670–830 nm, where the wavelength value is being controlled by suitable variations of the bias current.

Physically, the GaAs semiconductor laser is quite small. However, it is a very efficient device, since up to half of all the electrons injected into the active GaAs material produce photons that are output in the form of laser radiation. Besides this high efficiency, it is very easy to control the output power through small changes to the supply current or voltage. Semiconductors are produced in very large quantities for other (nonsurveying) applications, so they are relatively inexpensive, very reliable, and have a long life with low energy consumption. Negative points are that the output from a semiconductor laser has a rather poor spectral purity—which is not too important a characteristic in ranging applications—and has a rather wide divergent angle. When operated at the temperatures normally encountered in the field, the semiconductor material can reach quite high temperatures in order to generate suitable output power—for example, when operated in a pulsed mode over longer ranges. On the other hand, semiconductor diode lasers are often used in a continuous (nonpulsed) mode over shorter distances. In which case, it is easier to keep the power dissipation and heat levels at an acceptable level within the device. With continuous wave (CW) operation of this type of laser, the phase measurement technique allows range measurements to be made at a much higher rate than with the pulse-ranging technique.

1.6 Laser Ranging

The laser ranger or rangefinder is the device or instrument that is used to measure the slant range or distance to a reflective object such as the ground, a building, or a tree—that is often referred to in rather abstract terms as a "non-cooperative target." Besides being used purely as a distance measuring device for surveying, military, or even sporting purposes, the laser rangefinder forms the basis of all laser profilers and scanners. As mentioned in the introduction to this chapter, there are two measurement technologies and methodologies that are in use for topographic applications: (1) the TOF or timed pulse or pulse echo method; and (2) the multiple-frequency phase comparison or phase shift method for CW operation.

1.6.1 Timed Pulse Method

As discussed in the introductory section above, the TOF method that is used to determine the slant range from the measuring instrument to an object involves the very accurate measurement of the time required for a very short pulse of laser radiation to travel to the target and back. The method has been used to measure distances of a few tens or hundreds of meters in the case of terrestrial or ground-based rangers, profilers, and scanners. Obviously, the distances that need to be measured from an airborne platform will be much greater—in the order of hundreds of meters to several kilometers—while the ranging devices mounted on spaceborne platforms such as the Space Shuttle and satellites will be measuring distances of several hundred kilometers.

The design of a generic type of laser ranger or rangefinder using the timed pulse method is presented in Figure 1.11. The actual design and the specific components that will be used in a particular instrument will depend on the particular application, for example, whether it is being operated from a ground-based, airborne, or spaceborne platform. The high-energy laser will have a lens placed in front of it to expand and collimate the laser pulse so that it has a minimum divergence. When a pulse is emitted by the laser, a tiny part of its energy is diverted by a beam splitter on to a photodiode whose signal or start pulse triggers the timing device (Figure 1.12); the triggering is usually accomplished by simple thresholding. In practice, this timer is a high-speed counter that is controlled by a very stable oscillator. To achieve a 1 cm ranging resolution, the timer should be able to measure a 66 ps interval, which would require a clock rate of about 15 GHz.

After its reflection from the object, the returning pulse that is being measured is picked up by the receiving lens or mirror optics. With the long ranges of hundreds of kilometers that need to be measured from a spaceborne platform, large diameter mirror optics will need to be used to receive and detect the quite faint signals that will be returned over these huge distances. Besides, even at shorter ranges, when the emitted laser pulse reaches the object being measured, it will have diverged and spread. In addition, it will have been

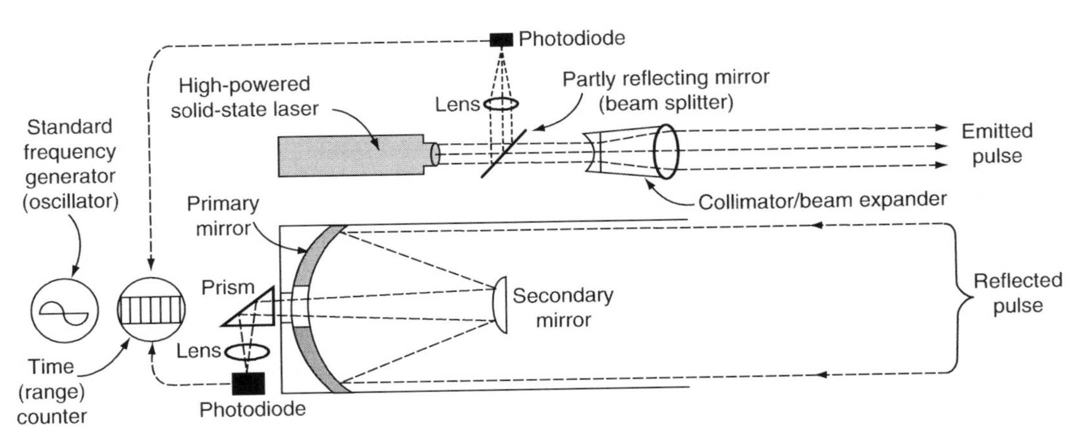

FIGURE 1.11
Layout of the main components of a pulse-type laser rangefinder and their interrelationship.

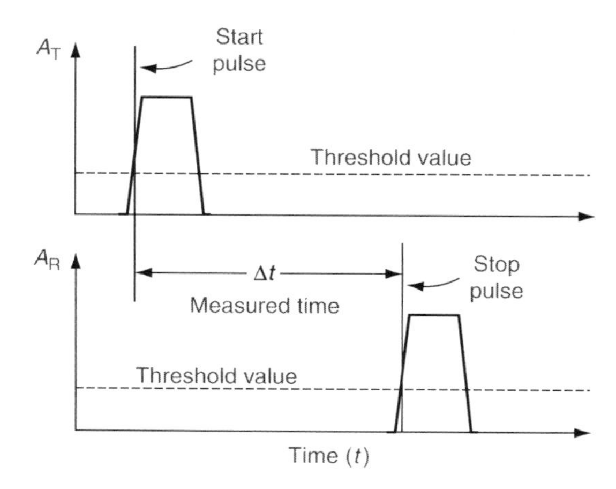

FIGURE 1.12
Diagrammatic representation of the action of the start and stop pulses used to control the time counter.

scattered in various directions by tiny particles that are present in the atmosphere, both on its outward path and on its return path to the ranger. So even at short ranges, there is a need to amplify the weak return signal electrically.

Whether large or small in terms of its size, the receiving lens or mirror optical system will focus the returning (reflected) pulse on to another very sensitive photodiode. Typical of those in use in laser rangefinders are silicon avalanche photodiodes (APDs) or InGaAs PIN (positive–intrinsic–negative) semiconductor photodiodes. Different types or models of these devices are used depending on the wavelength of the laser radiation being used. Besides which, narrowband optical filters will usually be placed over the photodiode in order to cut down or eliminate the effects of sunlight or other sources of optical noise that may cause spurious signals. In the simplest type of ranger, when the reflected pulse is received, the diode delivers a stop pulse to the time counter (Figure 1.12). Usually, the specific point on the leading edge, where the voltage generated by the diode reaches a predetermined threshold value, is used to start and stop the time counter. It should be noted that in real state-of-the-art systems much more sophisticated triggering is implemented. The time-to-digital converter process is rather simple as the measured time in terms of timer counts is converted to the corresponding range or distance value by using the known speed of light. If a series of reflections is being returned from a particular object such as a tree or an electric pylon from a single pulse, then the time (epoch) of each of the returned pulses—e.g., the first and last pulses—will be measured. Alternatively the complete waveform or shape of the whole of the return signal can be measured, digitized, and recorded on a time base for later analysis (Figure 1.13), as will be discussed in subsequent chapters.

The actual length of the pulse emitted by the laser is an important characteristic of any laser system, since it has a major impact on how multiple returns can be detected or differentiated. If the pulse lasts 10 ns (where 1 ns is 10^{-9} s) then the pulse length will be 3 m at the speed of light (v). Thus the measuring resolution of a ranging device utilizing timed pulses is governed, in the first place, by the length of the emitted pulse and the degree to which either the center or the leading or the trailing edge of the return pulse can be determined. In general, this can be a difficult task to perform since the return pulse will often be quite distorted after its reflection from the ground targets. Since the speed of electromagnetic energy (v) is 300,000 km/s, a resolution of 1 m in a single measured range requires the timing of the pulse to be accurate to

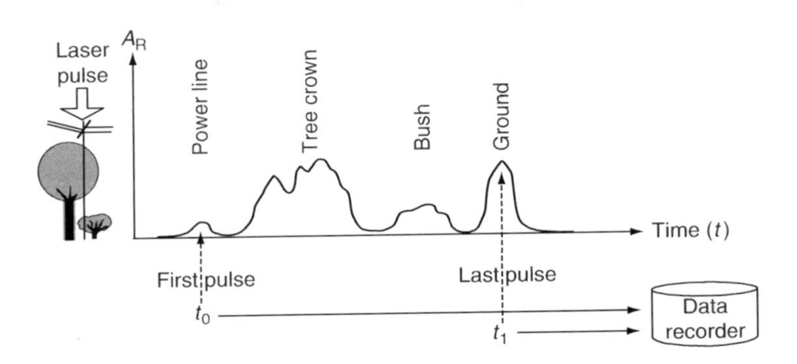

FIGURE 1.13
Showing the shape of the complete waveform of the returned (reflected) pulse that can be used for further analysis. (From Brenner, C., Aerial laser scanning. International summer school on "Digital Recording & 3D Modeling", Aghios Nikulaos, Crete, Greece, 2006, April 24–29. With permission.)

1/300,000,000 (3×10^{-9}) of a second. A resolution of 1 cm in the measured range requires a time measurement of 3×10^{-11} s, while a 1 mm resolution requires 3×10^{-12} s, equal to 3 ps (picoseconds). As mentioned above, a timing device based on a very accurate quartz-stabilized oscillator is required to carry out the measurement of the elapsed time (or round trip transit time) to the required level. In fact, the quality of the timer, including the start/stop triggering components, determines the achievable accuracy of the laser ranging instrument.

When short distances are being measured, as is the case with a terrestrial or ground-based ranging instrument, it is possible to make a relatively large number of measurements over a short period of time, that is, the pulse repetition frequency (PRF) can be high. Thus, for example, if a distance of 100 m has to be measured with a ground-based laser ranger, then the travel time over the 200 m total distance out and back is 200/300,000,000 = 0.67 µs. With longer distances, such as those required to be measured from airborne and spaceborne platforms, a longer time interval will elapse between the emission and the reception of individual pulses. For a distance of 1000 m that is typical for an airborne laser ranger, the elapsed time will be 6.7 µs, in which case, the maximum PRF will be 1/6.7 µs, which is 150 kHz. Thus, in general terms, a longer time interval between successive pulses will usually be required when longer distances need to be measured. However, this remark needs to be tempered by the recent introduction of the technique of having multiple laser pulses in the air simultaneously (Figure 1.14). This gives the capability for a laser ranger to emit a new pulse without waiting for the reflection from the previous pulse being received at the instrument. Thus, more than one measuring cycle can be taking place at any moment of time. Note that the concept of using multiple pulses was introduced in RADAR systems much earlier (Edde, 1992).

In the multiple pulse case, the limit of the number of pulses is primarily imposed by the laser source; in other words, how frequently can a pulse be emitted by the source. Note that systems from the early 2000s could increase their PRF at the price of reducing the energy of the emitted pulse and, in this way, the ranging accuracy was decreased due to the less favorable SNR (signal-to-noise ratio). Current laser supplies are more powerful, so this phenomenon does not impact current state-of-the-art systems. The fact that there are multiple pulses traveling back and forth, however, requires additional processing to resolve the ambiguity (to properly pair an emitted pulse to a received one). Since the number of pulses is low and the object distance is not changing too much between two pulses, this task is almost trivial and requires limited effort.

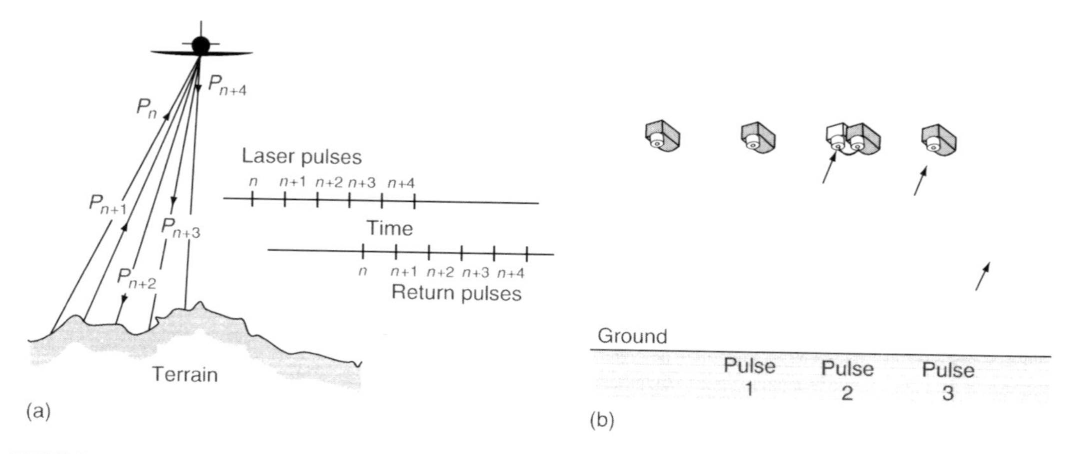

FIGURE 1.14
(a) Pulse interleaving concept for a five-time frequency rate is being applied in the cross-track direction from an airborne laser scanner (From Toth, C.K., Future Trends in Lidar. ASPRS Annual Conference, Denver, CO, May 23–28, 2004, Paper No. 232.), (b) Multiple pulses in the air are being emitted in the along-track direction to measure a longitudinal profile from a satellite.

1.6.2 Phase Comparison Method

The phase comparison method is used with those rangers where the laser radiation is emitted as a continuous beam—often referred to as a CW laser—instead of discrete pulses. In practice, due to the limited power of the CW laser, there are very few laser rangers, profilers, or scanners of this type in actual operation from airborne or spaceborne platforms as compared with those rangers using the timed pulse or pulse echo technique. (An exception is the ScaLARS research laser scanner of the University of Stuttgart that will be described later in Chapter 2.) By contrast, the phase comparison technique is much used in short-range terrestrial or ground-based laser scanners where the distances that need to be measured are often much shorter—typically <100 m. In those devices using the CW approach, the transmitted beam comprises a basic (laser) carrier signal on which a modulation signal has been superimposed for the purpose of measurement (Figure 1.15). This modulation signal or measuring wave is derived from and held at a constant value using a stable frequency oscillator. Thus, the measuring wave is used to control the amplitude of the carrier wave—a process known as amplitude modulation.

The reflection of the signal for a CW laser is similar to that for a pulsed one, except it is continuous. The transmitted beam then strikes the object being measured, and a small part of the signal is reflected back along the identical path to the instrument where it is detected using a silicon photodiode. The weakened reflected signal is then amplified and subjected to demodulation—which is the separation of the measuring and carrier waves. Next, the reflected signal is compared with the signal of the original transmitted beam (or reference signal). As mentioned above, the difference in phase (or phase angle, φ) between the two signals is then measured and represents the fractional part ($\Delta\lambda$) of the total slant range that is required (Figure 1.2). The integer number of wavelengths, $M\lambda$, cannot be derived from this single measurement; the task of determining M is the so-called ambiguity resolution problem. There are several ways to solve the ambiguity, depending on using various modulation frequencies, or simply following the range

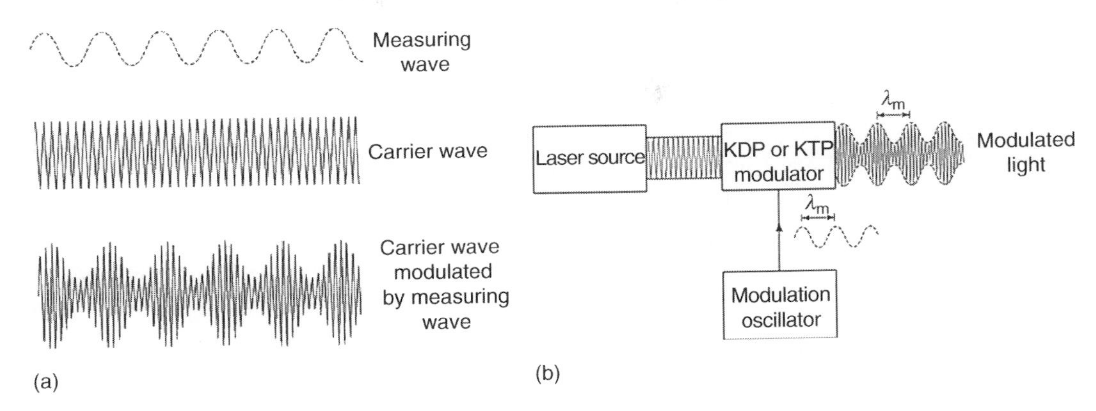

FIGURE 1.15
(a) Amplitude modulation with the measuring wave being used to vary the amplitude of the carrier wave, and (b) amplitude modulation of the emitted beam is performed by passing it through a KDP or KTP crystal to which a voltage has been applied. (From Price, W.F. and Uren, J., *Laser Surveying*, Van Nostrand Reinhold, 1990.)

changes, etc. The most widely applied method to resolve the ambiguity is to make a number of changes to the wavelength (λ) and therefore to the frequency (f) of the emitted beam very rapidly in succession—hence the description of this being a multifrequency phase measuring system. The required changes in wavelength are carried out automatically and very rapidly within the instrument without any human interaction. Thus, there is no need for this to be done manually in modern instruments. The resulting measurements of the series of phase angles φ lead to a set of simple equations having the general form:

$$
\begin{aligned}
R &= M_1 \cdot \lambda_1/2 + (\varphi_1/2\pi)\cdot(\lambda_1/2) \\
R &= M_2 \cdot \lambda_2/2 + (\varphi_2/2\pi)\cdot(\lambda_2/2) \qquad (1.6) \\
R &= M_n \cdot \lambda_n/2 + (\varphi_n/2\pi)\cdot(\lambda_n/2)
\end{aligned}
$$

where n is the number of different frequencies that are being employed. Solving the set of simultaneous equations will give the final value of the slant range (R).

It is important to note that, when using a modulated laser beam for ranging, the characteristics of the modulation signal, including the wavelength and phase measurement resolution, will determine the ranging precision of the system. When this type of multifrequency CW laser ranging system is being used, the lowest frequency associated with the longest wavelength, λ_{long}, will determine the longest slant range, R_{max}, that can be measured. By the same token, the highest frequency with the shortest wavelength, λ_{short}, will determine the range resolution and therefore the accuracy with which the slant range can be measured by such a device or system. For example, a 10 MHz modulation signal has a wavelength of 30 m, which with a 1° phase resolution could result in a precision of about 8 cm (30/360).

1.6.3 Power Output

Usually, the output power as it relates to a laser is expressed in watts (W) or milliwatts (mW), where 1 W (the SI unit of power) = 1 J/s. The joule is the SI unit of energy or capacity for doing work, equal to the work done when a current of 1 A is passed through a resistance of 1 Ω for 1 s. The output for pulsed lasers is often expressed as the radiant exposure, which is the concentration of the laser energy on a given area, expressed in terms of joules per square centimeter (J/cm^2).

Very different output power levels are encountered depending on the type of laser—pulsed or CW—that is being considered. Thus, at one end of the power spectrum are the terrestrial or ground-based laser scanners using CW lasers—such as those manufactured by Zoller & Fröhlich and Faro—that measure over short ranges (up to 70 m). Using the phase measuring technique, they employ CW semiconductor lasers with a relatively low power—between 10 and 20 mW for slant ranges up to 25 m and between 20 and 40 mW for distances up to 50–70 m. By contrast, the pulse-type lasers being used in airborne laser scanners measuring ranges of a few kilometers typically operate at peak power levels of 1–2 kW or even more (Flood, 2001b). In the latter case, if the pulse duration (t_p) is 10 ns, then the energy (E) generated per pulse is $E = P_{peak} \cdot t_p = 20\,\mu J$. The average power for a pulse repetition rate of 50 kHz will be $P = E\,F = 0.000020 \times 50{,}000 = 1\,W$.

The actual measurement of the laser power in a laboratory or a manufacturing facility is normally carried out using instruments based on one or other of two different types of detector: quantum detectors and thermopile detectors. The quantum detector is based on a semiconductor material such as indium gallium arsenide and measures laser power by counting photons. It converts the incoming photons into charge carriers (electrons and holes), which are then summed as a voltage or current using an amplifying circuit. By contrast, the thermopile detector acts essentially as a calorimeter. The incident radiation heats the device and a circuit measures the heat differential between the detector and an attached heat sink. Both detectors provide their final measurements to the user in units of watts. The quantum detector is more sensitive at low power levels—at the microwatt level or below. Both types of instruments can be used at higher power levels, but the thermopile detector is best if very high power levels need to be measured since the quantum detector can suffer from saturation effects (and possible damage) at these levels.

1.6.4 Power and Safety Concerns

Given the fact that many of the devices used in laser ranging, profiling, and scanning utilize powerful pulse-type lasers, the matter of safety and of safety standards is a matter of prime importance, especially during the actual use of these instruments in the field and in the air. For most of the world, the applicable safety standards are those set by the International Electro-technical Commission (IEC). These are known as the International IEC 60825 Standards. The exception to the adoption of this standard is the United States, which, so far, has not adopted this standard. Instead, it has its own standards set by the Center for Devices & Radiological Health (CDRH) and administered by the Food & Drug Administration (FDA). The American standard is ANSI Z136.1, while the manufacturer's standard is called CDRH 21 CFR, parts 1040.10 and 1040.11. Within Europe—where many of the manufacturers of laser rangers, profilers, and scanners are located—the IEC standard has been adopted and is known as the EN 60825 Standard. Each European country has its own version of this standard; for example, in the United Kingdom, it is known as the British Standard BS EN 60825. Recently the CDRH in the United States has decided to harmonize its standards with the IEC 60825 standard, although this process has not yet been completed. Reference may be made to the following web page http://en.wikipedia.org/wiki/Laser_Safety for details of the current classification.

1.6.5 Laser Hazard Classification

Of most importance to the present discussion is that both the International and U.S. standards divide lasers into more or less the same four broad categories on the basis of their hazards, especially concerning the risk of them causing damage to human eyes or skin, as follows:

1. Class I: These comprise those lasers and laser-based systems that do not emit radiation at known hazard levels and are exempt from radiation hazard controls during their operation in the field, though not during servicing. Thus the CW lasers used in ground-based laser scanners such as the Z+F and Faro scanners mentioned above operate with Class I lasers.
2. Class IA: This is a special group with an upper power of 4.0 mW, where the emission level does not exceed the Class I limit for an emission of 1000 s. In practice, it applies only to lasers that are "not intended for viewing" such as a supermarket laser scanner.
3. Class II: These are low-power visible light lasers that emit their radiation above Class I levels, but at a radiant power not above 1 mW. The concept is that the human aversion to bright light will protect a person. Only limited controls are specified regarding their operation.
4. Class IIIA: These comprise lasers with intermediate power levels—typically CW lasers in the range 1–5 mW that are only hazardous for intrabeam viewing. So only certain limited controls are specified for their operation.
5. Class IIIB: These comprise lasers with moderate powered emissions—for example, CW lasers operating in the range 5–500 mW; or pulsed lasers with outputs up to 10 J/cm^2. There are a set of specific controls recommended for their operation.
6. Class IV: This category comprises those lasers with power levels above those specified for Class IIIA lasers. They are hazardous to view under any circumstances; besides which, they are a skin hazard; and potentially a fire hazard. In the context of the present discussion, many airborne laser scanners, including the ALTM series from the leading supplier, Optech, use Class IV lasers (Flood, 2001a). So the appropriate safety measures must be followed both during their manufacture and servicing and also during their operation in the field or in the air.

It is important to note also that certain lasers emitting radiation in the short-wave infrared part of the electromagnetic spectrum—at wavelengths greater than 1400 nm are labeled as being eye-safe. An example is the Optech ILRIS-3D terrestrial laser scanner, which uses a laser rangefinder that emits its radiation at $\lambda = 1550$ nm. This designation arises from the fact that the water content of the cornea of the human eye absorbs radiation at these wavelengths. However, this label can be misleading since it really applies only to low-powered CW beams. Any high-powered Q-switched laser operating at these wavelengths can still cause severe damage to the eye of an observer.

1.6.6 Beam Divergence

No matter how well collimated the laser beam or pulse is when it leaves the ranging instrument, it will have spread to illuminate a circular or elliptical area when it reaches the ground or the object on the ground. For a given angular spread of the beam, the greater the range, the larger the diameter of the area that is being covered on the ground or at the object. Thus, if the ground is irregular in shape or elevation, the return

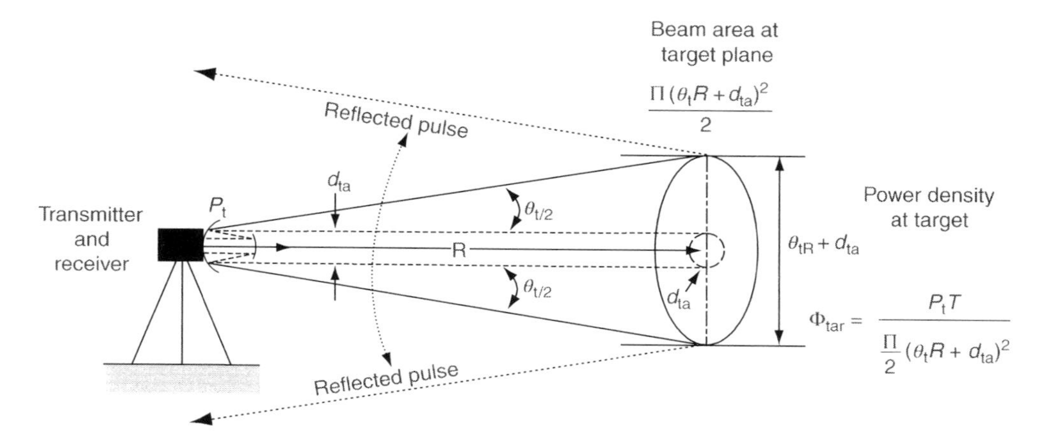

FIGURE 1.16
Spread of the beam or pulse that has been emitted by the laser rangefinder and its divergence on its path to the ground object or target and on its return from the object or target.

signal will be the average of the mixture of reflections occurring within the circular or elliptical area illuminated by the incident laser radiation.

For a terrestrial or ground-based laser rangefinder, the geometry of the beam traveling between the rangefinder and the target is shown in Figure 1.16. The area (A) covered by the diverging beam when it reaches the target is equal to $\pi(\theta/2R + d)/2$, where θ is the angle of divergence in radians; d is the diameter of the aperture; and R is the range.

Brenner (2006) gives further information about this matter. If the beam divergence is γ, then the theoretical limit of the divergence caused by diffraction is $\gamma \geq 2.44\ \lambda/d$, where d is the diameter of the aperture of the laser. In this case, using a pulsed laser emitting its radiation at $\lambda = 1064$ nm and having an aperture of 10 cm, the value of γ will be 0.026 mrad. Brenner quotes the typical values of γ for an airborne laser scanner as being in the range 0.3 to 2 mrad. The Optech ALTM 3100 and the Leica ALS50-II airborne laser scanners both have beam divergences at the smaller end (0.3 mrad) of this range. The Riegl LMS-Q560 laser scanner and the IGI LiteMapper and TopoSys Harrier airborne laser scanners that use the Riegl device as the laser engine in their systems have a beam divergence of 0.5 mrad. At a typical flying height of 1000 m, the laser footprint is about 30 and 50 cm for the beam divergences of 0.3 and 0.5 mrad, respectively.

1.6.7 Reflectivity

The reflectance or reflectivity of a ground target is another important matter to be considered in terms of the ranging performance of laser profiling and scanning systems. Reflectance is formally defined as the ratio of the incident radiation on a particular surface to the reflected radiation from that surface. Obviously, if the strength of the reflected signals that are received by the ranger is very weak, the range over which the required measurements can be made will be reduced since no detectable return signal will reach the receiver. The backscattering properties of the particular piece of ground or object being measured are therefore matters of considerable importance. For a diffusively reflective target such as a building, rock, or tree trunk—also called a hard surface—the reflected radiation can be envisaged (or idealized) as having been scattered into a hemispherical pattern with the

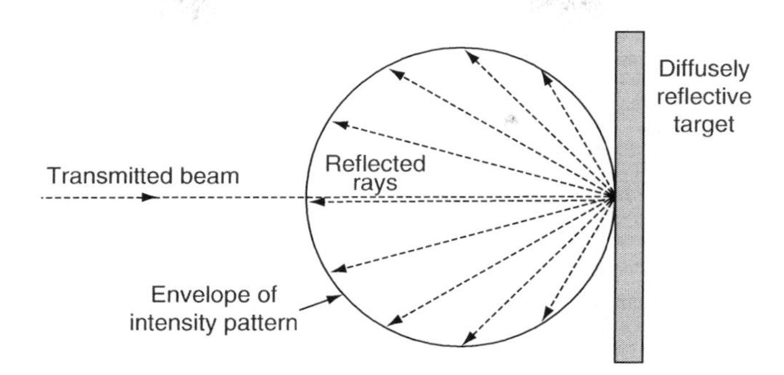

FIGURE 1.17
Reflectivity of a diffuse target.

maximum reflection taking place perpendicular to the target plane and the intensity diminishing rapidly to each side (Figure 1.17).

Furthermore, the reflective properties of the target will vary according to the wavelength of the radiation being emitted by the laser. In practice, it is often difficult to obtain information on these properties, since much of the published information on the subject of reflectivity and backscattering comes from remote sensing. In many of these reports on reflectivity, the incident radiation comes from the Sun and is incoherent in nature. Thus, the reflectivity may be very different when the radiation comes from a coherent source such as a laser emitting its radiation at a very specific wavelength. Wehr and Lohr (1999) have published a table based on information provided by the Riegl company that manufactures both airborne and terrestrial laser rangers, profilers, and scanners. This shows the typical reflectivity of various diffuse reflectors for laser radiation with a wavelength (λ) of 900 nm (Table 1.1).

TABLE 1.1

Typical Reflectivity of Various Reflecting Materials at a Wavelength of 900 nm

Material	Reflectivity (%)
Lumber (pine, clean, dry)	94
Snow	80–90
White masonry	85
Limestone, clay	Up to 75
Deciduous trees	Typ. 60
Coniferous trees	Typ. 30
Carbonate sand (dry)	57
Carbonate sand (wet)	41
Beach sand; bare areas in desert	Typ. 50
Rough wood pallet (clean)	25
Concrete, smooth	24
Asphalt with pebbles	17
Lava	8
Black neoprene (synthetic rubber)	5

Source: From Wehr, A. and Lohr, U., *ISPRS J. Photogram. Rem. Sens.*, 54, 68, 1990. With permission.

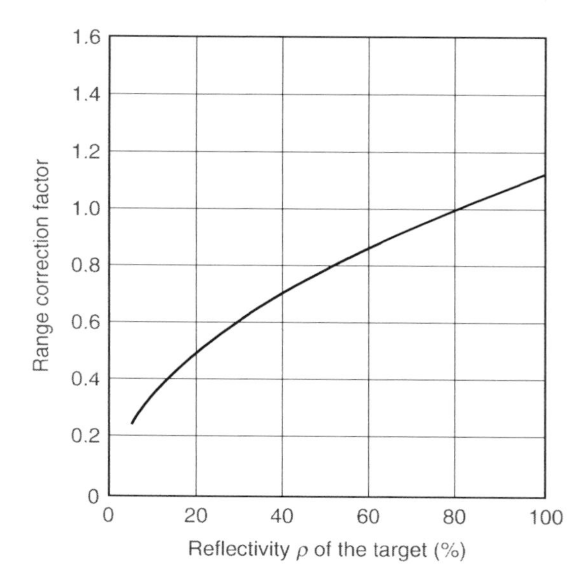

FIGURE 1.18
Correction factors for maximum range depending on the reflectivity of diffuse targets. (From Wehr, A. and Lohr, U., *ISPRS J. Photogram. Rem. Sens.*, 54, 68, 1990. With permission.)

Further information provided to the same two authors (Wehr and Lohr, 1999) by the Riegl company with regard to the variation in the maximum range with target reflectivity is given in Figure 1.18.

Still further information on the effects of variations in surface reflection comes from various manufacturers of ground-based laser scanners, particularly with regard to their effects on the range over which measurements can be made. Thus, the Riegl company quotes a maximum measuring range for its LMS-Z210i pulse-type terrestrial laser scanner of 350 m for natural targets having a reflectivity ≥50%. However, this is reduced to 150 m for natural targets having a reflectivity of ≥10%. Further information on this matter is provided by the i-site company from Australia in respect of its 4400LS laser scanner. The company quotes a maximum range of 150 m for measurements made with this instrument to a surface of black coal (with 5%–10% reflectivity); 600 m for measurements made to rock or concrete surfaces (with 40%–50% reflectivity); and up to 700 m for surfaces having a still higher reflectivity. Both the Riegl and the i-site instruments use the pulse-based ranging technique with their lasers emitting their radiation at the wavelength (λ) of 905 nm.

Turning next to those terrestrial short-range high-speed laser scanners that utilize CW lasers and the phase-based measuring technique, Leica Geosystems provides information in its literature on the effects of reflectivity on the precision with which different surfaces can be modeled using its HDS4500 scanner. This utilizes a CW semiconductor laser emitting its radiation at a wavelength (λ) of 780 nm. At a shorter range of 10 m, the precision is quoted as ≤1.6 mm for dark gray surfaces with 20% reflectivity, but ≤1.0 mm for white surfaces with 100% reflectivity. At a longer range of 25 m, the respective values of precision were ≤4.4 mm for the dark gray surface and ≤1.8 mm for the white surface.

Another important factor that governs the reflectivity of laser radiation is the angle that the ground or object makes with the incident pulse or beam (Figure 1.19). With an airborne laser scanner illuminating an area of smooth sloping terrain having a high (specular)

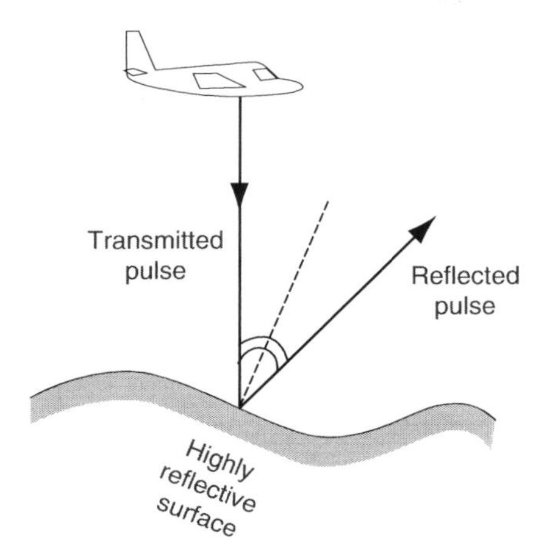

FIGURE 1.19
Highly reflective surface on the terrain that is not at right angles to the incident laser pulse or beam will reflect its radiation off to the side and will not return a signal to the laser rangefinder.

reflectance, a very large part of the incident radiation from the scanner will be reflected off to the side and will not be returned to the laser scanner. In the case of a forest canopy, the laser pulse emitted from the airborne scanner in the vertical or nadir direction will be able to penetrate the spaces in the canopy and produce a reflection from the ground below the canopy. However, the ability to penetrate the canopy and reach the ground will decrease greatly with increasing scan angles away from the nadir direction.

1.7 Power Received after Reflectance

Besides the loss of power in the emitted pulse through its spread caused by beam divergence and the limited reflectances from the ground and its objects, the radiation will also be subjected to scattering by the particles of dust and water droplets that are present in the air between the rangefinder and the target. Furthermore, it can be affected by scintillation effects in the atmosphere through which it is traveling. These effects will also be present during its return path after reflection. This results in a further diminution of the strength of the signal that will be returned to the measuring instrument. Indeed the power that will be reflected and received back at the rangefinder, profiler, or scanner will be a tiny fraction of that originally emitted by its transmitter.

If P_T is the power of the pulse transmitted by the airborne scanner, then the following expressions can be derived (Brenner, 2006):

1. Power that has been transmitted is P_T.
2. Power received at the ground will be $M \cdot P_T$ where M is the transmission factor through the atmosphere.
3. Power that is reflected from the ground (assuming a Lambertian surface) will be $(\psi/2\pi) \cdot \rho \cdot M \cdot P_T$.
4. The power received after reflectance from the ground and the return path through the atmosphere will be $P_R = A/(2\pi R^2) \cdot M \cdot \rho \cdot M \cdot P_T = \rho \cdot \{(M^2 \cdot A)/2\pi R^2\} \cdot P_T$ (Figure 1.20).

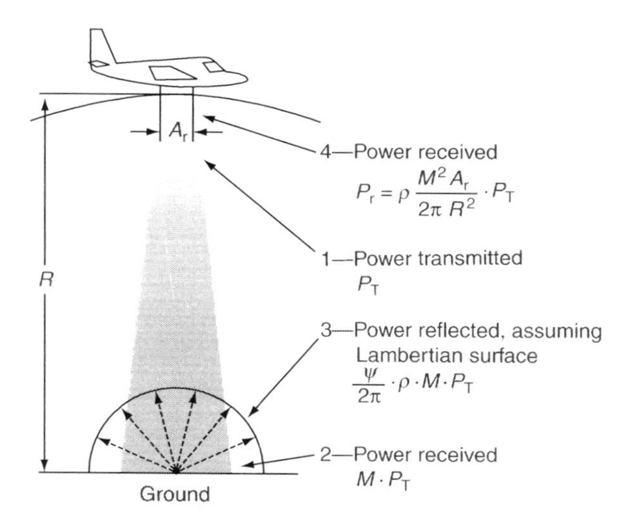

FIGURE 1.20
Power transmitted, reflected, and received, is shown graphically for an airborne laser scanner. (From Brenner, C., Aerial laser scanning. International summer school on "Digital Recording & 3D Modeling", Aghios Nikulaos, Crete, Greece, 2006, April 24–29. With permission.)

For example, if P_T = 2000 W, Atmospheric transmission (M) = 0.8, A = 100 cm², where D (aperture) = 10 cm, H (Flying Height) = 1 km, and Reflectivity (ρ) = 0.5, then $P = 4.10^{-10}$. P_T = 800 nW.

Thus, in terms of its power, only the tiniest fraction of the power of the emitted pulse is being received back at the laser rangefinder, profiler, or scanner after its travel to and from the ground target. As already mentioned, the received signal needs to be amplified and any accumulated noise needs to be filtered out before the final range value can be determined from the elapsed time.

1.8 Conclusion

This chapter has provided an introduction to lasers and laser ranging in the very specific context of their application to those rangefinders that form the basis of the laser profilers and scanners that are now being used extensively in topographic mapping and modeling. Based on this introduction, the two chapters that follow will describe the laser profilers and scanners that are currently in use from airborne and spaceborne platforms (Chapter 2) and in the terrestrial or ground-based applications of the technology (Chapter 3). The authors are grateful to Mike Shand who has provided all the drawings for this chapter.

References

Brenner, C., 2006 Aerial laser scanning. International summer school on "Digital Recording & 3D Modeling", Aghios Nikulaos, Crete, Greece, April 24–26.

Edde, B., 1992 *RADAR; Principles, Technology, Applications*, Prentice Hall, p. 816.

Flood, M., 2001a Eye safety concerns in airborne Lidar mapping. http://www.asprs.org/society/committees/lidar/Downloads/Flood%20–%20Eye%20safety.zip.

Flood, M., 2001b Laser altimetry: From science to commercial Lidar mapping. *Photogrammetric Engineering and Remote Sensing*, 67(11), pp. 1209–1217.

Miller, B., 1965 Laser altimeter may aid photo mapping. *Aviation Week & Space Technology*, March 29, 1965, 4 pp.

Petrie, G., 1990 Laser-based surveying instrumentation & methods, in *Engineering Surveying Technology*, Kennie, T.J.M. and Petrie, G. (Eds.), Chapter 2, Blackie, Glasgow & London and Halsted Press, New York, pp. 48–83.

Price, W.F. and Uren, J., 1990 *Laser Surveying*. Van Nostrand Reinhold (International). 256 pp.

Shepherd, E.C., 1965 Laser to watch height. *New Scientist*, April 1, 1965, 33 pp.

Toth, C.K., 2004 *Future Trends in LiDAR*. ASPRS Annual Conference, Denver, CO, May 23–28, 2004. Paper No. 232, 9 pp.

Wehr, A. and Lohr, U., 1990 Airborne laser scanning—an introduction and overview. *ISPRS Journal of Photogrammetry and Remote Sensing*, 54(2/3), pp. 68–82.

Further Reading

Baltsavias, E.P., 1999a Airborne laser scanning: Existing systems, firms, & other Resources. *ISPRS Journal of Photogrammetry and Remote Sensing*, 54(2–3), 164–198.

Baltsavias, E.P., 1999b Airborne laser scanning: Basic relations & formulas. *ISPRS Journal of Photogrammetry & Remote Sensing*, 54(2–3), 199–214.

Jensen, H. and Ruddock, K.A., 1965 Applications of a Laser Profiler to Photogrammetric Problems. Brochure published by Aero Service Corporation, Philadelphia, 22 pp.

Thiel, K-H. and Wehr, A., 2004 *Performance Capabilities of Laser Scanners—An Overview & Measurement Principle Analysis*. Proceedings of ISPRS WG 8/2 Workshop—"Laser Scanners for Forest & Landscape Assessment", Freiberg, Germany, October 3–6, 2004.

2

Airborne and Spaceborne Laser Profilers and Scanners

Gordon Petrie and Charles K. Toth

CONTENTS

2.1 Introduction

As already outlined in the introductory section of Chapter 1, the direct measurement of terrain from an airborne platform using laser rangefinders began with the laser profiling technique. This involved the measurement of successive ranges continuously from an aircraft vertically downwards toward the terrain along the ground track of the aircraft. Essentially these measured ranges gave the successive values of the ground clearance between the aircraft and the terrain. If the corresponding values of the successive positions and altitude of the aircraft in the air from which the laser ranges were made could be measured accurately, for example, using an EDM system, then this allowed the determination of the profile of the terrain along the ground track. The development of the profiling technology and methodology was an important first step that led eventually to the development of the airborne laser scanners that are now in widespread use for topographic mapping purposes. However it should be noted that a number of airborne laser profilers are still in use for scientific research purposes. In the case of spaceborne platforms, given the great distances (flying heights) and very high speeds (29,000 km/h) at which these platforms orbit the Earth, until now, only laser profilers can be operated from them. Laser scanners cannot be used from spacecraft as yet—given the present limitations in the laser ranging technology that is currently available.

Valuable though the terrain profiles were (and are), they have obvious limitations when large areas of terrain have to be measured either to form a terrain elevation model or to form part of a mapping project. However, once suitable scanning mechanisms had been devised successfully in the early to mid-1990s, airborne laser scanners were soon designed and built and they quite quickly started to come into operational use for the determination of terrain elevation values over large swaths of terrain. The practical implementation of the airborne laser scanning technique was also made possible through the successful concurrent development of integrated GPS/inertial measuring unit (IMU) systems that could determine the successive positions, altitudes and attitudes of the scanner systems while they carried out their scanning/ranging measurements. After their introduction in the mid-1990s, airborne laser scanners were at first adopted gradually and rather cautiously. However, since 2002, as the technology has improved and experience was gained, they have been brought into service in ever increasing numbers—in spite of their considerable cost of purchase. At the time of writing this chapter at the end of 2007, the authors estimate that over 200 airborne laser scanners are in operational use world wide. Thus they are now a major element in the current topographic mapping scene.

The objective of this chapter is to present to the reader an account of the technology of airborne and spaceborne laser profilers and scanners as it exists at the end of 2007. The arrangement of the chapter will be to first cover airborne and spaceborne laser profilers. After which, airborne laser scanners will be dealt with in much greater detail and at greater length—as befits their relative importance in carrying out mapping operations. The coverage of airborne laser scanners will be subdivided into two parts—those used for topographic mapping and bathymetric mapping operations respectively.

2.2 Airborne Laser Profilers

Soon after the first lasers had been constructed and demonstrated, they began to be used for rangefinding purposes. In particular, laser altimeters were devised for use in aircraft—for example, in the United Kingdom, an experimental laser altimeter using a GaAs semiconductor laser was demonstrated as early as 1965 (Shepherd, 1965). This device could measure flying heights of 1000 ft (300 m) above ground level (AGL) to an accuracy of 5 ft (1.5 m). Shortly after that, the first airborne laser profiler was introduced for use in commercial topographic mapping operations (Miller, 1965; Jensen and Ruddock, 1965). The instrument had been developed jointly by the Spectra Physics Company—which had built the laser—and the Aero Service Corporation—then a major aerial survey and mapping company that was owned by Litton Industries. The rangefinder part of the system was based on a helium-neon gas laser operating as a continuous wave (CW) device at the wavelength of 632.8 nm. The output signal was modulated using a KDP crystal that allowed the radiation to be emitted simultaneously at three different frequencies −1, 5, and 25 MHz. In each case, the return signals reflected from the ground were received and compared with the transmitted reference signal to determine the respective phase differences, so determining the actual range. To complete the laser profiling system, since GPS did not exist at that time, first of all a sensitive barometric pressure sensing device was added to measure the absolute value of the aircraft altitude while the laser rangefinder was measuring the range from the aircraft to the ground. Next, a strip camera, boresighted and mounted integrally with the laser rangefinder, was added to provide continuous linescan imagery of the profile line over the ground. Quite surprisingly, the system could still receive the return signals from the ground at altitudes up to 15,000 ft (4.5 km), though it was usually operated at much lower altitudes—in the range 1000–5000 ft (300–1500 m). In view of later developments, it was interesting to note that a rangefinder based on the use of a pulsed ruby laser was also tested during the development of the system. However, it was not utilized in the final system due to the low pulse repetition rate—since only a few pulses per second could be generated at that time.

2.2.1 Development of Airborne Laser Profilers

By the mid-1970s, the situation had changed entirely and pulsed laser rangefinders based both on Nd:YAG solid-state lasers and on GaAs semiconductor lasers had been developed that could be used to form the basis of airborne laser profiling systems. As an example, the Avco Everett company from Everett, Massachusetts produced its Avco Airborne Laser Mapping System in 1979 (McDonough et al., 1979). This utilized an Nd:YAG laser as the basis for its rangefinder and used a two-axis gyro to measure the aircraft attitude and to help locate the position of the laser spot on the ground. The position and attitude of the profiling system in the air was tracked continuously using a microwave ranging system mounted on the aircraft, which continuously measured the ranges to transponders located at three known base stations on the ground.

Another slightly later example from the 1980s was the PRAM III laser profiler, which was built by Associated Controls & Communications (later Dynatech Scientific) of Salem, Massachusetts. The laser rangefinder used in this profiler was based on a GaAs semiconductor diode laser operating in a pulsed mode at $\lambda = 904\,\text{nm}$ with a pulse width of 30 ns and pulse repetition rates of 1, 2, or 4 kHz. Various examples of the PRAM III had either Honeywell or Litton or Ferranti IMUs integrated into the system; for example, Nortech Surveys of Calgary, Canada used a Ferranti Inertial Land Surveyor (FILS-II) in combination with its PRAM III. Various types of microwave ranging systems were also utilized by different operators to provide the continuous accurate positioning of the system in the air. During the 1980s, the partnership of LeSchack Associates and Photo Science Inc. were prominent suppliers of commercial terrain profiling services in the United States using a PRAM III system.

Yet another well known airborne profiling system from the 1980s was the Model 501 SX built by the Optech company based in Toronto, Canada, which is now such a major force as a system supplier of airborne laser scanners. The Model 501 SX used a GaAs semiconductor diode laser operating at the wavelength of 904 nm with a 15 ns pulse width and measuring rates up to 2 kHz. One of these profilers was used as the basis for the Airborne Laser Profiling System (ALPS) devised by Dr. Joachim Lindenberger (1989), then of the University of Stuttgart, to carry out extensive tests of the system in collaboration with the State Land Survey Office of Baden-Württemberg. The system used a Sercel GPS receiver to provide the positioning of the airborne platform (a Dornier Do 28) using the differential GPS technique in combination with a Delco Carrousel IMU to provide the attitude measurements of the platform and its ALPS system. It will be seen later that this system, with its early use of GPS, contained most of the elements of a modern airborne laser scanning system. All it lacked was a scanning mechanism. Two or three years later, this additional mechanism was indeed provided by Optech, resulting in the company's ALTM 1020 system introduced in 1993 and sold to TopScan, a German company that had been set up by Lindenberger and his colleague, Peter Friess (Wever and Lindenberger, 1999).

2.2.2 Current Types of Airborne Laser Profilers

It may well seem that, given the extra coverage of the ground provided by the cross-track scanning method, the airborne laser profiler would by now (in 2007) have disappeared. However, a number still remain in operational service, being used typically for the mapping of barely changing surfaces such as water bodies, ice-covered terrain and flat land areas. A notable example is the GGL Airborne Laser Profiler of the Geodesy and Geodynamics Laboratory (GGL) of the ETH, Zürich (Geiger et al., 2007). This system is also based on an Optech Model 501 profiler which is used in conjunction with Trimble or NovAtel GPS receivers; a Sperry gyro platform to give attitude values; and a flux gate compass to provide values of the aircraft heading. Currently, the GGL system is mounted on a Twin Otter aircraft utilizing four GPS receivers to provide attitude data in combination with the gyro platform, besides generating the usual positional data. For this particular application, two of the GPS antennas are wing mounted, one on each of the engine covers on the wing,

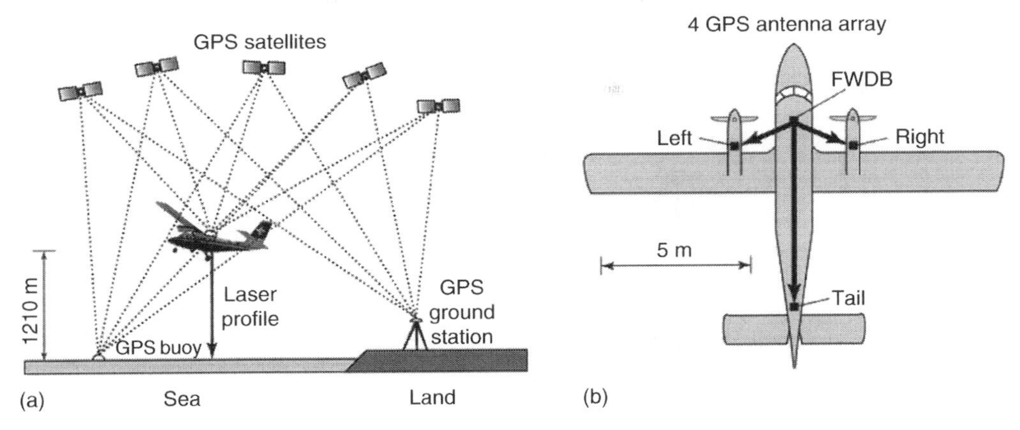

FIGURE 2.1
(a) Shows the overall concept of the airborne profiling being carried out by the GGL airborne laser profiler of the ETH, Zürich. (b) Shows the arrangement of the four GPS receivers that are mounted on the Twin Otter aircraft to provide positional and attitude data. (From Institute of Geodesy and Photogrammetry, ETH, Zürich; redrawn by M. Shand.)

while the third is mounted on the front of the aircraft's fuselage above the cockpit and the fourth is placed on top of the tail of the aircraft (Figure 2.1). The system has been used extensively to measure the sea surface topography of parts of the Mediterranean, Ionian and Aegean Seas around Greece in connection with regional studies of the geoid in this area and to carry out the calibration of the radar altimeters carried on the TOPEX/Poseidon and Jason-1 satellites. It should be noted that a number of the flights have been made to carry out simultaneous measurements of the local gravity field using a LaCoste/Romberg airborne gravimeter—which is also a profiling type of measuring instrument.

For a number of users, a big attraction of the airborne laser profiler is its very much lower cost, even though its coverage is more limited than that of the airborne laser scanner. Thus a number of American research groups have continued to use them for scientific research studies, for example to measure longitudinal profiles over glaciers, as well as executing geophysical research work simultaneously—in a similar manner to that being carried out by the ETH, Zürich group mentioned above. An early version of NASA's Airborne Topographic Mapper (ATM) from the late 1980s was a profiling laser altimeter complete with a laser ring gyro and a GPS receiver that was mounted on a Lockheed P-3 aircraft. This was used by NASA to measure longitudinal profiles of active temperate glaciers, including Breidamerkurjokull in south Iceland. More recently, a group from the Byrd Polar Research Center at the Ohio State University has carried out airborne laser profiling of ice streams occurring within the ice cap of West Antarctica (Spikes et al., 1999). For this task, they employed a profiler based on an Azimuth LRY laser rangefinder, which utilizes a diode-pumped Nd:YAG pulsed laser; a dual-frequency Ashtech GPS receiver and a Litton LTN-92 IMU unit. Yet another group from the Geophysical Institute of the University of Alaska has used a similar airborne laser profiling system that they have developed in-house for use in small aircraft to measure a series of profiles over more than 60 glaciers located in Alaska, British Columbia and the Cascades and Olympic Mountains in Washington State.

Forest research is another area where the use of airborne profilers continues. Still another system—called Portable Airborne Laser System (PALS)—has been developed recently for this application by NASA's Goddard Space Flight Center (GSFC) using low-cost commercial-off-the-shelf (COTS) components (Nelson et al., 2003). This system

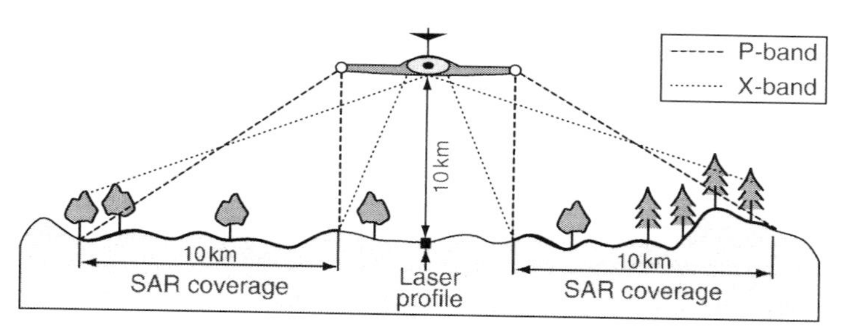

FIGURE 2.2
Cross-sectional diagram showing the ground coverage of the GeoSAR airborne radar imaging system that has now been equipped with a laser profiler. (From Fugro EarthData; redrawn by M. Shand.)

is based on a Riegl LD90-3800-VHS laser rangefinder with a 20 ns wide laser pulse and having a pulse repetition rate of 2 kHz. This is being used in combination with a hand-held Garmin III GPS receiver, together with a beacon receiver that receives the correction data being broadcast by U.S. Coast Guard beacons for the differential corrections of the positional measurements made by the GPS receiver. A Pulnix color video camera is also integrated into the PALS system. According to Nelson et al., the total cost of the hardware of the whole system was in the region of $30,000—which may be compared with the costs in the region of $850,000–$1,300,000 for a full-blown airborne laser scanning system from one of the mainstream system suppliers. The PALS system has been used extensively both in the United States and abroad, operating from low altitudes for the measurement of forest height along each profile line.

Another interesting recent development is the incorporation of an airborne laser profiler into the GeoSAR synthetic aperture radar system operated by the Fugro EarthData survey and mapping company. The GeoSAR features a dual-sided SAR, comprising two X-band and two P-band systems, collecting IfSAR image data on either side of the aircraft's flight line. This is supplemented by the nadir-pointing laser profiler that provides additional high-accuracy elevation data of the ground directly under the aircraft's flight path (Figure 2.2). This data can be used for sensor calibration or for quality control purposes by providing reference data for neighboring strips (Figure 2.3).

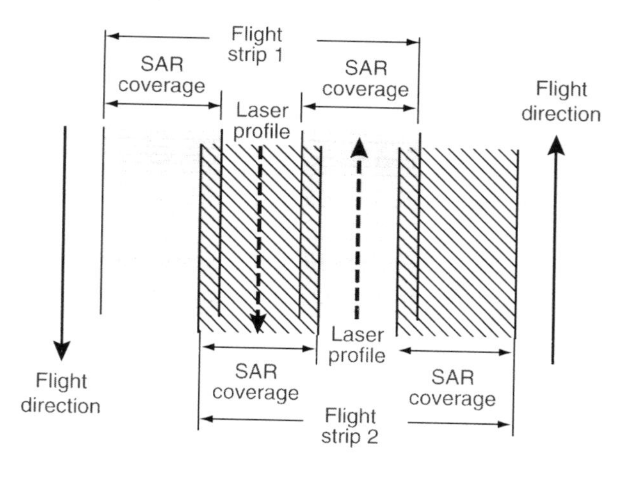

FIGURE 2.3
Plan view showing the ground coverage of both the SAR imagery and the laser elevation profiles being acquired by the GeoSAR system for two adjacent flights. (Drawn by M. Shand.)

2.3 Spaceborne Laser Profilers

A number of microwave (radar) profilers have been mounted in satellites and have been used very successfully to measure sea surface topography over large areas of the world's oceans. They include the well known TOPEX/Poseidon and Jason-1 missions operated jointly by NASA and the French CNES space agency and the various radar altimeters operated by the European Space Agency on board the ERS-1 and-2 and Envisat satellites. These microwave profilers work less well over areas of land and ice that are characterized by large variations in elevation. This is due in large part to the comparatively large angular beam widths of these microwave radar instruments and their resulting large area footprints on the ground. Thus a substantial group of scientists concerned with research studies of large ice sheets, such as those covering Greenland and Antarctica, campaigned for the development and use of laser altimeters (profilers) from satellites, since they appeared to offer much narrower angular beams and smaller ground footprints over the ice-covered terrain than those produced by microwave (radar) profilers. In turn, this would lead to much finer resolution (from the smaller footprints) and higher accuracies in the resulting elevation data. Besides which, the recording of the returned waveform would allow studies of surface roughness to be undertaken. Other scientists concerned with research into desert topography and the movement of sand-based geomorphological features were also interested in the improved elevation data that could be generated by spaceborne laser profilers. Furthermore, atmospheric scientists also pressed to have a laser profiler operated from space to measure cloud structures and aerosol layers. NASA responded to these various pressures by the scientific community by conducting a series of experimental missions using laser profilers mounted on various Space Shuttle spacecraft. This led ultimately to the development and deployment of a dedicated satellite (Ice, Cloud and Land Elevation Satellite [ICESat]) equipped with a specially designed laser profiler, as will be discussed later.

2.3.1 LiDAR In-Space Technology Experiment (LITE)

The characteristics of a laser profiler suitable for operation from space needed of course to be substantially different to those of an airborne profiler. In particular, the operational height over which the distance had to be measured would be in the order of 250–800 km instead of the few kilometers of an aircraft flying height, while the speed over the ground would be 29,000 km/h instead of the 200–300 km/h of an aircraft (Figure 2.4). NASA's first experimental mission was the LITE (LiDAR In-space Technology Experiment) which was orbited on board the Shuttle flight STS-64 in 1994. This used large diameter (1 m) optics that were very heavy, together with a very powerful flashtube-pumped Nd:YAG laser that required considerable electrical power (3 kW)—all of which could be coped with by the Shuttle spacecraft. In the event, the mission concentrated mainly on satisfying the requirements of atmospheric, weather and climatic research scientists through the measurement of cloud height and aerosol and dust layers rather than the Earth's topography.

2.3.2 Shuttle Laser Altimeter (SLA)

However the two follow-up missions that were accommodated on Shuttle flights in 1996 and 1997 were much more oriented toward topographic applications, including the measurement of Earth surface relief and vegetation canopies. For these applications, both

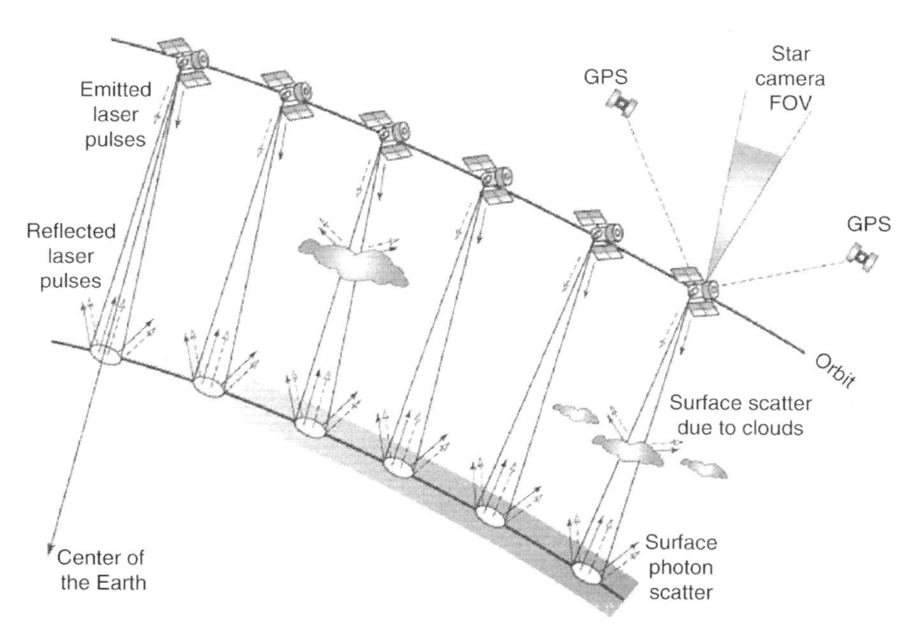

FIGURE 2.4
Diagram showing the operational mode of a spaceborne laser profiler with the transmitted pulses being reflected both from the ground and from the cloud cover. (From NASA; redrawn by M. Shand.)

missions used the SLA. This profiler employed a much smaller, lighter and less power-hungry laser rangefinder based on a Q-switched diode-pumped Nd:YAG laser (with $\lambda = 1064\,\text{nm}$) and a much smaller (38 cm) diameter optical telescope. The pulse energy of this laser was 40 mJ and the pulse repetition rate was 10 Hz. The first of these two missions—called SLA-01—was flown in January 1996 on Shuttle flight STS-72 and lasted 10 days. During this period, the rangefinder emitted 3 million pulses toward the Earth's surface during the 80 h of its operation. The area of its coverage lay between 28.45° North and South latitude, corresponding to the Shuttle's orbital inclination (i) of 28.45°. Only 475,000 of the pulses resulted in geolocated laser returns from land surfaces. A major problem with the SLA profiler that was experienced during this first mission was the deep saturation of the detector and its electronics arising from the strong levels of reflection that occurred over deserts and other flat surfaces. However, notwithstanding these difficulties, the SLA-01 mission was judged to be successful.

Based on the experience gained during the SLA-01 mission, the profiler was modified extensively for the SLA-02 mission that was flown on the Shuttle STS-85 flight which took place over a 12 day period in August 1997 (Carabajal et al., 1999). In particular, the laser rangefinder was fitted with a Variable Gain-state Amplifier (VGA) that allowed the signal intensity that was passed to the detector to be controlled from the mission ground control station to prevent the detector becoming saturated. Furthermore, for the SLA-02 mission, the rangefinder was fitted with a waveform digitizer instead of the simple first-return ranging scheme that had been employed during the SLA-01 mission. The time interval counter that was used to measure the round-trip travel time of the emitted pulses was coupled to the waveform digitizer. This allowed a better recognition of the leading edge of the reflected pulse that was used to establish the range to the highest detected surface within the profiler's 100 m diameter footprint on the Earth's surface. The pulses from the SLA profiler generated these 100 m footprints of the ground at a rate of 10 pulses per second (10 Hz) giving a spacing of 700 m between successive footprints. Nearly 3 million pulses were emitted from the profiler during the SLA-02 mission, yielding 590,000 geo-located returns from the land and more than 1.5 million returns from the ocean surfaces. With the Shuttle having an orbital inclination of 57°, much more of the Earth's surface was covered by the SLA-02 mission than was achieved with the previous SLA-01 mission (Harding et al., 1999).

2.3.3 Geoscience Laser Altimeter System (GLAS)

Based on the experience gained from the SLA missions, the GLAS was developed. It was launched in January 2003 as the sole instrument on board the ICESat. This has the primary mission of monitoring the Earth's ice sheets with a view to determining their mass balance and their contribution to sea level change. Changes in the elevations of regional ice sheets are being determined by comparison of the successive sets of elevation profile data measured using the GLAS instrument. A secondary objective of the mission is to obtain detailed information on the global distribution of clouds and aerosols. GLAS is equipped with three separate laser rangefinders that are fitted to a common optical bench. These rangefinders were designed to be operated independently so that, if one failed, another could then be used to replace it. In the event, this foresight has proven to be invaluable since the first of the three rangefinders did in fact fail after only 38 days of operation and the other two were then brought into operation to replace it.

Each of the GLAS rangefinders uses a Q-switched diode-pumped Nd:YAG laser with a pulse width of 6 ns that operates at the dual wavelengths of $\lambda = 1064$ and 532 nm in the infrared and green parts of the spectrum respectively (Abshire et al., 2005). These lasers are operated with a pulse repetition frequency (PRF) of 40 Hz and produce pulse energies of 74 and 36 mJ, respectively at the two different wavelengths. The lightweight beryllium mirror of the receiver telescope has a diameter of 1 m. An analogue detection scheme using an SiAPD detector is employed in the case of the reflected radiation of the infrared pulses—which are used primarily for the measurement of ground elevation. However, photon counting is used in the case of the reflections of the green pulses, which are used mainly to determine the heights of the optically thin cirrus clouds and aerosol layers that are also being observed and measured from the satellite. The pulses emitted from the laser rangefinder at its orbital height of 600 km illuminate a ground area that is 70 m in diameter, spaced at 170 m intervals along the orbital track (or profile line) over the Earth's surface (Figure 2.4). The accuracy of the ranging measurements is estimated to be better than ±10 cm; note however that, given the 70 m footprint size, this is hardly an achievable value even for flat surfaces. The SLA missions obtained their positional and attitude data from the ground tracking of the Shuttle spacecraft and from its internal navigation system. By contrast, the ICESat obtains the corresponding data in a rather different manner. In the first place, the positional data is generated using on-board GPS receivers carrying out differential measurements in conjunction with an extensive network of GPS receivers located at base stations on the ground. These measurements of the satellite's position are supplemented by the measurements made by a network of satellite laser rangers located at base stations on the ground. These track the satellite and measure the ranges to an array of nine corner cube reflectors mounted on a hemispherical surface on the ICESat (Figure 2.5). The attitude of the satellite and its laser rangefinders is measured continually using star tracking cameras mounted on the upper (zenith) side of the ICESat that constitute its Stellar Reference System (SRS) (Figure 2.6).

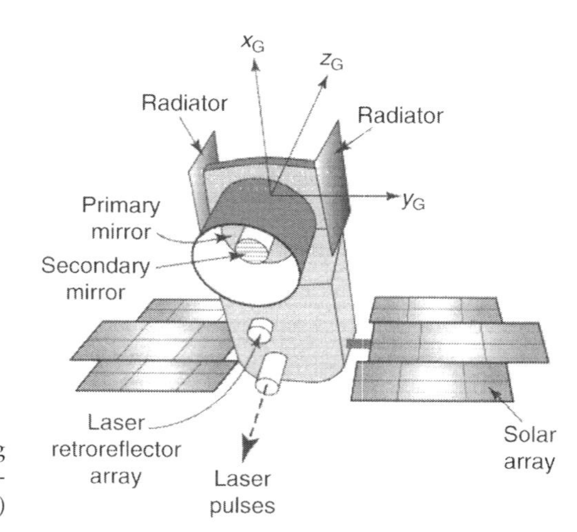

FIGURE 2.5
The diagram provides a nadir view of the ICESat, showing the main features of the satellite and its GLAS laser profiler. (From NASA; simplified and redrawn by M. Shand.)

After the launch of the ICESat in January 2003, the first of the three GLAS laser rangefinders was brought into operation during the following month. After a few weeks of operation, its performance began to drop and, after 38 days, the laser stopped emitting pulses. A very detailed investigation of this failure was carried out by NASA. A thorough and painstaking analysis of the housekeeping telemetry records reached the conclusion that the failure was caused by manufacturing defects in the array of diodes that pumped the laser. Since the other two lasers had been fitted with the same type of diode arrays, it was decided to operate them over shorter time periods—i.e., for three 33 day periods per year or 27% of the year. Over the first 3 years of the operation of the laser profiler, the GLAS profiles from the ICESat mission have dramatically improved the amount and quality of elevation data available for the Antarctic and Greenland ice sheets—with dense coverage over Antarctica and northern and central Greenland (Abshire et al., 2005). In addition, as a secondary activity, the GLAS instrument has measured numerous profiles over many of the Earth's vegetated areas (Harding and Carabajal, 2005).

2.3.4 Multi-Beam Laser Altimeter (MBLA)

In parallel with the development of GLAS and ICESat, NASA, in cooperation with the University of Maryland, developed its Vegetation Canopy Lidar (VCL) mission as part of

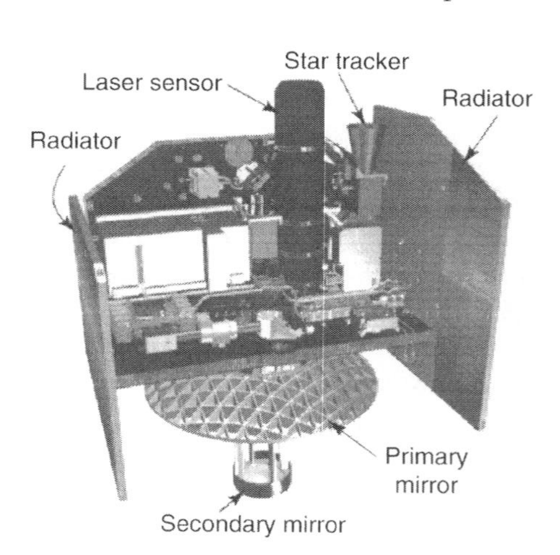

FIGURE 2.6
A CAD drawing of the main body of ICESat when viewed from the side and showing the main elements of the GLAS laser profiler. (From NASA; additional annotation by M. Shand.)

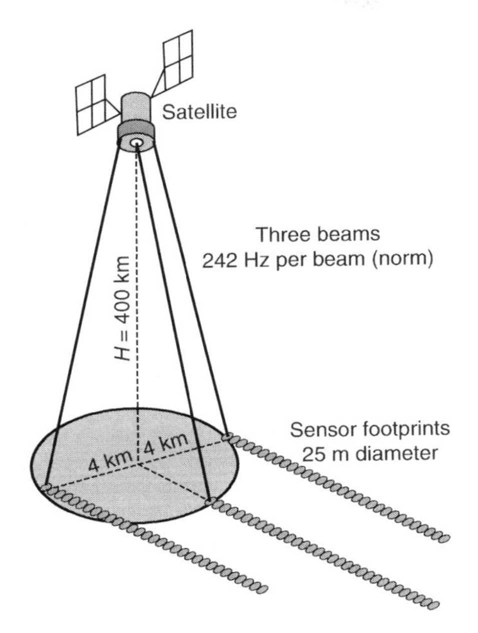

FIGURE 2.7
Diagram showing the ground coverage of the three spaceborne laser profilers of the MBLA as proposed for use in the NASA VCL mission (From NASA; redrawn by M. Shand.)

its Earth System Science Pathfinder (ESSP) Programme. As the title suggests, its main goal was the measurement of the vertical and horizontal structures of vegetation canopies and of their underlying topography. The MBLA that was developed to undertake these measurements comprised three laser rangefinders that were placed in a circular configuration that covered 8 km diameter on the ground. Each of the three laser rangefinders was designed to generate a profile of contiguous measured points in the along-track direction that were spaced 4 km apart on the ground (Figure 2.7). Each instrument utilized a diode-pumped Q-switched Nd:YAG laser, generating pulses that were 5 ns in width, had a pulse energy of 10 mJ and featured a pulse repetition rate of 290 Hz. The ground footprints of each emitted pulse would be 25 m. The diameter of the optical telescope receiving the reflected signals from the ground was 0.9 m, while the receiver itself comprised a detector with a waveform digitizer. Like the ICESat, the VCL satellite was to use a combination of a GPS receiver and satellite laser ranging for the measurement of its position. It was also to use a fiber-optic gyro (FOG) to control the orientation and direction of the laser pulses, together with a stellar camera to measure the actual attitude of the laser rangefinder. Unfortunately, although the VCL satellite and the MBLA hardware were built, the VCL mission was first delayed and then cancelled in 2003.

2.4 Airborne Laser Scanners

2.4.1 Overall System Configuration

A typical airborne laser scanner system comprises the following main components:

1. The basic laser ranging unit, including its transmitter and receiver optics, that has already been discussed in Chapter 1.
2. An optical scanning mechanism such as a rotating mirror (together with its angular encoder) that is used to carry out the scanning of the terrain in the cross-track direction.

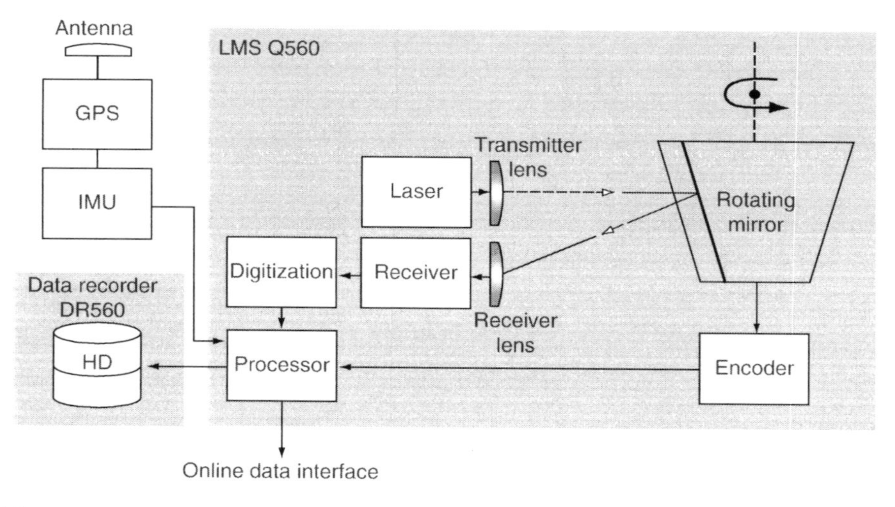

FIGURE 2.8
This diagram shows the relationship between the main hardware components of an airborne laser scanner system—using the Riegl LMS-Q560 as the model. (From Riegl; redrawn by M. Shand.)

3. An electronics unit that comprises various hardware elements that provides many of the control and processing functions of the overall system.
4. A positioning and orientation system comprising an integrated differential GPS/IMU forms an essential element of the total airborne laser scanning system—without which it cannot be operated in any really practical or useful manner.
5. The software that will be used to control and coordinate the operation of each of the main elements of the system and to carry out the recording, storage and preliminary processing of the measured data collected in-flight is a further essential component of the overall system. Furthermore, substantial processing is needed postflight in order to create the LiDAR point cloud.
6. An imaging device such as a digital camera, video camera or pushbroom line scanner often forms an integral part of the overall airborne laser scanning system.

The overall configuration and the relationship of the major hardware components or units within the overall system are given in Figure 2.8.

2.4.2 System Components

Next a more detailed discussion of the role that each of these system components or elements plays in the overall laser scanning system will be undertaken.

1. The technology and operation of laser ranging units has already been discussed in Chapter 1. This rangefinder unit used in the airborne laser scanner will include the actual laser; the transmitting and receiving optics; and the receiver with its detector, time counter and digitizing unit. In practice, because of the relatively large distances over which the slant ranges to the ground have to be measured from an airborne platform, almost invariably, powerful pulsed lasers are used for the

required time-of-flight measurements of range (Pfeiffer and Briese, 2007). An exception to this is the ScaLARS laser ranger of the Institute of Navigation of the University of Stuttgart, which uses a CW laser, and the phase comparison method of measuring the range.

2. On optical scanning mechanism is attached rigidly in front of the output end of the laser ranging unit. As the name suggests, it uses an optical element such as a rotating plane or polygon mirror or a fiber-optic linear array to send a stream of pulses of laser radiation at known angles and at high speed along a line crossing the terrain in the lateral or cross-track direction relative to the airborne platform's flight path. This allows the sequential measurements of a series of ranges and the corresponding angles to be made to successive points along this line allowing a profile of the ground to be constructed along that line. The forward motion of the aircraft or helicopter on which the laser scanner is mounted allows a series of range measurements to be made along successive lines in the cross-track direction. The combination of the successive sets of range and angular measurements made in both the along-track and cross-track directions allows the elevations of the ground surface and its objects to be determined for a wide swath of the terrain.
3. The electronics unit is normally computer-based and is equipped with a display and an operator interface through which commands can be given to the system to execute specific actions. Besides the dedicated computer, certain electronic hardware components or cards may also be incorporated into the unit. In addition to controlling the operation of the laser rangefinder and the scanning mechanism, this unit controls the data recording system that collects and stores the measured time and waveform data from the laser rangefinder and the angular data from the scanning mechanism, together with the corresponding data from the GPS receiver and the IMU. The storage of all of this data is normally carried out using the hard disk or solid state drive of the computer, although earlier systems often recorded and stored their data on digital tape drives.
4. As its title suggests, the position and orientation system records various items of raw navigation data—including GPS observations, linear and angular accelerations and angular encoder values—that are measured continuously in-flight. After the mission has been completed, the accurate position, altitude and attitude of the laser rangefinder and its attached scanning mechanism will be determined during a postflight computer processing operation. The hardware side of the system comprises a closely integrated combination of a GPS receiver and an IMU. In many cases, the GPS receiver takes the form of a specially designed card (or cards), which is mounted on a rack within the electronics unit. However, stand-alone and/or additional receivers can also be used. The differential GPS processing requires ground reference GPS data that can be provided in any one of the four ways: (1) via a separate GPS receiver that is being operated at a base station on the ground; (2) using a wide-area service such as OmniSTAR that employs satellite broadcast technology to provide corrections in real-time, based on a worldwide network of reference stations; (3) utilizing data provided by a local network of reference stations, a solution that is increasingly becoming popular in the industrialized world due to the widespread availability of CORS (Continuously Operating Reference Stations) networks; and (4) employing PPP (Precise Point Positioning) technology that is based on the use of precise orbits and clock corrections for the

satellites provided by IGS, based on worldwide network of reference GPS stations. If a wide-area service such as OmniSTAR is used, then either an OmniSTAR-ready GPS receiver or an additional OmniSTAR L-band receiver is needed. Wide-area services allow for in-flight processing, while all the other solutions are based on postmission processing; note that, in some cases, the precise corrections may be available only after longer time periods, such as days or weeks.

5. Software is required for various purposes and for use in different stages of the overall airborne laser scanning operation. Initially a package or module will be provided for mission planning and for the implementation of the planned flight lines by the pilot. This latter operation will be carried out in conjunction with the data from the GPS receiver using a suitable cockpit display for the pilot who is flying the mission. This allows him to make the necessary adjustments to the actual path being flown by the aircraft to ensure that it is as close as possible to the planned flight line. A further module will allow the settings of various parameters of the laser rangefinder such as its scan rate, pulse rate and scan angle to be controlled by the system operator. Additional software will also be provided for the collection and storage of all the data being measured by these different parts of the overall system. The synchronization and time-tagging of this data in-flight is normally provided by the GPS receiver. Finally software needs to be provided to execute any preliminary processing and display of the raw data that needs to be carried out in-flight as well as that required for postflight processing (Figure 2.9).

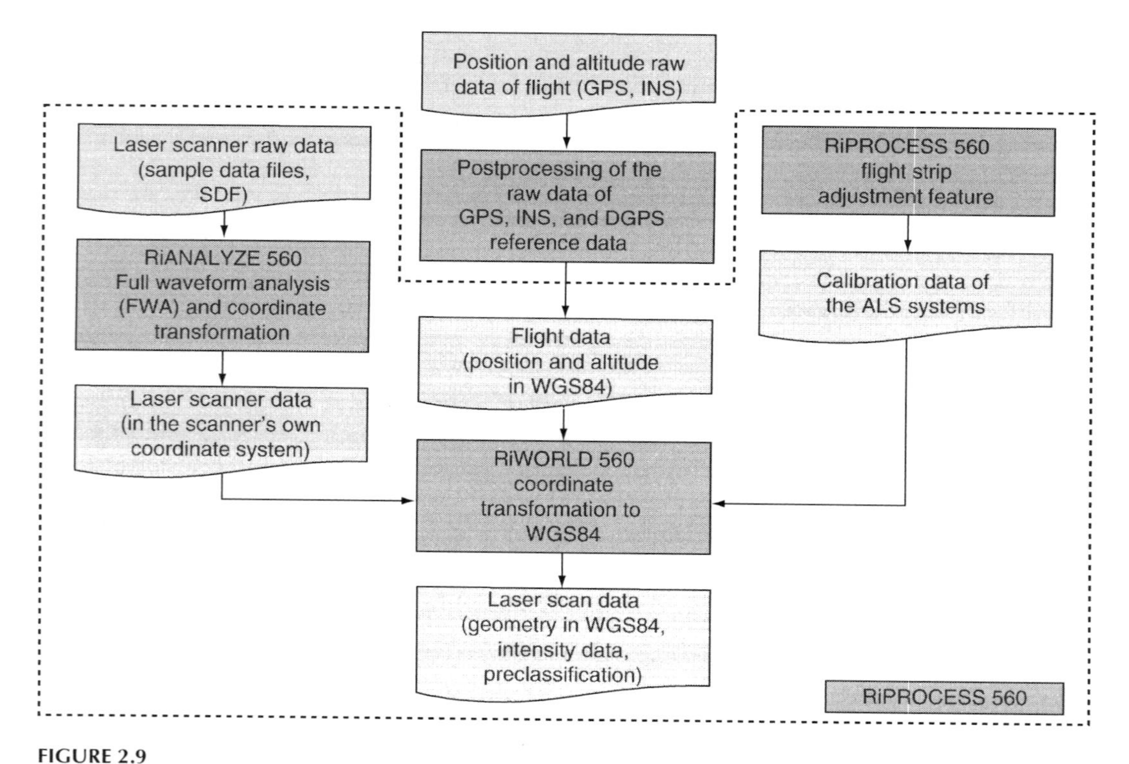

FIGURE 2.9
Diagrammatic representation of the processing operations being carried out by a representative software package—the Riegl RiPROCESS 560. (From Riegl; redrawn by M. Shand.)

6. The most commonly used imaging device that will form an essential component of the airborne laser scanning system is a small- or medium-format digital frame camera. Usually it will be attached rigidly to the same mount as the laser ranging unit and its operation will be closely integrated with that of the rangefinder and the scanning mechanism. However, occasionally, as an alternative, either a digital video camera or even a pushbroom line scanner—as with the TopoSys Falcon II system—may be incorporated into the system to generate the required imagery.

2.5 Airborne Topographic Laser Scanners

Taking the broadest view, airborne laser scanners may be divided into two distinct groups. The first of these comprises those systems that are designed solely for topographic mapping operations. The second group consists of systems that are designed principally to carry out measurements of water depth in the bathymetric mapping operations undertaken for marine or lacustrine charting applications. However these charting operations will often include some topographic mapping of the coast or shoreline areas.

2.5.1 Primary, Secondary, and Tertiary Classifications

The basis for the primary classification of the airborne laser scanning systems used for topographic mapping purposes that has been adopted here is to divide them into three distinct categories:

1. Scanner systems that have been manufactured in substantial numbers for commercial sale by the major system suppliers.
2. Custom-built systems that have been built and operated in-house by service providers, usually in relatively small numbers.
3. Airborne laser scanners that have been developed as technology demonstrators, principally by NASA and, on the application side, have been used mainly for scientific research purposes.

It is useful also to note another secondary classification based on the different types of scanning mechanisms that are being utilized to scan the ground and measure its topography. The scanning action of a particular mechanism defines the pattern, the spacing and the location of the points being measured both on the ground surface and on the objects that are present on it.

Besides this classification, it should be borne in mind that a further tertiary classification can also be envisaged on the basis of the range of flying heights over which the system can be used. In practice, this is based largely on the maximum distance or slant range that can be measured by the laser ranging unit. Thus a certain number of airborne laser scanning systems with relatively short-range capabilities are designed primarily for operation from relatively low altitudes and at slow flying speeds, most often using a helicopter as the airborne platform. Often such systems are used primarily to carry out the corridor mapping of linear features such as roads, railways, rivers, canals, power lines, etc. At the other end of the operational range are those systems designed for commercial operation over a much greater range of flying heights at higher altitudes up to 6 km (20,000 ft). These tend to be used in the topographic mapping of substantial areas of the Earth's terrain.

2.5.2 Scanning Mechanisms and Ground Measuring Patterns

With regard to the secondary classification outlined above, four main scanning mechanisms, each with its own distinctive ground measuring pattern can be distinguished, as follows:

1. Either a single mirror or a pair of oscillating plane mirrors is used in the systems that have been constructed by the two largest commercial suppliers of airborne laser scanning systems—Optech with its ALTM series and Leica Geosystems with its Aeroscan and ALS scanners. The precise angle that the mirror makes with the direction of the vertical is measured continuously using an angular encoder. The use of this type of bidirectional scanning mechanism results either in a Z-shaped (saw-toothed) pattern or in a very similar sinusoidal pattern of points being measured on the ground (Figure 2.10).
2. An optical polygon that is continuously spinning in one direction providing a unidirectional scanning motion is used in the various systems such as the IGI LiteMapper, TopoSys Harrier, iMAR,... etc.—that utilize the Riegl laser ranging and scanning engine, besides the complete airborne laser scanning systems that are supplied by the Riegl company (Figure 2.11). Similarly, the Terrapoint company uses multi-faceted polygons in the several ALTMS systems that it has constructed in-house to form the basis of its early airborne laser scanning services. The use of this type of scanning device results in a series of parallel lines of measured points being generated over the ground (Figure 2.12). The constant rotational velocity of the optical polygon means that there is no repetitive acceleration or deceleration of

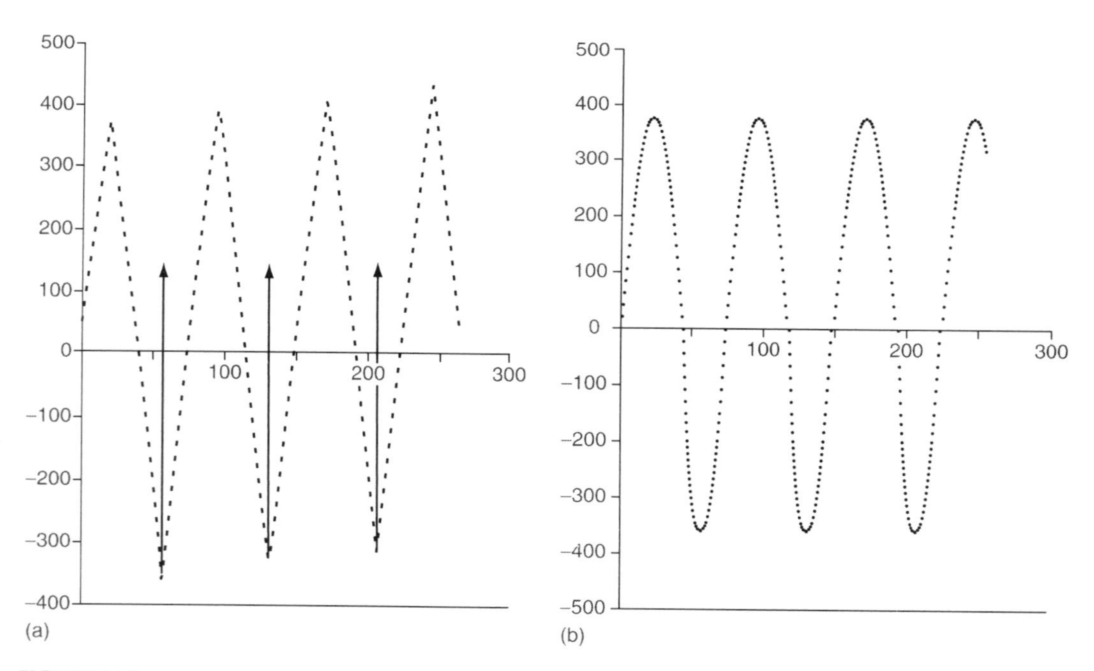

FIGURE 2.10
(a) Saw-toothed pattern over the ground that is produced by the Optech ALTM series of laser scanners; and (b) Sinusoidal pattern produced by the Leica ALS laser scanners—in both cases, using oscillating mirrors as the scanning mechanisms. (From Optech; redrawn by M. Shand.)

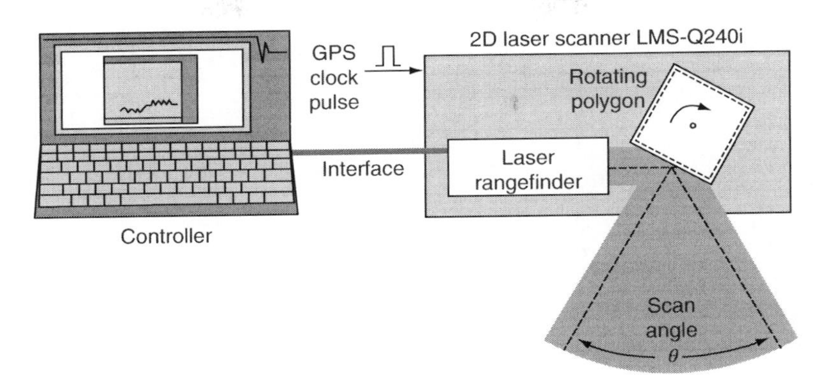

FIGURE 2.11
Diagram showing the unidirectional rotating optical polygon that forms the scanning mechanism that is used in Riegl laser scanning engines. (From Reigl; redrawn by M. Shand.)

the mirror. This provides for high frequencies such as the 160 Hz reached by the Riegl LMS-Q560 laser scanner. In turn, this offers better control over the spacing of the LiDAR points. The RASCAL airborne laser scanner constructed and operated by NASA also uses a polygon mirror. However this does not spin continuously in one direction. Instead it oscillates in a unidirectional mode over a range of ±16° from the nadir. Thus it falls into the previous category.

3. A nutating mirror producing an elliptical scan pattern—the so-called Palmer scan (Figure 2.13)—over the ground is used in NASA's ATM and Airborne Oceanographic Lidar (AOL) series of scanners; in the ScaLARS system of the University of Stuttgart; in the Geographia Survair systems; and in the later models in the TopEye series of airborne laser scanners. This produces a series of overlapping elliptical scans over the ground (Figure 2.14).
4. A pair of linear fibre-optic arrays is employed in the Falcon systems that have been built and operated in-house by the TopoSys company (Figure 2.15). A pair of tilted mirrors driven by a motor distributes and collects the pulses being sent to and from the arrays. The smaller size of the mirror allows a higher scan rate to be

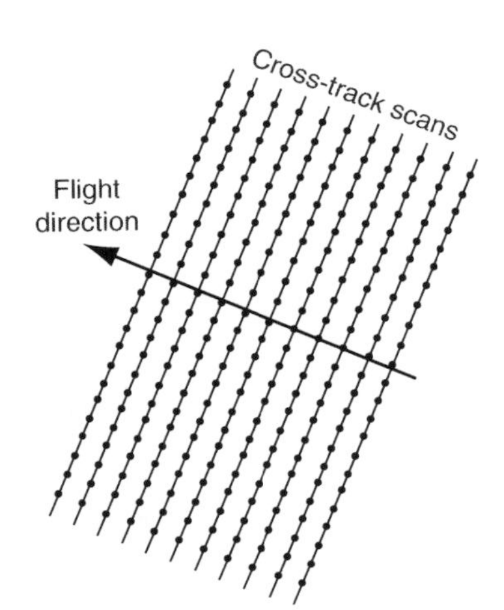

FIGURE 2.12
The resulting raster scanning pattern that covers the ground. (Drawn by M. Shand.)

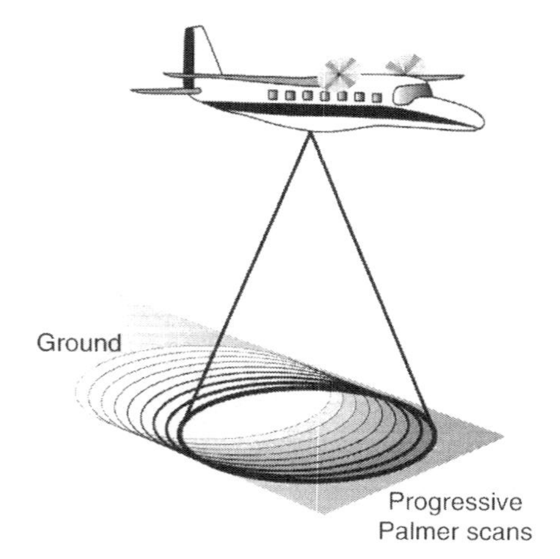

FIGURE 2.13
Diagram showing the elliptical ground scanning pattern and coverage of an airborne laser scanning system that is utilizing progressive Palmer scans. (Drawn by M. Shand.)

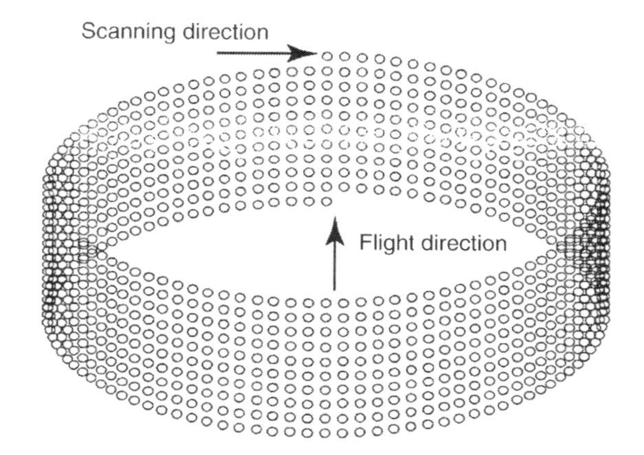

FIGURE 2.14
LiDAR point distribution on the ground of an airborne laser scanner using progressive Palmer scans. (Drawn by M. Shand.)

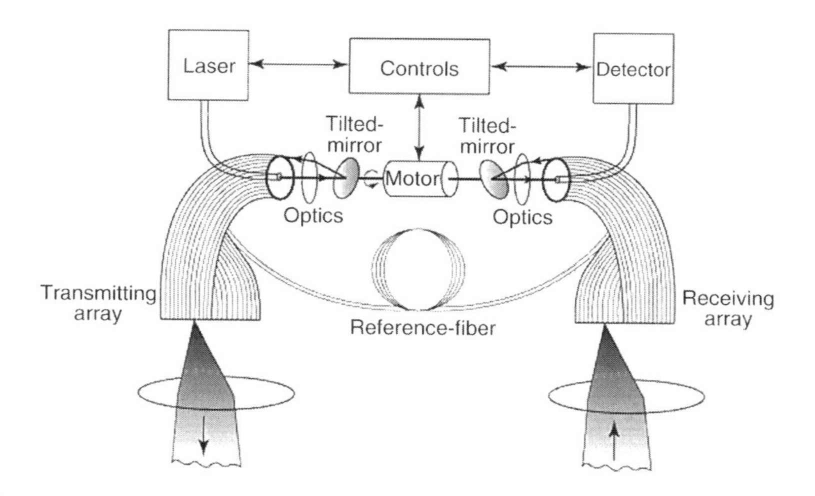

FIGURE 2.15
The optical and mechanical elements of the TopoSys Falcon series of airborne laser scanners are based on the use of linear fiber-optic arrays. (From TopoSys; redrawn by Mike Shand.)

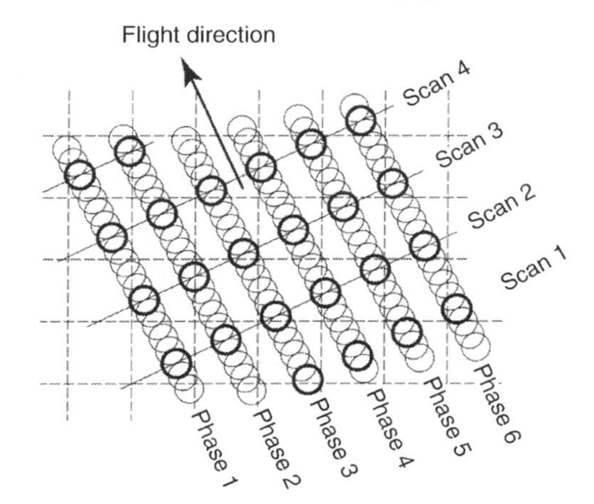

FIGURE 2.16
The resulting pattern of points measured over the ground by the Falcon system is a series of lines parallel to the fight direction. (From TopoSys; redrawn by M. Shand.)

implemented. This arrangement results in a series of scan lines that run parallel to the flight line as the measuring pattern (Figure 2.16). However, as will be seen later, this basic pattern has been modified somewhat in the later models in the series through the use of an additional swing mirror.

2.5.3 Commercial System Suppliers

At the present time, there are three main commercial suppliers of airborne laser scanner systems. The two principal suppliers in terms of volume are Optech International Inc. and Leica Geosystems. The third supplier, Riegl, also supplies complete systems. However, the company has also supplied quite a large number of laser scanning engines (comprising a laser rangefinder and a scanning mechanism) to a number of German system suppliers on an OEM basis. Besides which, Riegl has also sold a number of these laser scanning engines to several service providers in North America. They have then added the other required components that are available on a COTS basis to build up the final complete systems in-house.

2.5.3.1 Optech

This company, based in Toronto, Canada, is by far the largest commercial supplier of airborne laser scanners, having supplied over 100 of its Airborne Laser Terrain Mapper (ALTM) topographic scanner systems worldwide up to the time of writing this chapter. The first model in the series—ALTM 1020—was introduced in 1993. For some years prior to this, the Optech company had been involved in a number of laser research and development projects, including the construction and supply of laser rangefinders and airborne laser profilers such as the Model 501 profiler discussed above. Since its introduction, the basic design of the ALTM has stayed much the same, but with each new model in the series, the main elements of the design have been steadily improved and refined. The result of these developments has been a marked improvement in almost all of the operational characteristics of the ALTM series with a very substantial increase in performance and accuracy. A typical operational system showing its major components and operational parameters is shown in Figure 2.17.

The basic design of the ALTM series features a rangefinder employing a powerful Class IV pulsed laser emitting its radiation at $\lambda = 1047\,\text{nm}$ (Figure 2.18). In late 2006, Optech introduced its multiple pulse technology which allows the firing of a second laser pulse by

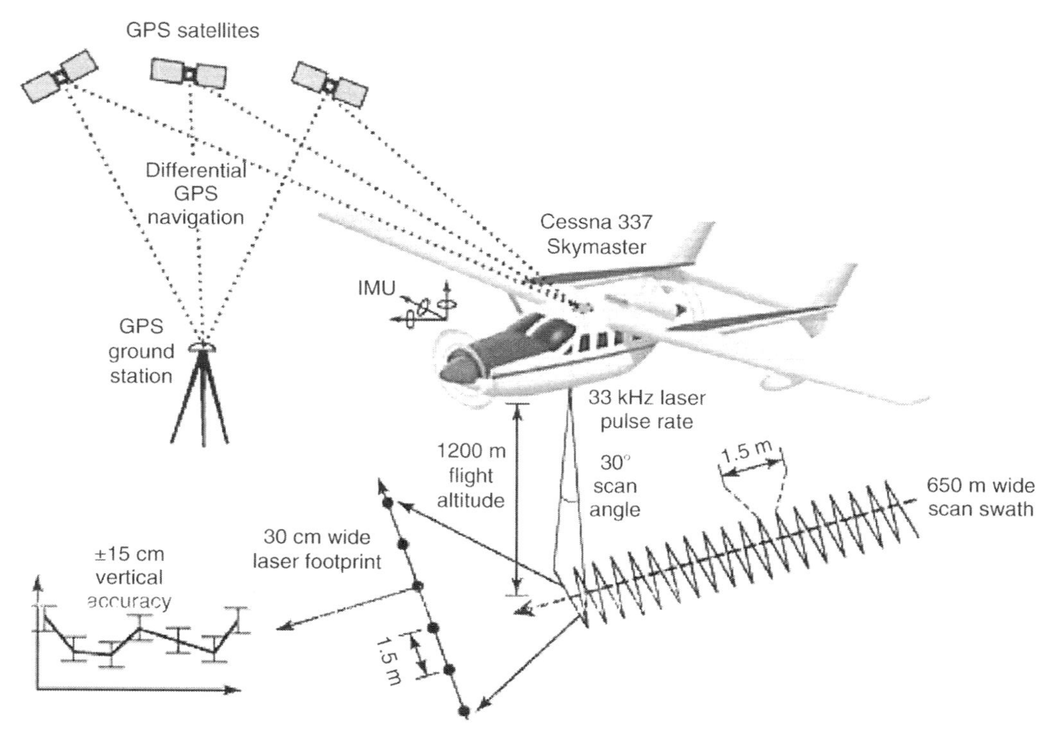

FIGURE 2.17
Diagram showing the operational concept of an Optech ALTM airborne laser scanner, including the use of GPS in the aircraft and at the ground base station. Note the zigzag pattern of the points being measured on the ground. (From Florida International University; redrawn by M. Shand.)

the rangefinder before the reflected signal from the previous pulse has been received and detected by the system (Toth, 2004). This has allowed the use of a much higher pulse repetition rate—in this case, 167 kHz—to be reached in the latest ALTM Gemini model. The ALTM scanners all employ oscillating plane mirrors as the mechanism for scanning

FIGURE 2.18
The laser head of an Optech ALTM 3033 airborne laser scanner installed in an observation hatch located in the floor of a helicopter. (From Helica, Italy.)

FIGURE 2.19
The principal elements of an Optech ALTM laser scanning system with the laser scanner unit located at right and the case containing the control electronics at left. In front of these two units are a small display screen that is used in the navigation of the aircraft in-flight (left) and a laptop computer that is used for flight management and data storage. (From Optech.)

the ground in the cross-track direction. This results in a saw-tooth pattern of measured points over the ground (Jenkins, 2006). The position and orientation systems used throughout the ALTM series are the POS/AV systems supplied by the Applanix company, a near neighbor of the Optech company in the Toronto area. A variety of dual-frequency GPS receivers—from Trimble, Ashtech, NovAtel, etc.—that are suitable for use in aircraft have been utilised in the POS/AV systems supplied for incorporation in the different ALTM models. Similarly, with the IMUs, different units—from Litton (now Northrop Grumman), Honeywell and Sagem—have been used, depending on the specified accuracy and the particular market into which the ALTM is being sold. In this last respect, licences for the sale of American-manufactured IMUs from Litton and Honeywell will not be granted by the relevant U.S. government agencies—the State Department or the Department of Commerce—if the scanner is to be sold to certain countries. The accuracies offered by the different versions of the POS/AV system that have been used in conjunction with the ALTM scanners are summarized in Table 2.1.

TABLE 2.1
POS/AV Absolute Accuracy Specifications—RMSE Values

Model Number	210	310	410	510	610
Position (m)	0.05–0.3	0.05–0.3	0.05–0.3	0.05–0.3	0.05–0.3
Velocity (m/s)	0.01	0.075	0.005	0.005	0.005
Roll and pitch (°)	0.04	0.015	0.008	0.005	0.025
True heading (°)	0.08	0.035	0.015	0.008	0.005

Source: From Optech and Applanix.
Notes: The lower-end POS/AV 210, 310, and 410 systems all use MEMS quartz gyros; the POS/AV 510 uses FOGs; while the POS/AV 610 uses ring laser gyros.

The steadily improving performance of the different models of the ALTM scanners over the years, can be seen, for instance, in the substantial rise in the PRF—from 5 kHz in the initial ALTM 1020 model from 1993 to 10 kHz in the ALTM 1210 model and 25 kHz in the ALTM 1225 model from 1999. Since then, the frequency has risen first to 33 kHz in the ALTM 2033 and 3033 models dating from the first 3 years of the twenty-first century to the 70 kHz of the ALTM 3070 model; the 100 kHz rate of the current ALTM 3100 models and the 167 kHz rate of the newly introduced ALTM Gemini model with its multiple-pulse technology (Petrie, 2006). Similarly the scan rates have risen from 28 Hz in the early ALTM models to 70 Hz in the current models. In operational terms, the maximum operational altitude has also risen from 1000 m with the ALTM 1020 model and 1200 m in the ALTM 1210 and 1225 models to 3500 and 4500 m, respectively with the current ALTM 3100 and Gemini models. While the quoted range resolution of the rangefinder has remained throughout at ±1 cm, a big improvement has been achieved in terms of the actual point positioning accuracy that is achieved in object space. This is mainly due to better scanning and improved geo-referencing performance. The resulting improvements are especially noticeable in the horizontal accuracy which has steadily improved from 1/1000 of the flying height (H) in the initial ALTM 1020 model to 1/5500. H in the current ALTM 3100 EA and Gemini models. The elevation accuracy will also vary with flying height and the pulse repetition rate. The elevation accuracy values that can be achieved for the current ALTM 3100EA and Gemini models, as quoted by Optech, are given in Table 2.2.

A very large proportion of the ALTM scanners from Optech have been supplied with a digital frame camera. This is fitted rigidly to the mount or base plate of the scanner and is used to acquire imagery of the ground simultaneously during the operation of the scanner. While some of the earlier ALTM systems utilized small-format digital cameras from the Kodak DCS series, in recent years, Optech has offered medium-format digital cameras in the form of the DSS (Digital Sensor System) series from Applanix and the various AIC (Aerial Industrial Camera) models produced by Rollei (now Rolleimetric). The earlier versions of both these cameras produced colored frame images via Bayer interpolation that were 4k × 4k pixels (16 megapixels) in size. The later DSS 322 model from Applanix produces either color or false-color images that are 4k × 5.4k pixels (22 megapixels) in size, while the latest DSS 439 model in the series features a 39 megapixel back. The corresponding model of the AIC camera from Rolleimetric is equipped with the P45 digital back from Phase One which also generates colored images that are 7.2k × 5.4k pixels (39 megapixels) in size.

2.5.3.2 Leica Geosystems

This major supplier of instrumentation and systems to the whole of the surveying and mapping industry entered the airborne laser scanning field through its purchase of the Azimuth Corporation based in Massachusetts in January 2001. Azimuth was a small specialist company that originally supplied laser rangefinders that were incorporated into

TABLE 2.2

Elevation Accuracy—RMSE Values

Laser Repetition Rate (kHz)	500 m Altitude (cm)	1000 m Altitude (cm)	2000 m Altitude (cm)	3000 m Altitude (cm)	4000 m Altitude (cm)
33	<5	<10	<15	<20	<25
50	<5	<10	<15	<20	N/A
70	<5	<10	<15	N/A	N/A
100	<10	<10	<15	N/A	N/A

Source: From Optech.
Note: The quoted accuracies do not include GPS errors.

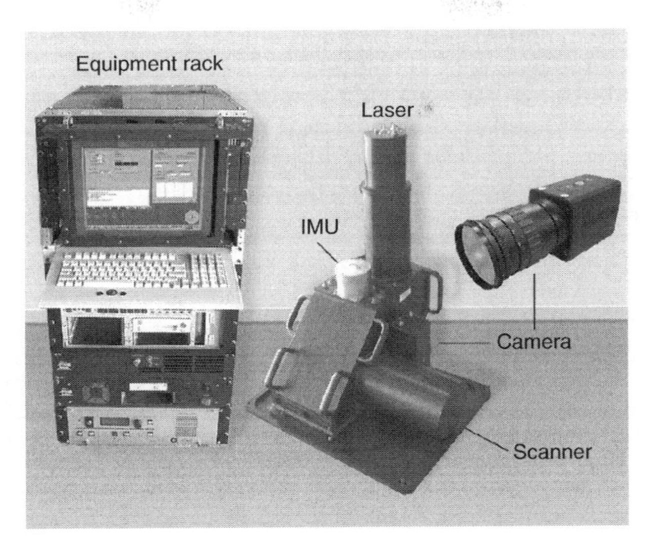

FIGURE 2.20
An Azimuth laser scanner forms part of the RAMS system built in-house by the Enerquest company, Denver, Colorado, now part of Spectrum Mapping. (From Spectrum Mapping.)

various custom-built airborne laser scanners such as the RAMS systems that was built in-house by the Enerquest (later Spectrum Mapping) company based in Denver, Colorado (Figure 2.20). However, in 1998, the Azimuth company started to build complete airborne scanner systems, under the brand name AeroScan. After the company's acquisition by Leica Geosystems, this system was re-branded and sold as the ALS40 in parallel with Leica's ADS40 pushbroom scanner producing optical linescan imagery. In 2003, a new and much more compact model called the ALS50 was introduced (Figure 2.21). In May 2006, an improved second generation model, called the ALS50-II, was introduced (Petrie, 2006). This features multiple-pulse laser ranging technology. Recently the production of the ALS50 series was moved from Massachusetts to Leica Geosystem's main manufacturing plant located in Heerbrugg, Switzerland.

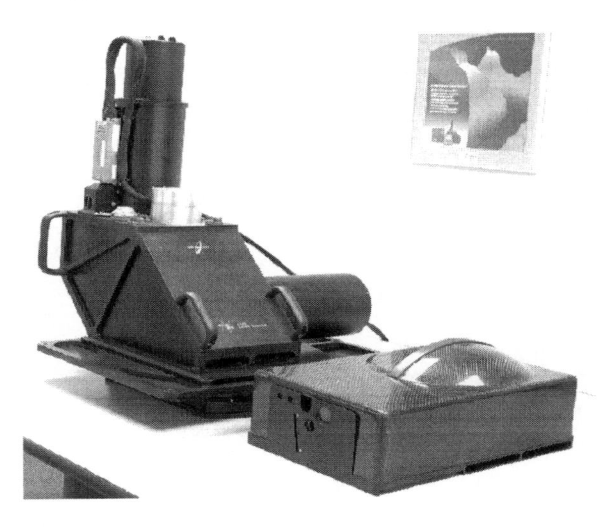

FIGURE 2.21
The difference in size (and weight) between the older Leica ALS40 laser scanner (left) and the newer ALS50 model (right) is quite striking. (From Spencer B. Gross.)

The original AeroScan and ALS40 systems used a large and powerful laser rangefinder emitting its pulses at the wavelength of 1064 nm. Indeed the special customized versions of the AeroScan built for the Fugro EarthData and 3001 aerial mapping companies in the United States could be operated at altitudes up to a maximum of 6 km. By contrast, the ALS50 has used a much more compact laser. With regard to the optical scanning mechanism, the ALS40 and ALS50 models both feature a bidirectional oscillating mirror that produces a sinusoidal pattern of measured points over the ground.

Regarding the position and orientation system, until recently, the various models of airborne scanners produced by Leica Geosystems all used one of the models in the POS/AV range of GPS/IMU systems from Applanix. However, in 2004, the Applanix company was acquired by Trimble, one of Leica's main commercial rivals in the surveying instrument field. This led Leica Geosystems to purchase (in 2005) the small specialist Terramatics company based in Calgary, Canada. This company had already developed its IPAS (Inertial Position and Attitude System) and this was quickly adapted for use in the ALS50 airborne laser scanning system (and also in Leica's ADS40 airborne pushbroom line scanner). As with the Applanix POS/AV system, a range of different IMUs with different capabilities can be utilized in an IPAS system according to the user's needs. These IMUs are supplied either by Northrop Grumman (formerly Litton) and Honeywell in the United States or from the European suppliers iMAR (Germany) and Sagem (France). Since the geo-referencing accounts for the largest part of the error budget, the accuracy values specified for each of these units are summarized in Table 2.3.

As with the Optech ALTM scanners discussed above, there has been a steady improvement in performance with the Leica ALS models over the years. Thus the PRF has risen steadily from the 15–25 kHz rate used in the AeroScan units and the initial models of the ALS40 to 38 kHz in the later models of the ALS40 from 2002. With the introduction of the ALS50 in 2003, the rate had increased to 83 kHz, while the introduction of the ALS50-II (Figure 2.22) with its multiple pulse technology in 2006 saw the maximum rate rise to 150 kHz. Similarly the maximum value of the scan rate has risen steadily from 26 Hz in the case of the original AeroScan units to 90 Hz with the ALS50-II. The ALS models have all had high values for their maximum operational altitude—with altitude values of 6 km with the AeroScan and ALS40 models;

TABLE 2.3

Specification and Accuracy Values for the Leica IPAS10 System

		NUS4	DUS5	NUS5	CUS6
Absolute accuracy after postprocessing (RMS)	Position	0.05–0.3 m	0.05–0.3 m	0.05–0.3 m	0.05–0.3 m
	Velocity	0.005 m/s	0.005 m/s	0.005 m/s	0.005 m/s
	Roll and pitch	0.008°	0.005°	0.005°	0.0025°
	Heading	0.015°	0.008°	0.008°	0.005°
Relative accuracy	Angular random noise	<0.05 deg/sqrt (h)	<0.01 deg/sqrt (h)	<0.01 deg/sqrt (h)	<0.01 deg/sqrt (h)
	Drift	<0.5 deg/h	<0.1 deg/h	<0.1 deg/h	<0.01 deg/h
IMU	High performance gyros	200 Hz FOG	200 Hz FOG	256 Hz Dry-tuned gyro	200 Hz Ring laser gyro
GPS receiver	Internal in IPAS10 control unit	12-channel dual frequency receiver (L1/L2) low noise, 20 Hz raw data, DGPS ready			

Source: From Leica Geosystems.

Notes: NUS4 is the iMAR FSAS unit; DUS5 is the Litton LN-200 unit; NUS5 is a Sagem unit; and CUS6 is the Honeywell MicroIRS unit.

FIGURE 2.22
Latest Leica ALS50-II airborne laser scanner is shown (right) together with its control electronics cabinet (left) and its aircraft-certified LCD displays that are used for flight management and system control purposes. (From Leica Geosystems.)

4km with the ALS50 and a return to 6km with the latest ALS50-II model. As one would expect, the continual improvements in the technology have also resulted in improved accuracies. Both the horizontal and elevation accuracy figures are directly related to flying height and are not too dissimilar to those quoted above for the latest Optech ALTM laser scanners.

A small-format digital frame camera with a 1024 × 1280 pixel format size is available as an optional item that can be fitted to the ALS50 model. However, many users of the ALS40/50 series have preferred to use larger-format digital frame cameras, especially the medium-format models available from Applanix (DSS), Rollei (AIC) and Spectrum (NexVue) (Figure 2.23), all of which have been integrated by Leica to operate in conjunction with its ALS50 scanners (Roth, 2005). In the summer of 2007, Leica started to supply its own RCD105 medium-format digital frame camera with a 39 megapixel back produced by an OEM supplier based in northeastern United States.

FIGURE 2.23
An ALS50-II laser scanner and a Spectrum NexVue medium-format digital camera mounted rigidly together on a base plate that sits an a set of anti-vibration and shock absorbing feet. (From Leica.)

2.5.3.3 Riegl

This company is based in the town of Horn, Austria. For many years, it has been engaged in the manufacture of laser measuring instrumentation such as distance meters, speed meters, levels, altimeters and anticollision devices for a wide range of industrial applications. In recent years, it has also developed a range of laser scanners for ground-based surveying applications. Within this particular field, according to a market survey made in 2005, the Riegl now occupies second place (after Leica Geosystems) in terms of sales of terrestrial 3D laser scanners. With all this relevant background allied to its extensive experience, Riegl entered the field of airborne laser scanners in 2003. However, it has done this in a very different manner to that of Optech and Leica Geosystems discussed above. Principally, it has developed a range of laser scanning engines, each comprising a laser rangefinder and a scanning mechanism together with the associated timing circuits and control electronics. It has supplied these laser scanner engines on an OEM basis to a number of system suppliers who have then added a position and orientation system, developed the appropriate software and integrated all these components to form a complete airborne laser scanning system. Besides which, Riegl has also developed its own complete system, which was first shown publicly at the end of 2006.

The Riegl company's laser scanning engines that are used in these various airborne systems come in two main flavors. The first of these is designed for relatively low-altitude applications—e.g., for detailed corridor mapping or power line surveys, typically using a helicopter as the airborne platform. The second is designed for operation from much higher altitudes for large-area surveys. Earlier models were the LMS-Q140 for low-altitude applications (Figure 2.24) and the LMS-Q280 for use at higher altitudes. The current low-altitude scanner engine is the LMS-Q240 model, which can be operated from flying heights up to 450 m with targets having an 80% reflectivity. The higher altitude model is the LMS-Q560 (Figure 2.25) which can be used at flying heights up to 1500 m with targets having an 80% reflectivity. This latter engine can also be fitted with a unit carrying out the recording of the full intensity waveform of the signal that has been reflected from the ground.

The Riegl laser rangefinder units feature powerful Class I pulsed lasers emitting their radiation at near infrared wavelengths ($\lambda = 900$ or 1550 nm) with repetition rates of up to 30 kHz in the case of the LMS-Q240 and up to 100 kHz with the LMS-Q560. All of these units utilize a scanning mechanism featuring a continuously rotating polygon block with a number of reflective surfaces giving a unidirectional scan of the terrain (Figure 2.11). The rotational speed of the reflective polygon is adjustable—in the case of the LMS-Q560 model to speeds between 5 and 160 Hz. The angular value of the rotating polygon block

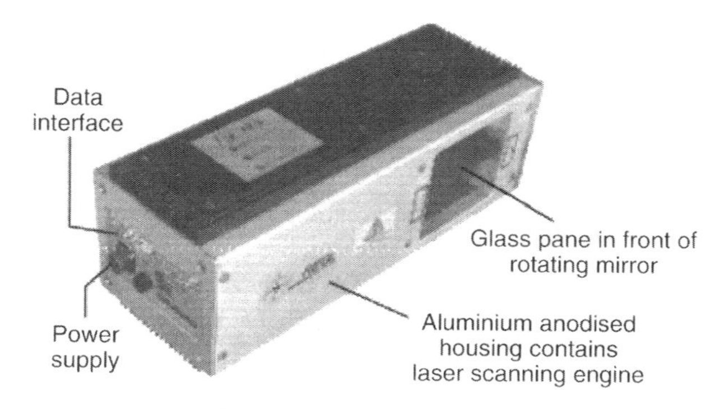

FIGURE 2.24
Riegl LMS-Q140 laser scanning engine designed for low-altitude operation. (From Riegl.)

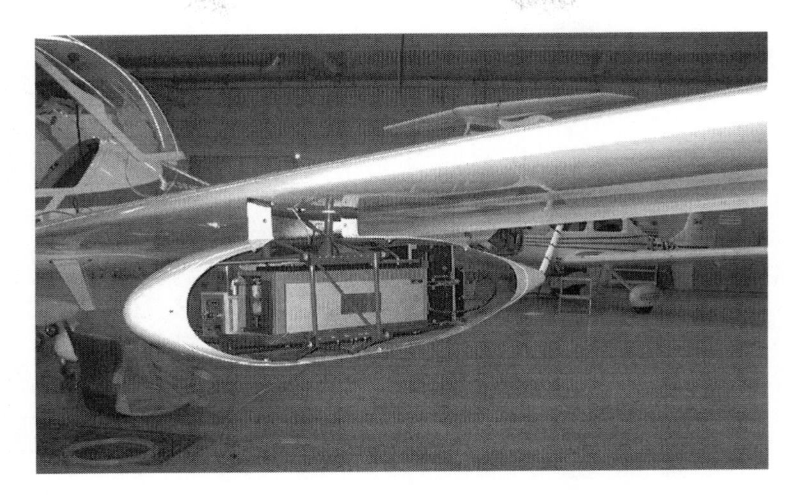

FIGURE 2.25
Riegl LMS-Q560 laser scanner unit fitted inside a pod attached under the wing of a Diamond HC36 Super Dimona single-engined aircraft. (From Riegl.)

can be read out continuously to a resolution of 0.001°. Based on these laser scanning engines, a number of airborne laser scanning systems have been developed by various German system suppliers—IGI, TopoSys and iMAR—who offer their systems for sale to service providers. They also form the basis of a number of systems built by two Canadian companies and at least one American company who use them for their operations in-house as service providers. It is interesting to note the variety of different solutions that have been developed using the same basic laser scanning engines from Riegl. In total, they amount to a substantial part of the market for commercially supplied systems.

Besides supplying the various laser scanning engines discussed above, late in 2006, Riegl also entered the market on its own account as a supplier of a complete airborne laser scanning system with the introduction of its LMS-S560 product. As shown in Figure 2.26, the version of this system that has been supplied to the Diamond Airborne Sensing company features two of Riegl's LMS-Q560 scanner engines, each able to operate with a PRF of

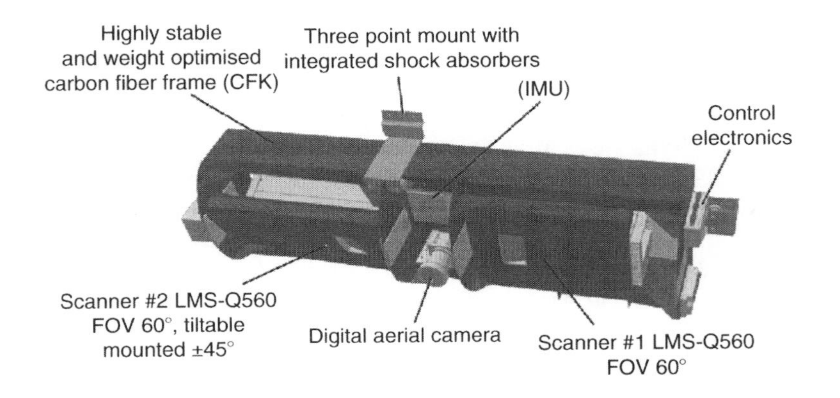

FIGURE 2.26
The carbon-fiber frame of this Riegl BP560 system is equipped with dual LMS-Q560 laser scanners separated by a medium-format digital frame camera and the system's IMU unit. The overall frame is placed on a three-point mount fitted with integrated shock absorbers and placed in a purpose-built belly pod. (From Riegl; redrawn by M. Shand.)

FIGURE 2.27
Diamond DA42 twin-engined multipurpose platform (MPP) aircraft is fitted with a belly pod containing the Riegl BP560 airborne laser scanning system. (From Diamond Aircraft.)

up to 200 kHz, resulting in a 400 kHz combined system pulse rate. These two scanner engines are mounted together on a specially built carbon fiber reinforced plastic (CFRP) frame. This frame can also accommodate the IGI AEROcontrol GPS/IMU unit used for position and orientation purposes and a medium-format digital frame camera, in this case, from Rolleimetric. One of the scanner engines sits in a separate tilt mount within the frame. This allows it to be tilted through a range of ±45° to improve the data acquisition possibilities in mountainous areas. The whole system including the CFRP frame is enclosed in a very stiff and aerodynamically optimized CFRP cover to form a pod that is fitted to the underside of the twin-engined Diamond DA42 Multi-Purpose Platform (MPP) aircraft (Figure 2.27).

2.5.3.4 IGI

This company, based in Kreuztal, Germany has developed its LiteMapper airborne scanner systems based on the Riegl engines discussed above. To these basic rangefinder/scanning units, IGI has added its own CCNS (Computer Controlled Navigation System) and its AEROcontrol GPS/IMU system. The latter uses IMUs based on FOGs that are constructed by the German Litef company, while the 12-channel dual-frequency L1/L2 GPS receivers are from Ashtech or NovAtel. These component systems are integrated together with a purpose-built, computer-based LMcontrol unit for the control and operation of the scanner (Figure 2.28). To these hardware components of the overall system, IGI has added its own software modules developed in-house—WinMP (for use with CCNS); AEROoffice (for use with AEROcontrol); and LMtools (for in-flight data registration). The systems resulting from the integration of all these components are the older LiteMapper 1400 and 2800 models for low- and high-altitude operations respectively, based on Riegl's earlier LMS-Q140 and LMS-Q280 engines and the newer LiteMapper 2400 and 5600 models which are based on Riegl's later LMS-Q240 and LMS-Q560 engines (Figure 2.28). The LiteMapper models can all be supplied with a digital frame camera—in this case, IGI's own DigiCAM model based on a Hasselblad camera with its digital backs from Imacon producing either 22 megapixel or 39 megapixel frame images.

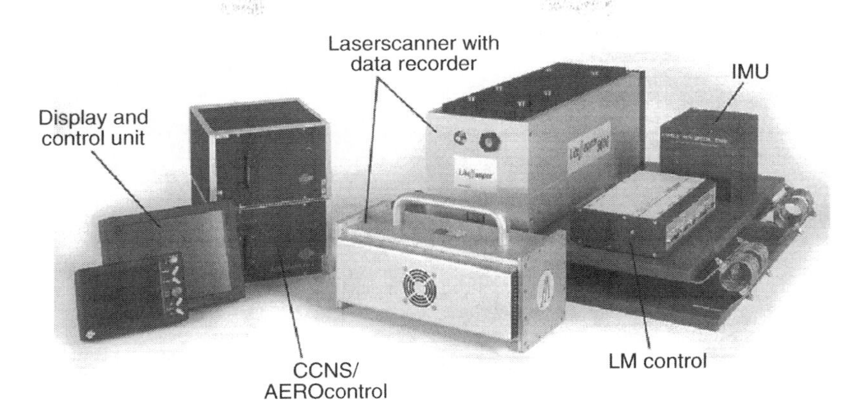

FIGURE 2.28
This IGI LiteMapper airborne laser scanning system is mounted on a purpose-built anti-vibration platform. The Riegl laser engine is the large white box sitting at the back with the AEROcontrol IMU unit placed alongside and to the right of it. The waveform data recorder and digitizer sit in front of the platform. At the left side of the picture are the CCNS airborne navigation and guidance system and the display screens. (From IGI; annotation by M. Shand.)

2.5.3.5 TopoSys

This is another German company based in Biberach located in the southern part of the country. TopoSys is best known for its Falcon airborne laser scanners based on the use of linear fiber-optic arrays (that will be discussed later). However the company's Falcon scanners have mainly been used in-house for the provision of airborne laser scanning services. In 2005, TopoSys decided to move into the business of building and supplying systems to service providers, based on the Riegl LMS-Q240 and LMS-Q560 laser scanning engines. The two resulting products are the Harrier 24 and 56 systems respectively. The position and orientation systems that are being utilized in these scanner systems are the POS/AV

FIGURE 2.29
An helicopter fitted with a specially built Helipod box that is mounted on an outrigger frame and contains an IGI LiteMapper system. (From IGI.)

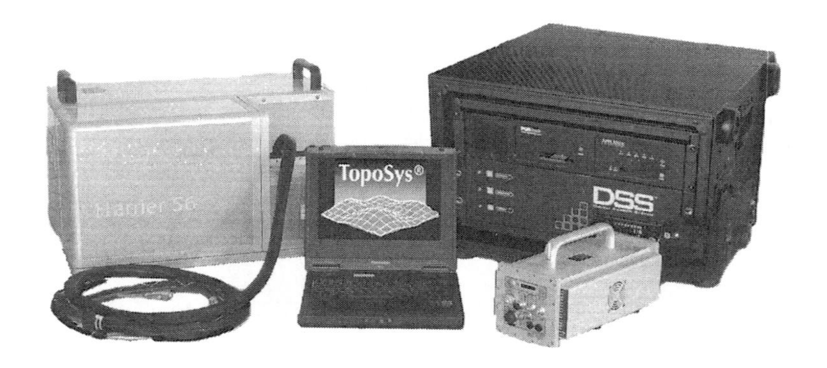

FIGURE 2.30
The TopoSys Harrier 56 system is based on a Riegl laser scanning engine, shown on the left of the picture. On the right is the control electronics unit for the Applanix DSS camera that is also offered optionally as part of the system. In front of this electronics unit is the waveform digitizer and recorder, which is also supplied by Riegl. (From Applanix.)

units from Applanix—which also supplies its DSS 322 medium-format digital frame camera for incorporation into those Harrier systems as an optional item (Figure 2.30), if requested by customers. An alternative camera of this type that can be supplied together with the Harrier is the Rolleimetric AIC (Figure 2.31).

2.5.3.6 iMAR

The iMAR company is located in St. Ingbert in the western part of Germany. It is very well known as a constructor of inertial systems having different capabilities that are used extensively in military, commercial and industrial applications. The company has also built a number of inertial systems for the surveying of roads and railways. In 2005, iMAR entered the airborne laser scanning field by introducing its iAIRSURV-LS1000 system. Again, this is based on the Riegl laser scanning engine, which, in this case, is used in conjunction with a GPS/IMU unit based on a ring laser gyro of iMAR's own manufacture and a dual-frequency GPS receiver from Javad or NovAtel. The system can also have a medium-format digital frame camera integrated with it in the form of a 22 megapixel camera supplied by Rollei.

FIGURE 2.31
Example of a Harrier 56 system that is enclosed in a special frame with the Riegl scanning engine located underneath and a Rolleimetric AIC medium-format digital frame camera bolted rigidly to the side of the frame. (From Rolleimetric.)

2.5.3.7 Helimap

Yet another European airborne laser scanner that makes use of a Riegl laser scanning engine is the Helimap system, developed originally by the Photogrammetry and Topometry laboratories of the Federal Institute of Technology of Lausanne (EPFL). This is a highly unusual hand-held laser scanning system that has been developed specifically for the mapping of areas of steep slopes in mountainous terrain that constitute a natural hazard—such as avalanche slopes, rock falls, landslides and glaciers—which are usually located in inaccessible areas (Vallet and Skaloud, 2004, 2005). The system is operated from low altitudes from a helicopter. In its original form, the Helimap system comprised the following four main elements—a Riegl LMS-Q140 laser scanner engine; a Javad dual-frequency L1/L2 GPS receiver; a Litton LN-200 IMU; and a Hasselblad medium-format camera equipped with a Kodak 16 megapixel digital back. The measured data from the GPS receiver and the IMU were passed via a specially built interface-cum-logger to be recorded on a portable PC. The system has since been upgraded to feature a Riegl LMS-Q240 scanner; an IMU from iMAR and a Hasselblad H1 camera equipped with an Imacon digital back generating 22 megapixel images. All of these components are mounted rigidly together on a light and compact carbon-fiber and aluminum frame so that they maintain a constant stable relationship with one another (Figure 2.32). The frame with its instrumentation is hand-held by an operator seated on the side of the helicopter, who, if required, can point the whole system obliquely to the side to measure steep slopes (Figure 2.33). Although the system was built and operated originally for scientific research purposes, it has also been operated commercially for the corridor mapping of roads, railways, power lines, pipelines, etc. by the Swiss UW+R engineering survey company (Skaloud et al., 2005).

2.5.3.8 Terrapoint

Originally, the TerraPoint company was founded in 1998 in Houston, Texas, utilizing airborne scanning technology developed by NASA. In 2004, it was purchased by a Canadian company,

FIGURE 2.32
Hand-held Helimap system comprises a Riegl laser scanning engine and a Hasselblad medium-format digital frame camera, mounted side-by-side on a rigid carbon-fiber frame. There are two handles for the operator to hold which are located on either side of the frame. The system IMU is enclosed in the box situated directly below the camera. (From EPFL, Lausanne.)

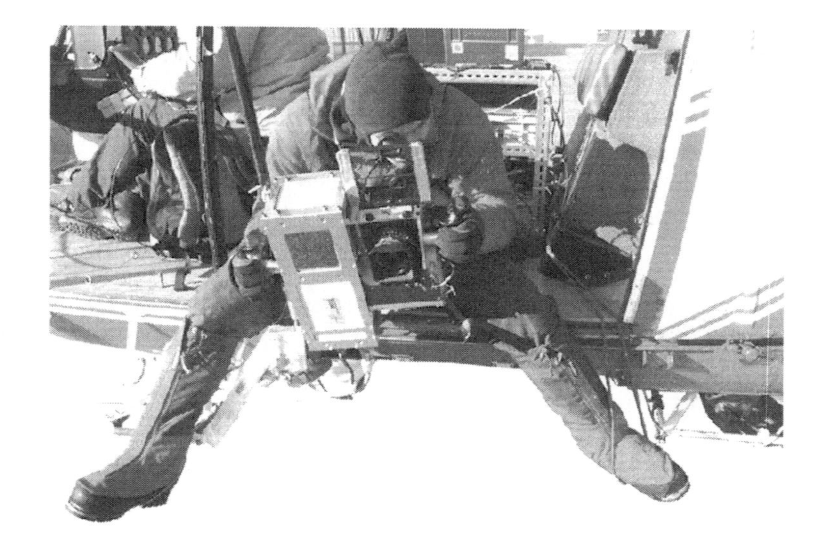

FIGURE 2.33
The Helimap system is being operated from the side of an Alouette III helicopter to measure the steep sides of a valley. (From UW + K.)

Mosaic Mapping Systems. A few months later, the combined company was purchased by another Canadian company, Pulse Data, which retained the title Terrapoint (but without the capital "P") for its marketing and operational activities. While the original TerraPoint company used its four NASA-developed ALTMS systems mainly for high-altitude large-area work using fixed wing aircraft, the later Canadian-based company developed a complementary range of laser scanner systems primarily for lower-altitude corridor-type work using helicopters as the airborne platforms. These systems, called the ALMIS series are based on the Riegl scanning engines. The first two systems use the LMS-Q140 laser scanning engine, while a third utilizes the LMS-Q560 engine. These have all been combined with dual-frequency GPS receivers and Honeywell IMU units and integrated with control systems and software that have been developed in-house by Terrapoint.

2.5.3.9 *Lidar Services International*

This Canadian company is based in Calgary, Alberta. Its Helix airborne laser scanning systems are used mainly for low-altitude applications, especially for the survey of electrical power transmission lines using helicopters as the airborne platforms. For this particular application, the company has formed a strategic alliance with Manitoba Hydro, which is a major Canadian electricity supply company. For its low-altitude work, Lidar Services International (LSI) has developed three Helix systems based on the Riegl LMS-Q140 (x2) and LMS-Q240 laser scanning engines, which have been integrated with GPS/IMU units in-house. Two newer systems that can be operated from higher altitudes from fixed-wing aircraft are based on the LMS-Q560 model complete with its digital waveform recording (Figures 2.34 and 2.35).

2.5.3.10 *Tuck Mapping*

The Tuck Mapping company based in Virginia has developed its "eagleeye" system in-house. Although it has been developed wholly independently, the eagleeye resembles

FIGURE 2.34
LSI Inc. operates this Cessna 185F aircraft, which is fitted with a carbon-fiber pod that is attached to the side of the aircraft. The pod houses a Helix laser scanning system. (From LiDAR Services International Inc.)

to some extent the Harrier 24 system already described above in that it is based on the Riegl LMS-Q240 scanner engine which is integrated with a POS/AV 410 GPS/IMU system and a DSS 322 medium-format digital camera from Applanix. The system also includes a Track'Air navigation system and a Sony digital video camera producing frame images that are 1280 × 960 pixels in size. These images are recorded using the sDVR

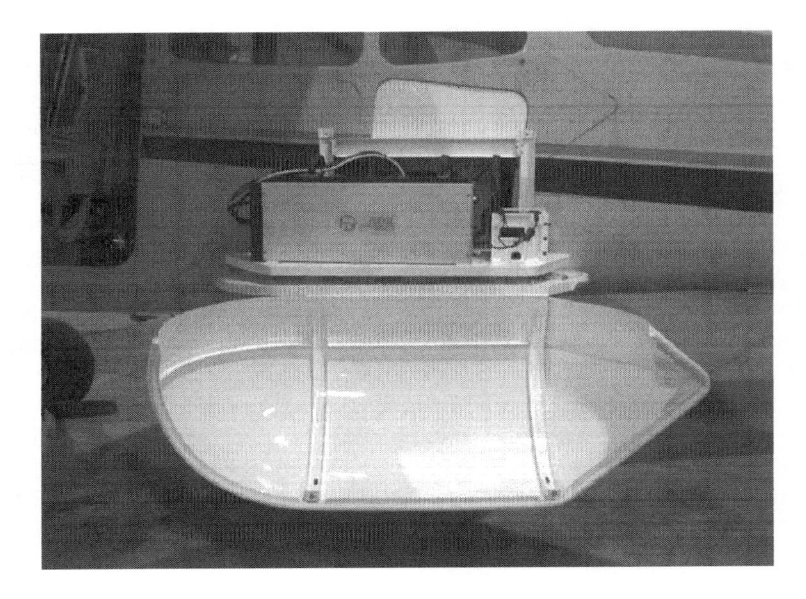

FIGURE 2.35
The cover of the pod is hinged down to show the Helix system with its Riegl LMS-Q560 laser scanning engine sitting on a rigid vibration-isolated base plate. (From LiDAR Services International Inc.)

video encoding system from Red Hen Systems. So far, two examples of the eagleeye have been built, both mounted on Bell helicopter platforms. A new system is at present under development, based on the Riegl LMS-Q560 laser engine.

2.5.4 Custom-Built and In-House Operated Systems

Besides the commercially available systems described above, a large number of systems have been built in-house by service providers using COTS components bought from a wide variety of suppliers. In the case of four of these service providers—TopEye, Fugro, TopoSys, and Terrapoint—their in-house produced systems have been built in some numbers. However, nearly all the other systems built by service providers have been custom-built and produced as individual one-of-a-kind systems.

2.5.4.1 TopEye

The origins of the TopEye airborne laser scanning systems lie with the Saab Dynamics Company from Sweden, which had previously developed the laser-based HawkEye airborne bathymetric mapping system. Starting in 1993, Saab Combitech (another company in the Saab aerospace group), in combination with Osterman Helicopters, formed the Saab Survey Systems company and developed an airborne topographic laser scanning system, which it called TopEye. In 1999, Osterman Helicopters purchased all the shares in the partnership and re-named the company TopEye AB. In 2005, the Blom mapping company from Norway bought TopEye AB, which is still based in Gothenburg, Sweden. It provides airborne laser scanning services all over Europe, while still undertaking the development and maintenance of the TopEye scanners.

The original TopEye Mk I scanners that came into service in the mid-1990s used a laser rangefinder with a maximum PRF of 8kHz in combination with twin oscillating mirrors producing a Z-shaped pattern of measurements of the ground—similar to that produced by the Optech ALTM series. The system could record four returns from each transmitted laser pulse (Al-Bayari et al., 2002). Over the last 3 years (2004–2007), these instruments have been radically re-built using components supplied by the AHAB company formed by three former Saab employees. The resulting TopEye Mk II scanners are equipped with new fiber-based lasers having a PRF of 50kHz and nutating mirrors that produce a series of overlapping elliptical (Palmer) patterns of measured points over the ground (Figure 2.36). The rangefinder's receiver is equipped with avalanche photo diodes that can record simultaneously both the first and last echoes and the full intensity waveform (with 128 samples) of the signal received after the reflection of the emitted pulse from the ground. The positioning and orientation system used in the TopEye scanners is the POS/AV system from Applanix. The TopEye scanners have been designed for operation from relatively low altitudes using helicopters as the airborne platforms. Six of these TopEye Mk II units are operated by TopEye AB; a further two are being operated by the Aerotec company in the United States. A completely new scanner, called the TS 3.5, is at present under development by AHAB based on its experience of helping to re-build the TopEye systems. Indeed the TS 3.5 design is quite similar to that of the TopEye Mk II system. For its high-altitude wide-area laser scanning operations, the Blom organization uses three Optech ALTM 3100 scanners.

2.5.4.2 Fugro

The Fast Laser Imaging Mobile Airborne Platform (FLI-MAP) is the name given to a series of airborne laser scanning systems produced in-house by the Fugro surveying and

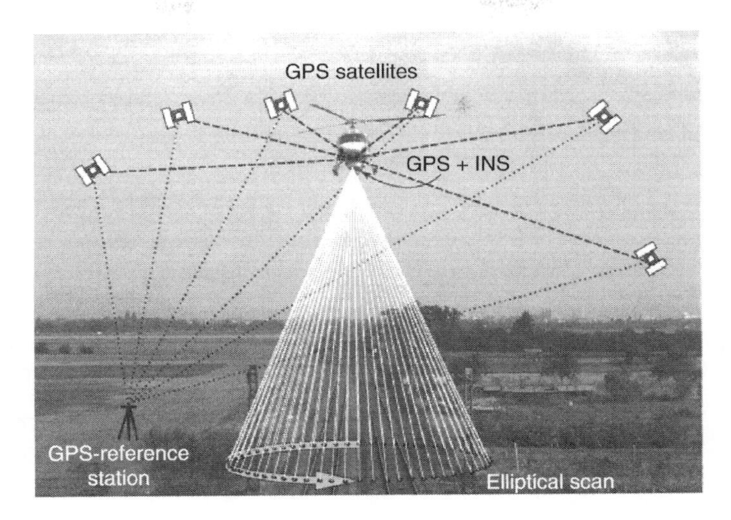

FIGURE 2.36
Diagram showing the overall concept of the TopEye Mk. II airborne laser system with its elliptical (Palmer) scanning of the ground and its use of GPS both in the air (in conjunction with an INS) and at the ground reference station. (From TopEye AB; redrawn by M. Shand.)

mapping company and operated principally by two of its subsidiaries—John E. Chance & Associates based in Lafayette, Louisiana and Fugro-Inpark based in the Netherlands. The development of the FLI-MAP airborne laser scanner by Fugro began in 1993 with the first system entering commercial service in 1995. The initial systems, named FLI-MAP I and II, all date from the late 1990s. Since then, a new system—FLI-MAP 400—has entered service. All of the FLI-MAP systems are mounted in helicopters and are used primarily for corridor surveys from low altitudes.

The original FLI-MAP I system was based on the use of a rangefinder utilizing a Class I semiconductor laser that emitted its pulses at the wavelength of 900 nm and at a PRF of 8 kHz. The maximum operating altitude of the system was 200 m. The scanning mechanism made use of a continuously rotating mirror giving a scan rate of 40 Hz. With the FLI-MAP II system, while the basic configuration remained the same, the system performance was upgraded substantially. The PRF of the laser rangefinder was increased to 10 kHz; the maximum altitude to 300 m; and the scan rate to 60 Hz. Both systems used a pair of video cameras, the one pointing vertically downwards, while the other was mounted obliquely in the forward pointing direction (Figure 2.37). These produced color S-VHS video images with a format size of 352 × 288 pixels.

The position and orientation system used in both models is of particular interest. It comprises four Trimble 4000SE dual-frequency GPS receivers and a vertical gyro, all of which are mounted on-board the helicopter platform (Figure 2.38). In combination with additional GPS receivers used as base stations on the ground, they are used to determine the position, altitude, heading, pitch and roll of the platform and its FLI-MAP scanner system. One of the four on-board GPS receivers is designated as the primary navigation receiver. After postprocessing its data in combination with the data collected by the ground-based stations, it generates the accurate X, Y, Z coordinates of its antenna. The primary navigation receiver also produces real-time positional values that are used as the basis for a flight navigation system that provides information to the pilot regarding deviations from the planned flight path and altitude using a system of LED light

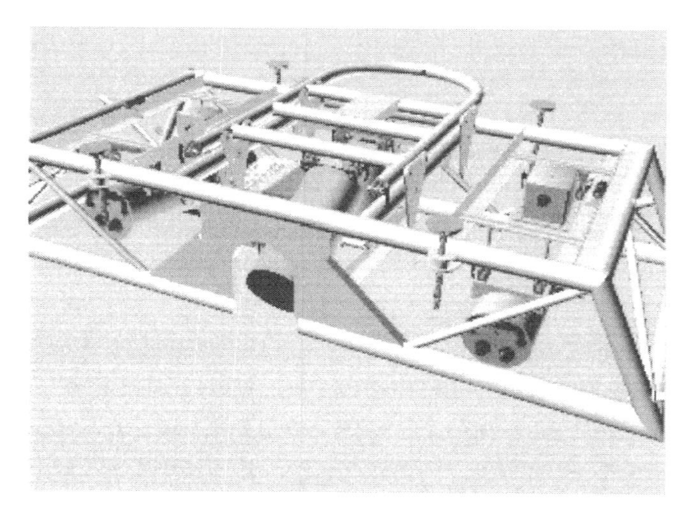

FIGURE 2.37
A CAD drawing of the frame that carries the FLI-MAP laser scanners and the system's video and digital cameras. (From Fugro.)

bars. The other three (secondary) GPS receivers that are mounted on-board the helicopter platform are used together with the primary receiver and the vertical gyro to determine the heading, roll and pitch of the platform and system every half-second. For this purpose, the antennas of two of the GPS receivers are mounted on outrigger pylons attached to each side of the helicopter (Figure 2.38). This multiple GPS antennae-based attitude determination system provides an adequate solution, since the flying height is limited. Thus the accuracy requirements for the attitude angles are less stringent compared with those of the fixed-winged aircraft with its longer ranges.

The latest model in the series is the FLI-MAP 400, which has been further upgraded as compared with its predecessors. In particular, it uses a more powerful laser (which is still

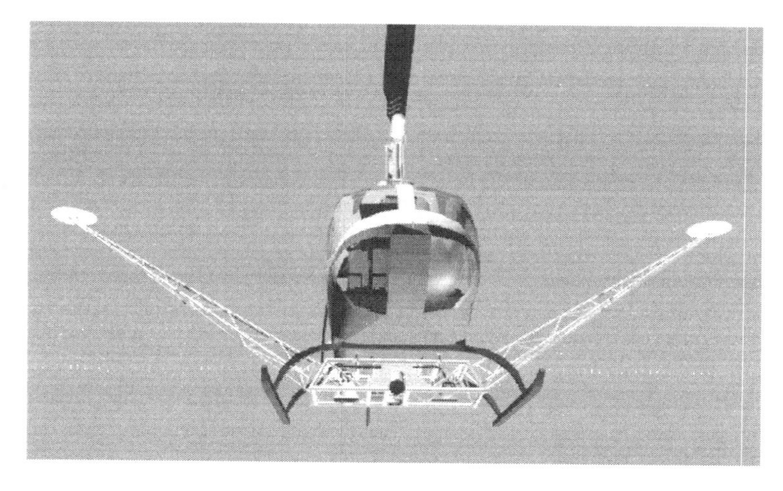

FIGURE 2.38
An helicopter equipped with a FLI-MAP system, which is mounted on the frame that is attached to the underside of the aircraft; note also the two outrigger pylons, each supporting a GPS antenna. (From Fugro.)

Class I eye-safe) and has the much faster maximum PRF of 150 kHz. These features enable the system to operate at higher altitudes—up to 350 m—which enables it to cover corridors having a much wider swath over the ground. While the earlier FLI-MAP I and II models only recorded a single (first) return from the reflected pulse from the ground, the FLI-MAP 400 allows the recording of four returns per emitted pulse. Furthermore, as well as the two video cameras used in the earlier systems, the FLI-MAP 400 features twin small-format digital still cameras, each producing images of the ground that are 11 megapixels in terms of their format size. While the FLI-MAP systems are used extensively for low-altitude surveys, Fugro, like several of the other service providers, uses a system from one of the main system suppliers for the high-altitude surveys of large areas—in Fugro's case, an ALS-50 laser scanner from Leica Geosystems.

2.5.4.3 TopoSys

This German company was founded in 1995 when it took over the quite unique laser scanner technology that was being developed by the Dornier aerospace company (now part of EADS Astrium). Its first system, called Falcon I, became operational in 1996 and was used by TopoSys to provide airborne laser scanning services in various countries in western and central Europe, usually in partnership with a local survey and mapping company within the country concerned. In 2000, TopoSys introduced an upgraded model called Falcon II (Lohr, 1999) and followed this by bringing a second version of the Falcon II into service in 2004, giving it three systems for its own use as a service provider. Another Falcon II was built and sold to an Egyptian agency. Most recently, a further upgraded model called Falcon III has been developed and placed into service in-house.

The Falcon series of scanners feature a unique and very interesting technology in the form of two linear fiber-optic arrays (Figure 2.15). The first of these arrays is used to transport and distribute the stream of laser pulses through an optical system toward the ground in the cross-track direction, while the second array carries out a similar action in reverse after the signals reflected from the ground have been picked up by the receiving optics. The laser rangefinder uses an erbium-based laser, which emits its pulses with a pulse repetition rate of 83 kHz, and the very high scan rate of 630 Hz at the wavelength (λ) of 1540 nm. The resulting stream of pulses is transmitted along a central fiber on to a nutating mirror, which reflects each pulse in sequence into a circular path into the successive adjacent fibers of the array. The fibers transport the pulses into the focal plane of the transmission optics where they emerge successively along a linear path set in the cross-track direction. They are then transmitted by the optics toward the ground. After striking the ground, the reflected signals are transmitted back toward the aircraft and its scanner. Here they are picked up by the receiving optics and are passed into the appropriate fiber in the receiving array. After being transported along the fiber, they are then reflected by a nutating mirror (similar to that used with the transmitting fibers) into the central pipe of this second array. After which, they are then passed through an optical filter to the receiver diode. Essentially the transmitting and receiving optical systems are identical but work in reverse to one another.

In the original Falcon I model, the arrays used in the transmission and receiving optical systems each comprised 128 fibers. This gave a comparatively narrow angular coverage of 14° over the terrain with quite a wide spacing between the points measured on the ground (Figure 2.15). In the Falcon II (Figure 2.39), each of the linear fiber-optic arrays was intended to be increased to 256 fibers (Wehr and Lohr, 1999). However it is not too clear whether this proposed increase was actually implemented. Because of the wide spacing between the parallel lines of measured points—especially as compared with the dense overlapping

FIGURE 2.39
TopoSys Falcon II laser scanning system mounted on board a Piper Seneca aircraft. (From TopoSys.)

coverage of points in the along-track (flight) direction, a rotating "swing" mirror was introduced into the optical system of the Falcon II (Wiechert, 2004). This caused a regular systematic sideways displacement or "swing" of the measuring pattern across the terrain, resulting in wriggly or "snake-like" coverage of the ground (Figure 2.40). In the latest Falcon III scanner (Artes, 2006), the angular coverage has been increased to 27°; the pulse repetition rate to 125 kHz at altitudes up to 1000 m; the maximum scan frequency to 415 Hz; and the maximum operating altitude to 2500 m. Full waveform digitizing of the reflected signals is also available as an optional feature on the Falcon III.

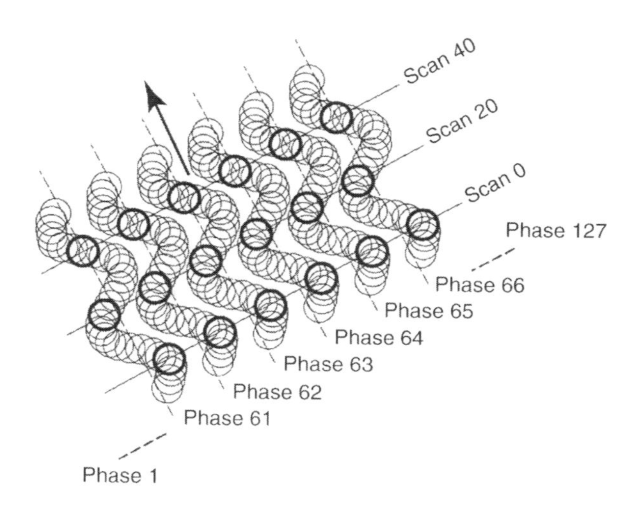

FIGURE 2.40
The Falcon II system incorporates a swing mirror that causes systematic displacements of the laser scan pulses in the cross-track direction resulting in a "snake-like" pattern of measurements across the terrain. (From TopoSys; redrawn by M. Shand.)

With regard to the position and orientation system, an Applanix POS/AV system featuring a Honeywell H-764G gyro and a NovAtel GPS receiver has been used in each of the different Falcon models. Turning next to the imagers that are integrated into the systems, the Falcon I used a video camera, while the Falcon II features a pushbroom line scanner with four parallel lines (RGB + NIR), each 682 pixels in length, from which a continuous narrow strip of color or false-color imagery of the ground can be generated. In the Falcon III, the scan width of the RGB linescan imagery has been increased to 2048 pixels. The Applanix DSS 322 medium-format digital frame camera is also available as an option that can be integrated into the overall Falcon III system.

2.5.4.4 Terrapoint

The original TerraPoint company, based in Houston, Texas, was formed in 1998 as part of a drive to commercialize the airborne laser scanner technology that had been developed at NASA's Wallops Flight Facility (WFF). The technology was developed further by the Houston Advanced Research Center (HARC) in cooperation with NASA, resulting in the Airborne Laser Topographic Imaging System (ALTMS). The TerraPoint company was then founded, owned partly by the HARC, but mostly by a subsidiary of the Transamerica Corporation. The ALTMS uses a powerful rangefinder employing a Class IV diode-pumped, Q-switched Nd:Vanadate laser together with a rotating multi-faceted polygon mirror which is used as the scanning mechanism (Figure 2.41), resulting in a regular grid pattern of points being measured on the ground. Both the laser pulse repetition rate and the mirror rotation speed are adjustable depending on the aircraft speed and altitude. In the earlier ALTMS 2536 models, the maximum laser pulse rate is either 20 or 25 kHz, whereas in the later ALTMS 4036 model, the maximum rate is 40 kHz. The maximum scan rate is 100 Hz. The ALTMS scanners record up to four returns from each emitted laser pulse and they can also be set to record the full intensity data from each reflected pulse. The ALTMS position and orientation system uses a Honeywell H-764G IMU based on a high-quality ring laser gyro for its attitude measurements, together with an internal (embedded) GPS receiver that is tightly coupled to the IMU. This produces a blended GPS/IMU data stream. A second Trimble GPS

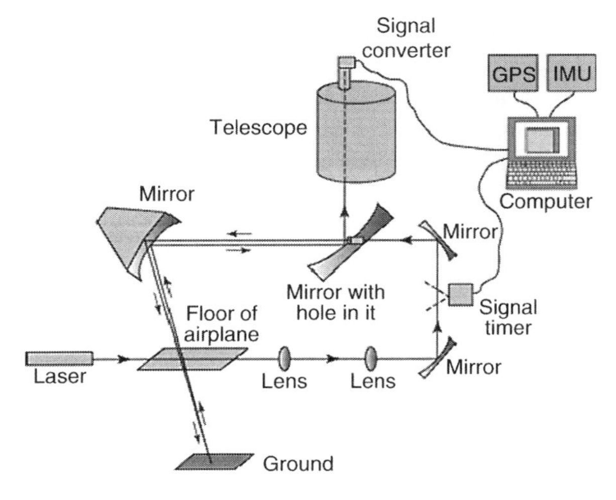

FIGURE 2.41
Optical design of the Terrapoint ALTMS laser scanner. (From Terrapoint; redrawn by M. Shand.)

FIGURE 2.42
A Terrapoint ALTMS laser scanner mounted together with a Canon EOD-1D small-format digital frame camera on a rigid base plate for use in a light aircraft. (From Terrapoint.)

receiver is also provided for real-time navigation purposes. The four ALTMS scanners have been used by TerraPoint from medium altitudes over the range 600–2500 m above AGL, mainly for large-area topographic surveys. A Canon small-format digital camera is used in conjunction with the laser scanner (Figure 2.42).

In November 2003, the TerraPoint company was purchased by the Mosaic Mapping company from Canada. Shortly afterwards, in May 2004, the combined company was acquired by another Canadian company, Pulse Data Inc., which then operated it under the Terrapoint brand (without the capital "P" of the original American company). At the time of writing in 2007, after 9 years' service, the four ALTMS systems are gradually being de-commissioned and replaced by Optech ALTM 3100 scanners for Terrapoint's high-altitude airborne laser scanning operations. For low-level scanning using helicopters, the company continues to use the ALMIS systems based on the Riegl laser scanning engines discussed above.

2.5.4.5 Nortech Geomatics

This company, one of several in the Nortech group, was based in Houston, Texas. As discussed previously, it was much involved in the development of airborne laser profiling and operated various profilers of its own manufacture during the 1980s and early 1990s.

Later, during the 1990s, Nortech developed its ATLAS-SL as a full-blown airborne laser scanning system that came into service in 1997. Two examples of this system were acquired and are still being operated by Advanced Lidar Technology Inc., which is also based in the Houston area. The ATLAS-SL uses a laser rangefinder that has a pulse repetition rate of 12 kHz operating at a scan rate of 20 Hz over a 60° angular field. The system is designed specifically for linear or corridor surveys from low altitudes—up to a maximum flying height of 260 m. It has a position and orientation system with a GPS/IMU of an unspecified type. The system also features two cameras—the first is a forward-pointing S-VHS video camera; the second is a small-format digital camera producing 2k × 2k = 4 megapixel frame images.

2.5.4.6 Other Custom-Built Systems

Besides these various custom-built systems that have been constructed in some numbers, there are a substantial number of other individual custom-built airborne laser scanning systems that have been constructed and operated in North America, Western Europe, and Japan. Thus, for example, in the United Kingdom, the QinetiQ defence research company has produced its ATLAS system (which has no relation and is quite different to the ATLAS systems built by Nortech). This is operated commercially by Survey Inspection Systems. Another example from the United Kingdom is that which has been built in-house and operated by Precision Terrain Surveys Ltd. In Sweden, the GeoMapper airborne laser scanning system has been built by Laseroptronix, while, in Japan, the Nakanihon Air Service mapping company has also built and operated its own home-built ALSS system.

2.5.5 Research Systems

Besides the various systems described above that are being used commercially to provide airborne laser scanning services to customers, a number of interesting systems have been built by U.S. government research agencies, principally NASA, and used primarily for scientific research purposes. A single university-built system developed by the University of Stuttgart in Germany also falls within this category.

2.5.5.1 NASA

NASA has been in the forefront of the organizations developing airborne laser profilers and scanners over a long period—since the mid-1970s. Much of its initial development took place within the AOL program, which was sponsored jointly by NASA and NOAA. Essentially this acted as a test-bed both of the technology and of its applications. The AOL system was continually modified and upgraded over a 20 year period. On the one hand, research was conducted into the use of airborne lasers to carry out the measurement of elevations and depths for both topographic and bathymetric mapping applications. In parallel with this, extensive research was undertaken into the fluorescence induced in the plankton that is present in the surface layers of the ocean when illuminated by an airborne laser. These two parts of the research program were separated in 1994 when two new systems were introduced. The fluorescing system retained the AOL title, while the topographic mapping oriented system was given the title ATM.

2.5.5.2 ATM

The ATM-I system was first flown in 1994, while the ATM-II was brought into service in 1996. The rangefinders of the two systems use a Spectra Physics Nd:YLF laser generating

7 ns wide pulses at λ = 1046 nm. The output pulses have been frequency doubled to be emitted simultaneously at λ = 523 nm in the green part of the spectrum. The rangefinder can be operated at pulse repetition rates between 2 and 10 kHz. The scanning mechanism of the ATMs uses a nutating mirror generating an elliptical (Palmer) scan pattern over the ground with a scan rate of 20 Hz. From 1997 onwards, the ATM-II system has been equipped with an imaging device that records passive panchromatic image data following the same elliptical scanning path as the active laser scanner. The ATM-II has been equipped with a dual-frequency GPS receiver allied to an IMU based on a laser ring gyro to act as its position and orientation system. The two ATM systems have been used extensively by NASA for the mapping of coastal beaches on both the Atlantic and Pacific coasts of the United States and in the Gulf of Mexico. The systems have also been used for the surveys of large areas of ice sheets and glaciers in Greenland, Northern Canada, Svalbard (Spitzbergen) and Iceland.

2.5.5.3 RASCAL

In parallel with its ATM systems, NASA also developed its RASCAL (RAster SCanning Airborne Laser) system at the GSFC in 1995 for the airborne mapping of surface topography. Like the ATM systems, the scanner's rangefinder utilized a Spectra Physics Q-switched, diode-pumped and frequency-doubled Nd:YLF laser to generate its pulses at 5 kHz, each of which is 6–10 ns long. However the scanning mechanism is very different to that of the ATM in that it utilizes a rotating aluminum scan mirror that is polygonal in shape with 45 individual sectors or faces. Each sector occupies 8° of the scan mirror wheel. On reflection, each sector gives a 16° coverage of the ground. The wheel with its mirror surfaces is rotated at a constant speed of 80 rpm giving a scan speed of 60 Hz—45 faces multiplied by 80 rpm. As each new sector is reached, it causes an instantaneous reset of the scan pattern over the ground. The result is a series of parallel scans in the cross-track direction producing a grid or raster pattern of measurements over the terrain—hence the RASTER name of the system. The position and orientation system uses a dual-frequency GPS receiver allied to an IMU based on a Litton LTN-92 ring laser gyro. Like its sister ATM systems, the RASCAL has been used extensively to generate the surface topography of specific geological and geomorphological features that are subject to changes in their surface. These have included volcanic calderas; active fault zones that are prone to earthquakes; and dam structures in California. The commercialization of the ATM and RASCAL technology was undertaken by HARC in cooperation with NASA. It resulted in the ALTMS scanners that have been operated by the TerraPoint company—as described above in Section 2.5.4.4.

2.5.5.4 SLICER

The SLICER (Scanning Lidar Imager of Canopies by Echo Recovery) is yet another airborne laser scanner that was developed by NASA in the mid-1990s. The system was based on the ATLAS (Airborne Topographic Laser Altimeter System), an earlier airborne laser profiler that had been built by NASA. The ATLAS was first modified into the SLICER through the installation of a more powerful laser and the addition of a waveform digitizer. The rangefinder in the SLICER uses a Q-switched, diode-pumped Nd:YAG laser that outputs a short-duration (4 ns) pulse at the wavelength (λ) of 1064 nm. The scanning mechanism comprises a simple oscillating mirror that is rotated rapidly to the successive positions of a set of fixed scan angles using a computer-controlled galvanometer. Usually, five fixed angular positions are used for each cross-track scan, so only a very narrow scan width is achieved. The detector used in the receiver is a silicon avalanche diode that converts the energy received from the reflected pulse into an output analogue voltage.

The waveform digitizer is an analogue-to-digital (A/D) converter that samples and records the output voltage values in digital form. The on-board position and orientation system utilizes twin Ashtech Z-12 dual-frequency GPS receivers. The first of these provides the required positional data, while the second set provides the time reference for the recorded data stream. The IMU is a Litton LTN-92 unit based on a laser ring gyro.

From this description, it will be seen that the SLICER scanner produces a very narrow swath of three-dimensional waveform data. So it is not used to cover large areas. Instead it is used to carry out transects across the forests or areas of vegetation that are being studied. The SLICER system has been used in a large number of research projects concerned with studies of forest canopies. These include the BOREAS project carried out over the Taiga boreal forests of Northern Canada; projects concerned with studies of the canopies of tropical forests in parts of the West Indies (Puerto Rico, Dominican Republic and the Lesser Antilles); and temperate forests in northwestern United States (Washington, Oregon, and California). Other projects involving the SLICER scanner have been research studies of Arctic vegetation carried out in Iceland and on Jan Mayen Island, lying between Iceland and Norway.

2.5.5.5 Laser Vegetation Imaging Sensor (LVIS)

As its title suggests, the LVIS has been developed by NASA for the mapping of topography and of the height and structure of the vegetation growing on it, based on the experience gained with the SLICER scanner (Blair et al., 1999). In particular, it has been designed to be operated from high altitudes for the coverage of a fairly wide swath over the ground (Figure 2.43). Furthermore it is unique in recording the active time history of both the emitted and reflected pulses. The LVIS rangefinder uses a powerful Q-switched, diode-pumped Nd:YAG laser supplied by Cutting Edge Optronics. This emits pulses having a power of 5 mJ (permitting high-altitude operations) and a pulse length of 10 ns at the wavelength (λ) of 1064 nm. In its earliest form, the pulse repetition rate of the laser could be varied over the range 100–500 Hz. In its later upgraded form, the rate was increased to 5 kHz. Furthermore, after the installation of the more powerful laser, the flying height has been extended to 12 km (40,000 ft), being operated from a Sabreliner business jet aircraft equipped with

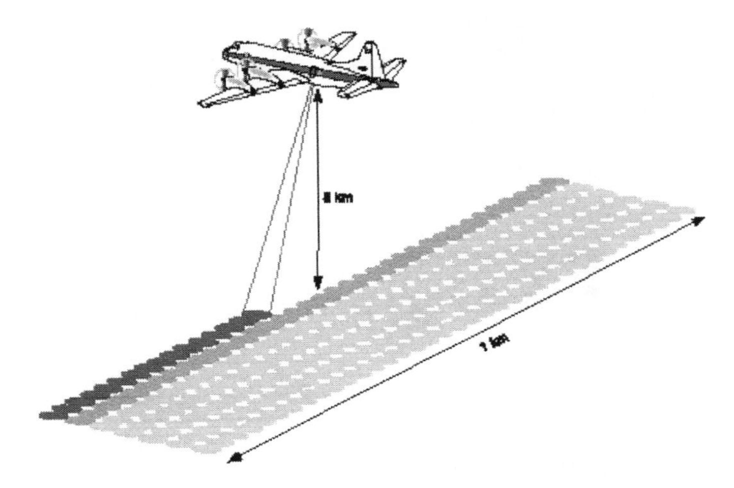

FIGURE 2.43
Stepped pattern of the coverage of the ground and its vegetation canopy that is implemented by the LVIS. (From NASA; redrawn by M. Shand.)

a high-quality optical window. From this altitude, the swath width has been extended to 2 km. The ranging and the waveform recording are all carried out using a single detector, digitizer and oscillator, the last acting as a clock or time base.

The scanning of the ground with its vegetation canopy is carried out in a stepped pattern with the galvanometer-driven mirror being stationary while the laser emits its pulse and the reflected energy is received from the ground (Figure 2.43). This solution is necessary since the travel time of the pulse is rather long at high altitudes. The scanning mechanism uses a very stiff but lightweight mirror that remains flat during the severe accelerations and decelerations that are experienced during the rapid starts and stops of this stepped scanning action with the mirror remaining stationary during the measurement process. The position and orientation system is similar to that of the SLICER, using a Litton LTN-92 IMU in combination with two Ashtech Z-12 dual-frequency GPS receivers. Initially the LVIS was regarded principally as an airborne prototype or simulator for the VCL spaceborne laser profiling mission featuring the MBLA that was discussed in Section 2.3.4 above. As such, it formed part of NASA's ESSP program. However the VCL mission was first delayed and then cancelled in 2002. After which, the LVIS was upgraded and re-packaged to fit within a standard Wild RC10 aerial camera mount (Figure 2.44). The LVIS scanner has been used extensively both in the United States (in California and in the eastern states) and in Central America (in Costa Rica and Panama). It has also been proposed for use in a major mapping project in the Amazonian forests of Brazil to give direct measurements of vegetation height and topography and provide estimates of aboveground biomass for the mapped regions.

FIGURE 2.44
LVIS airborne laser scanner. (From NASA.)

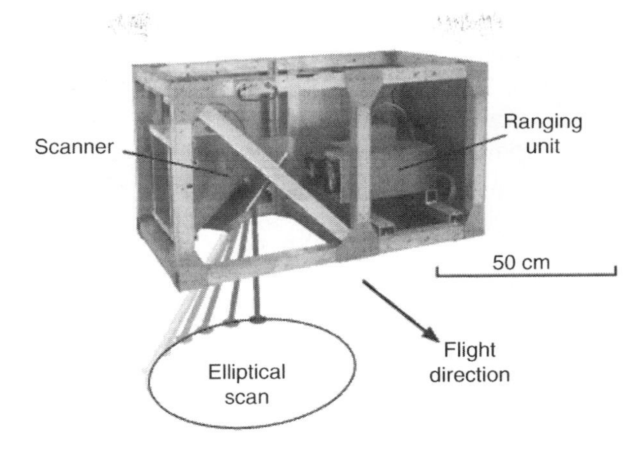

FIGURE 2.45
ScaLARS airborne laser scanner. (From Institute of Navigation, University of Stuttgart.)

2.5.5.6 *ScaLARS*

The name ScaLARS is an acronym formed from the full title of the Scanning Laser Altitude and Reflectance Sensor that was developed in the mid-1990s by the Institute of Navigation of the University of Stuttgart in Germany (Hug, 1994; Thiel and Wehr, 1999, 2004). As will have become apparent from previous references that have been made about this instrument, ScaLARS has a fairly unique design and construction among airborne laser scanners (Figure 2.45). In particular, it uses a CW diode laser as the basis for its rangefinder and employs the phase measuring technique (with measuring frequencies of 1 and 10 MHz, respectively) for the determination of the slant range from the airborne platform to the ground (Figure 2.46). The semiconductor material used as

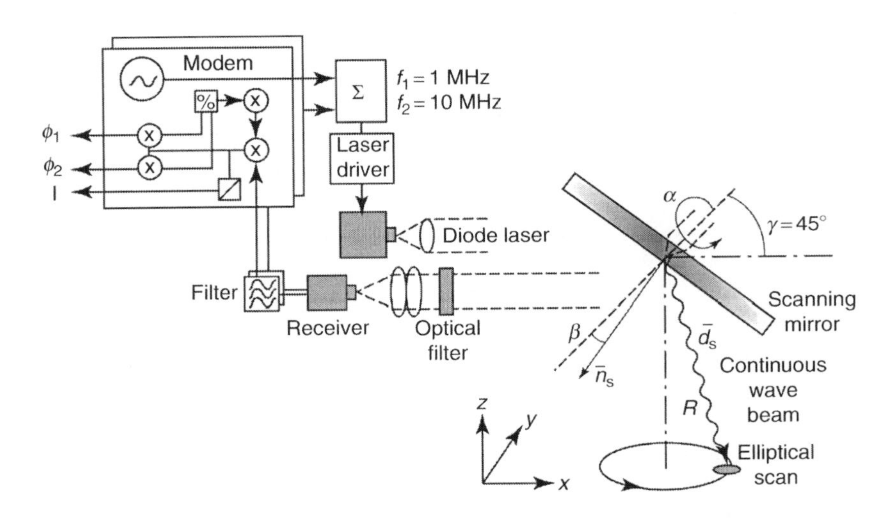

FIGURE 2.46
Design and major components of the ScaLARS airborne laser scanner, including the optical elements that produce the elliptical (Palmer) scan pattern over the ground. (From Institute of Navigation, University of Stuttgart; redrawn by M. Shand.)

the basis of its laser diode is In Ga(Al)As. The laser emits its radiation at $\lambda = 810$ nm with an average power of 0.8 W. The beam divergence is 1 mrad; the sampling rate for the measurements is 7.5 kHz; while the actual scan rate over the ground is ≤20 Hz. The ScaLARS instrument also features a Palmer scan mechanism generating overlapping elliptical scan patterns over the ground surface. The maximum range for ScaLARS is quoted as being 750 m for a ground reflectivity value of 20%, the swath width being 340 m from a flying height of 700 m. From this flying height, the elevation accuracy is quoted as being ±20 cm, while the positional accuracy is ±1 m.

The position and orientation system that was used in the original ScaLARS instrument, used a dual frequency GPS receiver in conjunction with a DTS-FP model IMU from iMAR that is based on a set of FOGs. The later ScaLARS-II instrument has utilized an Applanix POS/DG 310 IMU. Some of the main application areas for the ScaLARS are similar to those carried out by NASA's research laser scanners—namely glacier, coastal and forest surveys. For the first of these applications, detailed accounts of the mapping of the Unteraar Glacier in Switzerland have been published by Favey et al. (2000) and Baltsavias et al. (2001), including detailed comparisons with the results of mapping of the glacier carried out using both manual and automated photogrammetric methods using aerial photography. However ScaLARS has also been used to carry out the acquisition of elevation data of substantial areas for topographic mapping purposes on behalf of the official survey organizations of individual German states such as Thuringia.

2.6 Airborne Bathymetric Laser Scanners

As the title of this section suggests, airborne laser scanners have been devised to carry out mapping of the seabed and the adjacent land in coastal areas. The early systems that have been used for this type of bathymetric mapping were all laser profilers. Their pulse-type laser rangefinders measured the vertical distances from the aircraft to a series of successive points both on the sea surface and on the seabed, that was located directly below the aircraft using the reflected signals from these surfaces. These early and largely experimental profiling systems were built in the United States, Sweden, Canada, and Australia; a very detailed account of the development of these systems is given in the paper by LaRoque and West (1990). During the 1980s, these early profiling systems were modified and new systems were built that had a scanning capability added to them. They included the AOL developed jointly by NASA, NOAA and the U.S. Navy and the WRELADS-2 built by the Weapons Research Establishment (WRE) of the Royal Australian Navy (RAN). Optech were also involved in the development of these early scanner systems, building the LARSEN-500 system under the sponsorship of the Canadian Hydrographic Service (CHS) and the Canadian Centre for Remote Sensing (CCRS). The Optech company was also involved in the development of the FLASH-1 system for the Swedish Defence Research Establishment (FOA). In the early 1990s, all of these early efforts came to fruition with the development and the delivery of the first SHOALS system by Optech to the U.S. Army Corps of Engineers (USACE) and its entry into full operational service. Optech was also involved, together with Saab, in the development of the Hawk Eye bathymetric laser scanner systems that were built for the Swedish Department of Defence and entered operational service with Swedish maritime agencies. At the same time, the Laser Airborne Depth Sounder (LADS) system, built by an Australian partnership, entered service with the RAN. The same four countries—the United States,

Canada, Sweden, and Australia—that were involved in these early developments are still those among Western countries that are in the forefront of developing airborne bathymetric laser scanners today (Pope et al., 2001). Outside these four countries, experimental systems have also been built in Russia and China (LaRoque and West, 1990). Although all of these systems mentioned above were built to satisfy the requirements of government defence agencies and hydrographic surveying services, there are now three commercial operators of airborne bathymetric laser scanning systems offering their services to government agencies and commercial companies on a worldwide basis.

2.6.1 Laser Bathymetric Measurements

The basic principle of measuring the depth of the seabed (or lakebed) below the sea (or lake) surface with an airborne laser scanner usually involves the use of two laser rangefinders emitting pulses simultaneously at different wavelengths—in the infrared and green parts of the electromagnetic spectrum (Guenther et al., 2000). The infrared radiation is reflected from the water surface, whereas the pulse of green radiation passes into and through the water column and is reflected by the seabed back toward the rangefinder (Figure 2.47). The actual depth that can be measured is limited to 25–70 m, depending on the system that is being used and the clarity or turbidity of the water column through

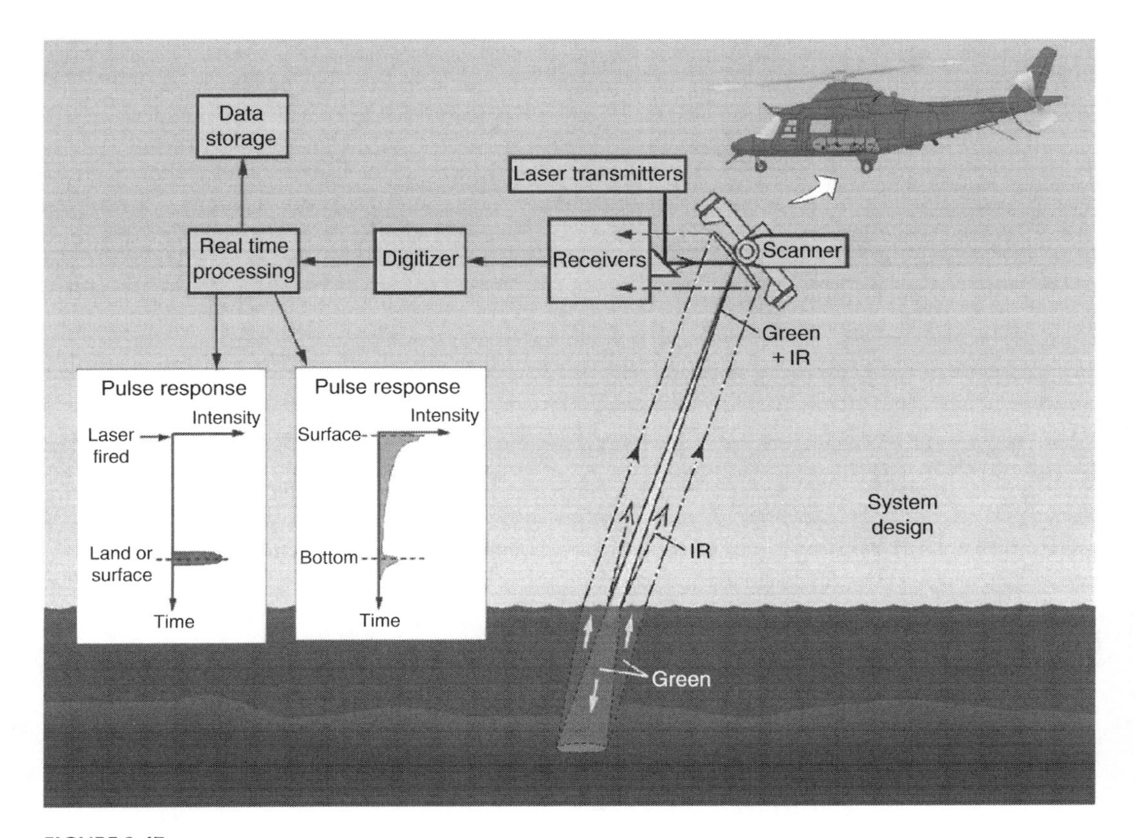

FIGURE 2.47
Overall concept of an airborne bathymetric laser scanner showing how the pulses from a red/NIR laser are reflected from the sea surface or the land—while the pulses from a green laser penetrate the water and are reflected back from the sea floor. (From Airborne Hydrography AB.)

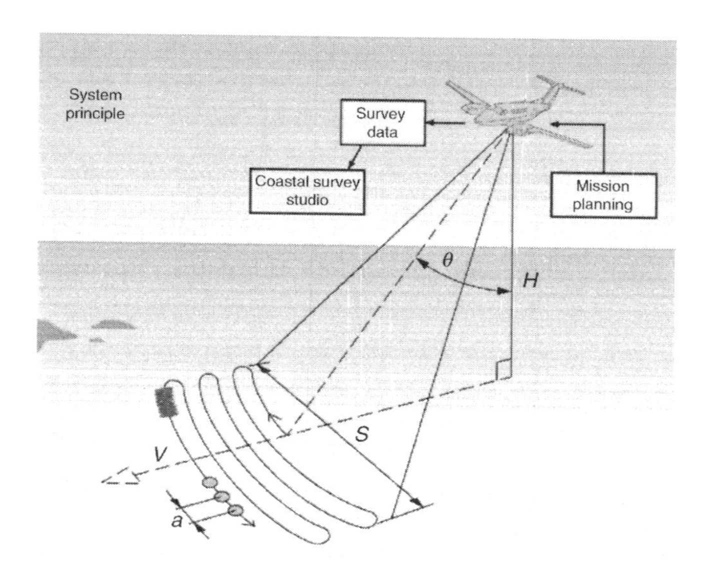

FIGURE 2.48
The operating principle of the Hawk Eye Mk. II airborne bathymetric and topographic laser scanning system showing the scanning pattern that is used to survey the land and the sea floor of coastal areas. (From Airborne Hydrography AB.)

which the radiation is passing. The reflected radiation at both wavelengths is then gathered by the appropriate detectors located within the rangefinder, and the elapsed time between the emitted and received signals is measured for both pulses. From these measurements of the elapsed time, the corresponding ranges can then be derived, knowing the speed of the radiation passing through air and water respectively. While this information provides depth values and creates a simple depth profile along the flight line, the addition of a cross-track scanning mechanism provides coverage of a swath or area of the sea surface and seabed (see Figure 2.48). As with an airborne topographic laser scanner, the position and attitude of the airborne platform and the bathymetric laser scanner mounted on it is provided by an integrated GPS/IMU system.

2.6.2 System Suppliers

As noted above, currently there are only a few constructors and suppliers of airborne bathymetric laser scanners.

2.6.2.1 Optech

In the introduction given above, it was noted that Optech had played a leading part in the initial development of airborne bathymetric laser scanners, having either constructed or been the supplier of major components of several early systems built for government agencies in the United States, Canada, and Sweden. The several complete bathymetric laser scanner systems that have been constructed by Optech since 1990 all bear the name SHOALS, which is an acronym for the full title of Scanning Hydrographic Operational Airborne Lidar Survey system (Irish and Lillycrop, 1999). There has been a steady development of these systems with successive models being labeled SHOALS-200, SHOALS-400, SHOALS-1000, and SHOALS-3000 respectively.

The original SHOALS system supplied to the USACE in 1993 used two pulsed lasers operating at infrared ($\lambda = 1064$ nm) and green ($\lambda = 532$ nm) wavelengths respectively, the frequencies having been chosen to optimize the detection of the air/water interface and water penetration respectively (Irish and Lillycrop, 1999). The system had five receiver channels. Two of these measured and recorded the waveforms (energy in time) for each reflected green pulse, while a second pair recorded the waveforms of each reflected infrared pulse. In each case, the use of two receiver channels (each recording the appropriate waveforms) was to help discriminate the actual surface from which the reflections were taking place. The remaining (fifth) channel measured and recorded the Raman energy that results from the excitation of the surface water molecules by the green laser. The power levels of the two lasers were 15 mJ in the case of the infrared laser and 5 mJ in the case of the green laser.

The original SHOALS system had a laser pulse repetition rate of 200 Hz—hence the designation SHOALS-200. It was flown using a Bell helicopter as the airborne platform. In 1998, the system was modified, including the fitting of a new laser that doubled the pulse rate to 400 Hz—hence the name SHOALS-400. The system was mounted in a Twin Otter aircraft. The positioning of the SHOALS system was carried out using a dual-frequency GPS receiver carrying out differential measurements, either in conjunction with local base stations or using the Fugro OmniSTAR system, in combination with an IMU that provided the required attitude measurements. The horizontal accuracies that were achieved were quoted as being ±1 to 3 m, while the vertical accuracy was stated to be ±15 cm. The scan pattern was defined by the lasers being pointed forward at an angle of 20° to the vertical, with the scan mechanism producing an arc of coverage in front of the aircraft. The laser range measurements were supplemented by a video camera that continuously recorded video frame imagery of the area being covered.

In 2003, Optech introduced its SHOALS-1000 model, the first example of which was delivered to the U.S. Navy. While, in principle, the basic system and the measuring principle remained the same as in the previous SHOALS models, this new model featured lasers operating at the much higher rate of 1000 Hz (infrared) and 400 Hz (green). The former was designed to measure the elevation data of the topography of the coastal land area as well as the surface of the water. The new system was also much reduced in terms of the weight −205 versus 405 kg—and power requirements −60 A at 28 V instead of the 150 A and 120 A at the same voltage—of the previous -200 and -400 models. The SHOALS-1000 formed a major element of the CHARTS (Compact Hydrographic Airborne Rapid Total Survey) program that was being implemented by the Joint Airborne LiDAR Bathymetry Technical Center of Expertize (JABLTCX), a partnership comprising the USACE, NOAA and the U.S. Navy's NAVOCEANO organization, which is based at the Stennis Center in Bay St. Louis in Mississippi (Heslin et al., 2003). Further examples of the SHOALS-1000 have been supplied to the Japanese Coast Guard's hydrographic and oceanographic survey organization and to the Fugro Pelagos company which had previously operated a commercial airborne laser bathymetric service using a SHOALS-1000 in partnership with Optech (Figure 2.49). Fugro had also operated the SHOALS-200 and -400 models on behalf of the JABLTCX through its John E. Chance subsidiary company.

In 2006, a further development took place in the form of the SHOALS-3000 (Figure 2.50) that was delivered to the U.S. Navy (Heslin et al., 2003). Essentially this was an upgrade to the SHOALS-1000 producing an improved performance with the measuring rate of the infrared (topographic) laser being increased to 3000 Hz. This has allowed higher data collection rates and a larger swath width (300 vs. 215 m) for a laser spot area of 4 × 4 m from an altitude of 400 m. This new version of the SHOALS system again forms a major component of the overall CHARTS system. This not only includes the improved SHOALS system but

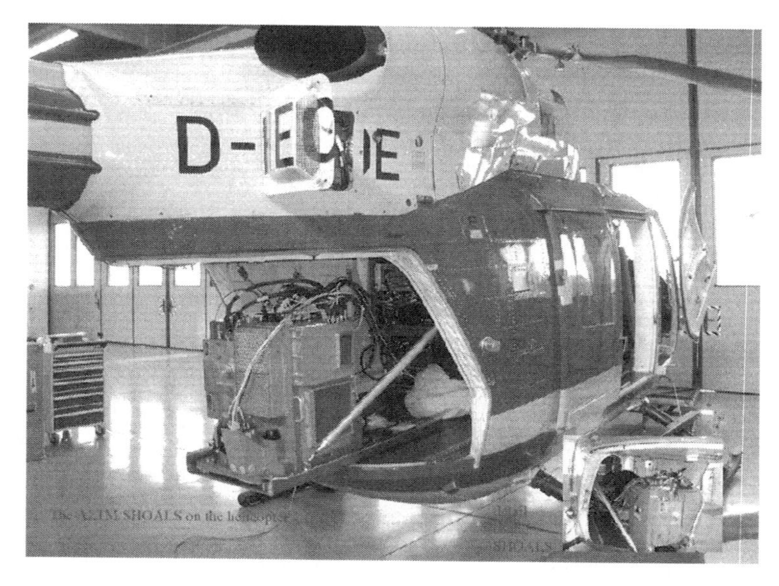

FIGURE 2.49
SHOALS system is being loaded into a large Mil helicopter prior to its operation over the coastal waters of the Adriatic Sea. (From Helica.)

it also incorporates a CASI-700 pushbroom line scanner supplied by another Canadian company, ITRES Research. The resulting hyperspectral linescan imagery allows the construction of classified (thematic) maps and charts showing the areas of sand, mud, rock, coral, sea grass, etc. that are present on the sea floor.

2.6.2.2 AHAB

The Airborne Hydrography AB (AHAB) company maintains the long-held Swedish interest in airborne laser bathymetry. As mentioned in the introduction to this Section 2.6,

FIGURE 2.50
SHOALS 3000 airborne bathymetric laser scanning system. (From Optech.)

FIGURE 2.51
Hawk Eye Mk. II system is seen in operation with Admiralty Coastal Surveys. (From Airborne Hydrography AB.)

this interest resulted initially in the Flash-1 system of the 1980s. At the beginning of the 1990s, the Saab group received the contract to develop a replacement for the Flash system and, with the help of Optech as its major subcontractor, two Hawk Eye systems were delivered to the Swedish Navy and the Swedish Maritime Administration in 1994 and 1995 respectively. The first of these two systems was later sold to the Indonesian Navy, on whose behalf it was operated by the Blom mapping organization to carry out extensive surveys and chart production of Indonesian coastal waters. The second system was in use for surveys around the Swedish coast carried out by the Hydrographics Unit of the Maritime Administration until 2003. Given the substantial part played by Optech in their development, these two Hawk Eye systems were, in many ways, quite similar to the original SHOALS system.

In 2002, Saab sold the product rights to the Hawk Eye system to three former Saab employees who had been involved in the Hawk Eye scanner development. They formed the AHAB company and started up the development of a new Hawk Eye II system as well as undertaking part of the upgrading of the TopEye II systems mentioned above. The first Hawk Eye II system was delivered in late 2005 to Admiralty Coastal Surveys AB, a company that is owned jointly by the United Kingdom. Hydrographic Office (49%), Blom ASA (21%) and AHAB (30%). The Admiralty company has been offering the system for bathymetric surveys on a worldwide basis and has already carried out a number of survey projects successfully (Figure 2.51). However the Blom company has just bought out its partners and is now operating the system on its own behalf.

The infrared ($\lambda = 1064$ nm) and green ($\lambda = 532$ nm) pulsed lasers that are used in the rangefinder of the Hawk Eye II are bought from a German company. They are then customized and integrated with the optics, receivers and electronics manufactured to AHAB's own specification (Figure 2.52). The laser pulse rates are 64 kHz for the infrared (topographic) laser and 6 kHz for the green (bathymetric) laser. The rangefinder is tilted 15°–20° from the nadir pointing forwards in the direction of the line of flight. The position and orientation system that is used in the Hawk Eye II is an Applanix POS/AV 410 GPS/IMU system. A small-format (2 megapixel) digital frame camera is used in conjunction with the laser scanning operations. The total weight of the system is 180 kg. The Hawk Eye II

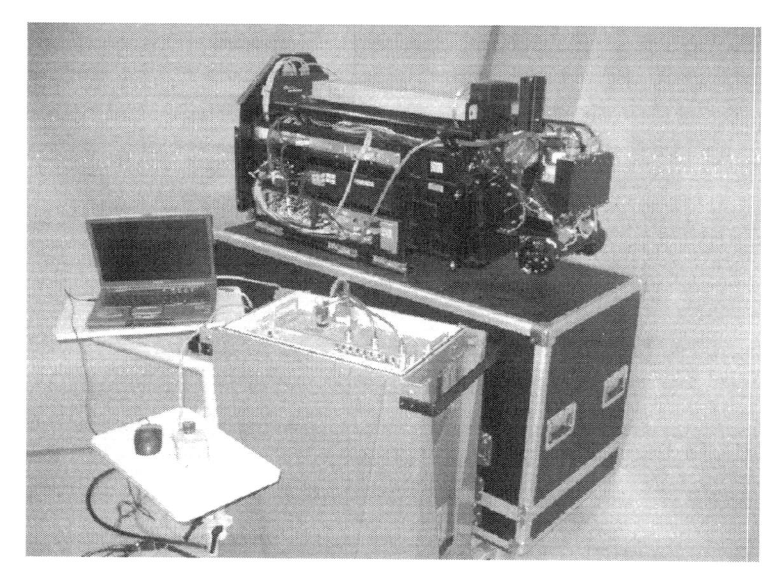

FIGURE 2.52
Hawk Eye Mk. II airborne bathymetric laser scanner sitting on its transport box, together with its control electronics unit located in front of it. (From Airborne Hydrography AB.)

is normally operated from a height of 200–300 m for IHO Class I surveys. The measured data can be displayed in real time on the system monitor and can, if required, be transferred in real time by a wireless communication link to a shore station or a survey ship. The data processing is carried out using AHAB's own Coastal Survey Studio (CSS) software, after which, the data can be passed to industry standard hydrographic chart production software such as CARIS for the generation of the final chart.

2.6.2.3 Tenix

Following on from the initial WRELADS systems of the 1980s, in 1989, the RAN awarded a contract to a partnership of two Australian companies—BHP Engineering and Vision Systems—for the construction of its LADS. The system was completed and brought into operational use by the RAN in 1993. The LADS system is mounted on a Fokker Friendship F27–500 aircraft equipped with twin turbo-prop engines, which is based at Cairns in Northern Queensland. It has been used continuously for bathymetric surveys, principally of the northern and eastern coasts of Australia, ever since its introduction into service in 1993. In 1998, a second system—LADS-II—entered service mounted on a De Haviland Dash-8 twin turbo-prop aircraft (Figure 2.53). This second system has been operated commercially by a division of the Tenix Corporation, which is a large Australian defence contractor that had bought out the original partnership that had constructed the LADS systems. The Tenix LADS company also provides the support for the original RAN LADS system, which is operated jointly by a team of Tenix and RAN personnel. The Tenix LADS-II system has been operated worldwide, including the execution of extensive bathymetric surveys of Alaskan coastal waters for NOAA; surveys of coastal areas for official agencies in the United Kingdom, Ireland and Norway; and the mapping and charting of the sea around Qatar and Dubai, as well as surveys in Australian waters, supplementing those being carried out by the RAN's LADS system (Figure 2.54).

FIGURE 2.53
De Haviland Dash-8 aircraft belonging to the Tenix LADS company undertaking a survey of coastal waters. (From Tenix LADS.)

Like the SHOALS and Hawk Eye systems, the LADS systems use infrared and green lasers for the measurements of depth. However, the LADS systems have the infrared laser pointing vertically downwards while the green laser points forward, both being mounted on a stabilized platform. The laser pulse rate that is used in the rangefinder of the LADS-II system is 900 Hz. As usual, the LADS systems incorporate a GPS/IMU

FIGURE 2.54
Control station of the LADS-II system that is installed in the Dash-8 aircraft. (From Tenix LADS.)

system for the measurement and generation of the required position and attitude data. Operated from a flying height of 500 m, the LADS-II scanner covers a swath width of 240 m with a sounding density of 5 × 5 m. At the time of writing, the original RAN LADS system is being upgraded to a higher specification "including a more capable laser system with greater positional capability, increased depth range and more detailed seabed coverage" according to the RAN Web site.

2.6.2.4 NASA

NASA's WFF designed and built its EAARL (Experimental Advanced Airborne Research Lidar) system primarily for the purposes of marine science research rather than the systematic collection of water depth data that the previous systems described above in this section have been used for (Figure 2.55). The EAARL system has been operational since 2001. It differs from the previously described systems in that it utilizes a single (green) laser rather than the twin (infrared + green) lasers of these previous systems. The single Nd:YAG laser, which emits its pulses at $\lambda = 532$ nm, is comparatively low-powered (70 μJ); has a high pulse rate (3000 Hz); and generates a well collimated pulse, giving a 15–20 cm diameter illumination spot at the water surface. However, the lower power also means that the maximum depth that can be measured is 20–25 m. The laser rangefinder is pointed vertically downwards, the scan mechanism with its oscillating mirror producing the corresponding pattern of depth measurements. When operated from a flying height of 300 m (1000 ft), the EAARL system covers a 240 m wide swath, with each scan providing 120 measured depths, having a 2 m spacing between these points in the cross-track direction. The scan rate is 20 Hz. The total weight of the system is comparatively low at 114 kg (250 lb), as are its power requirements –400 W at 28VDC.

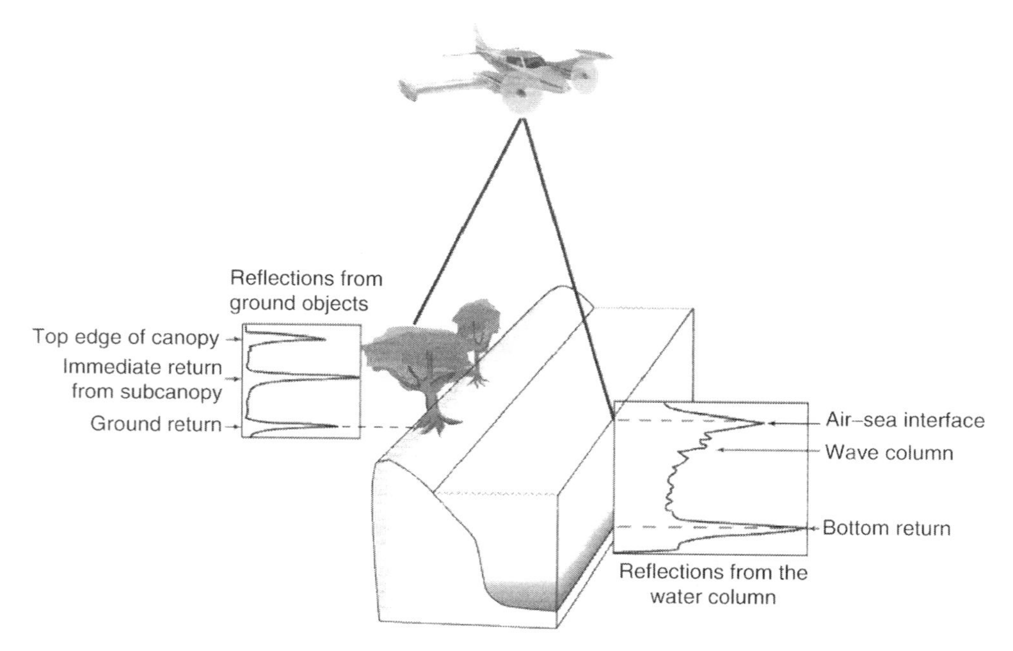

FIGURE 2.55
Overall concept of the EAARL system. (From NASA; Redrawn by M. Shand.)

The receiver side of the EAARL system is quite unique, comprising four subnanosecond photodetectors connected directly to four high-speed waveform digitizers. Each of the four photodetectors receives a fraction of the photons reflected from the sea and seabed surfaces directly below the aircraft. The most sensitive channel receives 90% of the photons, while the least sensitive channel receives 0.9%. The two middle channels share the rest. All four channels are digitized synchronously to an 8-bit level with the digitization commencing a few nanoseconds before the laser pulse is triggered and ending over 16,000 ns later. A small portion of each emitted laser pulse is sampled by a fiber-optic probe and injected in front of one of the photodetectors to capture the actual shape, timing and amplitude of the pulse as it is generated. The reflections are also digitized (Figure 2.55) and the resulting waveforms—which have been sampled at a 1 ns interval—are then analyzed in real time to locate key features such as the digitized transmitted pulse and the first and last returns. The reflections from the water surface are strong and will usually saturate the most sensitive channel in the receiver. In this case, the range will be resolved by the least sensitive channel. The reflections from the sea floor will be weak and the range will therefore be resolved using the most sensitive channel. A waveform having a complex shape will be produced from the multiple reflections from the different vegetation layers lying within the footprint of the laser pulse. The real-time processor automatically adapts to each returned waveform and records only the relevant portions of the waveform for further processing.

Besides the actual laser ranger, the EAARL system also comprises two digital frame cameras, both of which acquire their images at a 1 Hz rate. The first of these is an RGB color camera generating images with a Ground Sampled Distance (GSD) of 70–90 cm. The second camera generates false-color (green + red + near infrared) images with a GSD of 20 cm. The system also has a position and orientation system comprising two dual-frequency, carrier phase GPS receivers and an integrated miniature digital IMU, which together ensure submeter geo-referencing of each laser measurement. The 1 ns temporal resolution of the waveform digitizer equates to 13.9 cm in the air and 11.3 cm in water. The actual ranging accuracy of the system is quoted as being ±3–5 cm, while the horizontal positioning accuracy is quoted as being ≤1 m. The EAARL system has been used extensively for research mapping purposes in many coastal areas of South Florida, Puerto Rico and various American islands in the Caribbean Sea. Numerous other surveys have been undertaken in the immediate aftermath of the many devastating hurricanes that have hit the coastlines of the southern and eastern coasts of the United States in recent years. These have provided many spectacular images that have been displayed on a number of Web sites showing the resulting breaches in the barrier islands and coastal defences and the damage that has been wrought in these coastal areas.

References and Further Reading

Abshire, J.B., Sun, X., Riris, H., Sirota, J.M., McGarry, J.F., Palm, S., Yi, D., and Liiva, P., 2005, Geoscience laser altimeter system (GLAS) on the ICESat mission: On-orbit measurement performance. *Geophysical Research Letters*, Vol. 32, L21S02, 4 pp.

Al-Bayari, O.A., Al-Hanbali, N.N., Barbarella, M., and Nashwan, A., 2002, Quality assessment of DTM and orthphoto generated by airborne laser scanning system using automated digital photogrammetry. Proceedings ISPRS Commission III Symposium, Photogrammetric Computer Vision, Graz, September 9–13, 2002, 5 pp.

Artes, F., 2006, Fiber-based laser technology: On the cutting edge of today's Lidar systems. *GeoInformatics*, Vol. 9, No. 7, pp. 24–27.

Baltsavias, E.P., Favey, E., Bauder, A., Bosch, H., and Pateraki, M., 2001, Digital surface modelling by airborne laser scanning and digital photogrammetry for glacier monitoring. *Photogrammetric Record*, Vol. 17, No. 98, pp. 243–273.

Blair, J.B., Rabine, D.L., and Hofton, M.A., 1999, The laser vegetation imaging sensor: A medium-altitude, digitisation-only, airborne laser altimeter for mapping vegetation and topography. *ISPRS Journal of Photogrammetry and Remote Sensing*, Vol. 54, No. 2–3, pp. 115–122.

Carabajal, C.C., Harding, D.J., Luthcke, S.B., Fong, W., Rowton, S.C., and Frawley, J.J., 1999, Processing of shuttle laser altimeter range and return pulse energy data in support of SLA-02, Proceedings ISPRS Workshop Mapping Surface Structure and Topography by Airborne and Spaceborne Lasers, La Jolla, CA, *International Archives of Photogrammetry and Remote Sensing*, Vol. 32, Part 3-W14, pp. 65–72.

Favey, E., Pateraki, M., Baltsavias, E.P., Bauder, A., and Bösch, H., 2000, Surface modelling for alpine glacier monitoring by airborne laser scanning and digital photogrammetry, 19th ISPRS Congress, Amsterdam, The Netherlands, *International Archives of Photogrammetry and Remote Sensing*, Vol. 33, Part 4A, pp. 269–277.

Flood, M., 2001, Lidar activities and research priorities in the commercial sector. *International Archives of Photogrammetry and Remote Sensing*, Vol. 34–3/W4, pp. 3–7.

Geiger, A., Kahle, H.-G., and Limpach, P., 2007, Airborne laser profiling. ETH Research Database, ETH, Zurich, 15 pp.

Guenther, G.C., Cunningham, A.G., LaRoque, P.E., and Reid, D.J., 2000, Meeting the accuracy challenge in airborne Lidar bathymetry. Proceedings, 20th EARSeL Symposium—Workshop on LiDAR Remote Sensing of Land and Sea. Dresden, Germany, June 16–17, 2000, 28 pp.

Harding, D.J. and Carabajal, C.C., 2005, ICESat waveform measurements of within-footprint topographic relief and vegetation vertical structure. *Geophysical Research Letters*, Vol. 32, L21S010, 4 pp.

Harding, D.J., Gesch, D.B., Carabajal, C.C., and Luthcke, S.B., 1999, Application of the shuttle laser altimeter in an accuracy assessment of GTOPO30, a Global 1 km Digital Elevation Model, Proceedings ISPRS Workshop Mapping Surface Structure and Topography by Airborne and Spaceborne Lasers, LaJolla, CA, *International Archives of Photogrammetry and Remote Sensing*, Vol. 32, Part 3-W14, pp. 81–85.

Heslin, J.B., Lillycrop, J., and Pope, R.W., 2003, CHARTS: An evolution in airborne lidar hydrography. U.S. Hydro 2003 Conference, Biloxi, Mississippi, 4 pp.

Hug, C., 1994, The scanning laser altitude and reflectance sensor—An Instrument for Efficient 3D Terrain Survey. *International Archives of Photogrammetry and Remote Sensing*, Vol. 30, Part 1, pp. 100–107.

Irish, J.L. and Lillycrop, J., 1999, Scanning laser mapping of the coastal zone: the SHOALS System. *ISPRS Journal of Photogrammetry and Remote Sensing*, Vol. 54, No. 2–3, pp. 123–129.

Jenkins, L.G., 2006, Key drivers in determining lidar sensor selection. 5th FIG Regional Conference—Promoting Land Administration and Good Governance, Accra, Ghana, March 8–11, 2006.

Jensen, H. and Ruddock, K.A., 1965, Applications of a laser profiler to photogrammetric problems. Brochure published by Aero Service Corporation, Philadelphia, 22 pp.

LaRoque, P.E. and West, G.R., 1990, Airborne Lidar hydrography: An introduction. Proceedings ROMPE/PERSGA/IHB Workshop on Hydrographic Activities in the ROPME Sea Area and Red Sea, Kuwait, October 24–27, 1990.

Lindenberger, J., 1989, Test results of laser profiling for Topographic Terrain Survey. *Photogrammetric Week'89*, Special Publications of the Institute of Photogrammetry, University of Stuttgart, Vol. 13, pp. 25–39.

Lohr, U., 1999, High resolution laser scanning, not only for 3D-City Models. *Photogrammetric Week'99* (Eds. Fritsch, D. and Spiller, R.), Wichmann Verlag: Heidelberg, pp. 133–138.

McDonough, C., Dryden, G., Sofia, T., Wisotsky, S., and Howes, P., 1979, Pulsed laser mapping system for light aircraft. Proceedings. ASPRS Annual Conference, 10 pp.

Miller, B., 1965, Laser altimeter may aid photo mapping. *Aviation Week and Space Technology*, March 29, 1965.

Nelson, R., Parker, G., and Hom, M., 2003, A portable airborne laser system for forest inventory. *Photogrammetric Engineering and Remote Sensing*, Vol. 69, No. 3, pp. 267–283.

Petrie, G., 2006, Airborne laser scanning: New systems and services shown at INTERGEO 2006. *GeoInformatics*, Vol. 9, No. 8, pp. 16–23.

Pfeiffer, N. and Briese, C., 2007, Geometrical aspects of airborne laser scanning and terrestrial laser scanning. *International Archives of Photogrammetry and Remote Sensing*, Vol. 36, Part 3-W52, pp. 311–319.

Pope, R.W., Johnson, P., Lejdebrink, U., and Lillycrop, W.J., 2001, Airborne lidar hydrography: vision for tomorrow. U.S. Hydro Conference, Norfolk, Virginia, May 22–24, 2001.

Roth, R., 2005, Trends in sensor and data fusion. *Photogrammetric Week 05* (Ed. Fritsch, D.), Wichmann Verlag, Heidelberg, pp. 253–261.

Shepherd, E.C., 1965, Laser to watch height. *New Scientist*, April 1, 1965, p. 33.

Skaloud, J., Vellet, J., Keller, K., and Veyssiere, G., 2005, Rapid large scale mapping using hand held lidar/CCD/GPS/INS sensors on helicopters. ION GNSS 2005, Long Beach, California, September 13–16, 7 pp.

Spikes, V.B., Csatho, B., and Whillans, I.M., 1999, Airborne laser profiling of Antarctic ice stream for change detection. *International Archives of Photogrammetry and Remote Sensing*, Vol. 32, Part 3-W14, La Jolla, CA, 7 pp.

Thiel, K.-H. and Wehr, A., 1999, Advanced processing capabilities with imaging laser altimeter ScaLARS. SPIE Conference on Laser Radar Technology and Applications—IV, Orlando, Florida, April 1999. SPIE Vol. 3707, 0277–786X/99, pp. 46–56.

Thiel, K.-H. and Wehr, A., 2004, Performance capabilities of laser scanners—An overview and measurement principle analysis. *International Archives of Photogrammetry and Remote Sensing*, Vol. 36–8/W2, pp. 14–18.

Toth, C.K., 2004, Future trends in LiDAR. ASPRS Annual Conference, Denver, Colorado, May 23–28, 2004. Paper No. 232, 9 pp.

Vallet, J. and Skaloud, J., 2004, Development and experiences with a fully digital handheld mapping system operated from a helicopter. *International Archives of Photogrammetry and Remote Sensing*, Vol. 35, Part B, Commission V, 6 pp.

Vallet, J. and Skaloud, J., 2005, Helimap: Digital imaging/lidar handheld airborne mapping system for natural hazard monitoring. Sixth Geomatic Week Conference, Barcelona, 10 pp.

Wehr, A. and Lohr, U., 1999, Airborne laser scanning—An introduction and overview. *ISPRS Journal of Photogrammetry and Remote Sensing*, Vol. 54, pp. 68–82.

Wever, C. and Lindenberger, J., 1999, Experiences of 10 years laser scanning. *Photogrammetric Week '99*, pp. 125–132.

Wiechert, A., 2004, Linking laser scanning to surveying and visualization: Integrating geo-technologies with a new scanning method. *GeoInformatics*, Vol. 7, No. 5, pp. 48–52.

3

Terrestrial Laser Scanners

Gordon Petrie and Charles K. Toth

CONTENTS

3.1 Introduction

As discussed in the introduction to Chapter 1, there has been a widespread use of lasers in land and engineering surveying for the last 30 years. This can be seen in the incorporation of lasers into standard surveying instruments such as total stations and the laser rangefinders, profilers, levels and alignment devices that are in daily use within these fields of activity. Given this background, it would appear to be a natural development for scanning mechanisms to be added to total stations that were already equipped with laser rangefinders and angular encoders and for a terrestrial or ground-based laser scanner to be developed for use in field surveying. In this way, instead of individual points being measured on very specific ground features to a high degree of accuracy by a field surveyor using a total station—as required for large-scale cadastral, engineering and topographic maps—the laser scanner would allow the automated measurement and location of tens or hundreds or thousands of nonspecific points in the area surrounding the position where the instrument had been set up, all within a very short time-frame. Besides this favorable technological background against which terrestrial laser scanner development could have been expected to take place, there was no requirement for the supporting geo-referencing technology of integrated GPS/IMU systems that were essential for the development of airborne profiling and scanning. Indeed the locations of the ground stations where a terrestrial laser scanner had been set up on a tripod could easily be carried out using the traditional traversing and resection methods that are used by land surveyors.

However, for whatever reasons, the development of terrestrial or ground-based laser scanners has tended to lag behind that of airborne laser scanners. As Pfeifer and Briese (2007) have pointed out, terrestrial laser scanning has matured and been accepted by the overall geoinformatics industry rather later than its airborne equivalents. It may be that this is partly the result of the fact that a large segment of the professional surveying community comprises individual practitioners or small partnerships who operate purely on a local basis. Many of these do not possess the financial resources needed to acquire laser scanners costing much more than the total stations, GPS receivers, theodolites and levels that are the traditional forms of instrumentation being used in surveying. In fact, in many cases, the first users came from the surveying departments of large airborne mapping companies who had already been using laser scanners and thus had a good understanding of the technology. In this particular context, it is also interesting to note that, when it did take place, much of the early development of terrestrial laser scanners, especially over short ranges, was carried out by companies that were active in the manufacture and supply of measuring instruments and equipment for use in industrial metrology rather than the mainstream suppliers of surveying instruments—though the latter have been quick to adopt the technology once it had been fully developed.

Another interesting consequence of the developments that have come from the metrology side has been the introduction and widespread adoption of phase measuring techniques for the laser rangefinders that are used in short-range terrestrial laser scanners. This is in complete contrast to the situation in airborne laser scanning where the pulse-ranging technique is totally dominant. Now that terrestrial laser scanners have become sufficiently developed and have been accepted by the surveying profession, they have proliferated quite quickly and in large numbers, even though the investment is by surveying standards considerable—although it is only around one-tenth of that required for the purchase of an airborne laser scanner.

Turning next to the vehicle-mounted laser scanners that form part of certain mobile mapping systems, the technology has, in many respects, more similarities to that of airborne laser scanning systems than that used in terrestrial laser scanners. In particular, the requirements for the continuous geo-referencing of the moving vehicle have resulted in the use of integrated GPS/IMU systems as essential components of the overall mobile mapping system. However a special problem with these systems concerns their application to the mapping of built-up urban areas where the interruptions in the GPS signals caused by high buildings result in poor satellite configurations or even a complete loss of signal (Bohm and Haala, 2005). Besides which, the reflections of signals from nearby buildings can give rise to multipath effects which can reduce the positional accuracy of the GPS measurements. These deficiencies in the GPS measurements have resulted, in some cases, in their supplementation by distance measuring devices attached to the wheels of the vehicles on which the laser scanners have been mounted. In this context, it should also be noted that many terrestrial laser scanners are permanently installed on vehicles to improve their production efficiency—basically, to allow for their faster deployment in the field and, in general, to give more flexibility. In most cases, the laser scanner is attached to a mast that can be erected to higher positions, providing a better observation potential as compared with tripod-mounted installations. Obviously, these "stop-and-go" systems do not fall into the category of mobile mapping systems.

However terrestrial laser scanners, whether tripod- or vehicle-mounted, do have the great advantage of providing the details of building facades as required for the production of realistic 3D city models. In this respect, the data derived from terrestrial laser scanners complements the data acquired using airborne stereo-photogrammetric and laser scanning methods. The latter can produce the overall 3D city model at roof or ground level, but lack detailed information on the facades. Thus a combination of the two methods (airborne and terrestrial) is becoming increasingly common for the production of these 3D city models (Bohm and Haala, 2005).

As noted above, the terrestrial laser scanners that are being utilized for topographic mapping and modelling operations can be operated either from a static (or stationary) position—for example, being mounted on a tripod over a ground mark—or they can be operated from a dynamic (or moving) platform such as a van, truck or railcar. Quite different instrumental or system designs will result, depending on which of these two main operational modes has been adopted for the acquisition of the terrestrial laser scan data for mapping operations (Ingensand, 2006). Thus the coverage of terrestrial laser scanner technology that will be given in this chapter has been organized to fall into these two distinct categories—those covering static and dynamic laser scanners respectively.

3.2 Static Terrestrial Laser Scanners

The types of laser scanner that will be considered in this section of the chapter are those that measure the topographic features that are present on the ground in the area around the fixed (static) position that has been occupied by the instrument. They do this through the simultaneous measurement of slant range by a laser rangefinder and the two associated angles by angular encoders in the horizontal and vertical planes passing through the centre of the instrument (Figure 3.1). In most cases, prior to the actual scanning process, the angular increments in both directions, comprising the azimuth and vertical rotations, can be set by the user. Typically the angular step sizes are set to identical values; thus the scanner provides an equal spatial sampling in an instrument-centered polar coordinate system.

These simultaneous measurements of distance and angle are carried out in a highly automated manner using a predetermined scan pattern often at a measuring rate of 1000 Hz or more. As discussed in Chapter 1, the distance measurements that are made by the laser rangefinder will utilize either the pulse ranging or the phase difference measuring

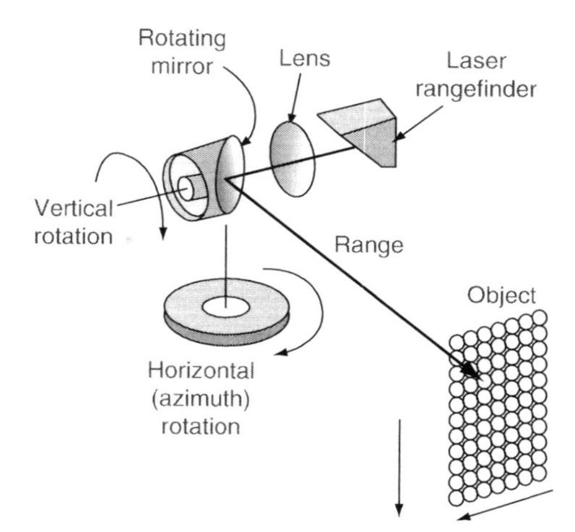

FIGURE 3.1
The range and the horizontal and vertical angular rotations that are measured toward objects in the terrain using a terrestrial laser scanner. (Drawn by M. Shand.)

technique. However, in this introductory discussion, it should be mentioned that there also exist a number of very short-range laser scanners, which are based on other measuring principles and which operate over ranges of a few meters, often to accuracy values of a fraction of a millimeter. A comprehensive review of the principles and technologies utilized in these ultrahigh accuracy, very short-range scanners is given in Blais (2004). These instruments are much used in metrology, industrial applications and reverse engineering; in body-scanning and medical research; and in the recording of objects by museum staff and archaeologists. A representative example of such an instrument—one among very many—is the Konica Minolta VIVID laser scanner. This operates on the basis of optical triangulation with the target or object being scanned with laser stripes whose reflected images are being recorded simultaneously by a digital camera. The maximum measuring range of such an instrument is only 2.5 m. These very short range laser scanners will not be considered here where the emphasis is exclusively on topographic applications.

3.2.1 Overall Classification

The primary classification that has been adopted widely in the published literature on static terrestrial or ground-based laser scanners differentiates between those instruments that utilize the pulse ranging or time-of-flight (TOF) measuring principle and those that employ the phase measuring technique—as set out in Chapter 1. Within the specific context of static terrestrial laser scanners, it can be said that the phase difference method produces a series of successive range measurements at a very high rate and to a high degree of accuracy, but only over distances of some tens of meters. Whereas the pulse-based TOF method allows much longer distances of hundreds of meters to be measured, albeit at a much reduced rate and a somewhat lower (though quite acceptable) accuracy as compared to those of the phase measurement method. The lower rate of the TOF method will be increased in the future using the multipulse technique.

However this simple classification based on the measuring technique that is used to measure distance does not take account of the angular scanning action that is used to ensure the required coverage of the ground and of the objects that are present on it. Therefore, it is essential to have a secondary classification which will account for both the scanning mechanism and the pattern or coverage that this produces over the ground.

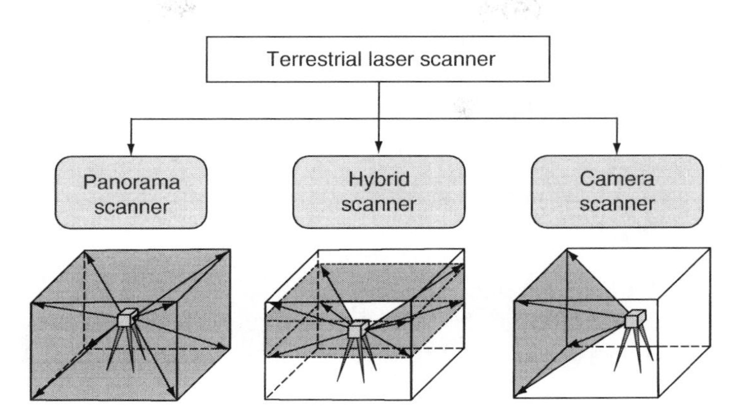

FIGURE 3.2
The classification of terrestrial laser scanners based on their respective scanning mechanisms and coverages. (From Staiger, R., Terrestrial laser scanning—Technology, systems and applications. Second FIG Regional Conference, Marrakech, Morocco, December 2–5, 2003; redrawn by M. Shand.)

The classification that has been adopted here is that introduced by Staiger (2003)—which differentiates between three types of static terrestrial or ground-based laser scanners—(i) panoramic-type scanners; (iii) hybrid scanners; and (iii) camera-type scanners (Figure 3.2).

1. Within the first of these three categories, panoramic-type scanners carry out distance and angular measurements in a systematic pattern that gives a full 360° angular coverage within the horizontal plane passing through the instrument's center and typically a minimum 180° coverage in the vertical plane lying at right angles to the horizontal plane—thus giving hemispheric coverage. However a still greater vertical field-of-view (FOV) of 270° or more is not uncommon—which means that a substantial coverage of the ground lying below the instrument's horizontal plane can be achieved. Indeed, the only gap or void in the coverage of a full sphere on a number of instruments is that produced by the base of the scanner instrument and its supporting tripod. While this panoramic scanning pattern is very useful in the context of topographic mapping, it is even more desirable, indeed often obligatory, in the measurement of complex industrial facilities, large quarries and open-cast mines and the facades of buildings within urban areas—or even indoors in large halls, churches, rooms, etc.
2. The instruments falling within the second category of hybrid scanners are those where the scanning action is unrestricted around one rotation axis—usually the horizontal scanning movement in the azimuth direction produced by a rotation of the instrument around its vertical axis. However, the vertical angular scan movement in elevation around the horizontal axis of the instrument is restricted or limited—typically to 50°–60°. This reflects the situation that is commonly encountered in medium- and long-range laser scanning carried out for topographic mapping purposes where there is no requirement to measure objects overhead or at steep angles, as will be needed within buildings.
3. The camera-type scanners that make up the third category carry out their distance and angular measurements over a much more limited angular range and within a quite specific FOV. Typical might be the systematic scanning of the surrounding

area over an angular field of 40° × 40° in much the same manner as a photogrammetric camera—at least in terms of its angular coverage, though obviously not in terms of the actual measurements that are being made and recorded.

As with airborne laser scanners, a third or tertiary classification can be envisaged based on the range or distance over which the static terrestrial or ground-based laser scanners can be used. The first group that can be distinguished comprises those laser scanners that are limited to short ranges up to 100 m—indeed some are limited to distances of 50–60 m. They mostly comprise those instruments that employ the phase measuring principle for distance measurement using the laser rangefinder—although, as will be seen later, there are instruments that employ pulse (TOF) ranging over these short ranges. Usually the limitations in range of these instruments are offset by the very high accuracies that they achieve in distance measurement—often to a few millimeters. A second group, almost entirely based on pulse ranging using TOF measurements of distance, can measure over medium ranges with maximum values from 150 to 350 m at a somewhat reduced accuracy. While a third long-range group, again using the pulse ranging technique, can measure still longer distances—up to 1 km or more in the case of the Optech ILRIS-3D instrument. However, the gain in range is normally accompanied by a reduction in the accuracy of the measured distances and in the pulse repetition rate—though this is still appropriate to the applications and is very acceptable to the users of these instruments.

3.2.2 Short-Range Laser Scanners

Both the main measuring principles outlined above—pulse ranging and phase measurement—are in use in those ground-based laser scanners that measure over short ranges—typically with maximum ranges from 50 to 100 m. It will be seen that all of the instruments that have been included in this group fall into the category of panoramic scanners. This enables them to be used indoors within buildings as well as outdoors where their operational characteristics make them well suited for use in the surveys and mapping of urban areas where the measurement of short distances and high vertical angles is required.

3.2.2.1 Short-Range Scanners Using Phase Measurement

Two of the most prominent manufacturers of this type of scanner are Zoller + Fröhlich (Z + F) and Faro—both of which are based in Germany. A third supplier is Basis Software Inc., which is based in the United States.

3.2.2.1.1 Zoller + Fröhlich

The current Z + F ground-based laser scanner is called the Imager 5006; a previous model from this manufacturer is the very similar Imager 5003. Both of these instruments have a servo motor and angular encoder that implements the angular scan movement in the horizontal plane (in azimuth) around the instrument's vertical axis through a rotation of the upper part of the instrument against its fixed base—which is normally mounted on a tripod so that it can be set accurately over a ground mark. The scan movement in the vertical direction is implemented using a lightweight, fast-rotating mirror placed on the horizontal trunnion axis of the instrument which is supported on two vertical standards or pillars (Figure 3.3). The horizontal rotation in azimuth covers the full circle of 360°, while the rotational movement of the mirror in these Imager scanners allows a scan angle of 310° within the vertical plane. The manufacturer's claimed accuracy in both horizontal and

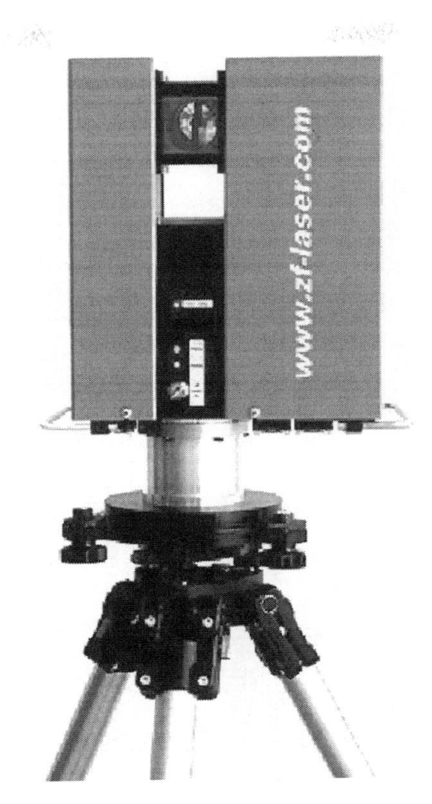

FIGURE 3.3
The Z+F Imager 5003 phase-based high-speed laser scanner. (From Zoller + Fröhlich. With permission.)

vertical angular measurement is ±0.007°—which is equivalent to ±6 mm accuracy at the measured points in both directions in the plane that is perpendicular to the laser direction at 50 m object distance. The maximum scan rate in the vertical (elevation) plane is 50 Hz; however a more typical rate in actual operational use is 25 Hz. The LARA laser rangefinder that is used in the Imager scanner instruments employs a Class 3R continuous wave (CW) laser that operates in the near infrared part of the spectrum at $\lambda = 780$ nm. In the earlier Imager 5003 model, the rangefinder was available in two alternative versions giving maximum ranges of 25.2 and 53.5 m, respectively (Mettenleiter et al., 2000). The current Imager 5006 model has a maximum range of 79 m. The accuracy in distance quoted by the manufacturer is ±6.5 mm over a range of 25 m. A detailed investigation into the accuracies of the distance and angular measurements of the Imager 5003 has been made by Schulz and Ingensand (2004) of the ETH, Zürich.

The processing of the measured range and angular data from the Imager scanners is carried out using the Z + F Light Form Modeller software. Besides measuring the distance and angles to the ground objects, the Z+F Imager scanners also measure the intensities of the signals that are reflected from these objects. This allows gray-scale reflectance images of the object to be formed. Data is stored either on the internal hard disk drive of the instrument or it can be transferred via an Ethernet interface to a laptop computer. An integrated control panel for the operation of the instrument is located on the side of one of the vertical standards of the instrument. The instrument can also be controlled remotely from a PDA using a wireless interface and connection. The Imager scanners can deliver in excess of 500,000 measured points per second, though in normal operation, the rate is likely to be somewhat lower.

FIGURE 3.4
The Leica Geosystems HDS6000 laser scanner which has the same performance as the Z + F Imager 5006. (From Leica Geosystems.)

The Z + F Imager 5003 instrument has also been marketed and sold very widely for mapping applications by Leica Geosystems as its HDS4500 laser scanner, where HDS is an acronym for "High Definition Surveying." This instrument has been available in the same two versions—with ranges of 25.2 and 53.5 m, respectively—as the Imager 5003 and with the same angular coverages of 360° in azimuth and 310° in the vertical plane. Similarly, the newer Z + F Imager 5006 instrument is now being offered by Leica as its HDS6000 scanner with the same product performance as the Z + F instrument (Figure 3.4).

3.2.2.1.2 Faro

The Faro company, which is mainly based in North America, is involved in the manufacture of a wide range of portable computer-based measuring instruments, including laser trackers, gauges and measuring arms, for use in industrial plants and factories. Its range of terrestrial or ground-based laser scanners was developed originally by the IQvolution company, based in Stuttgart, Germany which Faro purchased in 2005. The IQvolution company's laser scanner was called the IQsun 880. The instrument was renamed LS 880 after the take-over by Faro. Since then, two new shorter-range models—the LS 420 and LS 840—have been introduced to the market.

The basic design, construction and operation of all three models of this panoramic-type scanner is very similar. Furthermore, they all use a Class 3R continuous wave (CW)

FIGURE 3.5
The Faro LS 420 laser scanner that measures its ranges employing phase differences. (From Faro.)

semiconductor laser operating at $\lambda = 785\,\text{nm}$ in the near infrared part of the spectrum as the basis of their rangefinders. The output power of the laser is 20 mW in the case of the shortest-range LS 420 model (Figure 3.5) which has a maximum range of 20 m. The output powers of the lasers used in the two longer-range LS 840 and LS 880 models are 10 and 20 mW, respectively, allowing maximum ranges of 40 and 76 m to be measured respectively. With regard to the phase measuring technique used to measure distances, taking the LS 880 model as an example, the output laser beam is split and amplitude modulated to operate at three different wavelengths—76, 9.6, and 1.2 m—as shown in Figure 3.6. This allows the measured range to be determined to 0.6 mm in terms of its resolution value. The claimed accuracy of the measured ranges is ±3 mm at a distance of 10 m. The obvious advantage of using the phase measurement technique over the TOF or pulse ranging method is its speed of measurement. In the case of the LS 880 model, the laser rangefinder can measure distances at rates up to 120,000 points per second.

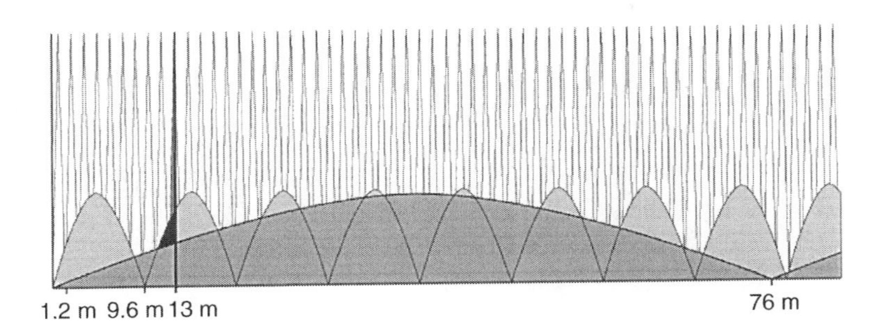

FIGURE 3.6
The measurement scheme used in the Faro LS 880 laser scanners showing the three different frequencies corresponding to distances of 76, 9.6, and 1.2 m, respectively at which the phase differences are measured. An example is shown for a measured distance of 13 m. (From Faro; redrawn by M. Shand.)

In all three models, the laser rangefinder is mounted in the horizontal plane and aligned with the horizontal (trunnion) axis of the instrument. The output beam from the laser passes into the centre of a continuously rotating (motor-driven) mirror that deflects it through a fixed angle of 90° to produce a vertical profile scan of the laser beam giving an angular coverage of 320° in the vertical plane. This places it in the category of panoramic scanners. The reflected signal that is returned from each point along the profile that is being scanned in the object field is then compared with the reference output signal to measure the phase differences, so determining the measured range. Besides which, the intensity of the return signal from each pulse that hits the object can also be measured to a 9-bit level. The 360° azimuth scan is implemented using a motor whose power is derived from a 24 V DC battery pack. The measured range, intensity and angular data can be recorded either directly on the instrument's internal hard disk or remotely via an Ethernet interface to an external laptop computer. The processing of the measured data is carried out using the Faro Scene software. Improved versions of the different models of the Faro scanners with upgraded electronics and better positional accuracy were introduced in March 2008 at the SPAR 2008 Conference under the name Photon and are now called the Photon 20 and Photon 80.

3.2.2.1.3 Basis Software Inc.

This American company is based in Redmond, Washington. Its main laser scanner product that can be used in topographic applications is the Surphaser (Figure 3.7), which again is a panoramic-type scanner. The company also manufactures short-range laser scanners for industrial use that will not be discussed here. The Surphaser is available in two models—the 25HS (Hemispheric Scanner), which has a maximum range of 22.5 m, and the 25HSX which has an extended range of 38.5 m. The basic design and construction of the Surphaser instruments is somewhat similar to that of the Faro scanners described above in that they both employ a laser diode rangefinder that emits its continuous beam of laser radiation

FIGURE 3.7
The Surphaser 25HS laser scanner produced by Basis Software. (From Basis Software.)

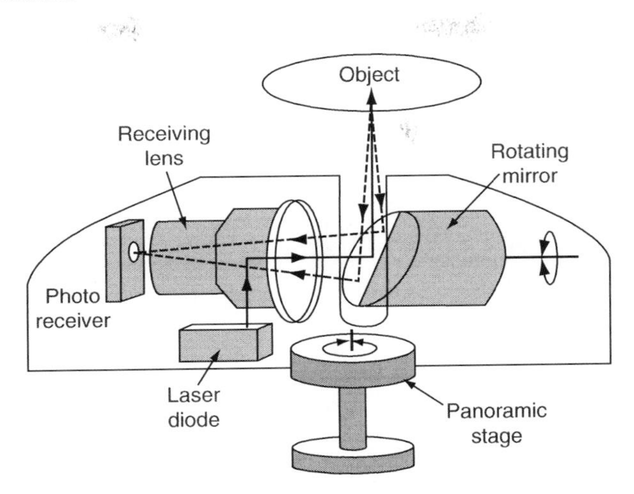

FIGURE 3.8
Diagram showing the design and the main features of the Surphaser laser scanner. (From Basis Software; redrawn by M. Shand.)

along the instrument's horizontal axis on to a continuously rotating mirror located at the end of a motor-driven shaft (Lichti et al., 2007). This turns the beam through a right angle to produce the required scanning motion in the vertical plane giving an angular coverage of 270° (Figure 3.8). The receiving lens focuses the reflected radiation from the object on to the photodiode that acts as the receiver allowing the continuous measurement of the phase differences and the intensity values. The maximum rate of range measurement of the laser rangefinder is 190,000 points per second, though often, in practice, a lower rate will be used. The semiconductor laser diode that is used in this instrument emits its radiation at the wavelength of 690 nm on the red edge of the visible part of the spectrum with a power of 15 mW. The 360° angular scan in the horizontal plane is implemented through the rotation of the upper part of the instrument which is driven in azimuth by a stepping motor and gearbox located in the lower (fixed) part of the instrument which can be fitted into a standard Wild/Leica tribrach. The instrument's power requirements are supplied by a standard 18/24 V DC battery. The measured data is transferred via a FireWire interface to be recorded on a laptop computer.

In November 2007, a version of the Surphaser was introduced by Trimble as the FX Scanner, the instrument being supplied by Basis Software to Trimble under an OEM agreement. The FX scanner is used in conjunction with Trimble's FX controller software, while Trimble's Scene Manager software is used to locate the positions where the scanner has been set up for its scanning and measuring operations. The measured data can then be used in Trimble's LASERGen suite of application software.

3.2.2.2 Short-Range Scanners Using Pulse Ranging

Only a single company—Callidus from Saxony in Germany—has adopted pulse ranging based on TOF measurement technology for its short-range laser scanners.

3.2.2.2.1 Callidus

Callidus has manufactured a short-range laser scanner based on pulse ranging for over 11 years—since 1996. Originally its scanner instrument was simply called Callidus 1.1. Later it was re-christened as the Callidus 3D Laser Scanner and, as such, it was also sold

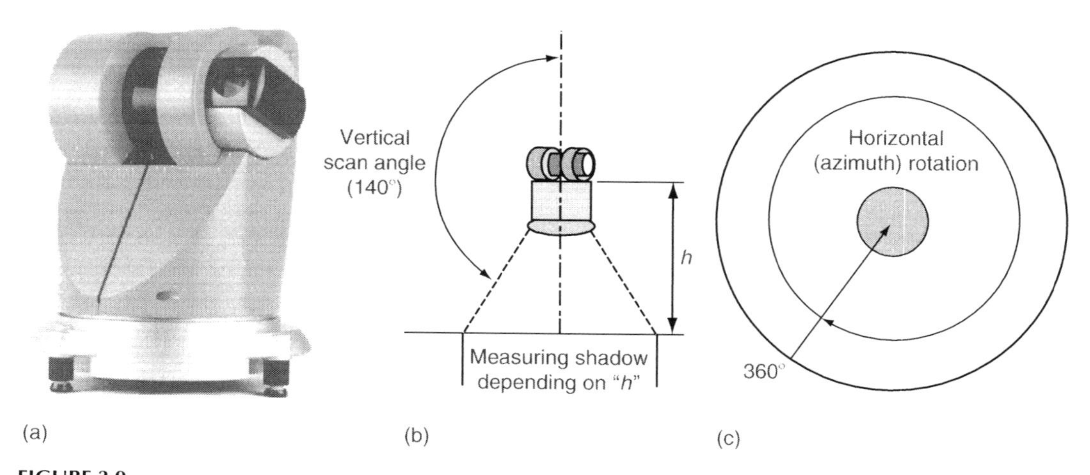

FIGURE 3.9
(a) The Callidus CP 3200 scanner that uses pulsed (TOF) based laser ranging technology, (b) vertical angular coverage, and (c) horizontal angular coverage of the CP 3200 of the instrument. (From Callidus; redrawn by M. Shand.)

by the Trimble company. Now it is known as the Callidus CP 3200 scanner (Figure 3.9a). In this context, it should be noted that the Callidus company also builds other very short-range laser scanners designed for the measurement of comparatively small objects over very short distances indoors—e.g., in factories or museums. These are the CT 900 and CT 180 instruments, both of which utilize laser triangulation for the measurement of object up to 1.5 m and 30 cm, respectively. As such, they will not be considered any further in this account which is concerned with the scanners that are used for topographic applications.

The CP 3200 instrument is a panoramic-type scanner providing a full rotation of 360° in azimuth (Figure 3.9c) in the horizontal plane driven by a servo motor that provides a range of angular step sizes between 0.0625° and 1.0° that are selectable by the operator. A rotatable mirror attached to an angular encoder is used to establish each successive step position over an arc of 280° in the vertical plane (Figure 3.9b). The angular step interval over this vertical scan is also selectable between 0.25° and 1.0° with a scan rate of up to 77 Hz. For each step position, the laser rangefinder measures the appropriate range in the vertical plane. The laser rangefinder that is used in the CP 3200 scanner utilizes a semiconductor diode laser with quite a large (0.25°) beam width operating at $\lambda = 905$ nm in the near infrared part of the spectrum. Different measuring ranges can be selected—up to 8 m; 32 and 80 m, respectively—according to the reflectivity of the surfaces and objects that are being measured. The quoted accuracies that can be achieved with the instrument are ±5 mm in distance at a range of 50 m and ±0.005° and ±0.009° in terms of its horizontal and vertical angles respectively. A typical measuring rate that is used with the instrument is 1750 points per second—although this rate is reduced if multiple measurements of each point are made. The resulting data is processed using Callidus's own 3D Extractor software. The CP 3200 also incorporates a software-controlled small-format (768 × 576 pixels) CCD camera with variable focal settings that can be used to generate digital frame images to supplement the range data being provided by the laser rangefinder.

Callidus has introduced its new CPW 8000 laser scanner at the Intergeo trade fair held in Leipzig in September 2007 (Figure 3.10). The instrument combines the two main mensuration methods—those of TOF pulse ranging and continuous wave phase measurement.

FIGURE 3.10
The Callidus CPW 8800 scanner utilizes a combination of pulsed (TOF) and phase-based laser ranging technologies. (From Callidus.)

The technique involves the basic measurement of the distance using the pulse ranging method. However the pulses are also modulated with a high frequency, which allows the phase difference between the emitted and received pulses to be measured. With this elegant solution, no attempt needs to be made to resolve the ambiguity that is inherent in a single phase measurement. Instead, the fine measurement of the distance given by the single phase measurement is combined with the overall distance value given by the pulse ranging measurement to provide the final range value to a high precision, ±2 mm at a range of 30 m. The maximum range of the instrument is quoted as being 80 m. A Class 3R laser emitting its pulses at $\lambda = 658$ nm in the red part of the spectrum is used as the basis of the instrument's rangefinder. A panoramic-type configuration ensures a full horizontal angular coverage of 360°, while the coverage of the scan in the vertical plane is 300° with the smallest angular resolution of 0.002° in both directions. Using this novel dual-measuring technique, a measuring rate of 50,000 measurements per second can be achieved—e.g., resulting in 54 min of scan time at an angular resolution of 0.02° in both directions.

3.2.3 Medium-Range Laser Scanners

As defined above, this group of scanners can measure distances over medium ranges with maximum values lying between 150 and 350 m. While most of the manufacturers of short-range laser scanners, such as Faro, Callidus and Basic Software, have strong interests in metrology—and indeed manufacture various other mensuration products that fall into that area—the situation is quite different with the manufacturers of medium-range laser scanners. In this area, several of the principal system suppliers of these scanners, such as Leica, Trimble and Topcon, are major manufacturers of surveying instrumentation such as GPS receivers, total stations and laser levels. The medium-range ground-based laser scanners that are manufactured and supplied by each of these companies are all based on the pulse ranging technique.

3.2.3.1 Leica Geosystems

Leica's entry into the field of terrestrial or ground-based laser scanners was made through its acquisition of the Cyra Technologies Inc. company in January 2001. This purchase was made at the same time as Leica's acquisition of another quite separate and independent American company, Azimuth, which ensured its entry into the field of airborne laser scanning—as already discussed in Chapter 2. The Cyra company had been set up originally in California in 1993. It developed its Cyrax terrestrial laser scanning instrument that finally entered the market in 1998. Leica first invested in the Cyra company as a minority shareholder in March 2000 before purchasing the rest of the shares in the company a year later. At first, the Cyra company kept its name, operating as a division of Leica Geosystems. However, in April 2004, its name was formally changed to Leica Geosystems HDS Inc.—HDS being an acronym for "High Definition Surveying."

The original model that was built and sold by the Cyra company was the Cyrax 2400. Its rangefinder used a Class 2 semiconductor diode laser operating at $\lambda = 532$ nm in the green part of the spectrum. This allowed a maximum speed of measurement of 800 points per second with a maximum range of 100 m—though 50 m was more realistic with objects having a moderate reflectivity. The stated accuracy in range was ±4 mm over a distance of 50 m. The Cyrax 2400 was a camera-type scanner that could scan a 40° × 40° FOV or window using a twin mirror optical scanning system. The scanner's main-body sat in a simple nonmotorized pan and tilt mount that allowed it to be pointed manually in steps over an angular range of 360° in azimuth and 195° in the vertical plane (Figure 3.11). It was followed by the improved Cyrax 2500 model (later called the Leica HDS2500) with

FIGURE 3.11
The Cyrax 2500 camera-type laser scanner with its manually operated tilt mount allowing an angular rotation around instrument's horizontal axis.

a similar specification (Sternberg et al., 2004). The Cyrax instruments were supplied with the accompanying Cyclone software.

In 2004, the popular Cyrax models were replaced by the Leica HDS3000 which had a very different design and specification. In particular, the camera-type layout of the previous Cyrax instruments was replaced by a dual-window design that gave a fully panoramic coverage of 360° in azimuth and 270° in the vertical plane. However the two windows are not in use simultaneously; thus two separate horizontal scans may be required to complete the required coverage or they may be used sequentially as required. The HDS3000 scanner used servo motors to rotate the scan mirror in the vertical plane and for the azimuth drive, resulting in a much higher scan rate than its Cyrax predecessors. The rangefinder used in the HDS3000 was based on a Class 3R laser, again operating in the green part of the spectrum (at $\lambda = 532\,\text{nm}$) and with a maximum operating range of 100 m—though the normal operational range over which high accuracy may be achieved was from 1 to 50 m. The accuracy of the measured distance was stated to be ±4 mm at a range of 50 m, while the maximum measuring rate was 1800 points per second. The HDS3000 scanner also incorporated a calibrated high-resolution digital video camera that was located internally within the instrument. This generated digital image data that could be overlaid on the scanned range and angular data using the Cyclone software.

In 2006, the HDS3000 was superseded by the ScanStation (Figure 3.12). This retained the overall design and construction of the HDS3000 panoramic-type instrument. However it had a much greater maximum operational range of 300 m; a maximum measuring rate of 4000 points per second; and the incorporation of a number of additional features such as a dual-axis compensator to allow conventional surveying operations such as resection and traversing to be carried out using the instrument. Most recently, in July 2007, the latest model in the series, called ScanStation 2, was introduced. This has the same overall design and appearance of the HDS3000 and the original ScanStation.

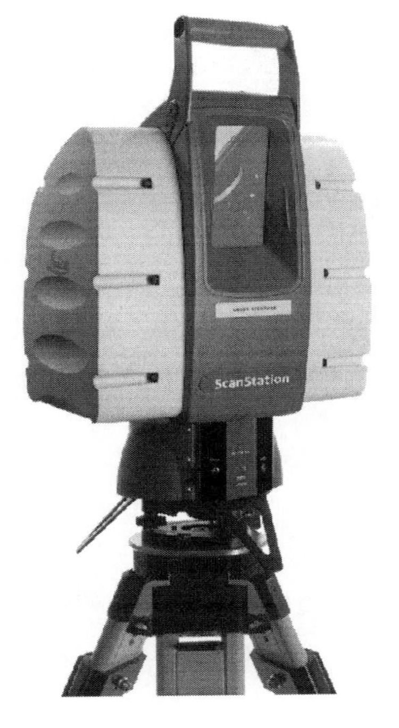

FIGURE 3.12
The Leica Geosystems ScanStation laser scanner with its scan mirror providing "panoramic" coverage of the surrounding area. (From Leica Geosystems.)

However the ScanStation 2 instrument has a new laser rangefinder and timing electronics that allows it to carry out its range measurements at a very much higher speed—with a maximum rate of 50,000 points per second—while still retaining the same maximum range of 300 m, with a reflectivity of 90% that had been achieved in the previous ScanStation model. Recently (in 2007) the production of the Scan Station 2 instruments has been moved from the factory in San Ramon, California to the main Leica manufacturing plant in Heerbrugg in Switzerland.

3.2.3.2 Trimble

Trimble is another of the major suppliers of surveying instrumentation that has entered the field of terrestrial or ground-based laser scanning through the acquisition of a much smaller company that specialized in this area. It did so in September 2003 through its purchase of the Mensi company based in Fontenay-sous-Bois in France—which had been one of the pioneering developers of laser measurement technology. This company had been founded in 1986 and had designed and built a number of short range laser scanners—in particular, the S-series (including the S10 and S25 models) based on laser triangulation—that were used principally in metrology applications carried out within industrial facilities.

In 2001, the Mensi company introduced a longer-range laser scanner in the form of its GS 100 instrument, based on pulse ranging (Figure 3.13). This was an early example of a hybrid type of scanner with a motorized 360° rotation in azimuth and a 60° angular rotation in the vertical plane using a lightweight scanning mirror. The angular resolution in the horizontal direction was 0.0018°, while that in the vertical direction was 0.0009°. The GS 100 instrument also used a rangefinder based on the use of a Class 2 laser operating at $\lambda = 532$ nm in the green part of the spectrum. The user could control the laser focus to produce a very small spot of 3 mm at 50 m to allow very precise measurements to be made. Alternatively an autofocus mode could be set, which allowed the laser spot to be refocused automatically according to the measured distance. The instrument could measure up to 5000 points per second with a measuring precision of ±5 mm. The data processing was carried out using Mensi's RealWorks Survey and 3Dipsos software packages that had been developed in-house.

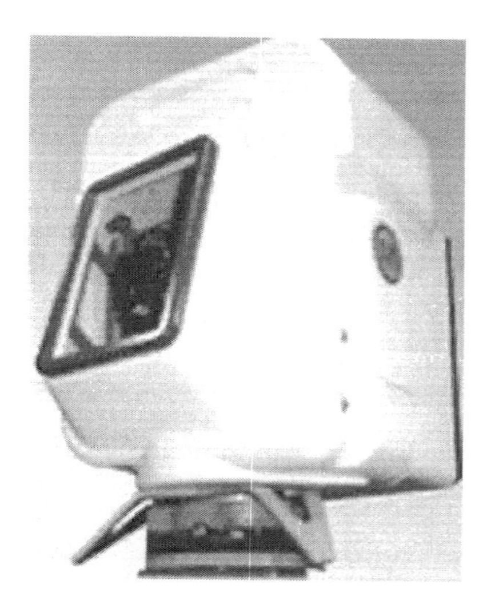

FIGURE 3.13
The original Mensi GS100 laser scanner.

A second model in the series, called the GS 200, followed soon after (Kersten et al., 2005). This had a similar design and angular coverage to the GS 100, but utilized more sophisticated receiver circuitry that allowed the maximum range to be doubled to 200 m and a higher precision of ±2 mm to be achieved. It also featured an overscan technique that allowed data to be captured at ranges up to 350 m. Furthermore the GS 200, like the GS100, incorporated a calibrated video camera with a zoom capability that was fitted internally within the instrument. This produced color video images with a format size of 768 × 576 pixels. The video images could be mosaiced and overlaid over the range and angular data that was being produced by the laser scanner. The GS 200 instrument could also be integrated together with a GPS receiver or a total station prism, these items being mounted on top of the laser scanner using a standard adapter.

The latest models in the series are the Trimble GX 3D scanners (Figure 3.14). These have a quite similar specification to that of the GS 200 model—with a hybrid-type angular coverage of 360° × 60° and a measuring rate of 5000 points per second. However, the instrument has been re-designed with a multishot capability to give an improved accuracy—with an absolute accuracy of ±3 to 8 mm depending on the range and the reflectivity of the objects being measured. It also incorporates automatic calibration of the zero index error together with a dual-axis compensator to correct any dislevelment of the instrument. The instrument also has the capability of polygonal framing—which allows the operator of the instrument to define an irregular shaped box, within which the instrument will carry out its scanning and measurements. The instrument also has a hand-held control unit which allows it to be operated remotely. In October 2007, Trimble introduced the GX Advanced model featuring its SureScan technology. This uses real-time data analysis to "regularize" the point density over the surface being measured. The user can set the desired grid interval and the algorithm changes the horizontal and vertical angles to provide an even density of points over the measured surface. Since March 2007, the production of the GX instruments has been shifted from France to the former Spectra Precision factory in Danderyd, a suburb

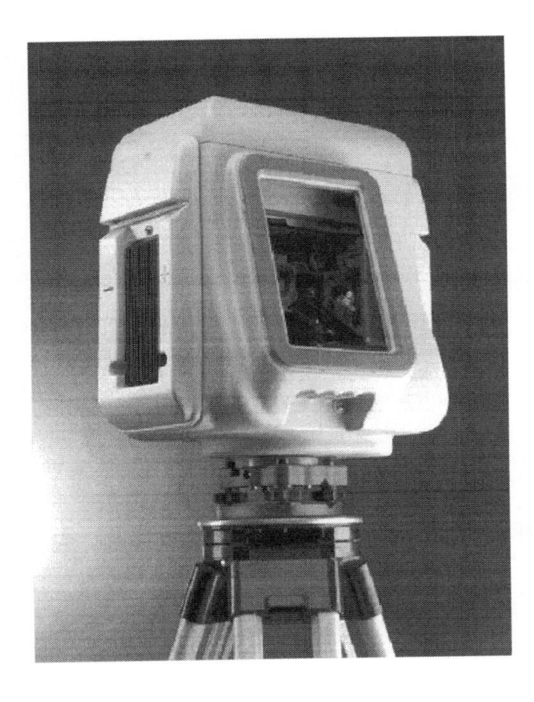

FIGURE 3.14
The latest Trimble GX 3D laser scanner. (From Trimble.)

of Stockholm, the Swedish capital city. Comparative tests of the latest Trimble GX and Leica ScanStation instruments employing the TOF method with the Z+F Imager 5006 and the Faro LS 880 scanners that use the phase measurement technique are reported in the paper by Mechelke et al (2007).

3.2.3.3 Topcon

Another of the major manufacturers of surveying instrumentation, Topcon, has announced that it will enter the field of terrestrial laser scanners early in 2008 with the introduction of its GLS1000 model (Figure 3.15). The preliminary information that has been released about this instrument is that it will be a hybrid-type instrument with a vertical angular coverage of 70°, besides the normal horizontal rotation in azimuth of 360°. The instrument's pulse-based rangefinder will feature a Class 1 laser that has a maximum range of 330 m to objects having a high (90%) reflectance and 150 m to those objects having a low reflectance (18%). The maximum rate of measurement is quoted as being 3000 points per second. The GLS1000 scanner will have an integral camera and will feature an integrated control panel located on the side of the instrument. The data measured by the scanner will be recorded on a removable storage card.

3.2.4 Long-Range Laser Scanners

Within this group of laser scanners, the maximum measuring range to highly reflective targets has been raised to more than 500 m; indeed, in some cases, it can be very substantially more. All of the instuments use pulse-ranging based on the TOF measuring principle. It will be

FIGURE 3.15
The newly announced Topcon GLS1000 terrestrial laser scanner. (From Topcon.)

noted that two of the principal suppliers—Optech and Riegl—are also the manufacturers of airborne laser scanners that measure over similar distances in terms of the flying heights at which they are operated.

3.2.4.1 Optech

The Optech company's activities encompass laser ranging and scanning devices operating in many different environments—spaceborne, airborne and terrestrial. Within this context, the Intelligent Laser Range Imaging System (ILRIS) was developed originally by Optech for the Canadian Space Agency (CSA) in the late 1990s, as a combined ranging and imaging device for use on-board spacecraft. In its original form, as the ILRIS-100, it used pulse ranging operating in both the visible (green) and infrared parts of the spectrum to scan a scene and produce simultaneous range, angular and intensity data at ranges up to 500 m. When processed, this data was combined to form a range and intensity data set that could be produced either in gray scale or color coded form that was 2 k × 2 k pixels (4 MB) in size.

In June 2000, Optech introduced its tripod-mounted ILRIS-3D version of the instrument (Figure 3.16). This had been developed specifically for topographic and open-cast mining applications on the one hand, and for industrial applications, especially the measurement and modelling of industrial plants and facilities, on the other. Thus the instrument was designed and constructed with long-range capabilities from the outset. Using a Class 1 laser rangefinder emitting its infra-red radiation at $\lambda = 1550$ nm, it had a maximum range of 800 m with a range resolution of 1 cm, even with a target having only 20% reflectance and a range of 350 m with a very low-reflectance (4%) target such as a coal stockpile. A measuring rate of 2000 points per second could be utilized while scanning over a 40° × 40° FOV using two internal deflection mirrors, producing a camera-type

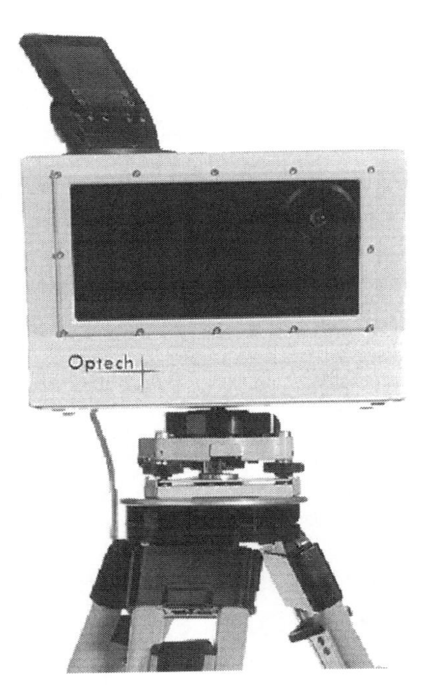

FIGURE 3.16
An early model of the Optech ILRIS-3D laser scanner. (From Optech.)

configuration. The measured data was written on interchangeable flash cards which were then transferred to a PC for postprocessing. An infrared interface allowed a palmtop computer to be used for set-up and control purposes. The main case of the ILRIS-3D instrument also contained a bore-sighted small-format digital camera giving a 640 × 480 pixel image. This camera also included an LCD viewfinder fitted to the back of the instrument that could be used both for set-up purposes and for display of the captured data. Options included a differential GPS unit and an attitude measurement system.

In 2004, a substantially upgraded model of the ILRIS-3D instrument was introduced. This provided an increased range (beyond 1000 m to highly reflective targets); an improved accuracy; an integrated CMOS-based camera giving a 6 megapixel image; and an integrated handle for carrying purposes. As noted above, the ILRIS-3D was a camera-type instrument with a fixed FOV of 40° × 40°. In order to cover a much larger area, Optech also introduced the ILRIS-36D version of the instrument, the first examples of which were shipped in May 2005. This instrument was equipped with a motorized pan-and-tilt base that allowed the scanner to cover a 360° × 230° FOV (Figure 3.17). For this to be implemented, the motorized base unit moves the scanner unit of the ILRIS-3D with its 40° × 40° FOV in a series of steps that are measured by angular encoders. Each 40° × 40° scan patch or window overlapped on its neighbors by 5°. However, in September 2007, a new profiling feature eliminated the need for these overlaps using multiple scan windows.

The basic ILRIS-3D instrument continues to be developed. In October 2006, two new optional features were introduced. The Enhanced Range (ER) option increases the range still further by 40% for use in open-cast mining and large-area topographic surveys.

FIGURE 3.17
The latest ILRIS-36D laser scanner with its motorized pan and tilt base and its carrying handle. (From Optech.)

The Motion Compensation (MC) option is designed to allow the ILRIS-3D instrument to be used from a moving or dynamic platform such as a boat or a vehicle and is based on the use of an additional integrated GPS/IMU system to provide the position and orientation data required for the motion compensation. Finally in March 2007, Optech introduced a so-called Value Package (VP) version of the ILRIS-3D which retains the main capabilities of the instrument in respect of its range and accuracy of measurement, but in a less expensive form. For the processing of the measured ILRIS-3D data, Optech utilizes the Quick Terrain Modeler from the Applied Imagery company based in Maryland, and modules from the Polyworks software package developed by another Canadian company, InnovMetric Software from Quebec.

3.2.4.2 Riegl

This company builds a number of long-range laser scanners based on pulse ranging that are entitled the LMS-Zxxx series. Currently this series comprises four different models—the LMS-Z210ii, LMS-Z390i, LMS-Z420i and LMS-Z620—having maximum measuring ranges of 650, 400, 1000 and 2000 m, respectively with objects having a reflectance of 80%. They will be treated here together since they all have similar design characteristics. For objects with a lower reflectance, the ranges measured by the four instruments will be much less—200, 100, 350, and 650 m, respectively for a reflectivity value of 10%. The manufacturer's claimed accuracy of distance measurement is ±4 to 15 mm for a range of 50 m. All four LMS-Zxxx instruments utilize the same type of laser engine with a continuously rotating optical polygon that is placed in front of the laser rangefinder to produce the basic optical scan—as used also in Riegl's airborne laser scanning engines. The rangefinder itself uses a Class 1 laser operating in the near infrared part of the spectrum either at $\lambda = 905$ or 1550 nm. However, with the ground-based systems, the rangefinder is placed in a vertical position pointing upward instead of the horizontal position that is used with the airborne version (Figure 3.18c).

The pulses are then directed toward the ground and its objects at the appropriate vertical angle by the rotating optical polygon. Thus the angular scan takes place within the vertical plane, instead of the plane containing the cross-track scan of the airborne versions of the laser scanning engine. The angular scanning range of the ground-based laser engine is 80°—which is normally implemented to provide ±40° above and below the horizontal plane when the rangefinder is set pointing vertically upward in its normal operating position. For the required horizontal angular rotation, the upper part of the instrument containing the rotating optical polygon head is moved in steps in azimuth through the full 360° rotation against the fixed lower part of the instrument that contains the laser rangefinder. This angular coverage of 360° × 80° places the instrument in the hybrid class of laser scanners—as defined above in Section 3.1.1. The rotational steps in azimuth within the horizontal plane—which define the intervals between successive vertical scan lines—can be set at different angular intervals, for example, between 0.004° and 0.075° in the case of the LMS-Z420i instrument. The maximum measuring rates that can be achieved using pulse ranging with these LMS-Zxxx models lies between 8,000 and 12,000 points per second, although often lower rates will be used in actual practice. The measured data will be transmitted via the instrument's built-in Ethernet interface and recorded on a laptop computer. It can then be processed using Riegl's own RiSCAN Pro software package that comes bundled with the laser scanners. A GPS-Sync option is available for the LMS-Zxxx series which allows each set of measurements to be time-stamped with respect to GPS time.

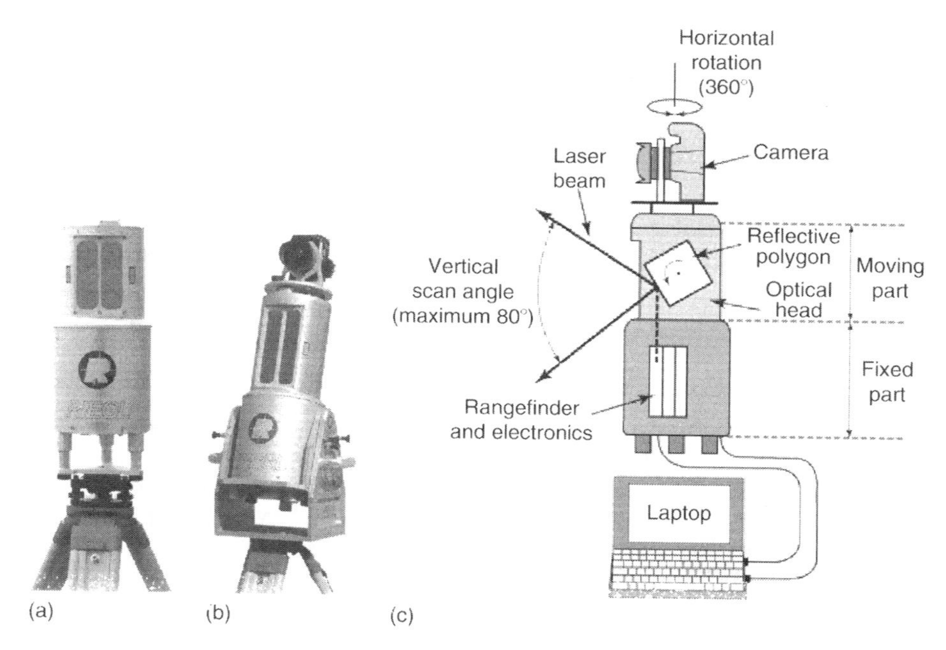

FIGURE 3.18
(a) A Riegl LMS-Z210i laser scanner sitting on a tribrach. (b) This example of a Riegl LMS-Z390i laser scanner is shown with an additional mount having a horizontal axis around which the main part of the instrument can be rotated and operated at an oblique angle to the vertical—if that is required. This particular example is also fitted with a calibrated small-format digital frame camera producing colored images. (c) Diagram showing the general layout of the Riegl LMS-Zxxx series of terrestrial laser scanners. (From Reigl; redrawn by M. Shand.)

This allows the integration of the laser scanner with a GPS/IMU system to provide position and orientation data, as required for dynamic scanning operations.

The Riegl laser engine with its rotating optical polygon head can be mounted on a conventional survey tribrach and operated with its main optical axis set in a vertical position using the tribrach footscrews (Figure 3.18a). However, the four LMS-Zxxx models can all be fitted optionally with a mount, with two vertical posts that provides an additional horizontal rotation axis that allows the scanner engine as a whole to be inclined at an oblique angle to the vertical axis (Figure 3.18b). The various models can also be equipped optionally with a small-format CCD digital frame camera that sits on a mount on top of the main scanner engine (Figure 3.18b and c). Currently, users are offered a choice between the Nikon D70s or D100 models (providing a 6.1 megapixel image), the D200 model (giving a 10.2 megapixel image) or the D300 model (giving a 12.3 megapixel image). If a still larger-format image is required, then the Canon EOS-1Ds Mark II camera with its 16.7 megapixel frame image is offered as an alternative. The resulting color images are transmitted via a USB interface to the laptop computer where the image data can be fused with the range data from the laser scanner.

Besides the LMS-Zxxx series of scanners, over the last decade, Riegl has also manufactured a series of long-range laser profile measuring systems. In 2008, it introduced the latest model in the series, which is called the LPM-321 profiler. This has a very different mechanical and optical design to that of the LMS-Zxxx scanners with the laser rangefinder unit mounted on the horizontal (trunnion) axis of the instrument, supported by a single pillar (Figure 3.19a). The rangefinder can measure ranges up to a maximum distance of

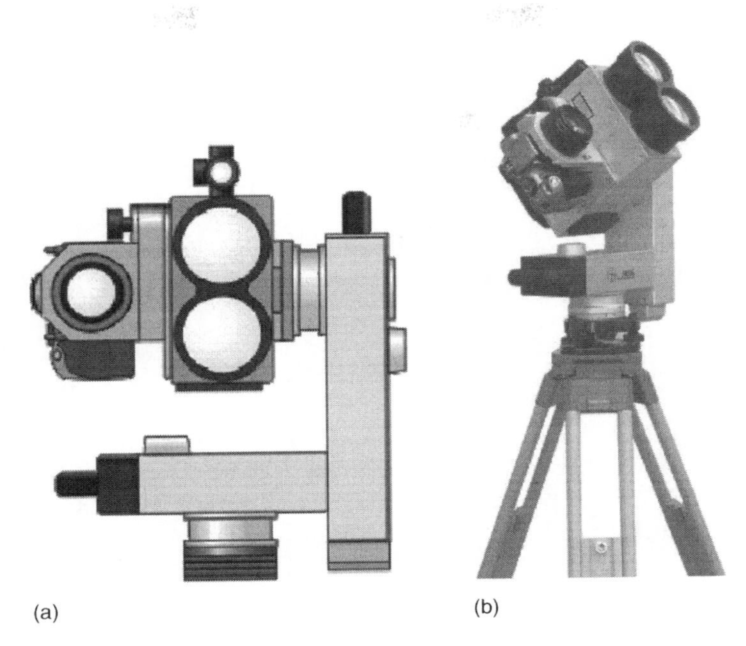

FIGURE 3.19
(a) The external design arrangement of the Riegl LPM-321 laser profiler with the double transmitting and receiving optics of the rangefinder attached to the vertical pillar; the sighting telescope mounted on top of the rangefinder unit; and the camera attached to the side of the rangefinder. (b) Photograph of the instrument. (From Riegl.)

6000 m to targets with 80% reflectivity (without the use of a reflector) and can be operated to measure profiles either manually or in an automated mode. It also offers a full waveform digitizing capability in the same manner as an airborne laser scanner. In its automated mode of operation, the LPM-321 can measure up to 1000 points per second. The instrument has a sighting telescope with up to 20× magnification that sits on top of the rangefinder and can have a calibrated digital camera fitted to the side of the rangefinder (Figure 3.19b).

3.2.4.3 I-SiTE

The I-SiTE company was founded in 1999 as a subsidiary of the Maptek company based in Adelaide in South Australia. Maptek is the developer of the Vulcan 3D geological modelling software that is used extensively in the mining industry world-wide. The I-SiTE company was formed specifically to support and develop Maptek's early interest and activity in laser scanning. The I-SiTE Studio software for use with ground-based laser scanned data was an early product from the company, often sold to users together with Riegl terrestrial laser scanners as a bundled package.

However, after 2 or 3 years of development, I-SiTE brought out its own Model 4400 laser scanner (Figure 3.20). This is a hybrid-type scanner having a motorized rotation of 360° in azimuth and 80° angular coverage in the vertical plane. The stated angular accuracy is ±0.04°. The instrument's rangefinder is based on a Class 3R laser which emits its pulses at $\lambda = 905$ nm with a power of 10 mW and measures its ranges at a maximum rate of 4400 points per second. The maximum range with highly reflective surfaces is 700 m.

FIGURE 3.20
An example of an I-SiTE model 4400 laser scanner. (From I-SiTE.)

This value will of course decrease somewhat to 600 m with rock and concrete surfaces with 40%–50% reflectivity, and go down to 150 m in the case of black coal surfaces with only 5%–10% reflectance. The claimed accuracy in range is ±5 cm at a range of 100 m. The intensity values of the return signals after reflectance can also be measured and recorded. Furthermore, the I-SiTE 4400 instrument also incorporates a digital panoramic line scanner equipped with a Nikon $f = 20$ mm lens that produces a 37 megapixel linescan image that is acquired concurrently during the laser scanning/ranging operation. The measured data is transferred via an Ethernet interface to be recorded on an external PC tablet computer. The corrected image is automatically rendered on to the 3D surface produced by the range and angular data using the I-SiTE Studio software. The instrument also incorporates a viewing telescope with 14× magnification that can be used for alignment and back-sighting operations. Slightly different versions of the instrument are available: (i) the 4400LR for use in the longer ranges encountered in topographic and open-cast mining applications; and (ii) the 4400CR for use in the close ranges encountered in police and forensic applications.

3.2.4.4 Measurement Devices Ltd.

This company, based in Aberdeen and York in the United Kingdom, has specialized in the manufacture of laser-based measuring systems since its foundation in 1983. Its products include a wide variety of hand-held laser distance measuring instruments for use in forestry, agriculture, etc.; various fan-beam laser systems for positioning ships and ensuring collision avoidance at sea; and laser scanning devices for use underground within cavities and voids. Nearer to the subject of this chapter, the company has also produced its Quarryman instrument that has been used extensively for many years in quarries and open-cast mines world-wide, to measure profiles across rock faces employing laser ranging techniques.

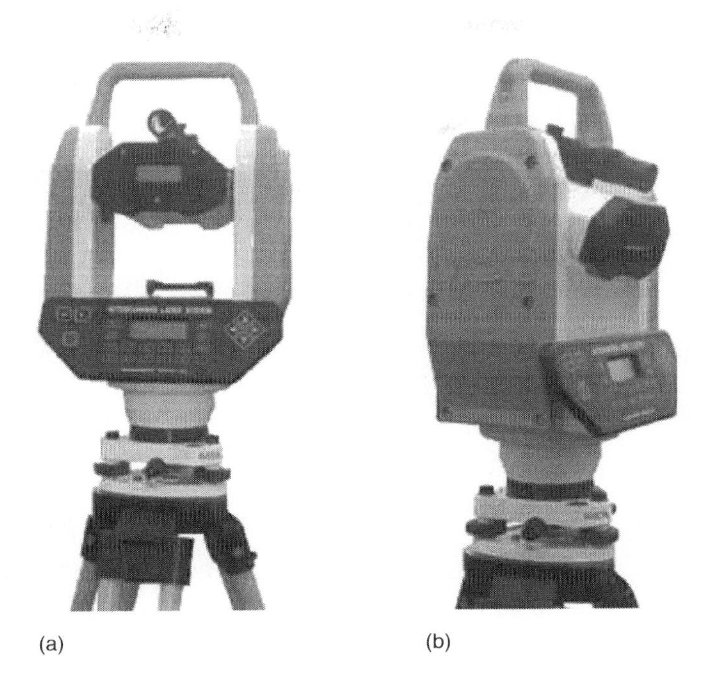

(a) (b)

FIGURE 3.21
(a) MDL LaserAce laser scanner. (b) Quarryman model. (From MDL.)

A redeveloped version of this instrument was introduced in 2004 in the form of the LaserAce Scanner, (Figure 3.21), which is designed to generate point cloud data in the manner of the other laser scanners being discussed in this section. The LaserAce Scanner uses a rangefinder that is mounted on the instrument's horizontal (trunnion) axis supported by two pillars or standards in the manner of a theodolite telescope. The rangefinder utilizes a Class 1 semiconductor laser emitting its pulses at $\lambda = 905$ nm in the near infrared part of the spectrum. This can measure a maximum range of 700 m to a moderately reflective target and 5 km with a prism reflector with a measuring resolution of 1 cm and an accuracy of ±5 cm. The instrument can be operated manually as a total station using a telescope that is mounted on top of the rangefinder. Alternatively it can be operated in an automated (robotic) mode as a laser scanner using its built-in motors, in which case, it can measure objects at a rate of 250 points per second. The corresponding angles are measured using horizontal and vertical angular encoders. The full angular coverage provided by the instrument is 360° × 135°. However, specific areas can be measured through the prior definition of a rectangle or polygon by the operator using the instrument's built-in numeric keyboard. The measured range and angular data is stored on a standard interchangeable flash card. The customized version of this instrument that is oriented toward MDL's traditional quarrying and open-cast mining market is called the Quarryman Pro. The instruments use the MDL Logger software on Psion PDA and Pocket PC computers to record the measured data and the company's Laser Cloud Viewer for the processing of the data. Further processing to form contours and sections and to compute volumes can be carried out using MDL's Face 3D Modeller software.

3.2.4.5 Trimble

In June 2007, Trimble introduced its VX Spatial Station (Figure 3.22). This instrument is constructed and operates on an entirely different concept to that of the company's GS and GX

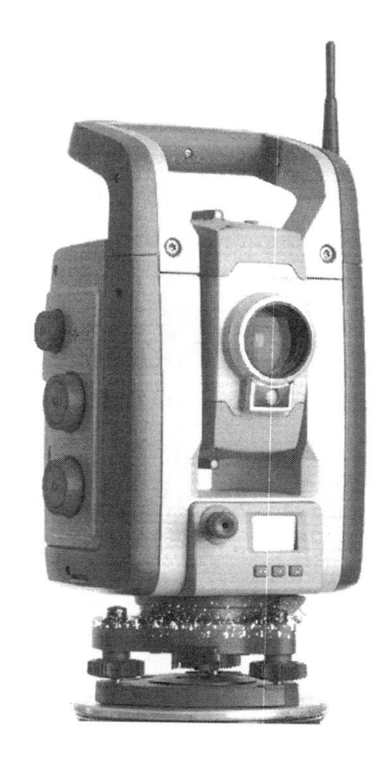

FIGURE 3.22
The Trimble VX Spatial Station. (From Trimble.)

series of scanners described above. Basically it can be viewed as an automated (robotic) total station equipped with an imaging and scanning capability. The instrument is equipped with a conventional surveyor's telescope rather than a scanning mirror. The total station capability of the instrument that is used for positioning and traversing provides a ±1 s angular accuracy and a ±3 mm distance accuracy with a maximum range of between 300 and 800 m without a prism and 3 km with a single prism. The scanning is carried out at a rate of 5–15 measured points per second with a scanning range of up to 150 m. The tiny on-board digital frame camera is fitted to the underside of the telescope with its optical axis parallel to it. This calibrated camera produces HDTV images that are 2048 × 1536 pixels in size and can be recorded in JPEG format at a maximum 5 FPS on the instrument's removable data collector unit. Obviously, the instrument is basically a total station that is designed to provide surveyors with laser scanning and imaging capabilities over limited areas or specific objects such as buildings, rather than carrying out the systematic scanning of large areas or objects that the Trimble GS/GX series is designed specifically to perform.

3.3 Dynamic Terrestrial Laser Scanners

The subject of vehicle-borne terrestrial laser scanners is an exciting one to explore. It forms only a part of a much wider subject area—namely that of conducting surveying and mapping operations from dynamic vehicular platforms. Up till now, this wider subject area has been concerned mainly with the imagery acquired from these platforms using multiple video and digital cameras in combination with the data acquired concurrently for direct geo-referencing purposes by integrated GPS/IMU units. This is then followed by the photogrammetric evaluation of the imagery on a highly automated basis to deal with the enormous number of small-format images that are collected during these operations. Initially this activity was a pioneering research effort carried out in the late 1980s by the Center for Mapping of the Ohio State University using its GPSVan (Bossler and Toth, 1996; Toth and Grejner-Brzezinska, 2003, 2004). It was joined somewhat later by an independent but parallel project at the University of Calgary using its similar VISAT van. Other similar projects were undertaken by various European universities and institutions such as the EPFL, Lausanne (Photobus) and the Institut Cartogràfic de Catalunya (ICC), Barcelona (GeoVan). These various research efforts have resulted in the establishment of a number of commercial companies carrying out mapping from vehicles using the purely camera-based technology, especially in North America, with companies such as Lambda Tech, Transmap, Facet, DDTI, Mandli (the United States) and Geo3D (Canada). These are concerned principally with the collection of geospatial data for road inventory purposes.

The development and use of laser scanners on vehicular platforms either instead of, or in conjunction with, cameras has taken place somewhat later. An early research effort was the Vehicle-borne Laser Mapping System (VLMS) of the Centre for Spatial Information Science of the University of Tokyo (Manandhar and Shibasaki, 2001, 2003). The camera-based GeoVan of the ICC, Barcelona was transformed into the Institute's Geomobil project (Talaya et al., 2004a,b). Both of these projects featured multiple cameras and laser scanners for data collection purposes. Only very recently, over the last 2 or 3 years, have these largely research-oriented projects resulted in the introduction of vehicular-based systems using laser scanners that operate on a commercial basis. Now vehicular-based systems are available for purchase that can be used routinely by service providers. There appears little doubt that this is merely the start of what should become a commonly available service—at least in the more advanced industrialized countries.

In this introduction to this first section, the use of laser scanners on vehicles for non-topographic purposes should also be mentioned—in particular, their use in navigation and collision avoidance systems. If these come to be mass-produced and used widely, then undoubtedly this will have a strong influence, especially with regard to the costs of laser scanners that can be mounted on vehicles. A further mention should also be made of the influence of the DARPA Grand Challenge (DGC) and DARPA Urban Challenge (DUC) contests that have hastened the development of these navigation and collision avoidance technologies for use in unmanned vehicles. There is a striking difference between the imaging sensor configurations used in the first DGC held in 2004 and in the recent DUC held in 2007, respectively. In 2004, digital cameras represented the primary sensors to observe the object space around the vehicles and stereo techniques were widely used for object space reconstruction. Laser profilers, predominantly SICK models, were only used to supplement the image sensing capabilities. The situation was reversed by the 2005 DGC, when the winner, the Stanley vehicle from Stanford University, used five SICK roof-mounted laser scanners to map the area in the front of the vehicle and cameras were only used at high speed to look ahead for objects that were farther away than 40 m.

Recognizing the potential of laser scanning, manufacturers developed dedicated laser scanners for the 2007 DUC. The Ibeo system provided laser profiling capabilities in four planes, thus replacing four SICK units. Then, more importantly, the introduction of the Velodyne laser scanner, which was used by the winner, the Boss, a fully autonomous Chevy Tahoe vehicle from Carnegie Mellon University, and by most of the top ten participants, represented a major technological breakthrough in 2007. The Velodyne HDL-64E High Definition Lidar (Figure 3.23) is based on using 64 laser units covering a 26.8° vertical spread, thus eliminating the need for any vertical mechanical

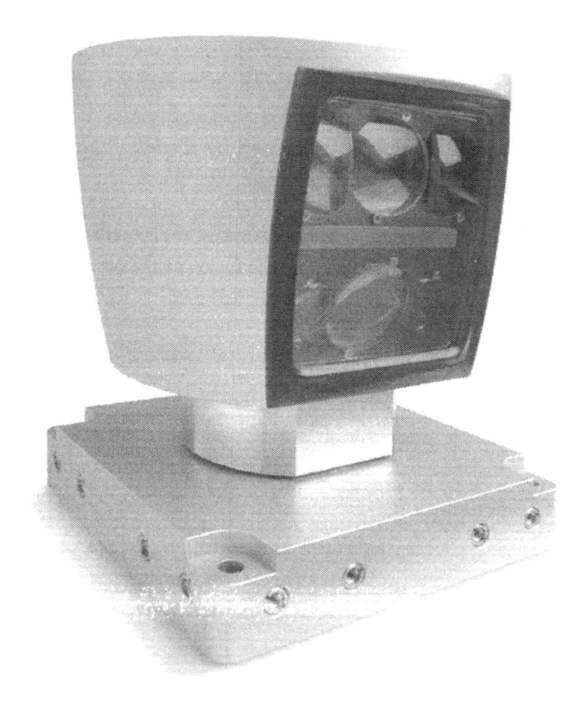

FIGURE 3.23
The Velodyne HDL-64E laser scanner. (From Velodyne.)

motion (Figure 3.24). The system sports high horizontal rotation rates of the laser sensors around the vertical axis of the unit, at up to 15 Hz, with an angular resolution of 0.09°. The Class 1 laser operates at the wavelength of 905 nm with a 10 ns pulse width. The ranging accuracy is claimed to be less than 5 cm for 50 and 120 m with reflectivities of 10% and 80%, respectively. The data collection rate of more than 1 million points per

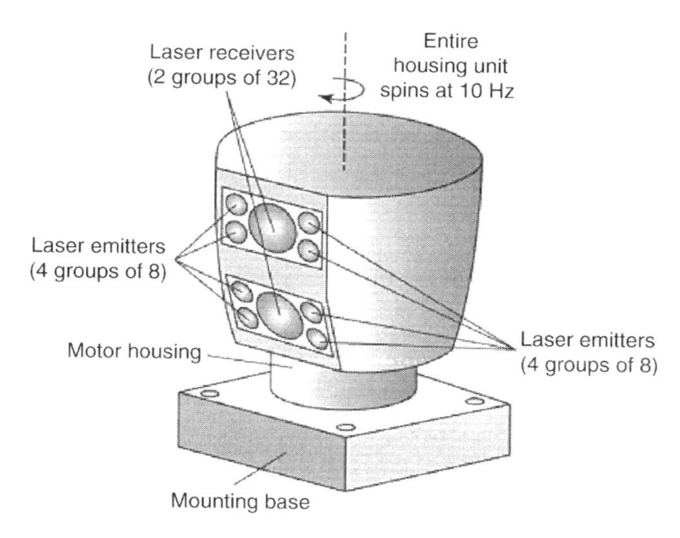

FIGURE 3.24
Diagram showing the main external features of the Velodyne HDL-64E spinning laser scanner. (Drawn by M. Shand.)

FIGURE 3.25
A Velodyne HDL-64E spinning laser scanner (circled at top) together with a Riegl scanner placed in front and to the left of it, both mounted on the Stanford University Junior vehicle (a Volkswagen Passat) that finished second in the DARPA 2007 Urban Challenge. (From Stanford Racing Team.)

second is simply amazing. In summary, these unmanned vehicles (Figures 3.25 and 3.26) have made extensive use of the laser scanners and integrated GPS/IMU systems, which are the building blocks of the vehicular systems that are being developed for topographic applications (Ozguner et al., 2007).

The account that follows this introduction will discuss: (1) commercially available systems; (2) custom-made systems; and (3) some representative research systems that utilize vehicle-mounted laser scanners for topographic applications—in much the same way as has been done with the airborne laser scanning systems covered in Chapter 2, with which they share many features in common.

FIGURE 3.26
A combination of a Velodyne scanner and twin Riegl scanner mounted on another DUC vehicle.

3.3.1 Commercial System Suppliers

3.3.1.1 Optech

As already mentioned above, Optech has developed a special MC version of its ILRIS-3D laser scanner for use on dynamic or moving platforms. This system has been used, for example, by the Sineco company in Italy for surveys of an open cast mine and for road surveys conducted from a moving van (Zampa and Conforti, 2008). Besides the ILRIS-3D laser scanner unit that generates the range, intensity, angle and time data, the overall system includes an Applanix POS/LV 420 GPS/IMU subsystem that provides the position and orientation data against time. The Polyworks software package is used by Sineco to carry out the subsequent processing of all the captured data.

Toward the end of 2007, Optech has also released a completely new product, called the LYNX Mobile Mapper (Figure 3.27). This is a purpose-built spinning laser profiling system designed specifically for attachment to standard vehicle roof racks with mounts for two laser scanners and two (optional) calibrated frame cameras as well as the system IMU and GPS antenna. The third dimension comes from the vehicle motion. Although the scanners used in the LYNX system also utilize a Class I laser as the basis for their laser rangefinders, they are very different units to those used in the ILRIS-MC, having a maximum range of 100 m; a full 360° angular coverage; a pulse measuring rate of 100 kHz; and a scan rate of 9000 rpm (150 Hz). The system control unit with its embedded navigation solution is based on the Applanix POS/LV 420 subsystem and can control up to four laser scanners simul-taneously using the laptop computer attached to the unit. The Applanix POSPAC software is used to process the POS/LV data while Optech supplies its own LYNX-Survey and LYNX-Process software for final postprocessing. Optech has

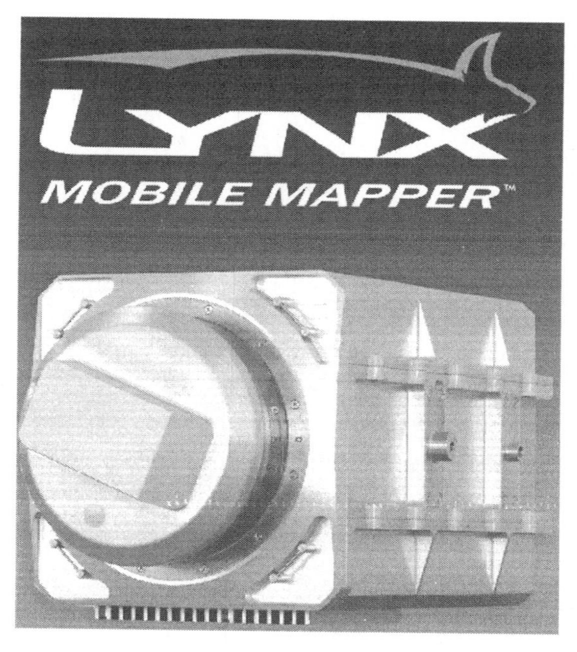

FIGURE 3.27
The LYNX Mobile Mapper scanning unit. (From Optech.)

announced that the LYNX systems are already being supplied to the Infoterra mapping company based in the United Kingdom and to the Italian Sineco company which already operates an ILRIS-MC system as described above.

3.3.1.2 3D Laser Mapping

This company, which is based in Nottingham in the United Kingdom, has developed its portable StreetMapper system specifically for use on moving vehicles in close collaboration with the German systems supplier, IGI, which produces the LiteMapper airborne laser scanning system (Hunter et al., 2006; Kremer and Hunter, 2007). For the StreetMapper, IGI supplies its TERRAcontrol GPS/IMU system, which is derived from its AEROcontrol unit, together with its own hardware and software solutions for the control of the laser scanners and data storage. The control unit is housed in a cabinet that is mounted inside the vehicle. IGI also contributes its TERRAoffice software (derived from its AEROoffice package) for the processing of the IMU data, while the differential GPS data is processed using the GrafNav package from the Waypoint division of NovAtel based in Canada. The TerraScan/TerraModeler/TerraMatch suite of programs from Terrasolid in Finland is utilized for the processing of the laser scan data and its transformation into the final 3D model data. The multiple laser scanner engines are supplied by Riegl, between two and four of its LMS-Q120 units (with their 150m range) being fitted on to a roof rack together with the IMU and the GPS antenna (Figures 3.28 and 3.29). Either video or digital still cameras can be

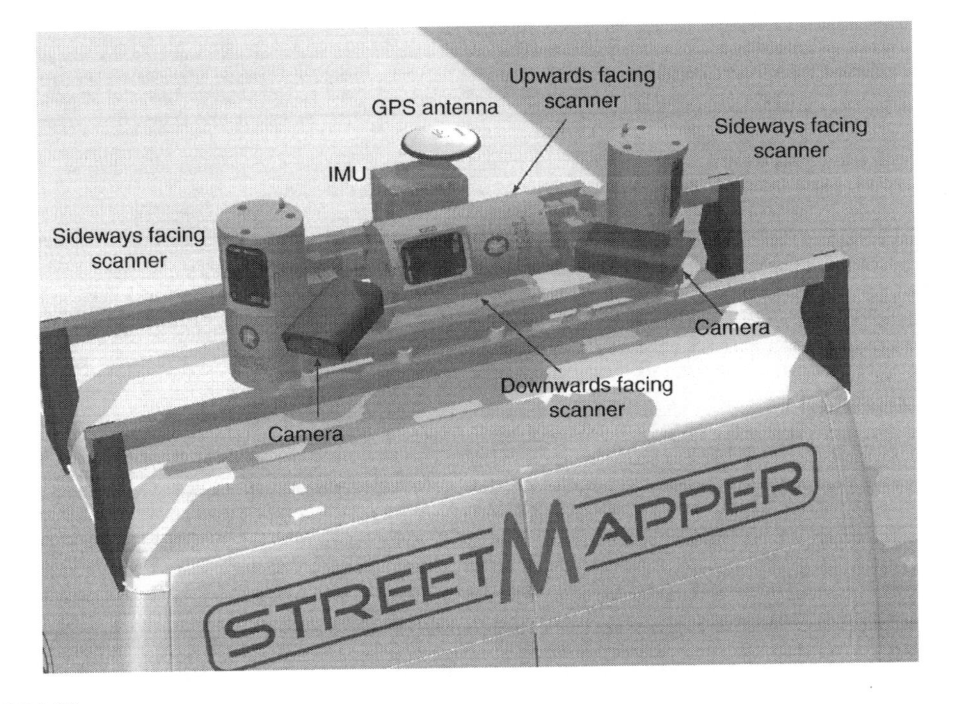

FIGURE 3.28
The StreetMapper van is equipped with its roof rack supporting four Riegl laser scanners—two sideways-facing; one upwards-facing and one downwards-facing—and two small-format cameras; together with the IMU and GPS antenna of the position and orientation system supplied by IGI. (From 3D Laser Mapping.)

FIGURE 3.29
Another example of a StreetMapper van with a protective cover mounted over its roof rack and contents. The windows allow the pulses from the laser scanners to be transmitted and received and the camera images to be exposed. (From 3D Laser Mapping.)

supplied to generate higher quality images that supplement the laser scanned data. Touch screen displays installed on the dashboard of the vehicle are used for the display of the captured data (Figure 3.30). A StreetMapper system has been used extensively by Reality Mapping, a service company based in Cambridge, England, to carry out corridor surveys along roads for highway asset management and to capture street level data in city centers in the United Kingdom.

FIGURE 3.30
Two touch screen displays are mounted on the dashboard of this StreetMapper van. (From 3D Laser Mapping.)

3.3.2 Custom-Built and In-House Operated Systems

3.3.2.1 Tele Atlas

This company, which is based in Ghent, Belgium, is a leading supplier of digital road map data for use in vehicle navigation and location-based systems. It operates fleets of vans that continually acquire data for the revision of its digital map database. In this context, between 15% and 20% of the information on roads contained in the database needs to be revised annually. For this purpose, Tele Atlas has a fleet of 22 camper vans (in an eye-catching orange color!) operating throughout Europe. Each van is equipped with six digital cameras, a GPS receiver that uses the Fugro OmniSTAR service; a gyro unit; and a distance measuring device attached to the rear wheel of the van for use when the GPS signals are lost. The resulting data is processed and analyzed in data centers located in Poland and India.

In North America, a fleet of nine smaller vans is used. Each van is equipped with a roof rack containing the cameras, GPS antenna, etc. as before. However each system also features twin laser scanners that are pointed to the side of the van to continuously collect street-level range data as the van travels forward (Figures 3.31 and 3.32). The laser scanners are supplied by the Swiss SICK company, which is well known as a supplier of short-distance laser scanners that are used for navigation and safety (warning) purposes on vehicles (such as fork-lift trucks) and potentially dangerous equipment (such as cranes and cutting and bending machines) being operated in factories. However, SICK and its two German subsidiaries and partners, LASE GmbH and Ibeo, also manufacture laser scanners that can measure over longer distances and some of these models are used in the

FIGURE 3.31
A Toyota mini-van of Tele Atlas North America equipped with a roof rack on which the system's cameras, laser scanners and GPS antenna are mounted. (From Tele Atlas. With permission.)

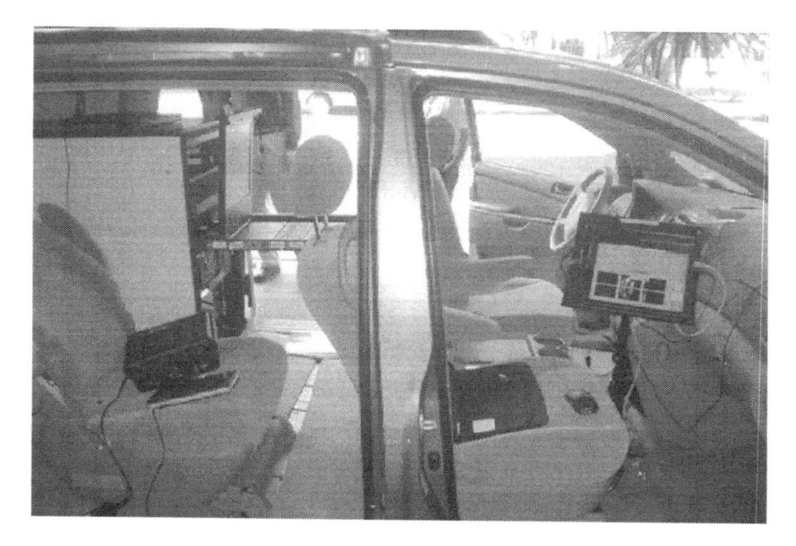

FIGURE 3.32
The inside of the mini-van showing the system control unit mounted at the rear and the display monitor at the front beside the dashboard. (From Tele Atlas.)

Tele Atlas mobile mapping vans that are being operated in North America (Figures 3.33 and 3.34). As discussed above, these SICK and Ibeo laser systems have also been used extensively on the unmanned vehicles taking part in the DGC and DUC competitions (Ozguner et al., 2007).

3.3.2.2 Terrapoint

Another interesting custom-built vehicle mapping system employing a laser scanner is the so-called SideSwipe system devised by the Canadian Terrapoint company for use in road surveys and mapping in Afghanistan (Newby and Mrstik, 2005). This system is a modified

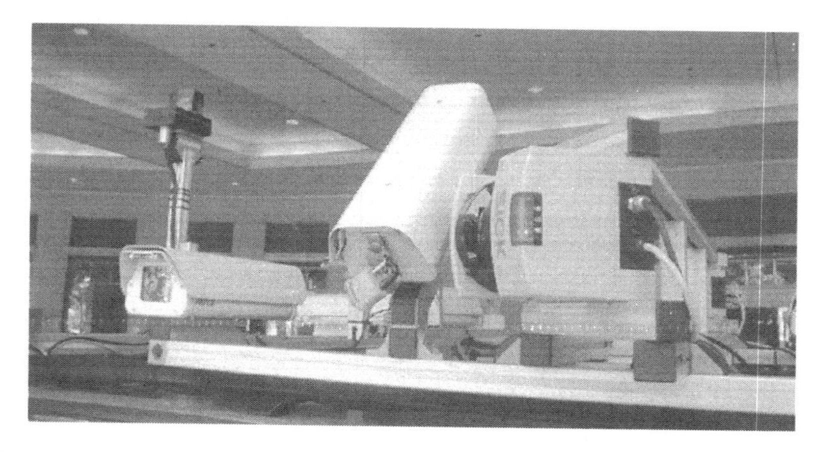

FIGURE 3.33
Showing cameras and a laser scanner mounted on the roof rack of the Tele Atlas mini-van that is being used as a mobile mapping system. (From Tele Atlas.)

FIGURE 3.34
A close-up photo of one of the SICK laser scanners that is being used in the mobile mapping system. (From Tele Atlas.)

version of one of the company's ALMIS-350 airborne laser scanners already discussed in Chapter 2. The laser scanner was mounted rigidly on a pole attached to the side of an open-backed pick-up truck (Figure 3.35) which also contained the system's two computers; the one used for route navigation, the other for system control and logging purposes. The system also included a high-resolution video camera that was integrated with the GPS/IMU system. The latter comprised an HG1700 tactical-grade IMU from Honeywell that is connected to a NovAtel GPS receiver. The operational procedure involved first driving down the road with the laser scanner pointing forward and tilting slightly downward,

FIGURE 3.35
The truck-mounted SideSwipe laser scanner system of Terrapoint has been used for mapping purposes along main highways in Afghanistan. (From Terrapoint.)

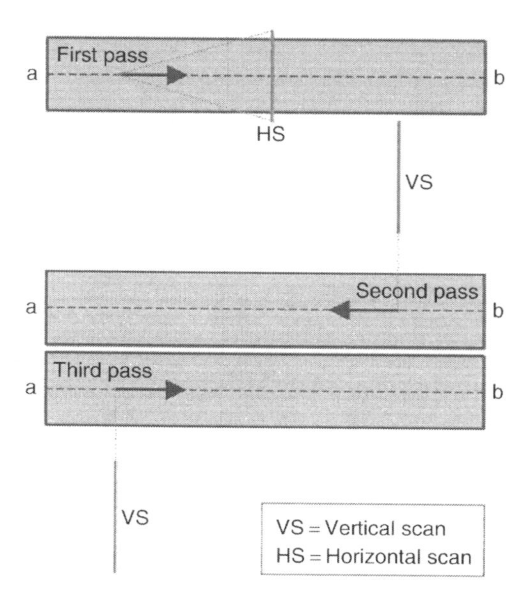

FIGURE 3.36
A diagrammatic representation of the three-pass solution that is used to provide coverage both of the road and the terrain on both sides of the highway. (From Terrapoint.)

so providing a horizontal scan or swath with a 60° angular coverage. During a second pass along the same stretch of road, the scanner was rotated and reset to scan vertically from the side of the vehicle producing side-scan coverage that extended from 5 to 100 m to the side of the vehicle. The same set-up was used on a third pass along the same section of road, but pointing to the other side (Figure 3.36). The three pass solution provided the data for the mapping of the highway and the corridor on both sides of the road.

Based on this successful initial activity, Terrapoint has developed its Tactical Infrastructure and Terrain Acquisition Navigator (TITAN) system (Glennie, 2007), designed to overcome the limitations of the multiple passes that were required with the SideSwipe system and the limited accuracy of its tactical-grade IMU. It features an equipment pod containing the system's laser scanners, IMU, GPS receivers and digital cameras that is mounted on a hydraulic lift attached to the floor of an open-back pick-up truck (Figure 3.37). The system comprises an array of four Riegl laser scanners; a higher-grade IMU from iMAR coupled to a NovAtel GPS receiver; and up to four digital video or frame cameras (Figure 3.38). The data collected by these various instruments is passed via cables to the data logging computers installed within the truck's cabin. Terrapoint uses its own software package called CAPTIN (Computation of Attitude and Position for Terrestrial Inertial Navigation) to carry out the postprocessing of the measured GPS/IMU data. Terrapoint has entered into a partnership with the Neptec Design Group based in Ottawa, Canada, under which Neptec will support the further development of the TITAN system through the supply of its analytical software algorithms. Neptec will also license the TITAN technology to produce vehicle-mounted laser scanner systems designed specifically for use by military and homeland security agencies.

3.3.3 Research Systems

As noted above, several vehicle-based laser scanning systems have been constructed and operated by university departments for research purposes. Two of these will be discussed in some more detail as exemplars of such systems.

FIGURE 3.37
The open back of this pick-up truck forms the base of the hydraulic lift supporting the equipment pod of the Terrapoint TITAN mobile mapping system. (From Terrapoint.)

3.3.3.1 University of Tokyo

The Centre for Spatial Information Sciences at the University of Tokyo, has been undertaking research into VLMS under the leadership of Prof. Shibasaki and with the collaboration of the Asia Air Survey Co. Ltd. since 1999 (Inaba et al., 1999). The first prototype system comprised four laser scanners, three of which scanned vertically while the fourth carried out a horizontal scan. The resulting laser scan data was supplemented by the image data captured by four CCD cameras, each producing small-format (640 × 480 pixels) images. A GPS receiver allied to an IMU and a shaft-driven precision odometer, produced the

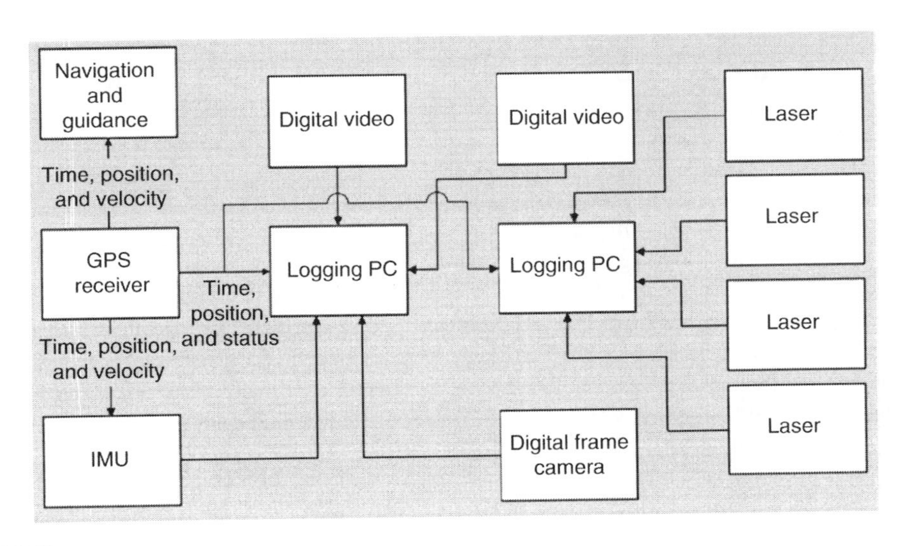

FIGURE 3.38
Block diagram of the TITAN system that shows the relationship of its various hardware components to one another. (From Terrapoint.)

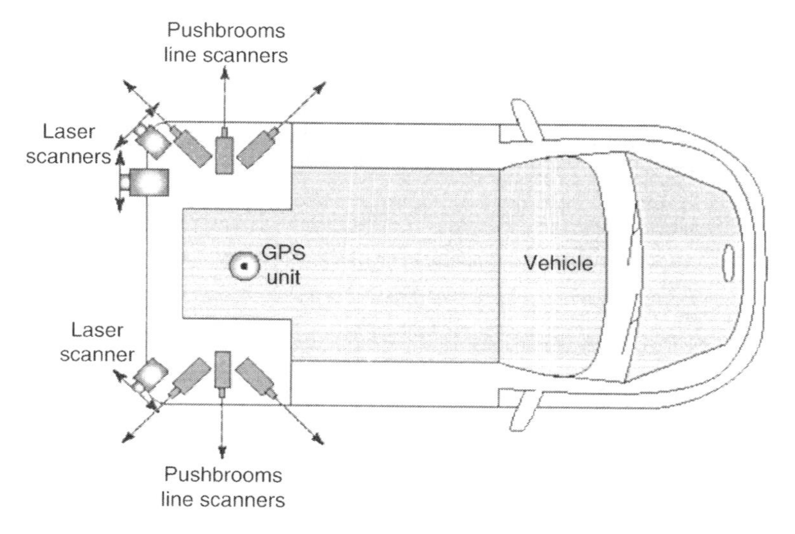

FIGURE 3.39
Diagram showing the layout and orientation of the six pushbroom line scanners, three laser scanners and the GPS antenna mounted on the roof of the GeoMaster vehicle used by the Centre for Spatial Information Sciences of the University of Tokyo. (From University of Tokyo; redrawn by M. Shand.)

required position and orientation data (Manandhar and Shibasaki, 2000). The system has been steadily improved, using laser scanners with higher frequencies and greater angular ranges. Furthermore, the digital frame cameras have been replaced by pushbroom line scanners that use the forward motion of the vehicle to produce continuous linescan imagery (Manadhar and Shibasaki, 2003; Zhao and Shibasaki, 2003). More recent developments are reported in detail in a paper by Zhao and Shibasaki (2005). The current laser scanners are LD-A models produced by Ibeo. These have a maximum range of 100 m, a scan frequency of 20 Hz and an average error in range of ±3 cm. Each of the pushbroom line scanners is equipped with a fish-eye lens giving an angular field of 180° and captures its image data at the rate of 80 Hz. The laser scanners and pushbroom line scanners are all mounted on the roof rack of the GeoMaster vehicle with their scanning planes oriented at different angles to reduce the occlusions caused by trees and other obstacles (Figure 3.39). Tests were carried out to map building facades in the Ginza district of Tokyo.

3.3.3.2 ICC

The ICC is the organization that is responsible for the official topographic mapping and cartographic activities in Catalonia, Spain. The Institute has been developing its own mobile mapping system since 2000. Initially this comprised its Geovan equipped with two small-format (one Megapixel) digital frame cameras producing monochrome images together with an Applanix based GPS/IMU position and orientation system (Talaya et al., 2004). In September 2003, a laser scanner was added to the system, after which it was known as the Geomobil system (Figure 3.40). The system's laser scanner is the Reigl Z-210 model that can collect data at rates up to 10 kHz with a vertical angular coverage of 80° and a horizontal coverage of ±166.5° (Alamus et al., 2004; Talaya et al., 2004). As reported in a further paper by Alamus et al. (2005), the system now has one IMU and two GPS receivers, the one a dual-frequency model for position determination, the other a single-frequency model to help improve the heading angle (azimuth) determination—forming part of

FIGURE 3.40
The ICC Geomobil van with its roof rack on which the cameras, laser scanner, and GPS antennas are mounted. (From ICC.)

the POS subsystem. The Applanix POSPAC software is used to process the GPS data. The Geomobil vehicle also has a distance measurement indicator (DMI) attached to one of the van's rear wheels to provide measurements of the distance traveled by the van. A further development is the installation of new digital color cameras pointing both forward and backward from the van (Figure 3.41). The Geomobil system is being used to make a

FIGURE 3.41
The specially constructed roof rack of the Geomobil with the digital cameras mounted at the rear corners of the frame, video cameras at the sides, the GPS antennas in the middle of the frame and at the front right, while the Riegl Z-210 laser scanner is placed at the front centre. (From ICC.)

geo-referenced inventory of the ground floor facades (in color) of all the buildings in the city of Barcelona, covering 3800 individual streets amounting to a linear distance of 1291.5 km.

References and Further Reading

Static Terrestrial Laser Scanners

Blais, F., 2004, Review of 20 years of range sensor development. *Journal of Electronic Imaging*, Vol. 13, No. 1, pp. 231–240.

Bohm, J. and Haala, N., 2005, Efficient integration of aerial and terrestrial laser data for virtual city modelling using lasermaps. Proceedings ISPRS WG III/3, III/4, V/3 Workshop Laser Scanning 2005, Enschede, The Netherlands, September 12–14, 2005, pp. 192–197.

Ingensand, H., 2006, Metrological aspects in terrestrial laser-scanning technology. Third IAC/12th FIG Symposium, Baden, Germany, May 22–24, 2006, 10 pp.

Kersten, T., Sternberg, H., and Mechelke, K., 2005, Investigations into the accuracy behaviour of the terrestrial laser scanning system Trimble GS100. *Optical 3D Measurement Techniques VII*, Vol. 1, Gruen, A. and Kahmen, H., Eds., pp. 122–131.

Lichti, D., Gordon, S., and Stewart, M., 2002, Ground-based laser scanners: Operation, systems and applications. *Geomatica*, Vol. 56, No. 1, pp. 21–33.

Lichti, D., Brustle, S., and Franke, J., 2007, Self calibration and analysis of the Surphaser 25HS 3D scanner. Strategic Integration of Surveying Services, FIG Working Week 2007, Hong Kong SAR, China, May 13–17, 2007, 13 pp.

Mechelke, K., Kersten, T., and Lindstaedt, M., 2007, Comparative investigations into the accuracy behaviour of the new generation of terrestrial laser scanning systems. *Optical 3-D Measurement Techniques VIII*, Vol. I, Gruen, A. and Kahmen, H., Eds., Zurich, July 9–12, 2007, pp. 319–327.

Mettenleiter, M., Hartl, F., and Fröhlich, C., 2000, Imaging laser radar for 3-D modelling of real world environments. International Conference on OPTO/IRS2/MTT, Erfurt, Germany, May 11, 2000, 5 pp.

Pfeifer, N. and Briese, C., 2007, Geometrical aspects of airborne laser scanning and terrestrial laser scanning. *International Archives of Photogrammetry and Remote Sensing*, Vol. 36, Part 3, W52, pp. 311–319.

Schulz, T. and Ingensand, H., 2004, Influencing variables, precision and accuracy of terrestrial laser scanners. INGEO 2004 and FIG Regional Central and Eastern European Conference on Engineering Surveying, Bratislava, Slovakia, November 11–13, 2004, 8 pp.

Staiger, R., 2003, Terrestrial laser scanning—Technology, systems and applications. Second FIG Regional Conference, Marrakech, Morocco, December 2–5, 2003, 10 pp.

Sternberg, H., Kersten, T., Jahn, I., and Kinzel, R., 2004, Terrestrial 3D laser scanning—Data acquisition and object modelling for industrial as-built documentation and architectural applications. *International Archives of Photogrammetry, Remote Sensing and Spatial Information Sciences*, Vol. 35, Commission VII, Part B2, pp. 942–947.

Dynamic Terrestrial Laser Scanners

Alamús, R., Baron, A., Bosch, E., Casacuberta, J., Miranda, J., Pla, M., Sànchez, S., Serra, A., and Talaya, J., 2004, On the accuracy and performance of the Geomobil system. *International Archives of Photogrammetry and Remote Sensing*, ISPRS Comm. IV, Vol. 35, Part B5, pp. 262–267.

Alamús, R., Baron, A., Casacuberta, J., Pla, M., Sánchez, S., Serra, A., and Talaya, J., 2005, Geomobil: ICC land-based mobile mapping system for cartographic data capture. Proceedings 22nd International Cartographic Conference, Corunna, Spain, 9 pp.

Bossler, J. and Toth, C., 1996, Feature positioning accuracy in mobile mapping: Results obtained by the GPSVan™. *International Archives of Photogrammetry and Remote Sensing*, ISPRS Comm. IV, Vol. 31, Part B4, pp. 139–142.

Glennie, C., 2007, Reign of point clouds: A kinematic terrestrial lidar scanning system. *InsideGNSS*, Vol. 2, No. 7, pp. 22–31.

Hunter, G., Cox, C., and Kremer, J., 2006, Development of a commercial laser scanning mobile mapping system—StreetMapper. Second International Workshop, The Future of Remote Sensing, Antwerp, Belgium, October 17–18, 2006, 4 pp.

Inaba, K., Manandhar, D., and Shibasaki, R., 1999, Calibration of a vehicle-based laser/CCD sensor system for urban 3D mapping. Proceedings 20th Asian Conference on Remote Sensing (ACRS), Hong Kong, China, 7 pp.

Kremer, J. and Hunter, G., 2007, Performance of the streetmapper mobile LiDAR mapping system in "real world" projects. *Photogrammetric Week '07*, Fritsch, D., Ed., pp. 215–225.

Manandhar, D. and Shibasaki, R., 2000, Geo-referencing of multi-range data for vehicle-borne laser mapping system (VLMS). Proceedings 21st Asian Conference on Remote Sensing (ACRS), Taipei, Taiwan, pp. 974–979.

Manandhar, D. and Shibasaki, R., 2003, Accuracy assessment of mobile mapping system. Proceedings 24th Asian Conference on Remote Sensing (ACRS), Busan, South Korea, 3 pp.

Newby, S. and Mrstik, P., 2005, LiDAR on the level in Afghanistan: GPS, inertial map the Kabul Road. *GPS World*, Vol. 16, No. 7, pp. 16–21.

Ozguner, U., Redmill, K., Toth, C., and Grejner-Brzezinska, D., 2007, Navigating these mean streets: Real-time Mapping in Autonomous Vehicles. *GPS World*, Vol. 18, No. 10, pp. 32–37.

Talaya, J., Bosch, E., Alamus, R., Serra, A., and Baron, A., 2004a, Geovan: The mobile mapping system from the ICC. Proceedings 4th International symposium on mobile mapping Technology, Kunming, China, March 29-31,2004, 7 pp.

Talaya, J., Alamus, R., Bosch, E., Serra, A., Kornus, W., and Baron, A., 2004b, Integration of terrestrial laser scanner with GPS/IMU orientation sensors. *International Archives of Photogrammetry and Remote Sensing*, ISPRS Comm. V, Vol. 35, Part B5, 6 pp.

Toth, C. and Grejner-Brzezinska, D., 2003, Driving the line: Multi-sensor monitoring for mobile mapping. *GPS World*, Vol. 14, No. 3, pp. 16–22.

Toth, C. and Grejner-Brzezinska, D., 2004, Redefining the paradigm of modern mobile mapping: An automated high-precision road centerline mapping system, *Photogrammetric Engineering and Remote Sensing*, Vol. 70, No. 6, pp. 685–694.

Zampa, F. and Conforti, D., 2008, Continuous mobile laser scanning. *GIM International*, Vol. 22, No. 1.

Zhao, H. and Shibasaki, R., 2003, A vehicle-borne urban 3D acquisition system using single-row laser range scanners. *IEEE Transactions*, SMC Part B: Cybernetics, Vol. 33, No. 4, pp. 658–666.

Zhao, H. and Shibasaki, R., 2005, Updating a digital geographic database using vehicle-borne laser scanners and line cameras. *Photogrammetric Engineering and Remote Sensing*, Vol. 71, No. 4, pp. 415–424.

4

LiDAR Systems and Calibration

Aloysius Wehr

CONTENTS

4.1 Laser Scanning Systems

Today several different laser scanners are available that can be distinguished, based on their intended application, as close range and airborne surveying systems. Close range means that laser scanning is carried out over ranges between 0 and 200 m and is typically conducted from fixed locations on the ground. These systems are sometimes referred to as terrestrial laser scanners. Commercial airborne systems typically operate over ranges of several hundred meters to several kilometers from helicopters or fixed-wing airplanes. Laser scanners belong to the family of active sensors, like the well-known radar, because the target is illuminated by the sensor itself. Therefore, the measurement is independent of external illumination. As similar ranging techniques are applied as for the first microwave radars, laser scanners are also called LiDARs which stands for light detection and ranging. For understanding the functioning of the different LiDARs, for evaluating the performance of the instrument components and for selecting the right components appropriate for specific surveying tasks, the principles of LiDAR systems are described here by subdividing the whole system into functional subunits. By explaining these units the reader will learn what the measurement potential of such systems is and will make him better able to select the optimum instrument implementation for a given surveying task. In the following we will focus the discussion on airborne LiDAR systems.

4.1.1 General Setup of Topographic LiDAR Systems

In the following I discuss a LiDAR system, because a valid surveying result is impossible if a LiDAR is used as an isolated device. In this context a surveying result means geocoded laser measurements. Figure 4.1 shows that LiDAR systems consist of an airborne and ground segment. The airborne segment includes the

- Airborne platform
- LiDAR
- Position and orientation system (POS)

The ground segment is comprised of

- Global positioning system (GPS) reference stations
- Processing hardware and software for synchronization and registration which is carried out off-line

During flight a LiDAR samples line-of-sight slant ranges referenced to the LiDAR coordinate system, and a POS stores GPS data including carrier phase information and orientation data of an inertial measurement unit (IMU). LiDAR and POS sample data independently. At the same time, on-ground GPS stations gather GPS data and GPS carrier phase data at known earth fixed positions for later off-line computing of differential GPS (DGPS) positions of the airborne platform. Using DGPS and inertial data, the position of the laser scanner can be computed with centimeter to decimeter accuracy and its orientation is determined to better than one-hundredth of a degree. This position and orientation data is stored as a function of the GPS time. As the laser scanner data are also stored with timestamps generated from the received GPS signal, the scanner and POS data sets can be synchronized. After synchronization the laser vector for each sampled ground point can be directly transformed into an earth fixed coordinate system (e.g., WGS84). By solving the vector geometry shown in Figure 4.2, geocoded laser data are obtained. $\vec{r}_L$ is the vector from Earth center to the origin of the laser beam given by the position data and $\vec{s}$ symbolizes the laser beam. Its direction is given by the orientation of the laser scanner unit determined by POS in 3D space and the direction of the laser beam defined by the instantaneous angular position of the laser beam deflection device defined in the scanner coordinate system. If vector $\vec{G}$ is calculated in WGS84, it represents the geocoded laser measurements. The process of calculating $\vec{G}$ is called registration. Today registered laser scanner data with accuracy better than 10 cm in 3D space are possible and state of the art. The accuracy is primarily determined by the accuracy of POS.

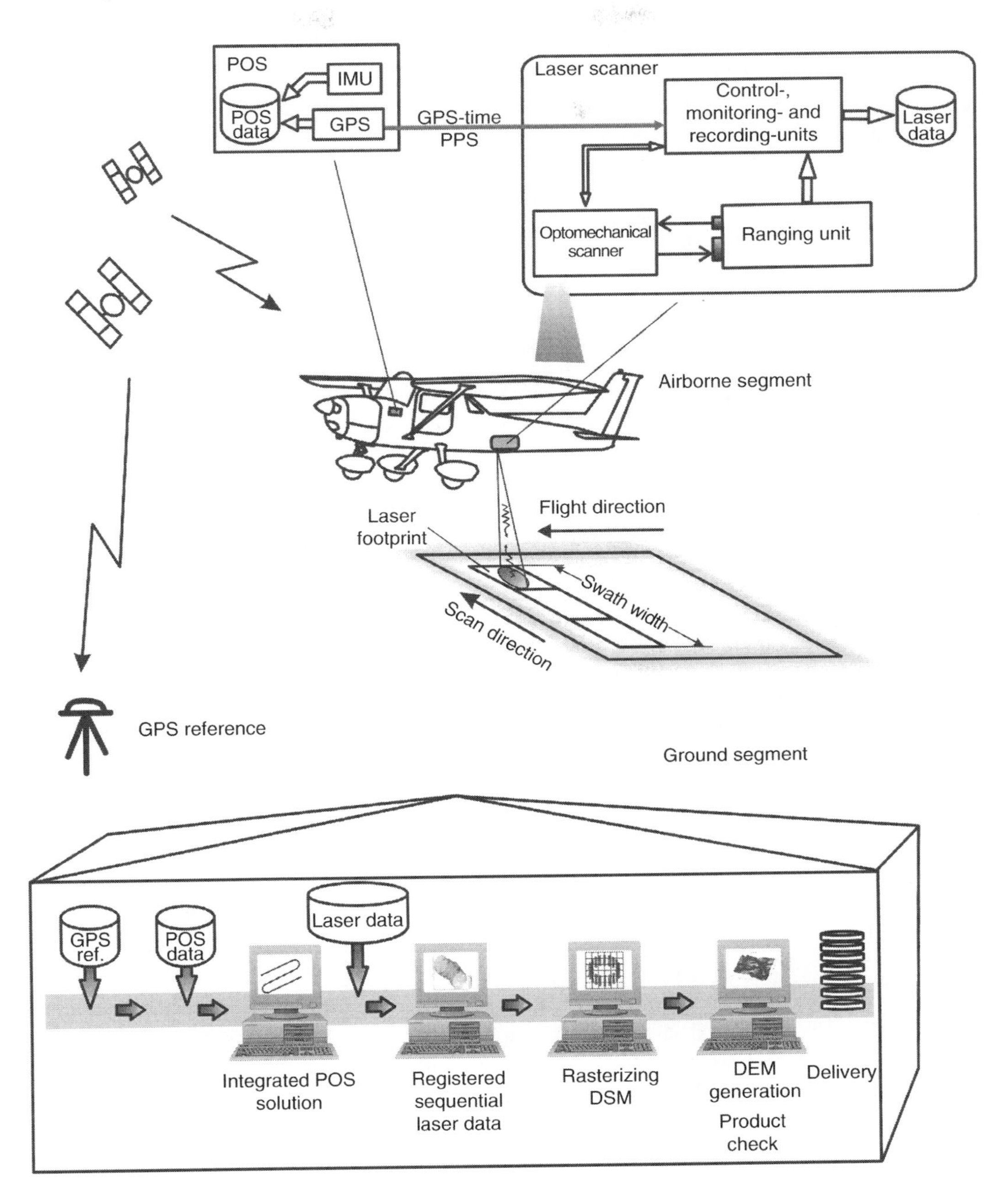

FIGURE 4.1
LiDAR system (airborne and ground segment).

In the following we will look at the airborne segment first. Figure 4.1 depicts that scanning LiDARs are comprised of the following key components excluding POS:

- Laser ranging unit (LRU)
- Opto-mechanical scanning device (OMSD)
- Controlling and data sampling unit

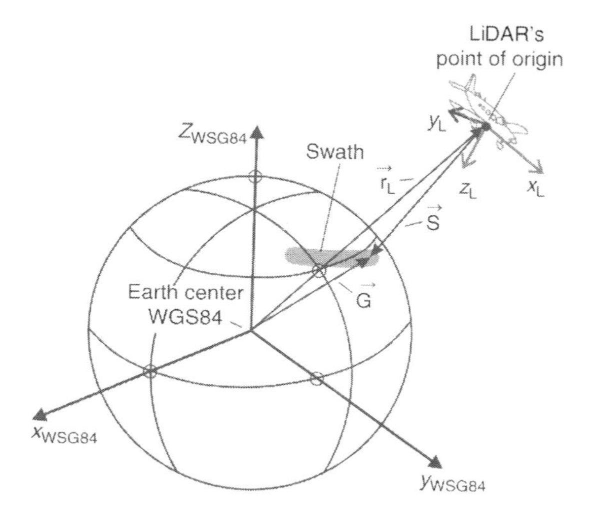

FIGURE 4.2
Vector setup for geocoded LiDAR data.

LRU measures the slant range between the sensor and the illuminated spot on ground. It comprises an emitting laser, an electro-optical receiver and a ranging electronics box. The ranging electronics box drives the laser either with pulsed or continuously modulated signals and computes the slant range from the traveling time of the laser signal from LRU to ground point and back. The transmitting and receiving apertures are mounted so that the transmitting and receiving paths share the same optical path. This assures that object surface points illuminated by the laser are always in the instantaneous field of view (IFOV) of the optical receiver. The angular divergence w of the laser, typically 0.3–2.5 mrad, is smaller than the receiver IFOV so that all of the laser spot is fully contained within the receiver field of view (FOV). To obtain surface coverage on the earth the laser spot and co-aligned receiver IFOV have to be moved across the flight path during flight. This is realized by the OMSD for which a moving mirror is commonly used.

4.1.2 Laser Ranging Unit

The ranging unit of a LiDAR is also known as an optoelectronical range finder in geodesy. In this chapter, the applied ranging principles and the setup are explained and discussed, and the achievable ranging accuracy is estimated by link budget calculations. A detailed performance analysis was published by Abshire (2000) and Gardner (1992).

4.1.2.1 Ranging Principles

Figure 4.3 shows a typical laser ranging setup. Transmitter and receiver are at the same location. The laser transmitter aperture normally is smaller than the receiver aperture. In airborne laser scanning, radar measurement principles are applied, because high range dynamics have to be covered. The most direct ranging measurement is determining the time-of-flight of a light pulse, i.e., by measuring the traveling time between the emitted and received pulse (see Figure 4.4).

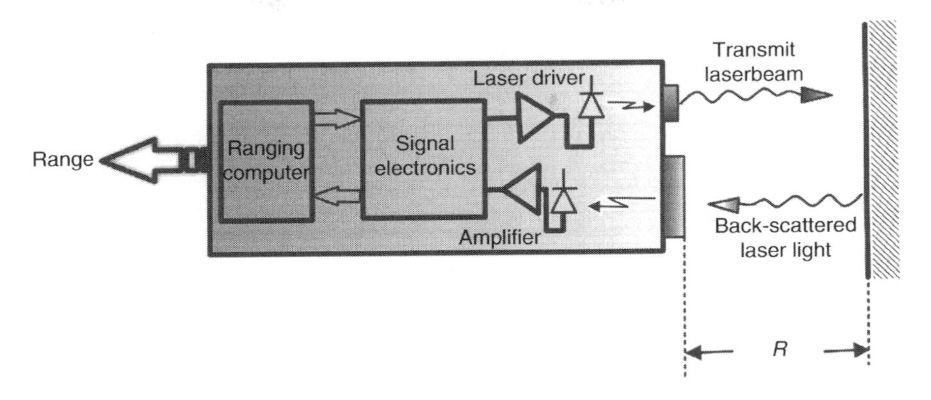

FIGURE 4.3
Two-way ranging setup.

According to Figure 4.4 the traveling time t_L of a light pulse is

$$t_L = 2 \cdot \frac{R}{c} \tag{4.1}$$

where
R is the distance between the ranging unit and the object surface
c is the speed of light

As the signal travels the distance two times (see Figures 4.3 and 4.4), one talks of two-way ranging. If the traveling time t_L is measured, the range can be directly computed by

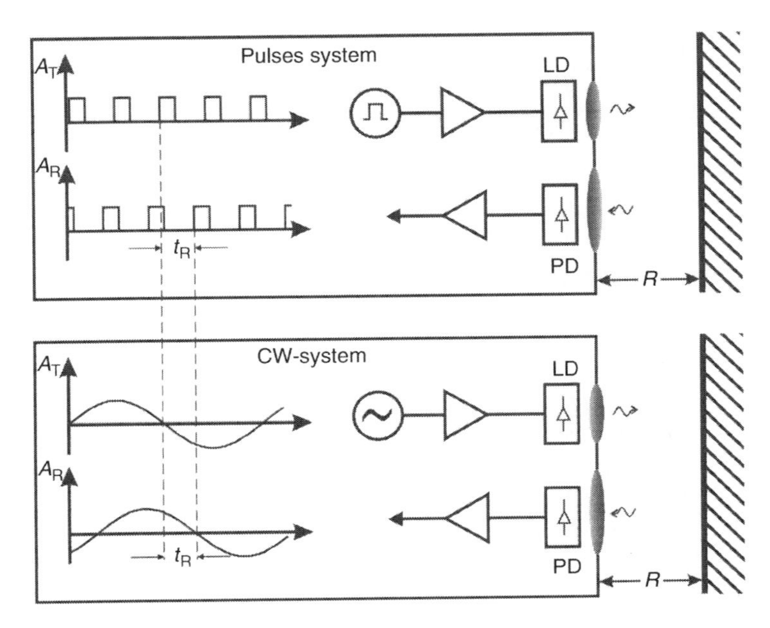

FIGURE 4.4
Traveling time ranging (upper image pulse ranging, lower image phase difference ranging).

$$R = \frac{c}{2} \cdot t_{\mathrm{L}} \tag{4.2}$$

The range resolution ΔR is determined by the obtainable time resolution Δt_{L} of the traveling time measuring instrument and is given by

$$\Delta R = \frac{c}{2} \cdot \Delta t_{\mathrm{L}} \tag{4.3}$$

It is also possible to determine the range, if the laser emits light continuously. Such a signal is called continuous wave (cw) signal. However, ranging can only be carried out, if the light has a deterministic intensity structure. This means that the intensity of laser light is modulated. Assuming the laser light is intensity modulated with a sinusoidal signal (Figure 4.4) which has the period T, the ratio traveling time t_{L} to period T equals the ratio phase difference between transmitted and received signal to 2π:

$$\frac{t_{\mathrm{L}}}{T} = \frac{\phi}{2\pi} \tag{4.4}$$

Putting Equation 4.4 into Equation 4.2 the range for two-way ranging is given by

$$R = \frac{c}{2} \cdot \frac{T}{2\pi} \cdot \phi \tag{4.5}$$

As the period is the reciprocal of the intensity modulation frequency f, R can be calculated by

$$R = \frac{1}{4\pi} \cdot \frac{c}{f} \cdot \phi \tag{4.6}$$

The ratio c/f is equivalent to the wavelength λ of the ranging signal, so that

$$R = \frac{\lambda}{4\pi} \cdot \phi \tag{4.7}$$

Measuring the range by the phase difference the achievable range resolution ΔR is not only dependent on the maximum phase resolution ϕ but also on wavelength λ:

$$\Delta R = \frac{\lambda}{4\pi} \cdot \Delta\phi \tag{4.8}$$

In comparison to pulsed systems (Equation 4.3), Equation 4.8 makes clear that by using cw signals one has an additional physical parameter that can be used to design a system with a desired resolution. This means, even if the phase resolution is kept constant, the range resolution can be improved by applying ranging signals with shorter wavelength. This is not possible with pulsed systems, because in Equation 4.3 the constant of proportionality is the speed of light which cannot be varied. Therefore cw signals are applied if very high range resolutions are required. For example, using a 1 GHz ranging signal, which corresponds to a wavelength λ of 30 cm, and assuming a phase resolution of 0.4°, the range resolution is calculated as 0.2 mm. Achieving this resolution with a pulsed system requires a time resolution of 1 ps which is very demanding and needs very sophisticated time interval counting electronics. This simple calculation makes clear, why cw systems are used for high precision ranging applications. However, they are usually only applied in close range applications, due to the high power required to continuously emit cw laser light and the limitations imposed by range ambiguity. The unambiguous range R_{unamb}, the maximum distance between targets that can be uniquely differentiated in the cw ranging signal, is given by

$$R_{\text{unamb}} = \frac{\lambda}{2} \tag{4.9}$$

If longer ranges are to be surveyed additional cw signals are used with lower frequencies than the frequency which determines the system resolution. In this case the highest frequency determines the resolution and the accuracy of the ranging system and the lowest frequency the maximum unambiguous range. To assure a well functioning cw LiDAR, at least three frequencies are required for airborne applications. Due to this fact and as a ranging accuracy of better than 5 cm is usually not required, CW-systems are rarely used for airborne surveying. Hence, only pulsed laser systems are on the market for airborne surveying and the remainder of this chapter deals only with pulsed laser systems.

Concerning pulsed laser ranging systems, the number of system parameters has to be extended to fully characterize their performance. Up to now only the range resolution ΔR has characterized the accuracy potential of a pulsed ranging system. Looking at Figure 4.5, we recognize that several targets at different ranges can be illuminated by a single laser pulse. Their different ranges can be resolved, if the laser pulse length is less than half the distance between the two targets. If T_P is the pulse length, the minimum resolvable distance ΔR_{tar} between targets is computed by

$$\Delta R_{\text{tar}} = \frac{c}{2} \cdot T_P \tag{4.10}$$

Let be T_P = 10 ns, then ΔR_{tar} = 1.5 m. This means targets must be more than 1.5 m apart in slant range to be identified as separate targets. Today also shorter laser pulses are possible. 10 ps are state of the art, which theoretically allows a target resolution of 1.5 mm. Here the limiting factor is determined by the processing speed of the ranging electronics, because the range measurement for one target must be completed during this short pulse or parallel gates are necessary. Today's laser scanners commonly generate pulses with pulse length of about 4 ns.

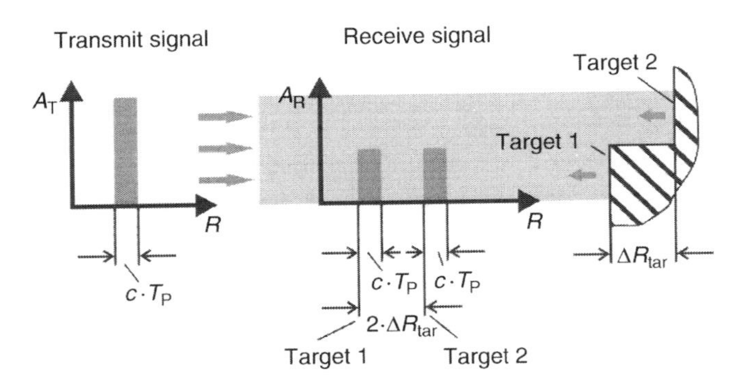

FIGURE 4.5
Target resolution.

4.1.2.2 Laser Transmitter

In laser scanning, the device which offers the electromagnetic signal in the optical spectrum is the laser. Normally semiconductor lasers or solid-state lasers which are pumped by semiconductor lasers are applied in airborne laser scanning. These lasers are optimized, so that very short pulses with high peak power levels and high pulse repetition rates are possible. Peak powers realized in commercially available laser scanners are up to 15 kW with pulse widths of 10 ns. The laser beam is collimated, so that narrow beam divergence and resulting small footprints are possible. Typical diameters for footprints are 30 cm to 1 m at flying altitudes of about 1000 m. The footprint diameter d is dependent on the ranging distance and laser beam divergence which is itself a function of the size of the transmitting aperture and the transmitted wavelength λ. A good estimate for the minimal possible footprint is obtained by calculating a diffraction-limited beam. For spatially coherent light at the wavelength λ, the beam divergence is

$$w = 2.44\frac{\lambda}{D} \tag{4.11}$$

where D is the aperture diameter (Young, 1986). Therefore, the diameter d of the illumination spot on ground is

$$d = D + \frac{\lambda}{D} \cdot R = D + w \cdot R \tag{4.12}$$

where R is the slant range. As D is small compared to the second term a good approximation for the footprint d is

$$d = w \cdot R \tag{4.13}$$

At nadir it corresponds with the flying altitude, because R equals the flying height H. Typical laser beam divergences vary between 0.3 and 2.7 mrad. This means that spot diameters on ground between 15 cm and 1.35 m are realized for a flying altitude of 500 m.

All airborne laser scanners available on the market emit light in the near infrared. Typical wavelengths are 900, 1064, and 1550 nm. The wavelengths 900 and 1550 nm are emitted from semiconductor lasers, whereas the 1064 nm signal is generated by diode pumped solid state Nd:YAG lasers.

The maximum possible emitted laser power is limited by eye safety regulations. Generally speaking, the longer the wavelength and the shorter the pulse width the more a system is eye safe. Therefore, 1550 nm laser systems can use higher power levels than a comparable system working at 900 and 1064 nm, respectively having the same divergence and pulse width.

4.1.2.3 Receiver

The optical receiver is composed of a receiving optic, an optical detector and a transimpedance amplifier (Keiser, 1983). The optical detector is realized with a semiconductor photodiode that converts the optical signal into an electrical current. The photodiode performance depends on the laser wavelength used. For wavelengths up to 1100 nm photodiodes made of silicon (Si) can be used and for wavelengths in the 1000–1650 nm range germanium (Ge) are commonly used. Instead of Ge, semiconductor alloys such as InGaAsP, GaAlSb, InGaAs, GaSb and GaAsSb can also be utilized. A primary criterion in the selection of an appropriate photodiode is its responsivity which should be as high as possible. The unity gain responsivity is given by

$$\Re_0 = \frac{\eta \cdot q}{h \cdot \nu} \tag{4.14}$$

where
η is the quantum efficiency
q is the electron charge
h is the Planck's constant (6.625×10^{-34} J s)
ν is the frequency of a photon

The responsivity specifies the photocurrent generated per unit optical power incident on the photodetector. The photocurrent I_{PH} can be computed by

$$I_{PH} = \Re_0 \cdot P_{opt} \tag{4.15}$$

where P_{opt} is the received optical power. The quantum efficiency η is given by the selected photodiode, the material it is made of, and the wavelength of the detected light. In Figure 4.6 typical quantum efficiencies for Si, Ge, and InGaAs photodiodes are plotted.

Often in LiDAR systems avalanche photodiode detectors (APDs) are used which feature higher responsivities (Sun et al., 1992), because they offer an internal amplification, referred to as the avalanche gain M. For APDs, the responsivity $\Re$ becomes

$$\Re = \frac{\eta \cdot q}{h \cdot \nu} \cdot M \tag{4.16}$$

For Si photodiodes M is as large as 300. However, for Ge diodes the avalanche gain is limited to 40. The avalanche gain is achieved by building up a region with a high electric field. As soon as the photogenerated electrons or holes traverse this region they gain so much energy that they ionize bound electrons in the valence band by collision. Also, these additional carriers experience accelerations by the high electric field and set free more ionized

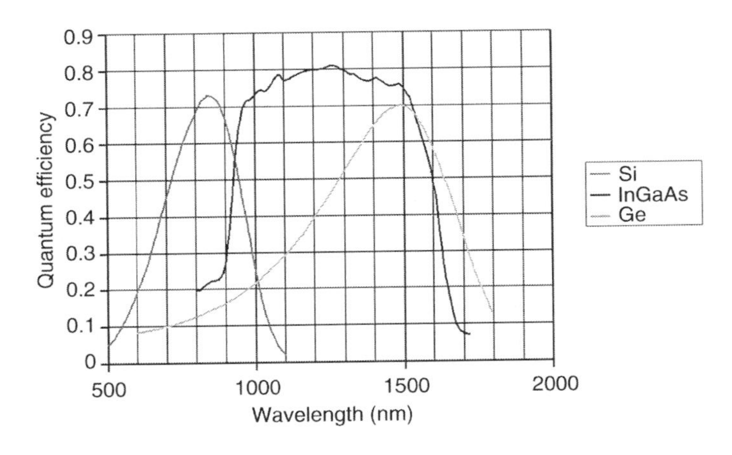

FIGURE 4.6
Quantum efficiencies for different photodiodes.

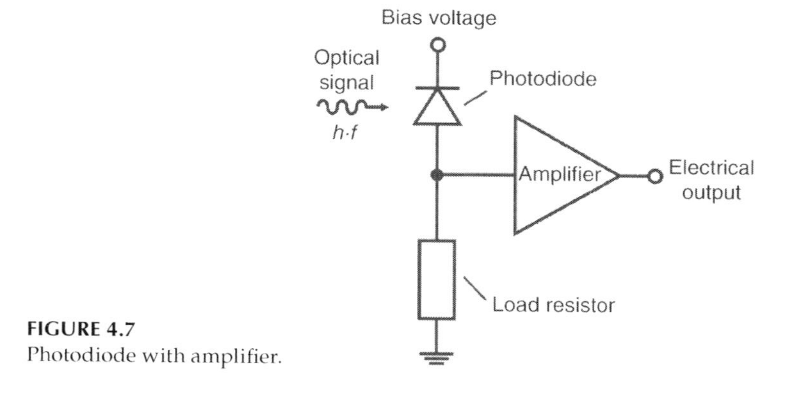

FIGURE 4.7
Photodiode with amplifier.

bound electrons. This physical process is called avalanche effect. As APDs need a high electric field, high bias voltages of several hundreds of volts have to be applied between the anode and cathode of the photodiode. If the bias voltage is lower than the breakdown voltage, a finite total number of carriers is created. APDs used in LiDARs normally work in this mode. Here, the avalanche gain M is a function of the bias voltage. Working above the break down voltage causes infinite number of carriers which may destroy the photodiode, because it is not a regular operation mode. However, there are special APDs available which can be used above the break down voltage. In this case they are functioning as photon counters.

For further signal processing the photocurrent must be converted into a voltage. This is accomplished by a transimpedance amplifier. Figure 4.7 shows a typical photodetector amplifier configuration. The transimpedance amplifier should exhibit a high conversion gain and a low noise figure. The combination of photodiode and transimpedance amplifier determines the sensitivity of the receiver. The higher the signal-to-noise ratio (SNR) at the output of the transimpedance amplifier the more sensitive is the receiver and the lower is the standard deviation for ranging, and the more precise is the ranging result.

4.1.2.4 Optics

Today, most LiDARs are realized with conventional optics. This means in the transmitting and receiving path optical lenses, mirrors, beam splitters, and interference filters are applied. Figure 4.8 shows a standard setup for pulsed ranging systems. The laser radiation is

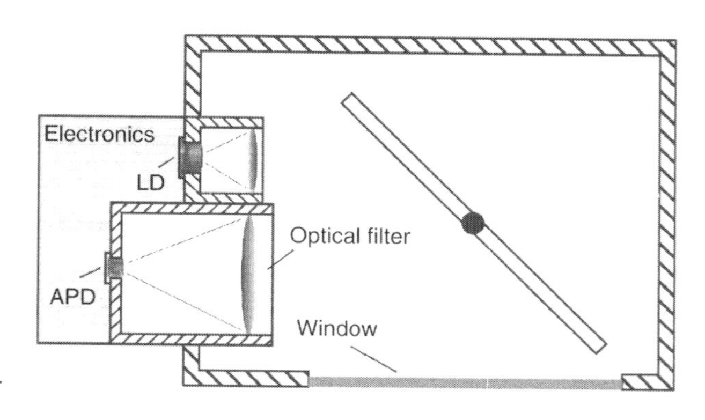

FIGURE 4.8
Optical setup with conventional optics.

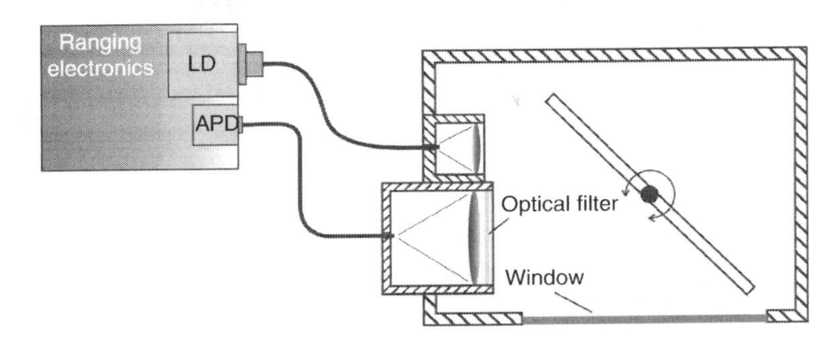

FIGURE 4.9
Principle setup with optical fibers.

collimated by a lens controlling the beam divergence. The backscattered signal from the target passes through an interference filter first and is then focused on the detector chip by a focusing lens. The interference filter is required to suppress background light seen within the receiver IFOV which causes additional noise in the receiver and degrades ranging precision. Most background light is sunlight backscattered from Earth's surface or from objects, such as clouds, within the IFOV. The typical bandwidths for interference filters are about 5–10 nm. Interference filters are also used in cw LiDARs. Here, the purpose is first to circumvent saturation of the detector by background light and second to reduce the received background power down to power levels that make optimum avalanche amplification possible. This results in reduced noise within the photo current. The still remaining background light, which can well be seen in the photocurrent directly at the photodiode (Wehr, 2007), is filtered out finally by further processing the electrical signal. Here synchronous demodulation or correlation receivers are applied.

As an alternative to a conventional optical design, fibers can be used to relay the optical signal. A configuration using fibers is shown in Figure 4.9. The laser light is directly coupled into a transmit fiber and the received light is focused onto a fiber that is coupled to the detector. Using such a setup, the electronical parts are separated from the optical parts. This makes the adjustment of the optics more simple and robust. Figure 4.10 shows examples of a fiber-coupled laser and APD receiver.

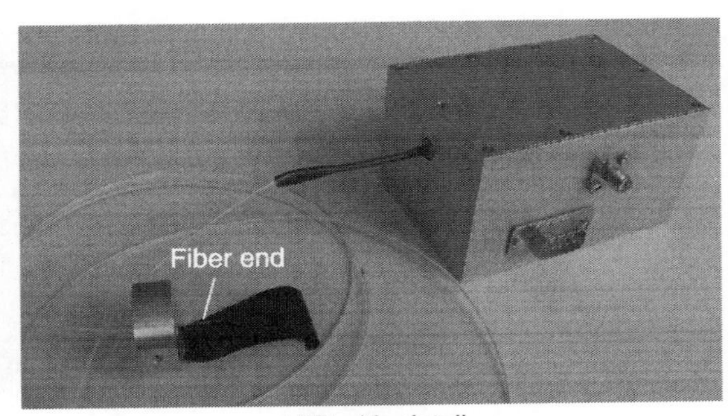

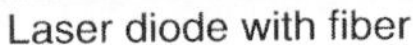
Laser diode with fiber

APD with pigtail

FIGURE 4.10
Laser and receiver with fibers.

Typically in the transmitter and receiver, the focusing objectives are realized with non-imaging biconvex lenses, because monochromatic light is transmitted and received and the goal is to gather the optical energy as efficiently as possible. A distortion free image is not required.

4.1.2.5 Link Budget

In Section 4.1.2.1 the principles of laser ranging were explained. However, degradation of the signal along the round-trip travel path from the laser to the photodetector was not considered. Calculations concerning the signal quality at the output of the receiver are known as a link budget. The signal quality is defined as the SNR at the receiver's output. This ratio determines the achievable standard deviation of the slant range measurement, which is inversely proportional to the square root of SNR:

$$\sigma_{\mathrm{R}} \sim \frac{1}{\sqrt{\mathrm{SNR}}} \tag{4.17}$$

SNR is mostly determined by the optical detector and the signal power incident on the detector element. According to Keiser (1983) the SNR for a photodiode is given by

$$\mathrm{SNR} = \frac{\frac{1}{2}(\Re_0 \cdot m \cdot P_{\mathrm{r}})^2}{2q\left(\Re_0 \cdot (P_{\mathrm{r}} + P_{\mathrm{B}}) + I_{\mathrm{D}}\right) \cdot F(M) \cdot B + \frac{4k_{\mathrm{B}} \cdot T \cdot B}{R_{\mathrm{eq}} \cdot M^2} \cdot F_{\mathrm{AMPL}}} \tag{4.18}$$

where

P_{r} is the received optical power
P_{B} is the received optical power of background signal
m is the modulation index
M is the avalanche gain
I_{D} is the primary bulk dark current
$F(M)$ is the excess photodiode noise factor $\cong M^x$ with $0 < x \leq 1$
B is the effective noise bandwidth
k_{B} is the Boltzmann's constant
T is the absolute temperature
R_{eq} is the equivalent resistance of photodetector and amplifier load
F_{AMPL} is the noise figure of the amplifier

This formula looks quite complicated and I will not go into much detail. But I think it is necessary to present the total formula for better understanding of the following assumptions and resulting reductions. Regarding commercial pulsed laser systems, the received power levels can be regarded as high and are well above the power of the background signal. In this case, the dark current I_{D} and the thermal noise term which is reduced by the high avalanche gain M, can be neglected so that Equation 4.18 can be simplified to

$$\mathrm{SNR} = \frac{\frac{1}{2} \cdot \Re_0 \cdot m^2 \cdot P_{\mathrm{r}}}{2q \cdot F(M) \cdot B} \tag{4.19}$$

Furthermore, $m = 1$ for pulsed signals, because the laser is switched on and off. This means SNR is directly proportional to the received optical power P_{r} and the responsivity $\Re_0$ and inversely proportional to the bandwidth B. Bandwidth B is a function of the sample rate and pulse duration T_P, respectively; the shorter the pulse the higher is the required bandwidth. However, a larger bandwidth means lower SNR if P_r is kept constant. This formula makes clear that here we face a typical engineering problem, that needs an engineering trade-off for each system. Introducing the discussed parameters of Equation 4.19 into proportionality (Equation 4.17), the resulting relation for the ranging standard deviation

$$\sigma_R \sim 2 \cdot \sqrt{\frac{q \cdot B}{\Re_0 \cdot P_r}} \tag{4.20}$$

underlines this fact. Looking at Equation 4.20, one may conclude that the bandwidth should be as small as possible. But in this proportionality the signal form (the pulse shape in a plot of pulse energy versus time) is not regarded. Jelalian (1992) considered the signal form and obtained the following equation for leading edge detection:

$$\sigma_R = \frac{c}{2} \cdot \sqrt{\frac{T_P}{2 \cdot B \cdot (E_r/N_0)}} \tag{4.21}$$

where

- T_P is the pulse length
- B is the effective bandwidth
- E_r is the received optical energy
- N_0 is the noise power per cycle

Leading edge detection means that the rising edge of the pulse waveform is used as a reference for time delay measurements. As the receiver has a limited bandwidth B, the slope of the received signal has a finite gradient (Figure 4.11a). The exact moment at which the received pulse is acknowledged as arrived is dependent on the detection method. If for example, threshold detection is applied the receive time varies with the threshold level (Figure 4.11a). Even if the threshold level is fixed the measured receive time of the pulse varies in dependence of the gradient of the leading edge (Figure 4.11b). The slope of the leading edge may vary by signal noise, amplitude variations of the received signal or spreading of the pulse due to elevation differences within the illumination spot of the laser pulse (Figure 4.11a and b). The first one introduces stochastic errors where as the other two cause systematic errors. The amplitude dependent systematic errors can be compensated by applying a constant fraction discriminator. Here the received signal is split into two channels. In one channel, the signal is delayed by a certain delay time t_d and in the other one, its amplitude is reduced by a factor m. Then, the delayed signal is subtracted from the reduced signal. The time at which the resulting signal crosses zero is independent on the amplitude and this moment is used for ranging. Such signal processing carries out ranging not on a fixed threshold but on a constant fraction of the total peak power of the received pulse.

The most precise pulse ranging measurements can be expected from full waveform systems, which today are the most advanced commercially available systems. They sample the transient of the return pulse with sampling rates of about 1 GHz, so that no analogue information gets lost (Figure 4.11c). The stored data can be processed by different sophisticated algorithms (Jutzi and Stilla, 2005). So it is possible to compensate systematic errors by, for example, calculating the centriod of the return signal and even different elevations can be resolved within the transient of one return pulse. This means that different vegetation elevation levels can be determined in dense forest

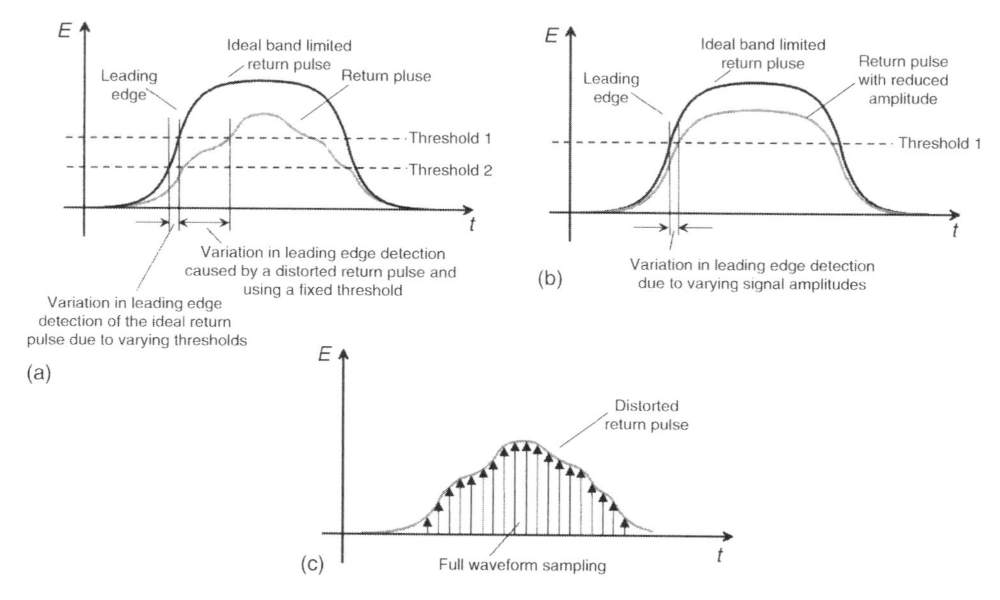

FIGURE 4.11
Pulse time delay measurement. (a) Timing variations caused by distorted signals and different thresholds, (b) timing variations caused by varying amplitudes and resulting different leading edge slopes, and (c) full waveform sampling.

areas. More details concerning full waveform systems and their data processing are presented in Chapters 3, 5, and 6.

In the following derivations and calculations, an ideal pulse detection is assumed, meaning there are no systematic errors in the measurement caused, for example, by amplitude variations in the return pulses. Assuming now perfect rectangular pulses of width T_P and a matched filter which means that

$$B = \frac{1}{T_P} \tag{4.22}$$

σ_R becomes

$$\sigma_R = \frac{c}{2\sqrt{2}} \cdot \frac{1}{\sqrt{\text{SNR}}} \cdot T_P \tag{4.23}$$

This corresponds to proportionality (Equation 4.17), because T_P is normally constant. Introducing Equation 4.19 into Equation 4.23 σ_R amounts to

$$\sigma_R = c \cdot \frac{1}{\sqrt{2}} \sqrt{\frac{q \cdot F(M)}{\Re_0}} \cdot \sqrt{\frac{T_P}{P_r}} \tag{4.24}$$

Equation 4.24 depicts that σ_R is proportional to the square root of T_P and inversely proportional to the square root of the received optical power P_r.

For small optical signals onto the detector, this is especially the case in CW-systems, the thermal noise term in the numerator of Equation 4.18 dominates so that

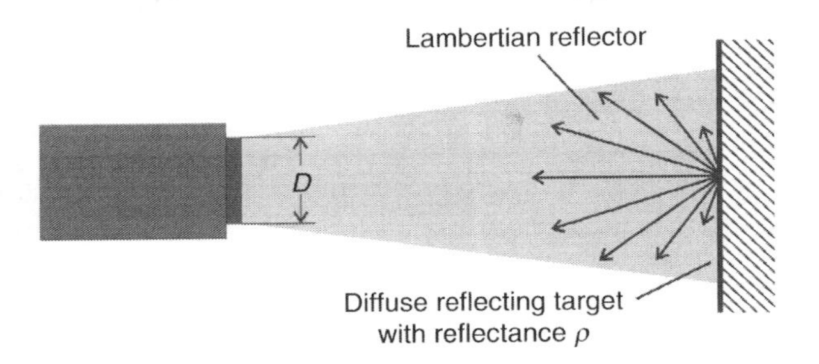

FIGURE 4.12
Ranging setup for diffuse a reflecting target.

$$\sigma_R = c \cdot \sqrt{\frac{k \cdot T \cdot F_{\text{Ampl}}}{\Re_0 \cdot R_{\text{eq}} \cdot M^2}} \cdot \frac{\sqrt{T_P}}{P_r} \tag{4.25}$$

This means also here the ranging standard deviation is proportional to the square root of pulse width T_P. However, it is inversely proportional to the received optical signal power P_r.

Equations 4.24 and 4.25 make clear that P_r is the primary parameter which determines the accuracy during flight. Therefore, the determination of P_r. is worked out in the following. Figure 4.12 shows the two-way ranging setup.

Assuming a diffuse reflecting target which is equal or larger than the laser footprint, then the received optical power P_r is computed by

$$P_r = \frac{1}{4} \cdot P_T \cdot \tau_{\text{total}} \cdot \rho \cdot \left(\frac{D}{R}\right)^2 \tag{4.26}$$

where

- P_T is the transmitted optical power
- ρ is the reflectivity
- D is the diameter of receiving aperture
- R is the distance from LiDAR to target
- τ_{total} is the total transmission the beam experiences along the two-way path from the transmitting laser aperture to the detector element

In Equation 4.26 diffuse reflection is approximated by considering Lambertian reflectors (Figure 4.12). This means the laser light is backscattered into the half sphere and its radians is independent of the viewing angle. For a pulsed system, the peak power P_{peak} is used as P_T. The total transmission τ_{total} is comprised of the transmission of the receiver objective, of the optical interference filter, of the atmosphere and of the scanning device. It should be noted that the atmosphere path is traveled twice and also that the scanning device has a transmit and receive path. P_T and D are constant determined by the instrument design. τ_{total} can be separated into a constant part τ_{system} defined by the LiDAR design and the transmission of the atmosphere, which is a function of the slant range R:

$$\tau_{\text{total}} = \tau_{\text{system}} \cdot 2 \cdot \tau_{\text{atmos}}(R) \tag{4.27}$$

Therefore, P_r can be described by a constant parameter C_{design} defined by the LiDAR design, the target dependent reflectivity ρ, the range dependent parameter τ_{atmos} and the slant range itself:

$$P_r = C_{design} \cdot 2 \cdot \tau_{atmos} \cdot \frac{\rho}{R^2} \tag{4.28}$$

Neglecting the range dependent influence of the atmosphere, P_r is proportional to the reflectivity of the target and inversely proportional to the square of slant range R. Applying Equation 4.28 in Equation 4.24 one sees clearly that for pulse systems ranging standard deviation σ_R becomes

$$\sigma_R \sim \frac{R}{\sqrt{\rho \cdot \tau_{atmos}}} \tag{4.29}$$

whereas for CW-systems with low receiving power levels the following proportionality is observed:

$$\sigma_R \sim \frac{R^2}{\rho \cdot \tau_{atmos}} \tag{4.30}$$

However, real LiDAR measurements show that the functional relationship between σ_R and slant range R is located somewhat between proportionality (Equations 4.29 and 4.30). Furthermore, it should be mentioned that there are special detection cases, in which this very general approach does not apply, e.g., detection of powerlines. Here the fundamental form of the radar equation should be used, which contains the radar cross section σ_{cross} as a parameter. σ_{cross} describes the property of a scattering object. σ_{cross} is a function of the wavelength, the size, the directivity of the backscattered signal and the reflectivity of the target. This parameter is used to model specular reflection and retro reflecting targets. Regarding σ_{cross} Equation 4.26 becomes

$$P_r = P_T \cdot \tau_{total} \cdot \frac{1}{w} \cdot \frac{D^2}{R^4} \cdot \sigma_{cross} \tag{4.31}$$

where w is the divergence of the laser beam, which is measured in radians. As w is an instrument constant, it is clear that P_r is inversely proportional to the fourth power of the slant range. This maximum possible accuracy in slant range degrades with $\sigma_R \sim R^2$ as far as pulsed systems are concerned. A more detailed analysis was carried out by Litchi and Harvey (2002) and Jutzi et al. (2002).

The LiDAR link budget calculation presented in this chapter was reduced to the simpler forms in order to highlight the key parameters that govern instrument performance. These parameters are those with greater relevance for people who will be configuring, using or procuring a LiDAR system. Calculating the precise ranging accuracy of a system is often difficult because the transmitted power levels are usually not mentioned. Therefore, the equations were reduced to proportionalities so that the influences of the different parameters are seen clearly. Performance of a particular system can be extrapolated to different flying altitudes based on the ranging accuracy achieved at one altitude. Profound calculations are published by Abshire (2000), Bachman (1979), Skolnik (2001), and Sun et al. (1993).

4.1.3 Scanning Devices

Figure 4.1 depicts that the transmit beam and the co-aligned receiver IFOV must be deflected across the flight line to obtain surface coverage. In all commercial LiDARs, optomechanical

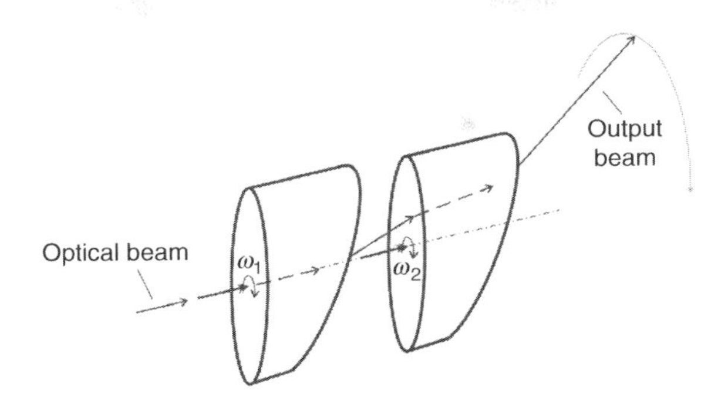

FIGURE 4.13
Principles of rotating prisms.

scanners are implemented to accomplish this. Using conventional optics two physical principles can be exploited: refraction and reflection. Refraction scanners are realized with two rotating prisms (Figure 4.13).

Such scanners have been used in military LiDARs for power line detection in helicopters. Very fast scans can be realized, because the continuous, uni-directional rotation of each prism can be performed at high speeds. Figure 4.14 depicts a typical setup with two prisms without drives. With such a setup all kind of Lissajous figures are possible by changing the phase between the two prisms. A spiral scanning pattern (Figure 4.15), for example, has been used for power line detection. However, they are not routinely applied in airborne surveying. Mirrors are typically used in commercial airborne LiDARs to carry out the scanning.

4.1.3.1 Reflecting Scanning Devices

This section presents and discusses the different scanner setups exploiting the physical effect of reflection. Typical mirror scanners with the associated scanning patterns are compiled in Table 4.1. For all airborne LiDAR scanners, the primary design goal is to scan sufficiently fast to compensate for the forward velocity of the platform while achieving the desired sampling density in the flight path direction. Oscillating mirrors rotate back and forth to achieve a cross-track scan pattern. For oscillating mirrors the achievable scanning speed is dependent on the inertia of the moving parts and the available power of the scanner

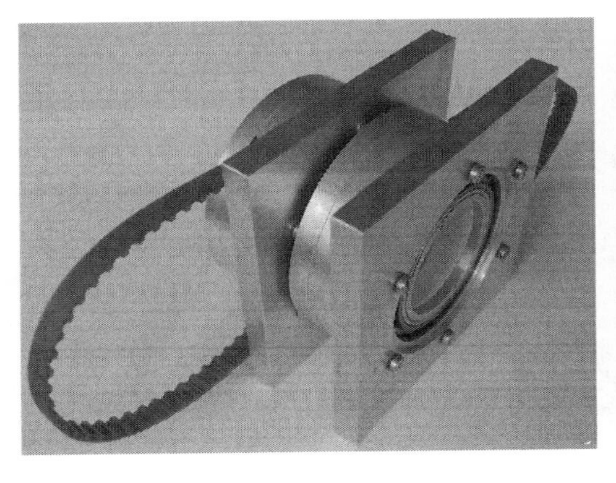

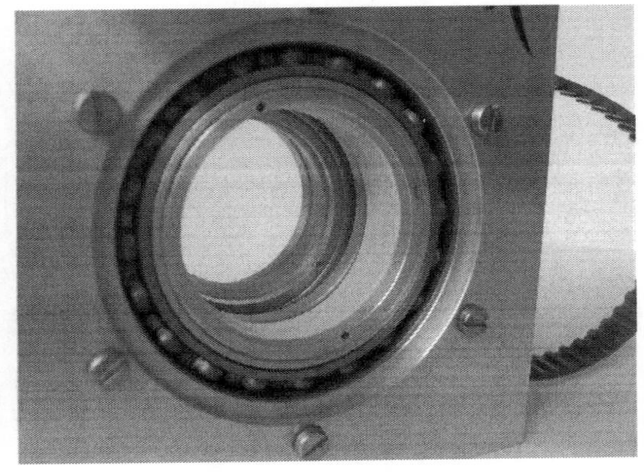

FIGURE 4.14
Rotating prisms without drives.

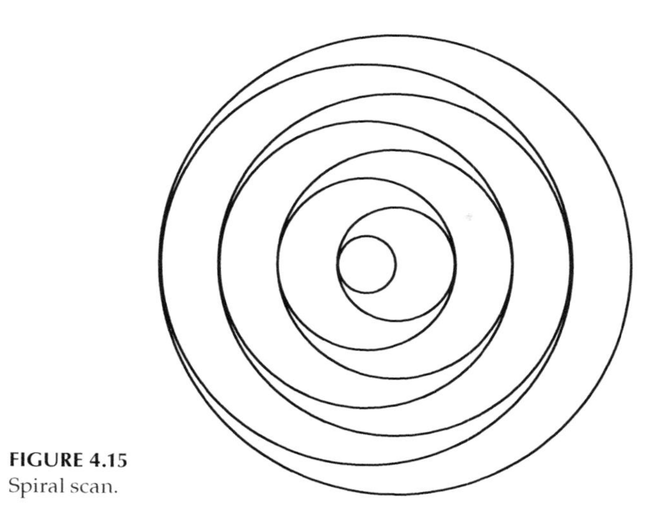

FIGURE 4.15
Spiral scan.

TABLE 4.1
Overview of Mirror Scanners

Name	Principle	Scanning Pattern
Oscillating mirror scanner	α	
Oscillating mirror scanner with cam drive	Cam drive	
Polygon scanner		
Palmer scanner	α ω Mirror	Flight direction Forward scan Backward scan

drives, because the mirrors experience accelerated movements. The inertia is a function of the mirror's size and shape. The required size of mirrors is determined by the size of the transmitting and receiving aperture. Here, we again face an engineering problem which means optimizing the parameters scanning speed, accuracy and laser power, because accuracy can only be improved for a given laser power by enlarging the receiving aperture. Larger apertures mean larger mirrors and higher inertias. Higher inertias of moving parts require more powerful drives. If the drive power cannot be improved the mirror must move more slowly.

Faster scans are possible with the polygon scanner that employs a multifacet mirror, and the Palmer scanner, because these mirrors do not experience any acceleration once the working rotation speed is reached.

Table 4.1 makes evident that the laser beam deflected by facet mirrors always moves in one direction, either from left to right or vice versa. Figure 4.16 presents typical facet mirror scanners. A rotating multi-facet mirror, composed of four facets, is used in the scanners manufactured by Riegl. For oscillating and faceted scanners, the obliquity of the scanning lines with respect to the flight direction and the line spacing in the flight direction are a function of the ratio of the scanning speed to the flying speed. The spacing of the laser spots along the scanning track is dependent on the scanning speed and laser pulse rate.

The Palmer scanning mechanism is the most simple one as far as the opto-mechanical bearing and drive setup are concerned, if large apertures are required. The ideal Palmer scanning pattern is generated on Earth's surface by the translation of a pure circular scan (Table 4.1). A circular scan is realized by mounting a mirror at the phase side of a rotating shaft. If the normal of the mirror has an angle α with the rotation axis and assuming a narrow laser beam radiated in the direction of the rotation axis, an ideal circle is projected back into the direction the laser light is originating from. The circle size increases by angle α. Normally, the shaft is mounted so that the shaft's rotation axis exhibit 45° with regard to incident laser beam. Using such a configuration, the scanning pattern is deflected by 90° (Table 4.1). A detailed ray tracing analysis shows, that the deflection of 90° causes a deformation of the ideal circular scanning pattern to an egg-shaped one. However for flight planning, the scanning pattern is approximated with an ellipse. During flight a translated ellipse is observed on Earth's surface. Such scanning devices were first applied in airborne and spaceborne remote sensing missions in the early seventies. For example, the mulispectral scanner flown on Skylab uses this scanning concept. In the nineties, this scanning principle was used for laser scanning (Hug, 1994; Krabill et al., 1995, 2002).

A typical Palmer scanner assembly, built at the Institute of Navigation, University of Stuttgart and used in the ScaLARS LiDAR, is shown in Figure 4.17. The TopEye Mk II

FIGURE 4.16
Facet mirror scanners.

FIGURE 4.17
Palmer scanner.

(Blom Swe AB & TopEye AB, Stockholm, Sweden) also uses the Palmer scanning principle. The helical pattern produced by a Palmer scanner is a redundant one, because the same area on ground is sampled twice. Thiel and Wehr (1999) show that the sampling cross-overs formed by the intersecting forward and backward scans can be exploited for intrinsic calibration of the external orientation parameters of POS and LiDAR.

All discussed scanners are designed so that loss of transmit and received laser energy upon reflection is minimized and the direction of the deflected beam is known with sufficient accuracy to achieve the required position accuracy of the laser spot on Earth's surface. Regarding for example the oscillating mirror scanner, the angular accuracy of rotation about the main axis should typically be better than a hundredth of a degree. Flying at an altitude of 1000 m and assuming a scan angle of 20° off-nadir, an angular accuracy of a hundredth of a degree yields a position accuracy of 20 cm.

4.1.3.2 Fiber Scanning Technology

The company TopoSys (Topographische Systemdaten GmbH, Biberach, Germany) is the only one worldwide which manufactures and offers airborne laser scanners using a so-called fiber scanner. They are sold under the product name Falcon (Topographische Systemdaten GmbH, Biberach, Germany). Their latest product in this category is Falcon III. Fiber scanners were first designed and applied for reconnaissance tasks in military jet planes. Due to the high flying speed, very fast scanning was required which could not be realized with conventional techniques. Therefore, a solution with small moving parts was developed. Figure 4.18 shows that here also mirror scanners are applied. However due to the small aperture of the fibers, small mirrors can be used. As the mirrors are small and rotated constantly in one direction, high scanning speeds up to 415 Hz are possible.

Besides the high scanning speed, the main advantage of this configuration is that transmit and receive optics are designed and manufactured identically. They do not have any moving parts. Therefore, they are very robust and stable during surveying missions. Incredibly remarkable is the design idea to transform a circular scan pattern into line scan by using fibers. Identical fiber line arrays are mounted in the focal plane of the receiving and transmitting lenses. By means of two rotating mirrors moved by a common drive, each fiber in the transmitting and receiving path is scanned sequentially and synchronously. In the transmitting path, the laser light is relayed from the central fiber to a fiber of the fiber array mounted in a circle around the central fiber. The optics between mirror

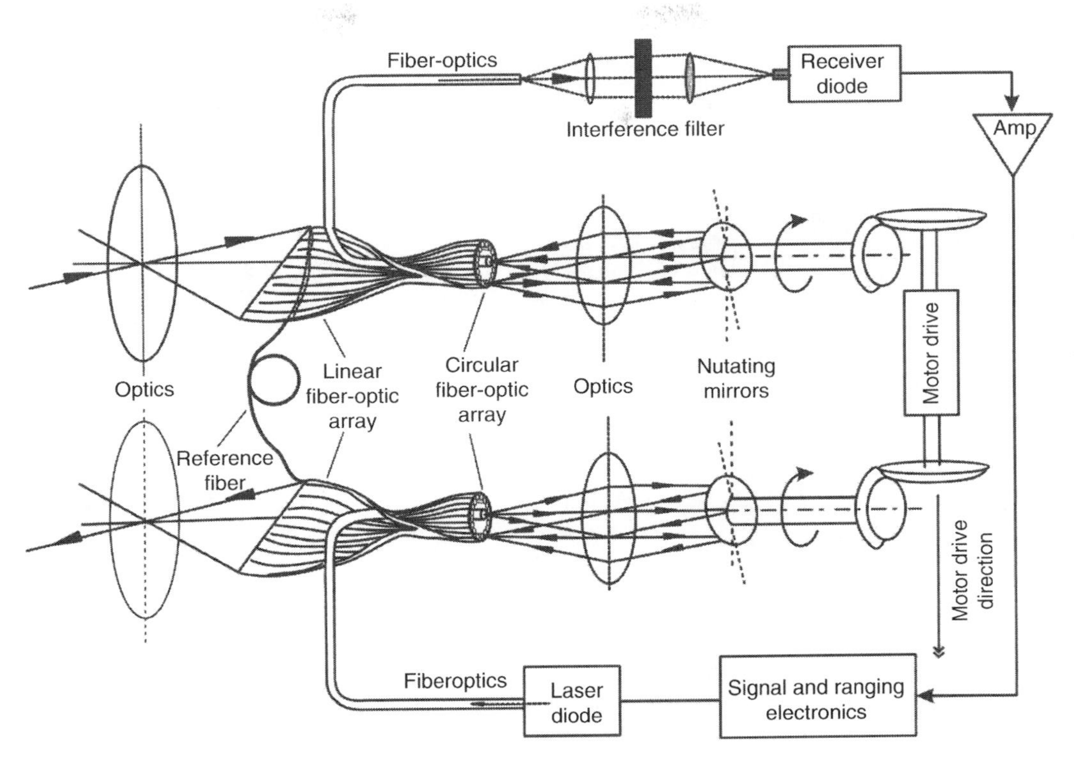

FIGURE 4.18
Principle of TopoSys fiber scanner.

scanner and fibers is adjusted in a way that the aperture of the laser fiber is imaged onto the aperture of the instantaneously illuminated fiber of the array. As the received part exhibits the identical optical path, the transmitted laser pulse backscattered from the target is seen by the corresponding fiber in the receiving fiber array. The receiver sees in the IFOV background light besides the laser signal, which must be filtered to improve SNR (Section 4.1.5.2) and protect the detector against saturation. This is carried out by an optical filter in the central fiber linked to the photodiode. High scanning speeds (up to 415 Hz) can be achieved. This is not possible with conventional mirror scanners. 127 fibers plus a fiber for the reference signal are realized in Falcon III. The maximum across-track spacing Δx_{across} is dependent on the scan angle θ across the flight direction, which is also called FOV, the flying altitude H above the ground and the number of fibers N. It is given by

$$\Delta x_{\text{across}} = \frac{\theta}{(N-1)\cdot \cos^2(\theta/2)} \cdot H \tag{4.32}$$

In the flight direction the spacing Δx_{along} is determined by the flying speed above ground v, and the number of revolutions per unit time f_ω:

$$\Delta x_{\text{along}} = \frac{v}{f_\omega} \tag{4.33}$$

Falcon III has 127 fibers and a FOV of 28°. Assuming a flying speed of 75 m/s and an altitude of 1000 m, the across-track spacing Δx_{across} equals 4 m and spacing along track is 18 cm

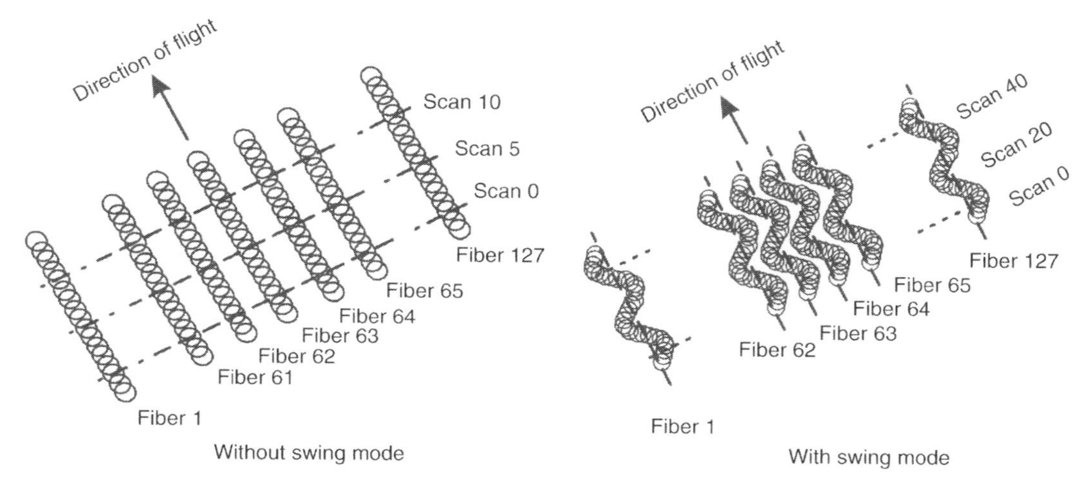

FIGURE 4.19
Scan pattern of TopoSys airborne laser scanner system.

with $f_\omega = 415\,\text{Hz}$. This example demonstrates that the spacing along-track is much smaller than acrosstrack. This is depicted in Figure 4.19. Ideal scan pattern designs try to achieve equally distributed laser points on ground. In the right drawing of Figure 4.19 the scanning pattern is dithered to fill the gaps between fibers by using a swing mode which is realized by controlled synchronous movements of the receive and transmit fiber arrays. Regarding in addition the stochastic movements of the airborne platform, laser spots appear more equally distributed on the ground. This technique improves for example the detection of powerlines and building edges. As the production of fiber arrays requires very sophisticated know-how, LiDARs using fiber scanners are only offered by one company.

4.1.4 Controlling and Data Sampling Unit

Figure 4.1 shows that the control, monitoring and recording unit is a key device of a LiDAR system. It synchronizes the ranging unit with the scanner, triggering the pulsed laser synchronously with the incremental scanner steps. In addition, this unit stores to hard disk the ranging data set, including the slant ranges of the return pulses, the return intensity, if available, the instantaneous scanning angles and high precision time stamps. The time stamps, required for later synchronization with the POS data, are derived from the GPS 1 pulse per second (pps) signal. Figure 4.20 presents graphically the information flow. The unit is realized by a powerful data handling device, because data have to be transferred to the disk at high data rates which can easily exceed 1 MB/s. Range calculations and decisions concerning the reliability of ranging are also carried out in the ranging unit.

Data handling for fullwave recording LiDARs is even more demanding, because the full receive pulse shape is digitized rather than simply ranging to its leading edge. First, a sophisticated A/D-board (analogue to digital conversion) is required which samples the transient of the return pulse at high rate, for example at 1 G samples per second. This value is equivalent to a range resolution of only 15 cm. Assuming an amplitude resolution of 8 bits and continuous data logging one has to deal with a data stream of 1 GB/s. As today's commercial data recorders typically handle data rates only up to 80 MB/s, the waveform data have to be reduced by special algorithms. For example, rather than continuously recording the A/D output from the time of transmission to reception of the return, transmit and receive waveforms can be recorded only when signal is detected above a threshold.

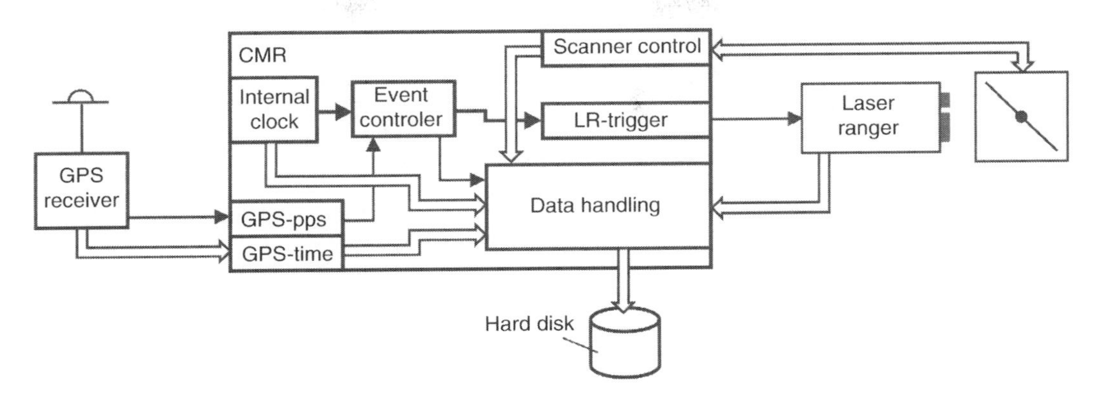

FIGURE 4.20
Information flow of control, monitoring, and recording unit.

In this case, the time interval between the transmitted and received pulse must also be measured and recorded. Applying such preprocessing filters and taking into account a data rate of 80 MB/s, surveying times of more than 2 h are possible with hard disk storage of 640 GB.

4.1.5 Position and Orientation System

Previously it was explained that the position and orientation of the LiDAR must be known very precisely for every instant. A so-called POS is used for this purpose. It is comprised of a DGPS and an IMU. DGPS requires reference stations on ground. To guarantee reliable phase solutions with centimeter level accuracy, using current GPS equipment and typical commercial processing methods, the reference stations must not be further away than 25 km from the airborne platform. DGPS data processing for LiDAR surveys is normally carried out off-line after the flight. Therefore, a telemetry link from the reference stations to the airborne LiDAR is not required. During surveying, the airborne POS data including GPS position and carrier phase information and the IMU data are stored together with GPS-time on memory cards in the airplane. Figure 4.21 shows the components of a typical POS.

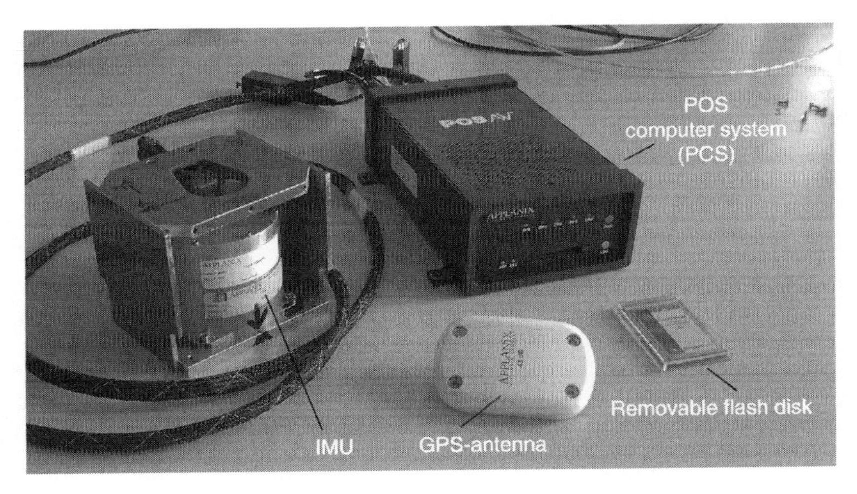

FIGURE 4.21
Applanix POS-AV. (From Applanix, Richmond Hill, Ontario, Canada.)

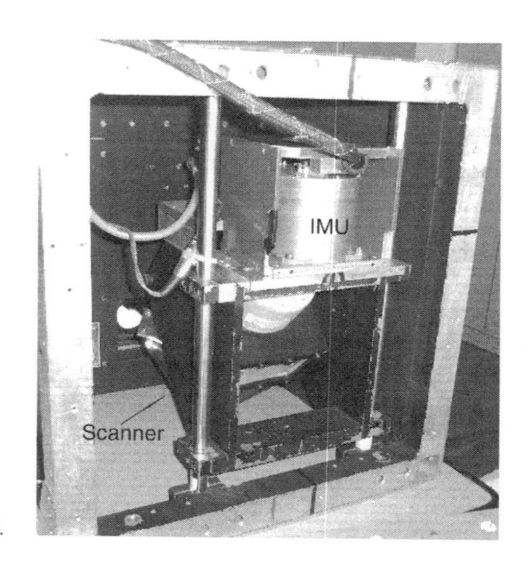

FIGURE 4.22
Mounting of IMU.

Typically the IMU is positioned as close to the scanner as possible in order to record orientation and aircraft vibrations at the location of the LiDAR. In Figure 4.22, the IMU is directly mounted on top of the scanner. With IMU systems recording at a bandwidth of 100 Hz, the vibration of the scanner and the carrying structure of the airplane can well be seen in the IMU raw data.

After the mission, first GPS-data collected by the POS receiver and the GPS-data of the reference stations are processed by special DGPS-software, which computes the LiDAR positions with cm- to dm-accuracy, typically in 0.5 s steps. In a second step, an integrated position and orientation solution is calculated with the DGPS-position data and the IMU-data by another software module, yielding positions determined with cm- to dm-accuracy in three dimensions and roll, pitch and yaw angles of the LiDAR determined to better than one hundredth of a degree. These POS data together with the instantaneous time at which the position and orientation were sampled are stored in a file.

4.1.6 Synchronization

In the preceding sections the sampling and the recording of the data were explained. In general, the LiDAR and the POS are independent units produced by different companies. The combination of LiDAR and POS units may vary from survey to survey, and for research purposes often several POS are flown together for comparative studies. Because little is published on the methods by which commercial surveying companies synchronize the LiDAR and POS data sets, I describe here an approach developed at the Institute of Navigation of the University Stuttgart that is independent of the LiDAR and POS units employed.

Figure 4.23 makes clear that the laser scanner measurements are controlled and stored by the LiDAR control unit (LCU), and the GPS and IMU data by the POS Computer System (PCS). LCU and PCS both work in their own time system. PCS is related to the GPS time where as the LCU time is defined by its internal computer clock. Assuming that the POS delivers synchronized GPS and IMU data, synchronization can be reduced to synchronizing the GPS time with the LCU time. According to Figure 4.23, a software module in the LCU stores the laser scanner data with additional time data controlled by interrupt IRQ2 during flight. At the start of each scan or scan line, the local LCU time is linked into the data stream. In parallel to this recording, the time tag of the pps-signal of the POS GPS-unit

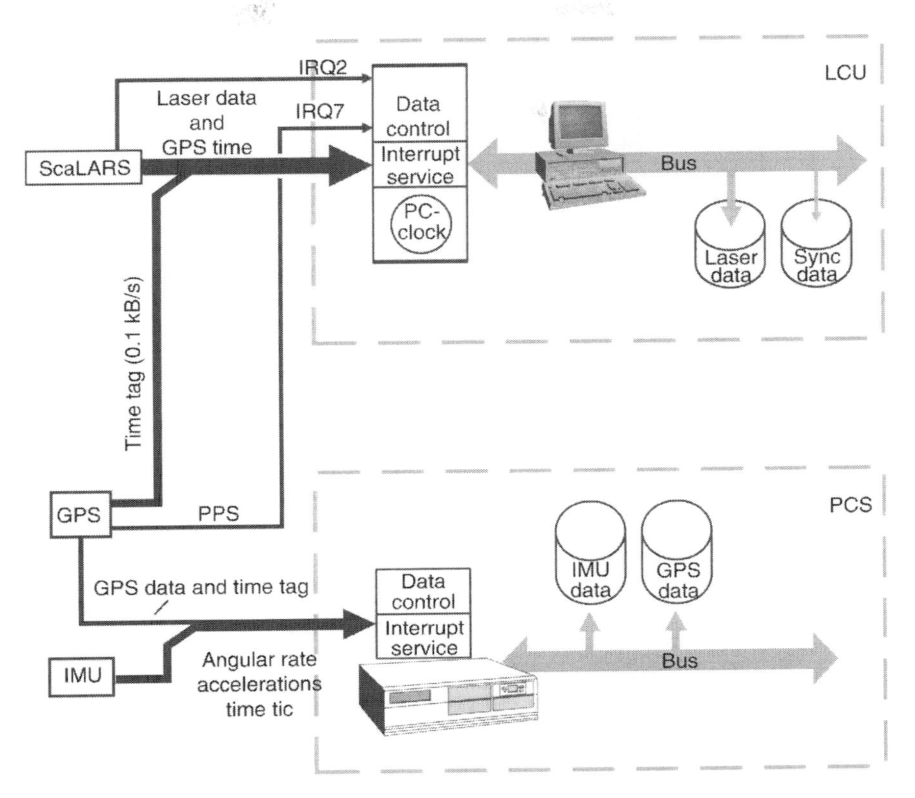

FIGURE 4.23
Synchronization.

and the local LCU time at the pps instant are stored in a separate protocol file. As the pps-signal of the onboard GPS receiver triggers the hardware interrupt IRQ7, the GPS-time and local LCU time are stored at the same instant. LCU time differences from the pps interval can be directly identified as timing errors between the highly accurate and stable GPS time and the LCU time. They are applied as correction for synchronization errors. This method achieves an operational synchronization of better than 10 μs. As the protocol file comprises the actual synchronization information, an offline synchronization of the POS and LiDAR data is possible. Before the synchronization the user has three files:

1. LiDAR raw data file containing LCU-time for each scanline start followed by a number of data sets with instantaneous scanning angles, slant range, and intensity
2. Protocol file
3. POS file containing the instantaneous three dimensional position and, the orientation angles with regard to GPS time

Normally, the LiDAR file comprises more data lines per time interval than the POS file, because the sampling rate, equivalent to the laser pulse rate, is much higher. This requires interpolation of the POS positions and orientation angles. In case of ScaLARS a linear interpolation is applied. After synchronization data are available in the following possible arrangement:

GPS time | x, y, z in (WGS84) | heading, pitch, role | slant range | intensity | scanning angle

These data are the input for calculating geocoded data in a process known as registration (Vaughn et al., 1996).

4.1.7 Registration and Calibration

The accuracy of the registered data is very dependent on calibration data of the LiDAR and POS sensors. Incorrect calibration not only introduces systematic errors but can also introduce random noise recognized by higher elevation standard deviations. The importance of calibration is emphasized by the improvement in LiDAR survey accuracy that has been achieved in recent years. This has been achieved primarily, without changing the LiDAR surveying hardware, but rather by improvements in calibration procedures and associated improvements in navigation data accuracy and data synchronization (Burman, 2002; Filin, 2003a,b; Filin and Vosselman, 2004; Huising and Gomes Pereira, 1998).

For better understanding the process of registration it is assumed first that the inner orientation of the LiDAR is known. In this case, calibration means determination of the exterior orientation described by bore sight misalignment angles in roll $\delta\omega$, pitch $\delta\varphi$ and yaw $\delta\kappa$.

In Section 4.1.7.1, registration of laser data is explained provided that the system is calibrated. Section 4.1.7.2 deals in detail with the problem of LiDAR calibration.

4.1.7.1 Registration

The registration process is best mathematically described by the simple vector approach already shown in Figure 4.2:

$$\vec{G} = \vec{r}_L + \vec{s} \tag{4.34}$$

where

$\vec{G}$ is the vector from the earth center to the ground point
$\vec{r}_L$ vector from the earth center to the LiDAR's point of origin
$\vec{s}$ slant ranging vector

The LiDAR's point of origin is a fixed point in 3D space from which the laser beam originates at the instantaneous scanning angle. The actual position of this point is very dependent on the scanning mechanism used, e.g., in case of ScaLARS it is the turning point of the mirror. To make a straightforward registration calculation with a precision as high as possible by applying Equation 4.34, the IMU is normally mounted on the support of the LiDAR instrument or as shown in Figure 4.22 on the suspension of the scanner. The POS data are transformed to LiDAR's point of origin. For this transformation, the user has to enter parameters for the so-called lever arms (Figure 4.24) into the POS processing software, which are two three-dimensional vectors. One vector is from the origin of the LiDAR to the center of the IMU and the other is from the origin of the LiDAR to the phase center of the GPS-antenna.

In the following, the origin of the LiDAR defines the origin of the LiDAR coordinate system L depicted in Figure 4.2. The x_L-axis points into the flight direction, the y_L-axis points to the right of the airplane and the z_L-axis points downwards perpendicular to the plane defined by the axes x_L and y_L. Vector $\vec{r}_L$ is measured and described in WGS84. The

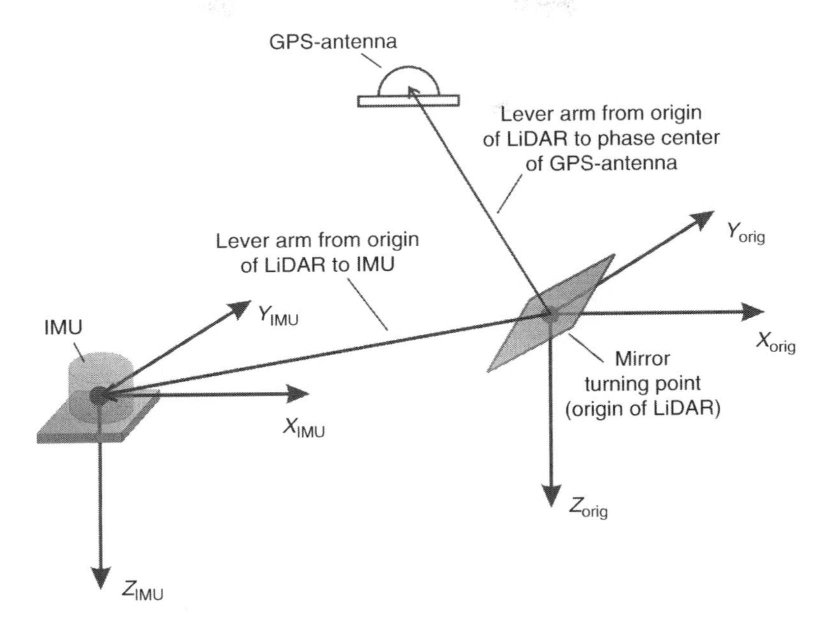

FIGURE 4.24
Lever arms.

registered laser measurement points $\vec{G}$ should also be described in WGS84. As $\vec{s}$ is measured in the coordinate system L it has to be transformed by transformation matrices into WGS84. Regarding the coordinate system used the general approach becomes

$$\underline{G}_{WGS84} = \underline{r_L}_{WGS84} + (\_\_)_H^{WGS84} \cdot (\_\_)_{IMU}^H \cdot (\_\_)_L^{IMU} \cdot \underline{s}_L \tag{4.35}$$

The product of the matrices $(\_\_)_L^{IMU}$ and $(\_\_)_{IMU}^H$ describes the orientation of the coordinate system L with respect to the horizontal coordinate system H. Assuming that the LiDAR is perfectly aligned with the IMU—this means the coordinate system L and the coordinate system of the IMU have the same orientation–, the matrix $(\_\_)_L^{IMU}$ becomes unity. In this case only the instantaneous orientation of the airplane or more precisely the orientation of the IMU, determines together with the instantaneous scanning angle the actual direction of the laser beam.

The orientation of the IMU in relation to the horizontal system H (Figure 4.25) is described by the angles roll ω (rotation about the x_L-axis), pitch φ (rotation about y_L-axis) and heading κ (rotation about the z_L-axis). If the rotations are carried out in the following sequence: first rotation about the x_L-axis (roll), second rotation about the y_L-axis (pitch) and finally rotation about the z_L-axis (heading), the matrix $(\_\_)_{IMU}^H$ can be set up as

$$(\_\_)_{IMU}^H = \begin{pmatrix} a_{11} & a_{21} & a_{31} \\ a_{12} & a_{22} & a_{32} \\ a_{13} & a_{23} & a_{33} \end{pmatrix} \tag{4.36}$$

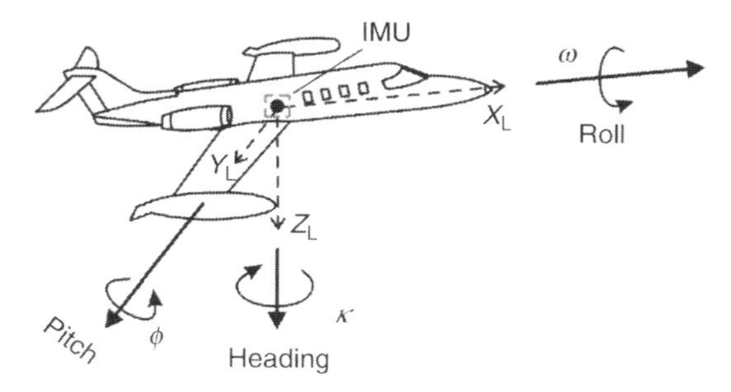

FIGURE 4.25
Orientation angles (IMU and LiDAR aligned).

with

$$\begin{pmatrix} a_{11} \\ a_{12} \\ a_{13} \end{pmatrix} = \begin{pmatrix} \cos(\kappa)\cdot\cos(\varphi) \\ \sin(\kappa)\cdot\cos(\varphi) \\ -\sin(\varphi) \end{pmatrix} \tag{4.37}$$

$$\begin{pmatrix} a_{21} \\ a_{22} \\ a_{23} \end{pmatrix} = \begin{pmatrix} \cos(\kappa)\cdot\sin(\varphi)\cdot\sin(\omega)-\sin(\kappa)\cdot\cos(\omega) \\ \sin(\kappa)\cdot\sin(\varphi)\cdot\sin(\omega)+\cos(\kappa)\cdot\cos(\omega) \\ \cos(\varphi)\cdot\sin(\omega) \end{pmatrix} \tag{4.38}$$

$$\begin{pmatrix} a_{31} \\ a_{32} \\ a_{33} \end{pmatrix} = \begin{pmatrix} \cos(\kappa)\cdot\sin(\varphi)\cdot\cos(\omega)+\sin(\kappa)\cdot\sin(\omega) \\ \sin(\kappa)\cdot\sin(\varphi)\cdot\cos(\omega)-\cos(\kappa)\cdot\sin(\omega) \\ \cos(\varphi)\cdot\cos(\omega) \end{pmatrix} \tag{4.39}$$

The transformation matrix $(\_\_)_H^{\mathrm{WGS84}}$ regards the orientation between the horizon system H and the WGS84. The orientation is defined by the geographical latitude Φ_0 and longitude Λ_0, so that $(\_\_)_H^{\mathrm{WGS84}}$ becomes

$$(\_\_)_H^{\mathrm{WGS84}} = \begin{pmatrix} -\cos\Lambda_0\cdot\sin\Phi_0 & -\sin\Lambda_0 & -\cos\Lambda_0\cdot\cos\Phi_0 \\ -\sin\Lambda_0\cdot\sin\Phi_0 & \cos\Lambda_0 & -\sin\Lambda_0\cdot\cos\Phi_0 \\ \cos\Phi_0 & 0 & \sin\Phi_0 \end{pmatrix} \tag{4.40}$$

Equation 4.35 and the matrices show that registered LiDAR measurement points can only be computed if the orientation and position of the LiDAR are known. These values are measured by POS. Up to now it has been assumed that LiDAR and POS have the same orientation. This is not the case in general. Due to tolerances in the mechanical setup one has to take into account a misalignment between POS and the LiDAR. The misalignment is described by the misalignment angles in roll $\delta\omega$, in pitch $\delta\varphi$ and in heading $\delta\kappa$. As the sequence of rotations is equal to the rotation order of matrix $(\_\_)_{\mathrm{IMU}}^H$, the matrix $(\_\_)_L^{\mathrm{IMU}}$ looks like $(\_\_)_{\mathrm{IMU}}^H$. However, the orientation angles ω, φ, and κ have to be exchange with the misalignment angles $\delta\omega$, $\delta\varphi$, and $\delta\kappa$. Therefore $(\_\_)_L^{\mathrm{IMU}}$ is given by

$$(\_)_{L}^{\text{IMU}} = \begin{pmatrix} b_{11} & b_{21} & b_{31} \\ b_{12} & b_{22} & b_{32} \\ b_{13} & b_{23} & b_{33} \end{pmatrix} \tag{4.41}$$

with

$$\begin{pmatrix} b_{11} \\ b_{12} \\ b_{13} \end{pmatrix} = \begin{pmatrix} \cos(\delta\kappa)\cdot\cos(\delta\varphi) \\ \sin(\delta\kappa)\cdot\cos(\delta\varphi) \\ -\sin(\delta\varphi) \end{pmatrix} \tag{4.42}$$

$$\begin{pmatrix} b_{21} \\ b_{22} \\ b_{23} \end{pmatrix} = \begin{pmatrix} \cos(\delta\kappa)\cdot\sin(\delta\varphi)\cdot\sin(\delta\omega) - \sin(\delta\kappa)\cdot\cos(\delta\omega) \\ \sin(\delta\kappa)\cdot\sin(\delta\varphi)\cdot\sin(\delta\omega) + \cos(\delta\kappa)\cdot\cos(\delta\omega) \\ \cos(\delta\varphi)\cdot\sin(\delta\omega) \end{pmatrix} \tag{4.43}$$

$$\begin{pmatrix} b_{31} \\ b_{32} \\ b_{33} \end{pmatrix} = \begin{pmatrix} \cos(\delta\kappa)\cdot\sin(\delta\varphi)\cdot\cos(\delta\omega) + \sin(\delta\kappa)\cdot\sin(\delta\omega) \\ \sin(\delta\kappa)\cdot\sin(\delta\varphi)\cdot\cos(\delta\omega) - \cos(\delta\kappa)\cdot\sin(\delta\omega) \\ \cos(\delta\varphi)\cdot\cos(\delta\omega) \end{pmatrix} \tag{4.44}$$

The misalignment angles are determined by the calibration.

The presented formulas document that the calculation of registered laser measurement points is straight forward, assuming that the system is calibrated and any misalign ment between the LiDAR and POS are constant. However, examining airborne LiDAR data georeferenced by the described method, it becomes obvious, that there are spatially varying errors present in the form of elevation differences (Bretar et al., 2003; Kager and Kraus, 2001) and tilts between adjacent scanning paths (Rönnholm, 2004). Various solutions for this problem are presented in the literature, e.g., Burman (2002), Crombaghs et al. (2000), Schenk (2001a,b), and Vosselmann and Maas (2001) and will also be explained in the next section. Most of the solutions are based on bundle block adjustment. The source of these errors are not fully understood and likely vary between different systems. The cause may be time-varying orientation changes in the LiDAR itself, of the IMU to LiDAR or LiDAR to GPS level arms, or time-varying errors in the POS position and attitude solutions. Also synchronization errors between the orientation angles computed by Kalman filters must be taken into account (Latypov, 2002, 2005; Schenk, 2001a,b).

4.1.7.2 Calibration

In Section 4.1.7.1, it was assumed that the bore sight misalignment angles $\delta\omega$, $\delta\varphi$, and $\delta\kappa$ were calibrated. Now, calibration is explained by studying the calibration procedures of ScaLARS (Schiele et al., 2005). ScaLARS uses the Palmer scanning mechanism (Section 4.1.3.1). Due to its special helical scanning pattern with a constant off-nadir scan angle which can be separated into forward and backward views (Table 4.1) an intrinsic system calibration is possible. Other scanning mechanisms do not offer this feature. However, the basic calibration steps of ScaLARS are also valid for these scanning mechanisms.

For precise geocoding the set of calibration data is extended by the wobble angle γ_M and an additive parameter s_{Add} of the slant range. The wobble angle corresponds to the maximum scanning angle of a moving or flipping mirror or in other words the FOV

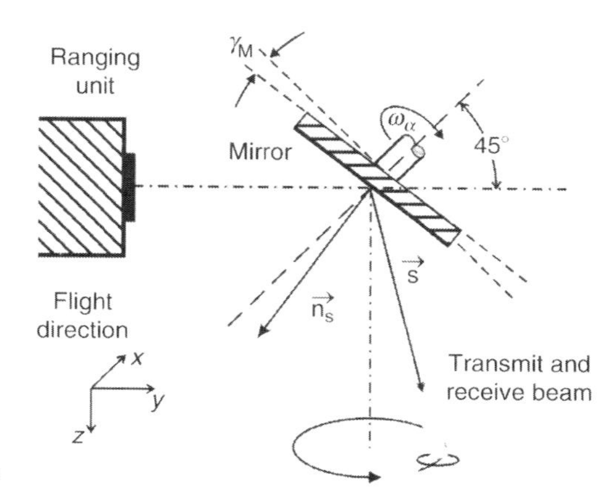

FIGURE 4.26
Palmer scanner of ScaLARS.

or swath width (SW). Therefore, in total five parameters have to be calibrated: the bore sight misalignment angles $\delta\omega$, $\delta\varphi$, and $\delta\kappa$, wobble angle γ_M and the additive parameter s_{Add}. According to Figure 4.26 the instantaneous slant range vector of ScaLARS is calculated in the LiDAR coordinate system L by

$$\underline{s}_L = (s + s_{\text{Add}}) \cdot \begin{pmatrix} \sin(\delta) \cdot \sin(\xi) \\ -\cos(\delta) \\ \sin(\delta) \cdot \cos(\xi) \end{pmatrix} \tag{4.45}$$

with

$$\delta = 2 \cdot \arccos\left[\cos(\alpha_\omega) \cdot \sin(\gamma_M) \cdot \sin(\vartheta) + \cos(\gamma_M) \cdot \cos(\vartheta)\right]$$

$$\xi = -\arctan\left[\frac{\sin(\alpha_\omega) \cdot \sin(\gamma_M)}{\cos(\alpha_\omega) \cdot \sin(\gamma_M) \cdot \cos(\vartheta) + \sin(\gamma_M) \cdot \sin(\vartheta)}\right]$$

where
- s is the measured slant range
- α_ω the instantaneous rotation angle of the motor shaft
- ϑ is the tilt of the motor shaft which is defined by the scanner design

It is valid 45°. Equation 4.46 makes clear that only two parameters of $\underline{s}_L$ are of interest for calibration. These are the wobble angle γ_M and the additive parameter for slant ranging s_{Add}. All other calibration parameters are located in the transformation matrix $(\_\_)_L^{\text{IMU}}$ which describes the misalignment between IMU and LiDAR. This fact is clearly seen by looking at the equation for registered laser points in WGS84, which is

$$\underline{G}^{\text{WGS84}} = \underline{r}_L^{\text{WGS84}} + \underline{(\Lambda_0, \Phi_0)}_H^{\text{WGS84}} \cdot \underline{(\omega, \kappa, \varphi)}_{\text{IMU}}^H \cdot \underline{(\delta\omega, \delta\kappa, \delta\varphi)}_L^{\text{IMU}} \cdot \underline{s}_L(s_{\text{Add}}, \gamma_M) \tag{4.46}$$

For better understanding only the used parameters are written in the symbols for the matrices. On the basis of Equation 4.46, a Gauß-Markoff model is set up to estimate the calibration parameters: bore sight misalignment angles $\delta\omega$, $\delta\varphi$, and $\delta\kappa$, wobble angle γ_M and the additive parameter s_{Add}. The calibration is carried out by selected flat areas. Calibration tests showed that white marks on runways are optimum targets for this task but other features like flat roofs are also possible. However, the main objective of these

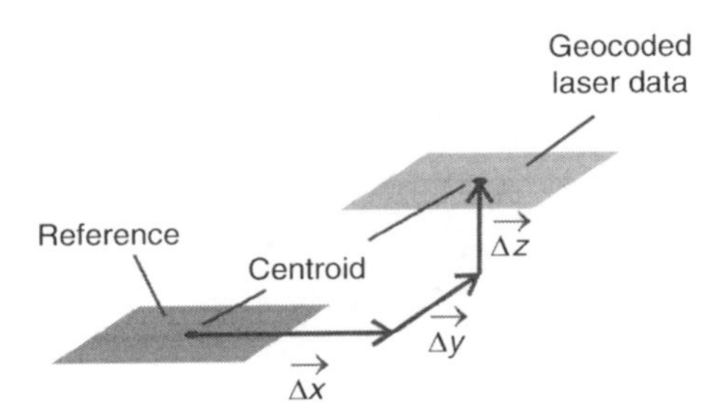

FIGURE 4.27
Control area.

targets is, that they exhibit a high contrast to the background and that a representative elevation point can be computed out of the laser data sampled from it. The high contrast is necessary to extract the targets from the surveyed area automatically. These chosen areas for calibration are called control area in the following. Figure 4.27 shows, how control areas are used for calibration. Regarding the centroid of the control area it is obvious that the centroid derived from geocoded LiDAR data is brought to coincidence with the real one if the displacements Δx, Δy, and Δz are zero which means the system is exactly calibrated. Therefore, the objective of the Gauß-Markoff model is to minimize Δx, Δy, and Δz by varying the calibration parameters.

The calibration procedure is going to be explained exemplarily by markers on runways. They can be easily identified out of the LiDAR intensity data (Figure 4.28).

Furthermore, they can be surveyed using GPS surveying instruments (Figure 4.29) and the reference centroid in 3D space can be computed by simple geometry calculations. The centroid determination from LiDAR data requires much more effort and sophisticated algorithms, because LiDAR data are not equally distributed over the target's surface (enlarged runway markers in Figure 4.28).

After extracting all laser measurements belonging to the control area by special software, the extracted surface is modeled by triangulation. A Delaunay triangulation algorithm was preferred for this task. As the target's surface is approximated by n triangles the coordinates of the centriod can be computed independently on the density of the laser measurements by

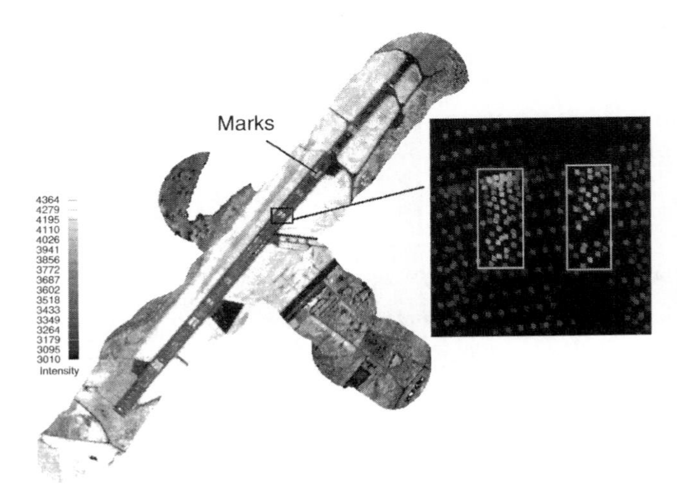

FIGURE 4.28
LiDAR intensity data of runway.

FIGURE 4.29
Surveying a control area with GPS.

$$\vec{x}_{cent} = \frac{1}{\sum_{i=1}^{n} F_i} \cdot \sum_{i=1}^{n} \left(\vec{x}_i \cdot F_i \right) \tag{4.47}$$

where
F_i is the ith triangle
$\vec{x}_i$ the corresponding vector to its centroid

The process of using $\vec{x}_{cent}$ and the location of the centroid derived from the GPS measurement for calibration is called calibration with coordinated control points. Applying this method the calibration parameters were determined with an accuracy listed in Table 4.2.

Noncoordinated control points are derived from surveyed features, where external reference information about shape, size, or position is not available. This means the location of the corresponding centroid is not known a priori and can only be determined by measured LiDAR data. In this case, the parameters wobble angle γ_M and bore sight misalignment angles $\delta\omega$, $\delta\varphi$, and $\delta\kappa$ can be relatively calibrated only. Calibration with non coordinated control points becomes important if areas are surveyed where in the field measurements are impossible, e.g., carrying out LiDAR surveys high in the mountains, in

TABLE 4.2

Accuracy of Calibration Parameters Determined for Two Surveying Campaigns

Estimated Parameter Accuracy	γ_M (°)	$\delta\omega$ (°)	$\delta\varphi$ (°)	$\delta\kappa$ (°)	Add (m)
July 2002 ($\gamma_M \approx 10°$)	0.004	0.006	0.006	0.024	0.03
April 2003 ($\gamma_M \approx 7°$)	0.003	0.005	0.005	0.028	0.03

primeval forests and over the poles. Furthermore this calibration technique permits to study system drifts along the flying path during the total mission.

4.1.8 From Flight Planning to Final Product

In the preceding chapters the components of a LiDAR system were presented and their performance attributes were discussed. Here, the different working steps required to obtain a digital elevation model (DEM) as a final product are explained.

Normally the customer defines the surveying area, the density of laser points on the ground and the elevation accuracy. In some cases special surveying objectives are defined for specific target types such as powerlines, building geometries including tilted roofs or embankments. For these special cases translation of measurement requirements into the necessary LiDAR performance attributes by the surveyor requires substantial practical experience that is not readily captured by a few system parameters. Therefore, in the following we will focus only on the first, more general kind of surveying project requested by a customer. Figure 4.30 presents an overview of the processing steps from flight planning to final product. Each step will be discussed in separate sections that follow.

4.1.8.1 Flight Planning

The flight planning stands at the beginning of an airborne surveying project. Several flight planning programs are available on the market. These products are normally sold with flight guidance software, which tells the pilot the flight course and elevation during flight and when he has to start and to stop LiDAR surveying along a flight line. However, the input of this software is not compatible with specifications listed in Section 4.1.8. Therefore flight planning for airborne LiDAR surveys needs additional working steps. Furthermore, the GPS reference station and calibration areas must be selected. That is why I prefer the term project or mission planning.

On the basis of the costumer's requirements, which are the surveying area size, the density of laser points on ground, and the elevation accuracy, the proper LiDAR system must be chosen first. This means selecting the appropriate LiDAR, POS and the carrier either an airplane or helicopter platform. Helicopters are used if high laser point densities are necessary and detailed surveys of small areas are essential. For extended surveys airplanes are preferred. The performance characteristics of the LiDAR and POS determine the point density across the flight direction, the elevation accuracy, the elevation resolution (Equation 4.10) and the pointing accuracy.

The point density in the flight direction Δx_{along} is given by the speed of the platform v and the scanning speed which is described by the scanning rate f_{sc}:

$$\Delta x_{\text{along}} = \frac{v}{f_{\text{sc}}} \tag{4.48}$$

If the airplane is not known, a speed of 75 m/s is a good choice to start planning. Helicopters can fly more or less with arbitrary speeds.

The point density across flight direction Δx_{across} can be approximated by

$$\Delta x_{\text{across}} = \frac{\theta}{N} \tag{4.49}$$

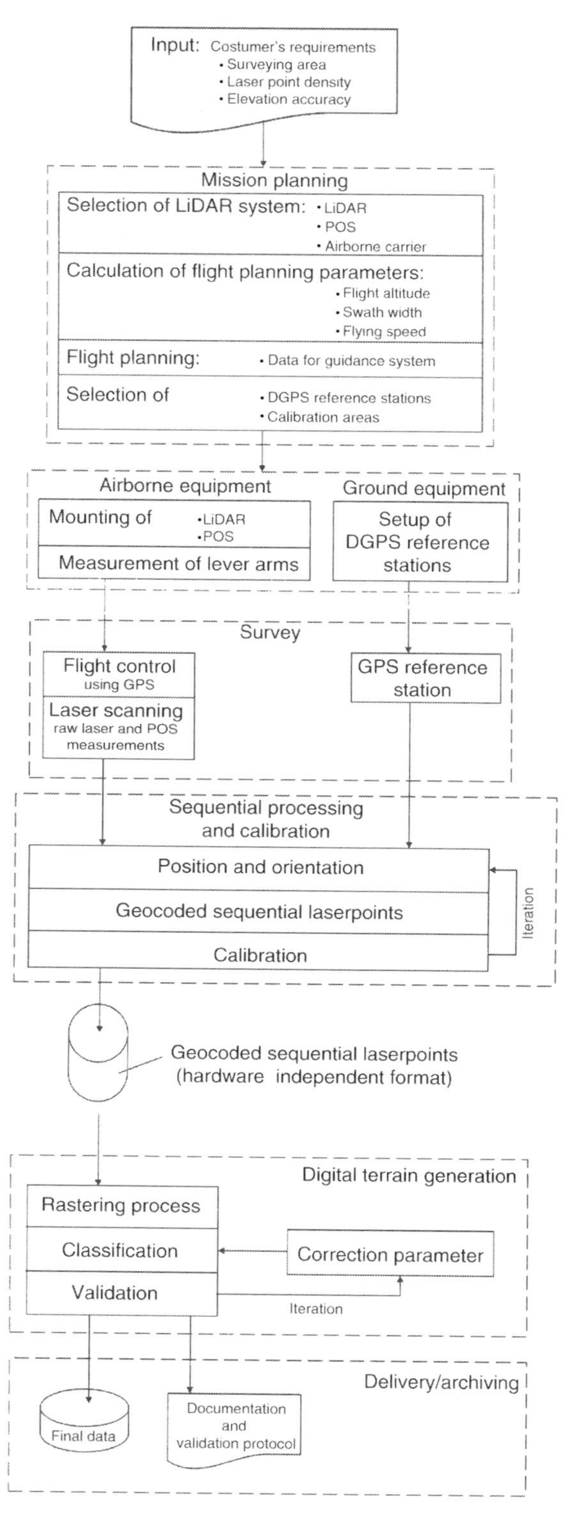

FIGURE 4.30
Flowchart processing steps.

where

θ is the SW (either expressed in meters or angular degrees)

N is the number of points per scan line and assuming a flat terrain and that the distances between points along the scan line are equal

N is calculated by

$$N = \frac{f_{\text{pulse}}}{f_{\text{sc}}} \tag{4.50}$$

where f_{pulse} is the laser pulse rate. Figure 4.31 depicts a dependence of the spacing between adjacent points on the instantaneous scanning angle α. The surveying parameters have to be dimensioned for the worst case within the scan line. This means for the maximum scanning angle which is half the angular SW θ. Therefore, a more precise formula for point density across flight direction Δx_{across} is derived from Figure 4.31:

$$\Delta x_{\text{across}} = \frac{\theta}{N} \cdot \frac{H}{\cos^2(\theta/2)} \tag{4.51}$$

where H is the flying altitude above ground. In case that terrain features a slope with angle i along the scanning line (Figure 4.32) the maximum spacing across the flight line becomes

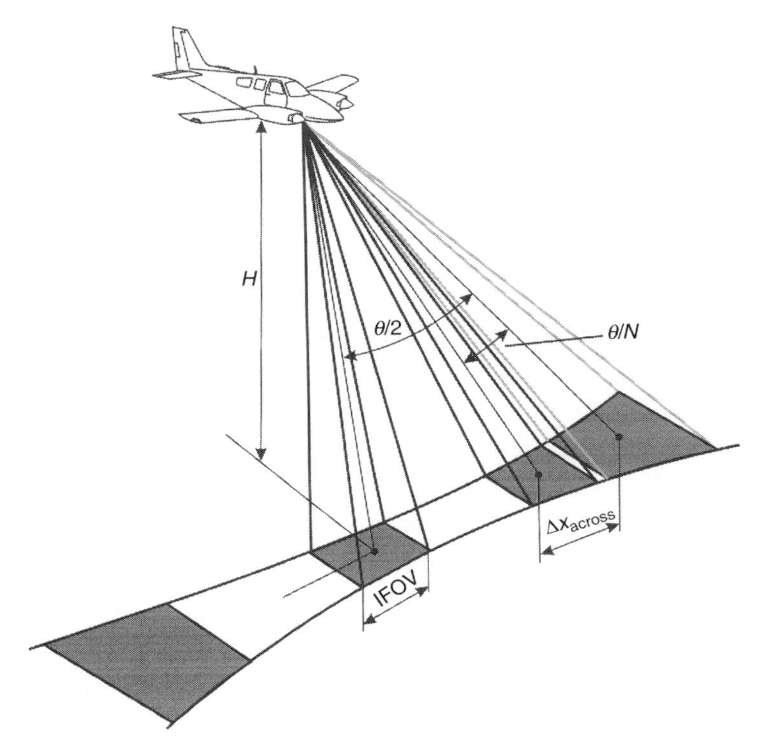

FIGURE 4.31
Geometry for sampling across flight direction.

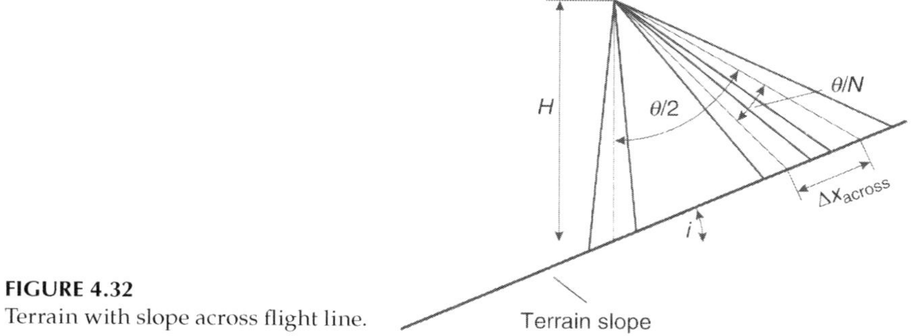

FIGURE 4.32
Terrain with slope across flight line.

$$\Delta x_{across} = \begin{cases} \dfrac{\theta}{N} \cdot \dfrac{H}{\cos^2\left(\dfrac{\theta}{2}\right) \cdot \cos(i) \cdot \left[1 - \tan\left(\dfrac{\theta}{2}\right) \cdot \tan(i)\right]^2}, & \text{if } i \geq 0 \\ \dfrac{\theta}{N} \cdot \dfrac{H}{\cos^2\left(\dfrac{\theta}{2}\right) \cdot \cos(i) \cdot \left[\tan\left(\dfrac{\theta}{2}\right) \cdot \tan(i) - 1\right]^2}, & \text{if } i < 0 \end{cases} \quad (4.52)$$

The values Δx_{along} and Δx_{across} are used to determine the minimum point density d_{min} which is given in number of laser spots per square meter.

$$d_{min} = \frac{1}{\Delta x_{along} \cdot \Delta x_{across}} \quad (4.53)$$

The parameters calculated with Equations 4.49 through 4.53 are not only dependent on LiDAR parameters. Rather they are a function of the flying altitude H. The flying height H or the resulting maximum slant range R_{max} which are related by

$$R_{max} = \frac{H}{\cos(\theta/2)} \quad (4.54)$$

determines the achievable accuracy (Equations 4.29 and 4.30). As the accuracy is set by the customer, H is limited. According to Equation 4.52 across track spacing of the laser points limits also H. This means, both limitations must be fulfilled and the most stringent one determines the SW which is

$$\text{SW} = 2 \cdot H \cdot \tan\left(\frac{\theta}{2}\right) \quad (4.55)$$

Now, the key parameters for flight planning are available which are flight altitude H and SW. In addition the overlapping factor ζ for neighboring paths has to be set. According to Figure 4.33 it is defined by

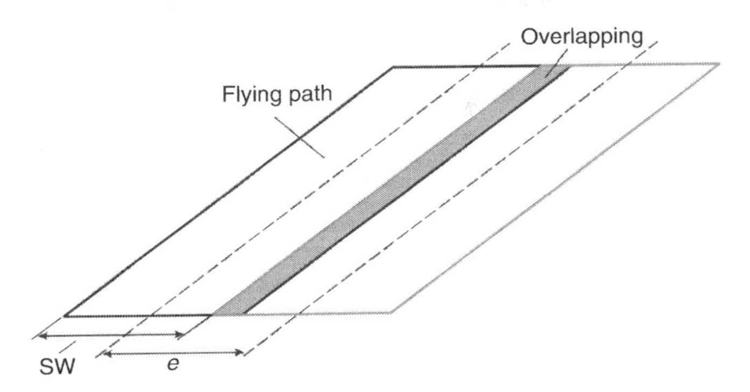

FIGURE 4.33
Overlapping of flight paths.

$$\frac{e}{\mathrm{SW}} = 1 - \xi \tag{4.56}$$

where e is the distance between the centerlines of adjacent paths. The overlapping is dependent on the pilot's flying precision. Planning surveys without any overlap is critical, because one has to face the problem that surveyed areas are not completely covered due to rolling of the aircraft and improper straight flight lines. At least a 20% overlap is advisable. For highly resolved surveys an overlapping of at least 50% may also be applied if high point densities are essential which cannot be achieved with the scanning parameters of LiDAR and the forward speed of the surveying carrier.

Table 4.3 compiles the relevant parameters for flight planning. Figure 4.34 shows a typical graphical output of a flight planning program. The numerical output is transferred to the guidance system of the surveying airplane. In addition the flight planning result can be used to estimate the expenditure of the project. The length and number of flying lines determines the flight duration. One has to keep in mind that long flight lines are preferable, because after each line the airplane has to fly turns into the next line where the survey is interrupted. On the other hand too long lines degrade the accuracy of POS caused by IMU drifts. Flying a straight line for ca. 15–20 min., some turns are required.

Based on the flight planning results the mission planning can be completed by positioning the GPS reference stations which should not be more than 20 km away from each surveying position of the airplane. Therefore, the number of reference stations is determined by the size of surveying area and the possibility to deploy a reference. In inaccessible areas are e.g., primeval forests, summit of glaciers, etc. Today in many countries networks of GPS reference stations exist. Paid GPS reference data are available to everybody. In all other cases the surveyor has to see that enough stations are available.

4.1.8.2 Installation of Airborne Equipment

In particular if the surveying company has not got an own airplane with fixed LiDAR installations, the LiDAR has to be mounted in the surveying airplane at mission start. The opto-mechanical assembly is typically mounted in openings which are also used by metric cameras etc. These holes and the mechanical interfaces are more or less standardized. Nevertheless during planning one should check if the LiDAR's FOV is oblique and the available electrical power satisfies the LiDAR's need.

TABLE 4.3

Complication of Key Parameters for Flight Planning

Sketch	Parameter		Formula	Variables
θ, R_{max}, H	Altitude above ground H		$H = R_{max} \cdot \cos\left(\frac{\theta}{2}\right)$	R_{max} max. slant range θ angular FOV
θ, H, SW	SW		$SW = 2 \cdot H \cdot \tan\left(\frac{\theta}{2}\right)$	H altitude θ angular FOV
Δx_{along}, Δx_{across}	Point spacing	Across Δx_{across}	$\Delta x_{across} = \frac{\theta}{N} \cdot \frac{H}{\cos^2\left(\frac{\theta}{2}\right)}$	H altitude θ angular FOV N number of points in one scanning line
		Along Δx_{along}	$\Delta x_{along} = \frac{v}{f_{sc}}$	v forward speed f_{sc} scanning rate or scan frequency
	Point density d_{min}		$d_{min} = \frac{1}{\Delta x_{along} \cdot \Delta x_{across}}$	Δx_{across}, Δx_{along} point spacings
SW, e	Overlapping factor ζ		$\xi = 1 - \frac{e}{SW}$	SW swath width e flight line separation
	Flight line separation e		$e = SW \cdot (1 - \xi)$	SW swath width ζ overlapping factor

After completing the installation, the so called lever arms (Section 4.1.7.1, Figure 4.24) have to be surveyed, which describe the displacements between IMU and GPS-antenna on top of the airplane. With additional geometrical information concerning the displacement between IMU and surveying origin of LiDAR all post processed POS data are related to this surveying origin.

4.1.8.3 Survey

During survey the LiDAR and POS data are stored on hard disk or memory cards in the airplane and the GPS reference stations save their data locally. The pilot is guided by the flight guidance software which uses data out of the flight planning program. He has to pay attention to circumvent roll angles of more than ±20°, because otherwise it could happen that the phase carrier of DGPS cannot be resolved continuously in post processing.

A so-called calibration area (e.g., a runway with markings) should be surveyed at the beginning and the end of a flight. This helps to identify system drifts. It is a must if the LiDAR and POS system were just new assembled or reassembled and mounted the first time or mounted again in the airplane. The area should exhibit special features which support the calibration precision.

4.1.8.4 Sequential Processing and Calibration

After the surveying flight back-up copies and working copies of the gathered LiDAR and POS data are stored on different standard hard disk. Normally the LiDAR and POS

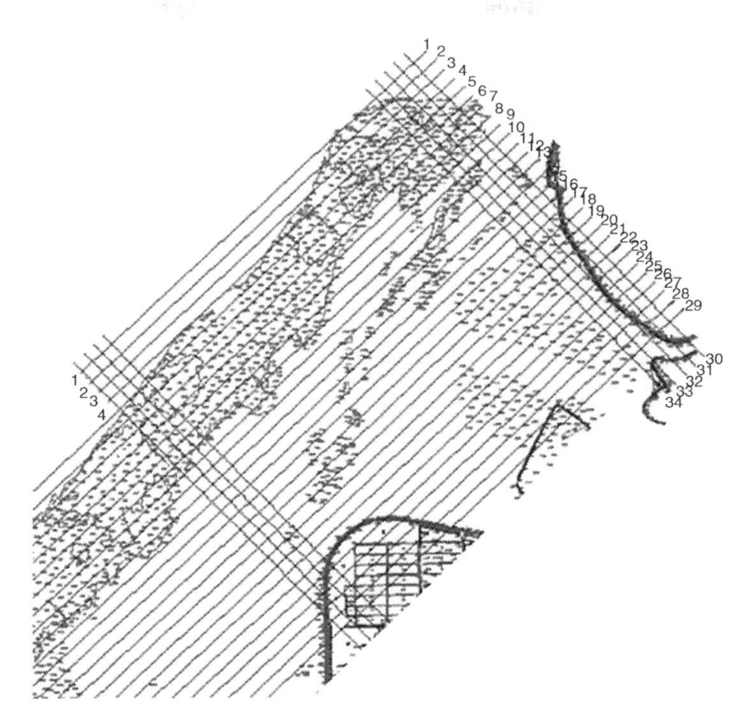

FIGURE 4.34
Flight lines out of a flight planning program.

data are stored in a compressed binary format. This means the data have to be uncompressed and converted to ASCII data format by special program routines. In the first processing step LiDAR and POS data are treated separately. The POS data processing is discussed first.

The POS data comprises IMU and GPS data. Before integrated POS data can be computed DGPS data using phase carriers have to be calculated by using the GPS data of the reference stations. The integrated solution with DGPS and IMU data offers the following principle data set:

$$\text{GPS time } x_{\text{WGS84}}, y_{\text{WGS84}}, z_{\text{WGS84}} \text{ roll, pitch, yaw}$$

The position coordinates are related to the LiDAR origin (Section 4.1.8.4) and the orientation angles are the rotations about the instantaneous local horizontal system.

In a second step LiDAR and POS data have to be synchronized by the procedure described in Section 4.1.6. After synchronization a file with the following principle content is generated for further processing:

$$\text{GPS time } x_{\text{WGS84}}, y_{\text{WGS84}}, z_{\text{WGS84}} \text{ roll, pitch, yaw scan angle, slant range, intensity}$$

This is the output of the processing step "Position and Orientation" presented in Figure 4.30.

Assuming a perfect inner orientation calibration the LiDAR geocoded laser measurement points on ground can be simply computed by Equation 4.35, if the misalignment angles in roll $\delta\omega$, pitch $\delta\varphi$, and yaw $\delta\kappa$ are known. These angles are caused by misalignment between IMU and LiDAR. Besides other calibration parameters discussed in Section 4.1.7.2 the misalignment can be determined by the LiDAR data gathered from the calibration areas. Thus, an iteration is foreseen in Figure 4.30. After a new set of calibration parameters is derived, the possible improvement has to be checked and validated.

The outputs of the processing step "Sequential Processing and Calibration" are files containing geocoded—also known as registered—laser measurements in a chronological order:

$$\text{GPS time } x_{\text{Laser\_WGS84}}, y_{\text{Laser\_WGS84}}, z_{\text{Laser\_WGS84}}, \text{ intensity}$$

This means that the geometry of the scanning pattern is still present. Figure 4.35 displays sequential laser data surveyed with a Palmer scanning pattern. This pattern is extreme with regard to geometry and time. Due to its egg-shaped pattern, ground points especially along the center line of the flight path are sampled twice, however at different times. Comparable images are obtained by other scanning mechanisms. In an operational processing environment the GPS time is not used and therefore normally not stored.

Up to this post processing step programs provided by the manufacturer of the LiDAR system are applied, because they must be programmed with a good knowledge of the hardware.

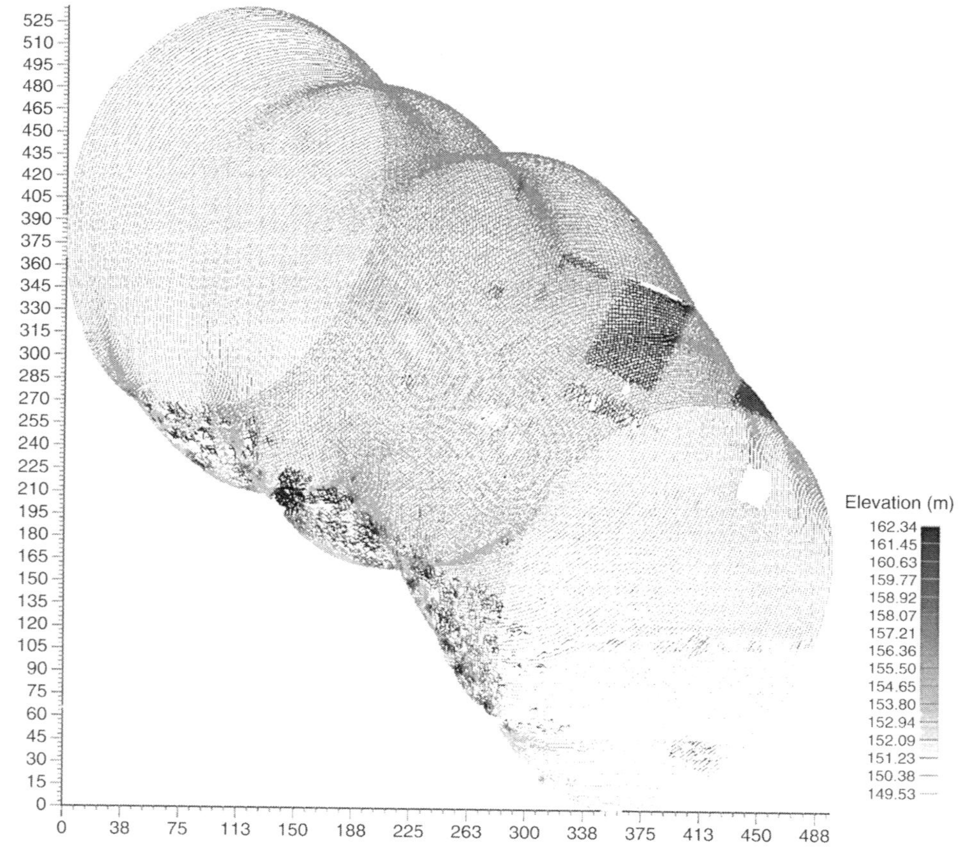

FIGURE 4.35
Sequential laser data.

4.1.8.5 Digital Terrain Model Generation

For generating final products defined by the customer the sequential data sets generated in Section 4.1.8.5 have be further processed. Final products are, e.g., digital surface models, DEMs or extracting other geoinformation, e.g., house geometries, shapes of roofs, tree heights, etc. In other words this work package must carry out some classification tasks. Today a lot of remote sensing and airborne surveying software is available on the market. But most of them work with rasterized data and cannot handle the so-called point cloud data delivered from LiDARs. Hence in the block "Digital Terrain Model Generation" the first task is rasterizing sequential data. The rasterizing process is best explained by looking at Figure 4.36. All laser data are sorted into raster cells which are defined by a grid. The sizes of the cells are determined by the resolution projected by the customer. Also the total image size is selected in agreement with the client. In general, a raster size should be larger or equal than the largest spacing between neighboring laser points to assure that all raster cells contain at least one laser point. The raster cells are also called pixels in the following. Having several elevation measurements in one pixel, sophisticated algorithms are applied to obtain the desired information, because the final output must have only one pixel value.

The generation of classified rasterized data is explained exemplarily for the generation of DEMs. Selecting a pixel size, so that several laser points are sorted in one pixel, one recognizes an elevation distribution. Even if it is not distinguished between first and last pulse, the user can easily realize from the way how the measurements spread if forest and vegetation or ground is measured (Figure 4.37).

In case of ground one line can be clearly identified. The spreading of the elevations is primarily determined by the measurement quality of the LiDAR, whereas in forest areas some measurements represent either backscattered signals from ground and from treetops. Now, applying sophisticated software tools ground points can be classified and the most likelihood value is stored for this pixel. Figure 4.38 shows a rasterized DEM. The scanning pattern is not apparent anymore. Such classifications are carried out automatically or semiautomatically with today's software tools. However, a good knowledge of setting filter parameters is necessary to achieve optimum classification results. Therefore, in the flowchart (Figure 4.30) an iteration for optimizing parameters is foreseen.

The most time consuming part in DEM production is the validation of, e.g., the DEMs. First accuracy checks have to be carried out. However, the calibration areas cannot be used for this test, because they fit per definition. Therefore, different, already well surveyed areas must be selected. By checking sequences of pixel lines the plausibility of ground lines in

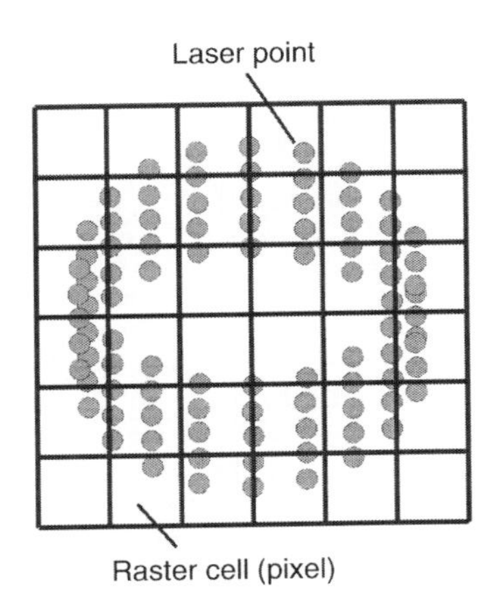

FIGURE 4.36
Rasterizing.

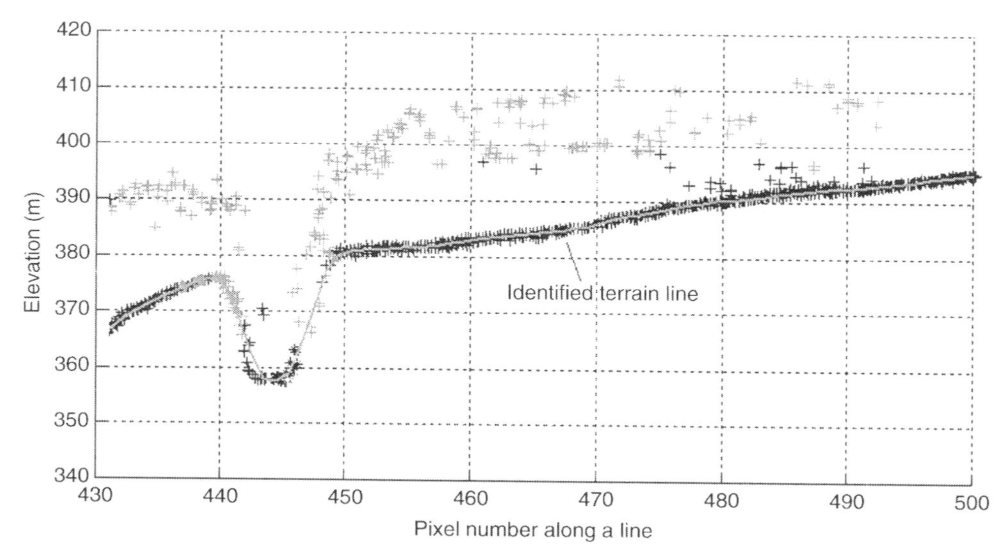

FIGURE 4.37
Pixel line with all laser measurements.

forest areas can be examined. Here an absolute check is impossible, because secured measurement data for comparison do not exist in many cases. If validation is completed successfully the data are stored on appropriate media, e.g., CDs or DVDs in agreement with the customer, who also defines the archiving nomenclature.

The different processing steps explained in the preceding chapters were not treated at great length, because the objective was to give an overview about what is needed to achieve a final product from a LiDAR survey. The remarks make clear that successful LiDAR surveying is mainly dependent on the quality and availability of the used software and the implemented algorithms. Looking at the technical data of the different LiDAR systems, the performance of today's LiDARs differ marginal. However, in practice very different accuracy values concerning geocoded data can be observed. This shows one has to regard the complete LiDAR System as defined in Section 4.1.1.

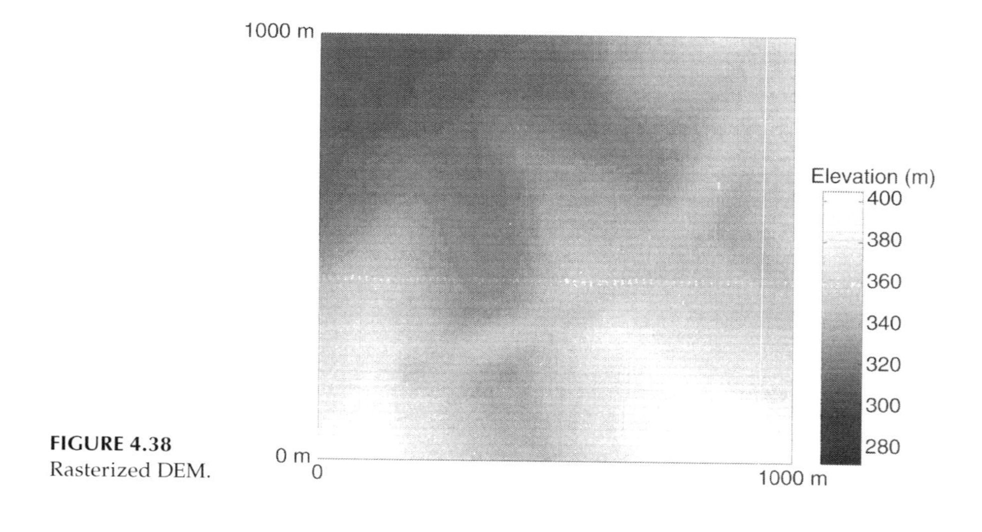

FIGURE 4.38
Rasterized DEM.

References

Abshire, J., Sun, X., and Afzal, R., 2000. Mars orbiter laser altimeter: Receiver model and performance analysis. *Applied Optics*, 39(15), 2449–2460.

Bachman, Chr. G., 1979. *Laser Radar System and Techniques*. Artech House, MA.

Bretar, F., Pierrot-Oeseilligny, M., and Roux, M., 2003. Estimating image accuracy of airborne laser data with local 3D-offsets. *International Archives of Photogrammetry and Remote Sensing*. 34(3-W13), 20–26.

Burman, H., 2002. Laser strip adjustment for data calibration and verification. *International Archives of Photogrammetry and Remote Sensing*, 34(part 3A), 67–72.

Crombaghs, M., Brügelmann, R., and de Min, E., 2000. On the adjustment of overlapping strips of laseraltimeter height data. *IAPRS*, 33(part 3A), 230–237.

Filin, S., 2003a. Recovery of systematic biases in laser altimetry data using natural surfaces. *Photogrammetric Engineering & Remote Sensing*, 69(11), 1235–1242.

Filin, S., 2003b. Analysis and implementation of a laser strip adjustment model. *International Archives of Photogrammetry and Remote Sensing*, 34(3-W13), 65–70.

Filin, S. and Vosselman, G., 2004. Adjustment of Airborne Laser Altimetry Strips. Proceedings ISPRS XXth Congress, Vol. XXXV Part B, Commission 3, p. 285 ff.

Gardner, C.S., 1992. Ranging performance of satellite laser altimeters. *IEEE Transactions on Geoscience and Remote Sensing*, 30(5), 1061–1072.

Hug, C., 1994. The scanning laser altitude and reflectance sensor: An instrument for efficient 3D terrain survey. *International Archives of Photogrammetry and Remote Sensing*, 30(Part 1), 100–107.

Huising, E.J. and Gomes Pereira, L.M., 1998. Errors and accuracy estimates of laser data acquired by various laser scanning systems for topographic applications. *ISPRS Journal of Photogrammetry and Remote Sensing*, 53(5), 245–261.

Jelalian, A.V., 1992. *Laser Radar Systems*. Artech House, Norwood, MA, p. 45.

Jutzi, B., Eberle, B., and Stilla, U., 2002. Estimation and measurement of backscattered signals from pulsed laser radar. In: Serpico, S.B. (Ed.) *Image and Signal Processing for Remote Sensing VIII*. SPIE Proceedings. Vol. 4885, pp. 256–267.

Jutzi, B. and Stilla, U., 2005. Measuring and processing the waveform of laser pulses. In: Gruen A, Kahmen H (Eds.) *Optical 3-D Measurement Techniques VII*. Vol. I, pp. 194–203.

Kager, H. and Kraus, K., 2001. Height discrepancies between overlapping laser scanner strips. *Proceedings of Optical 3D Measurement Techniques V*, October, Vienna, Austria, pp. 103–110.

Keiser, G., 1983. *Optical Fiber Communications*. McGraw-Hill series in electrical engineering, communications and information theory, McGraw-Hill, Japan.

Krabill, W., Abdalati, W., Frederick, E., Manizade, S., Martin, C.,Sonntag, J., Swift, R., Thomas, R., and Yungel, J., 2002. Aircraft laser altimetry measurement of elevation changes of the Greenland ice sheet: Technique and accuracy assessment. *Journal of Geodynamics*, 34, 357–376.

Krabill, W.B., Thomas, R.H., Martin, C.F., Swift, R.N., and Frederick, E.B., 1995. Accuracy of airborne laser altimetry over the Greenland ice sheet. *International Journal of Remote Sensing*, 16(7), 1211–1222.

Latypov, D., 2002. Estimating relative lidar accuracy information from overlapping flight lines. *ISPRS Journal of Photogrammetry and Remote Sensing*, 56(4), 236–245.

Latypov, D., 2005. Effects of laser beam alignment tolerance on lidar accuracy. *ISPRS Journal of Photogrammetry and Remote Sensing*, 59(6), 361–368.

Litchi, D. and Harvey, B., 2002. The effects of reflecting surfaces material properties on time-of-flight laser scanner measurements. *Symposium of Geospatial Theory, Processing, and Applications*, Ottawa.

Rönnholm, P., 2004. The evaluation of the internal quality of laser scanning strips using the interactive orientation method and point clouds. *Proceedings ISPRS XXth Congress*, Vol. XXXV Part B, Commission 3, p. 255 ff.

Schenk, T., 2001a. Modeling and analyzing systematic errors of airborne laser scanners. Technical Notes in Photogrammetry No. 19, Department of Civil and Environmental Engineering and Geodetic Science, The Ohio State University, Columbus, OH, 40 pp.

Schenk, T., 2001b. Modelling and recovering systematic errors in airborne laser scanners. *Proceedings of OEEPE workshop on airborne laser scanning and Interferometric SAR for Detailed Digital Elevation Models*, Stockholm, Sweden.

Schiele, O., Wehr, A., and Kleusberg, A., 2005. Operational calibration of airborne laserscanners by using LASCAL. In: Grün, A., Kahmen, H., Vienna (Eds.) *Papers presented to the Conference Optical 3-D Measurement Techniques VII*, Vol., Austria, Oct. 3–5. pp. 81–89.

Skolnik, M.I., 2001. *Radar Systems*. McGraw-Hill, New York.

Sun, X., Abshire, J., and Davidson, F., 1993. Multishot laser altimeter: Design and performance. *Applied Optics*, 32(24), 4578–4585.

Sun, X., Davidson, F., Boutsikaris, L., and Abshire, J., 1992. Receiver characteristics of laser altimeters with avalanche photodiodes. *IEEE Transactions of Aerospace and Electronic Systems*, 28(1), 268–275.

Thiel, K.-H. and Wehr, A., 1999. Operational data processing for imaging laser altimeter data. *Proceedings of the Fourth International Airborne Remote Sensing Conference and Exhibition*, ERIM, June, 21–24, Ottawa, Canada.

Vaughn, C.R., Bufton, J.L., Krabill, W.B., and Rabine, D.L., 1996. Georeferencing of airborne laser altimeter measurements. *International Journal of Remote Sensing*, 17(11), 2185–2200.

Vosselman, G. and Mass, H.-G., 2001. Adjustment and filtering of raw laser altimetry data. OEEPE Workshop on Airborne Laserscanning and Interferometric SAR for Detailed Digital Elevation Models, Stockholm, Sweden.

Wehr, A., Hemmleb, M., Thomas, M., and Maierhofer, C., 2007. Moisture detection on building surfaces by multi-spectral laser scanning. *Proceedings of Optical 3-D Measurement Techniques VIII*, Conference at ETH-Zuerich, Switzerland, July, 9–12. pp.79–86.

Young, M., 1986. *Optics and Lasers—Series in Optical Sciences*. Springer, Berlin, p. 145.

5

Pulsed Laser Altimeter Ranging Techniques and Implications for Terrain Mapping

David Harding

CONTENTS

5.1 Introduction

The emergence of airborne laser swath mapping (ALSM), initially as a research tool and in the last decade as a commercial capability, has provided a powerful means to characterize the elevation of the Earth's solid surface and its overlying covers of vegetation, water, snow, ice and structures created by human activity. Other terms used for ALSM include scanning laser altimetry, airborne laser scanning, and LIght Detection and Ranging (LIDAR). The rapid collection of georeferenced, highly resolved elevation data with centimeter to decimeter absolute vertical accuracy achieved by ALSM enables unprecedented studies of natural processes and creation of elevation map products for application purposes. Scientific uses of the data span many diverse disciplines, attesting to the value of detailed elevation information. Uses include, but are not limited to, evaluation of natural hazards associated with surface-rupturing faults, volcanic flow pathways, slope instability and flooding, monitoring of coastal change due to storm surges and sediment transport, characterization of ecosystem structure from which estimates of above-ground biomass and assessments of habitat quality are derived, and quantification of ice sheet and glacier elevation change and their contribution to sea level rise. In applied areas, the multilayered map products showing ground topography, even where densely covered by vegetation, as well as canopy density and height, and building footprints and heights make ALSM data uniquely suited for resource management and land use purposes.

In order to most effectively use the products generated from ALSM systems, for both scientific and applied purposes, their principles of operation and the resulting effects on the acquired data need to be well understood. In this chapter, laser ranging methods are examined focusing on the most commonly employed type of laser altimeter technology that uses pulsed laser light. Waveform recording systems and those that generate point clouds are described. For the latter type, discrete return and single-photon detection approaches are addressed. Relationships between ranging methods, instrument parameters, and the character of the resulting elevation data are considered. The emphasis is on vegetated landscapes and, in particular, the measurement of ground topography beneath vegetation cover and characterization of forest canopy structure.

Factors are discussed that affect the amplitude of the received signal, as that bears on the detection of surface returns. These factors also are important considerations in the interpretation of intensity images that are produced along with elevation data by more recent ALSM systems. When properly calibrated, the return intensity can be used to create monochromatic, single-band reflectance images at the laser wavelength. Because the laser return is range resolved, the return intensity can be ascribed to different levels within a scene as, for example, from overlying vegetation and the underlying ground surface. This added information has the potential to improve the value of ALSM mapping by providing better differentiation of surface types and features.

The fundamental observation made by pulsed ALSM systems is the range (i.e., distance) from the instrument to a target, determined by timing the round-trip travel time of a pulse of laser light reflected from a surface. Travel time is converted to distance using the speed of light. Combining range with knowledge of the orientation of the transmitted laser pulse, and the position of the instrument in a coordinate frame, yields a vector that defines the location of the reflecting target, usually expressed as a latitude, longitude and elevation. The laser beam orientation is determined using two devices consisting of an inertial measurement unit (IMU) and an angle encoder. The IMU defines the roll, pitch and yaw of the instrument aboard an airborne platform and the angle encoder defines the angular position of the scanner mechanism used to deflect the laser beam across the swath being mapped. The instrument position at the instant the laser pulse is fired is established using a combined solution derived from a differential Global Positioning System (DGPS) aircraft trajectory and the IMU acceleration data. Here the focus is on the range measurement and the implications for interpreting information about the elevation structure of a target.

5.2 Signal Strength

Before describing ranging methods, the factors that control the signal strength observed by a laser altimeter need to be addressed. The amplitude of the received signal depends on the wavelength and energy of the transmit pulse, the distance to the target, the target reflectance, the transmission of the atmosphere, the area of the receiver aperture, the throughput efficiency of the receiver, the sensitivity of the detector and, in the case of analog detection, the amplification gain applied to the detector output.

Of particular relevance here are those factors that are wavelength dependent: detector sensitivity, atmospheric transmission, and target reflectance. Figure 5.1 illustrates the sensitivity of several detector types used, or having potential for use, in laser altimeters. Wavelengths that are commonly used in ALSM systems are indicated: 900nm, produced by semiconductor

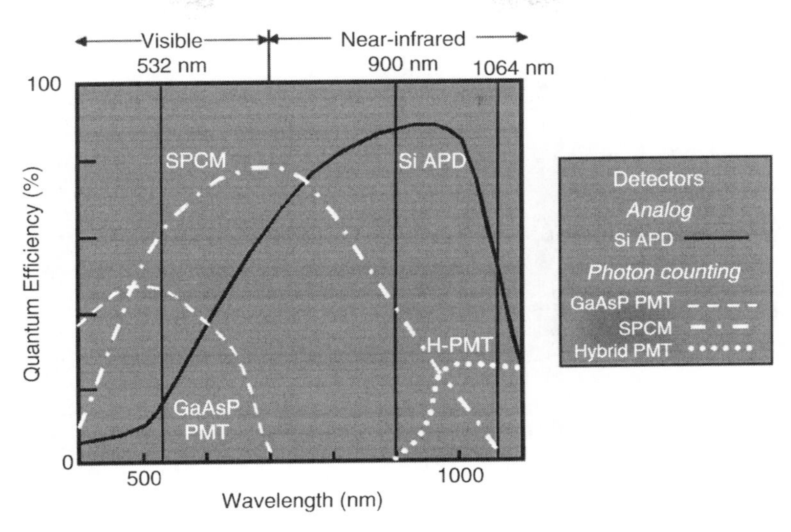

FIGURE 5.1
Representative curves of detector sensitivity as a function of wavelength, expressed as quantum efficiency. The Si APD example is for a particular detector with sensitivity preferentially shifted toward the near-infrared. Other Si APDs have higher visible sensitivity.

lasers, 1064nm, produced by Nd:YAG lasers, and 532nm (green visible light), generated by frequency doubling of Nd:YAG output. Other transmitter choices used in ALSM systems, not indicated in Figure 5.1, include 1560nm semiconductor lasers and 1470nm Nd:YLF lasers. Atmospheric transmission (Figure 5.2) depends on the density (i.e., optical depth) of aerosols and clouds. Normally ALSM missions are conducted with a cloud-free path between the instrument and the target, but haze due to water vapor and aerosols can significantly reduce

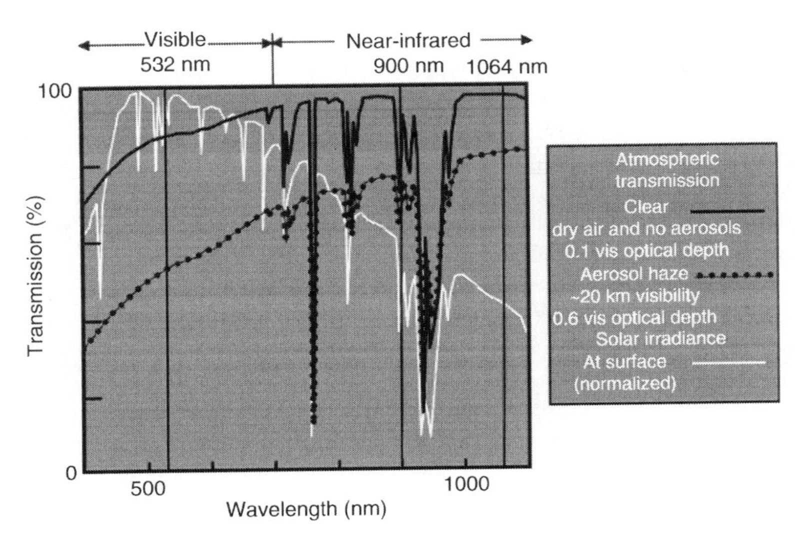

FIGURE 5.2
Wavelength dependence of atmospheric transmission shown for clear and hazy conditions. Also shown is down-welling solar irradiance at the Earth's surface which is a source of noise for laser altimeters. Minima in the transmission curves are due to water vapor absorption. Additional minima in the irradiance spectra are due to absorption occurring in the sun.

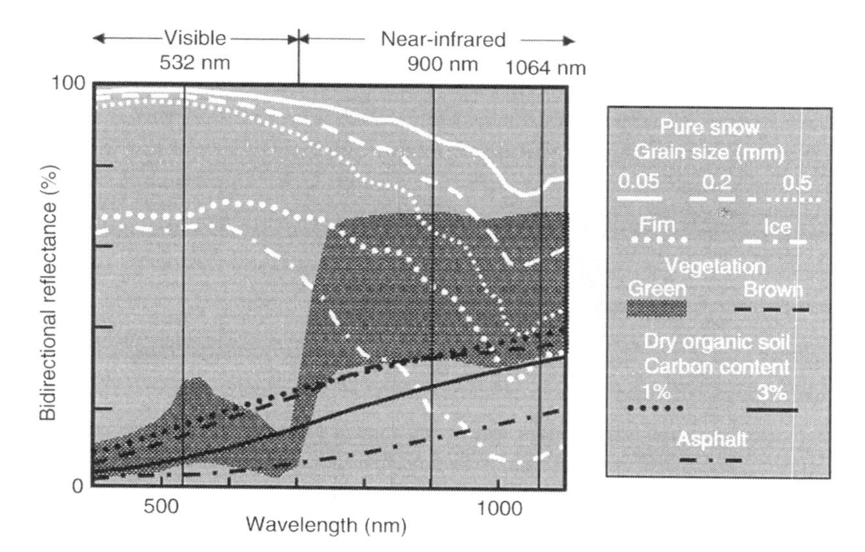

FIGURE 5.3
Wavelength dependence of bidirectional reflectance for representative surface materials. Green vegetation reflectance, a function of leaf structure and chlorophyll content, is represented as a broad band because of the variability observed as a function of species and environmental conditions. Firn is metamorphosing snow transitioning to ice. The spectra are compiled from multiple literature sources.

transmission, especially at visible wavelengths. Examples of typical reflectance curves for representative surface materials are illustrated in Figure 5.3, showing strong dependence on wavelength. These spectra show bidirectional reflectance data that are acquired with a phase angle (angular separation of the illumination source and the receiver) of several tens of degrees in order to be applicable to nadir-viewing, passive imaging systems viewing scenes illuminated at an angle by the sun.

Reflectance data acquired under bidirectional conditions are not directly applicable to assessing laser altimetry signal strength. Laser altimeters acquire data in a geometry unlike traditional passive optical imaging systems reliant on solar illumination. Specifically, the illumination and receiver view paths are either exactly parallel, for mono-static laser altimeters that use a common transmit and receiver aperture, or very nearly parallel, for bi-static designs that use adjacent apertures. Observing in this retroreflection "hot spot" geometry, requires consideration of opposition effects because most natural surfaces, although they are diffuse reflectors that reflect light in all directions, are not Lambertian (equal reflectance in all directions) as is commonly assumed. Enhancement of reflectance due to opposition effects, compared to a nominal bidirectional geometry, is typically a factor of two and can be as large as a factor of three. The increase is a function of surface composition, roughness and wavelength. As an example, Camacho-de Coca et al. (2004) studied the opposition hot-spot effect at four wavelengths using the airborne, multiangle, passive imaging POLarization and Directionality of Earth Reflectance (POLDER) instrument flown in the direction of the solar principal plane. Reflectance for corn, barley and bare soil increases with decreasing phase angle and peaks in the retroreflection orientation at 0° phase angle (Figure 5.4). The solid curves in Figure 5.4 are predicted reflectance based on a simple radiative transfer model that incorporates leaf reflectance, solar zenith angle and geometric factors that account for scene self-shadowing (casting of shadows by one part of the target onto another part). The model assumes the opposition effect is due to a reduction of observed shadowed surfaces (i.e., increased shadow hiding) as 0° phase angle is approached.

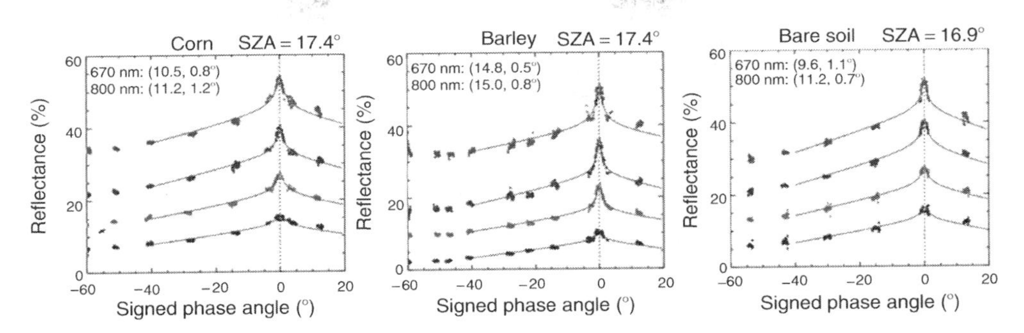

FIGURE 5.4
Phase angle dependence of solar reflectance illustrating the opposition hot-spot effect for corn, barley, and bare soil at 443, 550, 670, and 800 nm (plotted from bottom to top). (From Camacho de Coca, F.C., Breon, F.M., Leroy, M., and Garcia-Haro, F.J., *Rem. Sens. Environ.*, 90, 63, 2004. With permission.)

Several studies have documented the increase in laser reflectance from natural terrestrial materials at 0° phase angle (e.g., Kaasalainen and Rautiainen, 2005; Kaasalainen et al., 2006). The increase in reflectance with decreasing phase angle consists of two components: a linear increase over a large angular range, approximated by the dashed lines in Figure 5.5, and a narrow, nonlinear peak in reflectance beginning at about 2.5°. In the interpretation by Hapke et al. (1996, 1998) and Nelson et al. (1998, 2000), the opposition effect linear increase is due to increased shadow hiding with decreasing phase angle and the narrow peak is due to the combined effects of shadow hiding and light amplification caused by coherent interference of the transmitted and reflected light. However, uncertainty remains about the physical origin of the laser retroreflection peak, and models of the effect do not fully account for all its attributes (Helfenstein et al., 1997; Shkuratov and Helfenstein, 2001).

Shadow hiding and coherent interference are retroreflection effects present to varying degrees for all surfaces independent of the angle of incidence of the light onto the surface. They are not related to the more commonly known specular reflection phenomenon that controls the brightness of water observed by laser altimeters. A planar interface between transmissive media with differing refractive indices, such as the interface

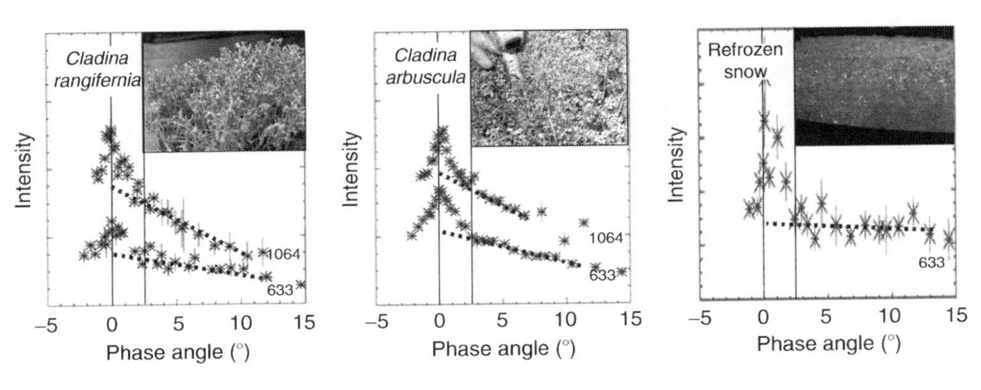

FIGURE 5.5
Phase angle dependence of laser reflectance illustrating the opposition effect at 633 and 1064 nm for lichen, and 633 nm for a snow surface after partial melting and refreezing. (Adapted from Kaasalainen, S. and Rautiainen, M., *J. Geophys. Res. Atmos.*, 110, 2005. With permission; Kaasalainen, S., Kaasalainen, M., Mielonen, T., Suomalainen, J., Peltoniemi, J.I., and Naranen, J., *J. Glaciol.*, 52, 574, 2006. Reprinted with permission from the International Glaciological Society.)

between air and smooth water, reflects and refracts the incident light. The reflection is specular, or mirror-like, and the angles of the incident and reflected rays with respect to the surface normal are equal. The intensity of light specularly reflected from an interface is governed by the Huygens-Fresnel Law, which defines reflectivity as a function of incident angle, the refractive index contrast across the interface and the polarization state of the incident light. Over the typical range of scan angles used by airborne swath mapping laser altimeters, up to 30° of nadir, the reflectance of a water surface is only 2% regardless of polarization state. Although this reflectance is very low, for collimated laser light the mirror-like reflection preserves a collimated beam and the light in the reflection direction is much more intense than is the diffuse reflection from nonspecular surfaces.

For the 2% of incident laser energy reflected from a water surface, the intensity of the laser return in the retroreflection direction observed by an altimeter receiver depends on the surface area of facets oriented perpendicular to the laser pulse vector (90° incident angle). For smooth, flat water, not roughened by wind, most surface facets are coplanar and horizontal, resulting in an intense return for laser pulses oriented near-nadir and a rapid fall-off of backscattered energy at small off nadir angles, accounting for the loss of signal from inland water bodies observed by airborne scanning laser altimeters at off-nadir angles. Bufton et al. (1983) describe how wind speed, and the resulting angular distribution of surface facets, is the controlling factor on laser backscatter intensity from roughened water surfaces. The remaining 98% of incident light not reflected from a water surface is refracted into the water column. The depth of penetration is defined by the absorption attenuation coefficient which is a function of wavelength. At visible wavelengths, the attenuation coefficient is small and light penetrates to several tens of meters depth through clear water, accounting for the use of 532 nm lasers for bathymetric mapping. At near-infrared wavelengths the attenuation coefficient of clear water is several orders of magnitude larger due to enhanced molecular absorption and light penetrates to only several centimeters to decimeters depth, thus acting to warm the uppermost layer of the water column and making 900 and 1064 nm lasers unsuitable for bathymetry purposes.

5.3 Analog Detection

Figure 5.6 illustrates commonly used laser ranging methods, depicting transmit pulses and received signals from a multistoried forest canopy as recorded by analog detection and photon counting approaches. In analog ranging, a detector converts received optical power into an output voltage yielding signal strength as a function of time. The signal is composed of reflected laser energy and noise sources, consisting of internal detector noise (i.e., dark counts) and background noise from solar illumination (Figure 5.2). In analog systems, high signal-to-noise ratio (SNR) performance per pulse is achieved (i.e., monopulse detection), using laser pulses with high peak power. The goal is reception of thousands of reflected photons per pulse in order to exceed the detector noise floor and have sufficient signal to accurately determine the range to the illuminated target.

Because the pulse energy must be high, the pulse width (i.e., its duration) is typically relatively broad, commonly being about 7 ns wide (full-width at half the maximum (FWHM) amplitude), equivalent to a FWHM pulse width of ~1 m. Semiconductor lasers or diode-pumped, Q-switched solid-state Nd:YAG laser transmitters operating at rates

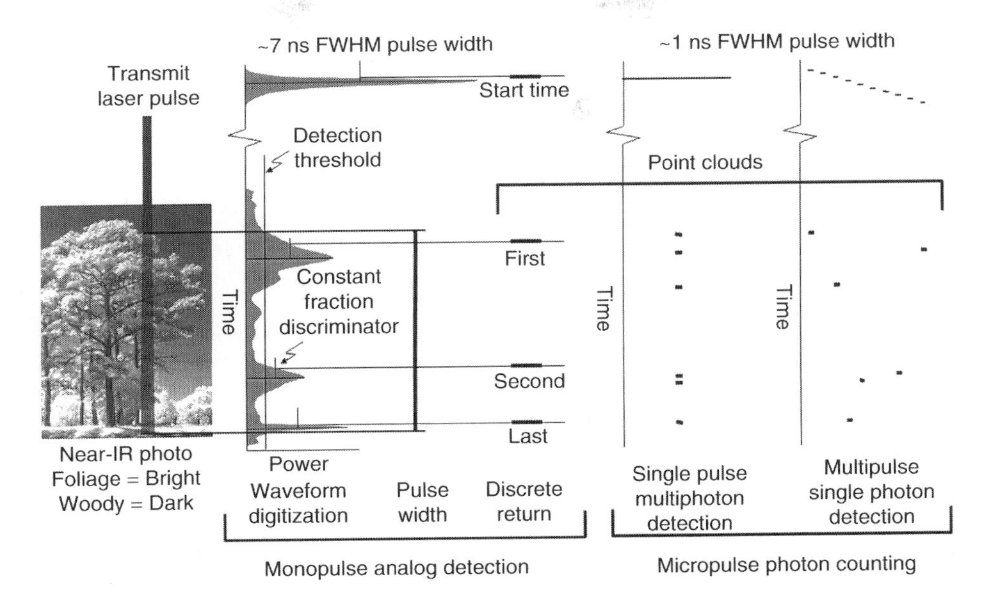

FIGURE 5.6
Illustration of laser ranging methods, depicting transmit pulses and received signals from a multistoried forest canopy as recorded by analog detection and photon counting approaches.

of hundreds to tens of thousands pulses per second are usually used for this purpose. The detector used for analog ranging is usually a silicon avalanche photodiode (Si APD), which has high sensitivity across visible to near-infrared wavelengths (Figure 5.1).

5.3.1 Waveform Digitization

Digitization of the detector output time series using an analog-to-digital converter yields a waveform that fully characterizes the vertical structure of the target (gray shaded signal in Figure 5.6). For a hard target such as bare ground or a building roof, a single return peak is normally observed. The broadening of the received peak, relative to that of the transmit peak, is a measure of the vertical relief of the target within the laser footprint due to surface roughness and/or slope. In some hard target cases, multiple return peaks are observed as for example when a laser footprint intersects a building edge, illuminating the top of the building and the adjacent ground. For distributed targets like vegetation canopies where multiple vertically distributed surfaces composed of leaves, stems and branches and underlying ground are illuminated by a single laser pulse, a complexly shaped received signal can result. The signal in that case is a measure of the height distribution of illuminated surfaces weighted by the spatial distribution of laser energy within the footprint and the retroreflectivity of the surfaces at the laser pulse wavelength (Harding et al., 2001). The pulse width of the received signal (Figure 5.6) is the duration from the first to last crossing of the reflected laser energy above a detection threshold, representing the height range of the target from signal start to end. Mapping vegetated surfaces with a scanning waveform-recording system that produces laser footprints contiguous across- and along-track within a swath produces a three-dimensional, volumetric image of reflected laser energy (Weishampel et al., 2000). As laser energy is intercepted by canopy surfaces, the pulse energy decreases with depth through vegetation.

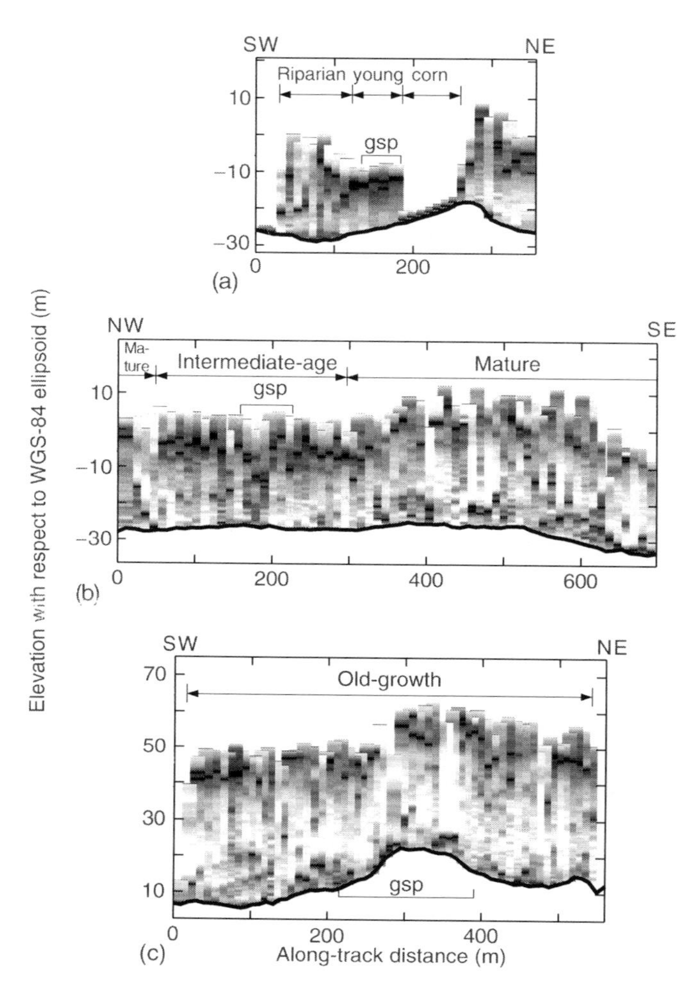

FIGURE 5.7
Examples of waveform-derived canopy height profiles showing normalized plant area from ALSM transects across deciduous forest stands of different ages. Each vertical bar corresponds to the diameter of one laser pulse (~10 m), and the gray shading corresponds to the relative amount of plant area in 66 cm height increments (darker = denser). (From Harding, D.J., Lefsky, M.A., Parker, G.G., and Blair, J.B., *Rem. Sens. Environ.*, 76, 283, 2001. With permission.)

Accounting for this extinction effect, height profiles of normalized canopy plant area (Figure 5.7) can be derived from the waveform data.

Normally in full-waveform systems both the transmit pulse shape and the received signal shape are digitized and recorded. The range to a specific feature within the received signal, such as the last peak in Figure 5.6 that is due to laser energy reflected from the ground surface, is usually defined as the distance between the center of the transmit pulse signal and the center of the received peak. Different methods for defining the peak center have been applied to waveform data, including use of the centroid (center of mass) or mean of the distribution or the center of a function, such as a Gaussian distribution, fit to the signal (Hofton et al., 2000; Harding and Carabajal, 2005). For waveform-recording systems, a well-defined and consistent transmit pulse shape is an important system attribute as is the signal-handling performance of the receiver, which must record the received pulse shape, free of distorting effects such as detector saturation.

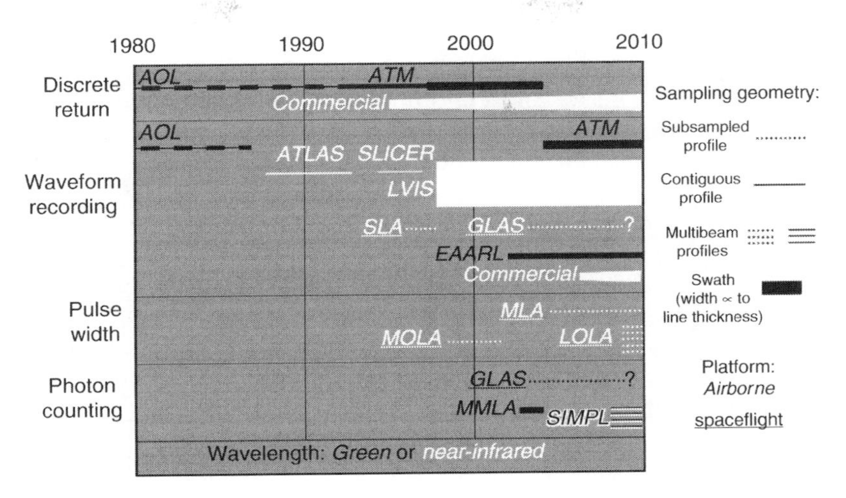

FIGURE 5.8
Timelines of laser altimeter operations for instruments developed by NASA and commercially, differentiating systems based on the ranging method employed, the transmit pulse wavelength, the sampling pattern, and the type of platform used. For swath mapping systems their swath width is proportional to the line thickness (LVIS width corresponds to its nominal 4 km swath acquired using 20 m diameter footprints).

The evolution of laser altimeter systems is shown in time-line form in Figure 5.8, grouped by ranging method and emphasizing research systems developed by NASA, many of which established measurement principles later used in commercial systems. Systems are characterized by operating wavelength (green or near-infrared), the platform on which they are flown (airborne or satellite), and the sampling geometry of the laser footprints. Early experiments in waveform sampling of land topography and vegetation structure were performed by the Airborne Oceanographic Lidar (AOL) (Krabill et al., 1984), acquiring in some cases single profiles and at other times using scanning to map a swath, and by the profiling Airborne Topographic Laser Altimeter System (ATLAS) (Bufton et al., 1991). Later, scanning airborne systems advanced the technology, including the Airborne Topographic Mapper (ATM) (Krabill et al., 2002) which is a follow-on to AOL, the Scanning Lidar Imager of Canopies by Echo Recovery (SLICER) (Harding et al., 2000), the wide-swath mapping Laser Vegetation Imaging Sensor (LVIS) (Blair et al., 1999), and the Experimental Advanced Airborne Research Lidar (EAARL) (Brock et al., 2002, 2006; Nayegandhi et al., 2006). The EAARL system, operating at 532 nm, simultaneously acquires bathymetry and topography data using a novel design that splits the received signal into three channels consisting of 90%, 9%, and 0.9% of the return energy. This is done in order to accommodate large variations in signal strength from diffuse surfaces (land, vegetation and the bottom of water bodies), and specular water surfaces that normally cause saturation and distortion of the waveforms from smooth water. Recently, several versions of near-infrared, waveform-recording ALSM systems have become available for purchase from commercial vendors.

Waveform-recording systems intended for characterization of canopy structure (e.g., SLICER, LVIS) normally use large diameter footprints (>10 m) in order to sample a representative canopy volume in a single footprint, while also illuminating the entire scene with adjacent, contiguous footprints. These are sometimes referred to as large-footprint systems. Other systems geared to high-resolution terrain mapping (e.g., ATM, EAARL, commercial) employ small diameter footprints (<1 m). In forested regions, these small-footprint systems usually yield simpler, but less representative waveforms than do large-footprint systems because the laser footprints intersect only a few distinct surfaces and are noncontiguous.

Satellite laser altimeters to date have not implemented scanning systems, instead acquiring single profiles with footprints that are separated along-track, because of the very high ground speeds of spacecraft and high laser power needed for analog ranging from orbital altitudes. Satellite-based waveform-recording systems in Earth orbit include the Shuttle Laser Altimeter (SLA) (Garvin et al., 1998) and the currently-operating Geoscience Laser Altimeter System (GLAS) (Abshire et al., 2005) aboard NASA's Ice, Cloud and land Elevation Satellite (ICESat) (Zwally et al., 2002; Schutz et al., 2005). SLA demonstrated techniques for space-based waveform measurements of canopy height and continental-scale topographic profiles that are now being used by ICESat to sample the biomass stored in forests and monitor elevation changes of the Earth's ice sheets and glaciers as they respond to global warming. The Mars Orbiter Laser Altimeter (MOLA), (Abshire et al., 2000; Neumann et al., 2003) is an analog detection system that recorded the received pulse-width, rather than the complete waveform, in order to reduce the data volume needing transmission. MOLA profiles acquired continuously over a period of several years were used to create a topographic map of Mars that has revolutionized understanding of that planet's evolution (Smith et al., 1999, 2001; Zuber et al., 2000). The MESSENGER Laser Altimeter (MLA) (Ramos-Izquierdo et al., 2005) is currently en-route to Mercury where it will provide the first topographic mapping of that planet using pulse-width recording. The Lunar Orbiter Laser Altimeter (LOLA) (Riris et al., 2007; Chin et al., 2007), a multibeam, pulse-width recording system will use five parallel profiles to map the Moon in unprecedented detail beginning in 2008.

5.3.2 Discrete Returns

Discrete return ranging identifies distinct peaks in the analog detector output time series that exceed a detection threshold. This is usually accomplished by means of a constant fraction discriminator that times the leading edge of the peak at an energy level that is some specified fraction of the peak amplitude. Fifty percent constant fraction discrimination is illustrated in Figure 5.6. This approach substantially reduces the data volume that must be recorded, as compared to full waveforms, while identifying the height of prominent surfaces illuminated by the laser pulse. A key attribute of systems that utilize discrete return detection is the number of returns that can be recorded per laser pulse. Earlier versions of such systems recorded only one return, with the timing electronics configured to record either the highest detected distinct peak or the lowest peak. Subsequent systems were configured to record two returns, the highest and the lowest, or multiple returns, up to a maximum of five. Figure 5.6 illustrates a case in which three discrete returns were identified. Most recently, systems have added detection and recording of the return amplitude associated with each discrete return, in some cases reporting the maximum amplitude of the peak and in others reporting an integrated signal strength (i.e., area of the peak).

For discrete return systems, the range is defined as the distance from the transmit pulse leading edge (start time in Figure 5.6) to the leading edge of the distinct peaks. Therefore, in these systems a well defined leading-edge on the transmit pulse is most important, and the trailing shape of the pulse is of lesser consequence for ranging accuracy. Each return, when geolocated, yields a discrete location defined by the latitude, longitude and elevation of a point. Combining the discrete returns from many laser pulses yields a three-dimensional distribution of geolocated points, commonly referred to as a point cloud. Small-diameter footprints (<1 m) are usually used in discrete-return systems because of the need, when a pulse intercepts multiple surfaces, to have distinct return peaks that are separated in range. In addition, the footprint size is a limiting factor on spatial resolution; small footprints enable high-resolution mapping of topography and canopy structure.

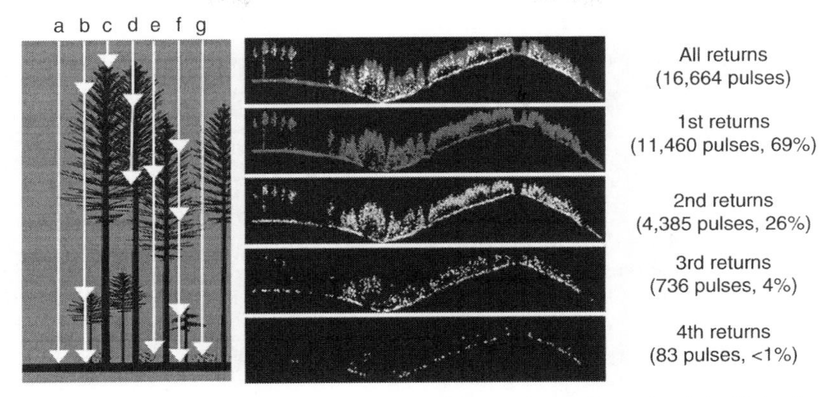

a: 1st = Ground b: 1st = Highest canopy, 2nd = Mid-story, Last = Ground c: 1st = Highest canopy
d: 1st = Within crown, Last = Within crown e: 1st = Within crown, Last = Under-story
f: 1st = Highest canopy, 2nd = Within crown, 3rd = Mid-story, Last = Ground g: 1st = Under-story

FIGURE 5.9
Schematic illustrating the detection of multiple discrete returns for seven laser pulses transmitted through a forest canopy (left) and a transect through discrete return point cloud data acquired across the Capitol Forest, WA showing data separated by return number. (Graphics courtesy of Bob McGaughey, USDA Forest Service PNW Research Station.)

Figure 5.9 is an illustration of point cloud data acquired by a multiple-return, discrete return, small-footprint laser ranging system, depicting acquisition of up to four returns per laser pulse reflected from a forest canopy. The number and height of returns detected for a pulse depends on the distribution of surfaces, and their reflectance at the laser wavelength, encountered along the path of the laser beam. Discrete returns are detected when the area and reflectance of an illuminated surface is sufficient to yield received energy that exceeds the detection threshold. First returns correspond to the leading edge of the first detected signal above the threshold level, and may be from the canopy top, from a layer within the canopy or from the ground. Last returns correspond to the leading edge of the latest detected peak, and may be from the ground or from a layer within the vegetation canopy. The detection of ground returns in areas of vegetation cover, depends on the spatial and angular distribution of open space (gaps) through the canopy, the scan angle of the laser beam, the divergence of the laser beam and the range to the target (defining the diameter of the laser footprint), the footprint sampling density, and the reflectivity of the ground at the laser wavelength. Unlike waveform recording systems that fully characterize vegetation structure, discrete-return representations of canopies are highly instrument-dependent. Measurements of canopy attributes such as height, crown depth or the distribution of under-story layers and gaps obtained using different discrete-return instruments are therefore generally not equivalent and comparisons for change detection purposes for example, must be made with care. Lefsky et al. (2002) reviews the use of waveform-recording and discrete-return systems for estimation of vegetation structural attributes.

The ability to detect separate returns from closely spaced surfaces is dependent on instrument parameters including the laser pulse width (shorter being better), detector sensitivity and response time (i.e., bandwidth), the system signal-to-noise performance, the detection threshold, and the ranging electronics implementation. This is especially relevant for detection of ground returns beneath short-stature vegetation. The elevation recorded by discrete return ranging is dependent on the shape of the received signal's leading edge, illustrated in Figure 5.10 in which waveforms represent four surface types: (a) highly reflective flat ground,

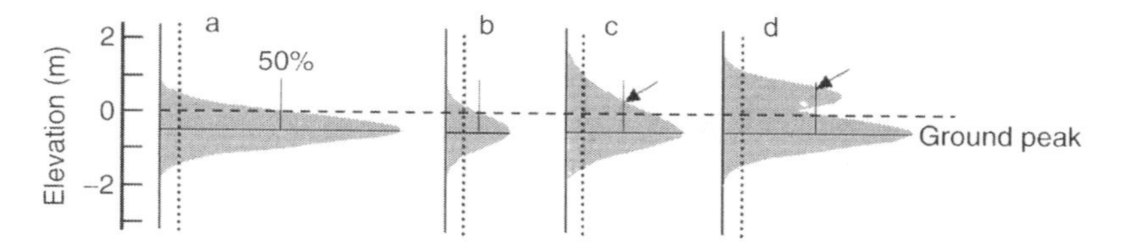

FIGURE 5.10
Schematic waveforms depicting returns from (a) highly reflective flat ground, (b) less reflective flat ground, (c) flat ground with low vegetation cover such as tall grass, and (d) flat ground with a low but distinct vegetation layer such as a forest under-story. The detection threshold is dotted, leading edge timing for the ground peak using 50% constant fraction discrimination is dashed, and arrows mark the height of upward biased returns for the vegetated cases.

(b) lower reflectivity flat ground, (c) ground covered by tall grass causing an upward skew in the received signal, and (d) ground overlain by a distinct layer of low vegetation. Flat surfaces reflect a signal having the same shape as the transmit pulse, in these examples a Gaussian distribution with a 7 ns (1 m) FWMH pulse width. For returns with energy exceeding a detection threshold (dotted lines), 50% constant fraction discrimination yields a range that is insensitive to peak amplitude, so, bright and dark surfaces should have unbiased elevations (a and b, dashed line). However, a laser pulse reflected from low vegetation cover and the underlying ground can yield a composite signal from which only one peak is identified above the threshold, yielding a single discrete return. The vegetation shields the leading edge of the ground return, with the upward-displaced leading edge of the composite signal yielding a return that is close to, but above, the ground (c and d, arrows). This shielding is of course enhanced when leaves are present, not only because this increases plant cover but also because of the brighter reflectance of green vegetation as compared to soil or leaf litter, especially in the near-infrared (Figure 5.3). Due to this shielding, discrete return analog systems can yield elevations that are biased high with respect to the ground, in areas with short-stature vegetation cover.

Derivation of gridded Digital Elevation Models (DEMs) from ALSM point cloud data is done to define surfaces using those points associated with specific landscape features. For example, the highest reflective surface, or Digital Surface Model (DSM), is produced from the first returns for each laser pulse (Figure 5.11). In vegetated areas, this surface is a representation of the canopy top. The bald Earth surface, or digital terrain model (DTM), is produced from those returns inferred to be from the ground based on spatial filtering that identifies lowest returns defining a continuous surface. Interpolation is applied to produce a DTM, because returns from the ground can be irregularly distributed and widely spaced especially where vegetation cover is dense. In one interpolation approach the terrain elevation is first represented as a triangulated irregular network (TIN) created from the returns classified as ground. The elevation of the TIN is then sampled at regularly spaced grid points to derive the DTM. The georeferenced DSM and DTM products, provide a simplified representation of the point cloud data in a form amenable for use in raster-based image processing software and Geographic Information Systems. Note, however, as illustrated in Figures 5.9 and 5.11 there is additional information in the point cloud data set, regarding the structure and density of vegetation within the canopy that is not described by the DSM and DTM surfaces. Vector-based analysis tools are needed to fully assess information concerning canopy structure contained within the irregularly distributed point cloud data.

Examples of the highly detailed images of ground topography and forest canopy structure that can be produced using discrete-return ALSM data are illustrated in Figure 5.12, depicting commercial data acquired for the Puget Sound Lidar Consortium (Haugerud et al., 2003).

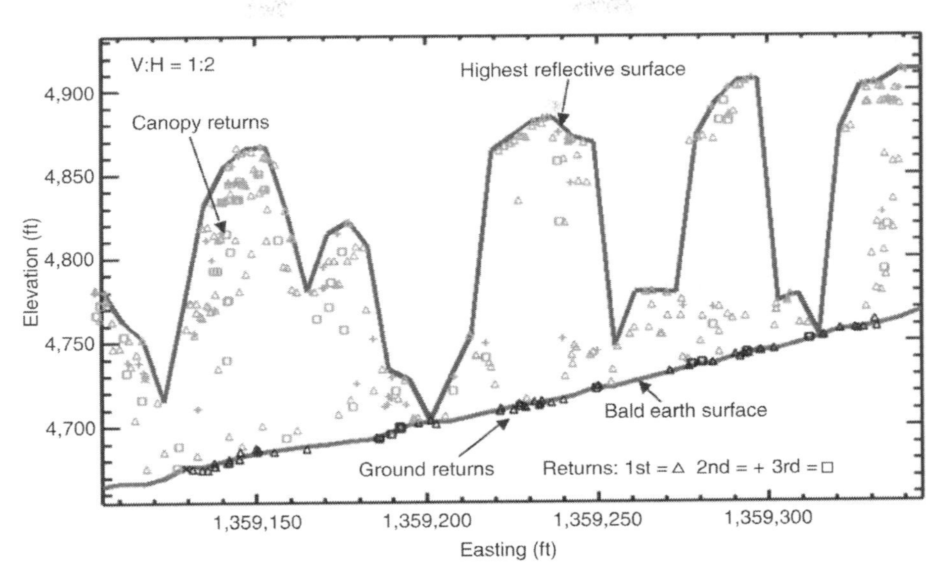

FIGURE 5.11
Derivation of DEM products from a multi-hit, discrete return point cloud, showing the highest reflective DSM produced from first returns and the "bald Earth" topography DTM produced from those returns inferred to be from the ground (black) based on spatial filtering. In this example, depicting returns in a six foot wide corridor from individual tree crowns in a forest stand above sloping ground, the surfaces are generated with six foot postings using data acquired for the Puget Sound Lidar Consortium. (From Haugerud, R., Harding, D.J., Johnson, S.Y., Harless, J.L., Weaver, C.S., and Sherrod, B.L., *GSA Today*, 13, 4, 2003.) Triangle, plus, and square symbols indicate first, second, and third returns, respectively.

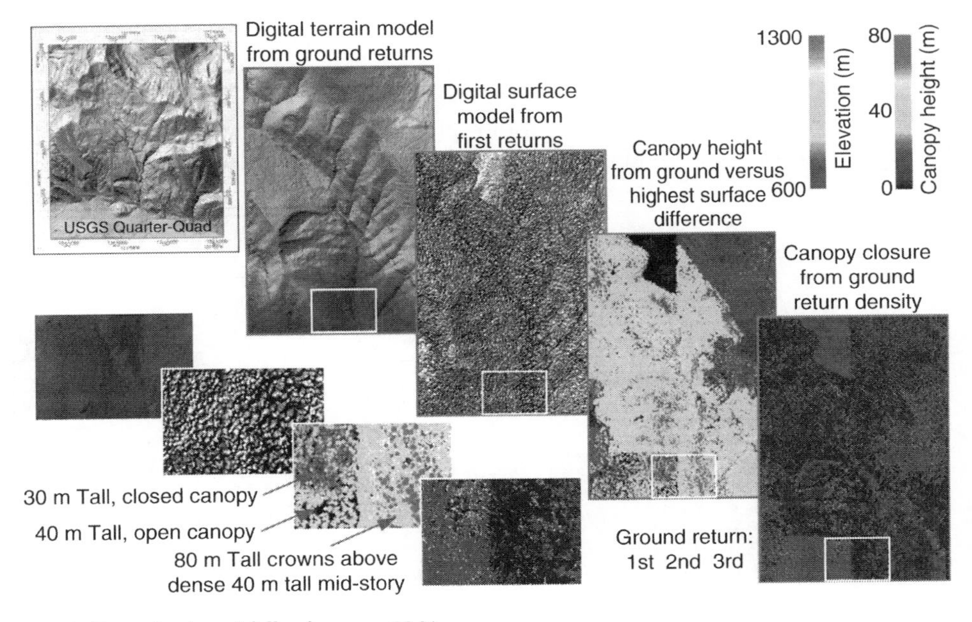

FIGURE 5.12 (See color insert following page 334.)
Discrete return point clouds and the DEMs generated from them provide highly detailed images of ground topography and forest canopy structure. Here commercial ALSM data acquired for the Puget Sound Lidar Consortium (From Haugerud, R., Harding, D.J., Johnson, S.Y., Harless, J.L., Weaver, C.S., and Sherrod, B.L., *GSA Today*, 13, 4, 2003) and gridded at six foot postings is used to depict several representations of topography and vegetation cover for the area indicated by the red rectangle in the upper left DTM shaded relief image corresponding to a USGS 7.5 in. quarter quadrangle. The area indicated by the white rectangles is enlarged in the insets to show image details.

The data were acquired with submeter diameter laser footprints having a nominal density of two footprints per square meter using a system that recorded up to four returns per pulse. A heavily forested, topographically rugged area in the Cascade Range, Washington State is portrayed as a color hill-shade image of the ground topography DTM, a gray-scale hill-shade image of the canopy top DSM, a color image of canopy height (DSM–DTM), and a measure of canopy closure represented by the distribution of ground returns (black areas lack ground returns due to dense canopy cover). The inset enlargements illustrate tree crowns in three stands with different canopy structures produced by varying forest management practices. The area on the right corresponds to a mature, closed-canopy forest with a low density of ground returns and isolated 80 m tall tree crowns above a 40 m tall mid-story canopy layer. The areas on the left are managed stands at different stages of regrowth following clear-cut logging. The upper area is a young, dense stand whereas the lower area is an older, taller stand that has undergone selective thinning.

Implications for the vertical accuracy of DTMs due to the upward bias of discrete-return, leading-edge ranging are apparent from histograms of elevation differences between ground control points and an ALSM ground topography DTM (Figure 5.13). The ground control was collected using differential global position system surveying at randomly selected locations uniformly distributed across the DTM. For nonvegetated areas, random ranging and geolocation errors yield a distribution of elevation differences that is symmetric and has a mean difference very close to zero. In these bare ground areas, the DTM is unbiased and has a vertical accuracy, expressed as a root mean square error (RMSE) of 19 cm. For areas with tall grass cover, the difference histogram is skewed, with the derived DTM surface sometimes occurring several decimeters above the ground control points yielding a surface that is, on average, too high by 7 cm. For forested areas, the skew can be more pronounced and the distribution can be bimodal. The forest example in Figure 5.13 has a peak at zero for true ground returns, and at −50 cm, due to returns from dense under-story vegetation being misclassified as ground. As a consequence, the mean upward bias of the DTM is significant and the RMSE vertical accuracy is substantially degraded. Instrumentation-dependent biases related to vegetation shielding of the ground can be a significant source of error quantifying topographic change by differencing discrete return DTMs acquired separated in time by different systems. Improvements in instrument parameters, such as a narrowed pulse-width to better detect closely space returns, decreased

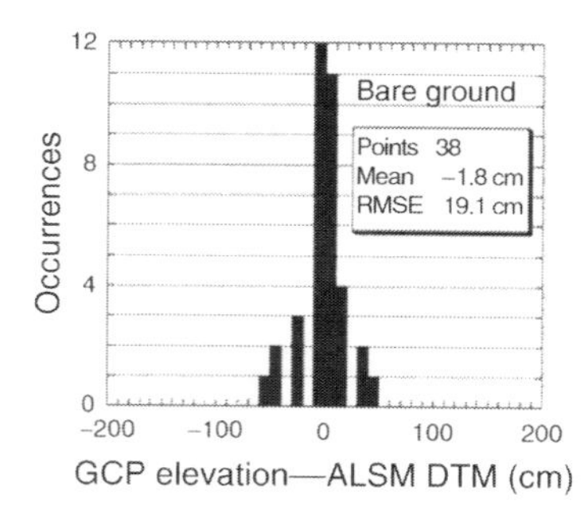

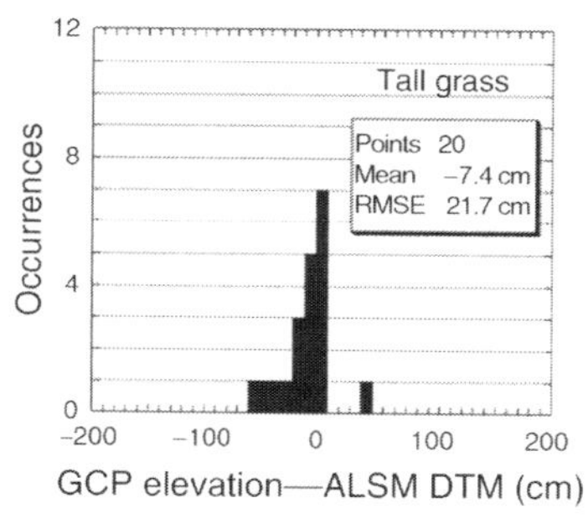

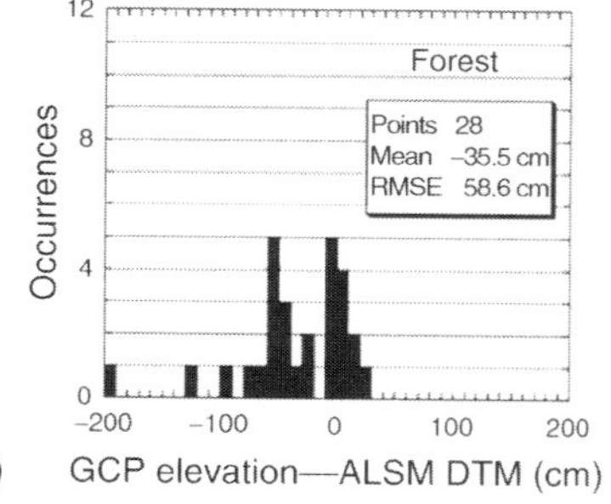

FIGURE 5.13
Cover-type specific histograms of elevation differences between a DTM, produced from ALSM discrete returns inferred to be from the ground, and ground control points measured by differential GPS surveying. Because the difference is computed as GCP–DTM, locations where the DTM is biased high yield negative values. The GPS and ALSM data were acquired for the Puget Sound Lidar Consortium. (From Haugerud, R., Harding, D.J., Johnson, S.Y., Harless, J.L., Weaver, C.S., and Sherrod, B.L., *GSA Today*, 13, 4, 2003. With permission.)

footprint size, increased footprint sampling density, or improved filtering to minimize misclassification of low vegetation as ground are means to lessen this source of error in discrete-return ALSM data.

The AOL system (Krabill et al., 1984), pioneered airborne topographic profiling and mapping using discrete-return analog ranging (Figure 5.8) and directly led to the development of the scanning ATM system (Krabill et al., 1995, 2002). ATM transitioned to a full waveform-recording system in 2004. ATM has conducted comprehensive surveying of Greenland and parts of Antarctic for more than a decade in order to monitor elevation change associated with changes in ice sheet mass balance (Krabill et al., 2000; Abdalati et al., 2002; Thomas et al., 2004, 2006). ATM has also conducted repeated surveys of U.S. coastlines to document beach elevation change related to storm surges, flooding and sediment transport (Brock et al., 2002; Sallenger et al., 2003). Given the need for repeatable surveying with high spatial resolution and absolute accuracy necessary for change detection, the development and evolution of the ATM instrumentation and associated geolocation processing (Krabill and Martin, 1987; Vaughn et al., 1996), established many of the principles and practices now used in the large number of commercial ALSM instruments operating worldwide. ATM includes a laser return intensity mapping capability that is unique in that the amplitude of the solar background reflectance, observed between laser pulse fires, is also recorded. Thus, solar bidirectional reflectance and laser retroreflectance at 532 nm can be directly compared, to gain a better understanding of the hot-spot opposition effect (Figure 5.14).

5.4 Photon Counting

Unlike analog detectors, which convert received power into an output voltage, photon counting detectors record the arrival of single photons. Their output can be thought of as digital, where there is either an absence of signal or discrete events upon detection of a photon (Figure 5.6). The principles and methods for photon counting laser ranging were first established in ground-based systems used to range to Earth-orbiting satellites and are now being extended to ALSM systems (Degnan, 2001). For laser ranging implementations, photon counting detectors are used that have very low dark counts, so that the rate of detector noise is very small as compared to the rate of received signal photons. Combining a photon counting detector with timing electronics, the time-of-flight between laser fire and reception of a single photon is recorded. The single photon is a sample from the full distribution of surfaces illuminated by the laser pulse, and accumulation of many single photon ranges can recreate the height structure of a target. During daytime operations, solar illumination reflected from the Earth's surface or from clouds is a significant source of noise for photon counting systems (Figure 5.2), more so than for analog systems in which the detector is the dominant noise source. The rate of solar background noise detection must therefore be controlled. This is usually accomplished by a combination of narrow bandpass filtering, to block detection of energy at all wavelengths other than that of the laser, and use of a small receiver field-of-view, restricting collection of light to the location illuminated by the laser.

Unlike analog systems, in which high energy laser pulses are used to achieve an adequate SNR, photon counting systems employ much lower pulse energies with the goal of detecting only a small number of photons per pulse (i.e., micropulse detection). At these low pulse energies, short transmit pulse widths of ≤1 ns FWHM (≤15 cm) can be achieved.

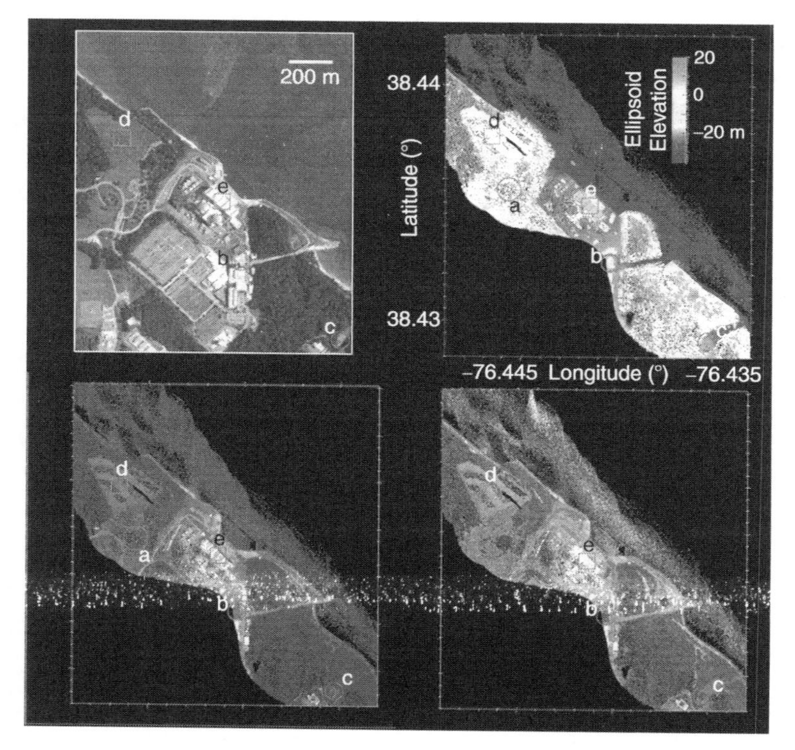

FIGURE 5.14 (See color insert following page 334.)
ATM mapping of the Chesapeake Bay western shoreline at Calvert Cliffs Nuclear Power Plant, Maryland acquired on September 17, 1997. Individual laser footprints from multiple flight passes are plotted showing elevation (upper right) and uncalibrated received 532 nm laser energy (lower right). Also shown is the ATM measurement of uncalibrated reflected 532 nm solar energy acquired between laser pulses (lower left) and a natural-color aerial photograph from Google Earth (upper left). Examples of intensity differences include areas where laser backscatter is lower, with respect to the surroundings, as compared to the reflected sunlight (circles: (a) parking lot, (b) building roof, (c) clay tennis court) and where the laser backscatter is higher (squares: (d) grass, (e) cooling tower roof). Also, intense specular reflections from offshore waves are observed in the laser backscatter image but not in the reflected sunlight image. (ATM images provided by Bill Krabill and Serdar Manizade, NASA Goddard Space Flight Center.)

This narrow pulse width, along with low-jitter detectors and high-resolution timing electronics, makes possible centimeter to decimeter range precision for single detected photons (Priedhorsky et al., 1996; Ho et al., 1999). Photon counting systems, therefore, have the potential to more efficiently acquire ALSM data than can analog systems, using less power, smaller receiver apertures, and/or higher pulse repetition rates (Degnan, 2002). Because very short pulses and single photons are the basis for ranging, biases due to the inability of distinguishing closelyspaced surfaces, that affects discrete-return leading-edge ranging, can be avoided. Microchip lasers, operating at thousands to tens of thousands of pulses per second, or fiber lasers, operating at hundreds of thousands of pulses per second, are well suited as short-pulse transmitter sources for photon counting systems. Analog discrete returns and photon counting both yield point clouds composed of many individual returns, each positioned at a specific location in three-dimensional space. Because of that commonality, the extensive body of software that has been developed for the processing, analysis and visualization of discrete return point clouds, and derivation of map products, is equally applicable to photon counting ALSM data.

5.4.1 Single Pulse Multi-Photon Detection

The specific manner in which a photon counting ALSM system is implemented depends on the performance characteristics of the detector used. Specifically, detector dead-time is a controlling factor. Dead-time is the time needed to recover after detection of one photon, returning to a state in which a second photon can be detected with equal sensitivity. Photomultiplier tubes (PMTs) commonly have very short dead times so that successive photons reflected from closely spaced surfaces, such as are encountered in vegetation canopies, can be differentiated. Using PMTs, single pulse multiphoton detection can be employed in which multiple photons are detected per pulse (Figure 5.6). To date, commercially available PMTs suitable for photon counting ranging are limited to operating at visible wavelengths, restricting their use to ALSM systems employing Nd:YAG lasers frequency-doubled to 532 nm (Figure 5.1). An additional limitation of PMTs is limited lifetime due to reduction in detection sensitivity as the total number of detected photons increases through time. An emerging detector technology needing further development, the hybrid PMT (Figure 5.1), offers the potential to conduct low dead-time, photon counting ranging at near-infrared wavelengths without the lifetime limitations of traditional PMTs (Sun et al., 2007).

The first ALSM system to implement PMT-based photon counting ranging (Figure 5.8) was the experimental Multikilohertz Micro-Laser Altimeter (MMLA) (Degnan et al., 2001) developed with funding from NASA's Instrument Incubator Program (IIP). MMLA employed a four-element PMT detector array, with each channel of the array capable of detecting up to four individual photon returns per laser fire. Thus, up to 16 single photon range measurements could be obtained per pulse. A rotating transmissive wedge produced a helical scan pattern thus mapping a narrow swath. Figure 5.15 illustrates an MMLA swath acquired during mid-day (e.g., high solar background) across the Wicomico Demonstration Forest, Maryland Eastern Shore. Map and cross-section representations of the elevation of individual photon returns depict variations in canopy height with distinct changes across stand edges. Accumulating single photon returns from areas approximately 20×30 m in size to form elevation histograms, yields representations of canopy vertical structure analogous to the waveforms of an analog detection system. The histograms, produced with 25 cm height bins, reveal differences in canopy height and layering related to stand maturity and forest management practices.

Even with the full leaf-on canopy cover conditions present at the time of data collection, a sufficient number of photons were returned from the ground to form a well-defined last peak beneath the canopy. The accumulation of ground returns in just one, or a few, 25 cm height bins is indicative of the range precision achieved using short-pulse photon counting. Solar background noise is present but at very low levels, as seen by the scarcity of detected photons below the ground surface and above the canopy top. A second-generation system based on MMLA, the Coastal Area Tactical-mapping System (CATS) (Carter et al., 2004; Slatton et al., 2005), is now being tested with the ultimate goal of deploying small-aperture, low-power, photon-counting ALSM systems on unmanned aerial vehicles. MMLA and CATS, operating at 532 nm, can simultaneously acquire topography and bathymetry data. MMLA has demonstrated single photon ranging that detects the water surface elevation, the vertical distribution of scattering within the water column, and the depth of the water bottom.

5.4.2 Multi-Pulse Single-Photon Detection

In contrast to short dead-time PMT detectors, single photon counting modules (SPCM) based on a Si APD detector (Sun et al., 2004) have long dead-times requiring a different approach

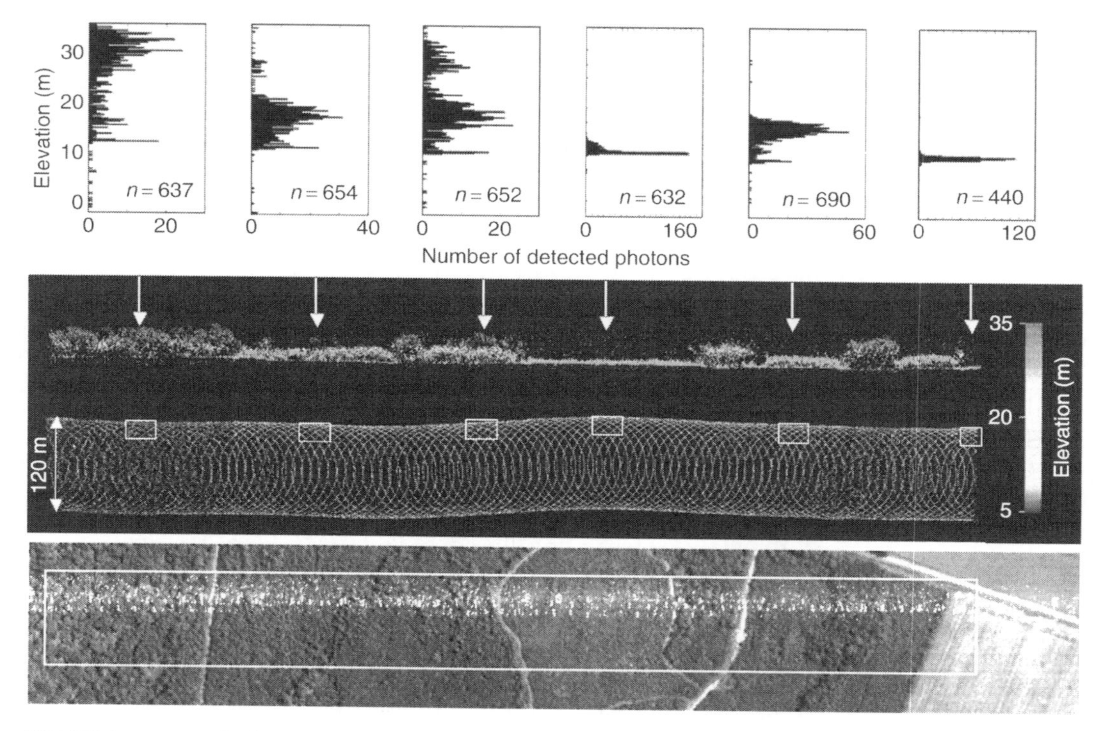

FIGURE 5.15 **(See color insert following page 334.)**
MMLA photon-counting, helical-scanning airborne data acquired in a 120m wide swath across the Wicomico Demonstration Forest, MD Eastern Shore on September 12, 2002. Middle: single photon returns color-coded by elevation in map and cross-section views (ground = cyan, canopy = green through red); the cross-section portrays laser pulse returns and solar background noise (no noise filtering has been applied) from a ~20m wide corridor along the top edge of the swath where returns are densest (~1 return per square meter). Top: height distributions (0.25m bins) of single photon returns accumulated over ~20 × 30m areas (white boxes) showing canopy vertical structure, distinct ground peaks, and low levels of solar background noise. Bottom: natural color aerial photograph showing approximate location of the MMLA swath (white rectangle). (Geolocated MMLA data was provided by Jan McGarry, NASA Goddard Space Flight Center.)

to photon counting ranging. In order to avoid a range bias due to photons being received but undetected during the dead-time, a very low energy per laser pulse is used so that the probability of detecting more than one photon from a target for each pulse is low. A detection probability of less than one is therefore desired, meaning, for some pulses no photon returns will be detected. By transmitting pulses at very high rates (e.g., several hundred kilohertz) individual returns can be rapidly accumulated, yielding return densities comparable to multiphoton detection conducted at lower pulse repetition rates (Figure 5.6). Although the requirement for a low probability of detection per pulse is a significant constraint, the sensitivity of SPCM detectors, unlike PMTs, is not degraded as the total dose of detected photons increases. Like PMTs, SPCM detectors are most sensitive at visible wavelengths, achieving even higher sensitivity than PMTs at 532nm, but their sensitivity falls off less rapidly with increasing wavelength (Figure 5.1), providing some capability for operation in the near-infrared. Single photon avalanche diode (SPAD) detectors that can compensate for range bias effects due to reception of multiple photons closely spaced in time have been demonstrated (Kirchner and Koidl, 1999), providing a potential alternative to the SPCM detectors.

Although primarily used to profile atmospheric clouds and aerosols, the GLAS atmospheric LiDAR channel operating at 532nm (Spinhirne et al., 2005) has demonstrated

SPCM photon-counting ranging to the Earth's surface from space (Figure 5.8). An airborne instrument, the Swath Mapping Multipolarization Photon-counting Lidar (SIMPL), based on multipulse single-photon ranging is currently in development using NASA IIP funding. SIMPL will be a multibeam instrument, operating simultaneously at 532 nm and 1064 nm wavelengths, using a 400 kHz fiber laser, a beam splitter and SPCM detectors. Received photons will be detected in channels with filters oriented parallel and perpendicular to the polarization plane of the transmit pulses in order to measure the depolarization of the reflected laser energy. Depolarization is a function of the amount of surface versus volume scattering that occurs during reflection of plane-polarized laser light. Acquiring depolarization data at visible and near-infrared wavelengths directly associated with the ranging data will contribute to classification and mapping of surface types based on their height and scattering properties. The nonranging, analog-detection Airborne Laser Polarization Sensor (ALPS) developed by NASA demonstrated differentiation of needle-leaf and broad-leaf tree species based on 532 nm and 1064 nm depolarization (Kalshoven and Dabney, 1993). With the addition of waveform-recording ranging, ALPS was renamed as the Multiwavelength Airborne Polarimetric Lidar (MAPL) and is being used to test new vegetation remote sensing applications (Tan and Narayanan, 2004; Tan et al., 2005).

5.5 Summary

ALSM is a uniquely capable method for collection of high resolution elevation data. In particular, in vegetated areas it provides measurements of ground topography and canopy structure unmatched by any other remote sensing technique. The resulting data products are used for scientific investigations and applied mapping purposes. The method used to acquire the fundamental ALSM measurement, the range to the surface, defines the characteristics of the elevation data. Scanning analog-detection systems that record full-waveforms provide three-dimensional representations of backscattered laser pulse energy at visible or near-infrared wavelengths. Analog discrete return systems detect distinct surfaces within that three-dimensional volume that reflect energy exceeding a threshold, yielding point cloud representations of surface elevations. Detection of a discrete return is dependent on instrument parameters and geometric and reflectance attributes of the target. Because laser altimeters acquire data at 0° phase angle, with parallel illumination and view angles, the received energy is a function of the retroreflectance of the target. Retro-reflectance is enhanced, compared to bidirectional reflectance, by shadow hiding and coherent backscatter opposition effects. Specular reflection from water surfaces yields a high amplitude return for data acquired near-nadir and weak or no signal strength from off-nadir angles. Photon-counting systems, using detectors that record the arrival of single photons and low-energy, short-pulse, high-repetition rate laser transmitters, have the potential to acquire data more efficiently than analog systems. Like discrete-return systems, they record individual returns that when combined form point clouds. Unlike discrete-return data, that can be biased upward due to the use of leading-edge ranging, photon counting when properly implemented yields unbiased range data that samples the height distribution of surfaces illuminated by the laser pulse. Research systems, many of which have been developed by NASA for airborne and spaceflight use, have established measurement principles and practices now used in commercially available systems that form the basis of a robust, worldwide mapping industry.

Acknowledgments

This chapter has benefited greatly from discussions over the past two decades with the many experts in laser altimetry at Goddard Space Flight Center. In particular, I would like to thank James Abshire, Bryan Blair, Jack Bufton, Phil Dabney, John Degnan, Jim Garvin, Bill Krabill, and Xiaoli Sun.

References

Abdalati, W., Krabill, W., Frederick, E., Manizade, S., Martin, C., Sonntag, J., Swift, R., Thomas, R., Wright, W., and Yungel, J., 2002. Airborne laser altimetry mapping of the Greenland ice sheet: Application to mass balance assessment, *J. Geodyn.*, 34, 391–403.

Abshire, J.B., Sun, X.L., and Afzal, R.S., 2000. Mars Orbiter Laser Altimeter: Receiver model and performance analysis, *Appl. Optic.*, 39, 2449–2460.

Abshire, J.B., Sun, X.L., Riris, H., Sirota, J.M., McGarry, J.F., Palm, S., Yi, D.H., and Liiva, P., 2005, Geoscience Laser Altimeter System (GLAS) on the ICESat mission: On-orbit measurement performance, *Geophys. Res. Lett.*, 32, L21S02, doi:10.1029/2005GL024028.

Blair, J.B., Rabine, D.L., and Hofton, M.A., 1999. The Laser Vegetation Imaging Sensor: A medium-altitude, digitisation-only, airborne laser altimeter for mapping vegetation and topography, *ISPRS J. Photogramm. Rem. Sens.*, 54, 115–122.

Brock, J.C., Wright, C.W., Sallenger, A.H., Krabill, W.B., and Swift, R.N., 2002. Basis and methods of NASA airborne topographic mapper lidar surveys for coastal studies, *J. Coast. Res.*, 18, 1–13.

Brock, J.C., Wright, C.W., Kuffner, I.B., Hernandez, R., and Thompson, P., 2006. Airborne lidar sensing of massive stony coral colonies on patch reefs in the northern Florida reef tract, *Rem. Sens. Environ.*, 104, 31–42.

Bufton, J.L., Hoge, F.E., and Swift, R.N., 1983. Airborne measurements of laser backscatter from the ocean surface, *Appl. Optic.*, 22, 2603–2618.

Bufton, J.L., Garvin, J.B., Cavanaugh, J.F., Ramos-Izquierdo, L., Clem, T.D., and Krabill, W.B., 1991. Airborne lidar for profiling of surface topography, *Opt. Eng.*, 30, 72–78.

Camacho-de Coca, F.C., Breon, F.M., Leroy, M., and Garcia-Haro, F.J., 2004. Airborne measurement of hot spot reflectance signatures, *Rem. Sens. Environ.*, 90, 63–75.

Carter, W., Shrestha, R., and Slatton, K.C., 2004. Photon-counting airborne laser swath mapping (PC-ALSM), *Proc. SPIE, 4th Int. Asia-Pacific Environ. Rem. Sens. Symp.*, 5661, 78–85.

Chin, G., et al., 2007. Lunar reconnaissance orbiter overview: The instrument suite and mission, *Space Sci. Rev.*, 129, 391–419.

Degnan, J.J., 2001. Unified approach to photon-counting microlaser rangers, transponders, and altimeters, *Surv. Geophys.*, 22, 431–447.

Degnan, J.J., 2002. Photon-counting multikilohertz microlaser altimeters for airborne and spaceborne topographic measurements, *J. Geodyn.*, 34, 503–549.

Degnan, J.J., McGarry, J., Zagwodzki, T., Dabney, P., Geiger, J., Chabot, R., Steggerda, C., Marzouk, J., and Chu, A., 2001. Design and performance of an airborne multikilohertz photon-counting, microlaser altimeter, *Proc. Land Surface Mapping and Characterization Using Laser Altimetry, Int. Arch. Photogramm. Rem. Sens.*, XXXIV3-W4, Annapolis, MD, pp. 9–16.

Garvin, J., Bufton, J., Blair, J., Harding, D., Luthcke, S., Frawley, J., and Rowlands, D., 1998. Observations of the Earth's topography from the Shuttle Laser Altimeter (SLA): Laser-pulse echo-recovery measurements of terrestrial surfaces, *Phys. Chem. Earth Solid Earth Geodes.*, 23, 1053–1068.

Hapke, B., DiMucci, D., Nelson, R., and Smythe, W., 1996. The cause of the hot spot in vegetation canopies and soils: Shadow-hiding versus coherent backscatter, *Rem. Sens. Environ.*, 58, 63–68.

Hapke, B., Nelson, R., and Smythe, W., 1998. The opposition effect of the moon: Coherent backscatter and shadow hiding, *Icarus*, 133, 89–97.

Harding, D.J., Blair, J.B., Rabine, D.L., and Still, K.L., 2000. SLICER airborne laser altimeter characterization of canopy structure and sub-canopy topography for the BOREAS Northern and Southern Study Regions: Instrument and Data Product Description, in *Technical Report Series on the Boreal Ecosystem-Atmosphere Study (BOREAS)*, F.G. Hall and J. Nickeson, Eds., NASA/TM-2000-209891, 93, 45 pp.

Harding, D.J., Lefsky, M.A., Parker, G.G., and Blair, J.B., 2001. Laser altimeter canopy height profiles—Methods and validation for closed-canopy, broadleaf forests, *Rem. Sens. Environ.*, 76, 283–297.

Harding, D.J. and Carabajal, C.C., 2005. ICESat waveform measurements of within-footprint topographic relief and vegetation vertical structure, *Geophys. Res. Lett.*, 32, L21S10, doi:10.1029/2005GL023471.

Haugerud, R., Harding, D.J., Johnson, S.Y., Harless, J.L., Weaver, C.S., and Sherrod, B.L., 2003. High-resolution topography of the Puget Lowland, Washington—A bonanza for earth science, *GSA Today*, 13, 4–10.

Helfenstein, P., Veverka, J., and Hillier, J., 1997. The lunar opposition effect: A test of alternative models, *Icarus*, 128, 2–14.

Ho, C., Albright, K.L., Bird, A.W., Bradley, J., Casperson, D.E., Hindman, M., Priedhorsky, W.C., Scarlett, W.R., Smith, R.C., Theiler, J., and Wilson, S.K., 1999. Demonstration of literal three-dimensional imaging, *Appl. Optic.*, 38, 1833–1840.

Hofton, M.A., Minster, J.B., and Blair, J.B., 2000. Decomposition of laser altimeter waveforms, *IEEE Trans. Geosci. Rem. Sens.*, 38, 1989–1996.

Kaasalainen, S., Kaasalainen, M., Mielonen, T., Suomalainen, J., Peltoniemi, J.I., and Naranen, J., 2006. Optical properties of snow in backscatter, *J. Glaciol.*, 52, 574–584.

Kaasalainen, S. and Rautiainen, M., 2005. Hot spot reflectance signatures of common boreal lichens, *J. Geophys. Res. Atmos.*, 110, D20102, doi:10.1029/2005JD005834.

Kalshoven, J.E. and Dabney, P.W., 1993. Remote-sensing of the Earth's surface with an airborne polarized laser, *IEEE Trans. Geosci. Rem. Sens.*, 31, 438–446.

Kirchner, G. and Koidl, F., 1999. Compensation of SPAD time-walk effects, *J. Optic. Pure Appl. Optic.*, 1, 163–167.

Krabill, W.B. and Martin, C.F., 1987. Aircraft positioning using global positioning system carrier phase data, *Navigation*, 34, 1–21.

Krabill, W., Abdalati, W., Frederick, E., Manizade, S., Martin, C., Sonntag, J., Swift, R., Thomas, R., Wright, W., and Yungel, J., 2000. Greenland ice sheet: High-elevation balance and peripheral thinning, *Science*, 289, 428–430.

Krabill, W.B., Abdalati, W., Frederick, E.B., Manizade, S.S., Martin, C.F., Sonntag, J.G., Swift, R.N., Thomas, R.H., and Yungel, J.G., 2002. Aircraft laser altimetry measurement of elevation changes of the greenland ice sheet: Technique and accuracy assessment, *J. Geodyn.*, 34, 357–376.

Krabill, W.B., Collins, J.G., Link, L.E., Swift, R.N., and Butler, M.L., 1984. Airborne laser topographic mapping results, *Photogramm. Eng. Rem. Sens.*, 50, 685–694.

Krabill, W.B., Thomas, R.H., Martin, C.F., Swift, R.N., and Frederick, E.B., 1995. Accuracy of airborne laser altimetry over the greenland ice-sheet, *Int. J. Rem. Sens.*, 16, 1211–1222.

Lefsky, M.A., Cohen, W.B., Parker, G.G., and Harding, D.J., 2002. Lidar remote sensing for ecosystem studies, *Bioscience*, 52, 19–30.

Nayegandhi, A., Brock, J.C., Wright, C.W., and O'Connell, M.J., 2006. Evaluating a small footprint, waveform-resolving lidar over coastal vegetation communities, *Photogramm. Eng. Rem. Sens.*, 72, 1407–1417.

Nelson, R.M., Hapke, B.W., Smythe, W.D., and Horn, L.J., 1998. Phase curves of selected particulate materials: The contribution of coherent backscattering to the opposition surge, *Icarus*, 131, 223–230.

Nelson, R.M., Hapke, B.W., Smythe, W.D., and Spilker, L.J., 2000. The opposition effect in simulated planetary regoliths. Reflectance and circular polarization ratio change at small phase angle, *Icarus*, 147, 545–558.

Neumann, G.A., Abshire, J.B., Aharonson, O., Garvin, J.B., Sun, X., and Zuber, M.T., 2003. Mars Orbiter Laser Altimeter pulse width measurements and footprint-scale roughness, *Geophys. Res. Lett.*, 30(11), 1561, doi: 10.1029/2003GL017048.

Priedhorsky, W.C., Smith, R.C., and Ho, C., 1996. Laser ranging and mapping with a photon-counting detector, *Appl. Optic.*, 35, 441–452.

Ramos-Izquierdo, L., Scott III, S., Schmidt, S., Britt, J., Mamakos, W., Trunzo, R., Cavanaugh, J., and Miller, R., 2005. Optical system design and integration of the Mercury Laser Altimeter, *Appl. Optic.*, 44, 1748–1760.

Riris, H., Sun, X., Cavanaugh, J.F., Jackson, G.B., Ramos-Izquierdo, L., Smith, D.E., and Zuber, M., 2007. The lunar orbiter laser altimeter (LOLA) on NASA's lunar reconnaissance orbiter (LRO) mission, in *Sensors and Systems for Space Applications, Proc. SPIE*, R.T. Howard and D. Richards, Eds., 6555, 1–8.

Sallenger, A.H. et al., 2003. Evaluation of airborne topographic lidar for quantifying beach changes, *J. Coast. Res.*, 19, 125–133.

Schutz, B.E., Zwally, H.J., Shuman, C.A., Hancock, D., and DiMarzio, J.P., 2005. Overview of the ICESat Mission, *Geophys. Res. Lett.*, 32, L21S01, doi:10.1029/2005GL024009.

Shkuratov, Y.G. and Helfenstein, P., 2001. The opposition effect and the quasi-fractal structure of regolith: I. Theory, *Icarus*, 152, 96–116.

Slatton, K.C., Carter, W.E., and Shrestha, R., 2005. A simulator for airborne laser swath Mapping via photon counting, in *Detection and Remediation Technologies for Mines and Mine-Like Targets, Proc. SPIE*, R.S. Harmon, J.T. Broach, and J.H. Holloway Jr., Eds., 5794, 12–20.

Smith, D.E. et al., 1999, The global topography of Mars and implications for surface evolution, *Science*, 284, 1495–1503.

Smith, D.E. et al., 2001. Mars Orbiter Laser Altimeter: Experiment summary after the first year of global mapping of Mars, *J. Geophys. Res. Plan.*, 106, 23689–23722.

Spinhirne, J.D., Palm, S.P., Hart, W.D., Hlavka, D.L., and Welton, E.J., 2005. Cloud and aerosol measurements from GLAS: Overview and initial results, *Geophys. Res. Lett.*, 32, L22S03, doi:10.1029/2005GL023507.

Sun, X., Krainak, M.A., Abshire, J.B., Spinhirne, J.D., Trottier, C., Davies, M., Dautet, H., Allan, G.R., Lukemire, A.T., and Vandiver, J.C., 2004. Space-qualified silicon avalanche-photodiode single-photon-counting modules, *J. Mod. Optic.*, 51, 1333–1350.

Sun, X., Krainak, M.A., Hasselbrack, W.B., and La Rue, R.A., 2007. Photon counting performance measurements of transfer electron InGaAsP photocathode hybrid photomultiplier tubes at 1064 nm wavelength, in *Photon Counting Applications, Quantum Optics, and Quantum Cryptography, Proc. SPIE*, I. Prochazka, A.L. Migdall, A. Pauchard, M. Dusek, M.S. Hillery, and W. Schleich, Eds., 6583, 1–14.

Tan, S.X. and Narayanan, R.M., 2004. Design and performance of a multiwavelength airborne polarimetric lidar for vegetation remote sensing, *Appl. Optic.*, 43, 2360–2368.

Tan, S.X., Narayanan, R.M., and Shetty, S.K., 2005. Polarized lidar reflectance measurements of vegetation at near-infrared and green wavelengths, *Int. J. Infrared and Millimet. Waves*, 26, 1175–1194.

Thomas, R., Frederick, E., Krabill, W., Manizade, S., and Martin, C., 2006. Progressive increase in ice loss from Greenland, *Geophys. Res. Lett.*, 33, L10503, doi:10.1029/2006GL026075.

Thomas, R. et al., 2004. Accelerated sea-level rise from West Antarctica, *Science*, 306, 255–258.

Vaughn, C.R., Bufton, J.L., Krabill, W.B., and Rabine, D., 1996. Georeferencing of airborne laser altimeter measurements, *Int. J. Rem. Sens.*, 17, 2185–2200.

Weishampel, J.F., Blair, J.B., Knox, R.G., Dubayah, R., and Clark, D.B., 2000. Volumetric lidar return patterns from an old-growth tropical rainforest canopy, *Int. J. Rem. Sens.*, 21, 409–415.

Zuber, M.T. et al., 2000. Internal structure and early thermal evolution of Mars from Mars Global Surveyor topography and gravity, *Science*, 287, 1788–1793.

Zwally, H.J. et al., 2002. ICESat's laser measurements of polar ice, atmosphere, ocean, and land, *J. Geodyn.*, 34, 405–445.

6

Georeferencing Component of LiDAR Systems

Naser El-Sheimy

CONTENTS

Airborne laser scanning (ALS) is becoming a popular choice to determine surface elevation models. ALS systems integrate several surveying technologies. While there are differences between commercial systems, the basic package remains the same: A global positioning system (GPS) receiver and an inertial navigation system (INS) as the georeferencing component, a laser range finder and a scanner as the remote sensing component. This chapter will mainly discuss the georeferencing component of an ALS.

6.1 Introduction

Aerial remote sensing, more specifically aerial photogrammetry, in its classical form of film-based optical sensors (analog) has been widely used for high accuracy mapping applications at all scales. Ground control points (GCPs) were the only required source of information for providing the georeferencing parameters and suppressing undesirable error propagation. In general, the necessity for GCPs was so evident that all operational methods relied on them. Even with the major changes in photogrammetry from analog to analytical and then to the digital mode of operation, it was taken for granted that GCPs were the only source for providing reliable georeferencing information. External georeferencing information was therefore considered as auxiliary data, indicating that it was only useful in minimizing the number of GCPs. The drawback of indirect georeferencing is the cost associated with the establishment of the GCPs. This usually represents a large portion of the overall project budget. In some cases, this cost can be prohibitive; especially when imagery is to be acquired and georeferenced in remote areas such as those found in many developing countries (for more details see Schwarz et al., 1993).

The use of auxiliary position and navigation sensor data in the georeferencing process has been extensively studied for several decades. The output of these sensors is used to determine the six parameters of exterior orientation, either completely or partially, and thus to eliminate the need for a dense GCP network. These sensors include airborne radar profile (ARP) recorders, gyros, horizon cameras, statoscope, Hiran, and Shiran. However, the use of these auxiliary data was intended only to support the georeferencing process by reducing the number of GCPs. The accuracy achieved with most of these auxiliary data was limited, so, during the last two decades the use of such auxiliary data in the georeferencing process has almost disappeared completely from photogrammetry, except for the use of statoscope (for more details see Ackermann, 1995).

This situation changed fundamentally when GPS locations of the aerial camera at the instant of exposure were included in the block adjustment of aerial triangulation. In principle, the use of GPS data made block triangulation entirely independent of GCPs. For the first time in the history of photogrammetry, the georeferencing process became autonomous, as GCPs were not necessarily required any more (Ackermann, 1995). However, this is only true in the case of the block triangulation scheme with over-lapping images. Other sensors cannot be fully georeferenced by GPS alone, examples of these sensors are pushbroom digital scanners, LiDAR systems, and imaging radar systems, which are important in kinematic mapping applications.

Georeferencing of light detection and ranging (LiDAR) data, for example, requires the instantaneous position and attitude information of each range measurement. GCPs alone are not sufficient to resolve all the three position parameters and the three orientations parameters associated with each range measurement, of which there may be thousands in a single image. The data rate of most available GPS receivers is not high enough to support the data rate required for LiDAR systems. Also, for LiDAR systems, the use of GPS in stand-alone mode is not feasible, since no support by block formation is easily realized.

Only recently has direct georeferencing become possible by integrating GPS and INS (termed GPS/INS), such that all the exterior orientation information has become available with sufficient accuracy at any instant of time (Schwarz et al., 1993). The integration of GPS and INS not only puts the georeferencing of photogrammetric data on a new level and frees it from operational restrictions, but also opens the door for new systems such as LiDAR, which would not have been possible without GPS/INS. Together with digital data recording and data processing; it allows the introduction of LiDAR mapping systems.

6.2 Kinematic Modeling—The Core of Direct Georeferencing

Kinematic modeling is the determination of a rigid body's trajectory from measurements relative to some reference coordinate frame. It combines elements of modeling, estimation, and interpolation. Modeling relates the observable positions to the abstract trajectory. Estimation uses actual observations, i.e., it adds an error process to the model and solves the resulting estimation problem in some optimal sense. Interpolation connects the discrete points resulting from the estimation process and generates a trajectory by formulating some appropriate smoothness condition.

A rigid body is a body with finite dimensions, which maintains the property that the relative positions of all its points, defined in a coordinate frame within the body, remain the same under rotation and translation (Goldstein, 1980). The general motion of a rigid body in space can be described by six parameters. They are typically chosen as three position and three orientation parameters. The modeling of rigid body motion in 3-D space can be described by Equation 6.1:

$$\boldsymbol{r}_{\mathrm{i}}^{\mathrm{m}} = \boldsymbol{r}_{\mathrm{b}}^{\mathrm{m}}(t) + \boldsymbol{R}_{\mathrm{b}}^{\mathrm{m}}(t)\boldsymbol{a}^{\mathrm{b}} \tag{6.1}$$

where

$\boldsymbol{r}_{\mathrm{i}}^{\mathrm{m}}$ are the coordinates of point (i) in the mapping frame (m-frame)

$\boldsymbol{r}_{\mathrm{b}}^{\mathrm{m}}(t)$ are the coordinates of the center of mass (b) of the rigid body in the m-frame at time (t)

$\boldsymbol{R}_{\mathrm{b}}^{\mathrm{m}}(t)$ is the rotation matrix between the body frame (b-frame) and the m-frame at time (t)

$\boldsymbol{a}^{\mathrm{b}}$ is the fixed distance between point (i) and the center of mass of the rigid body

The right-hand side of Equation 6.1 consists of a translation vector $\boldsymbol{r}_{\mathrm{b}}^{\mathrm{m}}(t)$ and a rotational component ($\boldsymbol{R}_{\mathrm{b}}^{\mathrm{m}}(t)\boldsymbol{a}^{\mathrm{b}}$). The vector $\boldsymbol{a}^{\mathrm{b}}$ can be any vector fixed in the rigid body with its origin at the center of mass of the rigid body. Its rotation is equivalent to rotation about the center of mass of the rigid body.

Figure 6.1 illustrates the basic concept. The coordinate b-frame is fixed to the body and rotates in time with respect to the coordinate m-frame in which the translation vector $\boldsymbol{r}_{\mathrm{i}}^{\mathrm{m}}$ is expressed. The m-frame is, in principle, arbitrary and can be chosen to simplify the problem formulation. The m-frame can be a system of curvilinear geodetic coordinates (latitude, longitude, height), universal transverse mercator (UTM) or 3° transverse mercator (3TM) coordinate systems, or any other earth-fixed coordinate system.

Determining the position and orientation of rigid bodies in 3-D space is, in principle, a problem of trajectory determination, which requires measuring systems with the capability to sense six independent quantities from which these parameters can be derived. Most notable among them are strapdown inertial measuring units (IMU) in an INS and differential GPS positioning and orientation systems (DGPS).

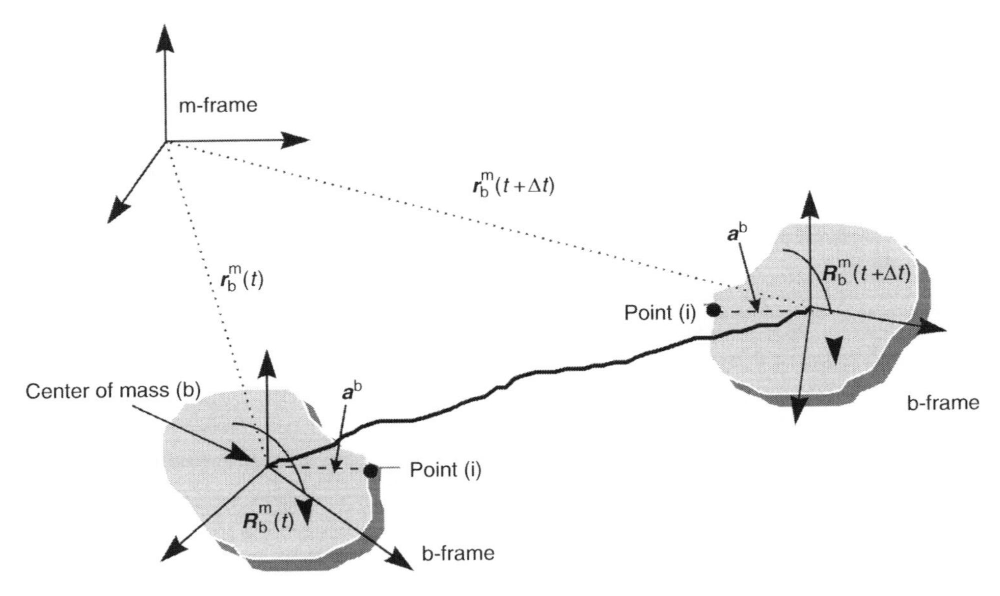

FIGURE 6.1
Modeling rigid body motion in space.

In general, the INS consists of three gyroscopes and three accelerometers. Gyroscopes are used to sense angular velocity ω_{ib}^{b} (which can be written in its skew symmetric matrix form as Ω_{ib}^{b}), which describes the rotation of the b-frame with respect to the inertial frame (i-frame), coordinated in the b-frame. The i-frame is a properly defined inertial reference frame in the Newtonian sense and, thus, can be considered as being nonaccelerating and nonrotating. Accelerometers are used to sense specific force f^{b} in the b-frame. The first set of measurements, the angular velocities ω_{ib}^{b}, are integrated with respect to time to provide orientation changes of the body relative to its initial orientation. The second data set, the specific force measurements f^{b}, are used to derive body acceleration which, after double integration with respect to time, give position differences relative to an initial position. Specific force and angular velocity can be used to determine all navigation parameters (r^{m}, v^{m}, R_{b}^{m}) required for trajectory determination by solving the following system of differential equations (see El-Sheimy, 2000):

$$\begin{pmatrix} \dot{r}^{m} \\ \dot{v}^{m} \\ \dot{R}_{b}^{m} \end{pmatrix} = \begin{pmatrix} D^{-1}v^{m} \\ R_{b}^{m} f^{b} - (2\Omega_{ie}^{m} + \Omega_{em}^{m})v^{m} + g^{m} \\ R_{b}^{m}(\Omega_{ib}^{b} - \Omega_{im}^{b}) \end{pmatrix} \tag{6.2}$$

To solve the system, the observables f^{b} and ω_{ib}^{b} are needed as well as the scaling matrix D^{-1}, the gravity vector g^{m}, the Earth rotation rate ω_{ie}^{m} and the dimensions of the implied reference ellipsoid. The gravity vector is normally approximated by the normal gravity field, while the Earth's rotation is assumed to be known with sufficient accuracy. The scaling matrix D^{-1} is obtained in the integration process using the implied reference ellipsoid.

GPS, on the other hand, is used for trajectory determination. The system outputs in this case are ranges and range rates between the satellites and receiver, derived from carrier phase data. The models that relate the position and velocity with the measurements are well known. In GPS stand-alone mode, a multiantenna system can be used to provide both position and attitude. The feasibility of attitude determination using multiantenna systems has been shown for applications not requiring the highest accuracy (see Cohen and Parkinson, 1992; El-Mowafy and Schwarz, 1994). Similar to the INS model, the GPS trajectory equation can be written in state vector form:

$$\begin{pmatrix} \dot{r}^{m} \\ \dot{v}^{m} \\ \dot{R}_{b}^{m} \end{pmatrix} = \begin{pmatrix} v^{m} \\ 0 \\ R_{b}^{m}\Omega_{mb}^{b} \end{pmatrix} \tag{6.3}$$

In Equation 6.3, the angular velocities in the body frame are obtained by differencing between antennas, satellites, and epochs. Note that the translation parameters of the trajectory are obtained by differencing between the master station receiver and the rover receiver, while the rotational parameters are obtained by differencing between the rover receivers only.

Both INS and GPS are, in principle, capable of determining position and attitude of the rigid body. In practice, due to the double integration of the INS acceleration data, the time-dependent position errors will quickly exceed the accuracy specifications for many trajectory determination applications. Frequent updating is, therefore, needed to achieve the required accuracies. GPS on the other hand, can deliver excellent position accuracy, but has the problem of cycle slips, which are, in essence, gross errors leading to a discontinuity in the trajectory. The combination of the two measuring systems, therefore, offers a number of advantages. In the absence of cycle slips, the excellent positioning accuracy of differential GPS can be used to provide frequent updates for the inertial system. The inertial sensors orientation information and the precise short-term position and velocity can be used for cycle slip detection and correction. In general, the fact that nine independent measurements are available for the determination of the six required trajectory parameters greatly enhances the reliability of the system. To optimally combine the redundant information, a Kalman filtering scheme is used whereby the inertial state vector is regularly updated by GPS measurement.

6.3 Development of Direct Georeferencing Technology

Direct georeferencing is the determination of time-variable position and orientation parameters for a mobile digital imager. The most common technologies used for this purpose today are satellite positioning by GPS and inertial navigation using an IMU. Although each technology can in principle determine both position and orientation, they are usually integrated in such a way that the GPS receiver is the main position sensor, while the IMU is the main orientation sensor. The orientation accuracy of an IMU is largely determined by the gyro drift rates, typically described by a bias (constant drift rate), the short-term bias stability, and the angle random walk. Typically, four classes of gyros are distinguished according to their constant drift rate, namely

1. Strategic gyros (0.0005–0.0010 deg./h or about 1 deg./month)
2. Navigation-grade gyros (0.002–0.015 deg./h or about 1 deg./week)
3. Tactical gyros (0.1–10 deg./h or about 1 deg./h)
4. Tow-accuracy gyros (100–10,000 deg./h or about 1 deg./s)

Only navigation and tactical grade gyros have been implemented in the georeferencing components of LiDAR systems. Operational testing of direct georeferencing started in the early 1990s (see, for instance, Cannon and Schwarz, 1990; Cosandier et al., 1993; Bossler, 1996; Toth, 1997) for airborne applications, and Bossler et al. (1993) and Lapucha et al. (1990) for land-vehicle applications). These early experiments were done by integrating differential GPS with a navigation-grade IMU (accelerometer bias: 2–3 × 10^{-4} m/s^2, gyro bias: 0.003 deg./h) and by including the derived coordinates and attitude (pitch, roll, and azimuth) into a photogrammetric block adjustment. Although GPS was not fully operational at that time, results obtained by using GPS in differential kinematic mode were promising enough to pursue this development. As GPS became fully operational, the INS/DGPS georeferencing system was integrated with a number of different imaging sensors. Among them were the Casi sensor manufactured by Itres Research Ltd. (Cosandier et al., 1993) and a set of CCD cameras (see El-Sheimy and Schwarz, 1993). By the end of 1993 experimental systems for mobile mapping existed for both airborne and land vehicles. The evolution of the georeferencing technology during the past decade was driven by ongoing refinement and miniaturization of GPS-receiver hardware and the use of low and medium cost IMUs that became available in the mid-1990s. Only the latter development will be discussed here.

The inertial systems used in INS/GPS integration in the early 1990s were predominantly navigation-grade systems, typically strapdown systems of the ring-laser type.

When integrated with DGPS, they provided position and attitude accuracies sufficient for all accuracy classes envisaged at that time. These systems came, however, with a considerable price tag (about US$130,000 at that time). With the rapidly falling cost of GPS-receiver technology, the INS became the most expensive component of the georeferencing system. Since navigation-grade accuracy was not required for the bulk of the low and medium accuracy applications, the emergence of low-cost IMU in the mid-1990s provided a solution to this problem. These systems came as an assembly of solid state inertial sensors with analog read-outs and a postcompensation accuracy of about 10 deg./h for gyro drifts and about 10^{-2} m/s^2 for accelerometer biases. Prices ranged between US$10,000 and 20,000 and the user had to add the A/D portion and the navigation software. Systems of this kind were obviously not suited as stand-alone navigation systems because of their rapid position error accumulation. However, when provided with high-rate position and velocity updates from differential GPS (1 s pseudorange solutions), the error growth could be kept in bounds and the position and attitude results from the integrated solution were suitable for low and medium accuracy applications (for details on system design and performance; see Bäumker and Mattissek, 1992; Lipman, 1992; Bader, 1993, among others).

With the rapid improvement of fiber optic gyro performance, the sensor accuracy of a number of these systems has improved by about an order of magnitude (to 1 deg./h and 10^{-1} m/s^2) in the past 5 years. Typical cost are about US$30,000. Beside the increased accuracy, these systems are more user-friendly and offer a number of interesting options. When integrated with a DGPS phase solution the resulting position and attitude are close to what is required for the high-accuracy class of applications. When aiming at highest possible accuracy these systems are usually equipped with a dual-antenna GPS, aligned with the forward direction of the vehicle. This arrangement provides regular azimuth updates to the integrated solution and bounds the azimuth drift. This is of particular importance for flights flown at constant velocity along straight lines, as is the case for photogrammetric blocks. Commercialization of direct georeferencing systems for all application areas has been done by the Applanix Corporation (now a subsidiary of Trimble, see www.applanix.com) and IGI mbH (Ingenieur-Gesellschaft fuer Interfaces (see http://www.igi-systems.com/index.htm). In general, the position and orientation accuracy achieved with these systems is sufficient for all but the most stringent accuracy requirements.

6.4 INS Equations of Motion

Inertial positioning is based on the simple principle that differences in position can be determined by a double integration of acceleration, sensed as a function of time, in a well-defined and stable coordinate frame, i.e., by

$$\Delta \boldsymbol{r}(t) = \boldsymbol{r}(t) - \boldsymbol{r}(t_o) = \int_{t_o}^{t} \int \boldsymbol{a}(\tau)\,\mathrm{d}\tau\,\mathrm{d}\tau \tag{6.4}$$

where

- $\boldsymbol{r}(t_o)$ is the initial point of the trajectory
- $\boldsymbol{a}(\tau)$ is the acceleration along the trajectory obtained from inertial sensor measurements in the coordinate frame prescribed by $\boldsymbol{r}(t)$

Note that the practical implementation of the concept is rather complex. It requires the transformation between a stable Earth-fixed coordinate frame, used for the integration, and the measurement frame defined by the sensitive axes of the IMU. The stable Earth-fixed coordinate frame is often chosen as a local-level frame (ℓ) (ℓ-frame—the z-axis is normal to the reference ellipsoid pointing upwards, the y-axis pointing towards geodetic north. The x-axis completes a right-handed system by pointing east and can be either established mechanically inside the IMU (stable platform concept) or numerically (strapdown concept). In the following, only the strapdown concept will be treated. Since inertial sensor measurements are always made with respect to an inertial reference frame, the rotational dynamics between the body frame of the IMU, the inertial reference frame, and different Earth-fixed frames is essential for deriving the acceleration $\boldsymbol{a}(\tau)$ used in Equation 6.4—this is typically achieved through integration of the gyro measurements.

The raw measurements from accelerometers and gyros are specific forces and angular velocities along and about the three axes of the body frame (b-frame), respectively. The navigation frame is where the data integration is performed. The local-level frame is often selected as the navigation frame for the following reason (El-Sheimy, 2000):

1. The definition of the local-level frame is based on the normal to the reference ellipsoid; as a result, the geodetic coordinate difference $\{\Delta\varphi, \Delta\lambda, \Delta h\}$ can be applied as the output of the system.
2. The axes of the local-level frame (NED) are aligned with the local north, east, and down directions. Therefore, the attitude angles (roll, pitch, and azimuth) can be obtained directly as an output of the mechanization equations.

According to Newton's second law of motion, the fundamental equation for the motion of a particle in the field of the earth, expressed in an inertial frame, is of the form

$$\ddot{r}^{\mathrm{i}} = f^{\mathrm{i}} + \bar{g}^{\mathrm{i}} \tag{6.5}$$

where
$\ddot{r}^{\mathrm{i}}$ is the acceleration vector
f^{i} is the specific force vector
$\bar{g}^{\mathrm{i}}$ is the gravitational vector

Equation 6.5 of motion can be transformed into local-level frame and can be expressed as a set of first order differential equation (see Shin, 2001).

$$\begin{bmatrix} \dot{r}^{\ell} \\ \dot{v}^{\ell} \\ \dot{C}_{\mathrm{b}}^{\ell} \end{bmatrix} = \begin{bmatrix} D^{-1}v^{\ell} \\ C_{\mathrm{b}}^{\ell} f^{\mathrm{b}} - (2\omega_{\mathrm{ie}}^{\ell} + \omega_{\mathrm{el}}^{\ell}) \times v^{\ell} + g^{\ell} \\ C_{\mathrm{b}}^{\ell}(\Omega_{\mathrm{ib}}^{\mathrm{b}} - \Omega_{\mathrm{i}\ell}^{\mathrm{b}}) \end{bmatrix} \tag{6.6}$$

$$D^{-1} = \begin{bmatrix} \dfrac{1}{M+h} & 0 & 0 \\ 0 & \dfrac{1}{(N+h)\cos\varphi} & 0 \\ 0 & 0 & -1 \end{bmatrix} \tag{6.7}$$

The specific force f^b is the raw output measured by the accelerometer and is defined as the difference between the true acceleration in space and the gravitational acceleration. The rotation matrix from b-frame to ℓ-frame, C_b^ℓ, is given as

$$C_\ell^b = R_x(\phi)R_y(\theta)R_z(\psi) \tag{6.8}$$

where ϕ, θ, and ψ are the three components of the Euler rotation angles roll, pitch, and azimuth, respectively, between the ℓ-frame and the b-frame. Similarly, the rotation matrix from the b-frame to the ℓ-frame can be obtained via the orthogonality criteria of direction cosine matrices (DCM):

$$\begin{aligned} C_b^l &= (C_\ell^b)^{-1} = (C_\ell^b)^T = R_z(-\psi)R_y(-\theta)R_z(-\phi) \\ &= \begin{bmatrix} \cos\psi & -\sin\psi & 0 \\ \sin\psi & \cos\psi & 0 \\ 0 & 0 & 1 \end{bmatrix} \begin{bmatrix} \cos\theta & 0 & \sin\theta \\ 0 & 1 & 0 \\ -\sin\theta & 0 & \cos\theta \end{bmatrix} \begin{bmatrix} 1 & 0 & 0 \\ 0 & \cos\phi & -\sin\phi \\ 0 & \sin\phi & \cos\phi \end{bmatrix} \\ &= \begin{bmatrix} \cos\theta\cos\psi & -\cos\phi\sin\psi + \sin\phi\sin\theta\cos\psi & \sin\phi\sin\psi + \cos\phi\sin\theta\cos\psi \\ \cos\theta\sin\psi & \cos\phi\cos\psi + \sin\phi\sin\theta\sin\psi & -\sin\phi\cos\psi + \cos\phi\sin\theta\sin\psi \\ -\sin\theta & \sin\phi\cos\theta & \cos\phi\cos\theta \end{bmatrix} \end{aligned} \tag{6.9}$$

M and N are radii of curvature in the meridian and prime vertical, respectively and can be expressed as follows:

$$N = \frac{a}{(1-e^2\sin^2\varphi)^{\frac{1}{2}}} \tag{6.10}$$

$$M = \frac{a(1-e^2)}{(1-e^2\sin^2\varphi)^{\frac{3}{2}}} \tag{6.11}$$

where a and e are the semi-major axis and linear eccentricity of the reference ellipsoid, respectively.

The position vector in the ℓ-frame is given by curvilinear coordinates that contain latitude, φ, longitude, λ, and ellipsoidal height, h:

$$r^l = [\varphi \quad \lambda \quad h]^T \tag{6.12}$$

The velocity vector in the ℓ-frame is given as follows:

$$v^l = \begin{bmatrix} v_N \\ v_E \\ v_D \end{bmatrix} = \begin{bmatrix} (M+h) & 0 & 0 \\ 0 & (N+h)\cos\varphi & 0 \\ 0 & 0 & -1 \end{bmatrix} \begin{bmatrix} \dot{\varphi} \\ \dot{\lambda} \\ \dot{h} \end{bmatrix} \tag{6.13}$$

where v_N, v_E, and v_D are north, east, and downward velocity components.

TABLE 6.1

Constant Coefficients for Normal Gravity

$a_1\,(\mathrm{m/s^2})$	9.7803267715	$a_4\,(\mathrm{m/s^2})$	−0.0000030876910891
$a_2\,(\mathrm{m/s^2})$	0.0052790414	$a_5\,(\mathrm{m/s^2})$	0.0000000043977311
$a_3\,(\mathrm{m/s^2})$	0.0000232718	$a_6\,(\mathrm{m/s^2})$	0.0000000000007211

The gravity vector in the local-level frame, g^{ℓ}, is expressed as the normal gravity at the geodetic latitude φ, and ellipsoidal height h (El-Sheimy, 2000)

$$g^{\ell} = [0 \quad 0 \quad g]^{\mathrm{T}}, \quad g = a_1(1 + a_2 \sin^2\varphi + a_3 \sin^4\varphi) + (a_4 + a_5 \sin^2\varphi)h + a_6 h^2 \tag{6.14}$$

where a_1–a_6 are constant values and are listed in the Table 6.1.

The rotation rate vector of the e-frame with respect to the i-frame projected to the e-frame is expressed as follow:

$$\omega_{\mathrm{ie}}^{\mathrm{e}} = \begin{bmatrix} 0 \\ 0 \\ \omega_{\mathrm{e}} \end{bmatrix} \tag{6.15}$$

Projecting the vector to the ℓ-frame utilizing is given as

$$\omega_{\mathrm{ie}}^{\ell} = C_{\mathrm{e}}^{\ell}\omega_{\mathrm{ie}}^{\mathrm{e}} = \begin{bmatrix} \omega_{\mathrm{e}}\cos\varphi \\ 0 \\ -\omega_{\mathrm{e}}\sin\varphi \end{bmatrix} \tag{6.16}$$

The transport rate represents the turn rate of the ℓ-frame with respect to the e-frame and is given using the rate of change of latitude and longitude which are given as follows:

$$w_{\mathrm{el}}^{\ell} = \begin{bmatrix} \dot{\lambda}\cos\varphi \\ -\dot{\varphi} \\ -\dot{\lambda}\sin\varphi \end{bmatrix} = \begin{bmatrix} v_{\mathrm{E}}/(N+h) \\ -v_{\mathrm{N}}/(M+h) \\ -v_{\mathrm{E}}\tan\varphi/(N+h) \end{bmatrix} \tag{6.17}$$

$\Omega_{\mathrm{i}}^{\ell}$ and $\Omega_{\mathrm{el}}^{\ell}$ are skew symmetric matrices corresponding to $\omega_{\mathrm{ie}}^{\ell}$and $\omega_{\mathrm{e}\ell}^{\ell}$, respectively.

The angular velocity, $\omega_{\mathrm{ib}}^{\mathrm{b}}$, is the raw output measured by the gyros and $\Omega_{\mathrm{ib}}^{\mathrm{b}}$is its skew-symmetric matrix.

$$\omega_{\mathrm{ib}}^{\mathrm{b}} = \begin{bmatrix} \omega_{\mathrm{x}} & \omega_{\mathrm{y}} & \omega_{\mathrm{z}} \end{bmatrix}^{\mathrm{T}} \tag{6.18}$$

The angular velocity $\Omega_{\mathrm{i}\ell}^{\mathrm{b}}$ is subtracted from $\Omega_{\mathrm{ib}}^{\mathrm{b}}$ to remove (1) earth rotation rate and (2) orientation change of the local level frame. As a result, $\Omega_{\mathrm{i}\ell}^{\mathrm{b}}$ is expressed as follows:

$$\Omega_{\mathrm{i}\ell}^{\mathrm{b}} = \Omega_{\mathrm{ie}}^{\mathrm{b}} + \Omega_{\mathrm{e}\ell}^{\mathrm{b}} \tag{6.19}$$

Thus, $\omega_{i\ell}^{b}$ can be obtained as follows:

$$\begin{aligned}\omega_{i\ell}^{b} &= C_{\ell}^{b}(\omega_{ie}^{\ell} + \omega_{e\ell}^{\ell}) = C_{\ell}^{b}\omega_{i\ell}^{\ell} \\ &= C_{\ell}^{b}\begin{bmatrix}\omega_{e}\cos\varphi + \dfrac{v_{E}}{(N+h)} & \dfrac{-v_{N}}{(M+h)} & -\omega_{e}\sin\varphi - \dfrac{v_{E}\tan\varphi}{(N+h)}\end{bmatrix}^{T}\end{aligned} \tag{6.20}$$

Consequently, the $\Omega_{i\ell}^{b}$ can be obtained through $\omega_{i\ell}^{b}$.

6.5 INS Mechanization Equations

Solving the vector differential Equation 6.6 will result in a time-variable state vector with kinematic subvectors for position, velocity, and attitude. In the literature, the integration algorithms are often called the mechanization equations. This term dates back to the time when stable platform systems were the norm and a specific platform orientation was maintained using mechanical systems, as for instance, in the local-level system or the space-stable system. The accelerometers mounted on the platform were isolated from rotational vehicle dynamics and the computation of the transformation matrix was replaced by a system of gimbals, supplemented by platform commands. Although the algorithms used for strapdown inertial systems can be considered as an analytical form of the platform mechanization, we will use the term integration equations in the following to distinguish between stable platform and strapdown systems.

The integration equations are applied to solve the equations of motion in order to obtain the necessary position, velocity and attitude increment. In practice, strapdown IMUs work in discrete form and provide angle and velocity increments, Δv_{f} and $\Delta\theta_{ib}^{b}$, respectively, over time interval t_k to t_{k+1} in the body frame. Combining these data with the initial condition of the system, it is possible to provide the navigation information. Figure 6.2 shows the algorithmic flowchart of the integration in the local-level frame.

The integration equations consist of three basic steps:

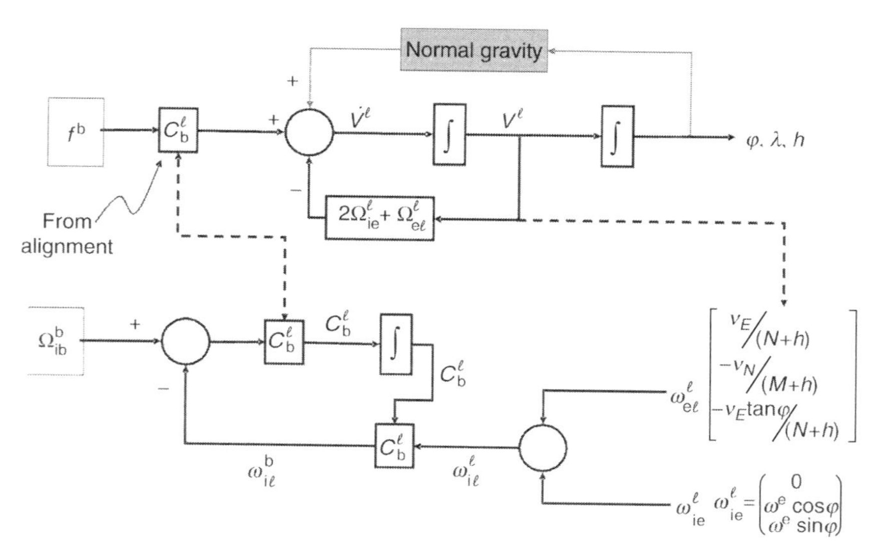

FIGURE 6.2
Flowchart of the integration process in the ℓ-frame.

1. Sensors Error Compensation: The accelerometer and gyros outputs are corrected for the constant sensor errors utilizing the following equations

$$\Delta v_{\mathrm{f}} = \begin{bmatrix} \frac{1}{(1+S_{\mathrm{gx}})} & 0 & 0 \\ 0 & \frac{1}{(1+S_{\mathrm{gy}})} & 0 \\ 0 & 0 & \frac{1}{(1+S_{\mathrm{gz}})} \end{bmatrix} (\Delta\tilde{v}_{\mathrm{f}} - b_{\mathrm{g}}\Delta t) \tag{6.21}$$

$$\Delta\theta^{\mathrm{b}}_{\mathrm{ib}} = \Delta\tilde{\theta}^{\mathrm{b}}_{\mathrm{ib}} - b_{\mathrm{w}}\Delta t \tag{6.22}$$

where

S_{gx}, S_{gy}, and S_{gz} are the scale factors of the accelerometer
b_g and b_w are the bias of the accelerometer and gyro, respectively
$\Delta\tilde{v}_f$ and $\Delta\tilde{\theta}^b_{ib}$ are the raw output of accelerometers and gyros, respectively
Δv_f and $\Delta\theta^b_{ib}$ are compensated output of accelerometers and gyros, respectively
$\Delta t = t_{k+1} - t_k$ = Time interval between two consecutive computation cycles t_k and t_{k+1}

2. Attitude Integration: The transformation matrix C^{ℓ}_{b} can be updated by solving the following set of differential equations

$$\dot{C}^{\ell}_{\mathrm{b}} = C^{\ell}_{\mathrm{b}}(\Omega^{\mathrm{b}}_{\mathrm{ib}} - \Omega^{\mathrm{b}}_{\mathrm{i}\ell}) \tag{6.23}$$

The body frame angular increment with respect to navigation frame (l-frame) is given

$$\begin{aligned} \Delta\theta^{\mathrm{b}}_{\ell\mathrm{b}} &= [\Delta\theta_{\mathrm{x}} \quad \Delta\theta_{\mathrm{y}} \quad \Delta\theta_{\mathrm{z}}] \\ &= \Delta\theta^{\mathrm{b}}_{\mathrm{ib}} - C^{\mathrm{b}}_{\ell}(\omega^{\ell}_{\mathrm{ie}} + \omega^{\ell}_{\mathrm{e}\ell})\Delta t \end{aligned} \tag{6.24}$$

The magnitude of the angular increment is given

$$\Delta\theta = \sqrt{\Delta\theta^2_{\mathrm{x}} + \Delta\theta^2_{\mathrm{y}} + \Delta\theta^2_{\mathrm{z}}} \tag{6.25}$$

Due to its computational efficiency, quaternion integration is again chosen for the transformation matrix update. Equations 6.24 and 6.25 are applied to update the quaternion (see El-Sheimy, 2000 for the detailed definition and properties of quaternions).

$$\begin{bmatrix} q_1(t_{k+1}) \\ q_2(t_{k+1}) \\ q_3(t_{k+1}) \\ q_4(t_{k+1}) \end{bmatrix} = \begin{bmatrix} q_1(t_k) \\ q_2(t_k) \\ q_3(t_k) \\ q_4(t_k) \end{bmatrix} + 0.5 \begin{bmatrix} c & s\Delta\theta_{\mathrm{z}} & -s\Delta\theta_{\mathrm{y}} & s\Delta\theta_{\mathrm{x}} \\ -s\Delta\theta_{\mathrm{z}} & c & s\Delta\theta_{\mathrm{x}} & s\Delta\theta_{\mathrm{y}} \\ s\Delta\theta_{\mathrm{y}} & -s\Delta\theta_{\mathrm{x}} & c & s\Delta\theta_{\mathrm{z}} \\ -s\Delta\theta_{\mathrm{x}} & -s\Delta\theta_{\mathrm{y}} & -s\Delta\theta_{\mathrm{z}} & c \end{bmatrix} \begin{bmatrix} q_1(t_k) \\ q_2(t_k) \\ q_3(t_k) \\ q_4(t_k) \end{bmatrix} \tag{6.26}$$

The parameters s and c are given as follows:

$$
\begin{aligned}
s &= 1 - \frac{\Delta\theta^2}{24} + \frac{\Delta\theta^4}{1920} + \cdots \\
c &= -\frac{\Delta\theta^2}{4} + \frac{\Delta\theta^4}{192} + \cdots
\end{aligned}
\tag{6.27}
$$

The initial value of the quaternion is obtained after determining the initial DCM using Equation 6.28 with the computed initial attitudes during alignment process.

$$
\begin{bmatrix} q_1 \\ q_2 \\ q_3 \\ q_4 \end{bmatrix} = \begin{bmatrix} 0.25(C_{32} - C_{23}) / 0.5\sqrt{1 + C_{11} + C_{22} + C_{33}} \\ 0.25(C_{13} - C_{31}) / 0.5\sqrt{1 + C_{11} + C_{22} + C_{33}} \\ 0.25(C_{21} - C_{12}) / 0.5\sqrt{1 + C_{11} + C_{22} + C_{33}} \\ 0.5\sqrt{1 + C_{11} + C_{22} + C_{33}} \end{bmatrix} \tag{6.28}
$$

The DCM is updated as follows:

$$
C_b' = \begin{bmatrix} (q_1^2 - q_2^2 - q_3^2 + q_4^2) & 2(q_1q_2 - q_3q_4) & 2(q_1q_3 - q_2q_4) \\ 2(q_1q_2 + q_3q_4) & (q_2^2 - q_1^2 - q_3^2 + q_4^2) & 2(q_2q_3 - q_1q_4) \\ 2(q_1q_3 - q_2q_4) & 2(q_2q_3 + q_1q_4) & (q_3^2 - q_1^2 - q_2^2 + q_4^2) \end{bmatrix} \tag{6.29}
$$

The Euler angle of the attitudes, roll, pitch, and azimuth, are then given as follows

$$
\theta = -\tan^{-1}\left[\frac{C_{31}}{\sqrt{1 - C_{31}^2}}\right] \tag{6.30}
$$

$$
\phi = a\tan 2(C_{32}, C_{33}) \tag{6.31}
$$

$$
\psi = a\tan 2(C_{21}, C_{11}) \tag{6.32}
$$

Where $C'_{i,j}$s, $1 \le i, j \le 3$ are the (i, j)th element of the DCM matrix and $a\tan 2$ is a four quadrant inverse tangent function.

3. Velocity and Position Integration: The body frame velocity increment due to the specific force is transformed to the navigation frame using Equation 6.33:

$$
\Delta v_f' = C_b' \begin{bmatrix} 1 & 0.5\Delta\theta_z & -0.5\Delta\theta_y \\ -0.5\Delta\theta_z & 1 & 0.5\Delta\theta_x \\ 0.5\Delta\theta_y & -0.5\Delta\theta_x & 1 \end{bmatrix} \Delta v_f^b \tag{6.33}
$$

The first order sculling correction is applied utilizing Equation 6.33. The velocity increment is obtained by applying the gravity and the Coriolis correction:

$$\Delta v^{\ell} = \Delta v_{\mathrm{f}}^{\ell} - (2\omega_{\mathrm{ie}}^{\ell} + \omega_{\mathrm{el}}^{\ell}) \times v^{\ell}\Delta t + g^{\ell}\Delta t \tag{6.34}$$

The velocity integration is then given as

$$v_{k+1}^{\ell} = v_k^{\ell} + \Delta v_{k+1}^{\ell} \tag{6.35}$$

The position integration is obtained using the second-order Runge-Kutta method:

$$r_{k+1}^{\ell} = r_k^{\ell} + 0.5 \begin{bmatrix} \dfrac{1}{(M+h)} & 0 & 0 \\ 0 & \dfrac{1}{(N+h)\cos\varphi} & 0 \\ 0 & 0 & -1 \end{bmatrix} (v_k^{\ell} + v_{k+1}^{\ell})\Delta t \tag{6.36}$$

6.6 GPS/INS Integration

6.6.1 Integration Strategies

The integration of GPS and INS has been investigated for several years in various applications including navigation, mobile mapping, airborne gravimetry, and guidance and control. Both systems are complimentary and their integration overcomes their individual limitations. In GPS/INS systems, the GPS provides position and velocity and the INS provides attitude information. In addition, the INS can provide very accurate position and velocity with a high data rate between GPS measurement fixes. Therefore, INS is used to detect and correct GPS cycle slips and also for navigation during GPS signal loss of lock. Finally, the GPS is used for the in-motion calibration of the INS accelerometer and gyro sensor residual errors. For all INS/GPS applications, navigation information parameters are obtained using kinematic modeling. Thus, the state-space representation is implemented in the mathematical modeling of INS, GPS and INS/GPS systems. In this context, the Kalman Filter (KF) has been commonly used as an optimal estimator and compensator of the INS/GPS system errors.

Two integration strategies can be implemented at the software level using the Kalman filter approach. In the first one, a common state vector is used to model both the INS and the GPS errors. This is often called the centralized filter or tightly coupled approach. It has been applied with good success in Knight (1997); Scherzinger and Woolven (1996), and Moafipoor et al. (2004). If double differenced carrier phases are used for GPS, the state vector is of the form:

$$\mathbf{x} = (\varepsilon, \delta\mathbf{r}, \delta\mathbf{v}, \mathbf{d}, \mathbf{b}, \delta\mathbf{N})^{\mathrm{T}} \tag{6.37}$$

where

- $\boldsymbol{\varepsilon}$ is the vector of attitude errors
- $\delta\mathbf{r}$ is the vector of position errors
- $\delta\mathbf{v}$ is the vector of velocity errors
- $\mathbf{d}$ is the vector gyro drift about gyro axes
- $\mathbf{b}$ is the vector of accelerometer bias
- $\delta\mathbf{N}$ the vector of double differenced ambiguities

All vectors, except $\delta\mathbf{N}$, have three components; $\delta\mathbf{N}$ has $(n - 1)$ component, where n is the number of satellites. INS measurements are used to determine the reference trajectory and GPS measurements to update the solution and estimate the state vector components. Cycle slips can be easily detected in this approach because they will show up as statistically significant differences between the predicted and the actual GPS measurements. Cycle slip correction is possible in various ways. One can either use a filtering approach as implemented in Wong (1988) or one can directly correct the cycle slip by using INS measurements. In the latter case, the ambiguity parameters do not show up in the state vector but appear as corrections to the phase measurements.

In the second approach, different filters are run simultaneously and interact only occasionally. This is often called the decentralized filter approach. For a discussion of the algorithm, see Hashemipour et al. (1988) and for the implementation, see Scherzinger (2002) and Shin (2005). It has advantages in terms of data integrity and speed but it is more complex in terms of program management. Again assuming only INS and GPS measurements, two different state vectors are defined, one for INS and one for GPS. They are of the form

$$\begin{aligned} \mathbf{x}_{\mathrm{I}} &= (\varepsilon, \delta\mathbf{r}, \delta\mathbf{v}, \delta\mathbf{d}, \mathbf{b})^{\mathrm{T}} \\ \mathbf{x}_{\mathrm{G}} &= (\delta\mathbf{r}, \delta\mathbf{v}, \delta\mathbf{N})^{\mathrm{T}} \end{aligned} \tag{6.38}$$

The interaction between master filter and local filter characterizes the decentralized filtering scheme. The local Kalman filter estimates a trajectory from GPS data only. In this case, the derived velocities are used for prediction and the carrier phase measurements for updates. The master filter estimates the trajectory from INS data only. At distinct epochs, the results from the local filter are fused with those of the master filter to obtain an optimal trajectory estimation. The approach has the advantage that each data stream is treated independently and blunders in one data set can be detected and corrected by using results from the other. The estimates of each local filter are locally best while the master filter gives a globally best estimate.

6.6.2 Filter Implementation Strategies

Since Kalman introduced his optimal estimation filter (Kalman, 1960), several KF approaches have been implemented for INS/GPS integration. In the early stages of INS and DGPS integration for high accuracy civil navigation applications (i.e., with benign dynamics), IMUs used were navigation-grade. At that time, the linearized KF (LKF) was commonly applied. With the recent advent in inertial sensor technologies, tactical-grade IMUs and consumer grade IMUs (gyro drift 100–10,000 deg./h) are currently used in several navigation applications. Therefore, for better error control with such systems, the LKF has been replaced by the extended KF (EKF). For low-cost IMUs such as micro-electro-mechanical systems (MEMS), the sensors have very low accuracy, which causes large navigation errors. This is especially true regarding the initial uncertainties.

Hence, the unscented KF (UKF) has been proposed lately for INS/GPS integration since it is able to cope with such large initial errors (Julier and Uhlmann, 2002).

In INS/GPS navigation, both the LKF and EKF algorithms are similar in that the INS position, velocity and attitude error models (i.e., the KF navigation error state vector) are obtained by linearizing the INS integration equations while the INS sensor error models (i.e., the KF sensor error state vector) are represented using a stochastic process. For more details about different possible INS sensor stochastic error models (see Nassar and El-Sheimy, 2005). In addition, a similar linearization procedure is performed on the observation update model to obtain the measurement update state vector. The error state vector (**x**) and the corresponding update equations are represented by linearized state-space models of the form:

$$\mathbf{x}_k = \mathbf{\Phi}_{k,k-1}\mathbf{x}_{k-1} + \mathbf{w}_{k-1} \tag{6.39}$$

$$\mathbf{z}_k = \mathbf{H}_k\mathbf{x}_k + \mathbf{e}_k, \tag{6.40}$$

where

k is a time epoch
$\mathbf{\Phi}$ is the system state transition matrix
$\mathbf{w}$ is the vector of the system input noise
$\mathbf{z}$ is the update measurements
$\mathbf{H}$ is the measurements design matrix
$\mathbf{e}$ is the update measurements noise

The LKF or EKF approaches can be then summarized as

$$\hat{\mathbf{x}}_k^- = \mathbf{\Phi}_{k-1}\hat{\mathbf{x}}_{k-1}^+ \tag{6.41}$$

$$\mathbf{P}_k^- = \mathbf{\Phi}_{k-1}\mathbf{P}_{k-1}^+\mathbf{\Phi}_{k-1}^{\mathrm{T}} + \mathbf{\Phi}_{k-1} \tag{6.42}$$

$$\hat{\mathbf{x}}_k^+ = \hat{\mathbf{x}}_k^- + \mathbf{K}_k(\mathbf{z}_k - \mathbf{H}_k\hat{\mathbf{x}}_k^-) \tag{6.43}$$

$$\mathbf{P}_k^{vv} = (\mathbf{H}_k\mathbf{P}_k^-\mathbf{H}_k^{\mathrm{T}} + \mathbf{R}_k) \tag{6.44}$$

$$\mathbf{K}_k = \mathbf{P}_k^-\mathbf{H}_k^{\mathrm{T}}(\mathbf{P}_k^{vv})^{-1} \tag{6.45}$$

$$\mathbf{P}_k^+ = (\mathbf{I} - \mathbf{K}_k\mathbf{H}_k)\mathbf{P}_k^- \tag{6.46}$$

where

$\mathbf{P}$ is the covariance matrix of the estimated state vector
$\mathbf{Q}$ is the covariance matrix associated with $\mathbf{w}$
$\mathbf{R}$ is the covariance matrix associated with $\mathbf{e}$
$\mathbf{P}^{vv}$ is called the covariance of the innovation sequence
$\mathbf{K}$ is the Kalman gain matrix
– and + (superscripts) are related to predicted and updated epochs, respectively

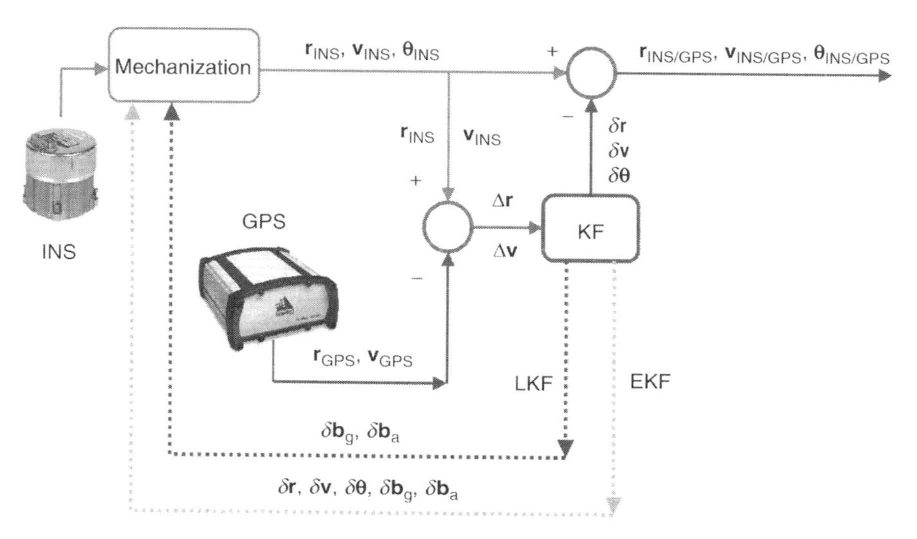

FIGURE 6.3
LKF and EKF in INS/GPS integration.

The derivation of such equations can be found in Gelb (1974) and Brown and Hwang (1997). As mentioned earlier, the update measurements (**z**) are the differences between the GPS position and velocity and the corresponding INS position and velocity.

The LKF differs from the EKF in that the LKF estimated errors are compensated for using a feedforward error control. In the INS/GPS integration case, the LKF estimated navigation errors are fed to the INS navigation output only and not to the INS mechanization. On the other hand, the EKF error compensation is performed using a feedback error control where the navigation errors are fed to both the INS navigation output and the INS mechanization. Figure 6.3 illustrates the LKF and EKF implementation in INS/GPS integration where ($\mathbf{r}$, $\mathbf{v}$, θ) represent position, velocity and attitude while ($\delta\mathbf{b}_g$, $\delta\mathbf{b}_a$) represent the gyros and accelerometers KF estimated errors. However, it should be noted here that the INS sensor error states $\delta\mathbf{b}_g$ and $\delta\mathbf{b}_a$ are fed back to the INS mechanization in both the LKF and EKF.

Due to the two different types of error control processes in of the LKF and EKF, the linearization process is different for each filter. The linearization for LKF is done about the INS mechanization trajectory (i.e., the predicted trajectory) while linearization is done about the updated trajectory in the EKF, and in theory, the performance of the EKF should be better than that of LKF. However, this will be only valid for high quality update measurements otherwise the EKF might diverge. Therefore, the EKF is sometimes riskier than the LKF especially when measurement errors are large (Brown and Hwang, 1997). On the other hand, in the case of using MEMS sensors, the predicted trajectory will be of low accuracy, and thus, the EKF performance will be better than the LKF. In addition, when GPS outages occur in MEMS/GPS navigation application, the LKF is expected to have divergence problems.

6.7 LiDAR Georeferencing—The Math Model

Georeferencing of LiDAR data can be defined as the problem of transforming the 3-D co-ordinate vector r^s of the laser sensor frame (S-frame) to the 3-D coordinate vector r^m of the mapping frame (m-frame) in which the results are required. The m-frame, as mentioned before, can be any earth-fixed coordinate system such as curvilinear geodetic coordinates

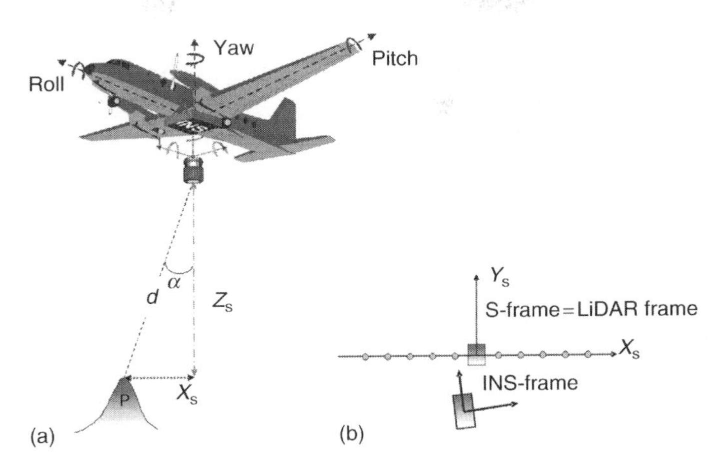

FIGURE 6.4
LiDAR scanner angle and distance measurement.

(latitude, longitude, height), UTM, or 3TM coordinates. The major steps in this transformation are depicted in Figure 6.4 for the airborne case, where the carrier could be an airplane or a helicopter.

The S-frame changes position and orientation with respect to the m-frame. Georeferencing is possible if at any instant of time (t) the position of the laser sensor center in the m-frame, i.e., $\mathbf{r}_s^m(t)$, and the rotation matrix between the S-frame and the m-frame $\mathbf{R}_s^m(t)$ have been determined. The georeferencing equation can then be written for any object point (i) as

$$\mathbf{r}_i^m = \mathbf{r}_s^m(t) + \mathbf{R}_s^m(t)\mathbf{r}_i^S \tag{6.47}$$

where

$\mathbf{r}_i^m$ is the position vector of an object (i) in the m-frame
(t) is the measurement epoch for measuring the distance d for point (i)
$\mathbf{r}_i^s$ is the position vector of an object (i) in the S-frame; which is given as

$$\mathbf{r}_i^s = \begin{pmatrix} -\mathrm{d}\sin\alpha \\ 0 \\ -\mathrm{d}\cos\alpha \end{pmatrix} \tag{6.48}$$

Equation 6.47 is however, only a first approximation of the actual situation. It implies that the coordinates of the laser sensor center can be directly determined. This is usually not the case because the navigation sensors—GPS antenna/INS body center—cannot share the same location in space with the laser sensor. Thus, small translations and rotations between the different centers have to be considered. The actual situation is shown in Figure 6.4b. It has been assumed that the laser sensor is mounted in the cargo area of the airplane, that the positioning sensor—a GPS antenna is mounted on top of the airplane, and that the attitude sensor an IMU is mounted in the interior of the aircraft, somewhere close to the laser sensor. In this case, aircraft position and attitude are defined by the INS center and the internal axes of the IMU (b-frame).

If the vector between the origin of the INS body frame (b-frame) and the laser sensor center is given in the b-frame as $\boldsymbol{a}^b$, $\mathbf{r}_s^m(t)$ can be written as

$$\mathbf{r}_s^m(t) = \mathbf{r}_{INS}^m(t) + \mathbf{R}_b^m(t)\boldsymbol{a}^b \tag{6.49}$$

where

$\mathbf{r}^{m}_{INS}(t)$ is the vector of interpolated coordinates of the INS in the m-frame at time (t) (resulting from the INS/GPS integration)

$\mathbf{R}^{m}_{b}(t)$ is the DCM, which rotates the b-frame into the m-frame

$\boldsymbol{a}^{b}$ is the constant vector between the laser sensor center and the center of the INS b-frame, usually determined before the mission by calibration

In addition to transformations between sensors, rotations between different sensor frames have to be taken into account. The INS b-frame cannot be aligned with the S-frame. The constant rotation $\mathbf{R}^{b}_{s}$ between the two frames is again obtained by calibration. In this case, $\mathbf{R}^{m}_{s}(t)$ can be written as

$$\mathbf{R}^{m}_{S}(t) = \mathbf{R}^{m}_{b}(t)\mathbf{R}^{b}_{S} \tag{6.50}$$

where $\mathbf{R}^{b}_{s}$ is the rotation between the S-frame and the INS b-frame as determined from a calibration process.

Applying Equations 6.49 and 6.50 to Equation 6.47, the final georeferencing formula can be written as

$$\mathbf{r}^{m}_{i} = \mathbf{r}^{m}_{INS}(t) + \mathbf{R}^{m}_{b}(t)\left(\mathbf{R}^{b}_{S}\mathbf{r}^{S}_{i} + \mathbf{a}^{b}\right) \tag{6.51}$$

where $\mathbf{R}^{b}_{s}$ transforms the vector $\mathbf{r}^{s}_{i}$ from the S-frame to the b-frame and $\mathbf{R}^{m}_{s}(t)$ transforms the vector $(\mathbf{R}^{b}_{s}\mathbf{r}_{i}^{s} + \mathbf{a}^{b})$ from the b-frame to the m-frame.

It should be noted that the vector $\mathbf{r}^{m}_{INS}(t)$, as well as the rotation matrix $\mathbf{R}^{m}_{s}(t)$ are time-dependent quantities while the vectors $\mathbf{r}^{s}_{i}$ and $\boldsymbol{a}^{b}$ as well as the matrix $\mathbf{R}^{b}_{s}$ are not. This implies that the carrier is considered as a rigid body whose rotational and translational dynamics are adequately described by changes in $\mathbf{r}^{m}_{INS}(t)$ and $\mathbf{R}^{m}_{s}(t)$. This means that the translational and rotational dynamics at the three sensor locations is uniform, in other words, differential rotations and translations between the three locations as functions of time have not been modeled. It also means that the origin and orientation of the three sensor systems can be considered fixed for the duration of the survey. These are valid assumptions in most cases, but may not always be true.

In georeferencing applications, inertial systems function mainly as precise attitude systems and as short-term interpolators for velocity, position, and attitude. Because of the time dependence of all major errors, the long-term velocity and position performance is not sufficient for precise georeferencing. Thus, regular position and/or velocity updates are needed to keep the overall errors within prescribed boundaries. The accuracy of inertial systems depends heavily on the quality of the sensors used which themselves are a function of the system costs. Noise levels in attitude can vary considerably, depending on the type of gyros used. We will therefore distinguish between high accuracy, medium accuracy, and low accuracy systems. High accuracy systems typically use navigation grade IMU, while medium and low accuracy systems typically employ civilian versions of tactical military IMU. Table 6.2 lists the achievable accuracies and the error characteristics of the sensors used in these systems. High- and medium-accuracy-grade systems can meet the accuracy requirements for LiDAR systems mounted on fixed-wing aircraft while low accuracy system can be adopted for low altitude helicopter based LiDAR systems.

TABLE 6.2

Characteristics and Achievable Accuracies of Inertial Systems for Direct Georeferencing Applications

Performance	High Accuracy	Medium Accuracy	Low Accuracy
Inertial system	Navigation grade	High end tactical grade IMU	Low end tactical grade IMU
Price (US$) of fully integrated GPS/INS systems	100–200 K	75–100 K	50–75 K
Gyro drift rate (deg./h)	≅0.015	0.1–1	3–10
Accelerometer bias (mg)	50–100	150–200	200–1000
Attitude accuracy	5–20 arcsec	0.1–1 arcmin	2–4 arcmin
Position accuracy (m)	0.05–0.3	0.05–0.3	0.05–0.3
Sensors/applications	Film based aerial cameras and Land DMM systems	High resolution digital aerial systems and LiDAR	Medium resolution digital aerial systems

References

Ackermann, F. 1995. *Sensor-and-Data Integration—The New Challenge, Integrated Sensor Orientation*, Eds. Colomina/Navaro, Wichmann, Heidelberg, pp. 2–10.

Bader, J. 1993. Low-cost GPS/INS. Proc. ION GPS-93, Salt Lake City, UT, pp. 135–244.

Bäumker, M. and Mattissek, A. 1992. Integration of a fibre optical gyro attitude and heading reference system with differential GPS. Proc. ION GPS-92, Albuquerque, NM, September 16–18, 1992, pp. 1093–1101.

Bossler, J.D. 1996. Airborne Integrated Mapping System, *International Journal of Geomatics*, Vol. 10, pp. 32–35.

Bossler, J.D., Goad, C., and Novak, K. 1993. Mobile mapping systems; new tools for the east collection of GIS information, *GIS'93*, Ottawa, Canada, March 23–35, 1993, pp. 306–315.

Brown, R.G. and Hwang, P.Y.C. 1997. *Introduction to Random Signals and Applied Kalman Filtering*. John Wiley & Sons, New York, USA.

Cannon, M.E. and Schwarz, K.P. 1990. A discussion of GPS/INS integration for photogrammetric applications. *Proc. IAG Symp. # 107: Kinematic Systems in Geodesy, Surveying and Remote Sensing*, Banff, September 10–13, 1990, pp. 443–452, Springer Verlag, New York.

Cohen, C.E. and Parkinson, B.W. 1992. Aircraft applications of GPS-based attitude determination, *Proc. ION GPS-92*, Albuquerque, NM, September 16–18, 1992.

Cosandier, D., Chapman, M.A., and Ivanco, T. 1993. Low cost attitude systems for airborne remote sensing and photogrammetry. *Proc. of GIS93 Conference*, Ottawa, March 1993, pp. 295–303.

El-Mowafy, A. and Schwarz, K.P. 1994. Epoch by epoch attitude determination using a multi-antenna system in kinematic mode, *Proc. KIS-94*, Banff, Canada, August 30 to September 2, 1994.

El-Sheimy, N. 2000. *Inertial Navigation and INS/GPS Integration*. Department of Geomatics Engineering, University of Calgary, Calgary, Alberta, Canada.

El-Sheimy, N. and Schwarz, K.P. 1993. Kinematic positioning in three dimensions using CCD technology. *Proc. IEEE/IEE Vehicle Navigation and Information System Conference (IVHS)*, October 12–15, 1993, pp. 472–475.

Gelb, A. 1974. *Applied Optimal Estimation*. MIT Press, Massachusetts Institute of Technology, Cambridge, MA.

Goldstein, H. 1980. *Classical Mechanics*, Addison Wesley, MA, 1980.

Hashemipour, H.R., Roy, S., and Laub, A.J. 1988. Decentralized structures for parallel Kalman filtering, *IEEE Trans. Automat. Contr.*, Vol. AC-33, 1988, pp. 88–94.

Julier, S.J. and Uhlmann, J.K., 2002. The scaled unscented transformation. *Proc. IEEE American Control Conference,* Anchorage, AK, 2002, pp. 4555–4559.

Kalman, R.E. 1960. A new approach to linear filtering and prediction problems. *ASME—Journal of Basic Engineering,* Vol. 82, series D, pp. 35–45.

Knight, D.T. 1997. Rapid development of tightly-coupled GPS/INS systems, *Aerospace and Electronic Systems Magazine,* IEEE, Vol. 12, Issue 2, February 1997, pp. 14–18.

Lapucha, D., Schwarz, K.P., Cannon, M.E., and Martell, H. 1990. The use of GPS/INS in a Kinematic Survey System, *Proc. IEEE PLANS 1990,* Las Vegas, NV, March 20–23, 1990, pp. 413–420.

Lipman, J.S. 1992. Trade-offs in the implementation of integrated GPS inertial systems. *Proc. ION GPS-92,* Albuquerque, NM, September 16–18, 1992, pp. 1125–1133.

Moafipoor, S., Grejner-Brzezinska, D., and Toth C.K. 2004. Tightly coupled GPS/INS integration based on GPS carrier phase velocity update, *ION NTM 2004,* January 26–28, 2004, San Diego, CA.

Nassar, S. and El-Sheimy, N. 2005. Accuracy Improvement of Stochastic Modeling of Inertial Sensor Errors. *Zeitschrift für Geodäsie, Geoinformation und Landmanagement (ZfV) Journal, Wißner, DVW, Germany,* Vol. 130, No. 3, pp. 146–155.

Scherzinger, B.M. 2002. Inertially aided RTK performance evaluation. *Proc. ION GPS-2002,* Portland, OR, September 24–27, 2002, pp. 1429–1433.

Scherzinger, B.M. and Woolven, S. 1996. POS/MV-handling GPS outages with tightly coupled inertial/GPS integration, OCEANS '96. MTS/IEEE. 'Prospects for the 21st Century' Conference Proceedings Vol. 1, No. 23–26, September 1996, pp. 422–428.

Schwarz, K.P., Chapman, M.A., Cannon, M.E., and Gong, P. 1993. An integrated INS/GPS approach to the georeferencing of remotely sensed data. *PE&RS,* Vol. 59, No. 11, pp. 1667–1674.

Shin, E. 2001. Accuracy improvement of low cost INS/GPS for land application, MSc thesis, December 2001, UCGE Report No. 20156.

Shin, E. 2005. Estimation techniques for low-cost inertial navigation. PhD thesis, Department of Geomatics Engineering, University of Calgary, Calgary, Alberta, Canada, UCGE Report No. 20156.

Toth, C.K. 1997. Direct sensor platform orientation: Airborne Integrated Mapping System (AIMS), *International Archives of Photogrammetry and Remote Sensing, ISPRS Comm. III,* Vol. XXXII, Part 3–2W3, pp. 148–155.

Wong, R.V.C. 1988. Development of a RLG strapdown inertial survey system, PhD thesis, Department of Geomatics Engineering, University of Calgary, Report No. 20027.

7

Waveform Analysis for Small-Footprint Pulsed Laser Systems

Uwe Stilla and Boris Jutzi

CONTENTS

7.1 Introduction

Aerial photogrammetry and airborne laser scanning (ALS) are the two most widely used methods for generating digital elevation models (DEMs), including digital terrain models (DTMs) that depict ground topography and digital surface models (DSMs) that depict the height of the ground, structures, and vegetation cover. In photogrammetry, the distance to a spatial surface is classically derived from a triangulation of corresponding image points from two or more overlapping images of the surface. These points are chosen manually or detected automatically by analyzing image structures. In contrast to photogrammetry, active laser scanner systems allow a direct and illumination-independent measurement of the distance to a surface, otherwise known as the range.

The interrelationship between aerial photogrammetry and ALS has been intensely discussed within the aerial surveying community in the last decade. Different comparison factors concerning data acquisition (e.g., coverage, weather conditions, costs, etc.) and surface reconstruction (e.g., accuracy, redundancy, post-processing time, etc.) have to be taken into account to choose the optimal method for a certain mapping campaign. An example of a study comparing photogrammetric image matching versus laser scanning for generation of high-quality DEMs for glacier monitoring (Lenhart et al., 2006) is given in Wuerlaender et al. (2004). In contrast to a decision to use one or the other technique, in some fields of applications a combined processing of laser data and stereo images is advantageous as shown in the generation of the extraterrestrial DTMs of Mars (Albertz et al., 2005; Spiegel et al., 2006) or DSMs for building characterization.

Conventional pulsed laser scanner systems for topographic mapping are based on time-of-flight ranging techniques to determine the range to the illuminated object. The time-of-flight is measured by the elapsed time between the emitted and backscattered laser pulses. The signal analysis to determine this time typically operates with analog threshold detection. For targets that have surfaces at different ranges illuminated by a single laser pulse, more than one backscattered pulse may be detected per emitted pulse. Most ALS systems are able to capture, at a minimum, the range for the first- and last-detected backscattered pulses. Some systems acquire ranges up to as many as five per emitted pulse for multiple backscattered pulses. First-pulse detection is the optimal choice to measure the hull of partially penetrable objects or the so-called volume scattering targets (e.g., canopy of trees). Last-pulse detection should be chosen to measure nonpenetrable surfaces (e.g., ground surfaces).

Currently, some commercial ALS systems not only capture the range for multiple pulse reflections but also digitize and record the received signal of the reflected laser energy, which allows for the so-called full-waveform analysis. This offers the possibility of analyzing the waveform off-line using digital signal processing methods in order to extract different surface attributes from the received signal based on the shape of the return pulses.

In the last decade, some waveform analysis investigations were carried out to explore the structure of vegetation and estimate aboveground biomass. For example, NASA developed the waveform-recording laser vegetation imaging sensor (LVIS) to measure vertical density profiles in forests (Blair et al., 1999). This experimental airborne system operates at altitudes up to 10 km and acquires waveforms for large diameter laser footprints (nominally 20 m) acquired across a wide swath. Another NASA system operating with a large footprint is the spaceborne geoscience laser altimeter system (GLAS) mounted on the ice, cloud, and land elevation satellite (ICESat). GLAS measures height distributions of atmospheric clouds and aerosols, and surface elevations of ice sheets, land topography, and vegetation (Brenner et al., 2003; Zwally et al., 2002). It is a profiling system that operates with a footprint diameter of 70 m and measures elevation changes with decimeter accuracy (Abshire et al., 2005; Schutz et al., 2005). In the analysis of data from both systems, surface characteristics are determined by comparing a parametric description of the transmitted and received waveforms (Hofton et al., 2000; Harding and Carabajal, 2005). Because the laser footprint is large and illuminates multiple surfaces, the resulting return waveform is an integrated, spatially nonexplicit representation of the range to illuminated surfaces separated both vertically and horizontally. The geometric organization of surfaces within a single footprint can therefore not be determined.

In contrast to large-footprint systems, small-footprint ALS systems illuminate only one or a few surfaces within the footprint, yielding waveforms with distinct return pulses corresponding to specific surfaces. One of the first developed small-footprint waveform-recording systems is the scanning hydrographic operational airborne LiDAR survey (SHOALS) instrument used for monitoring nearshore bathymetric environments. SHOALS has been in full operation since 1994 (Irish and Lillycrop, 1999; Irish et al., 2000). More recently commercial full-waveform ALS systems for terrestrial mapping have been developed (Jutzi and Stilla, 2006b; Wagner et al., 2006), which operate with a transmitted pulse width of 4–10 ns and allow digitization and acquisition of the waveform with approximately 0.5–1 GSample/s. Reitberger et al. (2006, 2007) have recently reported results that show clearly the potential of airborne, small-footprint, full-waveform data for the comprehensive analysis of tree structure and species classification. A set of key attributes have been defined and extracted based on the 3D distribution of the returns in combination with their characteristics in the full-waveform signal, providing information about tree microstructure such as the organization of the trunk and branches.

In this chapter, we focus not on a given application in the context of a data set from a given sensor rather than on general principles. Specifically, we describe the different approaches for designing a laser system, modeling the spatial and temporal properties of the emitted lasers pulses, detecting return pulses, and deriving attributes from the waveform. We emphasize aspects of the received waveform that are especially relevant for the newly available small-footprint, full-waveform commercial systems that yield distinct return pulses when multiple surfaces are illuminated by a laser pulse.

The design of a laser system impacts its measurement capabilities and the manner in which the signal has to be modeled and analyzed. Section 7.2 gives a brief overview of the features that characterize the design of laser ranging systems. Section 7.3 focuses on modeling of the temporal waveform and the spatial beam distribution. Different strategies for pulse detection are explained in Section 7.4. Section 7.5 describes the attributes that can be extracted from a single laser shot and presents an analysis of an entire scene that was recorded with an experimental small-footprint, full-waveform laser ranging system. A summary and outlook is given in Section 7.6.

7.2 Characterization of a Laser System

Depending on the application, laser systems can be designed in different ways. They may differ in techniques concerning the type of laser used, the modulation, type of measured features, detection method, or design of beam paths (construction). Figure 7.1 sketches a simplified overview of features characterizing a laser system. More detailed descriptions can be found in Kamermann (1993).

7.2.1 Laser Type

A laser works as an oscillator and an amplifier for monochromatic radiation (infrared, visible light, or ultraviolet). The operative wavelength of available lasers is located between 0.1 μm und 3 mm. For comparison, it should be mentioned that the visible domain is from 0.37 to 0.75 μm. To achieve a good signal-to-noise ratio (SNR) over long ranges, conventional scanning laser systems emit radiation with high energy. However, this could endanger the health of humans due to the focusing of laser radiation on the retina, which is most

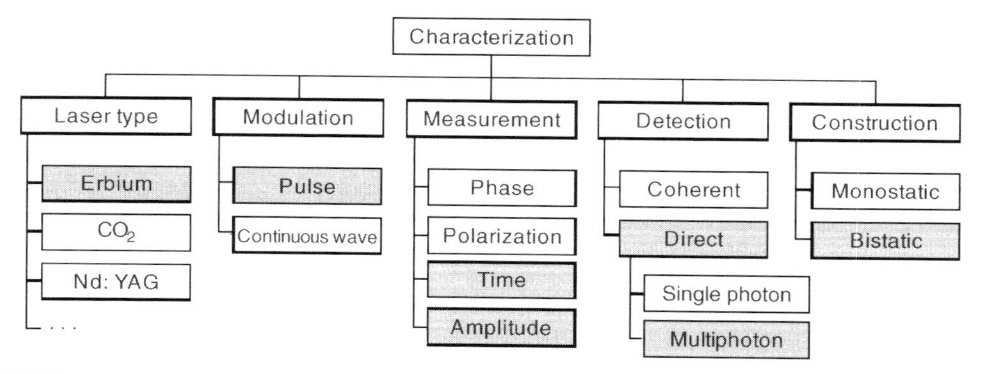

FIGURE 7.1
Features characterizing a laser system (with shaded boxes indicating the characteristics that are the focus of this chapter).

susceptible to damage at visible wavelengths. For this reason, most eye-safe laser systems used for mapping purposes operate with a wavelength outside the visible spectrum. This allows working with an emitted energy that is many times higher (up to 10^6) compared to the visible domain without the potential for retinal damage. Eye-safe lasers of greatest interest for long-range laser scanning are erbium-fiber lasers, carbon dioxide (CO_2) lasers, and neodymium:YAG (Nd:YAG) lasers.

Erbium-fiber lasers are optically pumped by a semiconductor diode and the active medium is an erbium-doped fiber. Their construction can be compact while still achieving a high output power. CO_2 lasers use carbon dioxide in gas form as the active medium. Although their construction is simple, their large size and mass are significant disadvantages. Solid-state Nd:YAG lasers can be pumped by various sources that define the characteristics of the emitted laser radiation. In this contribution we focus on erbium-fiber lasers (Figure 7.1).

7.2.2 Modulation Technique

Concerning modulation techniques, laser systems can be divided into two groups: continuous wave (CW) and pulsed lasers. A CW laser continuously emits electromagnetic radiation. The temporal energy distribution of the transmitted signal is influenced by amplitude modulation (AM) or frequency modulation (FM). Depending on the applied modulation technique, specific measurement techniques (Section 7.2.3) are required. A pulsed laser emits electromagnetic radiation in pulses of short duration. For laser ranging it is desirable to emit a pulse as short as possible and with a pulse energy as high as possible in order to obtain a precise range with a high probability of detection. However, design limitations on maximum peak power introduce a trade-off that requires a compromise between the length and the energy of the pulse. The length of the pulse (full width at half maximum, FWHM) is typically between 2 and 10 ns. For applications in remote sensing with long ranges, pulsed lasers with higher power density as compared to CW lasers are advantageous. In this contribution we focus on pulsed laser systems (Figure 7.1).

7.2.3 Measurement Technique

Measurement techniques using laser systems can be distinguished by the exploited signal properties such as phase, amplitude, frequency, polarization, time, or any combination of them. An amplitude modulated CW laser system is used to measure the range by exploiting the phase of a sinusoidal modulated signal. A phase difference Φ_d can be determined from a given phase of the transmitted signal and the measured phase of the received signal. With wavelength λ_m of the modulated signal, a corresponding range r can be calculated by $r = \lambda_m \cdot \Phi_d / 4\pi$. If the measurement of the phase difference Φ_d cannot be distinguished from $\Phi_d + n\pi$, the unambiguity interval of the range measurement will be limited to a maximum range $r_{max} = \lambda_m/2$. Assuming that the system is able to resolve an angle difference $\Delta\Phi_d$, the range resolution Δr corresponds according to $\Delta r = \lambda_m \cdot \Delta\Phi_d / 4\pi$. To increase the range resolution for a given $\Delta\Phi_d$, the modulation wavelength λ_m has to be decreased. However, this results in a reduction of the unambiguity interval of the range determination.

The problem of ambiguity can be solved by using multiple simultaneous offset sinusoidal modulation frequencies (multiple-tone sinusoidal modulation). In this case the maximum modulation wavelength defines the unambiguity and the minimum modulation wavelength defines the range resolution. In addition to this, partially illuminated surfaces with different ranges within the beam corridor result in a superimposed signal depending on the range and the reflectance of the surface. Because only a single phase value can be determined at the receiver, the ambiguities caused by the partially illuminated surfaces cannot be resolved (Thiel and Wehr, 2004). An incorrect intermediate value is measured.

The measurement of the amplitude value is feasible for CW lasers as well as for pulsed lasers. The amplitude is influenced by background radiation, the range of the object to the laser system, and the size, reflectance, slope, and roughness of the illuminated surface. In this chapter, we are interested in measuring and analyzing the received pulse waveform, i.e., the dependence of the intensity over time (Figure 7.1).

7.2.4 Detection Technique

Detection techniques can be divided into coherent detection and direct detection (Jelalian, 1992). Coherent detection is based on signal amplification due to constructive interference of the wave front of the received signal with that of the reference signal emitted from a CW laser. In direct detection laser systems, the received optical energy is focused onto a photosensitive element that generates an output signal that depends on the received optical power. Two direct detection techniques are appropriate for recording the temporal characteristics of the backscattered signal: multiphoton detection and single-photon detection.

7.2.5 Multiphoton Detection

The classical measurement technique for direct detection operates with a photodiode. For optical detectors, a positive intrinsic negative diode (PIN) or the more sensitive avalanche photodiode (APD) is used. The photodiode generates an electrical signal (voltage or current) that is directly proportional to the optical power of incident light composed of multiple photons. Figure 7.2 sketches a pulse resulting from a varying number of photons n over time t. For a detailed analysis of the analog signal a digitizing receiver unit is essential. To analyze the signal of the emitted short duration laser pulse with only a few nanoseconds pulse width, a high bandwidth receiver that resolves the signal at gigahertz rates and a correspondingly high digitizer sampling rate is required. Increasing bandwidth results in decreasing sensitivity of the photodiode, which can be compensated by increasing the power of the emitting laser source. An example of an Nd:YAG laser pulse sampled with 5 GSample/s is given in Figure 7.4a.

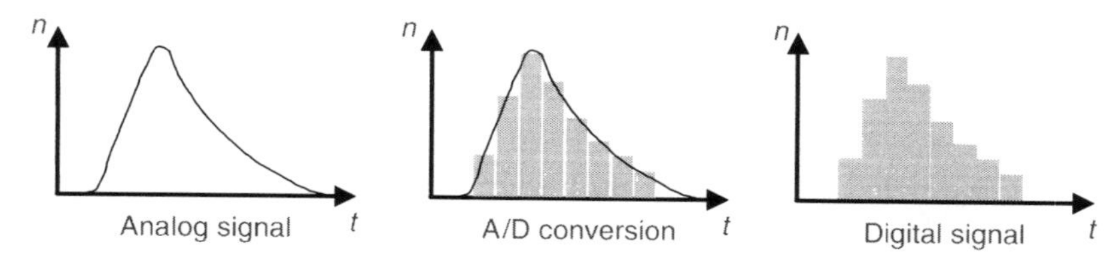

FIGURE 7.2
Digital recording of the signal with multiphoton detection.

7.2.6 Single-Photon Detection

The principle of single-photon detection is depicted in Figure 7.3. A short duration pulse is emitted by the laser source. A single photon of the backscattered pulse is detected by the receiver after time interval τ_1. This event blocks the receiver for a certain period of time during which no further photons are able to trigger the receiver. The time-of-flight of this single event is collected into a corresponding time bin of a histogram. After the period of blocking, the receiver is open to detect a new single-photon event. Multiple measurements are repeated and the time-of-flight of each single event ($\tau_2, \tau_3, \ldots$) is registered into the corresponding time bin of the histogram.

Let us assume a stationary scene and a stationary sensor platform. In this case, the statistical properties of the laser radiation do not change with the time and time-average quantities are equal to the ensemble quantities. Under these assumptions, the radiation ensembles are stationary and ergodic (Papoulis, 1984; Troup, 1972). The counting of single-photon events with assignment of their time-of-flight into time bins of a histogram is closely related to the integration of multiphotons over time (Alexander, 1997; Gagliardi and Karp, 1976; Loudon, 1973). In other words, the temporal waveform of the pulse can be reconstructed from a histogram of single-photon arrivals over time.

Many transmit pulses are necessary to obtain the waveform with single-photon detection. The quality of the sampled waveform depends on the number of photon counts. Various optical detectors can be used for this purpose, namely, PMT (photomultiplier tubes), MCP (microchannel plate), or APD detectors. Figure 7.4b shows a pulse plotted from a histogram containing the time-of-flight measurements from 16252 photons distributed in 50 bins, where the bin width is 40 ps. Note that the FWHM of the pulse

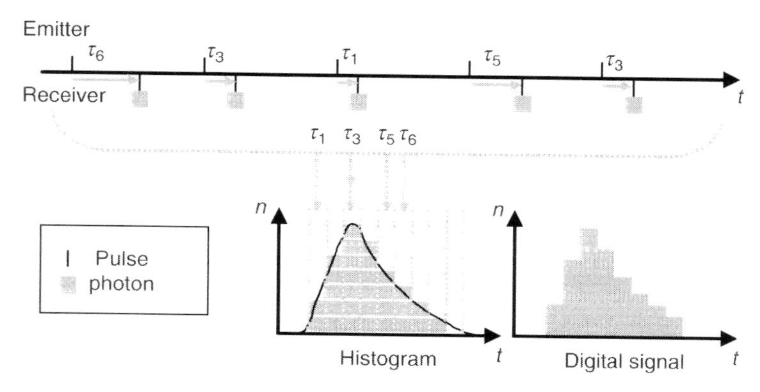

FIGURE 7.3
Digital recording of the signal with single-photon detection.

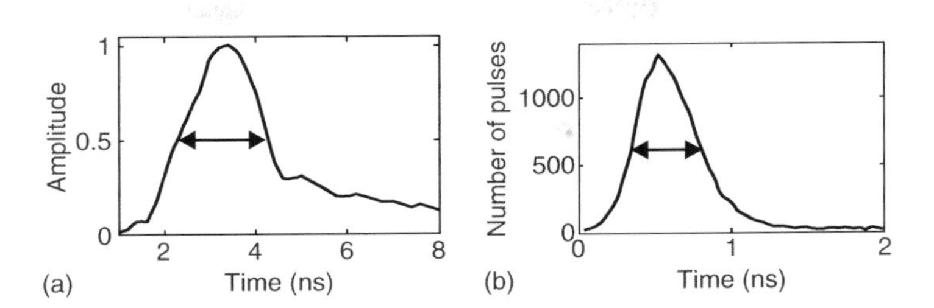

FIGURE 7.4
Examples for pulses backscattered from a diffuse surface. (a) Multiphoton detection (FWHM = 2.1 ns) and (b) single-photon detection (FWHM = 0.4 ns).

in Figure 7.4a is about five times of the pulse in Figure 7.4b. In this chapter, we focus on multiphoton detection (Figure 7.1).

7.2.7 Construction

Depending on the construction of the transmitter and receiver optics, monostatic and bistatic laser systems can be distinguished.

Monostatic laser systems have transmitter and receiver optics collocated on the same optical axis. A disadvantage of this construction is the higher number of components compared to a bistatic laser system, increasing the effort needed to optimally align the components. Advantages of this construction are the isogonal measurement of angles and the exact measurement of ranges, because the illuminated surface area and the observed field of view of the receiver are coincident for all ranges.

Bistatic laser systems have a transmitter and receiver optics that are spatially separated and thus the illumination and view angles are divergent. In general, both optics are close together and oriented in nearly the same direction. Objects are illuminated via a lens from the transmitter optic and the backscattered radiation is transferred via a separate lens to the receiver optic. An advantage of this measurement system is that the design can be easily constructed. A disadvantage of this design is that depending on the range to the illuminated surface the angle between the transmitter and the receiver optic varies. Furthermore, depending on the range, only a partial overlap is obtained between the illuminated surface and the observed field of view.

7.3 Modeling

The received waveform depends on the transmitted waveform of the emitted laser pulse, the spatial energy distribution of the beam, and the geometric and reflectance properties of the surface. In order to describe the temporal and spatial properties of a pulse, appropriate models that parameterize the pulse attributes have to be introduced.

7.3.1 Waveform of the Laser Pulse

Mathematical functions can be used to approximate the shape of laser pulse waveforms. Depending on the system, the shape may be best represented by a rectangular, exponential, or Gaussian distribution. A simple model is given by a rectangular distribution $s(t)$ with an amplitude a, pulse width w, and time delay τ

$$s(t) = a\text{rect}\left\{-\frac{(t-\tau)}{w}\right\} = \begin{cases} a & \text{for} \quad -\tau \le t \le w-\tau \\ 0 & \text{else.} \end{cases} \tag{7.1}$$

Especially for short laser pulses a rectangular model often differs from the measured shape. A waveform with an exponential distribution (e.g., for a Q-switched laser) is applied by Steinvall (2000)

$$s(t) = t^2 \exp\left\{-\frac{(t-\tau)}{w}\right\} \tag{7.2}$$

A temporally symmetric Gaussian distribution for modeling the waveform of the spaceborne GLAS is proposed by Brenner et al. (2003). The basic waveform $s(t)$ of the used laser system can be described by

$$s(t) = \frac{2a}{w}\sqrt{\frac{\ln(2)}{\pi}}\,\exp\left\{-4\ln(2)\frac{(t-\tau)^2}{2w^2}\right\} \tag{7.3}$$

The width of a pulse w is commonly defined as one-half of the pulse's maximum amplitude known as FWHM.

7.3.2 Spatial Energy Distribution

The spatial energy distribution of a laser (also known as the beam profile) depends on the pump source, the optical resonator, and the laser medium. In general, beam profiles are modeled by a cylindrical distribution (top-hat form) or by a 2D-symmetric Gaussian distribution (Kamermann, 1993). The measured cylindrical beam distribution of a pulsed

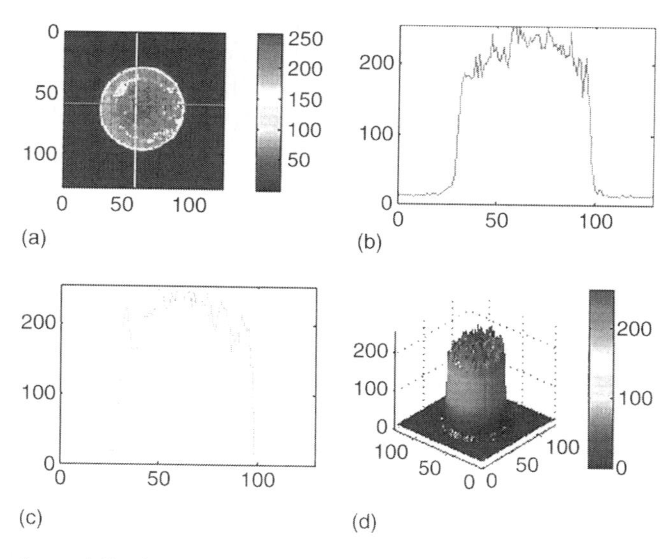

FIGURE 7.5 (See color insert following page 334.)
Measured cylindrical beam distribution (top-hat form). (a) 2D visualization, with the row (red) and column (green) of the maximum intensity indicated, (b) profile of the indicated row, (c) profile of the indicated column, and (d) 3D visualization.

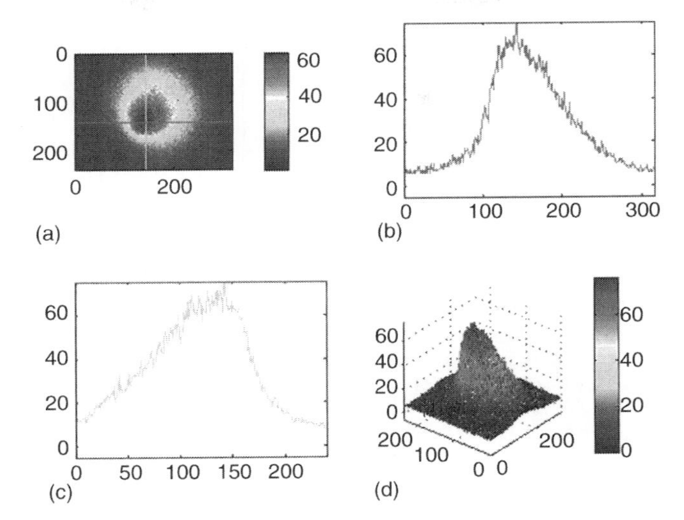

FIGURE 7.6 (See color insert following page 334.)
Measured Gaussian beam distribution. (a) 2D visualization, with the row (red) and column (green) of the maximum intensity indicated, (b) profile of the indicated row, (c) profile of the indicated column, and (d) 3D visualization.

erbium fiber laser that operates at a wavelength of 1550 nm is depicted in Figure 7.5. A Gaussian beam distribution of a Raman shifted Nd:YAG laser that operates at a wavelength of 1543 nm is depicted in Figure 7.6. Both measurements differ more or less from these idealized distributions.

7.4 Analyzing the Waveform

Various detection methods are used to extract attributes of the reflecting surface from the waveform. To obtain the surface attributes, each waveform $s(t)$ is analyzed. The surface within the beam corridor generates a return pulse. To detect and separate this pulse from the noise, a signal-dependent threshold is estimated using the signal background noise. For example, in one particular implementation if the intensity of the waveform is above three times the noise standard deviation ($3\sigma_n$) for a duration of at least the full-width-half-maximum of the pulse, a pulse is assumed to be found. The section of the waveform including the detected pulse is passed onto the subsequent processing steps.

Typical surface attributes to extract from a waveform are range, elevation variation, and reflectance corresponding to the waveform attributes of time, width, and amplitude.

A rough surface, i.e., a surface of a certain vertical extent, will widen the laser pulse upon reflection. Therefore, the width of the pulse is a measure of the elevation variation of the surface. In addition, the widening of the pulse causes the reflected photons to be spread over a greater amount of time, thus reducing the peak amplitude. Therefore, to estimate the elevation variation or reflectance attributes of a surface, the pulse width and amplitude have to be known. Estimating just the amplitude of a pulse without considering this dependency will lead to inaccurate and noisy reflectance values.

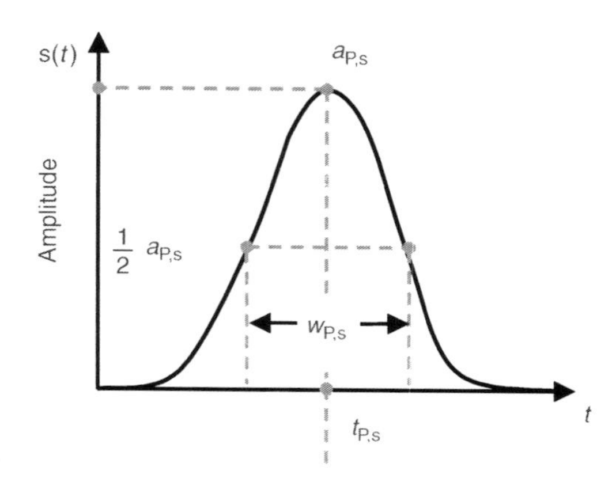

FIGURE 7.7
Attribute extraction with peak detection.

Determination of the range to a surface can be accomplished with different schemes that include peak detection, leading edge ranging, constant fraction detection, center of gravity detection, and Gaussian decomposition and deconvolution. A discussion of key elements of each approach follows.

7.4.1 Peak Detection

The values of range $r_{P,s}$ and amplitude are determined at the maximum pulse amplitude $a_{P,s}$ (Figure 7.7), and the width $w_{P,s}$ is estimated at full-width-half-maximum of the pulse. Local spikes on the pulse waveform strongly effect the attribute extraction. Therefore, for noisy signals, a smoothing filter is recommended to determine the global maxima.

7.4.2 Leading Edge Detection

A threshold crossing of the pulse waveform leading edge determines the range value $r_{LE,s}$ (Figure 7.8). The threshold value can be a predefined fixed value, but then the ranging detection strongly depends on the amplitude and width of the pulse waveform, introducing a ranging bias dependent on pulse shape referred to as range walk. The half of the maximum amplitude $a_{LE,s}$ of the pulse for a threshold is used for range determination.

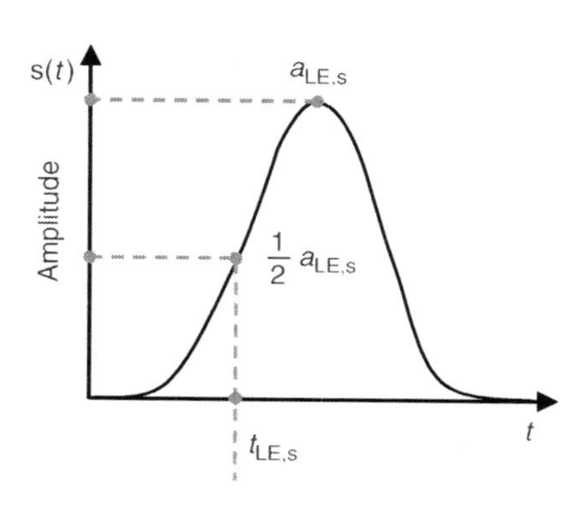

FIGURE 7.8
Attribute extraction with leading edge detection.

7.4.3 Constant Fraction Detection

A ranging implementation designed to be insensitive to amplitude-dependent biases applies a constant fraction detection circuit in which the pulse waveform $s(t)$ is inverted and delayed by a fixed time τ and added to the original pulse (Figure 7.9). The combined signal $s_{CFD}(t)$ gives a constant fraction signal with a zero crossing point at t_{CFD} (see Figure 7.10).

$$s_{CFD}(t_{CFD}) = 0 \quad \text{with} \quad s_{CFD}(t) = s(t) - s(t+\tau) \tag{7.4}$$

The determined t_{CFD} is insensitive to the pulse amplitude, but depends on the pulse waveform and width (Kamermann, 1993). A suitable value for the delay time τ is the FWHM of the waveform.

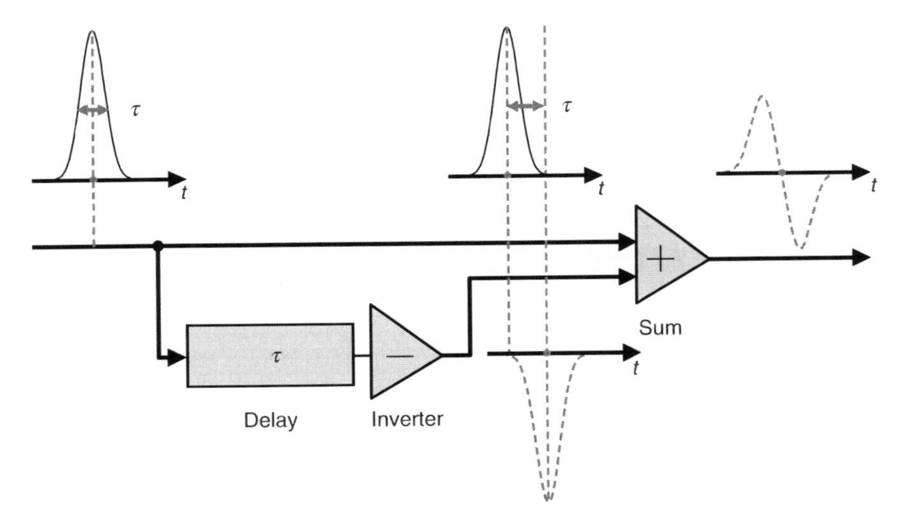

FIGURE 7.9
Simplified schematic visualization of the processing steps for the constant fraction detection.

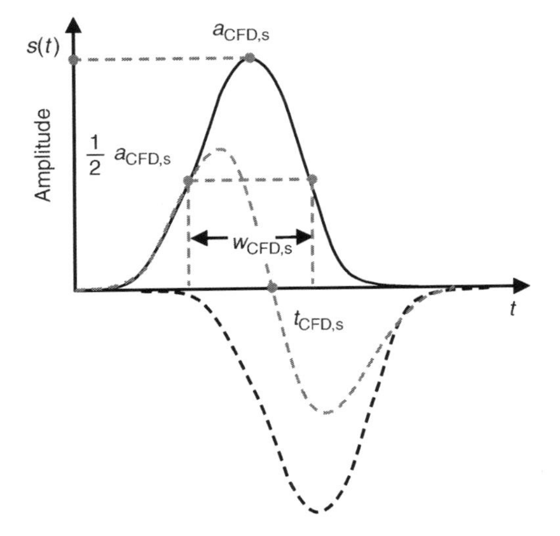

FIGURE 7.10
Attribute extraction with constant fraction detection.

For symmetric waveforms, the traditional constant fraction algorithm delivers unbiased ranging results. However, for an asymmetric noisy waveform the delayed signal should be reversed in time as well, to avoid ambiguities of the zero crossing point.

7.4.4 Center of Gravity Detection

The temporal center of gravity of the pulse waveform is $t_{CoG,s}$ (Figure 7.11). The time value (range) is determined by integrating the pulse waveform $s(t)$

$$t_{CoG,s} = \int_{t=t_{CoG_1,s}}^{t_{CoG_2,s}} t\, s(t)\,dt \Bigg/ \int_{t=t_{CoG_1,s}}^{t_{CoG_2,s}} s(t)\,dt \tag{7.5}$$

It delivers good results for returns with various pulse amplitudes and pulse widths that have low noise. For returns with an asymmetric pulse shape skewed to longer ranges, this method results in a detected range that is slightly longer than the range value obtained with the peak detection.

The following methods to further process the pulse properties are not part of the center of gravity algorithm, but are well suited to complement it. Generally, integration over a section of the signal has the advantage of reducing the noise dependence compared to the aforementioned methods relying on single samples. We call the integral of the waveform $s(t)$ shown in the denominator of Equation 7.4 as the pulse strength. From this, the value a_{CoG} can be calculated assuming a Gaussian and using the Inverse error function (erf^{-1}) and the width w

$$a_{CoG} = \frac{2\,\text{erf}^{-1}(0.5)}{\sqrt{\pi}\,w} \int_{t=t_{CoG_1}}^{t_{CoG_2}} s(t)\,dt \tag{7.6}$$

Furthermore, the width $W_{CoG,s}$ is approximated by the width of the central pulse area contributing 0.76 of this pulse strength with

$$\int_{t=t_{CoG}-\frac{w_0}{2}}^{t_{CoG}+\frac{w_0}{2}} s(t)\,dt = \text{erf}\left(\sqrt{\ln 2}\right) \int_{t=t_{CoG_1}}^{t_{CoG_2}} s(t)\,dt \tag{7.7}$$

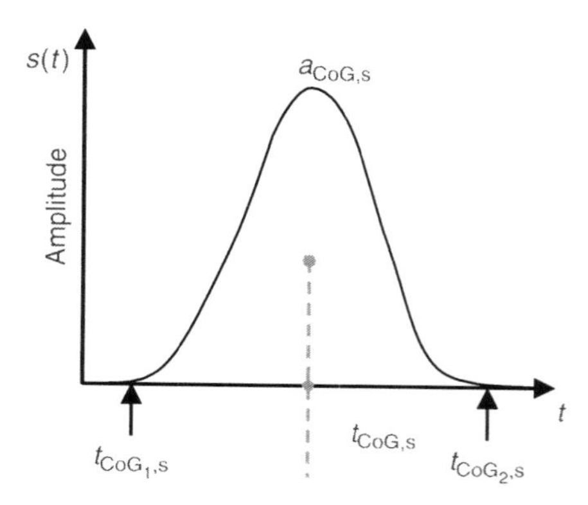

FIGURE 7.11
Attribute extraction with center of gravity detection.

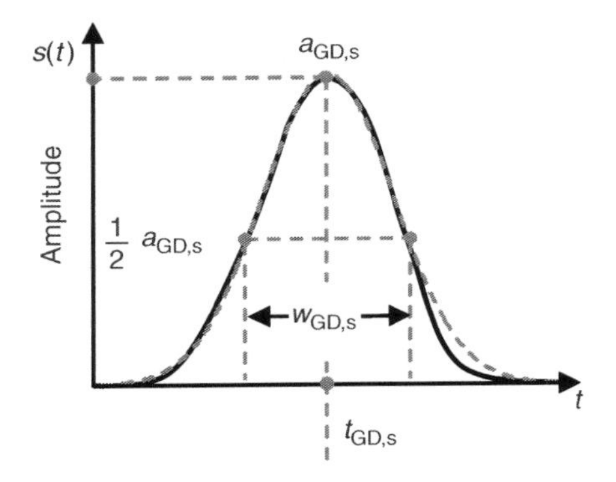

FIGURE 7.12
Attribute extraction with Gaussian decomposition algorithm.

7.4.5 Gaussian Decomposition

Assuming a Gaussian function for the waveform (Equation 7.3), the surface attributes can be extracted by estimating the parameters of the adapted function:

$$s(t) = a_{\mathrm{GD,s}} \exp\left\{-4\ln(2)\frac{(t - t_{\mathrm{GD,s}})^2}{(w_{\mathrm{GD,s}})^2}\right\} \tag{7.8}$$

For a parametric description of the pulse properties, a Gaussian decomposition on the waveform can be used. Different methods are known, for example, Expectation Maximation (EM) algorithm (Persson et al., 2005) and Gauss–Newton or Levenberg–Marquardt algorithm (Hofton et al., 2000; Jutzi and Stilla, 2005; Reitberger et al., 2006). In Figure 7.12, the estimated attributes of the received waveform are depicted.

7.4.6 Deconvolution

Analysis of a received waveform in order to extract the attributes of the illuminated surface is a difficult task because different processes impact the shape of the waveform. The received waveform $s(t)$ of a laser pulse depends on the transmitted waveform $r(t)$, the impulse response of the receiver, the spatial beam distribution of the laser pulse, and the geometric and reflectance properties of the illuminated surface. The impulse response of the receiver is mainly affected by the photodiode and amplifier, and the spatial beam distribution typically has a Gaussian distribution. Let us assume that a receiver consists of an ideal photodiode and that the amplifier has an infinite bandwidth with a linear response. In that case the received waveform depends mainly on the transmitted waveform $r(t)$ and the properties of the illuminated surface. The 3D characteristics of the surface can be captured by a time-dependent surface representation, referred to as the surface response $h(t)$. In this case the received waveform $s(t)$ can be expressed as

$$s(t) = h(t) * r(t) \tag{7.9}$$

where ($*$) denotes the convolution operation. By transforming $s(t)$ into the Fourier domain and solving the resulting equation for the spectral surface function $\underline{H}(f)$, we obtain

$$\underline{H}(f) = \underline{S}(f)/\underline{R}(f) \tag{7.10}$$

The surface response $h(t)$ is obtained by transforming $\underline{H}(f)$ into the time domain. By applying a Gaussian decomposition method to the surface response, surface attributes can be extracted. The deconvolution removes the characteristics of the transmitted waveform from the received waveform and enables a description of the observed surface.

For a reliable deconvolution a high SNR for the received waveform is essential. In addition to this, the waveform has to be captured with a high bandwidth receiver and with an adequate sampling rate of the analog-to-digital converter. Furthermore, it has to be mentioned that large numerical errors may appear depending on the receiver noise. A Wiener filter used for deconvolution reduces the noise of the determined surface response (Jutzi and Stilla, 2006a). This method allows discriminating differences in range, e.g., given by a stepped surface within the beam, which are smaller than the length of the laser pulse. Experiments have shown that a step smaller than 10 times of the pulse length can be distinguished.

7.5 Attribute Extraction

An example of a signal profile applied to multiple pulses is depicted in Figure 7.13. The waveform parameters for each detected pulse of this signal profile are estimated by a Gaussian decomposition method using the Levenberg–Marquardt algorithm. The extracted attributes are described in the table. The estimated waveform is shown below the original waveform in Figure 7.13.

By comparing the range values in the table, we can observe that the distance between the first and second pulse is about 10 m and between the third and fourth pulse about 2.5 m. The third pulse shows the highest maximum amplitude and the pulse width of the first and

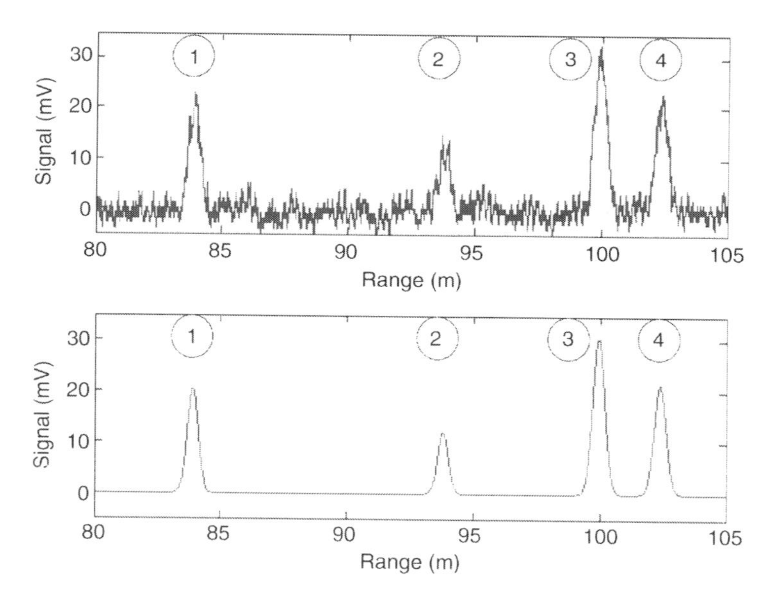

FIGURE 7.13
Signal profile with multiple reflections.

second pulse is slightly lower than the third and fourth pulse. It is not possible to classify the type of surfaces illuminated within the single-beam corridor using the amplitude, pulse width, and range properties alone. For assigning each return pulse to a specific surface type additional information is required, which can result from the 3D geometrical relationships of the returns within a point cloud.

As an example of information retrieval achieved by combining return pulse properties and the spatial relationship of the returns, a full-waveform data set of a test scene was captured by scanning along the azimuth and elevation and recording the return intensity sampled over time t. Neglecting angular variations of the scan, the measured intensities as a function of time t sampled over the azimuth and elevation can be interpreted as a 3D data set forming a cuboid with Cartesian coordinates x, y, and t. The sampling along the time axis can be recalculated into corresponding range values z. These data cuboid can be analyzed in several different ways.

Figure 7.14 shows a set of image slices ($y-t$ planes). The second slice from the left ($x = 4$) shows vegetation in the center (near range) and building structures on the right side (far range). The grey values correspond to the intensity of the signal. The intensity values along the marked solid line are the intensity values of the waveform shown in Figure 7.13.

Note that although this way of displaying the data suggests that a full 3D representation of the scene has been obtained, this is in fact not the case. Just as with point clouds measured by conventional laser scanners, the data cube represents only 2.5d information. This is because of occlusion effects that are dependent on the target size in relation to the beam footprint size. It is possible that the laser pulse is mostly intercepted by and backscattered from the first illuminated surface along the propagation direction and that the following surfaces along the laser vector are hidden, giving weak or no reflections. For instance, a tree with dense foliage may return only a single reflection response per laser pulse even though multiple surfaces are present after the first detected return along the path of the laser vector.

In the following, we use the Wiener filter method to extract attributes from received waveforms. The extraction is carried out without considering spatial neighborhood relations. The results of the extracted surface attributes from the data cuboid are shown in xy plane by

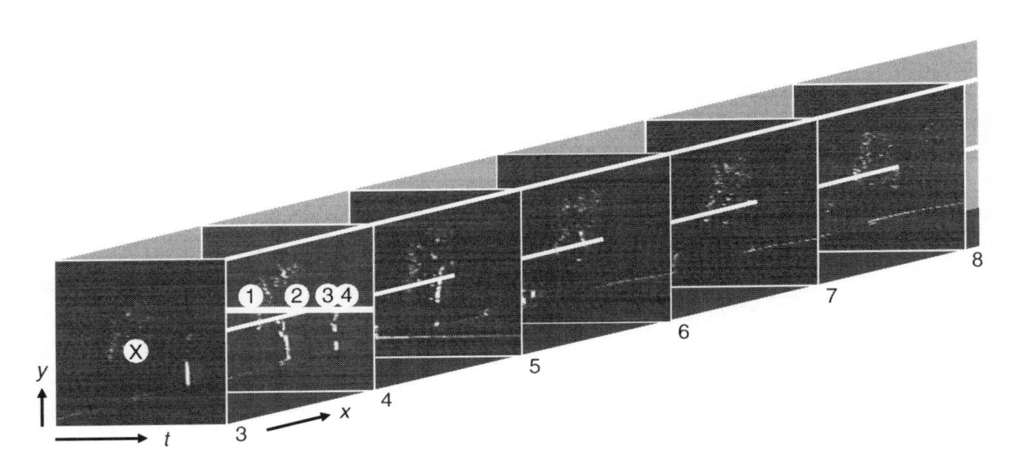

FIGURE 7.14
Vertical image slices with ground, vegetation, and building structures.

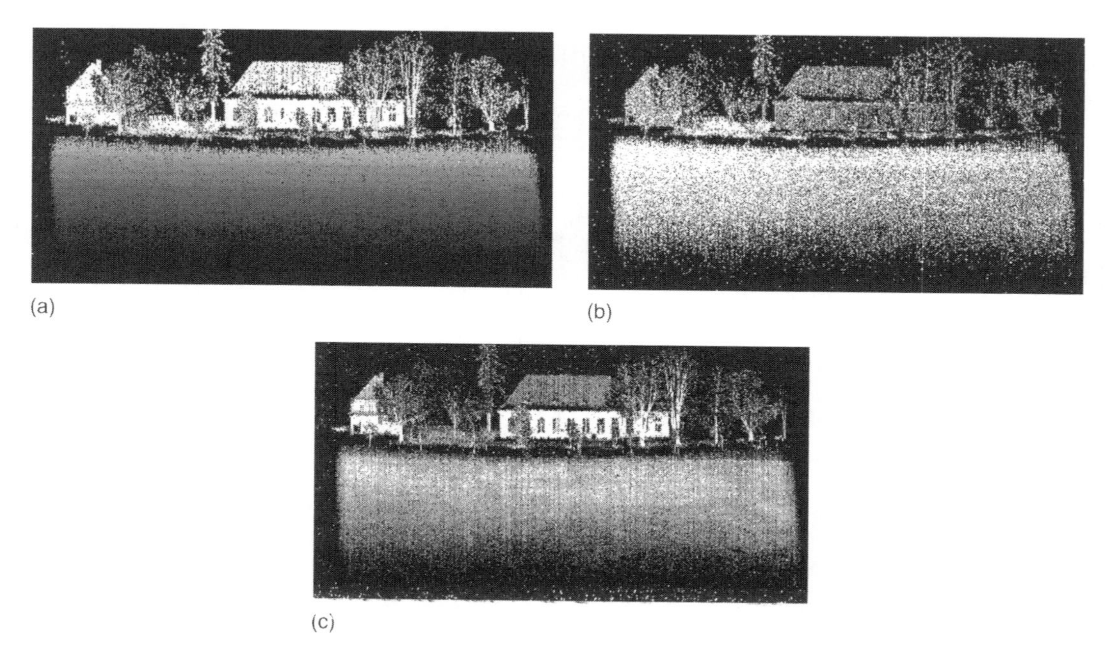

FIGURE 7.15
Extracted attributes of pulses: (a) range, (b) pulse width, and (c) intensity.

images of 320 × 600 pixels (Figure 7.15). Figure 7.15a shows the range image, Figure 7.15b shows the width of the pulse, and Figure 7.15c shows the intensity of the pulse. Larger values for an attribute are displayed by brighter pixels. Due to the fact that only a single value can be shown in the 2D images, only the first reflection is considered in cases where multiple reflections are present for a laser pulse.

The attributes maximum amplitude and pulse width were extracted using the Wiener filter and examined for their ability to discriminate different object classes. The entire scene and three objects classes, namely, buildings, trees, and meadow, are shown in Figure 7.16. Column A of the figure depicts range images of the entire scene (Figure 7.16a, column A) and of the selected objects classes (Figure 7.16b through d (column A)). Column B depicts scatter plots of maximum amplitude versus pulse width for the entire scene and the object classes (Figure 7.16, column B).

In Figure 7.16a, it can be observed that for the entire scene small values of the maximum amplitude have a large spreading of the pulse width. By decomposing the scene into the three object classes, it is apparent that the vegetation (trees and meadow) in most cases produces the signal returns with small maximum amplitudes but high values for the pulse width. The building in Figure 7.16b shows small values for the pulse width with little variation and a large range in the maximum amplitude values. Furthermore, the maximum amplitude of the building shows a cluster of higher values than that from the trees and meadow. These high values may result from high reflectance by the white façade. The trees are depicted in Figure 7.16c, where the pulse width shows high variation with mostly small maximum amplitude values. The meadow (Figure 7.16d) induces large variations of the pulse width and small maximum amplitude values. In general, the trees and the meadow produce larger pulse widths than the building.

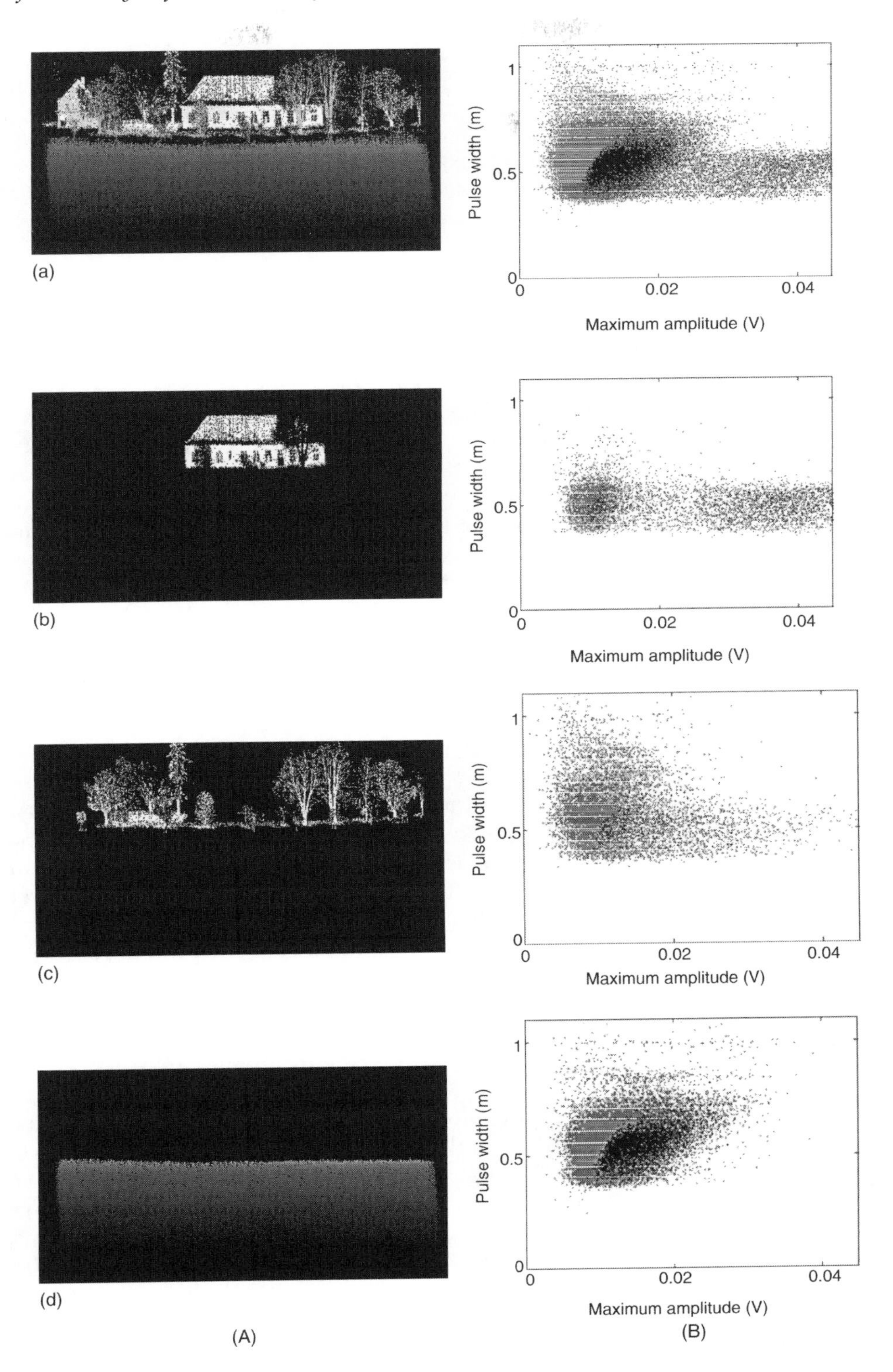

FIGURE 7.16
Comparison of maximum amplitude versus pulse width for selected objects of the measured scene. (a) Scene, (b) building, (c) trees, and (d) meadow. Column (A) denotes range images, and column (B) denotes maximum amplitude versus pulse width.

7.6 Summary and Outlook

It has been shown that full-waveform analysis enables extraction of more information compared to classical analogous pulse detection methods. First, recording of the received waveform offers the possibility for the end-user to select different methods to extract range information. The shape of the pulse and the entire signal can be considered for determining the range more accurately. Further improvements on reliability and accuracy can be derived by signal-processing methods based on the transmitted and the received waveform, e.g., deconvolution. Additionally, attributes of the surface can be derived from a parametric description of the waveform. The attributes maximum amplitude and pulse width may support the discrimination between volume scatterer (vegetation) and hard targets (man-made objects) (Kirchhof et al., 2007).

In the preceding description laser pulses were analyzed without considering the information of neighboring measurements. For reconstructing man-made objects, the introduction and test of hypotheses about the shape of the surface (e.g., plane, sphere) may efficiently support the analysis of single waveforms. Two different strategies assuming a planar shape in the local neighborhood of the surface and introducing this assumption into the signal analysis should be addressed. Both strategies combine information from top-down (surface primitives) and bottom-up (signal processing) for an extended analysis of full waveform laser data.

The first strategy (Kirchhof et al., 2007) uses an iterative processing of waveforms considering a predicted shape of the waveform from the local neighborhood. A presegmentation based on surface attributes is carried out to distinguish between partly penetrable objects (e.g., trees, bushes) and impenetrable surfaces (e.g., roof, wall). Derived range values from presegmentation of the impenetrable surfaces are used to automatically generate surface primitives (e.g., planes).

This allows a refinement of each range value, considering the surface geometry in a close neighborhood. Furthermore, partly occluded surface areas are extended by prediction of the expected range values. This prediction is further improved by considering the surface slope for the estimated received waveform. Expected pulses are simulated and correlated with the received waveforms. Accepted points that were missed in the first processing step due to weak signal response are associated to the point cloud. The procedure is repeated several times until all appropriate range values are considered to estimate the surface.

The second strategy (Stilla et al., 2007) uses a slope compensated stacking of waveforms. Weak pulses with a low SNR are discarded by classic threshold methods and get lost. In signal and image processing, different stacking techniques are used to improve the SNR.

For detection of weak laser pulses, hypotheses for planes of different slopes (e.g., angle difference 5°) are generated. According to the slope of the hypothesis, the waveforms in the local neighborhood are shifted in range. A superimposed signal is calculated from the stack of shifted waveforms. The maxima of superimposed signals from all hypotheses are compared to verify a hypothesis. Each signal is assessed by a likelihood value with respect to its contribution to the accepted hypothesis. Finally, signals are classified according to the likelihood values obtained using two thresholds and visualized by the traffic-light paradigm. The results contain detected pulses reflected from objects, which cannot be predicted by the previously detected point cloud.

Both strategies show promising results that encourage the continuity of work on the analysis of full-waveform laser pulses.

References

Abshire, J.B., Sun, X.L., Riris, H., Sirota, J.M., McGarry, J.F., Palm, S., Yi, D.H., and Liiva, P., 2005. Geoscience Laser Altimeter System (GLAS) on the ICESat mission: On-orbit measurement performance. *Geophysical Research Letters*, 32:L21S02, doi:10.1029/2005GL024028.

Albertz, J., Attwenger, M., Barrett, J., Casley, S., Dorninger, P., Dorrer, E., Ebner, H., Gehrke, S., Giese, B., Gwinner, K., Heipke, C., Howington-Kraus, E., Kirk, R.L., Lehmann, H., Mayer, H., Muller, J.-P., Oberst, J., Ostrovskiy, A., Renter, J., Reznik, S., Schmidt, R., Scholten, F., Spiegel, M., Stilla, U., Waehlisch, M., and Neukum, G., 2005. HRSC on Mars Express—Photogrammetric and Cartographic Research. *Photogrammetric Engineering and Remote Sensing*, 71(10):1153–1166.

Alexander, S.B., 1997. Optical communication receiver design. In: *SPIE Tutorial Texts in Optical Engineering*, TT22, SPIE Press, Bellingham, WA.

Blair, J.B., Rabine, D.L., and Hofton, M.A., 1999. The Laser Vegetation Imaging Sensor (LVIS): A medium-altitude, digitization-only, Airborne Laser Altimeter for mapping vegetation and topography. *ISPRS Journal of Photogrammetry and Remote Sensing*, 54(2–3):112–122.

Brenner, A.C., Zwally, H.J., Bentley, C.R., Csatho, B.M., Harding, D.J., Hofton, M.A., Minster, J.B., Roberts, L.A., Saba, J.L., Thomas, R.H., and Yi, D., 2003. Geoscience Laser Altimeter System (GLAS)—Derivation of range and range distributions from Laser Pulse Waveform Analysis for surface elevations, roughness, slope, and vegetation heights. Algorithm Theoretical Basis Document—Version 4.1. Available at http://www.csr.utexas.edu/glas/pdf/Atbd_20031224.pdf (accessed September 1, 2006).

Gagliardi, R.M. and Karp, S., 1976. *Noncoherent (Direct) Detection: Optical Communications*. John Wiley & Sons, New York.

Harding, D.J. and Carabajal, C.C., 2005. ICESat waveform measurements of within-footprint topographic relief and vegetation vertical structure. *Geophysical Research Letters*, 32:L21S10, doi:10.1029/2005GL023471.

Hofton, M.A., Minster, J.B., and Blair, J.B., 2000. Decomposition of laser altimeter waveforms. *IEEE Transactions on Geoscience and Remote Sensing*, 38(4):1989–1996.

Irish, J.L. and Lillycrop, W.J., 1999. Scanning laser mapping of the coastal zone: The SHOALS System. *ISPRS Journal of Photogrammetry and Remote Sensing*, 54(2–3):123–129.

Jelalian, A.W., 1992. *Laser Radar Systems*. Artech House, Boston, MA.

Jutzi, B. and Stilla, U., 2005. Measuring and processing the waveform of laser pulses. In: Gruen, A., Kahmen, H., Eds., *Optical 3-D Measurement Techniques VII*, Vol. 1, pp. 194–203. Vienna University of Technology.

Jutzi, B. and Stilla, U., 2006a. Range determination with waveform recording laser systems using a Wiener filter. *ISPRS Journal of Photogrammetry and Remote Sensing*, 61(2):95–107 [doi:10.1016/j.isprsjprs.2006.09.001].

Jutzi, B. and Stilla, U., 2006b. Characteristics of the measurement unit of a full-waveform laser system. Symposium of ISPRS Commission I: From Sensors to Imagery. *International Archives of Photogrammetry, Remote Sensing and Spatial Information Sciences*, 36(Part 1/A).

Kamermann, G.W., 1993. Laser radar. In: Fox, C.S., Ed., *Active Electro-Optical Systems, The Infrared and Electro-Optical Systems Handbook*. SPIE Optical Engineering Press, Ann Arbor, MI.

Kirchhof, M., Jutzi, B., and Stilla, U., 2008. Iterative processing of laser scanning data by full waveform analysis in close neighbourhood, *ISPRS Journal of Photogrammetry and Remote Sensing*, 63(1), pp. 99–114 [doi:10.1016/j.isprsjprs.2007.08.006].

Lenhart, D., Kager, H., Eder, K., Hinz, S., and Stilla, U., 2006. Hochgenaue Generierung des DGM vom vergletscherten Hochgebirge—Potential von Airborne Laserscanning. Arbeitsgruppe Automation in der Kartographie: Tagung 2005, Mitteilungen des Bundesamtes für Kartographie und Geodaesie, Band 36:65–78 (in German).

Loudon, R., 1973. *The Quantum Theory of Light*. Clarendon Press, Oxford.

Papoulis, A., 1984. *Probability, Random Variables, and Stochastic Processes*. McGraw-Hill, Tokyo.

Persson, Å., Söderman, U., Töpel, J., and Ahlberg, S., 2005. Visualization and analysis of full-waveform airborne laser scanner data. In: Vosselman, G., Brenner, C., Eds., *Laserscanning 2005. International Archives of Photogrammetry, Remote Sensing and Spatial Information Sciences*, Enschede, 36(Part 3/W19):109–114.

Reitberger, J., Krzystek, P., and Stilla, U., 2006. Analysis of full waveform lidar data for tree species classification. In: Förstner, W., Steffen, R., Eds., Symposium of ISPRS Commission III: Photogrammetric Computer Vision PCV06. *International Archives of Photogrammetry, Remote Sensing and Spatial Information Sciences*, Bonn, 36(Part 3):228–233.

Reitberger, J., Krzystek, P., and Stilla, U., 2007. Analysis of full waveform LIDAR data for the classification of deciduous and coniferous trees. *International Journal of Remote Sensing*, 29(5):1407–1431.

Schutz, B.E., Zwally, H.J., Shuman, C.A., Hancock, D., and DiMarzio, J.P., 2005. Overview of the ICESat Mission. *Geophysical Research Letters*, 32, L21S01, [doi:10.1029/2005GL024009].

Spiegel, M., Stilla, U., and Neukum, G., 2006. Improving the exterior orientation of Mars Express regarding different imaging cases. *International Archives of Photogrammetry and Remote Sensing and Spatial Information Sciences*, 36(4) (on CD).

Stilla, U., Yao, W., and Jutzi, B., 2007. Detection of weak laser pulses by full waveform stacking. In: Stilla, U. et al., Eds., PIA07: Photogrammetric Image Anlysis 2007. *International Archives of Photogrammetry, Remote Sensing and Spatial Information Sciences*, Munich, 36(3/W49A):25–30.

Steinvall, O., 2000. Effects of target shape and reflection on laser radar cross sections. *Applied Optics*, 39(24):4381–4391.

Thiel, K.H. and Wehr, A., 2004. Performance capabilities of laser-scanners—An overview and measurement principle analysis. *International Archives of Photogrammetry, Remote Sensing and Spatial Information Sciences*, 36(Part 8/W2):14–18.

Troup, G.J., 1972. Photon Counting and Photon Statistics. In: Sanders, J.H., Stenholm, S., Eds., *Progress in Quantum Electronics*, Vol. 2 (Part 1), Oxford, Pergamon.

Wagner, W., Ullrich, A., Ducic, V., Melzer, T., and Studnicka, N., 2006. Gaussian decomposition and calibration of a novel small-footprint full-waveform digitising airborne laser scanner. *ISPRS Journal of Photogrammetry and Remote Sensing*, 60(2):100–112.

Wuerlaender, R., Eder, K., and Geist, T., 2004. High quality DEMs for glacier monitoring: Image matching versus laser scanning. In: Altan, M.O., Ed., *International Archives of Photogrammetry and Remote Sensing*, 35(Part B7):753–758.

Zwally, H.J., Schutz, B., Abdalati, W., Abshire, J., Bentley, C., Brenner, A., Bufton, J., Dezio, J., Hancock, D., Harding, D., Herring, T., Minster, B., Quinn, K., Palm, S., Spinhirne, J., and Thomas, R., 2002. ICESat's Laser Measurements of polar ice, atmosphere, ocean, and land. *Journal of Geodynamics*, 34(3–4):405–445.

- Pairwise or Multiple Surface Matching: In computer vision, the general problem of matching 3D range data is defined for multiple surfaces with no restriction on the complexity of 3D surfaces. However, this universal model is rarely needed in a typical LiDAR strip adjustment, which is mostly concerned with matching two surfaces, which are less complex and usually represented in a 2.5D raster format (DEM). The sensor parameter-based strip adjustment methods usually impose certain requirements on the data, in terms of overlap configuration; thus, multiple surface matching techniques can be applied in these cases, such as for overlap areas in multiple cross-strips.
- Strip Deformation: Data driven techniques are generally based on 3D transformations of variable complexity. Many methods assume that the transformation between the LiDAR point cloud and the reference ground surface (DEM) is adequately modeled by a rigid body transformation. More general solutions allow for strip deformation.
- Use of LiDAR Intensity Data: Earlier methods could only use range data, as there was originally no intensity data available. Since intensity data has become a standard product, it can be used to support the strip adjustment process in various ways, such as providing clues for segmentation or establishing correspondence between strips, typically supporting matching in the horizontal domain.
- Grid-Based Processing: Methods vary as to whether they can directly handle irregular data, such as techniques that work on a TIN, or evenly spaced (gridded) raster data, such as a 2.5D DEM may be required, which is obtained by interpolating the LiDAR point cloud, which will potentially introduce additional errors in the data.
- Point or Feature-Based Techniques: There are methods that can directly work on the original LiDAR point cloud or on derived raster data, while others are based on using features such as planar patches.
- Object Space Requirements: Certain techniques impose restrictions on the object space in terms of shape, such as flat areas, buildings, and rolling terrains, where they can be applied.
- Ground Control Use: LiDAR strip adjustment by definition is a relative correction, as all the observations are differential, and hence the elimination or reduction of strip discrepancies does not mean that the absolute accuracy of the LiDAR point clouds will be better after the strip adjustment, regardless of whether data driven or sensor calibration-based methods are used. Therefore, the use of ground control is always necessary to assess overall point cloud accuracy. Most strip adjustment methods can accommodate the use of ground control.
- Manual or Automated Methods: While most of the strip adjustment techniques are self-contained and require no or little human interaction, there are still a large number of frequently used techniques that are partially based on operator measurements. This is mainly due to the use of ground control, which is less automated in general, although depending on the type of ground control, the need for human interaction can be totally eliminated, such as using LiDAR-specific ground targets or a calibration range.

LiDAR strip adjustment is a complex process, and to understand its performance potential as well as limitations several aspects should be considered, including surface representation, error characteristics of LiDAR, surface comparisons, modeling, and adjustment techniques. This chapter is structured to first review these fundamentals of the underlying theory, and then the most important techniques of LiDAR strip adjustment will be discussed. The term strip adjustment will cover both techniques: data driven and sensor calibration-based methods in the following discussion.

8.2 Surface Representation

The surface description obtained from LiDAR data is primarily given as a set of points called a point cloud, which is, in fact, a random or irregularly sampled representation of the actual surface, including objects (DSM). Ideally, a true surface elevation data, S_c with respect to a mapping plane can be described as a 2D continuous function:

$$S_c = f(x, y) \tag{8.1}$$

For practical reasons, the discrete representation of the surface should be considered, which is obtained by an evenly spaced 2D sampling of the continuous function and by converting the continuous elevation values to discrete ones:

$$E_{ij}^d = Q_p(S_c) = Q_p(f(x_i, y_j)) \tag{8.2}$$

where

E_{ij}^d is the discrete surface representation

Q_p is the quantization function (typically a regular step-function), which maps the continuous input parameter space to 2^p discrete levels

x_i and y_j are the 2D sampling locations

The fundamental question is how well the second format describes the first representation. Obviously, for simple surfaces that can be composed from planar patches, spheres, conical shapes, etc., the continuous surface representation can be restored from the discrete representation without any error based on the Shannon sampling theory (discussed later). In reality, however, the Earth's surface and other objects cannot be observed with unlimited precision. Some level of surface detail cannot be discerned and measurement errors exist; hence, the practical question is how to optimize the parameters of a discrete representation, such as sampling distance (sampling spatial frequency) and quantization levels, to achieve given accuracy requirements in terms of acceptable surface deviations. For example, what is the maximum sampling distance to keep the differences between the two representations below a predefined threshold? The answer is dependent on the application circumstances. For instance, for creating a topological map of a road, the surface roughness is irrelevant and small surface variations need not be considered. However, for road design or maintenance purposes, the details of the elevation changes are equally or even more important than the global nature of the surface, such as the road's absolute location in a mapping frame.

From a strictly theoretical point of view, the problem of how well a discrete representation describes the continuous case is well understood from Shannon's information theory (Shannon, 1948). Probably the most relevant and well-known expression is related to the sampling frequency, which is mostly known as the Nyquist frequency (Shannon, 1949). Rephrasing it for our case in one dimension and ignoring the impact of the signal quantization, it simply states that if a surface has a given maximum detail level (the surface changes are less than a predefined value), then there is an optimal sampling distance, and, thus, any discrete representation that has this optimal sampling distance (or shorter) can reconstruct S_c from E_{ij}^d without any degradation. In other words, if f_{max} is the highest spatial spectral frequency for a given surface, then the sampling distance d_s is sufficient for the complete representation of this surface and, consequently, the continuous surface can be restored without any error from the discrete representation in this ideal case. The Nyquist criterion for the 1D case is

$$d_s \le \frac{1}{2f_{max}} \tag{8.3}$$

Similarly, for 2D representations the directional spatial frequencies should be considered and the Nyquist criterion should be satisfied in both directions.

$$d_s^x \le \frac{1}{2f_{max}^x} \quad \text{and} \quad d_s^y \le \frac{1}{2f_{max}^y} \tag{8.4}$$

If the Nyquist criterion is satisfied, then the reconstruction of the continuous surface from the discrete values using the required or shorter sampling distances is described by

$$S_c(x,y) = \sum_{i=-\infty}^{\infty} \sum_{j=-\infty}^{\infty} E_{ij} \frac{\sin\left(\pi d_s^x \left(x - \frac{i}{d_s^x}\right)\right)}{\pi d_s^x \left(x - \frac{i}{d_s^x}\right)} \frac{\sin\left(\pi d_s^y \left(x - \frac{j}{d_s^y}\right)\right)}{\pi d_s^y \left(y - \frac{j}{d_s^y}\right)} \tag{8.5}$$

In this ideal case, the reconstruction introduces no errors, as the discrete representation provides a complete description of the surface function (see Figure 8.4). In reality, however, it is impossible to achieve this situation for several reasons. These include the characteristics of real surfaces and the inherent limitations of the measurement processes in LiDAR systems, discussed in Section 8.3.

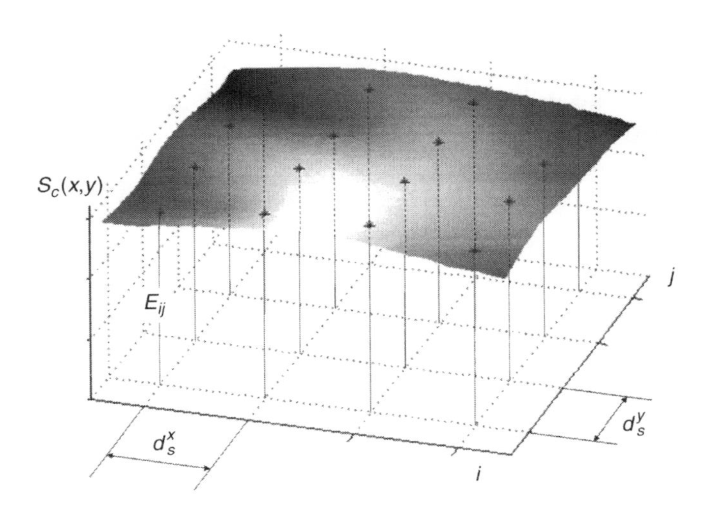

FIGURE 8.4
Surface reconstruction from discrete representation.

Another important aspect of the sampling process is the quantization of the elevation data. What is the smallest elevation difference that can be distinguished? Although the quantization is a nonlinear transformation, its impact in practice can be safely ignored, as in modern digital systems, the usual numerical representation provides fine representation of a wide signal range, in order that the error introduced by converting the continuous signal into a discrete one is usually negligible.

8.3 Characteristics of LiDAR Data

The most important characteristics of LiDAR data with respect to strip adjustment are the irregular point distribution, the laser point accuracy, the impact of beam divergence (footprint size), and the return signal dependence on the physical characteristics of the surface. Although these features are important in general to LiDAR data processing, they have a specific impact on the strip adjustment processes and, therefore, a short review is provided here.

8.3.1 Irregular Point Distribution

The LiDAR point cloud has an irregular spatial distribution and, therefore, Equations 8.3 through 8.5 describing the surface representation completeness cannot be applied directly. Although the theory can be extended to treat irregularly sampled data, in practice it is not used because of its high complexity. Since users try to set LiDAR sensor parameters to achieve nearly identical sampling distances in both directions, the widely accepted practice is to consider the average of the along and across flight point spacing values as the equivalent sampling distance. It is important to note that this simplified model works well only for smaller scan angles, as at the end of profiles the along track sampling distances will double for most LiDAR systems (obviously, depending on the scanning pattern, as discussed in Chapter 2).

One major problem with real surfaces is that vertical surfaces, such as buildings and walls, represent a discontinuity of the surface function and, in theory, a complete discrete representation would require an infinitely short sampling distance, as the spatial frequency would be unbounded. Therefore, in practice, discontinuities cannot be properly represented by a grid-based surface model and are almost always approximated with a smoothed transition between the two surface values, where the closeness of the approximation improves with decreasing sampling distance. If the terrain surface is separated from buildings and other man-made objects, different data formats and modeling techniques can be applied, which can properly describe vertical walls and similar features.

8.3.2 Laser Point Accuracy

The LiDAR systems are multisensory and dynamic systems with many potential error sources, including both systematic and random error components. Due to the complexity and highly interconnected structure of the system, not all the errors can be observed from strip discrepancies (more details to follow). LiDAR systems incorporate at least three main sensors: GPS and INS navigation sensors, and the laser-scanning device. Furthermore, there is a moving component in the laser system (e.g., oscillating or rotating mirrors) with its usual problems of position encoding, wear and mechanical hysteresis that can further degrade the accuracy of the acquired LiDAR data. In general, the errors in laser scanning data can come from individual sensor calibration (called measurement errors), lack of synchronization, and misalignment between the different sensors. Baltsavias (1999) presents an overview of basic relations and error formulae concerning airborne laser scanning, and Schenk (2001) provides an early error analysis on LiDAR. Even after careful system calibration, some errors could be present in the data, and navigation errors usually dominate. The errors are seen as discrepancies between overlapping strips and at ground control surfaces. Most of these systematic errors can be corrected using strip adjustment (with or without ground control) by eliminating the discrepancies between overlapping LiDAR strips using various strip adjustment methods developed over the years. The understanding of the error terms as well as the overall error budget is important and will be discussed next.

LiDAR is based on direct georeferencing and the position of a laser point measured at time t_p is computed by the LiDAR equation:

$$r_M(t_p) = r_{M,INS}(t_p) + R_{INS}^{M}(t_p) \cdot \left(R_L^{INS} \cdot \begin{bmatrix} 0 \\ \sin(\beta(t_p)) \\ \cos(\beta(t_p)) \end{bmatrix} \cdot d_L(t_p) + b_{INS} \right) \quad (8.6)$$

where

$r_M(t_p)$ —3D coordinates of the laser point in the mapping frame 3D INS coordinates (origin) in the mapping frame, provided by

$r_{M,INS}(t_p)$—GPS/INS (the navigation solution typically refers to the origin of the INS body frame)

$R_{INS}^{M}(t_p)$ —Rotation matrix between the INS body and the mapping frame

R_L^{INS} —Boresight matrix between the laser frame and the INS body frame

$d_L(t_p)$ —Range measurement (distance from the laser sensor reference point to the object point)

b_{INS} —Boresight offset vector (vector between the laser sensor reference point and the origin of INS) defined in the INS body frame

$\beta(t_p)$ —Scan angle defined in the laser sensor frame (x_L is flight direction, y_L to the right, and z_L goes down)

Based on Equation 8.6, the laser point accuracy in a mapping frame can be determined by applying the law of error propagation. Earlier discussions on the errors and model parameter recovery can be found in (Baltsavias, 1999; Schenk, 2001; and Filin, 2003a,b) and a recent comprehensive discussion on the various terms of the error budget, including analytical and simulation results, can be found in (Csanyi May, 2007). The difficulty of applying the theory in practical cases, however, is the time dependency of some of the parameters (note that several parameters are modeled as a function of time). In a simplified interpretation, there is no clear separation between the systematic and stochastic error terms for longer periods of time, as several components could have nonstationary behavior in terms of changing over intervals ranging from 10 to 30 min or even hours. For instance, the navigation part, which typically represents the largest terms in the error budget, may have components changing slowly in time and could be considered as drifts with respect to the other stochastic components with higher dynamics. These could subsequently be modeled as systematic error terms for shorter time periods, such as for the time it takes to fly a strip. In fact, these phenomena form the basis of the need for strip adjustments, as these error terms could change significantly between surveys as well as within surveys, similar to experiences with the GPS-assisted aerial triangulation with longer flight lines (Ackerman, 1994). This is also the reason why it is difficult to obtain reliable estimates of error terms. Table 8.1 lists some of the basic and generally accepted error sources and typical values, which can be used for an error propagation analysis for the

TABLE 8.1

Major Error Sources Affecting the Accuracy of LiDAR Point Determination

	Errors	Typical Values
Navigation solution (GPS/INS)	Errors in sensor platform position and attitude (shift and attitude errors)	σ_X, σ_Y: 2–5 cm; σ_Z: 4–7 cm σ_ν, σ_φ: 10–30 arcsec σ_κ: 20–60
Laser sensor calibration	Range measurement error	σ_r: 1–2 cm
	Scan angle error	σ_β: 5 arcsec
	Error in reflectance-based range calibration	[−20–10] cm
Inter-sensor calibration	Boresight misalignment between the INS body and laser sensor frames (shifts and angular errors)	σ_{Xb}, σ_{Yb}, σ_{Zb}: < 1 cm $\sigma_{\nu b}$, $\sigma_{\varphi b}$: 10 arcsec, $\sigma_{\kappa b}$: 20 arcsec
	Error in measured lever arm (vector between GPS antenna and INS reference point)	σ_{Xa}, σ_{Ya}, σ_{Za}: < 1 cm
Miscellaneous errors	Effect of beam divergence (footprint)	[0–5] cm
	Terrain and object characteristics	
	Time synchronization	
	Coordinate system transformations	
	Atmospheric refraction	
	Sensor mounting rigidity	

point cloud accuracy. In addition, the importance of these values is that they provide a lower limit on what can be realistically expected from a strip adjustment-based correction using sensor calibration-based methods.

8.3.3 Impact of Beam Divergence

The discussion above assumed an ideal sampling of the surface; in other words, the laser beam was modeled as a line with no width. In reality, there is always beam divergence of the laser pulse, which is usually rather small in topographic applications and can be controlled by system design. With increasing object distance, the beam divergence, however, can become relatively large, resulting in a footprint size that can be comparable to the surface variations. This can introduce additional errors in the reconstructed laser point position; in fact, it could produce substantial errors and, therefore, should be carefully considered in applications where high accuracy is required. In practice, the footprint of the laser pulse is used as a measure to assess the error introduced by the divergence of the laser beam. The footprint is defined by the intersection of the surface with the cone representing the divergence of the laser beam; for a vertical laser beam and horizontal surface case, the footprint size is

$$d_{\mathrm{fp}} = 2r \cdot \tan\left(\frac{\Delta\theta}{2}\right) \tag{8.7}$$

where

d_{fp} is the diameter of the footprint
r is the range (the distance between the laser sensor and the surface)
$\Delta\theta$ is the beam divergence

An additional correction can be introduced to model a nonvertical laser beam as well as sloped surface situations. Ignoring the cases of extreme parameter values and assuming small beam divergence and smooth terrain, the footprint can be approximated as

$$d_{\mathrm{fp}} = 2r \cdot \frac{\tan\left(\frac{\Delta\theta}{2}\right)}{\cos(\beta - \alpha)} \tag{8.8}$$

where
β is the scan angle of the laser beam
α is the angle between the surface normal and the vertical

Strictly speaking, the return pulse is formed from all the photons reflected from the footprint in the incoming pulse direction, and they jointly determine the shape or waveform of the return pulse (more details on waveform are in Chapter 7). If the reflection characteristics of the surface vary considerably within the footprint, such as having small very reflective objects, then a dominant return pulse can be generated from any such object within the footprint. For steep terrains, the uncertainty of the range measurement, which is close to the vertical coordinate accuracy of the point cloud, could be large even for small scan angles (see Figure 8.5). For instance, a 0.3 mrad beam divergence at 600 m flying height (AGL) represents an approximately 20 cm footprint, which could easily result in a 5 cm range error at an extreme case of 10° scan angle and 20° surface slope.

Concerning the relation between the horizontal accuracy and the footprint, the simplest model is if the footprint is considered as an error ellipse. More refined models attempt to model the energy distribution of the laser pulse within the footprint, which is usually

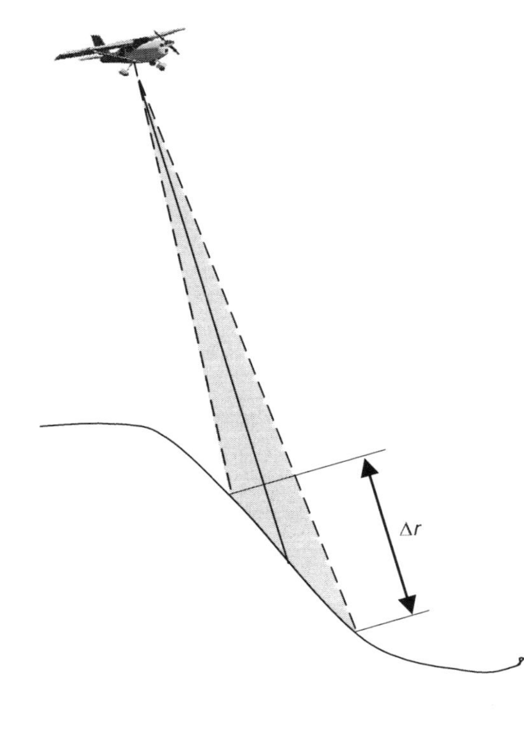

FIGURE 8.5
The impact of footprint (beam divergence) on range measurement uncertainty at sloped surface.

modeled by a 2D Gaussian function (the radius of the footprint is equal to the 2σ value) (Kukko and Hyyppä, 2007). However, for horizontal error estimation, a model based on uniform distribution is frequently preferred; the argument is that an area close to the perimeter of the footprint with objects of strong reflectance could easily produce a return signal even if the energy is relatively small as discussed above; obviously, it is a slightly conservative model.

8.3.4 LiDAR Intensity and Waveform

The ranging accuracy of the laser sensor is defined for hard surfaces with certain physical (reflectance) characteristics and, thus, the dependency on the surface material should not be generally ignored; even for man-made surfaces, which are usually hard, the ranging accuracy can vary significantly (see Figure 8.6). The pulse formation of a LiDAR system is controlled in terms of time and space. The distribution of the laser energy within a pulse, which is typically 5–10 ns, is normally modeled by a Gaussian as a function of time (Kukko and Hyyppä, 2007). The beam formation at the sensor is defined by the divergence angle, which is another (optical) design parameter of the LiDAR system; in most applications, a narrower divergence angle is required. As the laser pulse travels in air, the shape of the pulse as well as its energy distribution will change. As the pulse reaches the surface, photons will be reflected in many directions, depending on the surface geometrical (orientation and roughness) characteristics. In addition, the proportion of reflected photons to the absorbed photons varies, depending on the physical characteristics of the surface. Furthermore, the reflection is not instantaneous, and there could be a slight delay in the process, again, depending on the material characteristic of the surface.

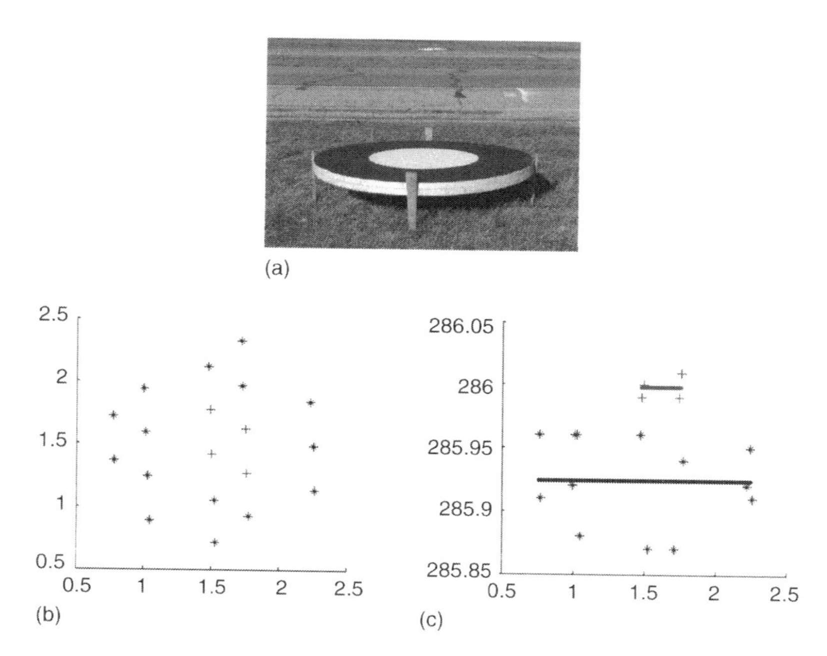

FIGURE 8.6
Ranging bias due to differences of the physical materials; bright surface areas produce shorter ranges compared to dark surfaces of the target; (a) LiDAR-specific target, (b) and (c) points reflected back from the target, top, and side view.

The photons reflected back from the footprint jointly form the shape of the return signal that is detected in the laser sensor, where, depending on the detection mechanism, various measurements can be derived. Considering the pulse's travel to and from the footprint and the interactions during the reflection, the shape of the return signal could be quite different from the emitted pulse at the laser source (the firing point). Obviously, the several-order difference in energy can be properly handled at the detector, and only the profile of the return signal needs special consideration. In the simplest case, only the arrival of the pulse is detected and a single range is computed from the travel time, or multiple ranges for the cases when multiple pulses are detected. If the detected pulse is characterized by a number, which relates to its energy content and shape, then an intensity signal is created. Ultimately, the entire returning pulse's waveform can be recorded and passed to the user; hence, the range determination as well as extraction of any other information is left to the user, as discussed in Chapter 7.

The understating of the above aspects of the range data is essential to strip adjustment. Although any anomaly in the range is likely to cancel out in overlapping LiDAR strips, an error caused by differences in surface material can easily lead to biases in the data and create strip degradation or deformation. While early LiDAR strip adjustment techniques were based on using only range data, current systems typically provide multiple return and intensity signal capabilities that can improve the LiDAR point cloud accuracy and support various strip adjustments processes. For example, the intensity-based range compensation became a standard on modern systems, resulting in improved vertical accuracy for several surface types. In addition, multiple returns or waveforms could offer the potential for direct measurement of surface discontinuities, such as if the footprint happens to include the upper and lower edges of a vertical wall (see Figure 8.7) and, thus, can improve strip adjustments by providing both vertical and horizontal control.

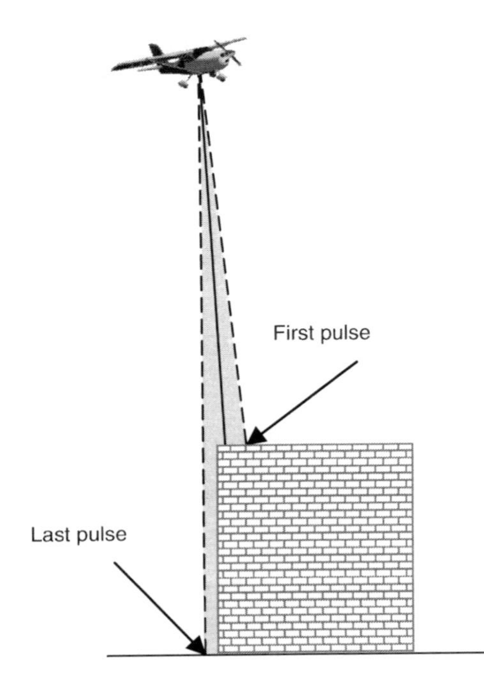

FIGURE 8.7
Multiple returns along vertical surfaces.

8.4 Concept of LiDAR Strip Adjustment

The purpose of this section is to introduce the fundamentals of the basic mathematical models and techniques on a conceptual level before the major strip adjustment methods are discussed in the next section. For the simplicity of the discussion, the case for two surfaces is considered, although the generalization to multiple surfaces is possible.

8.4.1 Strip Adjustment Models

The data driven methods are based on establishing a 3D transformation T between two 3D point sets $P = \{(x_i^p, y_i^p, z_i^p), i = 0,\ldots, n\}$ and $Q = \{(x_j^q, y_j^q, z_j^q), j = 0,\ldots, m\}$, which represent an irregular (random) spatial sampling of the same surface. As no correspondence is assumed between the two data sets, the determination of the transformation T described in Equation 8.9 by minimizing the Euclidean distance between one data set and the transformed other data set is not a trivial task.

$$\underset{T}{\mathrm{Min}} \left\| P - T(Q) \right\| \tag{8.9}$$

The transformation $T=[T_x\, T_y\, T_z]^T$ can be established between two overlapping LiDAR strips or more frequently between a reference surface and a LiDAR strip. In the first case, the strip discrepancies are reduced or removed by applying the transformation to one data set, but no improvement in terms of the absolute accuracy of the point cloud can be guaranteed, in general. In contrast, by using a reference surface, the correction applied to the LiDAR strip should and will result in better absolute accuracy, provided the reference surface is sufficiently discrete and accurate. This typically requires either a 3D test range or an area with precisely surveyed man-made objects, as the accuracy of older existing DEM datasets is not even close to the typical LiDAR vertical accuracy. For multiple strip overlap situations, a quasi-reference surface can be established by averaging the strips and then establishing and applying the transformation parameters for each strip individually (see Equation 8.10). Obviously, this approach only works if the distribution of the strip discrepancies is close to symmetrical.

$$\underset{T_k}{\mathrm{Min}} \left\| \frac{1}{s} \sum_{l}^{s} Q_l - T_k(Q_k) \right\| \tag{8.10}$$

where s is the number of overlapping strips.

The sensor model-based strip adjustment methods aim to calibrate the sensor parameters to minimize the strip discrepancies. As described in Equation 8.9, the Euclidean distance is minimized, except that the points forming the two sets are expressed in sensor parameters in this case as

$$\begin{aligned} P &= \left\{ (x_i^p, y_i^p, z_i^p), i = 0,\ldots, n^p \right\} \\ &= \left\{ \left(r_{\mathrm{M}}(t_i^p) = r_{\mathrm{M,INS}}(t_i^p) + R_{\mathrm{INS}}^{\mathrm{M}}(t_i^p) \cdot \left(R_{\mathrm{L}}^{\mathrm{INS}} \cdot \begin{bmatrix} 0 \\ -\sin(\beta(t_i^p)) \\ \cos(\beta(t_i^p)) \end{bmatrix} \cdot r_{\mathrm{L}}(t_i^p) + b_{\mathrm{INS}} \right) \right), i = 0,\ldots, n^p \right\} \end{aligned} \tag{8.11}$$

$$Q = \{(x_i^q, y_i^q, z_i^q),\ i = 0,\ldots,m^q\}$$

$$= \left\{ \left(r_{\mathrm{M}}(t_i^q) = r_{\mathrm{M,INS}}(t_i^q) + R_{\mathrm{INS}}^{\mathrm{M}}(t_i^q) \cdot \left(R_{\mathrm{L}}^{\mathrm{INS}} \cdot \begin{bmatrix} 0 \\ -\sin(\beta(t_i^q)) \\ \cos(\beta(t_i^q)) \end{bmatrix} \cdot r_{\mathrm{L}}(t_i^q) + b_{\mathrm{INS}} \right) \right),\ i = 0,\ldots,m^q \right\} \tag{8.12}$$

where, the notation of the various parameters is described above for Equation 8.6. Although the criterion to adjust the parameters is the same in both models, the sensor model-based computation is adversely affected by a few problems. Firstly, the functional correlation of the parameters should be mentioned; there is the typical problem of correlation between attitude and position of a sensor platform, such as error in roll and across track offset resulting in the same discrepancy of object point locations. Another problem is whether some parameters can be observed at all, or there exists a data set where sufficient observation can be obtained. Another complication is the stability of the sensor parameters, such as the slowly changing offset or drift of the georeferencing solution, expressed as time-dependent terms in Equations 8.11 and 8.12. In summary, there is no general solution for simultaneously calibrating all the sensor parameters; therefore, certain restrictions on sensor parameters and data should be introduced to recover a subset of the calibration parameters.

Most LiDAR systems used in topographic mapping provide only the point cloud as the primary sensor data that serve as the starting point for any further processing, such as bare earth filtration, road or building extraction, and biomass estimation. If the actual raw sensor data are made available to the strip adjustment process, then the sensor parameter recovery could be significantly improved. For example, having the six georeferencing parameters, including position and attitude data, and the scan angle and range measurements for every laser point (all of them are easily available during the LiDAR point cloud creation) can eliminate most of the parameter correlation and observation deficiencies of the sensor calibration-based methods.

Regardless of which model is used for strip adjustment, the main difficulty in implementing an adjustment, based on Equations 8.9, 8.11, and 8.12, is how to establish correspondence between the two 3D data sets. The characteristics of the LiDAR data, including irregular point distribution, point density, error budget, etc., make the determination of 3D discrepancies between overlapping LiDAR strips or between a LiDAR strip and control information (such as a reference surface) nontrivial. In contrast to traditional photogrammetry, establishing point-to-point correspondence between overlapping LiDAR strips is practically impossible. Therefore, either data must be resampled to allow for point- or area- and feature-based processing, rather than point-based algorithms have to be used; note that methods that can directly treat irregularly distributed data are also available. Since a large number of methods are based on using interpolated data, the following subsection will review the frequently used surface interpolation techniques.

8.4.2 Surface Interpolation Methods

The simplest way to establish point-to-point correspondence between two 3D data sets is if one of them is resampled to the other's grid. This can be done between two irregularly spaced data sets, such as two overlapping LiDAR strips, or, most commonly, the irregularly spaced LiDAR data are converted to an independent regular grid. From the perspective of strip adjustment, both cases are important and rely on using one of the major interpolation techniques. Before reviewing the relevant interpolation methods, the major data formats used to represent surfaces are briefly discussed.

Traditionally, digital elevation data are stored in three main structures: (1) triangulated irregular networks (TINs), (2) regular grid (raster format), and (3) lines of equal elevation (contours); an extensive discussion on digital elevation models can be found in Maune (2007) and El-Sheimy et al. (2005). Compared to past DEM data sets, LiDAR data are primarily different in the significantly larger number of surface points per unit area and by the dissimilar accuracy characteristics (the accuracy terms are practically reversed). LiDAR has significantly better vertical accuracy compared to the horizontal term, which is the opposite of the case for stereo photogrammetrically derived data.

At the introduction of LiDAR technology, the vast amount of data obtained presented a formidable challenge for the data processing and storing systems, but now the rapid technological advancements of computer technology mean that data can be easily processed and stored in any of the three major formats. However, preference appears to be given to raster format, as it is broadly supported on most computer platforms, offering easy visualization and basic processing tools. Furthermore, the raster format is typically the main distribution structure for national elevation models, such as the Nationa l Elevation Dataset in the United States (Gesch et al., 2002), where the DEM is available in the form of GeoTIFF files. It is important to note that TIN is also a widely used format, with several advantages for storing and processing surface data. Furthermore, it directly provides for interpolation and, thus, it frequently serves as an intermediate format from which the raster representation is derived.

Creating a gridded surface (regular or irregular) from irregularly spaced data requires the estimation of elevation values at the grid points, which is based on the interpolation of the known data. The interpolation methods can be grouped in several ways, but the most important aspect is the spatial extent from which points are used for interpolation. Some techniques use only a small number of surrounding sample points to determine an unknown elevation while other methods may consider a larger neighborhood, or even all sample points together are used to determine the unknown elevations. In the first case, the elevation data at the sample points is usually preserved. In the second case, some kind of surface fitting is applied to the sample data to determine the unknown elevations and, therefore, the sample point elevations may not necessarily be preserved. Mostly because of terrain characteristics and undersampled data, no single interpolation technique works best in every situation. This description provides a review of the existing, most commonly used interpolation methods for terrain data.

Inverse distance weighted (IDW) interpolation determines the unknown values as a weighted average of the surrounding sample points. The weights are a function of the inverse distances between the unknown and the measured points. This method works for both regular and irregular point distribution, and requires a parameter such as search radius to select sample points for the interpolation around an unknown point. Generally, it has drawbacks for irregular point distribution, as shown in Figure 8.8 for LiDAR data.

Natural neighbor interpolation is a more general form of IDW, as it uses area-based weighting to determine the unknown elevations. Consequently, this widely used method works for both irregular and regular point distributions. The algorithm is based on the Delaunay triangulation and Voronoi diagram (Okabe et al., 2000). The Voronoi neighbors of the unknown point are determined in a way that the insertion of the unknown point would result in the Voronoi neighborhood of the measured points. Then these Voronoi neighbors are used to determine the elevation of the unknown point. One advantage of the method is that since the method itself determines the surrounding measured points to be included in the determination of the unknown point, there is no need for additional parameters (search radius) that have drawbacks when the data points have an irregular distribution.

After the points to be included in the calculation are found, the unknown elevation is calculated as a weighted average of these elevations. The weights are based on the common

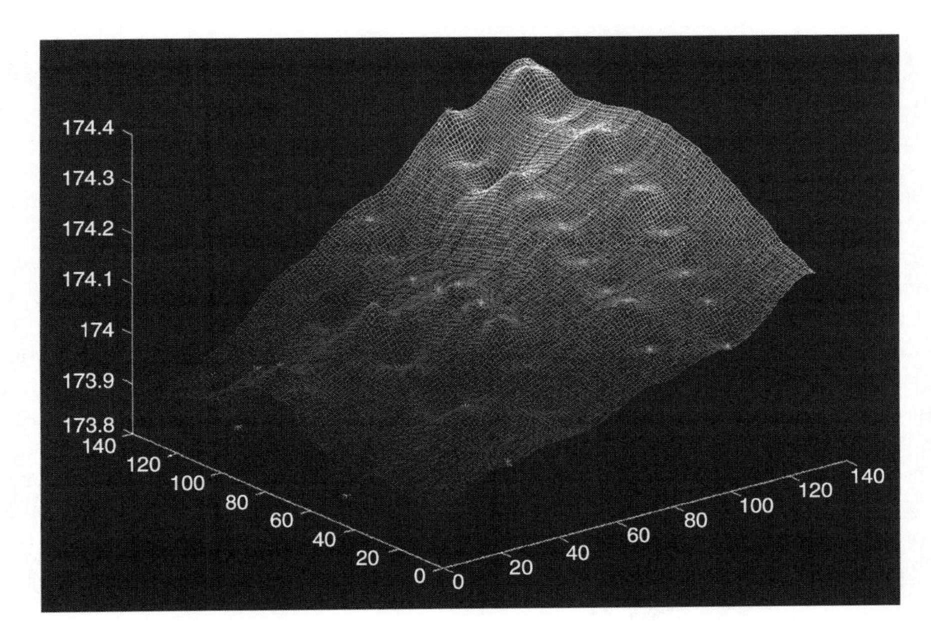

FIGURE 8.8
Dimpled effect around measured points in the case of IDW.

area of the Thiessen polygon of the unknown point (Heywood et al., 1998) and the Thiessen polygons of each selected point before the new point has been inserted to the triangulation. The interpolated surface passes through the sample points and it is constrained by the input data range, and, therefore peaks and valleys appear in the interpolated surface only if they were measured.

Spline interpolation fits a mathematical function, defined by piecewise polynomials of varying orders, to some neighboring measured points to determine the value at the unknown locations, which may be compared to bending a sheet of rubber through the measured points (Loan and Charles, 1997). The lowest order spline is equivalent to linear interpolation, or bilinear transformation for surfaces. Spline interpolation results in a smooth surface that passes through the sample points and, in general, gives a good representation of smoothly changing terrain (no sudden elevation changes, such as buildings and other man-made objects). This is a very useful method if the goal is to derive good quality contours.

Another technique is **kriging**, named after a South African mining engineer D.G. Krige who developed the technique (Oliver and Webster, 1990). Kriging estimates the unknown values with minimum variances if the sample data fulfills the condition of stationarity, which means that there is no trend in the data, such as main slope. Similar to some other methods, kriging calculates the unknown values as a weighted average of the sample values; however, the weights are based not only on the distance between the sample points and unknown point but also on the correlation among the sample points. The first step in kriging is the determination of the variogram, which is found by plotting the variances of the elevation values of each sample point with respect to each other sample point versus the distance between points. Once the experimental variogram is computed, the next step is to fit a mathematical function to the variogram that models the trend in it. This fitted polynomial is called the model variogram, and is then used to compute weights for the sample points for the calculation of unknown elevations. This method is called ordinary kriging, and if stationarity of the data is not fulfilled, the more universal kriging can be used. Kriging works with regular and irregular sample point distributions and is an exact interpolation method, as the interpolated surface passes through all the sample points.

Trend surface analysis approximates the surface by fitting a polynomial to the sample points. The proper order of the polynomial can be chosen considering the rule that any cross-section of an n-order surface can have at most $n - 1$ alternating maxima and minima. The coefficients of the fitted polynomial to the sample data can be determined by least squares adjustment, minimizing the square sum of the differences between the z values of measured points and the fitted surface at the sample points. Obviously, increasing the order means a better fit at the sample locations, but between the points it can result in large and sudden changes due to the lack of constraint. Therefore, in common practice, the order of the polynomial normally does not exceed five. This method works for both regular and irregular sample point distributions; however, it can change very rapidly between sample points at locations where the sample points are far from each other and especially near the borders of the interpolated surface. This method is appropriate to model the main trend in a surface, but it cannot model local irregularities. Therefore, it is not always applicable, and usually is applied to model the trend of smaller areas.

Orthogonal transformation-based methods reconstruct a surface function (2.5D) by using a linear combination of a set of orthogonal basis functions. For example, Fourier and wavelet transformations transform the data from time or space domain to frequency domain. By inverse transformation using the coefficients, the original surface can be reconstructed and the surface values can be calculated at the unknown locations too. The numerical methods of forward and backward transformation work only for regular point distribution; therefore, in their original form these series are not directly useable for the interpolation of irregularly distributed sample points. However, the coefficient determination can be formulated as a least squares problem, allowing the extension of the method to irregular point distribution (Toth, 2002). Figure 8.9 shows a LiDAR surface model that was created by a combination of trend surface analysis and Fourier harmonics-based modeling.

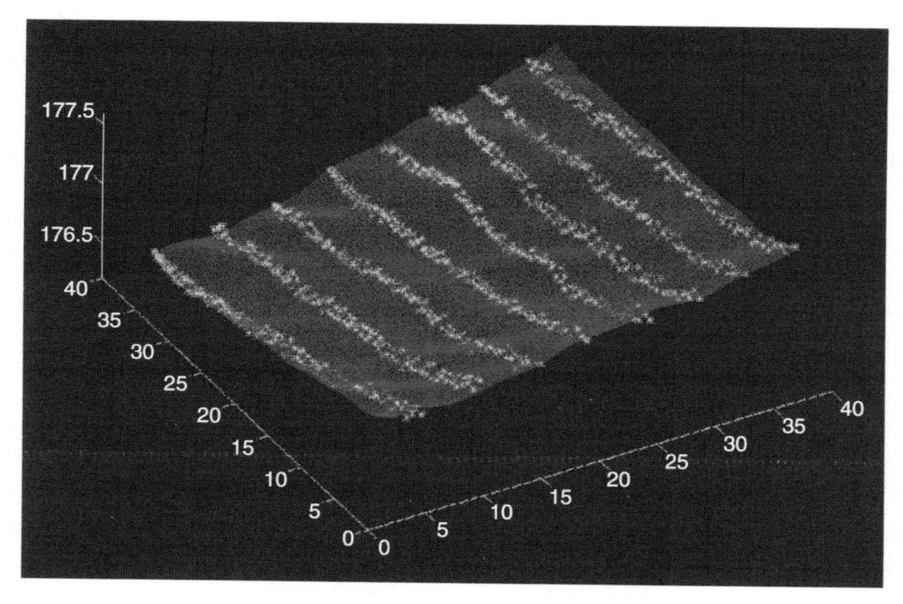

FIGURE 8.9
Fourier series approximation with a third-order polynomial extension.

The use of interpolation generally introduces error in the data, except for the rare case when the sampling criterion is met, which actually could be easily achieved for flat areas. However, in these cases, the strip adjustment suffers from the lack of the surface signal (terrain undulation), as no 3D discrepancies can be determined between the two surfaces (based on range data). In the general case, the LiDAR data tend to be undersampled and, consequently, if interpolated, errors will be introduced. Therefore, strip adjustment methods that can directly deal with irregular data should be preferred.

8.4.3 Strip Deformation Modeling

Another key aspect of a strip adjustment model is whether strip deformation is allowed or not. Strip deformation occurs if a rigid body model is not adequate to describe the transformation between the ground surface and the LiDAR data. Strips can be deformed for various reasons, but the changing quality of the georeferencing solution is frequently the main cause. Obviously, the concept of strip deformation assumes a spatial or time extent; sensor characteristics and environmental conditions could vary noticeably over longer distances and for extended time periods. For example, the errors of the georeferencing solution can be usually considered static for short time intervals, as discussed earlier. For example, during long and straight strips, such as for strip lengths of 50–100 km and flying time over several hours, the georeferencing solution can show slowly changing errors (drifts), frequently resulting in strip deformation; in particular, this is the case over terrain with significant elevation changes, such as high mountain ranges. Therefore, strips are frequently segmented into smaller sections, which are individually treated using a rigid body model.

Data driven strip adjustment methods are always based on a linear 3D transformation, which takes the general form:

$$\begin{bmatrix} x_\mathrm{p} \\ y_\mathrm{p} \\ z_\mathrm{p} \end{bmatrix} = \begin{bmatrix} t_0 & t_1 & t_2 & t_3 \\ t_4 & t_5 & t_6 & t_7 \\ t_8 & t_9 & t_{10} & t_{11} \end{bmatrix} \cdot \begin{bmatrix} x_\mathrm{q} \\ y_\mathrm{q} \\ z_\mathrm{q} \\ 1 \end{bmatrix} \tag{8.13}$$

Depending on the complexity of the applied model, which is determined by the data that can be observed and application-specific considerations, the actual number of the independent parameters varies from 1 to 12. The simplest strip correction is based on applying only a vertical shift, which was a typical procedure at the introduction of LiDAR, as the modest point density did not allow for accurate horizontal measurements of strip discrepancies. As technology improved, full 3D offset correction became feasible. The next step was when a similarity transformation was introduced, and since then it has been the most widely used model for data driven strip corrections. In many cases, the scale parameter is assumed to be unity, and then only the offset and rotation are considered for correction (6-parameter rigid body transformation). The use of the full 12-parameter, 3D affine transformation model is rather rare.

Strip deformation modeling can differ from the sensor calibration-based strip correction methods, where the deformation phenomenon is modeled through the sensor system parameters. In the general solution, all the terms, including the georeferencing solution and the laser sensor parameters, can be considered as time-dependent parameters, and thus would allow for optimal error modeling and correction. However, as discussed earlier, the implementation of this approach is simply not feasible because of parameter correlation and data dependencies. Therefore, restrictions and simplifications should be introduced to allow for the recovery of the sensor parameters, or more precisely, a subset of these parameters.

In a number of approaches, the georeferencing solution and the laser sensor calibration are separated, and thus the strip adjustment is formulated either for the georeferencing errors or for the sensor parameters. For example, laser sensor calibration parameters may slowly change over longer time periods while the georeferencing solution can drift or fluctuate at a much faster rate. Thus, once the laser sensor is calibrated, including range, scan angle, and boresight calibration, the errors in the georeferencing solution can be separately recovered. The georeferencing solution describes the sensor platform trajectory, including three position and three attitude angles as a function of time; for shorter time periods, these parameters could be sufficiently modeled by constant offsets, which leads to the rigid body transformation model discussed earlier. In many cases, this provides an adequate solution, as the georeferencing terms typically account for the largest part of the error budget.

The laser sensor calibration poses more difficulties because of the strong parameter correlation and inseparability of measurements. For example, the ranging error can be only calibrated to the combined accuracy level of the georeferencing solution and the ground control. The usual DEM accuracy is generally below the laser ranging accuracy, which is typically in the 1–3 cm RMS range, while the georeferencing positioning accuracy for airborne platforms is in the 5–10 cm range. If a reference DEM (test range) was precisely measured, it will account for at least 1–5 cm error. The removal of systematic scan angle errors is important, as this error results in strip deformation. Using flat surfaces, one can recover the scan angle error, frequently called smiley error, from profile measurements even from a single strip. Both analytical modeling and table look-up methods can be used. The calibration of the laser sensor boresight parameters is based on a rigid body model. To avoid the introduction of any additional error sources, the installation of the laser and georeferencing sensors is expected to provide a stable and rigid relationship between the sensors.

8.4.4 Surface Matching

Surface matching, the automatic coregistration of point clouds representing 3D surfaces, is an essential and rather difficult step of any strip adjustment technique. The purpose of this task is to provide the measurements for the identification and determination of strip discrepancies. During surface matching, the surfaces being compared are typically transformed into representations where the comparison of surfaces is straightforward. The model of the surface formation from LiDAR data, including data driven and sensor calibration-based techniques, was introduced earlier; hence, a general description of major surface matching methods and fitting solutions relevant to strip adjustment is provided here.

Techniques dealing with surface matching, including comparisons, fitting, and coregistration, can be grouped in several ways, depending on the data format, data characteristics, and transformation model used to describe the geometrical relationship between the two data sets. In almost all cases, the surface matching used for strip adjustment is accomplished using the least squares methods to minimize the Euclidean distances between the surfaces and then recover the transformation model parameters. In strip adjustment practice, closeness between the two surface representations is always assumed. Therefore, only a subproblem of the general surface matching, the refinement, or alignment task is addressed. In complex mapping processes, such as the extraction of various natural and man-made object features, the more general problem entails a global search, where either coarse or no orientation approximations for the two surfaces are available. The general problem of surface matching, also called range data matching, is an important field in computer vision, and a large variety of techniques have been developed over the years; a good review is provided in (Campbell and Flynn, 2001). Here, only the least squares surface matching (Gruen and Akca, 2005) is discussed because of its importance in the geospatial field.

The functional model of least squares surface matching of two surfaces can be formulated based on using implicit functions as

$$d(x,y,z)=p(x,y,z)-q(x,y,z) \tag{8.14}$$

where

p and q represent the two surface descriptions of the same object area

d represents the differences between the two samples, introduced by the various sensor errors and environmental conditions

The least squares adjustment minimizes the sum of the squares of the Euclidean distances between all the points of the surface p and the surface q. The computation of these distances requires an interpolation of the second surface, i.e., conjugate point determination, usually accomplished by piecewise surface modeling. The conjugate surface, q^0, must be determined in order that the partial derivatives can be estimated to provide the first-order Taylor series coefficients needed for the least squares solution. Thus, Equation 8.14 becomes

$$p(x,y,z)-d(x,y,z)=q^0(x,y,z)+\frac{\partial q^0(x,y,z)}{\partial x}\mathrm{d}x+\frac{\partial q^0(x,y,z)}{\partial y}\mathrm{d}y+\frac{\partial q^0(x,y,z)}{\partial z}\mathrm{d}z \tag{8.15}$$

The transformation T from Equation 8.9 that describes the spatial relationship between the two surfaces is modeled by n_T number of parameters, $m_i = (0,1,\dots, n_T-1)$. If the transformation T is nonlinear, then first it should be linearized by Taylor series expansion. Consequently, the differentials are expressed in model parameters as

$$\begin{bmatrix} \mathrm{d}x \\ \mathrm{d}y \\ \mathrm{d}z \end{bmatrix} = \begin{bmatrix} \frac{\partial T_x}{\partial m_0} \frac{\partial T_x}{\partial m_1} \cdots \frac{\partial T_x}{\partial m_{n_T-1}} \\ \frac{\partial T_y}{\partial m_0} \frac{\partial T_y}{\partial m_1} \cdots \frac{\partial T_y}{\partial m_{n_T-1}} \\ \frac{\partial T_z}{\partial m_0} \frac{\partial T_z}{\partial m_1} \cdots \frac{\partial T_z}{\partial m_{n_T-1}} \end{bmatrix} \cdot \begin{bmatrix} \mathrm{d}m_0 \\ \mathrm{d}m_1 \\ \vdots \\ \mathrm{d}m_{n_T-1} \end{bmatrix} \tag{8.16}$$

From this point, a conventional or generalized Gauss–Markoff estimation model can be built, where each point from surface p constitutes one observation equation. This solution is directly applicable to data driven methods, where the most typical transformation is the 7-parameter similarity model. When this technique is applied to the sensor calibration-based model, i.e., using Equation 8.6, certain restrictions on the model parameters should be introduced to achieve solutions. Finally, the least squares surface matching can be further generalized for multiple surface matching and can be extended with intensity matching and ground control (see Akca and Gruen, 2007).

The most widely used as well as the simplest surface matching technique is based on using raster format surface models, or DEMs, in which case the two data sets are available

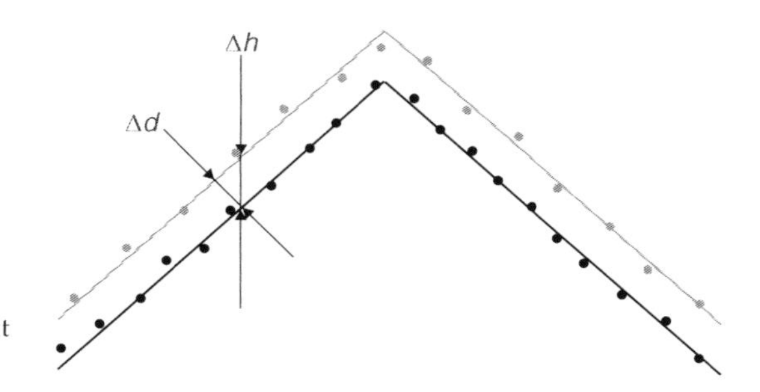

FIGURE 8.10
Distance along surface normal versus height difference.

in the same evenly spaced grid and, thus, vertical differences between them can be easily computed. This representation is also frequently called 2.5D data, referring to the fact that any location, (x, y) coordinate pair can have only one height value. The problem with this representation is that for terrains with surface normal vectors considerably deviating from the vertical, the simple vertical difference between the two data sets may not be the right measure for the surface discrepancy, as it does not account for horizontal differences and ignores their impact on the vertical differences. Considering the difference along the surface normal will result in the determination of correct 3D surface discrepancies, in which case, the volume between the two surfaces is minimized during the adjustment (see Figure 8.10).

The main advantage of using 2.5D surface data representation is that standard image processing techniques can be directly applied to the data, including area- and feature-based matching methods. Considering height as though it were image intensity is a very useful interpretation from a matching perspective, since in both cases the signal changes and patterns are needed to achieve successful matches. Note that basic image matching solutions usually result only in a 2D offset, and additional data and processing are needed to obtain 3D surface discrepancies. If LiDAR intensity image is used for matching, the vertical discrepancy can be directly obtained from the corresponding range data. A clear disadvantage of the DEM representation with respect to strip adjustment is that it is derived from the primary point cloud data by interpolation and, thus, errors are introduced, which could be further amplified by the fact that the LiDAR point density usually falls below the minimum spatial sampling distance required for complete surface representation.

TIN-based data representation seems to be the natural format for LiDAR, as it can preserve the original 3D information of the point cloud and provide an interpolated value for any location, based on the surface defined by the linked triangles. TIN offers an efficient way to reduce data size without compromising the surface representation; for example, the large number of surface points of a flat area can be suitably described by a few triangles. The determination of surface differences is clearly much more complex than it is for raster data. Surface matching techniques that can handle irregularly distributed points generally consider one data set in a TIN-based surface representation and then try to minimize the cumulative distance between every point of the other data set and the corresponding triangle along the normal of the triangle (Figure 8.11).

All the surface matching methods discussed so far impose no restrictions on the surfaces, except to avoid totally flat surfaces (no surface signal, i.e., no terrain relief). If one of the surfaces to be matched is formed from basic shapes, such as planar patches, the matching process can exploit the simple geometry that can be described analytically.

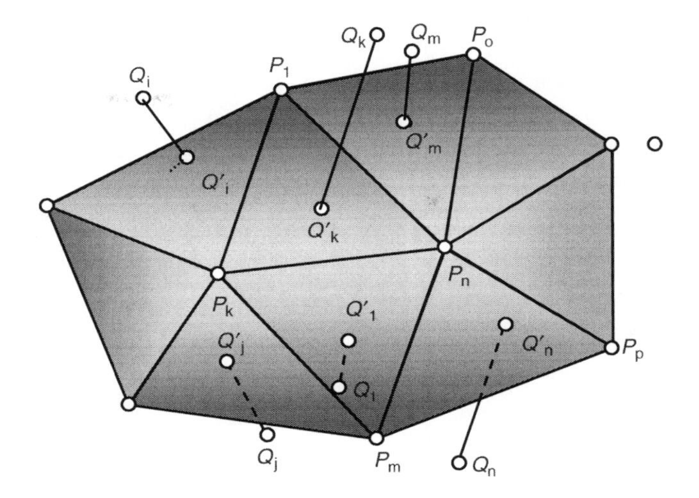

FIGURE 8.11
Distance definition between two point sets.

Although several shapes, such as linear, spherical, planar, and conical, have simple mathematical descriptions, they are rare in real life and, consequently, planar patches are used in most methods. Obviously, matching planes have a deficiency, as only the distance between the two data sets can be recovered. Therefore, several patches with reasonably varying surface normal vectors are needed to obtain 3D surface discrepancies. A good example is saddle roofs, which have quite different surface normal vectors and their flatness is generally good. Similarly, the intersection of planar patches can provide for linear features that can be matched. The advantage of featured-based matching is that there is no imposition on the primary data from where the features were extracted, and thus, it provides a mechanism to match surface data with high-level symbolically described object space data acquired by different sensors; Figure 8.12 shows roof scanned in two different strips.

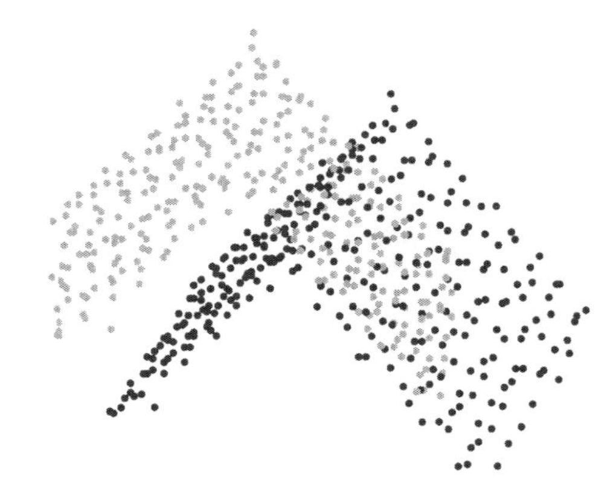

FIGURE 8.12
Matching linear features extracted from the roof intersection of planar roof surfaces.

Iterative registration algorithms are increasingly used in computer vision for registering 2D and 3D curves and range images. The well-known Iterative Closest Point (ICP) algorithm (Besl and McKay, 1992; Madhavan et al., 2005) has evolved significantly in recent years, and can provide robust solutions if modest approximations of the surface orientation are available, which is always the case in strip adjustment. The ICP algorithm finds the best correspondence between two point sets by iteratively determining the translations and rotations parameters of a 3D rigid body transformation:

$$M\,\mathrm{in}_{(R,T)} \sum_i \left\| M_i - (RD_i + T) \right\|^2 \tag{8.17}$$

where

R is a 3*3 rotation matrix
T is a 3*1 translation vector
Subscript i refers to the corresponding (closest) points of the sets M (model) and D (data)

In every iteration step, the ICP algorithm computes the closest point in M for each point in D, and then the incremental transformation (R, T) is computed and applied to D. If relative changes in R and T are less than a given threshold, the process terminates, otherwise another iteration is performed.

8.4.5 Major Strip Adjustment Processing Tasks

Despite the large variety of strip adjustment methods, there are several processing steps that are common to most techniques. As a precondition, the LiDAR data from an airborne survey are expected to be processed as a point cloud, with the entire available sensor and system calibration data applied before the strip adjustment can begin. The typical workflow consists of four steps:

- Selection of the strip overlap areas is concerned with finding appropriate areas and surface patches in the strip overlap regions. The determination of strip overlap in a data set is a simple task, which is generally well supported by most LiDAR data processing software. The selection of the areas in the overlap regions, however, is rather complex as several conditions should be satisfied. First, vegetated areas should be avoided as they provide less reliable surface points and structure. A general rule is that the selected areas should have no multiple returns. Next, there should be a good surface signal in the selected patches, such as moderately sloped terrain or buildings, depending on the type of strip adjustment approach. Preferably, the surface normal vectors over the selected areas should show diversity in terms of significantly deviating from the vertical in several directions. Finally, the selected surface patches should be evenly distributed in space to provide strong geometry for the adjustment. For example, for a cross-strip situation, if possible, the patches are selected in a similar pattern as the tie-points in aerial triangulation. The size of surface patches can vary on a larger scale and is mainly controlled by surface characteristics and the LiDAR point density; typically, it falls into the 10–50 m range. To simultaneously meet all the conditions is clearly not realistic in practice, yet enough attention must be paid to this crucial step, which is usually done in an interactive way. The selection of surface patches significantly reduces the amount of data that will be used in the subsequent processing steps.

- Determination of the strip discrepancies at the selected surface patches is based on surface matching and represents a basic part of all operator-based strip adjustment processes, as well as being essential to many automated techniques. Depending on the characteristics of the surface patch, vertical only, or horizontal, or 3D, discrepancies can be observed; for example, flat areas provide for height difference measurements based on range data, intensity data can determine horizontal shifts or a rolling terrain can produce 3D differences. As discussed earlier, a variety of surface matching techniques can be used, including direct and interpolation-based techniques, such as raster or irregular grid-based methods. Another method of categorizing the surface matching techniques is based on the type of primitives used in the matching process, such as point-to-point, TIN-to-TIN, surface-to-surface, line-to-line, etc. (Han et al., 2006). The performance of strip discrepancy determination is probably the most critical factor of the whole strip adjustment process.
- Selection of the surface-to-surface (data driven) transformation or the sensor and system calibration model to be used for the strip adjustment is based on a combination of performance expectation, application data specifics, and system parameters. For an application with modest accuracy requirements, a simpler data driven technique may be sufficient, while more demanding data quality calls for a sensor calibration-based method. For longer, corridor type of surveys, clearly the second approach is required, as the overlap area is relatively limited for a corridor and only a well-calibrated system can provide better overall data accuracy. In this case, additional calibration strips are advised, such as a few cross strips over an area with relatively good surface features.
- The final step of every strip adjustment is when the corrections, determined in the previous steps, are applied to the LiDAR data. For data driven methods, the corrections are applied to the original LiDAR point cloud typically by applying a 3D similarity transformation or simpler models. For the sensor calibration-based techniques, the general solution also could be the complete recreation of the LiDAR point cloud from the original sensor data using the refined sensor model. This step is usually a simple processing executed in batch mode.

8.5 Common Strip Adjustment Techniques

A large number of strip adjustment techniques have been developed in the last ten years, mostly in the academic environment. Some of them have remained in the research phase, while others have made it into mainstream LiDAR production. Since the various techniques typically share several similar processing components, there is no simple way to categorize them; therefore, the review below will follow the historical order of the strip adjustment techniques.

8.5.1 Early Strip Adjustments

The first widely referenced LiDAR strip adjustment method of transforming overlapping LiDAR strips to make them coincide with each other followed the practice of conventional photogrammetric strip adjustment of airborne imagery (Kilian et al., 1996). Using tie and ground control features (small DEM patches with a size of 10 m by 10 m), the process

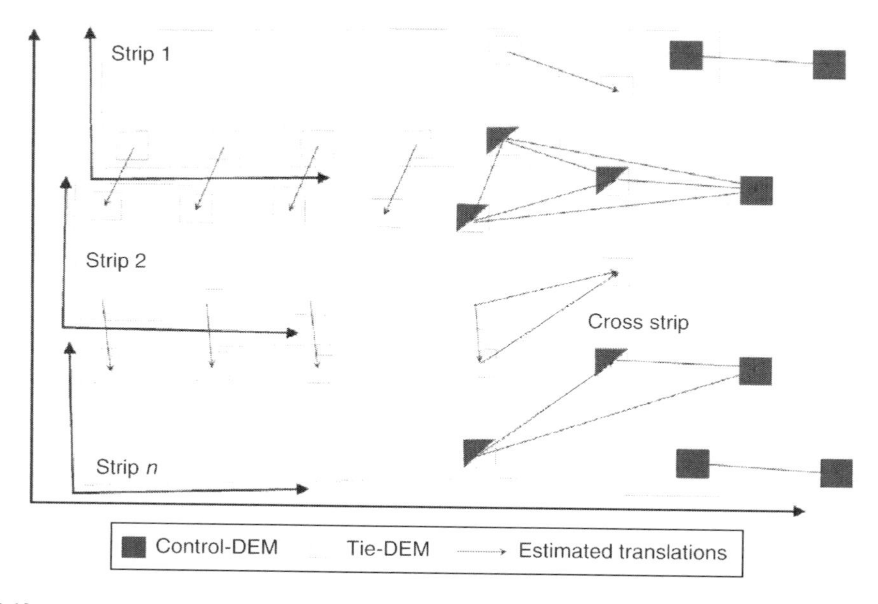

FIGURE 8.13
Tie and control point-based strip adjustment (From Kilian, J. et al., *Int. Arch. Photogram. Rem. Sens.*, 31 (Part B3): 383, 1996).

aimed at determining additional transformation parameters for each strip to transform the strips into a homogenous exterior coordinate system, as shown in Figure 8.13. In the first step, the tie and ground control information was extracted, which was then followed by matching the corresponding digital elevation models. The matching results provided three translation parameters for areas with sufficient terrain undulations or object content, or only height difference for flat areas. In the second step, using the tie and ground control information, a standard photogrammetric-style strip adjustment was performed based on a linear drift model that accounted for position and attitude corrections, including three position and attitude offsets with drift parameters, resulting in a 12-parameter model for each strip. The complexity of the model is related to the modest overall data quality of that time, as the LiDAR point spacing was rather sparse and the GPS/IMU solution was rather poor. The technique required substantial user interactions but was able to reduce strip discrepancies from the few meters range to the submeter level.

Several early strip adjustment methods minimize only the vertical discrepancies between overlapping strips or between strips and horizontal control surfaces. These strip adjustments can be referred as 1D strip adjustment methods (Crombaghs et al., 2000; Kager and Kraus, 2001; Kornus and Ruiz, 2003). Tie or absolute control features used for this adjustment are flat horizontal surfaces, where the differences in the vertical direction were estimated by simply averaging about 100+ points (Crombaghs et al., 2000). The problem with this kind of adjustment is that existing planimetric errors are likely to remain in the data.

8.5.2 3D Strip Adjustment

At the introduction of airborne laser scanning, LiDAR data was primarily perceived as a vertical or height measurement, and little attention was paid to the horizontal component. This misconception was mainly related to the low point density, as at a 0.1 pts/m^2 it is almost impossible to notice horizontal discrepancies while vertical differences can be easily observed. Therefore, the main concern was to minimize the height differences and simply ignore the often sizeable horizontal offsets. With the advancement of the LiDAR technology, better point density measurements and point accuracy improvements, however, changed the situation and users started to recognize the true 3D nature of the LiDAR data. In particular, the importance of the horizontal component was widely acknowledged. Vosselman and Maas (2001) showed that systematic planimetric errors are often much more significant than vertical errors in LiDAR data and, therefore, a 3D strip adjustment is the desirable solution for minimizing the 3D discrepancies between overlapping strips and at control points. The first 3D strip adjustment methods developed were predominantly based on using raster format LiDAR data and adjusted only for 3D offsets. Also, they were considered data driven techniques, as there was no attempt to rigorously model systematic sensor errors.

The first 3D strip adjustment technique that allowed for the direct use of the irregularly distributed LiDAR data was introduced by Maas (2000, 2002). The method, aimed at providing an efficient technique to determine strip discrepancies, is based on applying least squares matching (LSM) to the LiDAR data structured in a TIN representation. The LSM is performed on appropriately selected overlapping patches taken from different strips. The process tries to iteratively minimize the sum of height differences between the patches by adjusting the shift parameters in all the three coordinate directions. Observation equations are written for every data point of both patches, and the height difference is obtained by projecting a point to the closest triangle in the other patch (see Figure 8.11).

The technique requires surface patches with sufficient signal content to allow for the determination of the three shift parameters; basically, at least three noncoplanar surface areas should be in both patches. A rolling terrain with no vegetation is a good candidate for patch selection. Vegetated areas and buildings should be generally avoided because of occlusion. Obviously, the method fails on flat areas without any objects. Using circular patches with about 10 m radius, one can obtain 4–5 cm horizontal and 1 cm vertical precision for the shift parameters at a 0.3 pts/m^2 point density.

An alternative to area-based surface matching is feature-based matching, where features, typically higher-level data representations, are extracted from the LiDAR point cloud and the matching is performed in the feature domain. Vosselman (2002b) introduced a method to extract linear features, such as roof edge lines, gable roof ridgelines, and ditches, and then used them to determine strip discrepancies. The proper geometrical modeling of the linear features is essential to obtain suitable offset estimates.

8.5.3 Strip Adjustments Based on Sensor Calibration

The parameterization of the strip discrepancies in the physical sensor model and orientation measurement system (the introduction of self-calibration) represented a major milestone in the evolution of LiDAR strip adjustment techniques. Behan et al. (2000) proposed the first 9-parameter model that related to the georeferencing sensor by accounting for three offset, three attitude, and three drift parameters, which provided for GPS offset, initial IMU attitude bias, and IMU attitude drift. Burman (2000, 2002) treats the discrepancies between overlapping strips as positioning and orientation errors with special attention given to the alignment error between the INS and laser scanner. The first implementation needed raster data but was later extended to use a TIN structure. Besides providing a solution for stripwise elevation and planimetric differences, the boresight misalignment between the laser scanner and the IMU is determined based on multiple strip discrepancies. The more developed technique that formed the foundation for the TerraMatch product family (Burman, 2002; Soininen and Burman, 2005) can perform matching based on both range and intensity data, and applies corrections directly to the LiDAR point cloud.

Morin and El-Sheimy (2002) introduced a method that is also based on sensor and system error modeling but requires no ground control, as the assumption is that averaging overlapping strips will provide an unbiased estimate of the surface. Toth et al. (2002) presented a method that tries to make overlapping strips coincide, with the primary objective of recovering only the boresight misalignment (three angles) between the IMU and the laser sensor. The technique requires the rasterization of the LiDAR data to perform the strip discrepancy determination. Similar to the previous methods, it only works if a sufficient terrain signal exists.

Filin (2003a,b) presented a comparable method for recovering the systematic errors, which is based on constraining the position of the laser points to planar surfaces (Filin, 2001); linear features can also be used. The model initially considers a large number of error sources (Schenk, 2001), grouped as calibration and system errors. Because of functional correlation and the inseparability of several parameters, the recovery of all the systematic error terms for the total of 14 observations, including three position and three attitude for the GPS/IMU system, two for the range and scan angle measurements of the laser sensor (system measurements), and three offset and three misalignment angle errors for the IMU and the laser sensor, however, is not feasible. Therefore, certain assumptions, such as laser sensor scanner controls and object space characteristics, should be made to allow for the determination of a subset of the systematic error parameters. The planar surface patches connecting the LiDAR strips are introduced to the adjustment as ground control objects, with the four surface parameters known, or as tie objects, with surface parameters initially approximated from data and subsequently refined in the adjustment process are known. An extension of the surface-based strip adjustment method generalizes the control surfaces by allowing for natural and man-made nonplanar surface patches, which are locally approximated by a plane in a piecewise manner (Filin and Vosselman, 2004). The technique requires only the LiDAR point cloud data (no georeferencing data is needed), and is based on minimizing the distance between the laser points and the actual surface, as shown Figure 8.10. The segmentation is based on the clustering, where the feature vectors computed for each LiDAR point are classified in an unsupervised way (Filin, 2002). From the initial clusters, a complex validation process will select the tie surface patches based on the fitting accuracy of the analytical surface model to the points in the cluster with adequate upper and lower bound control to avoid under- and over-segmentation. Test results obtained from a block of 20 strips, flown as 10 strip subblocks in perpendicular directions, demonstrated that strip offsets can be reliably recovered and that they are not constant for a block. In addition, the analysis of results revealed that the horizontal offsets in the data are significantly larger than the vertical ones.

Another method introduced by Skaloud and Lichti (2006), which constrains the laser points to planar surface patches, aims at the calibration of the three boresight misalignment angles and the range-finder offset. Similarly to the method introduced by Filin (2003a), the system calibration parameters are modeled within the direct georeferencing equation, and then the laser points are conditioned to lie on a common planar surface patch without the need to know the true surface position and orientation. In contrast to previous techniques, the availability of the system level LiDAR data is assumed, including the sensor trajectory described by three position and three attitude parameters and the laser range and scan angle measurements. The planar surface patch selection and the determination of the surface normal are based on using principal component analysis, which offers computational advantages for larger data sets. Experience showed that the range offset recovery is somewhat limited, while the boresight misalignment angles can be determined to a high accuracy, provided a sufficient number of planar patches with different spatial orientations are available. Numerical results showed remarkably small RMS residuals, such as 2 arcsec for the boresight angles and 1 cm ranging accuracy, while height differences fell to the few centimeters range.

8.5.4 Using LiDAR Intensity Data

With the increasing availability of LiDAR intensity data, also frequently called reflectance values, LiDAR strip adjustment methods started to exploit this additional information primarily to support the matching between different strips. As discussed in the previous paragraphs, methods that are exclusively based on the use of the LiDAR point cloud (a mass of points defined by three coordinates) require adequate terrain characteristics, such as planar or smoothly changing surface areas with different orientations, to successfully recover systematic error terms. Large areas with no surface undulations or with limited slope cannot provide for sufficient strip discrepancy determination, in particular in the horizontal direction, and consequently any 3D adjustment will fail in such cases. Intensity data, now a standard output on modern LiDAR systems, complements the blind LiDAR point cloud with a conventional image-type of data, which is similar to an image produced by a single spectral band of a hyperspectral camera. Since LiDAR intensity generally provides more variation in terms of image texture or contrast, compared to elevation data, therefore, it can support matching in areas where the height differences are limited or nonexistent. A good example, as shown in Figure 8.14, is the transportation road network (which generally represents locally flat areas), where ubiquitous pavement markings are clearly visible in the intensity image and thus can be routinely matched. Obviously, the image domain matching only provides for the determination of horizontal offsets. As LiDAR intensity and range data are perfectly coregistered, the intensity domain matching results can be directly converted to 3D strip discrepancies.

One of the first methods exploiting the use of intensity data was reported by Maas (2001). In the proposed two-step implementation, the vertical and horizontal offsets are determined

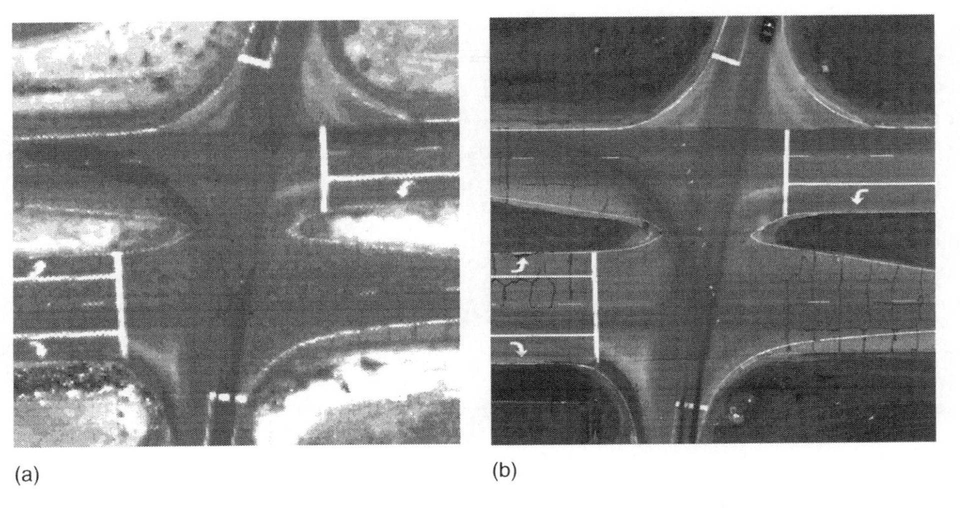

(a) (b)

FIGURE 8.14
Pavement marking appearance in (a) LiDAR intensity image and (b) reference optical image.

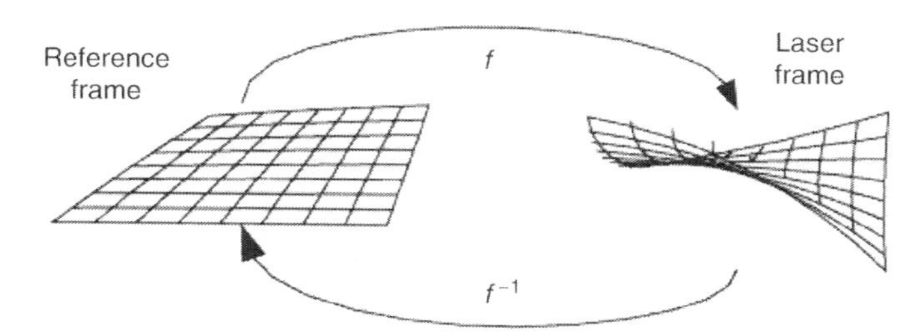

FIGURE 8.15
3D affine model used for surface fitting.

by two independent LSM processes of height and intensity data, and then the method described in (Maas, 2000) is used. Although the modest laser point density imposes limits on the matching performance in the image domain, experimental results indicated that a significant improvement can be achieved in the horizontal component, while the vertical performance remains unchanged. Vosselman (2002a) provides a comprehensive analysis of the LiDAR intensity signal, including (1) the formation of the signal at the sensor level, such as instantaneous or integrated sampling, (2) the problem of spatial undersampling, where point spacing is not sufficient to represent the surface, in particular, in urban settings, (3) the impact of multiple reflections at object boundaries, and (4) the impact of the footprint size. The combined effect of these errors results in relatively noisy characteristics of the intensity image and, therefore, feature-based matching, such as matching edges, is suggested for planimetric offset determination. To compensate for the undersampling and finite footprint size, a gray value edge modeling is introduced and the use of longer linear features is advised.

8.5.5 Miscellaneous Techniques

A modern data driven technique for LiDAR strip adjustment introduced by Bretar et al. (2004) provides a general solution for surface matching and is based on the formation of a homogenous 3D deformation model between two digital surface models (see Figure 8.15). The 12-parameter affine transformation is quite different compared to the conventional rigid body transformation, which is based on simple translation and rotation. The determination of the affine model is achieved by local 3D translations using a modified Hough transformation. The technique can be applied to coregister a reference surface to LiDAR data or LiDAR strips to each other or any two digital surface models.

Another data driven method that provides a methodology to avoid directly employing conjugate features in the strip adjustment process was introduced by Han et al. (2006). The technique is based on using the contour tree (CT) representation of the surface data and employs a 3D conformal transformation model. In an iterative process, the transformation parameters are sequentially refined until the leaves of the CT surface representation reach a minimum. The execution of the process is rather computationally intensive.

8.6 Summary

The success of strip adjustment depends on a lot of factors, including object space specifics, such as surface geometry and material characteristics, and ground control; sensor and system configuration and specifications; airborne survey parameters, including flight parameters and strip patterns; and the method and transformation model used. The combination of all of these will determine the improvement potential of the strip adjustment, i.e., to define the performance expectations in terms of error estimates and to assess the quality of the LiDAR product after corrections are applied. In other words, it is imperative to understand the limitations of the strip adjustment process.

First, the surface representation of the surveyed area should be considered, and the LiDAR point cloud should have an average sampling rate close to the sampling distance defined by the Nyquist criterion over the selected surface patches where the strip discrepancies will be determined. It is fair to say that current LiDAR data are typically undersampled and, therefore, care should be exercised when the patches are selected. Next the object composition of the surveyed area should be considered. Vegetated areas should be avoided and hard surface areas with comparable material signatures should be selected. An additional aspect in selecting the patches is the object shape requirement, if any, which is based on the methodology used for the strip adjustment. For example, some methods require man-made objects with hard surfaces, such as building and roads, while others need a sloping terrain with modest or no vegetation. In parallel, the reflectance characteristics of the surface materials should be also considered as they determine whether a good return signal can be obtained. The impact of the footprint size is also very critical; if it is relatively large then the extent to which the object space is restricted can be observed. Finally, the availability of ground control is important; otherwise, the validation of the data in absolute terms is not feasible.

The sensor parameters and configuration jointly define the performance potential of any LiDAR system, and the objective is to approach it as closely as possible under normal operational conditions. The laser sensing unit is generally characterized by its ranging and scanning accuracy. Modern systems support extremely good ranging accuracy, typically 1–2 cm (1σ), up to relatively high flying heights (1000 m AGL). As the moving parts in the laser sensor have shrunk over time, and these smaller components can be better controlled, this has resulted in faster and more even mirror motion, for example. In addition, the encoding performance has improved and, thus, the mirror position can be calibrated to a high accuracy.

In contrast to the laser sensor, the performance of the georeferencing component can vary a great deal and, generally, the georeferencing errors account for the largest contributor of the overall LiDAR error budget. Coincidentally, the georeferencing is a system component that can be significantly influenced by user control and, therefore, is a key part of flight planning. In the simplest approach, the quality of the differential GPS solution determines the overall georeferencing performance, provided that, at least, a medium-grade (i.e., tactical grade) IMU is used. The GPS data quality of both the airborne platform and ground reference are equally important to the DGPS solution, as well as their separation (base distance or virtual base for network solutions). With proper flight planning and execution, accurate georeferencing solutions can be obtained, resulting in smaller strip adjustment corrections, if any, and, consequently, better overall LiDAR product quality.

The selection of the strip adjustment technique for a particular application primarily depends on the object space characteristics and the expected performance of the strip adjustment, which is mainly determined by the sensor parameters. Additional operating aspects, including flight control, DGPS availability, processing resources and environment, and time requirements may also influence this selection process. From the large variety of the available strip adjustment methods, several perform relatively well in most situations, while specific conditions require more application-tailored solutions. In all cases, regardless of the actual strip adjustment method used, the object space and sensor parameters jointly define the quality of the LiDAR-derived surface, which is characterized by the completeness of the surface representation and its accuracy (defined by the point cloud) and, consequently, how well strips can be compared to each other and ultimately corrected for differences.

Strip adjustment techniques continue to evolve primarily driven by technological advances. Most importantly, as multipulse systems develop further, the higher pulse rates will result in better LiDAR point cloud densities. This will substantially improve surface representations, allowing for the observation of smaller object features. In a parallel development, georeferencing technology has made significant progress recently and further advances are expected, resulting in more robust and accurate solutions under almost any conditions.

From the algorithmic point of view, the current trend is to move toward sensor calibration-based strip adjustment techniques. In addition, due to improving feature extraction performance and fusion with optical imagery, strip adjustment methods will probably be combined with higher level feature extraction mapping processes in the future. As sensor performance, including laser and georeferencing sensors, continues to improve, the role of strip adjustment will slowly shift toward QC, becoming a primary tool for assessing relative data quality.

References

Ackerman, F., 1994. Practical experience with GPS supported aerial triangulation, *The Photogrammetric Record*, XIV(84): 861–874.

Akca, D. and Gruen, A., 2007. Generalized least squares multiple 3D surface matching, *International Society for Photogrammetry and Remote Sensing*, XXXVI: (Part 3/W52).

ASPRS LiDAR Committee, 2004. ASPRS Guidelines Vertical Accuracy Reporting for LiDAR Data. Available at http://www.asprs.org/society/committees/lidar/Downloads/Vertical_Accuracy_Reporting_for_Lidar_Data.pdf.

Baltsavias, E.P., 1999. Airborne laser scanning: Basic relations and formulas. *ISPRS Journal of Photogrammetry and Remote Sensing*, 54: 199–214.

Behan, A., 2000. On the matching accuracy of rasterized scanning laser altimeter data. *International Archives of Photogrammetry and Remote Sensing*, 33 (Part 2B): 75–82.

Behan, A., Maas, H.-G., and Vosselman, G., 2000. Steps towards quality improvement of airborne laser scanner data, *Proceedings of the 26th Annual Conference of the Remote Sensing Society*, Leicester, September 12–14, 9 pp, CD-ROM.

Besl, P.J. and McKay, N.D., 1992. A method for registration of 3-D shapes, *IEEE Transactions on Pattern Analysis and Machine Intelligence*, 14(2), 239–256.

Bretar, F., Roux, M., and Pierrot-Deseilligny, M., 2004. Solving the strip adjustment problem of 3D airborne Lidar data. *Proceedings of the IEEE IGARSS'04*, Anchorage, Alaska, September 2004.

Burman, H., 2000. Adjustment of laser scanner data for correction of orientation errors. *International Archives of Photogrammetry and Remote Sensing*, 33 (Part B3): 125–128.

Burman, H., 2002. Laser strip adjustment for data calibration and verification. *International Archives of Photogrammetry and Remote Sensing*, 34 (Part 3A): 67–72.

Campbell, R.J. and Flynn, P.J., 2001. A survey of free-form object representation and recognition techniques. *Computer Vision and Image Understanding*, 81(2): 166–210.

Crombaghs, M.J.E., Brügelmann, R., and de Min, E.J., 2000. On the adjustment of overlapping strips of laseraltimeter height data. *International Archives of Photogrammetry and Remote Sensing*, 33 (Part B3/1): 224–231.

Csanyi, May, N., 2008. A rigorous approach to comprehensive performance analysis of state-of-the-art airborne mobile mapping systems, PhD dissertation, The Ohio State University.

Csanyi, N., Paska, E., and Toth, C., 2003. Comparison of various surface modeling methods, terrain data: Applications and visualization—making the connection, *ASPRS/MAPPS*, Charleston, SC, October 27–30, 2003, CD-ROM.

Csanyi, N., Toth, C., Grejner-Brzezinska, D., and Ray, J., 2005. Improving LiDAR data accuracy using LiDAR-specific ground targets, ASPRS Annual Conference, Baltimore, MD, March 7–11, CD-ROM.

Csanyi, N. and Toth, C., 2007. Improvement of LiDAR data accuracy using LiDAR-specific ground targets, *Photogrammetric Engineering & Remote Sensing*, 73(4): 385–396.

Duda, R.O. and Hart, P.E., 1972. Use of the Hough transformation to detect lines and curves in pictures, *Graphics and Image Processing*, 15: 11–15.

El-Sheimy, N., Valeo, C., and Habib, A., 2005. *Digital Terrain Modeling: Acquisition, Manipulation, and Applications*, Artech House, Boston, MA.

Filin, S., 2001. Recovery of systematic biases in laser altimeters using natural surfaces, *International Archives of Photogrammetry and Remote Sensing*, 34, (3/W4): 85–91.

Filin, S., 2002. Surface clustering from airborne laser scanning data, *International Archives of Photogrammetry and Remote Sensing*, 34, (3A): 117–124.

Filin, S., 2003a. Recovery of systematic biases in laser altimetry data using natural surfaces, *ISPRS Journal of Photogrammetric Engineering and Remote Sensing*, 69(11): 1235–1242.

Filin, S., 2003b. Analysis and implementation of a laser strip adjustment model. *International Archives of Photogrammetry and Remote Sensing*, 34 (Part 3/W13): 65–70.

Filin, S. and Vosselman, G., 2004. Adjustment of laser altimetry strips. *International Archives of Photogrammetry and Remote Sensing*, 34 (Part 3/W13): 285–289.

Gesch, D., Oimoen, M., Greenlee, S., Nelson, C., Steuck, M., and Tyler, D., 2002. The National Elevation Dataset, *Photogrammetric Engineering & Remote Sensing*, 68(1): 5–13.

Gruen, A. and Akca, D., 2005. Least squares 3D surface and curve matching. *ISPRS Journal of Photogrammetry and Remote Sensing*, 59(3): 151–174.

Heywood, I., Cornelius, S., and Carver, S., 1998. *An Introduction to Geographical Information Systems*, New Jersey, Prentice Hall.

Han, D., Lee, J., Kim, Y., and Yu, K., 2006. Adjustment for Disrepencies Between ALS Data Strips Using Contour Tree Algorithm, ACIVS, LNCS 4179, pp. 1026–1036.

Hough, P.V.C., 1959. Machine analysis of bubble chamber pictures, *International Conference on High Energy Accelerators and Instrumentation*, CERN, Geneva, Switzerland.

Kager, H. and Kraus, K., 2001. Height discrepancies between overlapping laser scanner strips. *Proceedings of Optical 3D Measurement Techniques V*, October, Vienna, Austria: pp. 103–110.

Kilian, J., Haala, N., and Englich, M., 1996. Capture and evaluation of airborne laser scanner data. *International Archives of Photogrammetry and Remote Sensing*, 31 (Part B3): 383–388.

Kornus, W. and Ruiz, A., 2003. Strip adjustment of LiDAR data, Dresden: *International Archives of Photogrammetry and Remote Sensing*, 34 (3/W): 47–50.

Kukko, A. and Hyyppä, J., 2007. Laser scanner simulator for system analysis and algorithm development: A case with forest measurements, *International Archives of Photogrammetry and Remote Sensing*, 36 (Part 3/W52): 234–240.

Loan, V. and Charles, F., 1997. *Introduction to Scientific Computing*, New Jersey: Prentice Hall.

Maas, H.-G., 2000. Least squares matching with airborne laserscanning data in a TIN structure. *International Archives of Photogrammetry and Remote Sensing*, 33 (Part B3/1): 548–555.

Maas, H.-G., 2001. On the use of pulse reflectance data for laserscanner strip adjustment. *International Archives of Photogrammetry, Remote Sensing and Spatial Information Sciences*, 33 (Part 3/W4): 53–56.

Maas, H.-G., 2002. Methods for measuring height and planimetry discrepancies in airborne laserscanner data. *Photogrammetric Engineering & Remote Sensing*, 68(9): 933–940.

Maas, H.-G., 2003. Planimetric and height accuracy of airborne laserscanner data: User requirements and system performance, D. Fritsch (Ed.), *Proceedings 49. Photogrammetric Week*, Wichmann Verlag, 117–125.

Madhavan, R., Hong, T., and Messina, E., 2005. Temporal range registration for unmanned ground and aerial vehicles, *Journal of Intelligent and Robotic Systems*, 44(1): 47–69.

Maune, D. (Ed.), 2007. Digital elevation model technologies and aApplications: *The DEM Users Manual*, 2nd edn. American Society for Photogrammetry and Remote Sensing, Bethesda, MD.

Morin, K. and El-Sheimy, N., 2002. Post-mission adjustment of airborne laser scanning data, *Proceedings XXII FIG International Congress*, Washington, DC, April 19–26, 12 pp., CD-ROM.

Okabe, A., Boots, B., Sugihara, K., and Chiu, S.N., 2000. *Spatial Tessellations - Concepts and Applications of Voronoi Diagrams*, 2nd edn., New York, John Wiley.

Oliver, M.A. and Webster, R., 1990. Kriging: A method of interpolation for geographical information system, *International Journal of Geographical Information Systems*, 4(3): 313–332.

Pfeifer, N. and Briese, C., 2007. Geometrical aspects of airborne laser scanning and terrestrial laser scanning, *International Archives of Photogrammetry*, 36 (Part 3/W52): 311–319.

Renslow, M., 2005. The status of LiDAR today and future directions, 3D mapping from InSAR and LiDAR, ISPRS WG I/2 Workshop, Banff, Canada, June 7–10, CD-ROM.

Schenk, T., 2001. Modeling and analyzing systematic errors in airborne laser scanners, technical notes in photogrammetry, Vol. 19, The Ohio State University, Columbus, 46 pp.

Shannon, C.E., 1948. A mathematical theory of communication, *Bell System Technical Journal*, 27: 379–423, 623–656.

Shannon, C.E., 1949. Communication in the presence of noise, *Proceedings of Institute of Radio Engineers*, 37(1): 10–21.

Skaloud, J. and Lichti, D., 2006. Rigorous approach to bore-sight self-calibration in airborne laser scanning, *International Journal of Photogrammetry and Remote Sensing*, 61: 47–59.

Soininen, A. and Burman, H., 2005. TerraMatch for MicroStation. Terrasolid, Finland.

Toth, C.K., 2002. Calibrating airborne LiDAR systems, *International Archives of Photogrammetry and Remote Sensing*, Vol. XXXIV(part 2): 475–480.

Toth, C., Csanyi, N., and Grejner-Brzezinska, D., 2002. Automating the calibration of airborne multisensor imaging systems, *Proceedings ACSM-ASPRS Annual Conference*, Washington, DC, April 19–26, CD-ROM.

Toth, C., 2004. Future Trends in LiDAR, *Proceedings of ASPRS 2004 Annual Conference*, Denver, CO, May 23–28, CD-ROM.

Vosselman, G. and Maas, H.-G., 2001. Adjustment and filtering of raw laser altimetry data, *Proceedings OEEPE Workshop on Airborne Laserscanning and Interferometric SAR for Detailed Elevation Models*. OEEPE Publications No. 40, pp. 62–72.

Vosselman, G., 2002a. On the estimation of planimetric offsets in laser altimetry data. *International Archives of Photogrammetry and Remote Sensing*, 34 (Part 3A): 375–380.

Vosselman, G., 2002b. Strip offset estimation using linear features. 3rd International LiDAR Workshop, October 7–9, Columbus available at http://www.itc.nl/personal/vosselman/papers/vosselman2002.columbus.pdf.

9

Accuracy, Quality Assurance, and Quality Control of LiDAR Data

Ayman Habib

CONTENTS

9.1 Introduction

The improving capabilities of direct georeferencing technology Global Navigation Satellite System/Inertial Navigation System (GNSS/INS) are having a positive impact on the widespread adoption of Light Detection and Ranging (LiDAR) systems for the acquisition of dense and accurate surface models over extended areas. Unlike photogrammetric techniques,

LiDAR calibration is not a transparent process, and remains restricted to the system's manufacturer. Moreover, derived footprints from a LiDAR system are not based on redundant measurements, which are manipulated in an adjustment procedure. Consequently, one does not have the associated measures (e.g., variance component of unit weight and variance–covariance matrices of the derived parameters) that can be used to evaluate the quality of the final product. In this regard, LiDAR systems are usually viewed as black boxes that lack a well-defined set of quality assurance and quality control (QA/QC) procedures. This chapter introduces the concepts of QA/QC of LiDAR systems and derived data. The most important activity in QA is the system calibration. Therefore, we will present a conceptual approach for LiDAR calibration and the necessary prerequisites for such a calibration. On the other hand, the main focus of this chapter is the introduction of QC procedures to verify the accuracy of the LiDAR footprints. The main premise of the proposed QC procedures is that overlapping LiDAR strips represent the same surface if and only if there are no biases in the derived surfaces. Therefore, we will use the quality of coincidence of conjugate surface elements in overlapping strips as the basis for deriving the QC measures. The chapter will start with a brief discussion of LiDAR principles, which will be followed by general remarks regarding QA/QC procedures. Then, an analysis of error sources in a LiDAR system and their impact on the resulting surface will be presented. This analysis will be followed by explanations of several procedures for QA/QC of LiDAR systems and derived data.

9.2 LiDAR Principles

A typical LiDAR system consists of three main components: a GNSS system to provide position information, an INS unit for attitude determination, and a laser unit to provide range (distance) information from the laser beam firing point to the ground point. In addition to range data, modern LiDAR systems can capture intensity images over the mapped area. Figure 9.1 shows a schematic diagram of a LiDAR system, together with the coordinate

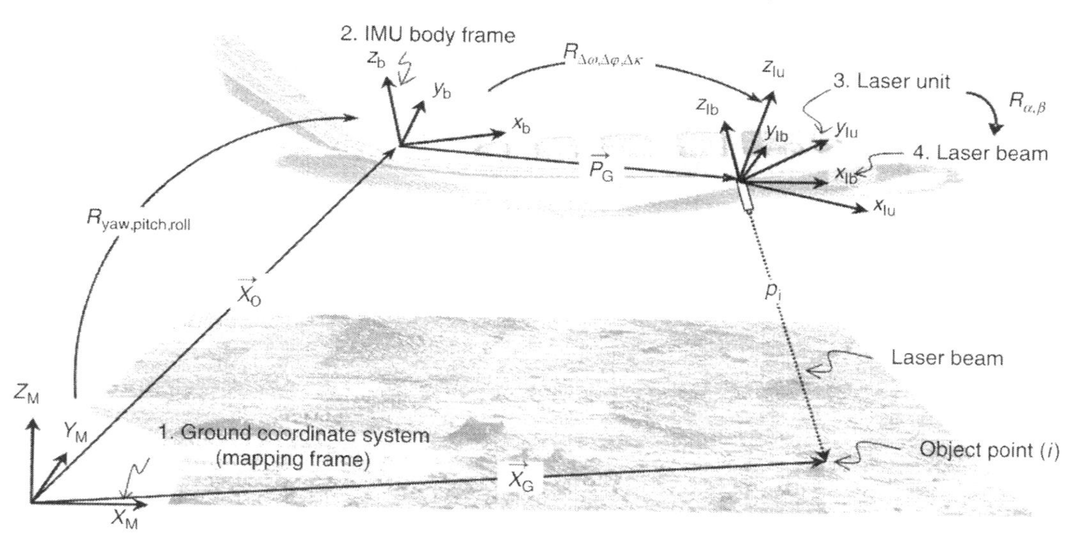

FIGURE 9.1
Coordinates and parameters involved in LiDAR data acquisition.

systems involved. Equation 9.1 is the basic model that incorporates the LiDAR measurements for deriving positional information for the laser beam footprint (El-Sheimy et al., 2005). This equation relates four coordinate systems: the ground coordinate system, the inertial measurement unit (IMU) body frame, the laser unit coordinate system, and the laser beam coordinate system. This equation is simply the result of a three-vector summation: $\vec{X}_o$ is the vector from the origin of the ground coordinate system to the IMU body frame, $\vec{P}_G$ is the offset between the laser unit and the IMU body frame with respect to the laser unit coordinate system, and $\vec{\rho}$ is the vector between the laser beam firing point and the target point. The summation of these three vectors, after applying the appropriate rotations ($R_{\text{yaw,pitch,roll}}$, $R_{\Delta\omega,\Delta\phi,\Delta\kappa}$, $R_{\alpha,\beta}$), yields the vector $\vec{X}_G$, which contains the ground coordinates of the object point under consideration. The quality of the derived surface depends on the accuracy of the involved sub-systems (i.e., laser, GNSS, and inertial measurement unit [IMU]) and the calibration parameters relating these components (i.e., bore-sighting parameters). In general, the system manufacturer provides a range of the expected accuracy of the derived point cloud. For example, Optech provides the following accuracies for their ALTM 2050 and ALTM 3100 systems (OPTECH ALTM 3100).

ALTM 2050: Horizontal accuracy: 1/2000 × altitude; 1 sigma
Elevation accuracy: <15 cm at 1.2 km; 1 sigma
<25 cm at 2.0 km; 1 sigma

ALTM 3100: Horizontal accuracy: 1/2000 × altitude; 1 sigma
Elevation accuracy: <15 cm at 1.2 km; 1 sigma
<25 cm at 2.0 km; 1 sigma
<35 cm at 3.0 km; 1 sigma

$$\vec{X}_G = \vec{X}_o + R_{\text{yaw,pitch,roll}} R_{\Delta\omega,\Delta\phi,\Delta\kappa} \vec{P}_G + R_{\text{yaw,pitch,roll}} R_{\Delta\omega,\Delta\phi,\Delta\kappa} R_{\alpha,\beta} \begin{bmatrix} 0 \\ 0 \\ -\rho \end{bmatrix} \tag{9.1}$$

Other than the expected accuracy, the quality of the LiDAR surfaces derived from LiDAR systems can be ensured and checked by the QA/QC procedures, which are the main emphasis of this chapter. QA encompasses management activities that are carried out prior to data collection to ensure that the raw and derived data are of the quality required by the user. These management controls cover the calibration, planning, implementation, and review of data collection activities. Quality control, on the other hand, takes place after data collection to determine whether the desired quality has been achieved. It involves routines and consistent checks to ensure the integrity, correctness, and completeness of the raw and derived data. To illustrate the nature of the QA/QC procedures, one can refer to the respective activities in photogrammetric mapping (Brown, 1966; Fraser, 1997; Habib and Schenk, 2001). The QA procedures for photogrammetric mapping include camera calibration, total system calibration, and flight configuration (e.g., flying height, overlap percentage, and side lap percentage). Quality control measures for photogrammetric mapping, on the other hand, include internal measures such as a posteriori variance component and the variance–covariance matrix resulting from the bundle adjustment procedure. External measures include the root mean square error (RMSE) derived from a check point analysis procedure using independently measured targets. The remaining sections of this chapter will focus on various methods of QA/QC of LiDAR after a brief discussion of the error budget of LiDAR systems.

9.3 LiDAR Error Budget

The error in the LiDAR-derived coordinates is affected by errors in the components of the LiDAR equation. These components, or input parameters, can either be measured or estimated from a system calibration procedure. In this section, we are interested in analyzing the effects of random noise and systematic biases in the measurements from the various LiDAR components on the final product. The purpose of such an analysis is to enable the estimation of the quality of the final product from the quality of the system's measurements. Moreover, knowing the expected accuracy of the final product, one might be able to interpret the outcome of the QC procedure as either acceptable or as an indication of the presence of systematic biases in the data acquisition system. Finally, by analyzing the effects of systematic biases, one might be able to offer some diagnostic tips regarding the origin of the discrepancies identified from the proposed QC procedures.

For any point measured by a LiDAR system, error propagation can be used to determine the error in the LiDAR-derived coordinates given the errors in the LiDAR input parameters. Another issue related to LiDAR error analysis is the nature of the errors resulting from random errors in the input system measurements. Usually, it is expected that random noise will lead to random errors in the derived point cloud. Moreover, it is commonly believed that random noise will not affect the relative accuracy. However, this is not the case for LiDAR systems. In other words, some of the random errors might affect the relative accuracy of the derived point cloud. Depending on the parameter being considered, the relative effect of the corresponding noise level in the output might not be the same. As an illustration, Figure 9.2 shows that a given attitude noise in the GNSS/INS-derived orientation (i.e., errors in the derive attitude from the GNSS/INS integration process) will

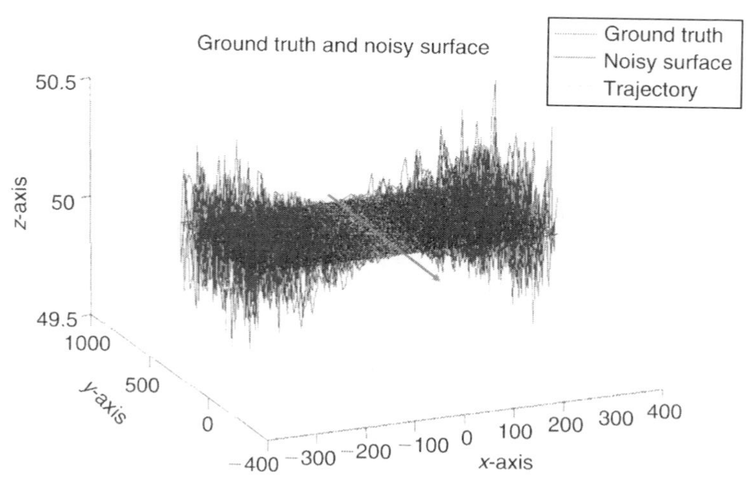

FIGURE 9.2
Effect of attitude errors on a simulated horizontal surface.

TABLE 9.1

Systematic Biases and Their Impacts on the Derived Surface

	Flying Height	Flying Direction	Look Angle
Bore-sighting offset bias	Impact is independent of the flying height	Impact is dependent on the flying direction (except elevation)	Impact is independent of the look angle
Bore-sighting angular bias	Impact increases with the flying height	Impact changes with the flying direction	Impact changes with the look angle (except in the across-flight direction)
Laser beam range bias	Impact is independent of the flying height	Impact is independent of the flying direction	Impact depends on the look angle (except in the along-flight direction)
Laser beam angular bias	Impact increases with the flying height	Impact changes with the flying direction (except in the along-flight direction)	Impact changes with the look angle (except in the across-flight direction)

Note: The table assumes a linear scanner flying over a flat horizontal terrain along a straight-line trajectory with constant attitude.

affect the nadir region of the flight trajectory less significantly than the off-nadir regions. Thus, the GNSS/INS attitude error will affect the relative accuracy of the LiDAR-derived point cloud. The following list gives some diagnostic hints regarding the impacts of noise in the system measurements on the derived point cloud.

- GPS noise: It will lead to a similar noise level in the derived point cloud. Moreover, the effect is independent of the system parameters (flying height and look angle).
- Angular noise (attitude or mirror angles): With this type of noise, the horizontal coordinates are affected more than the vertical coordinates. In addition, the effect is dependent on the system parameters (flying height and look angle).
- Range noise: It mainly affects the vertical component of the derived coordinates. The effect is independent of the system's flying height. However, the impact is dependent on the system's look angle.

Systematic biases in the system measurements (e.g., GNSS/INS-derived positions and attitudes, mirror angle measurements, measured ranges) and calibration parameters (e.g., bore-sighting parameters relating the system components) lead to systematic errors in the derived point cloud. Table 9.1 provides a summary of the various systematic biases and their impacts on the derived LiDAR coordinates.

9.4 QA of LiDAR Systems

The QA procedures for LiDAR systems are established prior to the mapping mission and include flight planning, setting up the GNSS base stations, selecting appropriate time for the flight mission to assure optimal satellite availability, and calibrating the system. Among these procedures, system calibration is very critical for ensuring the utmost quality of the derived LiDAR coordinates. Calibration procedures can be divided into two main categories: laboratory calibration and *in situ* calibration procedures. Laboratory calibration is usually carried out by the system manufacturer. On the other hand, *in situ* calibration can be carried out by the system's user. Due to the nontransparent nature of LiDAR systems, *in situ* LiDAR calibration of LiDAR systems has not been sufficiently addressed by prior literature. The calibration methodology developed by Morin (2002) uses the LiDAR equation to solve for the bore-sighting misalignment angles and the scanner angle correction. These parameters are either estimated using ground control points or by observing discrepancies between tie points in overlapping strips. However, the identification of distinct control and tie points in the LiDAR data is a difficult task due to the irregular nature of the collected point cloud. To alleviate this difficulty, Skaloud and Lichti (2006) presented a calibration technique using tie planar patches in the overlapping strips. The underlying assumption of this procedure is that systematic errors in the LiDAR system will cause the noncoplanarity of conjugate planar patches as well as bending effects in these patches. The calibration process uses the LiDAR equation to simultaneously solve for the plane parameters as well as for the bore-sighting misalignment angles. However, this approach requires having large planar patches, which might not be always available. In addition, systematic biases, which would not affect the coplanarity of conjugate planar patches, could still remain. To overcome such a problem, control patches can be used for *in situ* calibration of LiDAR systems, as shown in Figure 9.3. Using such control patches, the target function for system calibration should minimize the normal distance between the laser footprint (as derived from the LiDAR equation) and the control surface, as illustrated in Figure 9.3. Therefore, one only needs to determine the correspondence between the LiDAR footprint and the control surfaces. The LiDAR equation (Equation 9.1) can be used to estimate the systematic errors (e.g., bore-sighting parameters) that minimize this target function. Flight and control surface configurations (e.g., different flight heights in opposite and cross directions together with sloping control surfaces with varying aspects) must be carefully established in order to enable accurate estimation of these parameters (Habib et al., 2007).

The use of the above calibration methodologies, however, is only possible if we are dealing with a transparent system, that is, one in which the raw measurements (i.e., navigation data, mirror angles, and ranges) are available to the user. Unfortunately, current LiDAR systems do not furnish raw measurements; therefore, as far as the user is concerned, the data acquisition system is a black box. For this reason, explicit QA procedures such as this cannot be used to ensure that the derived LiDAR data meets the user's accuracy requirements. The quality of the LiDAR data can, however, be maximized through careful flight planning and

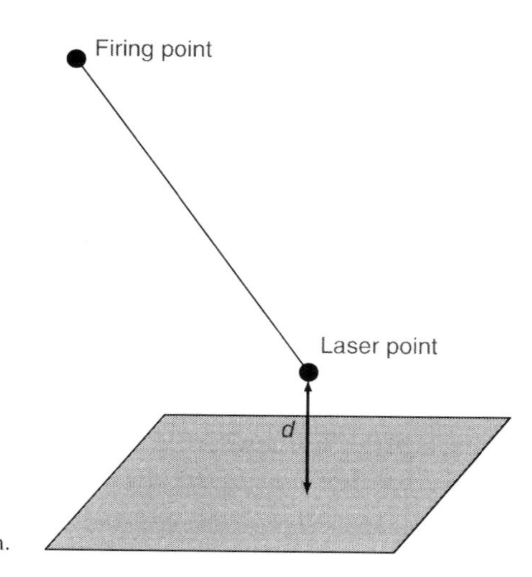

FIGURE 9.3
Target function for LiDAR calibration.

the use of sufficient control. One must then turn to QC procedures to assess the performance of the system and the quality of its output after the data has been collected.

9.5 Quality Control of LiDAR Data

The following sections will provide some tools for internal and external QC checks. The internal measures are used to check the relative consistency of the LiDAR data. This is usually conducted by checking the compatibility of the LiDAR footprints in the overlapping strips. On the other hand, the external QC measures verify the absolute quality of the LiDAR data by checking its compatibility with an independently collected and more accurate surface model.

9.5.1 Internal Quality Control of LiDAR Data

As it can be seen in Equation 9.1, there is no redundancy in the LiDAR measurements leading to the derivation of the coordinates of the laser footprint. Therefore, unlike with photogrammetric data, one cannot use explicit measures (e.g., a posteriori variance component and variance–covariance matrix of the derived ground coordinates of the LiDAR footprints) to assess the quality of the LiDAR-derived positional information. Hence, alternative QC methods are necessary for this type of data. The next paragraphs provide some measures for the internal QC of LiDAR data.

With the exception of narrow corridor mapping, LiDAR data is usually acquired from the overlapping strips from different flight lines, such as those shown in Figure 9.4. A common QC procedure is to assess the coincidence of conjugate features in the overlapping strips.

FIGURE 9.4
A pair of overlapping LiDAR strips.

Such a procedure ensures the internal quality of the available LiDAR data. There are two main approaches to ensure quality: (1) comparing interpolated range or intensity images from the overlapping strips and (2) comparing the conjugate features extracted from the strips. The degree of coincidence of the extracted features can be used as a measure of the quality of the data and to detect the presence of systematic biases. In other words, the conjugate features in the overlapping strips will coincide if and only if the LiDAR data is quite accurate. Therefore, the separation between conjugate features can be used as a QC measure.

9.5.1.1 Quality Control Using Interpolated Range and Intensity Images

This approach can be applied for either range or intensity measurements. When using range measurements, the data for two overlapping areas are interpolated onto a regular grid to create two range images. Image differencing is then performed, and the resulting image shows the deviations between the two range images. These deviations are used as a measure of QC; the smaller the deviations, the higher the quality of the datasets. Figure 9.5 shows two interpolated range images and their difference image.

Similarly, intensity measurements can also be interpolated onto a grid to obtain two overlapping intensity images of an area. The conjugate features in these images are then identified and their 3D coordinates are compared. Planimetric coordinates are

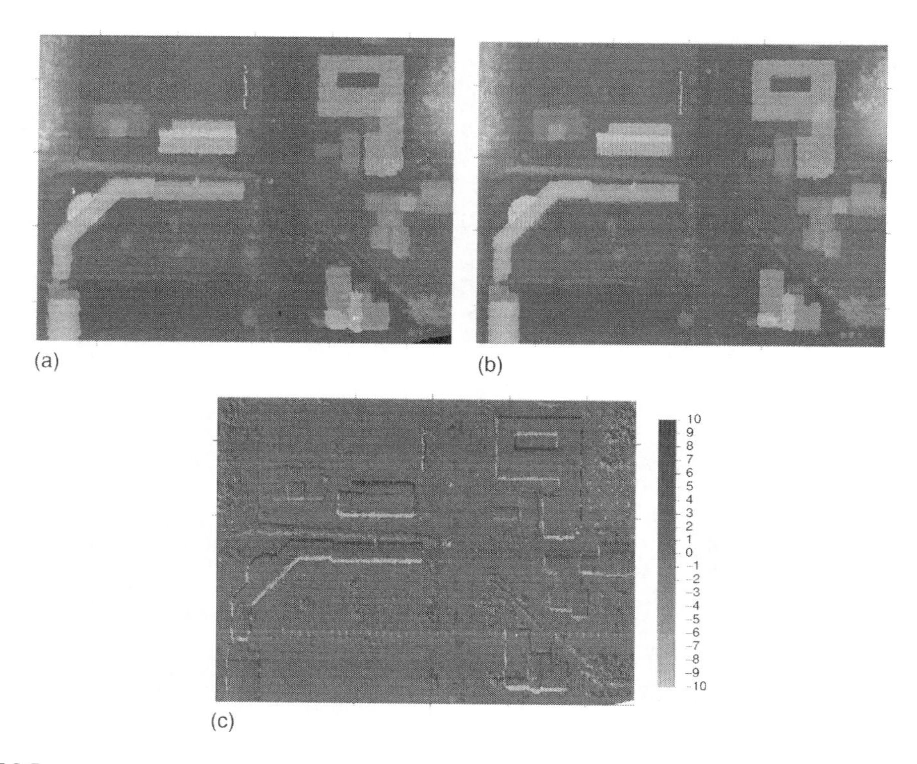

FIGURE 9.5
Image differencing of interpolated range images generated from overlapping strips: (a) interpolated range image from first strip, (b) interpolated range image from second strip, and (c) difference image.

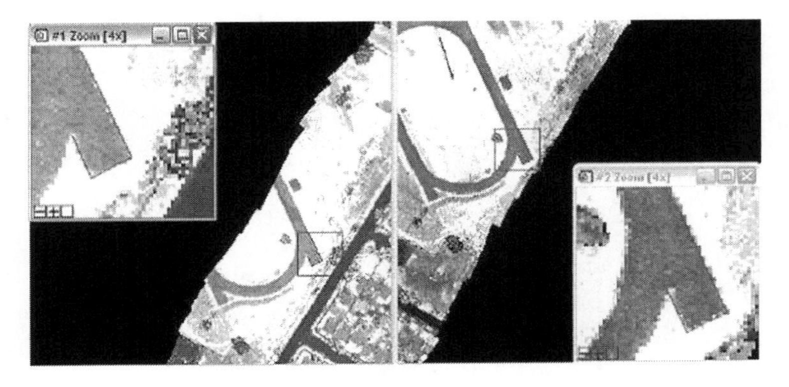

FIGURE 9.6
Comparison of conjugate features in interpolated intensity images.

derived from the intensity images while the vertical coordinates are determined from the corresponding range images. Differences in the derived coordinates of the conjugate features indicate the presence of biases in the data acquisition system. Figure 9.6 illustrates the identification and comparison of the conjugate features in interpolated intensity images.

A disadvantage of this method of QC is that the interpolation of the irregular LiDAR data leads to artifacts in the derived range or intensity images, especially at the vicinity of discontinuities in the range or intensity data. Because of these artifacts, incorrect conclusions may be made about the quality of the LiDAR data, especially in urban areas, where discontinuities in the data are quite common (refer to the significant differences in Figure 9.5 at the building boundaries). Therefore, an alternative method of QC that does not involve the interpolation of LiDAR footprints should be used.

9.5.1.2 *Quality Control by Checking the Coincidence of Conjugate Straight Lines in Overlapping Strips*

In this section, we will introduce a QC procedure that is based on linear features represented by their end points. It should be noted that this approach does not require lines that have conjugate end points in both LiDAR strips and, depending on the nature of the overlap, conjugate lines may arise from quite different parts of the same object. The quality of the coincidence of the conjugate linear features can be used to evaluate the internal quality of the LiDAR data. More specifically, the linear features extracted from the irregular LiDAR footprints in the overlapping strips can be used to compute estimates of the conformal transformation parameters that are needed to coalign these features, as shown in Figure 9.7. Deviations from the optimum values (zero shifts, zero rotations, and unit scale factor) can be used as indications of systematic biases in the LiDAR system. Before considering the mathematical details of using linear features, which are represented by nonconjugate end points in the overlapping strips, for parameter estimation, we will begin with the established procedure for their extraction. Using intensity images, one can develop an interface that extracts the LiDAR point cloud in the vicinity of selected points (i.e., extract the original LiDAR points within a given radius from an identified point in the intensity image). The extracted point cloud usually corresponds to buildings that contain linear features. Using an automated segmentation

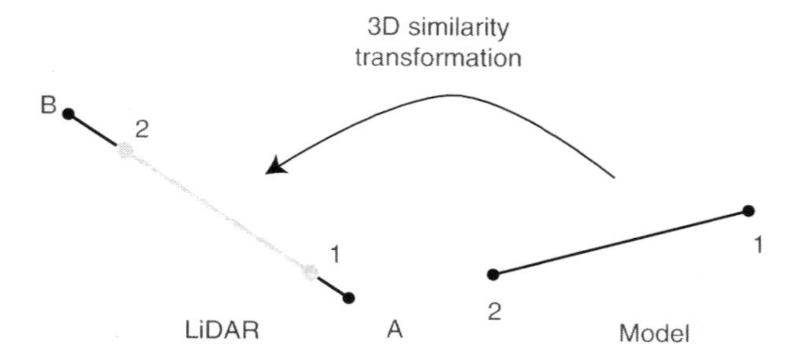

FIGURE 9.7
Conceptual basis of the use of linear features in a line-based approach for the determination of the conformal transformation parameters between two 3D datasets.

procedure, planar patches are identified in the extracted point cloud. Then, neighboring patches are intersected to produce infinite line segments. Finally, using the segmented patches and a given cylindrical buffer whose axis coincides with the infinite line, one can define the end points for the line of intersection. More specifically, LiDAR points within the segmented patches that lie within the given buffer are projected to the line of intersection, and extreme points along the line of intersection are used to define the end points of that line. This procedure should be applied to the overlapping strip. After the extraction of the linear features from the overlapping strips, an automated procedure can be applied to identify the conjugate linear features. The matching of the conjugate features should consider the normal distance between the lines, the parallelism of their direction vectors, and the percentage of their overlap.

The objective of this approach is to introduce the necessary constraints to describe the fact that the line segment from the first strip (12) coincides with the conjugate segment from the overlapping strip (AB) after applying the transformation, as illustrated in Figure 9.7. For these points, the constraint equations can be written as in Equation 9.2 (i.e., a line point in the first strip should be collinear with the conjugate line in the second strip).

$$\begin{bmatrix} X_T \\ Y_T \\ Z_T \end{bmatrix} = S\,R_{(\Omega,\Phi,K)} \begin{bmatrix} X_1 \\ Y_1 \\ Z_1 \end{bmatrix} = \begin{bmatrix} X_A \\ Y_A \\ Z_A \end{bmatrix} + \lambda_1 \begin{bmatrix} X_B - X_A \\ Y_B - Y_A \\ Z_B - Z_A \end{bmatrix} \tag{9.2a}$$

$$\begin{bmatrix} X_T \\ Y_T \\ Z_T \end{bmatrix} = S\,R_{(\Omega,\Phi,K)} \begin{bmatrix} X_2 \\ Y_2 \\ Z_2 \end{bmatrix} = \begin{bmatrix} X_A \\ Y_A \\ Z_A \end{bmatrix} + \lambda_2 \begin{bmatrix} X_B - X_A \\ Y_B - Y_A \\ Z_B - Z_A \end{bmatrix} \tag{9.2b}$$

where

$(X_T Y_T Z_T)^T$ is the translation vector between the strips
$R_{(\Omega,\Phi,K)}$ is the required rotation matrix for the coalignment of the strips
S, λ_1, and λ_2 are the scale factors

Through the subtraction of Equation 9.2a from b, and the elimination of the scale factors, Equation 9.3 can be written to relate the coordinates of the points defining the line segments to the rotation elements of the transformation.

$$\frac{(X_B - X_A)}{(Z_B - Z_A)} = \frac{R_{11}(X_2 - X_1) + R_{12}(Y_2 - Y_1) + R_{13}(Z_2 - Z_1)}{R_{31}(X_2 - X_1) + R_{32}(Y_2 - Y_1) + R_{33}(Z_2 - Z_1)}$$
$$\frac{(Y_B - Y_A)}{(Z_B - Z_A)} = \frac{R_{21}(X_2 - X_1) + R_{22}(Y_2 - Y_1) + R_{23}(Z_2 - Z_1)}{R_{31}(X_2 - X_1) + R_{32}(Y_2 - Y_1) + R_{33}(Z_2 - Z_1)} \tag{9.3}$$

These equations can be written for each pair of conjugate line segments, giving two equations, which contribute towards the estimation of two rotation angles, i.e., the azimuth and the pitch angle along the line. The roll angle across the line, on the other hand, cannot be estimated. Hence, a minimum of two nonparallel lines is needed to recover the three elements of the rotation matrix. To enable the estimation of the translation parameters and the scale factor, Equation 9.4 below can be derived by rearranging the terms in Equations 9.2a and b and eliminating the scale factors λ_1 and λ_2. A minimum of two noncoplanar lines is required to recover the scale and translation parameters. To recover all seven parameters of the transformation function, a minimum of two noncoplanar line segments is required. For more details regarding this approach, interested readers can refer to Habib et al. (2004).

$$\frac{(X_T + S\,x_1 - X_A)}{(Z_T + S\,z_1 - Z_A)} = \frac{(X_T + S\,x_2 - X_A)}{(Z_T + S\,z_2 - Z_A)}$$
$$\frac{(Y_T + S\,y_1 - Y_A)}{(Z_T + S\,z_1 - Z_A)} = \frac{(Y_T + S\,y_2 - Y_A)}{(Z_T + S\,z_2 - Z_A)} \tag{9.4}$$

where

$$\begin{bmatrix} x_1 \\ y_1 \\ z_1 \end{bmatrix} = R_{(\Omega,\Phi,K)} \begin{bmatrix} X_1 \\ Y_1 \\ Z_1 \end{bmatrix} \quad \text{and} \quad \begin{bmatrix} x_2 \\ y_2 \\ z_2 \end{bmatrix} = R_{(\Omega,\Phi,K)} \begin{bmatrix} X_2 \\ Y_2 \\ Z_2 \end{bmatrix}$$

9.5.1.3 Quality Control by Checking the Coplanarity of Conjugate Planar Patches in Overlapping Strips

Instead of using linear features, one can incorporate conjugate planar patches in the QC procedure. In this approach, planar patches are represented by equivalent sets of nonconjugate points (i.e., two sets with the same number of points), which can be derived using a planar patch segmentation procedure in the overlapping strips and arbitrarily truncating the segmented points in one of the strips to be of equivalent size to that derived from the other strip. The requirement for having an equivalent number of nonconjugate points along corresponding planar patches is dictated by the utilization of a point-based approach to estimate the conformal transformation parameters for the optimal coalignment of these patches. In other words, nonconjugate points along the corresponding patches will be used in a point-based approach that assumes that these are conjugate points. To compensate for the fact that the utilized points along the planar patches are not conjugate, one can expand the variance–covariance matrices for these points in the plane direction, as illustrated in Figure 9.8. The expansion can be carried out using Equation 9.5, where R is the rotation matrix relating the original *XYZ*-coordinate system and the *UVW*-coordinate system, which is defined with the U and V axes aligned along the plane (i.e., the W-axis is aligned along the normal to the plane). It should be noted that N and M in Equation 9.5 refer to

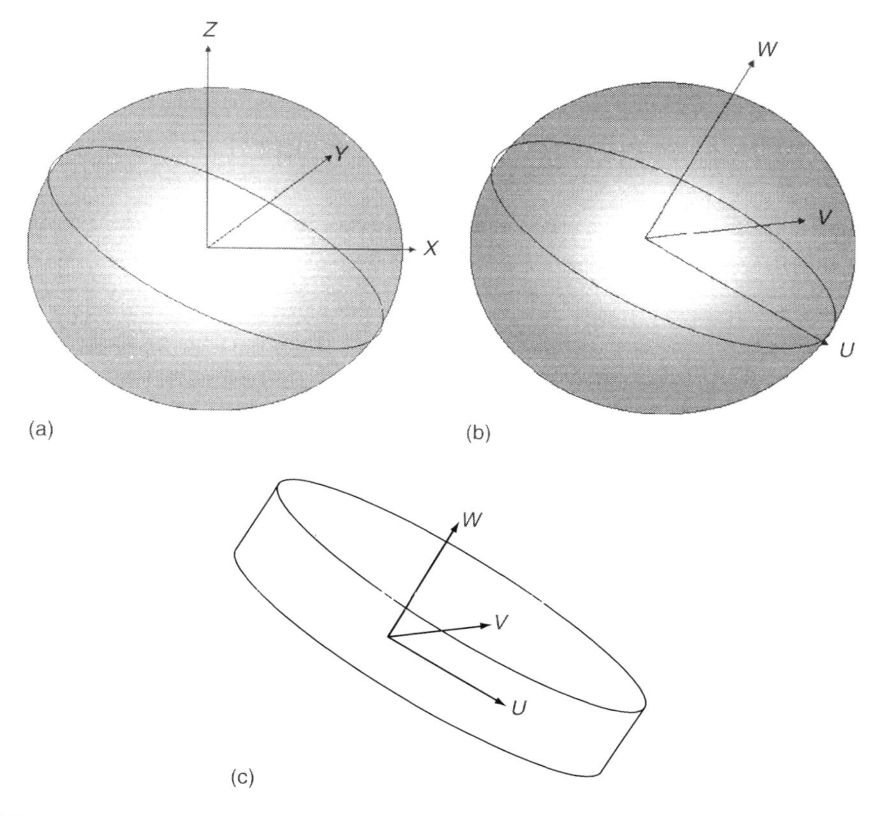

FIGURE 9.8
Conceptual basis for the use of planar patches in a point-based approach for the determination of the conformal transformation parameters between two 3D datasets: (a) the *XYZ* coordinate system, (b) the *UVW* coordinate system, and (c) the expansion of the error ellipsoid.

arbitrarily chosen large numbers. Finally, the expanded variance–covariance matrix in the *XYZ*-coordinate system can be derived according to Equation 9.6. The defined points along the conjugate planar patches, together with their expanded variance–covariance matrices, can be used in a point-based conformal transformation to estimate the necessary shifts, rotations, and scale to coalign these patches in the overlapping strips. Deviations from the optimum values (zero shifts and rotations and unit scale factor) indicate the presence of biases in the LiDAR unit. It should be noted that a sufficient number of patches with varying slope and aspect values should be available for reliable estimation of the transformation parameters.

$$\sum{}'_{UVW} = R\begin{bmatrix} \sigma_X^2 & \sigma_{XY} & \sigma_{XZ} \\ \sigma_{YX} & \sigma_Y^2 & \sigma_{YZ} \\ \sigma_{ZX} & \sigma_{ZY} & \sigma_Z^2 \end{bmatrix} R^T + \begin{bmatrix} N & 0 & 0 \\ 0 & M & 0 \\ 0 & 0 & 0 \end{bmatrix} \tag{9.5}$$

$$\sum{}'_{XYZ} = R^T \sum{}'_{UVW} R \tag{9.6}$$

9.5.1.4 Quality Control by Automated Matching of the Original LiDAR Footprints in Overlapping Strips

The above QC measures require preprocessing of the LiDAR point cloud (e.g., interpolation, planar patch segmentation, and intersection of neighboring planar patches). Therefore, it is preferred to develop alternative measures, which can deal with the original LiDAR footprints. The following paragraphs present an alternative approach for evaluating the degree of similarity between the overlapping LiDAR strips, which are represented by irregularly distributed sets of points. The approach presented does not assume one-to-one correspondence between the involved points. More specifically, the approach evaluates the transformation parameters relating the involved strips, determines the correspondence between conjugate surface elements, and derives an estimate of the degree of similarity between the two datasets.

To illustrate the conceptual basis of the suggested approach, let us assume that we are given two sets of irregularly distributed points that describe the same surface, as shown in Figure 9.9. Let $S_1 = \{p_1, p_2, \ldots, p_l\}$ be the first set and $S_2 = \{q_1, q_2, \ldots, q_m\}$ be the second set, where $l \neq m$ (i.e., there is no assumption of point-to-point correspondence). These points are randomly distributed and the correspondences between them are not known. Furthermore, it cannot be assumed that there is one-to-one correspondence between the points in the two datasets. Also, the two point sets might be given relative to two different reference frames. The transformation between these reference frames is modeled by a seven-parameter transformation involving three shifts, one scale, and three rotations (X_T, Y_T, Z_T, S, Ω, Φ, and K, respectively). Hence, the problem at hand is to determine the degree of similarity between the two point sets describing the surface, establish the correspondences between conjugate surface elements, and estimate the transformation parameters relating the respective reference frames. The proposed approach creates a TIN model using the points in S_1 to form a group of nonoverlapping triangles, shown in Figure 9.9. The individual triangles in the derived TIN are assumed to represent planar patches.

Now, let us consider the surface patch S_p in S_1, which is defined by the three points p_a, p_b, and p_c. The fact that a point q_i in the second set S_2 belongs to the surface patch S_p in the first set can be mathematically described by the determinant constraint in Equation 9.7, illustrated in Figure 9.9.

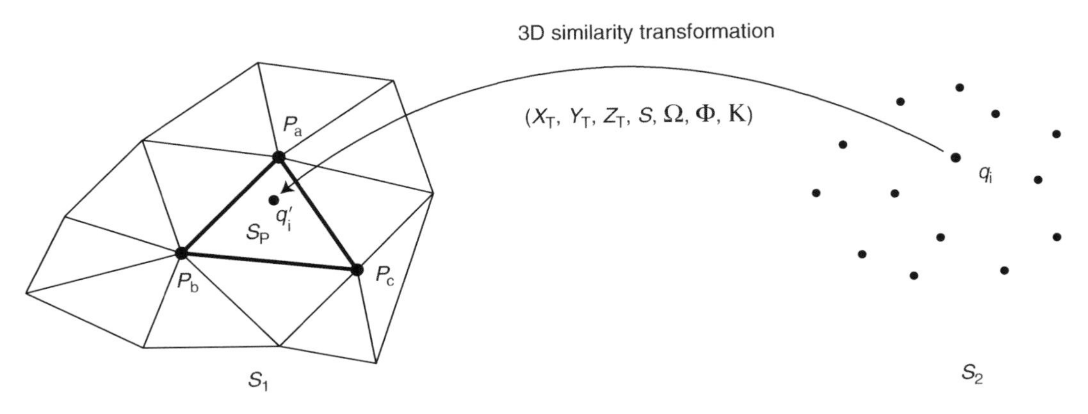

FIGURE 9.9
Comparing two datasets with irregularly distributed sample points.

$$\begin{vmatrix} x_{q'_i} & y_{q'_i} & z_{q'_i} & 1 \\ x_{p_a} & y_{p_a} & z_{p_a} & 1 \\ x_{p_b} & y_{p_b} & z_{p_b} & 1 \\ x_{p_c} & y_{p_c} & z_{p_c} & 1 \end{vmatrix} = 0 \tag{9.7}$$

The vector $(x_{q'_i}, y_{q'_i}, z_{q'_i})$ in the above equation represents the point coordinates in the second dataset transformed to the reference frame associated with the first one, using Equation 9.8. Equation 9.7 simply states that the volume defined by the points q'_i, p_a, p_b, and p_c is zero. In other words, these points are coplanar (i.e., the normal distance between q'_i and the surface patch S_p is zero). After establishing the correspondences between the points in S_2 and the patches in S_1, one can solve for the transformation parameters (implicitly considered in the first row of the determinant of Equation 9.7 through the substitution of Equation 9.8) using a least squares adjustment procedure. The estimated normal distances between point-patch pairs from the adjustment procedure represent the goodness of fit (degree of similarity) between the two point sets after the coalignment of their respective reference frames. Moreover, the deviation of the estimated transformation parameters from the optimal ones can be used as an indication of biases in the data acquisition system.

$$\begin{bmatrix} x_{q'_i} \\ y_{q'_i} \\ z_{q'_i} \end{bmatrix} = \begin{bmatrix} X_T \\ Y_T \\ Z_T \end{bmatrix} + S \times R_{(\Omega,\Phi,K)} \begin{bmatrix} x_{q_i} \\ y_{q_i} \\ z_{q_i} \end{bmatrix} \tag{9.8}$$

So far, we have established the mathematical model that can be used to derive the transformation parameters relating the reference frames associated with the two point sets. However, the derivation of these estimates requires knowledge of the correspondence between conjugate surface elements in the available datasets (i.e., which points in S_2 belong to which patches in S_1). Figuring out such a correspondence requires careful consideration since the two sets might be given relative to different reference systems. Therefore, the remaining problem is to establish a reliable procedure for the identification of such a correspondence. The proposed solution to this problem is based on a voting scheme, which simultaneously establishes the correspondence between conjugate surface elements and the transformation parameters relating the two datasets.

As mentioned earlier, the parameters of the 3D similarity transformation in Equation 9.8 can be estimated once the correspondences between seven points in S_2 and seven patches in S_1 are known, generating seven constraints of the form given in Equation 9.7. The suggested procedure can start with choosing any seven points in S_2 and matching them with all possible surface patches in S_1, generating several matching hypotheses. For each matching hypothesis, a set of seven equations can be written and used to solve for the transformation parameters. One should repeat this procedure until all possible matches between the points in S_2 and the patches in S_1 are considered. Throughout these combinations, correct matching hypotheses will lead to the same parameter solution. Therefore, the most frequent solution resulting from the matching hypotheses will be the correct set of transformation parameters relating the two datasets in question. Also, the matching hypotheses that led to this solution constitute the correspondences between the points in S_2 and the patches in S_1. A seven-dimensional accumulator array, which is a discrete tessellation of the expected solution space, can be used to keep track of the matching hypotheses and the associated solutions.

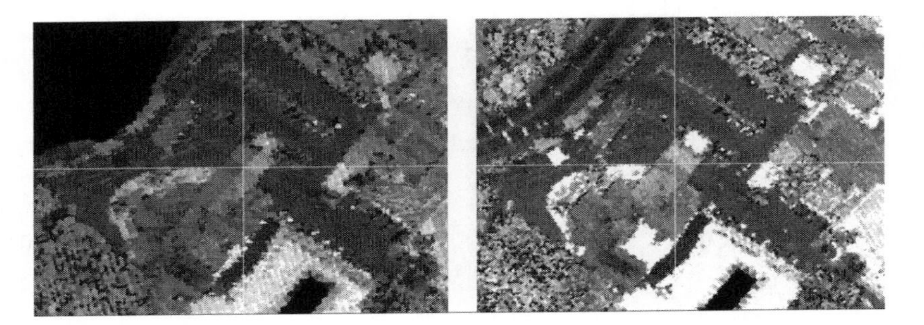

FIGURE 9.15
A sample pair of conjugate points identified in overlapping strips.

and Z directions are 1.65, 1.11, and −0.63 m, respectively. Therefore, one can conclude that the performance of a QC procedure, which is based on the manipulation of intensity and range images, is not reliable and is subjective to the ease of interpretation of the involved intensity images.

9.6.2 Checking the Coincidence of Conjugate Straight Lines in Overlapping Strips

To test this procedure, we manually collected conjugate lines in the three overlapping strips (Figure 9.14). Figures 9.17 through 9.20 illustrate the interface developed for the extraction of these features in two overlapping strips. Table 9.3 summarizes the conformal transformation parameters that were estimated, together with the average normal distance between conjugate linear features before and after applying the transformation. Figure 9.21 shows a sample pair of conjugate linear features in the overlapping strips before and after the application of the conformal transformation parameters. It is quite evident that the degree of coincidence among the conjugate features has significantly improved after applying the estimated transformation parameters (refer to the last two rows of Table 9.3 and Figure 9.21b). A closer look at the numbers reported in Table 9.3 reveals that the estimated discrepancies between the strips, which are mainly in the planimetric coordinates, depend on the flying direction.

9.6.3 Checking the Coplanarity of Conjugate Planar Patches in Overlapping Strips

As mentioned earlier, instead of using the conjugate linear features, one can determine the quality of LiDAR data by checking the coplanarity of the conjugate planar patches in the

TABLE 9.2
Averages and Standard Deviations of the Estimated Discrepancies between Overlapping Strips, Using 100 Points

	Average (m)	Standard Deviation (m)
X	0.45	0.36
Y	0.50	0.37
Z	0.22	0.28

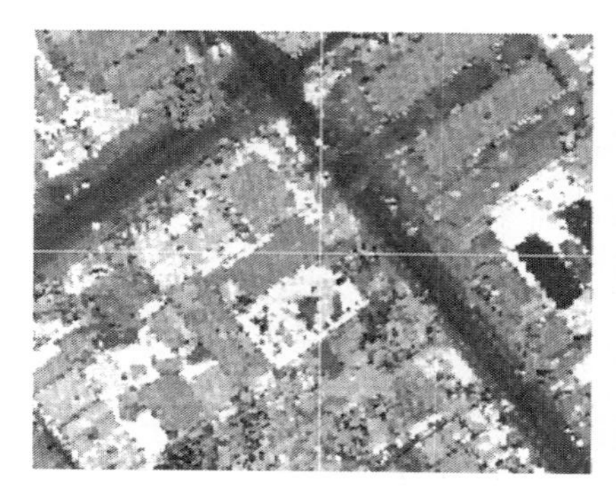

FIGURE 9.16
Another sample pair of conjugate points identified in overlapping strips.

overlapping strips. The coplanarity of these patches can be checked by estimating the conformal transformation parameters (shifts, rotations, and scale) that need to be applied to one strip to ensure the coalignment of the two strips. To test this procedure, we manually collected conjugate patches in the three overlapping strips (refer to Figure 9.22 for examples of these patches). These patches correspond to the roof of a building that appears in the overlapping LiDAR strips. Table 9.4 summarizes the conformal transformation parameters estimated using the extracted planar patches. A visual comparison of the results reported in Tables 9.3 and 9.4 shows that the parameters estimated from the line and patch procedures are quite compatible. However, the utilization of the planar patches eliminates the need for the intersection of neighboring planes to derive the linear features, thus saving some processing time.

FIGURE 9.17
An example of a building that has been used for the generation of linear features.

FIGURE 9.18
Selected areas in the intensity images that correspond to the building in Figure 9.17.

FIGURE 9.19
Segmented patches in the selected areas in the intensity images that correspond to the building in Figure 9.17.

FIGURE 9.20
Lines extracted from the segmented patches in the selected areas in the intensity images that correspond to the building in Figure 9.17.

TABLE 9.3

Transformation Parameters Estimated Using Conjugate Linear Features in Overlapping Strips, Together with the Normal Distances between the Linear Features before and after Applying the Transformation

	Strips 2 and 3	Strips 3 and 4	Strips 2 and 4
Transformation parameter/number of lines	24	36	24
Scale factor	1.0002	1.0006	1.0013
X_T (m)	−0.56	0.75	0.10
Y_T (m)	0.04	−0.17	−0.16
Z_T (m)	0.03	0.05	0.13
Ω (°)	0.0205	−0.0386	−0.0147
Φ (°)	0.0062	−0.0125	−0.0073
Κ (°)	0.0261	−0.0145	−0.0113
Normal distance (m) (before)	0.38 ± 0.22	0.49 ± 0.24	0.26 ± 0.14
Normal distance (m) (after)	0.18 ± 0.19	0.18 ± 0.18	0.16 ± 0.11

9.6.4 Automated Matching of the Original LiDAR Footprints in Overlapping Strips

To test this approach, we extracted several areas in the overlapping LiDAR strips. Figure 9.23 shows the selected areas in strips 3 and 4. The estimated parameters together with the average normal distances between the conjugate surface elements are provided in Table 9.5.

Based on a comparison of Tables 9.3 through 9.5, it is clear that the proposed approaches yield comparable results. Moreover, these results suggest the presence of systematic biases among the strips. The magnitude of these biases depends on the flying direction (i.e., large biases are detected for strips flown in opposite directions—compare the estimated biases for strips 2 and 3, 3 and 4, and 2 and 4). Also, in spite of the fact that these biases are in the range of the expected accuracy from the involved LiDAR system, this does not mean that they are acceptable. Since these measures are estimated from a large sample, the estimated

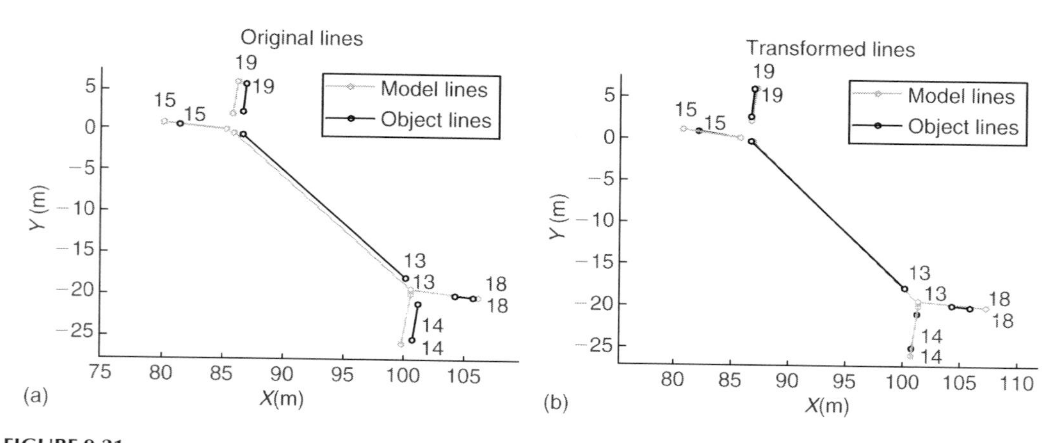

FIGURE 9.21
Sample pair of conjugate linear features in overlapping strips before (a) and after (b) applying the estimated conformal transformation parameters provided in Table 9.3.

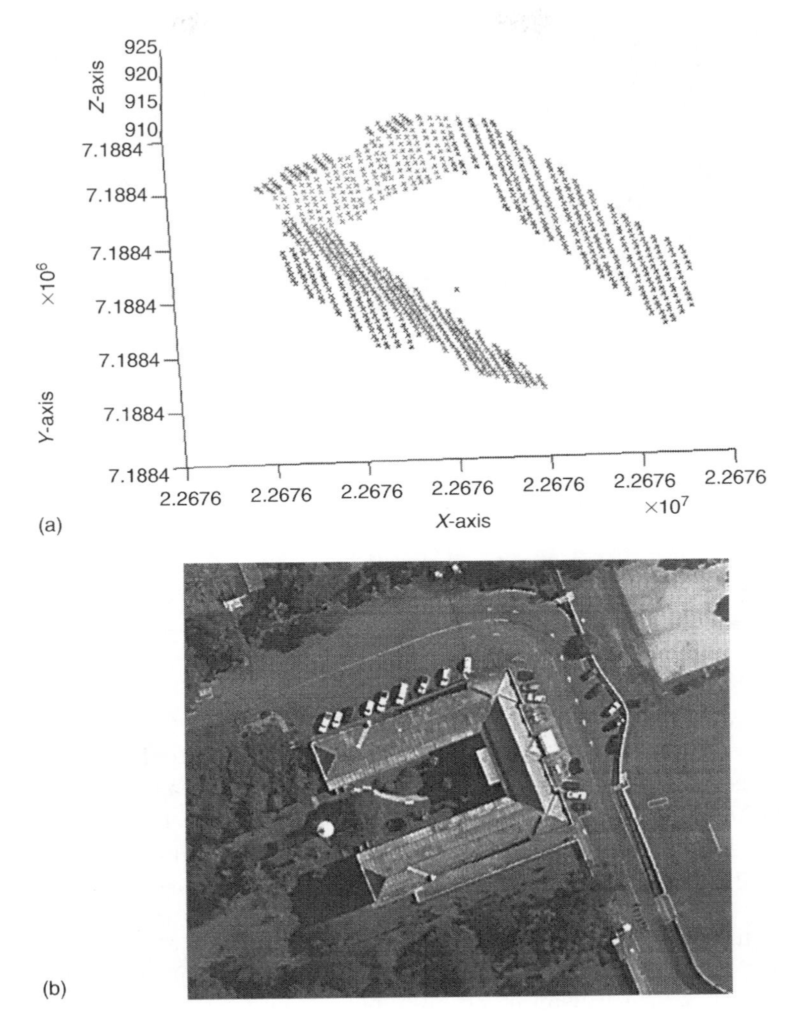

FIGURE 9.22
Planar patches extracted from a LiDAR strip (a) that correspond to the building in the optical image (b).

TABLE 9.4
Transformation Parameters Estimated Using Conjugate Planar Patches in Overlapping Strips

	Strips 2 and 3	**Strips 3 and 4**	**Strips 2 and 4**
Transformation parameter/number of patches	21	22	22
Scale factor	1.0,000	0.9996	0.9995
X_T (m)	−0.52	0.72	0.08
Y_T (m)	−0.13	−0.17	−0.21
Z_T (m)	0.05	0.09	0.14
Ω (°)	0.0289	−0.0561	−0.0802
Φ (°)	0.0111	−0.0139	−0.0342
Κ (°)	0.0364	0.0288	0.0784
Average normal distance (m) (after)	0.04	0.03	0.04

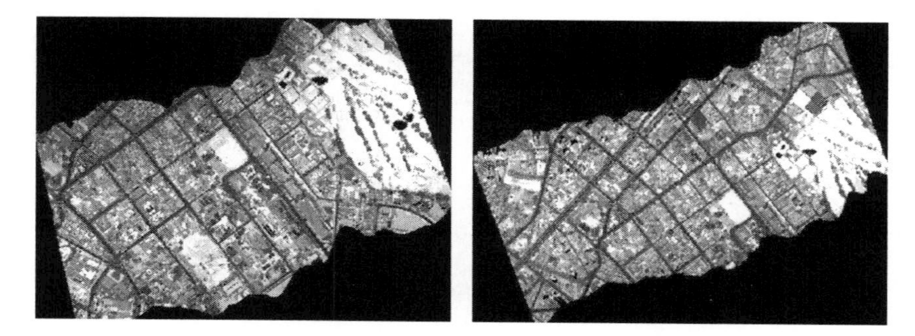

FIGURE 9.23
Locations of areas selected for the alignment and automated matching of original LiDAR footprints in strips 3 and 4.

biases should average to zero. In other words, the derived biases are much larger than their respective standard deviations.

9.7 Concluding Remarks

This chapter presented alternative procedures for the QA/QC of LiDAR systems and derived data. In general, the QA of LiDAR systems is restricted to flight planning, since the calibration procedures require the raw measurements (e.g., ranges, mirror angles, and navigation data) that are not usually available to the end user. Therefore, QC is an essential procedure to ensure that the data derived from a given system meets the users' requirements. The proposed procedures are based on evaluating the conformal transformation parameters needed to coalign conjugate surface elements in overlapping LiDAR strips. Deviations in the estimated transformation parameters from the theoretical ones (zero rotations and translations and unit scale factor) are used as QC measures to detect the presence of biases in the data acquisition system. When dealing with overlapping LiDAR strips, the deviations are considered as internal (relative) QC measures. On the other hand,

TABLE 9.5
Transformation Parameters Estimated through Automated Matching of Original LiDAR Footprints

	Strips 2 and 3	Strips 3 and 4	Strips 2 and 4
Scale factor	0.9996	0.9998	0.9993
X_T (m)	−0.55	0.75	0.19
Y_T (m)	−0.06	−0.13	−0.18
Z_T (m)	0.03	0.12	0.16
Ω (°)	0.0080	−0.0267	−0.0213
Φ (°)	0.0059	−0.0088	−0.0053
K (°)	−0.0009	−0.0003	0.0012
Average normal distance (m) (after)	0.09	0.09	0.10

external (absolute) QC measures can be derived by comparing the LiDAR surface with another version of the surface that has been independently and accurately acquired.

More specifically, we introduced one approach that utilizes linear features with nonconjugate end points. Additionally, the chapter introduced an approach that utilizes conjugate planar patches, which are represented by nonconjugate points. To compensate for the noncorrespondence of the selected points along the planar patches in the two strips, we artificially expanded their variance–covariance matrix in the plane direction. Finally, we introduced an automated approach that is based on the identification of conjugate surface elements while estimating the transformation parameters. Experimental results with real data have demonstrated the feasibility of the proposed algorithms for detecting biases in the horizontal directions between overlapping LiDAR strips. These errors are speculated to be due to spatial or angular bore-sighting biases. More specifically, by estimating the discrepancies between overlapping strips, which are flown in opposite directions, one can exclude some of the biases as possible sources of such discrepancies. Moreover, the different approaches have been shown to produce comparable results. When compared with traditional QC techniques (e.g., the utilization of intensity and range images), the proposed approaches delivered more reliable estimates. Moreover, the last approach can be applied to any coverage area, with no requirement for LiDAR targets or structures with linear features (e.g., urban areas).

Acknowledgments

This research work has been conducted under the auspices and financial support of the British Columbia Base Mapping and Geomatic Services (BMGS) and the GEOIDE Research Network. The author is grateful for the Technology Institute for Development–LACTEC–UFPR, Brazil for supplying the LIDAR data.

References

American Society of Photogrammetry and Remote Sensing LIDAR Committee, May 2004. ASPRS Guidelines—Vertical Accuracy Reporting for LIDAR Data and LAS Specifications. Ed. Flood, M. Retrieved April 19, 2007, from http://www.asprs.org/society/divisions/ppd/standards/Lidar%20guidelines.pdf.

Brown, D., 1966. Decentric distortion of lenses. *Photogrammetric Engineering and Remote Sensing*, 32(3), March 1966: 444–462.

Csanyi, N. and Toth, C., 2007. Improvement of LiDAR data accuracy using LiDAR-specific ground targets. *Photogrammetric Engineering and Remote Sensing*, 73(4), April 2007: 385–396.

El-Sheimy, N., Valeo, C., and Habib, A., 2005. Digital terrain modeling: Acquisition, manipulation and applications, *Artech House Remote Sensing Library*, 200 pp.

Fraser, C., 1997. Digital camera self-calibration. *ISPRS Journal of Photogrammetry and Remote Sensing*, 52(4), April 1997: 149–159.

Habib, A. and Schenk, T., 2001. Accuracy analysis of reconstructed points in object space from direct and indirect exterior orientation methods. *OEEPE Workshop on Integrated Sensor Orientation*. Institute for Photogrammetry and Engineering Surveying, University of Hanover, Germany, September 17–18, 2001.

Habib, A., Lee, Y., and Morgan, M., 2001. Surface matching and change detection using a modified Hough transformation for robust parameter estimation. *Photogrammetric Record*, 17(98): 303–315.

Habib, A., Ghanma, M., and Tait, M., 2004. Integration of LiDAR and photogrammetry for close range applications, *Proceedings of the ISPRS XXth Conference*, Istanbul, Turkey, 35B (7), 6 pp.

Habib, A., Cheng, R., Kim, E., Mitishita, E., Frayne, R., and Ronsky, J., 2006. Automatic surface matching for the registration of lidar data and mr imagery authors. *ETRI (Electronics and Telecommunication Research Institute) Journal*, 28(2): 162–174.

Habib, A., Bang, K., Shin, S., and Mitishita, E., 2007. LIDAR system self-calibration using planar patches from photogrammetric data. *Fifth International Symposium on Mobile Mapping Technology (MMT'07)*, May 28–31, 2007, Padua, Italy.

Morin, K., 2002. Calibration of airborne laser scanners. M.Sc. thesis, Department of Geomatics Engineering, University of Calgary, November 2003, UCGE Report No. 20179, 125 pp.

OPTECH. ALTM 3100: The Next Level of Performance. Retrieved September 29, 2007, from http://www.optech.ca/pdf/Specs/specs_altm_3100.pdf.

Skaloud, J. and Lichti, D., 2006. Rigorous approach to bore-sight self-calibration in airborne laser scanning. *ISPRS Journal of Photogrammetry and Remote Sensing*, 61: 47–59.

10

Management of LiDAR Data

Lewis Graham

CONTENTS

10.1 Uniform Gridded Data

Conventional two-dimensional image data or three-dimensional gridded elevation data are organized on uniformly spaced grids. This allows one to specify an origin and spacing (possibly different for x and y) and then simply list the attribute (such as height) values. For example, an elevation grid could be encoded as

x origin = 0.0
y origin = 0.0
x spacing = 0.5
y spacing = 1.0
x rows = 1000
y columns = 1000
$z_1 = 182.3$
$z_2 = 181.7$

etc.

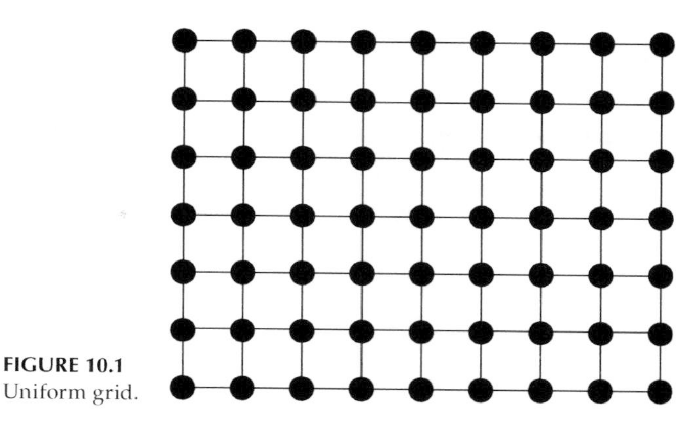

FIGURE 10.1
Uniform grid.

An example of a uniform grid is illustrated in Figure 10.1.

Generally, the header portion of the encoding contains information about the data themselves and is thus termed "metadata." In the example above, we have specified all of the parameters to locate, in a relative way, all of the elevation values (often termed "posts") in the file. Note that it is not necessary to specify the individual values of each x, y location; they are inferred from the point's location in the data file. We can quantify this by a simple equation (this example is for a row major file). If N is the point count in the file then $N = \text{rows} \times \text{columns}$. If we consider the first point index to be 0, then the $< x, y >$ coordinates of the ith point in the file are given by

$$x = (\text{i MOD Columns}) * x \text{ spacing} + x \text{ origin}$$

$$y = (\text{I DIV Columns}) * y \text{ spacing} + y \text{ origin}$$

where DIV is the integer value of the division operation. Of course, z is just the value read directly from the data file at position i. External file descriptions must specify if the data are row major or column major. A row major file organizes data by rows meaning that if a row is K elements long, the first K elements of the file will represent the first row, the second K elements the second row, and so forth. On the other hand, a column major file containing J elements per column will be organized such that the first J elements in the file represent column 1, the next J elements column 2, and so forth.

Additional attributes can be added by either considering each record in the file as a data structure (e.g., z value, intensity) or by storing sequential records in the file in

TABLE 10.1
A Data Interleaved File Format

Post n	Height
	Intensity
	Classification
Post $n + 1$	Height
	Intensity
	Classification
Post $n + 2$	...

TABLE 10.2
Block Organized File

Intensity block	Post n
	....
Height block	Post n
	Post $n + 1$
	.....
Classification block	Post n
	Post $n + 1$
	...

block fashion. The first strategy is generally called data interleaving and is illustrated in Table 10.1. The second approach is termed sequential data blocks and is illustrated in Table 10.2

The gridding of data can result as a natural consequence of the organization of the data formation system or as a result of a gridding procedure. A charged-coupled device (CCD) is a sensor that is organized as a grid and hence data collected from such a device naturally lend themselves to a gridded organization. Another common example from remote sensing is uniform scanning of a photograph using a photogrammetric scanner.

LiDAR data are, in general, not organized on a uniform grid due to the nature of the geometry of the scanning device. Nonuniform data can be resampled to a grid or kept in a storage format that supports nonuniform organization. The process of resampling data is in the domain of digital signal processing. Stringent processing procedures such as the Nyquist sampling criteria and frequency prefiltering must be followed to avoid adding false information (called aliasing) to the gridded data. It is generally not considered acceptable to convert point cloud LiDAR data to a gridded format except as a final delivery product.

10.2 Point Cloud Organization

Three-dimensional data that are not organized on a grid are often termed point cloud since they have ill-defined boundaries similar to a cloud. A comparison of a gridded organization to a cross section of a point cloud is depicted in Figure 10.2.

Since the X, Y locations in the point cloud are not known a priori as they are in a grid, all three coordinate values must be encoded for each point. Point cloud organizations are inherently more difficult to work with when performing operations such as searching for a particular point or interpolating a location that lies between points. However, point clouds have the advantage of being able to precisely encode data from a sensor that collects data in a nonuniform pattern such as most LiDAR systems. Point cloud organization is quite common in geographic information systems (GIS) when the individual points are connected via edges to their neighbors in a triangular organization.

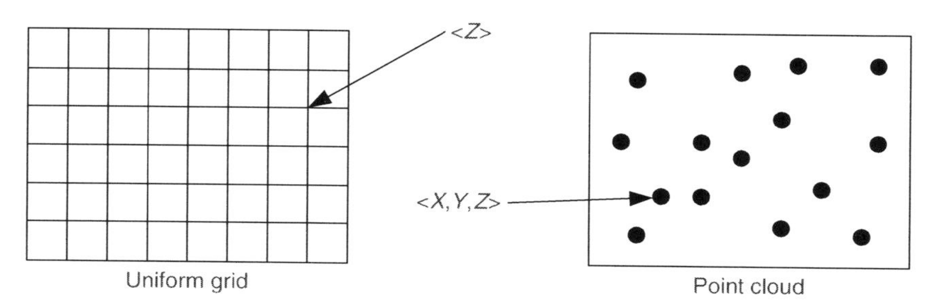

FIGURE 10.2
Comparing a grid to a point cloud.

Such a format is termed a Triangulated Irregular Network (TIN). A TIN has the distinct advantage of being able to accurately represent nonregular geomorphology (ridges, drainage, and so forth).

10.3 Encoding Approaches

Elevation data are commonly encoded as either binary or American Standard Code for Information Exchange (ASCII). Binary encoding is compact, using the minimum file space possible for each data item. For example, if the height values (z) are stored as 32-bit integers, then this is all the space that is occupied per height value (4 bytes per record). Generally, binary files cannot be read without agreement on both the writer's and reader's part of the specific binary format (i.e., 32-bit integer, signed, little-endian format).

To overcome this difficulty, elevation data are often delivered in ASCII encodings (or string). In this format the data are written out as text (either 8 bit or 16 bit per character). ASCII format elevation data can be read and edited in a simple word processor such as Notepad. ASCII format data are very inefficient for two reasons. The first is space. The elevation value"21,432,124" requires 4 bytes when encoded as a 4-byte integer. However, it requires at least 9 bytes when encoded as ASCII. Eight bytes are required for the actual digits of the value and one additional character is required as a "terminator." A terminator (usually a byte whose value is zero, called a null) allows a computer algorithm to distinguish between the end of one value and the start of the next. Occasionally one encounters other software (such as programs that use Hollerith encoding) that require fixed format ASCII values. These programs specify, for example, that each value must be 10 characters long with leading zeros. Of course, this is even more inefficient than null terminated ASCII in terms of storage requirements.

The second major inefficiency of ASCII as a storage format is that computers do not perform numerical computations in ASCII. This means that every ASCII elevation point (and other attribute data, if present) must be read in, parsed to find the end of the string, and then converted to an internal numeric format such as 32-bit integers.

An alternative text-based encoding is extensible markup language (XML). XML has the advantage of ensuring proper reading of data fields (assuming it is a properly formed file) but has the same disadvantages of bloated size and text-to-binary conversion.

A comparison of the various encoding formats for a 1 million point elevation file that requires 32-bit integers for storage is provided as Table 10.3. In this example the heights are encoded in centimeters.

TABLE 10.3

A Comparison of Encodings for a 1 Million Post File

Format	Encoding Example	Size, in Bytes
Binary (Int32)	iiii	4,000,000
Binary (float)	ffff	4,000,000
Binary (double)	dddddddd	8,000,000
ASCII (cm)	487600Ø	7,000,000
Fixed ASCII	0487600	7,000,000
XML	<ht>487600</ht>	15,000,000+

10.4 Exploitation versus Transport Formats

When selecting data formats it is necessary to consider the application goals for the data. Data can be classified as belonging to one or more states with the primary state often dictating the optimal format. Exploitation formats are used when data are actively being accessed by application software. Transportation formats are appropriate when moving data from one application domain to another. Perhaps the most well-known transportation format is XML. XML is a tagged data format that fully specifies each data element in a system neutral format. It is actually a rich, human and machine readable tagged ASCII format. The advantage of XML is that it can transmit information between disparate systems using an agreed upon tagging scheme. Many different schemas have been developed for different disciplines using XML. Geography markup language (GML) is an emerging standard for GIS.

Where formats such as XML are highly suited for transmitting data between systems, they are ill-suited for voluminous data such as LiDAR or image data and for exploitation of data.

Data exploitation involves the process of actually using the data in a presentation or algorithmic manner. An example might be perusing millions of LiDAR data points, locating all points that were acquired on a specific flight line. Such operations are repeated over and over during LiDAR data processing and therefore require a data organizational format that is optimized for accessing and processing the data.

The LiDAR community has settled on a compromise format that is very efficient for data transmission of very large data sets and moderately efficient for exploitation. The format is a binary point cloud organization called LAS.

10.5 Important Characteristics of LiDAR Data

Unlike most conventional image data sets, LiDAR data generally provides a significant amount of sensor characteristics to the exploitation software. This is primarily due to the emergent nature of data processing algorithms. Thus standard LiDAR sensor geocoding software emits LiDAR point data with a rich set of associated attributes. Examples of LiDAR data attributes are listed in Table 10.4.

A quick perusal of Table 10.4 reveals how much data are lost to processing analysis when LiDAR data are delivered in so-called *X, Y, Z* ASCII format. All of the information that are critical to classification of the data are lost. Thus it is important that LiDAR specifications always require the delivery of data with fully attributed points.

TABLE 10.4

Example LiDAR per-Point Data Attributes

Attribute	Description
X, *Y*	The planimetric ground location of the point
Z	The elevation of the point
Intensity	The laser pulse return intensity at the sensor
GPS time	The time (in GPS clock time) of the receipt of the return pulse
Number of returns	Number of returns detected for a given transmitted pulse
Return number	The return number of this pulse (e.g., return two of three returns)
Mirror angle	Angle of the scanner mirror at the time of this pulse (only applies to scanning sensors)
Classification	Surface (or other) attribute assigned to this point such as ground, vegetation, and so forth
Point source ID	A unique identifier to reference this point back to a collection source

10.6 Emergence of a LiDAR Data Standard

In 2002, representatives from Z/I Imaging Corporation (now a part of Intergraph Corporation), EnerQuest (now Spectrum Mapping), Optech Corporation, Leica Geosystems, and the U.S. Army Corps of Engineers Topographic Engineering Center (TEC) formed an informal committee to develop a standard interchange format for LiDAR data. LiDAR data were being produced by sensor vendors (e.g., Optech and Leica Geosystems) in proprietary formats. Software vendors such as Terrasolid Oy (Helsinki, Finland) had written special data readers to ingest these sensor-specific formats. EnerQuest, working with TEC, had also developed custom processing software that had to deal with the vendor-specific formats. It was becoming increasingly clear that high-performance processing systems would need to integrate hardware and software from a variety of vendors. This set of common needs led to the formation of this ad hoc format group (LAS Version 1.1).

EnerQuest had developed an internal format that they used to exchange data between a number of different application programs they had developed under a TEC Small Business Innovative Research (SBIR) grant. This format, called LAS (for LASer), was a binary format that was abstracted above any particular hardware vendor's implementation. EnerQuest and TEC agreed to donate this format to the public domain. The ad hoc committee used this original data specification as the genesis of what would become an American Society for Photogrammetry and Remote Sensing (ASPRS) standard for LiDAR exchange. The LAS name was retained for this format (this is a bit unfortunate since, unknown to the group at the time, LAS is the acronym for the oil industry standard file format Log ASCII Standard).

LAS was internally used by the members of the forming committee while simultaneously being offered as a standard to ASPRS. After slight modifications by the ASPRS LiDAR subcommittee, LAS Version 1.0 was approved as an ASPRS data standard on May 9, 2003. This standard was officially updated to LAS 1.1 (March 7, 2005), with minor changes in data fields. At the time of this writing, a major new release of LAS (Version 2.0) is nearing final ASPRS approval. LAS has been one of the more successful standardization efforts for creating a commercially viable data exchange. LAS 1.0 is currently supported by all commercial LiDAR data processing software vendors, allowing seamless interchange of data.

10.7 LAS LiDAR Data Standard

The LAS data format specification can be downloaded in pdf format from www.lasformat.org. This specification defines a binary file format for storage of LiDAR point data as a set of record blocks. The header section of the file is a fixed length record that contains information about the point records that are stored in variable length record blocks. While the standard supports a user-extensible point format, in reality only two point record types are commonly seen. These record types are dubbed Type 0 and Type 1. The only difference between the two is that Type-1 records include the time of detection of the point by the sensor in the GPS time format. Thus a Type-1 record is 8 bytes larger than a Type-0 record (this is because the GPS time is encoded as a double precision binary number) (Graham, 2005).

Like any data standard, LAS was a compromise between storage efficiency and richness of the data contained in the file. LiDAR-processing algorithms are still at the point when any metadata regarding the state of the sensor during acquisition can be useful in the classification process. For example, if the scan angle of the sensor for a particular point is zero (meaning the point was collected at nadir), then the point probably did not reflect from a vertical surface of a building. LAS is very rich in the collection of per-point metadata. However, the LAS format does not specify an order to the points in the data file and thus random access is not possible in the native LAS format. This organization was by design because it was meant to allow very rapid (real time) storage of data during collection (a so-called streaming format).

It is very important not to make *any* assumptions about the order of points in an LAS file. If a single file represents a single LiDAR flight line, the data will typically be ordered by GPS time. However, there is no guarantee that this will be the case. The LAS format does require that multiple returns be sequentially encoded in the file. For example, a pulse that had three returns will be in sequential order of pulse 1 of three, followed by pulse 2 of three, and finally pulse 3 of three. It is very important when writing software that creates LAS format data to ensure that this rule is followed as LAS does not provide an associativity mechanism that allows points to be related.

Of course, software applications are free to reorganize LAS data during processing. For example, it is quite common to read a LAS data file and then reorganize the data into a spatial arrangement to allow rapid indexing such as a Quad Tree representation. Some vendors spatially organize LiDAR data in the LAS file structure itself and use the user-defined records to store the spatial index. This allows software that understands the spatial index to rapidly access blocks of LiDAR points while not affecting software that does not use the index. The only drawback to this scheme is that software that relies on the indexing scheme must be able to detect whether the data have been randomized by software not using the index scheme. If randomization is detected, the indexing software must reord the file.

10.8 Processing Conventions

LiDAR data attributes should only be modified at points in the processing chain that specifically and reasonably deal with that attribute. For example, *geocoding* software is responsible for encoding the *X*, *Y*, and *Z* values of the data based on GPS/IMU input and LiDAR sensor modeling. Once the *X*, *Y*, and *Z* values have been encoded by this modeling (the software that performs these functions is usually supplied by the sensor manufacturer), the coordinates should not be modified without a rigorous, well-understood model. An example of a well-understood model would be changing WGS-84 ellipsoid heights to North American Vertical Datum, 1988 (NAVD-88), using a transformation model such as one supplied by the U.S. National Geodetic Survey (NGS). An example of modifying data outside the constraints of a well-understood model would be to attempt to determine the average elevation of vegetation that is not being penetrated by the LiDAR and then moving the LiDAR data points in these areas down by this average height.

Sometimes heuristic processing can be rationalized based on accuracy envelopes. A common example of this is data smoothing. Consider a situation in which a LiDAR dataset has an absolute vertical accuracy requirement of 20 cm with some specified sample standard deviation. Analysis of the data indicates a vertical accuracy of 8 cm with similar sample standard deviation. However, the noise in the vertical causes contours to be created with objectionable aesthetics (due to the in-specification vertical noise). Smoothing the vertical component of the LiDAR data within the vertical accuracy envelope would be deemed an acceptable practice (of course, the stated product sample standard deviation would have to be increased as a result of the smoothing).

Points should never be physically deleted from a LiDAR dataset. Instead, points that are to be excluded from consideration are flagged as withheld. LAS 1.0 required changing the Classification field to withheld. LAS 1.1 added a special attribute bit that can be used to indicate the withheld state. This addition in LAS 1.1 allows the original class of a withheld point to be maintained (hence, for example, a point in the Ground class can be flagged as withheld without losing the ground class tag). Points are typically withheld because they are outliers, are in the overlap regions where the desire is to maintain uniform data density, or are on the edge of flight lines or other similar reasons. It is generally a good idea to do an analysis of outliers to determine their cause. Reflection from a bird, cloud, or aircraft is normal, whereas many outliers with unknown cause can be indicative of an impending equipment failure.

10.9 Managing Data

Designing a data management scheme for LiDAR depends on the stage of the data and the application area. An overview of management schemes is provided as Table 10.5.

The choice of data format during LiDAR processing is dictated by the tools that will be employed. Nearly all software for processing LiDAR and elevation data will read and write tabular ASCII data. However, this is the poorest choice of format for exploitation (due to large data sizes and often the loss of attributes such as intensity) and will not be discussed in this section.

The primary consideration with respect to LiDAR data formats is to ensure that the production chain remains modular and open. Thus the best approach is to view production as a series of processing blocks with a neutral data format connection between the blocks. This approach to managing data will allow the processing chain to be upgraded in a modular form as various tools improve.

All commercially available LiDAR processing tools on the market today can read LAS 1.0 (some are still not compliant with the LAS 1.1 specification). Most of these tools will also write LAS 1.0. A number of tools read LAS format and then convert the data to a proprietary format for processing. This sort of scheme has no limitations as long as the data can be converted back to a neutral, open format, preferably during the write back to store operation from the processing tool.

TABLE 10.5

Processing Stages and Management Schemes

Stage	Scheme	Notes
Collection	Typically a vendor proprietary format	At this stage, data are not geocoded. They are typically in a raw form of time-encoded data packets
Geocoding	1 LAS file per flight line	Data are typically fully geocoded and classified to unclassified
Geometric analysis/correction	1 LAS file per test patch	Test patches are extracted from the flight lines and used for geometric analysis
Editing	Tiled LAS	Tiling scheme is generally determined by the maximum number of data points that the editing software can accommodate while remaining performant. This typically ranges from 200 thousand to 4 million points per tile
Quality check	Tiled LAS	Tiling schemes are occasionally different than editing due to different software or delivery requirements
Delivery	Tiled LAS or ASCII	Delivery tile schemes are usually dictated by ground gridding schemes (e.g., 5 × 5 km tiles). Often the primary delivery is gridded data with point cloud LiDAR data as ancillary deliverable

10.10 LiDAR and Databases

Spatial databases such as Oracle Spatial and ESRI ArcSDE are beginning to add native support for LiDAR point data (Oracle Spatial, 2007). With the advent of these storage technologies the question naturally arises as to when database storage is preferable over file storage (often called flat file storage). A general comparison of file versus database storage is provided in Table 10.6.

At the time of this writing, database storage for LiDAR data is very new, existing only as beta versions of released software. The general guideline should be to plan to use file-based storage for high throughput production operations (even as database storage matures) due to the high input–output requirements as well as the fact that data are quite transient during production.

TABLE 10.6

Database versus File Storage

Consideration	File	Database
Rapid data input–output	+	–
Transportable at the tile level	Yes	No
Directly compatible with exploitation and viewing tools	Yes	No
Simple data backup schemes	+	+
Random access based on geospatial criteria	No	Yes

There are really no advantages of having LiDAR data resident in a database during production operations and instead there are many disadvantages. However, for applications where point cloud data will be relatively static over long time periods (months, years), database storage should be investigated as an option. This is particularly true for situations such as removing arbitrary areas of interest for importation into analysis tools such as ESRI 3D Analyst.

10.11 Organizing Data for Production

The most effective current technique for organizing LiDAR data for high throughput production (where production includes primarily the process of extracting features such as bare earth, buildings, and other attributes from the data) is as a grid of tiles of a size optimized for the processing tool. A variety of commercial off-the-shelf (COTS) software applications are available for reordering flight line-based LiDAR data into tiles. Examples of software applications include the LiDAR 1 CuePac from GeoCue Corporation and TerraScan from Terrasolid. The LAS format maintains a field per point record that tracks the source of the points. This Source ID attribute allows the data to be filtered back into original source organization such as flight lines if this should prove necessary.

While many LiDAR processing companies organize data using a number scheme and file directory discipline, a graphical index will significantly aid maintaining the data inventory. In its simplest form, this index is a GIS layer with an attribute containing the storage location of the associated LiDAR data. More sophisticated management tools can arbitrate multiuser access to data tiles as well as retrieve copies of the data from the storage repository (Heller, 1990).

It is good practice to add tools to the processing array that can compute and display the convex hull of the LiDAR tile. The difference between a bounding rectangle display of LiDAR tiles and the convex hull display is depicted in Figure 10.3.

Analyzing LiDAR tiles based on a minimum bounding rectangle (MBR) rather than the Convex Hull can give a misleading picture of how well the tiles are edge-matched in terms of LiDAR data density. Obviously the data tile of Figure 10.3 has a void area in the lower right-hand corner of the tile that will cause a data gap problem when merged with data from adjacent tiles.

LiDAR data return intensity can be used to create a coverage map of the point data. Used in conjunction with the tile index graphic, the LiDAR ortho gives a rapid assessment of the quality of data coverage.

While LiDAR data are considered large as compared to vector-based GIS data, it is actually usually an order of magnitude smaller than associated image data. For example, a county-wide LiDAR project of 1000 km^2 with an average point spacing of 1 m will comprise

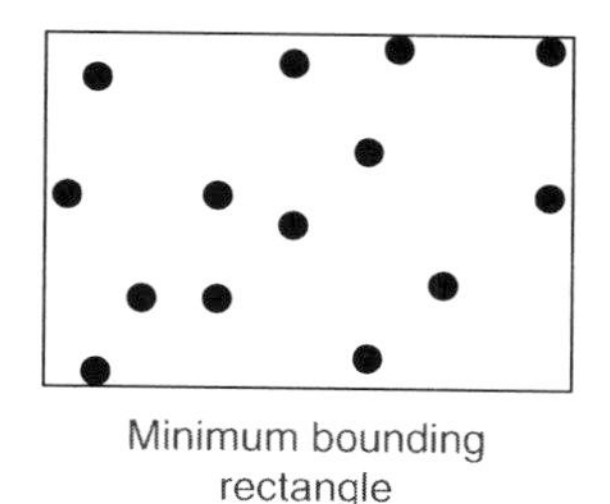

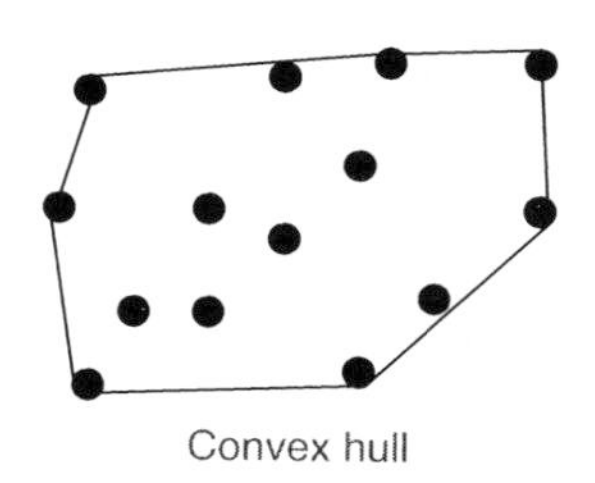

FIGURE 10.3
MBR compared to the convex hull.

about 26 GB of data (using 28 byte/point record type 1 LAS data storage). This same area covered by 15-cm ground sample distance (GSD) 8-bit per channel color imagery will comprise about 125 GB. Thus the challenges of managing LiDAR data are somewhat less than that of the image data typically associated with the project.

10.12 Workflow Considerations

It is a good idea to perform any necessary coordinate transformations at one point in the production workflow. LiDAR data are typically collected in the current epoch of WGS-84 with ellipsoid heights. Customers usually request the delivery of processed data in the local coordinate system with geoid heights. It is strongly recommended that these transformations be performed prior to geometric analysis of the data because this approach will consolidate all geometric testing in one part of the workflow.

LiDAR data should be copied at several steps in the processing chain to allow recovery in the event of an error or data disaster. The original flight lines in the collection coordinate system should always be maintained. The second phase of data duplication should occur at the point of classification of LiDAR tiles but prior to cutting out the customer delivery. These snapshots of the processed data are in addition to the normal system management disciplines.

It is usually not practical to back up LiDAR data to a tape archive (unless the archive system is an extreme performance device capable of 50 MB/s or greater transfer rates) due to the very dynamic nature of the data during processing. The best approach is to duplicate the data across two RAID systems at level 5 where the systems are in two separate physical locations. At the conclusion of the project, the original LiDAR flight line data (following the initial geocoding process) as well as the final fully classified and transformed data at the point that customer deliverables were produced should be archived. If the final product is a gridded data set that is to be derived from the point cloud data, the archive should include both the final classified point cloud data (in LAS format) and the gridded customer delivery.

10.13 Future Trends in LiDAR Data Storage

In the spring of 2008, the LAS format was on the verge of the next major release. Version 2.0 of the LAS specification will address several major shortcomings in the LAS 1.x versions. These include the following:

- LAS 1.x does not include a standard way to add additional data to the point attributes such as red-green-blue values from an ancillary camera.
- LAS 1.x supports only scaled integer storage for *X*, *Y*, and *Z*.
- Coordinate system specification in LAS 1.x has been problematic.

This new specification will have a relatively slow adoption curve with vendors because it represents a fairly significant modification to the 1.x specification. However, it adds a number of new capabilities to the current format to make it applicable to a much broader range of applications, particularly mixed sensor systems such as LiDAR units with digital cameras.

The most notable advances in data storage will be for analysis and exploitation (Van Kreveld, 1997). In these scenarios where the data set is populated and then remains essentially static, storage in a spatially indexed database offers significant advantages as compared to simple flat file storage. These systems will allow smooth integration between disparate data sets, enabling a rich array of heterogeneous exploitation tools. An example might be extracting building footprints from a two-dimensional vector file and attributing the height by extracting the associated LiDAR points classified as belonging to the Building class from the LiDAR database.

LiDAR data will continue to grow in dataset size due to two drivers. The first is the desire to collect data (whether it is LiDAR or imagery) at higher densities. The second is the trend of adding additional information to the LiDAR return data. Of the currently emerging trends, the collection and processing of full waveform data will contribute the most to increasing LiDAR dataset sizes. Fortunately, at the same time that LiDAR data sizes are increasing, the cost of large capacity storage devices are continuing to fall.

References

LAS Version 1.1 Specification. Available at www.asprs.org/las.

Graham, L. The LAS 1.1 Standard, Feature Article, *Photogrammetric Engineering and Remote Sensing*, July 2005, 777–780.

Heller, M. Triangulation algorithms for adaptive terrain modeling. In *Proceedings of 4th International Symposium on Spatial Data Handling*, pp. 163–174, 1990.

Oracle Spatial 11g: Advanced Spatial Data Management for Enterprise Applications, An Oracle White Paper, 2007. Available at http://www.oracle.com/technology/products/spatial/pdf/11g_collateral/spatial11g_advdatamgmt_twp.pdf.

van Kreveld, M. Digital elevation models and TIN algorithms. In M. van Kreveld, J. Nievergelt, T. Roos, and P. Widmayer, (Eds.), *Algorithmic Foundations of GIS, Lecture Notes in Comp. Science.* Springer-Verlag, 1997, 185–197.

11

LiDAR Data Filtering and DTM Generation

Norbert Pfeifer, Gottfried Mandlburger

CONTENTS

In this chapter, digital elevation models (DEM), also known as digital terrain models (DTM), will be discussed. After defining these terms we will first consider the specific advantages LiDAR offers for capturing the terrain surface, also in comparison to image-based photogrammetric techniques. The main topic in this chapter will be, however, the extraction of the terrain surface from the LiDAR measurements, either the original point cloud or the already preprocessed one with terrain elevations sorted into a raster. This process is a classification task, dividing the points or pixels into either ground or off-terrain, where some approaches derive more classes. The usage of full-waveform echo recording for terrain reconstruction will be also discussed within this scope. Computing a DTM from the results is the next step, but general discussion on strict interpolation and qualified approximation techniques will be very limited. However, in order to make a DTM or DEM usable for many disciplines, e.g., hydraulics, geology, etc., the so-called structure or feature lines should also be integrated. Likewise, the amount of data is typically very high and not adapted to the variations in terrain roughness. We will therefore continue with a section on the determination and exploitation of structure lines from the LiDAR data as well as with methods for reducing the data amount, i.e., compressing the elevation data, considering error bounds. The quality of DTMs from airborne LiDAR data will be treated in a separate section. Real-world examples, embedded in the text, will demonstrate the above. While DTMs can also be reconstructed from terrestrial laser scanning data, offering more detailed and accurate description, this is only possible for small areas in the order of 1 km^2 and below, whereas airborne laser scanning has been used to collect topographic information over more than 10,000 km^2 in dedicated projects. Therefore, only airborne laser scanning, also termed airborne scanning LiDAR, will be considered.

11.1 DTM and Laser Scanning

This section introduces digital terrain models (El-Sheimy et al., 2005; Ackermann and Kraus, 2004; Maune, 2001) and relates them to the data acquisition method laser scanning. It also compares the data acquisition methods laser scanning and photogrammetry (Kraus, 2007; McGlone et al., 2004) with respect to terrain derivation.

11.1.1 DTM Definition

From the many definitions of a digital terrain model (El-Sheimy et al., 2005), we choose one that does not prescribe the data structure, but concentrates on the geometrical aspects. A digital terrain model is a continuous function that maps from 2D planimetric position to terrain elevation $z = f(x,y)$. This function is stored digitally, together with a method on how to evaluate it from the geometrical and possibly explicitly stored topological entities. Sometimes DTM and DEM are distinguished by recognizing the difference that the first one includes break lines (and other topographic features), whereas the second one does not (El-Sheimy et al., 2005). In this chapter, the terms DTM and DEM will be used interchangeably. The inclusion or disregarding of break lines is rather a question of the level of precision, detail, and morphological quality. A more detailed discussion is necessary and will be given in the following paragraphs, as these definitions were coined when ground point data sets had a substantially lower density than those available from airborne laser scanning today.

For using the above definition, it is necessary to clarify the nature of the terrain (elevation). It is defined here as the boundary surface between the solid ground and the air. This definition (top soil, pavement, etc.) is also applied by Sithole and Vosselman (2004) for their filter comparison. This is typically also the surface of superficial water run-off. This definition is sufficient for well-defined surfaces, but a number of problems are encountered when looking at the details observed and the precision achieved by airborne laser scanning.

1. The concept refers to the so-called 2.5D approach, where for one ground position (x,y) only one height (z) may be expressed. For overhangs, which are occasionally recorded, the 2.5D assumption does not hold. Most algorithms for terrain extraction will therefore fail in this special circumstance (Pfeifer, 2005). However, it also plays a role when considering the digital surface model (DSM, defined below). From an application point of view, e.g., orthophoto production, a bridge may also be seen as integral DTM part, again violating the 2.5D assumption. In Sithole and Vosselman (2004), bridges are defined as objects, whereas ramps leading to bridges are considered as part of the terrain.
2. For hard surfaces (concrete, etc.), the boundary surface between ground and air is well defined. However, for natural surfaces, possibly featuring a dynamic aspect, it is less obvious. From lower flying altitudes and using high pulse repetition rate scanners the reconstruction of plough furrows becomes—theoretically—possible, although not interesting for a general purpose DTM. Additionally, elevation of surfaces collected by fieldwork over herbaceous vegetation typically refers rather to the soil elevation, whereas laser scanning records refer to some mean vegetation elevation. The effect of this method depends on the vegetation density and may easily reach, e.g., +15 cm for 30 cm high grass, and is thus notable given the state-of-the-art systems. Before the advent of high-precision terrain elevation capturing by airborne laser scanning, this question did not arise.
3. Large objects built on the terrain, e.g., houses, do not simply shadow the terrain surface, but cease its existence. To avoid holes, interpolation of the terrain surface under the houses becomes necessary. This situation is also visualized in Figure 11.7.

Eventually, the application determines which aspects are to be included in a digital terrain model.

From a geomorphological point of view, the terrain surface is structured by lines. At a so-called break line the first derivative of the surface is discontinuous, whereas it is smooth elsewhere. Additionally, form lines are sometimes defined as soft break lines. In a contour line plot, the form line is the connection of points with maximum curvature. So far, to the best of the authors' knowledge, no attempt to reconstruct form lines from airborne laser scanning data was published and they will not be discussed further.

11.1.2 Digital Canopy Models (DCM) and Digital Surface Model (DSM)

The DTM is one special surface that can be reconstructed from laser scanning. It is, however, not the only one. DSM and DCM are other surfaces, not necessarily thoroughly physically defined (Schenk and Cshato, 2007). The DCM represents the top surface that is visible from above. It is equivalent to the DTM in open areas and runs on top of the vegetation canopies in forested areas and over house roofs in the build-up environment. For these objects the DCM is—more or less—well defined, but other objects such as power lines or highly dynamic objects (cars, cattle, etc.) are not treated. A DCM is often computed by interpolating the points corresponding to the first echo of each emitted shot, although power lines and other elevated features make the result unpleasant to look at and harder to analyze. In the overlapping areas of adjacent strips, the dynamic objects typically show up twice, both times intermixed with points of the surface below them. The term DSM is used synonymously with DCM, although the "surface" in its acronym does not specify the surface to be modeled. The difference model of a DSM and a DTM is called a normalized DSM (nDSM); thus, $\text{nDSM}(x,y) = \text{DSM}(x,y) - \text{DTM}(x,y)$. The differences between DTM and

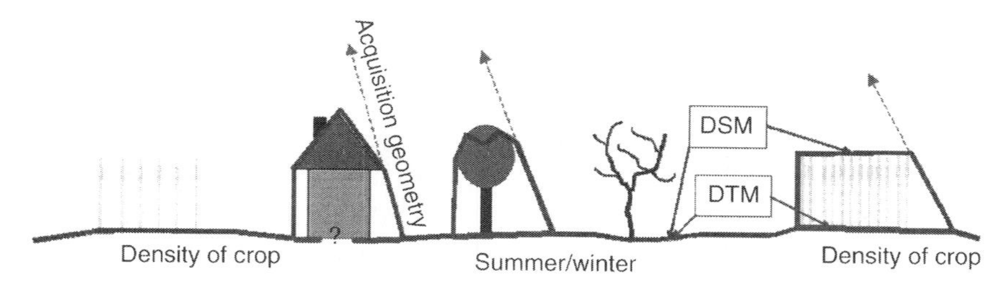

FIGURE 11.1
DSM runs over buildings and tree crowns, if they are contained in the data acquired by airborne laser scanning. Below houses the DTM is not properly defined. In open areas, the DTM and the DSM refer to the same surface.

DSM and also some of the factors influencing the DSM as an interpolated surface of all points are shown in Figure 11.1.

11.1.3 Data Structures

The terrain model function $z = f(x,y)$ is computed from the source data, i.e., 3D points $p_i = (x_i,y_i,z_i)$, $i = 1,\ldots, n$, with n the number of points. This function needs to be made persistent as the interpolation from the source data is unfeasible for each new evaluation of the terrain. The very large amount of data (millions to billions of points) makes the transformation into another format indispensable. Popular data formats are the raster, where an elevation is assigned to a small area, i.e., a rectangular pixel. The grid is the dual structure, heights are stored at discrete, regularly aligned points, and an interpolation method, typically bi-linear interpolation, is used to interpolate heights within a grid mesh. The pixel approach does, strictly speaking, not provide a DTM because it is not continuous. Therefore, grid evaluation is performed, even if the data is stored in a raster image format. A hybrid grid structure is obtained by intermeshing additional lines and special points into the grid structure (Ackermann and Kraus, 2004; Köstli and Sigle, 1986). Examples are given in Figures 11.10 and 11.11. The grid heights are obtained by interpolation methods as inverse distance weighting, moving least squares, and kriging. An alternative method is the triangulation, generating the so-called Triangular Irregular Network (TIN) data structure. The original points are used for reconstructing the surface. For large point sets, deriving the triangulation is time consuming, but the popularity of triangular meshes in computer graphics drives research also to overcome this problem (Isenburg et al., 2006; see, e.g., Figure 11.11).

11.1.4 Laser Scanning versus Photogrammetry for Terrain Capture

First commercial projects to acquire terrain data by laser scanning from airborne platforms were performed in Europe by TopScan in 1994/1995 (J. Lindenberger, personal communication), employing an Optech laser scanner. Alternative methods for terrain data capture are tachymetry, as ground technique only suited for very small areas, and aerial photogrammetry. From digital images (airborne and from satellite), the method of automatic image correlation can automatically produce point clouds. With digitally acquired images the overlap between consecutive images can become very high, allowing more robust multi-image matching (Grün and Baltsavias, 1988) and generating very dense point clouds. It therefore stands to reason to compare these two data acquisition methods a little more in detail with respect to terrain reconstruction (Kraus and Pfeifer, 1998).

1. Photogrammetry computes ground coordinates by forward intersection. Therefore, first the (automatic) finding of homologous points is necessary. Areas visible from one exposure position only cannot be reconstructed. This applies typically to the forest ground where ground points cannot be identified in adjacent images. Different gaps in the foliage seen from different positions. In laser scanning, one view to the ground is sufficient to record a point. It also applies to geometric shadows of buildings, which prevent the reconstruction of the ground surface or objects on it. Airborne laser scanning is not entirely free of that effect, but again the polar measurement method alleviates this drawback of airborne data acquisition.
2. In contrast to photogrammetry, laser scanning is an active measurement technique, and data acquisition is therefore independent of the sun position. Flying can be performed at night time, which is an important aspect next to airports or in high-latitude regions. In addition, laser scanning allows measurement of ground points independent of sun shadows, which are encountered on forest grounds as well as in city areas.
3. Both laser scanning and photogrammetry provide increased accuracy for decreasing flying heights. In laser scanning, however, the height component is the more precise one, whereas in photogrammetry planimetry can be measured more accurately. As camera opening angles are typically larger compared to laser scanner opening angles, a smaller number of flying strips is necessary in photogrammetry for covering the same area. However, strip distance and flying height are governed also by other factors such as minimum ground mapping units and point density. Furthermore, photogrammetry is a method based on measuring texture, which is high at object discontinuities, i.e., edges. Photogrammetrically acquired point clouds are therefore concentrated along distinct points and edges, whereas point clouds acquired by laser scanning are more reliable inside surfaces.
4. Laser scanning is, in first instance, a method without redundancy and, therefore, limited reliability. As no object surface point is measured twice, no tie points exist as in photogrammetry, and methods for checking and improving exterior orientation as well as system calibration are in photogrammetry.

It should also be mentioned that photogrammetrically acquired DTMs, which are a basic geodata product offered by national surveying authorities, are gradually being replaced by laser scanning DTMs featuring higher density and precision. Amongst these countries are the Netherlands, offering a nation-wide laser scanning DTM completed in 2003 (Brügelmann and Bollweg, 2004) and Switzerland for all areas with elevation below 2000 m a.s.l. (Artuso et al., 2003). In the United Kingdom, concentration is laid especially on flood plains. In other countries, including the United States, Germany, and Austria, DTMs are acquired by laser scanning not on a national but on a federal level.

11.2 Ground Point Extraction from Laser Scanning Point Clouds

In this section the classification of the acquired point cloud into ground points and off-terrain points, possibly split up in more classes, is discussed. The computation of digital terrain models, which is an integral part of some of these algorithms, will be treated in the next section. Before presenting an overview on filter algorithms, we will state the problem explicitly. Subsequently, approaches for filtering the airborne laser scanning data are presented. The filters are organized into four groups depending on the concept of the terrain surface used. Two filters in total will be described in more detail. They have been selected because they have reached certain popularity, they are very different and, thus, highlight the range of approaches and concepts applied.

11.2.1 Problem Definition

Formally, the problem can be stated as follows. Given is the entire set of points **P**, each point embedded in 3D space: $p_i = (x_i, y_i, z_i) \in \mathbf{R}^3$, $i = 1, \ldots, n$. There can be some additional attribute information a_i like echo number (first, intermediate, last echo), intensity (return amplitude) measure, color obtained from a digital image, echo width extracted from full-waveform analysis, or height precision. Those points p_i that lie on the terrain surface shall be identified considering the measurement precision. $\mathbf{P}_\mathrm{T}$ is the set of terrain points ($\mathbf{P}_\mathrm{T} \subseteq \mathbf{P}$), see also Figure 11.2.

This is the classification point of view on the subject, point-based or raster-based. In the latter case, the pixel position takes the role of x and y and the pixel value the role of z. Alternatively, the problem can be seen as directly obtaining the DTM from the set $\mathbf{P}_i$, without an interest in the classification itself. In any case, this task is not trivial, because

1. Objects slightly above the ground may have the same geometrical appearance as smaller terrain features. Thus, a criterion only based on points neighbouring p_i may fail in such cases.
2. The attribute information a_i typically cannot be used to classify the points into ground and off-terrain. Especially for multiple echos, the last echo does not always refer to the ground. It may originate from lower vegetation below a forest canopy. The last echo may also feature a multipath effect, effectively representing a point below the ground surface. However, using only last echoes as start point for ground extraction is an approach used frequently.
3. Human interpretation uses context knowledge significantly when looking at point clouds or a DSM in 3D perspective views, which is hard to incorporate in algorithms.

11.2.2 Rasterization: Pros and Cons

Most algorithms for ground extraction employ some geometrical reasoning to decide if a point is situated on the terrain surface. Rasterizing the data first, these algorithms can be introduced into the domain of digital image processing, where neighborhood operations run much faster. However, this results in a loss in precision (Axelsson, 1999). Additionally, depending on the algorithm, the interpolation of raster heights in areas without points may be necessary. There are a number of options for assigning a height to a cell, which will not be discussed in detail here. The possibilities include taking the height of the lowest/mean/closest to the center/median in height point, etc.

11.2.3 Simple Filters

Different concepts for filtering, with different complexity and performance characteristics, have been proposed in literature so far. The simplest filters operate by always taking the

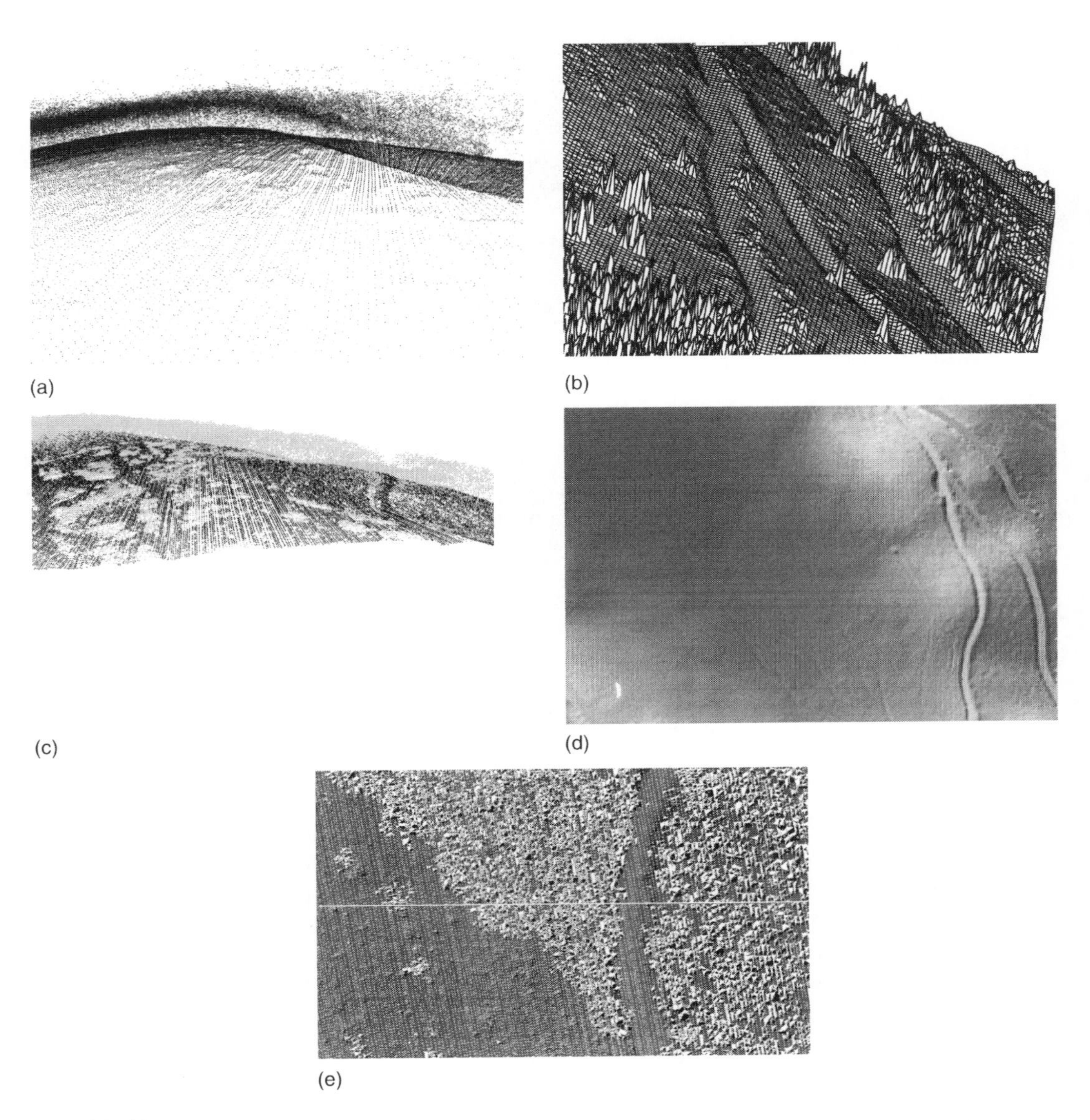

FIGURE 11.2
(a) Original point cloud in perspective view from the side, showing ground points, points on the tree crowns, at intermediate height, and points on features near the ground. (b) DSM of the same dataset, showing streets, bare ground, and wooded areas. (c) Classified point set: ground points (black) and off-terrain points (gray). Note that the features near the ground from (a) have been classified as off-terrain here. (d) Z-coding superimposed to shading of the DTM, plan view. Next to the streets, paths inside the forest also become visible. In (e) a shaded view of a DSM detail together with the original points and a profile of 85 m length are shown.

minimum elevation within a certain area. If this area is a raster cell, the lowest point falling within this cell defines the cell elevation, leading directly to a DTM in a regular grid structure. These block minimum filters have a number of properties that introduce systematic errors. First, there is an inherent assumption of a flat terrain. For inclined terrain, even without vegetation cover or buildings on top, the height will also be too low. This systematic error depends on both the terrain inclination and the neighborhood size.

11.2.4 Morphological Filter

The name for this group of filters is derived from mathematical morphology (Haralick and Shapiro, 1992), which is also explicitly stated in the "Morphological Filter" of Vosselman (2000). A structure element $\Delta h_{max}(d)$ describing admissible height differences as a function of the horizontal distance d is used in the erosion operation (see Figure 11.3a). The distance d is the planimetric distance; thus, $d(p_i,p_k)=\sqrt{(x_i-x_k)^2+(y_i-y_k)^2}$. The smaller the distances between a ground point and its neighboring points, the lesser the height difference accepted between them. The structure element is positioned at each candidate point and this point is identified as off-terrain point if one or more height differences to its neighbors are above the admissible height difference. This structure element has an effect up to the maximum distance d_{max}, thereby defining the neighborhood contributing to the point test. In the publications of Vosselman, d_{max} is typically in the order of 10 m. The structure element encodes information not only on the terrain but also on the points measured by laser scanning. It can be determined from assumptions on the maximum terrain slope found in the area and from the height precision of laser scanning points. With a maximum slope of $\tan(\gamma)$ and a vertical measurement precision of σ_z, it becomes (Vosselman, 2000):

$$\Delta h_{max}(d)=\sqrt{2}\sigma_z+d\tan(\gamma),\quad d\le d_{max}$$

The term $\sqrt{2}\sigma_z$ is the precision of the height difference of a point pair. The set of ground points is then defined as (Vosselman, 2000):

$$\mathbf{P}_\mathrm{T}=\{p_i\in\mathbf{P}\,|\,\forall\, p_k\in\mathbf{P}: z_i-z_k\le\Delta h_{max}(d_{ik})\},\quad d_{ik}=\sqrt{(x_i-x_k)^2+(y_i-y_k)^2}$$

The structure element can also be determined from terrain training data consisting of correctly classified ground points and vegetation points. The actual height differences between points of the same and of different classes can be used to derive a structure element that avoids omission errors (no real ground point is rejected as new off-terrain point) or commission errors (no real off-terrain point is classified as ground point). By averaging those two structure elements, the result is an optimal structure element to minimize commission and omission errors (Vosselman, 2000). In Figure 11.3b, the classification process, using the morphological filter, performed on a profile through the dataset of Figure 11.2 is shown.

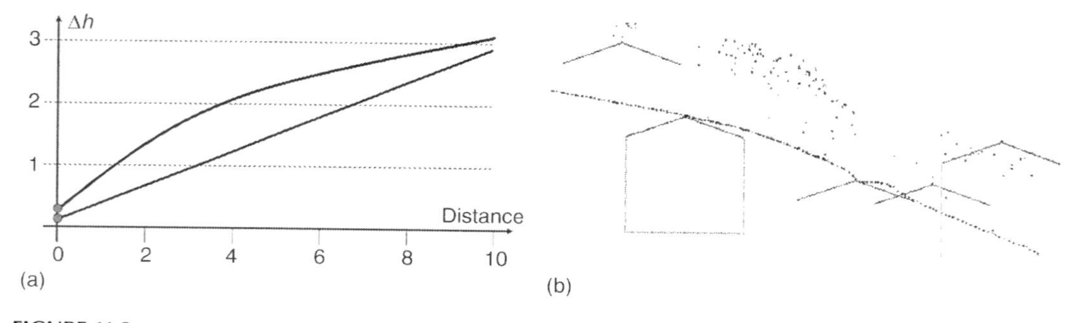

FIGURE 11.3
(a) Two different structure elements showing admissible height differences Δh_{max} versus distance, both axes are shown in meters. (b) Classification of a number of points in a profile by a structure element for an offset of 15 cm and a terrain slope of 20° is shown (straight structure element in left image). For the (two) ground points, no points are below the structure element, whereas for the (three) vegetation points, a number of points are below the structure element. The original data is shown in Figure 11.2e in plan view.

11.2.5 Extensions and Variants of the Morphological Filters

A morphological filter variant is described by Sithole (2001), where the rotationally symmetric structure element depends on the terrain inclination estimated from averaged height data. For steeper areas, larger admissible height differences are prescribed. Another extension is mentioned in the airborne laser scanning filter overview included in (Kobler et al., 2007) where the structure element is inclined as a whole in order to follow the terrain and the rotational symmetry is lost. The reason is that height differences downwards have different characteristics compared to the height differences upwards, which is not considered in the other morphological filters. Kilian et al. (1996) use multiple structure elements with different horizontal and vertical extents in the morphological opening operation. The structure element is, however, only horizontal. At different scales, spikes are thus removed, because for different structure element (i.e., window) sizes different objects (e.g., buildings and cars) are found. Weights are assigned to the points depending on the window size, which are finally used for the classification. A similar method is described by Zhang et al. (2003), augmented by a method for choosing the window parameters. In the "Dual Rank Filter," erosion and dilation is suggested by Lohmann et al. (2000) in order to replace raster terrain elevations with the filtered elevations. If erosion and dilation are applied with the same structure element it becomes the opening operation applied by Kilian et al. (1996) and Zhang et al. (2003).

Some of the above filters are described for point clouds (e.g., the morphologic filter of Vosselman) using planimetric distance as the neighborhood criterion, obtained by traversing a triangulation. The definition of neighborhood in airborne laser scanning data in general can be based on distance (fixed distance neighbors) or topology (*k*-nearest points in a tree-structure or tiers in a triangulation). Neighborhood systems have been investigated by Filin and Pfeifer (2005). Other approaches (e.g., by Kilian et al., 1996) work on digital images on which morphological operations can be applied very efficiently.

11.2.6 Progressive Densification

The filters in this group work progressively, rebuilding the ground by classifying more and more points as belonging to it. Axelsson (2000) uses the lowest points in large grid cells as the seeds of his approach. The current set of ground points is triangulated in order to form a reference surface. For each triangle one additional ground point is determined by investigating the unclassified points within the vertical column of that triangle. The offsets of these candidate points to the triangle surface are analysed.

These offsets are the angles α_1, α_2, and α_3 between the triangle face and the edges from the triangle vertices to the new point (Figure 11.4a). If a point is found with offsets below the threshold values, it is classified as a ground point and the algorithm proceeds with the next triangle. The algorithm may stop if no more points can be found in a triangle, if a certain density of ground points is achieved, or if all acceptable points are closer to the surface than another threshold.

Von Hansen and Vögtle (1999) describe a similar method with two differences. Firstly the starting points are the lower part of the convex hull of the point sets (Figure 11.4b), and secondly the offset is measured as the vertical distance Δh of a candidate point to the reference triangle (Figure 11.4a). Sohn and Dowman (2002) apply the progressive densification first with a downward step, where points below the current triangulation are added, followed by the upward step, where one or more points above each triangle are added. The initial triangulation is obtained from the corner points of the entire area. All of these filters use a triangulation for accessing the data and also as DTM.

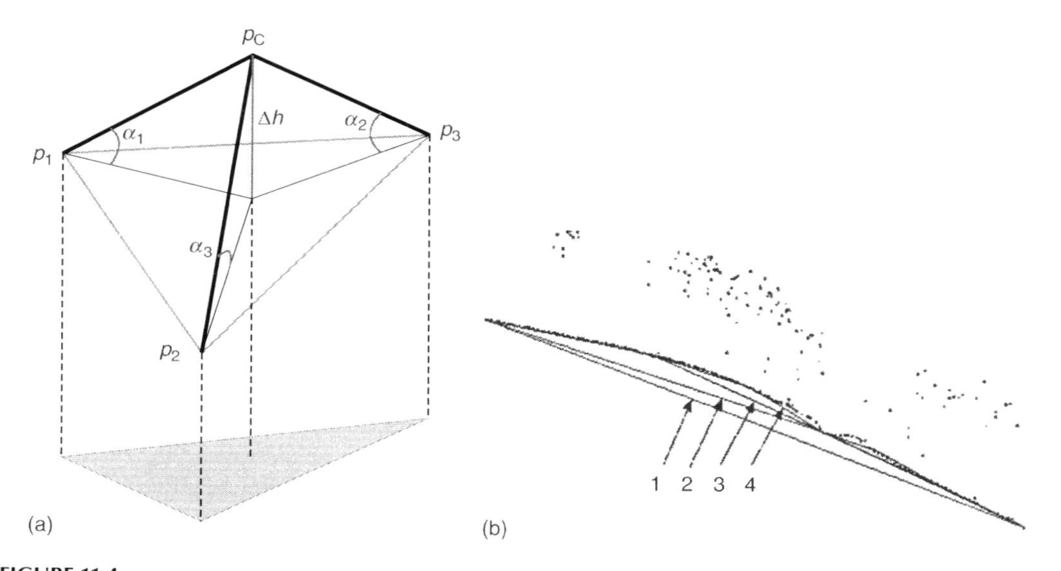

FIGURE 11.4
(a) Measures (α_1, Δh) to determine if a candidate point p_C is a ground point, as are the points p_1, p_2, and p_3 forming a triangle. (b) Lower part of the convex hull of a profile through the data of Figure 11.2 and a number of steps in the progressive densification.

11.2.7 Surface-Based Filter

While the progressive densifications rebuild the terrain surface, the group of filters presented now start by initially assuming that all points belong to the ground surface and then remove those points that do not fit. Therefore, a general surface model $s_l(x,y)$ is constructed to iteratively approach the terrain surface DTM(x,y), l being the iteration index. The method of Kraus and Pfeifer (1998), known as robust interpolation, will be described in detail now, especially in the context of ground points and points on the vegetation.

In this approach, linear least squares interpolation (also known as linear prediction or simple kriging) is used to describe the surface. The stochastic behavior of the terrain is described by the covariance function $C(p_i, p_k)$. with the variable planimetric distance d as defined above:

$$C(p_i, p_k) = C(0)e^{-d(p_i, p_k)^2/c^2}$$

The parameters c and $C(0)$ are computed from the data (Kraus and Pfeifer, 1997; Journel and Huijbregts, 1978), but estimates depending on the terrain type can also be used. The covariance function is used to set up a system of equations for computing the height z of one point $p = (x,y,z)$ from the given points $p_1, \ldots, p_n$. It has to be noted that a trend in the data is subtracted first, e.g., by determining an adjusting plane or low-order polynomial approximating all points. This will not be discussed further, and it is assumed that the trend has been removed.

$$z(x, y) = \mathbf{c}^{\mathrm{T}}\mathbf{C}^{-1}\mathbf{z}$$

$$\mathbf{c} = (C(p, p_1), \ldots, C(p, p_n))^{\mathrm{T}}, \quad \mathbf{z} = (z_1, \ldots, z_n)^{\mathrm{T}}$$

$$\mathbf{C} = \begin{pmatrix} V_{zzp1} & C(p_1, p_2) & \cdots & C(p_1, p_n) \\ C(p_1, p_2) & V_{zzp2} & & C(p_2, p_n) \\ \vdots & & \ddots & \\ C(p_1, p_n) & C(p_2, p_n) & \cdots & V_{zzpn} \end{pmatrix}$$

In linear least squares interpolation (Kraus and Mikhail, 1972), the surface does not pass exactly through the given points, but random errors can be filtered. The residual at a point is $r_i = z_i - z(x_i, y_i)$. The filtering strongly depends on the variances V_{zzpi} along the diagonal of the **C** matrix. In general, large variances cause large filter values and small variances force the surface to almost strictly interpolate the point. Expressed as weight, the most accurate points shall have the weight one, and it decreases for less accurate points. For the most accurate points, the difference $V_{zzpi} - C(0) = \sigma_z^2$ is the vertical measurement variance (square of precision, which is typically ±10 to ±30 cm). Thus, for points with weight w_i the variance V_{zzpi} becomes:

$$V_{zzpi} = \sigma_z^2 / w_i + C(0)$$

A method to compute a surface from a point set considering an individual weight per point has been described so far. This method will now be used for classifying ground versus off-terrain points.

The weights w_i are determined iteratively. In the first iteration, all points get the same weight $w_i = 1$, and all points have the same influence on the run of the surface. The values r_i are computed, and they typically show an asymmetric distribution. There is a cluster of negative values, corresponding to points below the surface and originating from the ground points. Because the initial surface is shifted upwards, the ground points form a group of points with negative residuals. The vegetation points do not form a clear cluster, because they are not necessarily on the canopy top surface only, but found at various height layers in the vegetation.

In robust estimation (Koch, 1999), a weight function is used to down-weight observations with larger residuals, which have presumably large errors. This is performed iteratively, so observations that only had large residuals in the first pass but actually fit well to the model (here: the ground surface) can be rehabilitated. However, observations that do not fit to the model obtain lower and lower weights in each iteration. The weight function must fulfil the property of weighing ground points higher (weight 1) and points on the vegetation lower. These properties are reached with an asymmetric weight function (Figure 11.5a).

$$w(r) = \begin{cases} r < g & : \quad 1 \\ g \le r \le g + h & : \quad \dfrac{1}{1 + (a(r-g))^b} \\ r > g + h & : \quad 0 \end{cases}$$

With the weight w, new variances V_{zzpi} are computed for each point. As noted above, points with large variances, i.e., those points with low weights, have less influence on the run of the surface, and the surface is more attracted to the (ground) points with higher weights. In the iterative process, the points on the vegetation get less and less weight. The parameters a and b determine how fast the weight function drops to zero, whereas the parameter g determines where this drop starts, and h is a parameter controlling the exclusion of (definite vegetation) points from the equation system. The results of the robust interpolation applied to the data of Figure 11.2 are shown in Figure 11.5b. For practical usage, a, b, and h

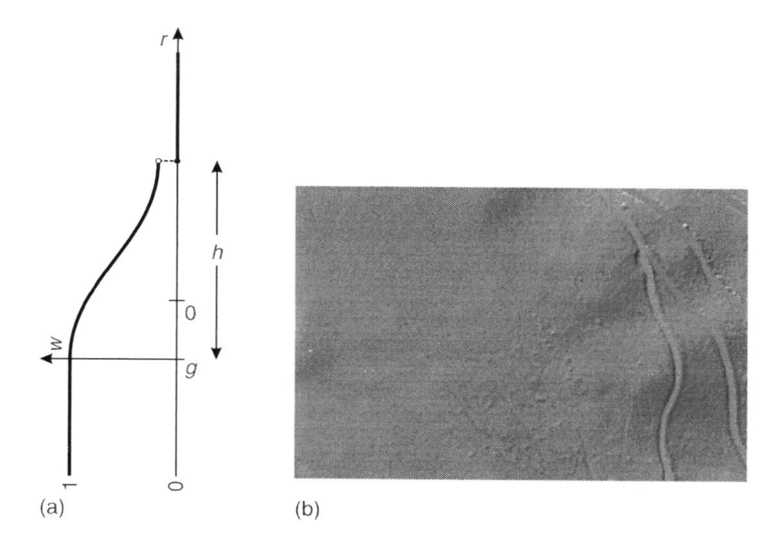

FIGURE 11.5
(a) Weight function of the robust interpolation, mapping a residual r to a weight w. (b) Shaded view of the DTM from the data set of Figure 11.2, filtered by robust estimation. Note that near-ground features have been classified as ground points incorrectly here (opposed to Figure 11.2c and d). There were no points below those features on the terrain surface and also height differences and inclination against the surrounding terrain is small.

can be set to 2 (m^{-1}), 2, and 5 m, respectively. This means that points with a residual of 0.5 m above g get the weight 1/2, and those with a residual of 5 m above g are ignored. The parameter g can be estimated from the residuals themselves. Methods are given in detail in (Kraus and Pfeifer, 1998). A robust method is to estimate the penetration rate, which need not be a very accurate estimation. If the estimation is, e.g., 40%, g then becomes the quantile of the residuals found at half of the estimated penetration rate (in the example the 20% quantile of residuals). The entire process is visualized for a profile in Figure 11.6.

As an equation system needs to be solved, the method is applied patch-wise. The functional model of surface interpolation could be replaced by other surface reconstruction techniques, e.g., moving least squares in place of Kriging, showing better computational performance, but giving less control over filter values.

11.2.8 Extensions and Variants of the Surface-Based Filters

In the work by Pfeifer et al. (2001), the robust interpolation method has been embedded in a hierarchical approach to handle large buildings and reduce computation time. At different levels, thinned out versions of the point cloud are used to compute surface models. The DTM from a coarser level is used to preselect probable ground points for the next layer. The process is illustrated in Figure 11.7.

Elmqvist (2001) uses a snake-approach (Kass et al., 1988), where the inner forces of the surface determine its stiffness and the external forces are a negative gravity. Iteration starts with a horizontal surface below all points that move, following the negative gravity, upwards to reach the points. Inner stiffness, however, prevents it from reaching up to the points on vegetation or house roofs. Brovelli et al. (2004) have a similar approach analyzing the residuals obtained from an averaging spline surface interpolation. In a second step, edges are extracted, connected to closed objects, and elevated areas are removed. All these filters work on the point clouds and no explicit neighborhood definition is necessary.

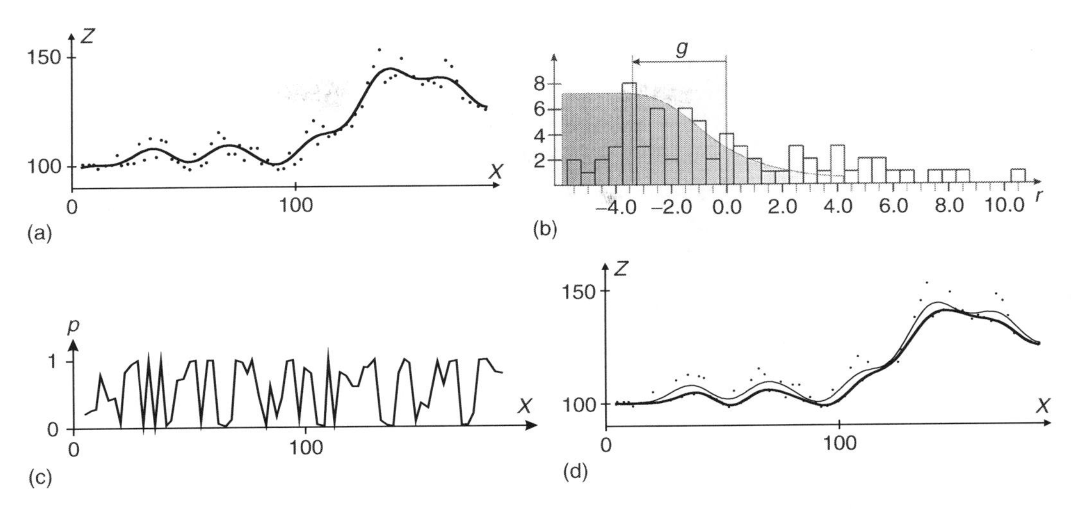

FIGURE 11.6
Profile of ALS data with a hilly terrain and points on the vegetation and the ground. Panel (a) shows the points and the averaging surface computed with equal weights for each point. Panel (b) shows the residuals, overlaid with the weight function giving smaller weights to points that lie higher above the surface. Panel (c) displays the distribution of weights between 0 (no influence on the surface run in the next iteration) and 1 (maximum influence). The surface computed under consideration of the weights is shown in panel (d). The previous (first) iteration is shown in gray.

It is given implicitly by the ground project plane and the interpolation method. A methodological strength of these filters is that surface behavior, e.g., smoothness, can be incorporated into the functional model of surface interpolation, whereas the stochastic properties of the laser points—including both random measurement errors and distribution of above ground objects—can be encoded in the weight function.

11.2.9 Segmentation-Based Filters

The fourth group of filters works on segments. As Filin and Pfeifer (2006) note, the processing of laser scanning point clouds can be strengthened by first aggregating information (i.e., building homogeneous segments) and then analyzing segments rather than individual points.

Segmentation can be performed directly in object space, using region-growing techniques. Often the normal vector or its change is used to formulate the homogeneity criterion, resulting in planar surfaces in the first case and smoothly varying surfaces in the second. Alternatively, homogeneity can be formulated in terms of height or height change only. In contrast to the region-growing techniques, segmentation can also be performed in a feature space, which has the advantage that the variability of the features is known before detecting clusters in feature space. In Figure 11.8a, the motivation for segmentation-based filtering is demonstrated. The original point cloud is shown with points from different segments in different colors. Each segment either belongs to the ground or contains only off-terrain points.

Sithole (2005) describes a method building segments by the rational that each point in a segment can be "reached" from each other point in the same segment, in the sense that it would be possible for a person to walk along such a path. The segmentation is performed first along two or more lines in different directions, and the height differences and jumps along this line as well as the horizontal distance are used to classify the segments as raised, lowered, terraced, and into other classes. The common points of line segments in different

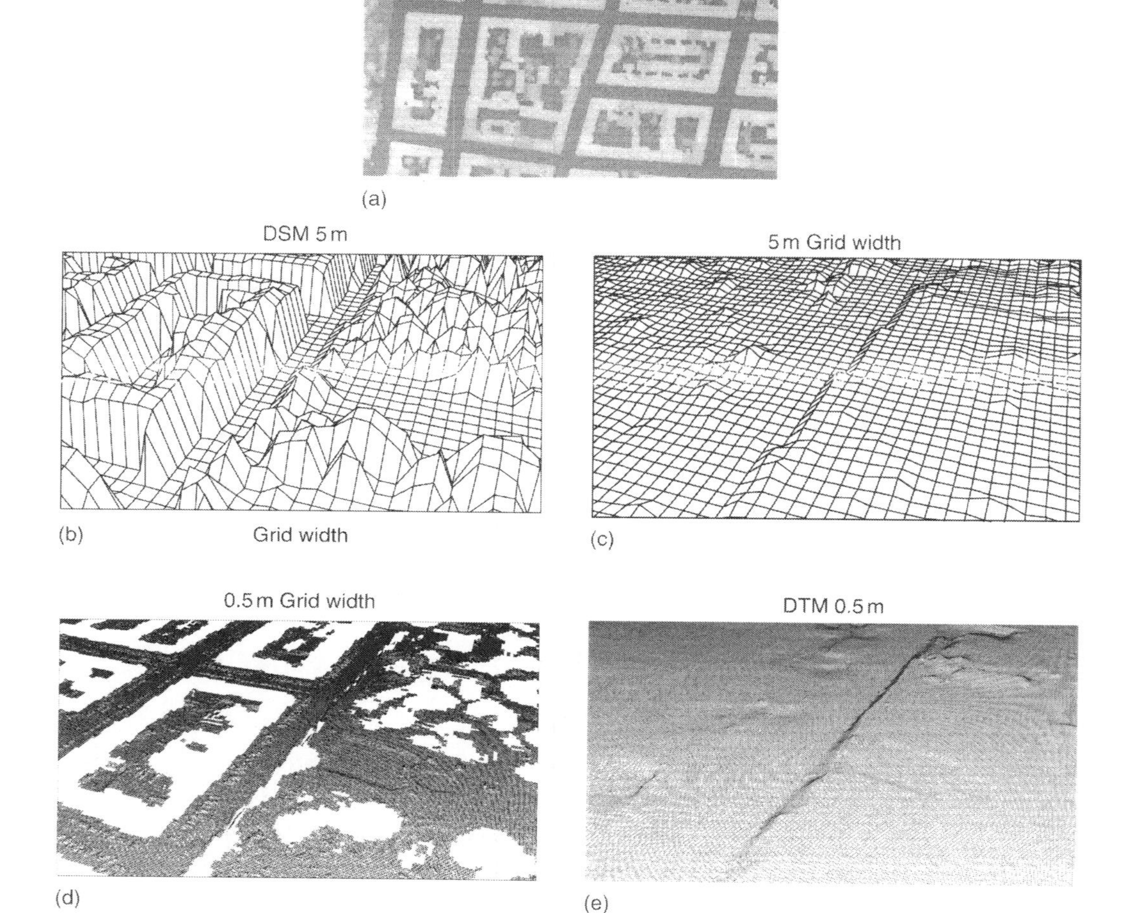

FIGURE 11.7
Counter clockwise: (a) Original DSM, (b) coarse DSM, filtered by robust interpolation to (c) coarse DTM, (d) DSM at finer level with points close to coarse level DTM, including cars, and (e) final DTM.

directions are used to form planimetrically extended segments. The relation to neighboring segments is used to classify the segments, where "ground" is one class.

Nardinocchi et al. (2003) apply a region-growing technique based on height differences to obtain segments. The geometric and topologic description of the regions can be expressed with graph theory, where the segments are nodes carrying information on segment size. The edges connect neighboring regions and carry information on height differences. Rules are applied to extract the segments representing the ground, vegetation, buildings, and other

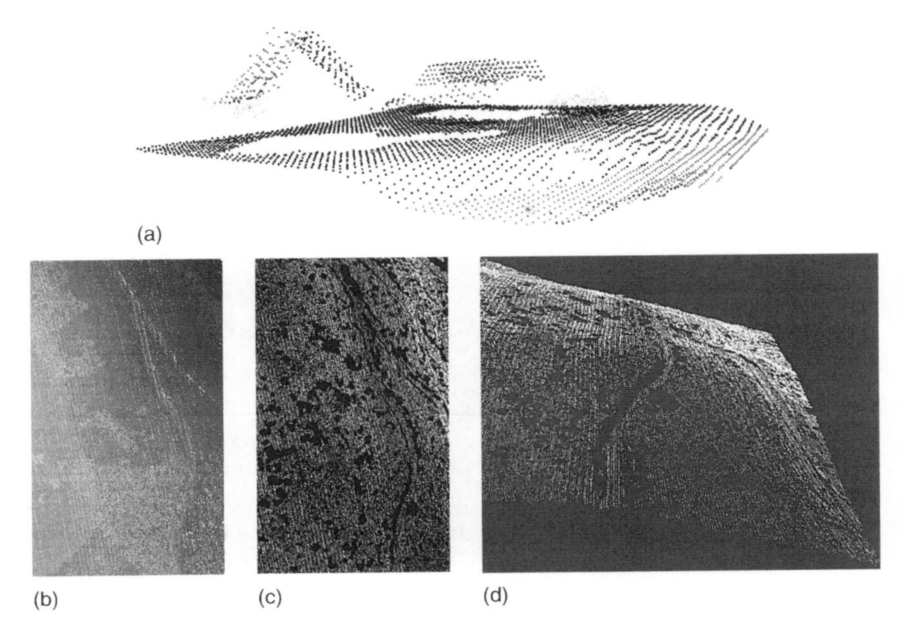

FIGURE 11.8 **(See color insert following page 334.)**
(a) Segmented point cloud of a small scene with two houses. Segments are distinguished by color. (b) Z-coded original point cloud of a forest and open area scene Figure 11.2b, white means higher). (c) Segmentation result, with points of the same segment are shown in the same color (only larger segments are displayed). (d) Perspective view of (c).

objects. Jacobsen and Lohmann (2003) proposed a comparable approach embedded in the eCognition software and, therefore, also on gridded data, and segments are obtained from region-growing. A number of criteria, e.g., compactness (not used by Nardinocci et al., 2003) and height differences to the neighboring segments, are used in order to detect different types of areas including terrain. Schiewe (2001) uses maximum and average gradients for classification of the data.

Tovari and Pfeifer (2005) use region-growing-based segmentation with the homogeneity formulated in terms of estimated normal vectors and spatial proximity. This segmentation was performed on the data of Figure 11.8b. Adjacent normal vectors were allowed to differ by 5°, and points had to have a maximum distance of 2 m to belong to the same segment. Segments obtained by this method with more than 10 points are shown in Figure 11.8c. As can be seen, segments below forest cover are less dense, but there are many holes due to small ground features and the surface next to the streets is also missing. In the approach of Tovari et al., the surface is interpolated, as in the robust interpolation, but a representative (e.g., mean) residual is computed for each segment, and depending on that residual the same weight is given to all points in that segment. Iteration continues and off-terrain segments are assigned lower and lower weights.

Many implementations require rasterized data, and also the segmentation methods are often borrowed from image processing. Generally, these filter methods work on larger entities, i.e., not on the single points or pixels, and are therefore less influenced by noise. This is demonstrated in Figure 11.8a, showing two large segments for the ground, one in the foreground and one in the background, a couple of segments for the roofs of the two connected houses, and many single or few point segments for the points on the vegetation.

11.2.10 Other Filter Approaches

There are a number of filters that cannot be assigned to one of the filter approaches above. The "Repetitive Interpolation" filter of Kobler et al. (2007) works on a prefiltered data set. It may contain off-terrain points but the points on the vegetation canopy shall be removed already. It proceeds by randomly picking points and computing a grid-based DTM by sampling the elevations from a triangulation. This process is repeated a number of times, each time choosing randomly different points. The final DTM is then computed from the individual DTMs in an averaging procedure.

11.2.11 Full-Waveform Exploitation for Ground Reconstruction

Full-waveform-recording laser scanners allow retrieving more information from an object surface than only its range to the sensor. The width of the echo holds information on the spatial (i.e., vertical) spread of the object surface. For small footprint scanners this is not influenced strongly by terrain slope because the height differences due to slope are typically much smaller than the length of the pulse. However, low vegetation and other non-ground objects feature a pronounced height variation within a footprint. Echo width can therefore be used as a preclassifier to exclude echoes that cannot originate from the ground (Doneus and Briese, 2006). Examples are shown in Figure 11.9. As full-waveform systems for airborne laser scanning became available only recently, research in the coming years will show the full potential for ground reconstruction.

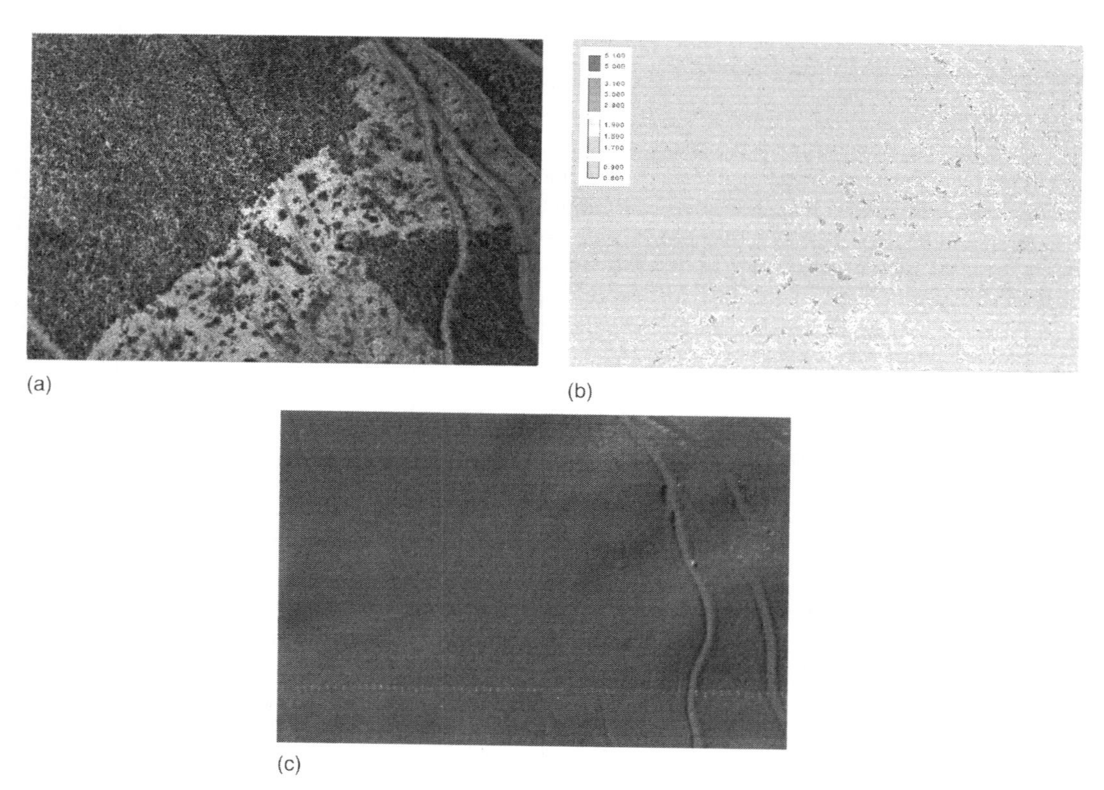

FIGURE 11.9
(a) Amplitude (the so-called intensity) of echoes extracted from the waveform, (b) **(See color insert following page 334.)** echo width, and (c) DTM obtained by robust interpolation excluding points with wide echoes (compare Figures 11.2 and 11.5b).

11.2.12 Comparison of Filter Algorithms

An experimental comparison of filter algorithms was published by Sithole and Vosselman (2004). The conclusions are that filters incorporating a concept of surface perform better, but it has to be mentioned that segmentation-based filters were not well developed at the time of this study. Problems were identified for all filters, typically along steep terrain edges und underpasses. We will return to the experimental filter comparison in Section 11.4 of this chapter, where the quality of DTMs from airborne laser scanning will be treated. The rest of this section is devoted to a comparison of different methodologies for solving the filtering problem.

Comparing the algorithms, it can be stated that the morphological filters mostly investigate height differences only, whereas surface-based filters consider the surface trend also. Therefore, they produce more reliable results. For the morphological filter, a problem faced is that a trade-off between erosion of ground on steep slopes on the one hand and the inclusion of off-terrain points in flat areas on the other hand has to be made.

The progressive densification and the surface-based filters use a surface in the generation of the classification. Progressive densification gradually adds points, building a DTM (or the classification) from a few points. The surface-based filters progress by removing points from the entire point cloud. A general advantage of these algorithms is that the geometrical terrain surface properties and the LiDAR data characteristics are, at least, to some extent separated. The structure element of the morphological filters on the other hand includes information on the terrain and on laser scanning characteristics as precision and vertical distribution of off-terrain objects. Each point is classified independently of its neighboring points. The progressive densification algorithms build a terrain model during execution, but the thresholds (angles or vertical distances) also encode terrain information and laser point vertical distribution aspects. In surface-based filters, especially in the robust interpolation, there is a separation between those two. The covariance function of kriging encodes the terrain characteristics, whereas the vertical distribution of ground and off-terrain points is mirrored in the weight function. Also, the iterative nature of the surface-based approaches, where a hard classification is only derived after some iteration steps, allows a better adaptation to the terrain features. As the surface-based approaches are built on a thorough terrain interpolation method, additional parameters, e.g., the range accuracy determined from full-waveform analysis, or knowledge about definite ground points can be considered.

The segmentation-based algorithms have advantages in areas strongly influenced by human building activities (houses, street dams, embankments, etc.). In these regions, distinct segments, often planes, are found. As these segments belong either entirely to the ground or to buildings, the classification is not affected by edge effects, which can be found in surface-based filters. In wooded areas, this does not hold and segment-based filters lose their advantage. A tight homogeneity criterion produces too many segments (approaching the case of point-based filtering) and a loose homogeneity criterion includes too many off-terrain features, which cannot be split in the segment classification step anymore (under-segmentation).

One general problem remaining for all filter algorithms is that the ground cannot be characterized by geometric properties only. Looking at a local neighborhood, similar characteristics may always be encountered when comparing small terrain features and small objects. To increase automation, information from other sensors or other information from the same sensor will have to be used. Spectral information may be one valuable source, especially if the data is acquired synchronously. This is a limiting factor as the combined use of laser scanning and imaging devices is not standard on one platform yet. Full-waveform systems offer an advantage here, but research in that direction is currently new.

11.3 DTM Derivation and Processing from Ground Points

11.3.1 DTM Interpolation

After the point cloud classification into terrain points and other points, the task of interpolation of the DTM remains. An overview of methods can be found in Maune (2001), El Sheimy et al. (2005), and Kraus (2000). As the number of points is typically very high and the density of points is also very high, it is not always necessary to apply the best predictors (e.g., kriging) but rather resort to methods exhibiting better computational performance. However, methods reducing the random measurement errors should be applied, making use of the redundancy in the measurements and thereby improving precision. The interpolated DTM can then be more accurate than a single point measurement.

11.3.2 Structure Line derivation

For the representation of models generated from ALS data, mostly rasters, grid models, or TINs are in use, which are determined from the irregularly distributed ALS point cloud. Due to the lack of structure information, it has to be considered that the description of break lines, which are crucial for a high quality delineation of a surface discontinuity, depend, next to the original sampling interval, also on the size of the stored raster, grid, or triangle surface cells, respectively. In contrast to these models without an explicit feature line description, it is essential for high quality terrain models to store these lines explicitly in the DTM (e.g., for hydraulic applications). For this aim, a 3D vector-based description of the lines is necessary. In addition, the line information can be helpful for the task of data reduction, because break lines allow describing surface discontinuities even in models with big raster or triangle cells (see next section).

The research in the area of break line extraction is focused on methods for the fully automated 2D detection in the so-called range images determined from the ALS point cloud, allowing the application of image processing techniques. Brügelmann (2000) presents one representative method, which introduces a full processing chain starting from an ALS elevation image leading to smooth vector break lines. The break lines are positioned at the maxima of the second derivation. The break line heights are retrieved by interpolation in the elevation image, which is a low pass filter. Borkowski (2004) presents two different break line modeling concepts. One makes use of snakes, whereas the second method utilizes a numeric solution of a differential equation describing the run of the break line. Brzank et al. (2005) present an approach for the extraction and modeling of pair-wise structure lines in coastal areas described by a hyperbolic tangent function.

In contrast to the approaches for break line modeling mentioned above, which rely on a previously filtered DTM, the subsequently presented approach uses the original irregularly distributed point cloud data as input (Briese, 2004). The break lines can then be used as additional information in the filtering of the ALS point cloud data. The basic concept of the approach is displayed in Figure 11.10a. Based on a local surface description on either side of the break line, the break line itself can be modeled by a sequence of locally intersecting surface patch pairs. For the estimation of the patches, the ALS point cloud in the vicinity of the line are used. In order to consider off-terrain points, the method of robust interpolation is integrated in the local patch determination in order to exclude the influence of off-terrain points on the run of the break line. One example of such a robustly estimated surface patch pair can be seen in Figure 11.10b. It can be seen clearly that after the robust estimation the off-terrain points do not affect the run of the right (green) patch and that the break line can be approximated by the intersection line between the two planar

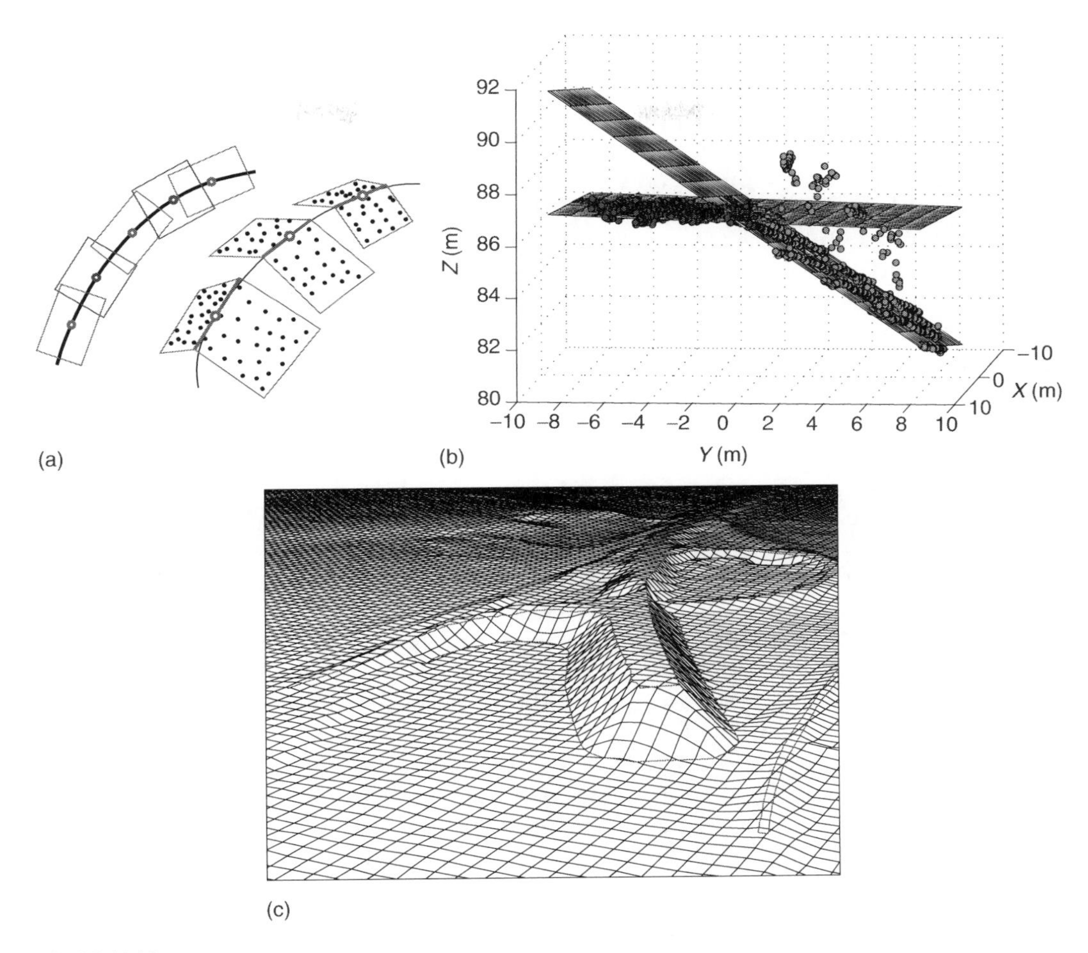

FIGURE 11.10
(a) Basic concept for the description of break lines by using overlapping, intersecting patch pairs. (From Kraus and Pfeifer, 2001. Advanced DTM generation from LiDAR data. IAPRS XXXIV 3/W4, Annapolis, MD.) (b) Practical example of a robustly estimated plane patch pair (10 by 10 m) based on unclassified ALS data (circles). The elimination of off-terrain points improved the sigma of the adjustment from 0.59 to 0.12 m. (From Briese, 2004. Three-dimensional modeling of breaklines from airborne laser scanner data. IAPRS XXXV B/3, Istanbul, Turkey.) (c) Extracted break lines meshed into the DTM.

patch pairs accurately. However, in order to estimate the surface in the vicinity of the break line, a 2D approximation of the break line is essential to delineate the patch pairs. To reduce the effort for a 2D approximation of the whole break line, line growing was introduced (Briese, 2004). Based on an initial 2D break line segment, the final break line is determined in a step-by-step expansion into forward and backward direction of previously determined break line segments. As long as the determination within the extrapolated patches is successful or a certain break-off point is reached, the growing procedure can continue. For the break-off point an evaluation criterion that describes the significance of the surface discontinuity (e.g., the intersection angle between the determined "left" and "right" surface) is necessary. Figure 11.10c shows the result of break line extraction and DTM reconstruction. At pronounced edges the break lines are meshed into the grid of the terrain model, whereas in flat and gently curved areas the terrain surface is represented by the grid only.

11.3.3 DTM Simplification (Thinning)

Mainly due to ALS the DTM point density has increased dramatically in the recent years. Modern ALS sensors allow mapping of topographic details at the price of a highly increased data volume. For that reason data reduction and surface simplification has become a major research topic not only in the field of geodesy but also in computer vision, computer graphics, and applied engineering technology. In any case the goal is to reduce the amount of data (vertices of a mesh, grid points, break lines, etc.) without losing relevant geometric details.

A survey of different simplification algorithms is published in (Heckbert and Garland, 1997). In this paper, curve as well as surface simplification are discussed. In the following, we concentrate on the simplification of bivariate functions as the DTM is commonly seen as 2.5D. In general, starting with a DTM description using m surface patches, the result of the thinning is a surface described by m_T patches, where $m_T \ll m$. The individual surface patches may either be linear, bilinear, or of higher degree. The surface approximation is achieved using a subset of the original points or any other points of the continuous surface. To describe the approximation error, the maximum error (L_∞) or the quadratic error (L_2) is in use.

Regular grid methods are the simplest reduction techniques. Hereby, a grid of height values equally spaced in x and y is derived from the original DTM either by interpolation or by resampling (low pass filter). While these methods work fast, they are not adaptive and therefore produce poor approximation results. Feature methods make a pass over the input points and rank them using a certain "importance" measure. Subsequently, the chosen features are triangulated. The quality of a feature-based surface approximation can be increased by adding break lines as constraints of the triangulation. Hierarchical subdivision methods build up a triangulation using a divide-and-conquer strategy to recursively refine the surface. They are adaptive and produce a tree structure, which enables the generation of different Levels of Detail (LoD). Better approximation quality can be achieved by applying refinement methods based on general triangulation algorithms like Delaunay triangulation. Refinement methods represent a coarse-to-fine approach starting with a minimal initial approximation. In each subsequent pass, one or more points are added as vertices to the triangulation until the desired approximation tolerance is achieved or the desired number of vertices is used.* In contrast, decimation methods work from fine-to-coarse. They start with the entire input model and iteratively simplify it, deleting vertices or faces in each pass. Due to the necessity of building a global triangulation, decimation methods are not suited for processing large DTMs, which arise when using high-density ALS data.

The performance of DTM data reduction is highly influenced by the existence of systematic and random measurement errors. For deriving thinned DTM models from original ALS point clouds, directly using refinement or decimation approaches as described above are not the first choice. Systematic errors have to be removed first by exact sensor calibration and fine adjustment of the ALS-strip data. The random measurement errors of the ALS points should rather be eliminated by applying a DTM interpolation strategy with measurement noise filtering capabilities. Good results can be achieved using linear prediction (Kraus, 2000) or Kriging (Journel u. Huijbregts, 1978), respectively. High reduction ratios can only be achieved for DEMs free of systematic and random errors.

* There is an interesting parallelity to progressive densification methods for ALS point cloud filtering for DTM extraction. One difference lies in the symmetry (thinning) versus asymmetry (DTM extraction) of the importance measure.

In contrast to the previously mentioned refinement approach, which rely on the original point cloud, the subsequently presented refinement framework uses the filtered hybrid DTM (regular grid and break lines, structure lines, and spot heights) as input (Mandlburger, 2006). The basic parameters for the data reduction are a maximum height tolerance Δz_{max} and a maximum planimetric point distance Δxy_{max}. The latter avoids triangles with too long edges and narrow angles. The algorithm starts with an initial approximation of the DEM comprising all structure lines and a coarse regular grid (cell size = $\Delta xy_{max} = \Delta_0$), which are triangulated using a Constrained Delaunay Triangulation. Each Δ_0-cell is subsequently refined by iteratively inserting additional points until the height tolerance Δz_{max} is achieved.

The additional vertices can either be inserted hierarchically or irregularly. In case of hierarchical breakdown, the grid cell is divided into four parts in each pass, if a single grid point within the regarded area exceeds the maximum tolerance. By contrast, only the grid point with the maximum deviation is inserted when using irregular division. Higher compression ratios (up to 99% in flat areas) can be achieved with irregular point insertion, whereas the hierarchical mode is characterized by a more homogeneous data distribution. The described framework is flexible concerning the reduction criterion. The decision to insert a point can be based on the analysis of the local surface slope and the curvature derived from the DTM or on the vertical distance between the DTM point and the approximated TIN. The results of a practical example are shown in Figure 11.11.

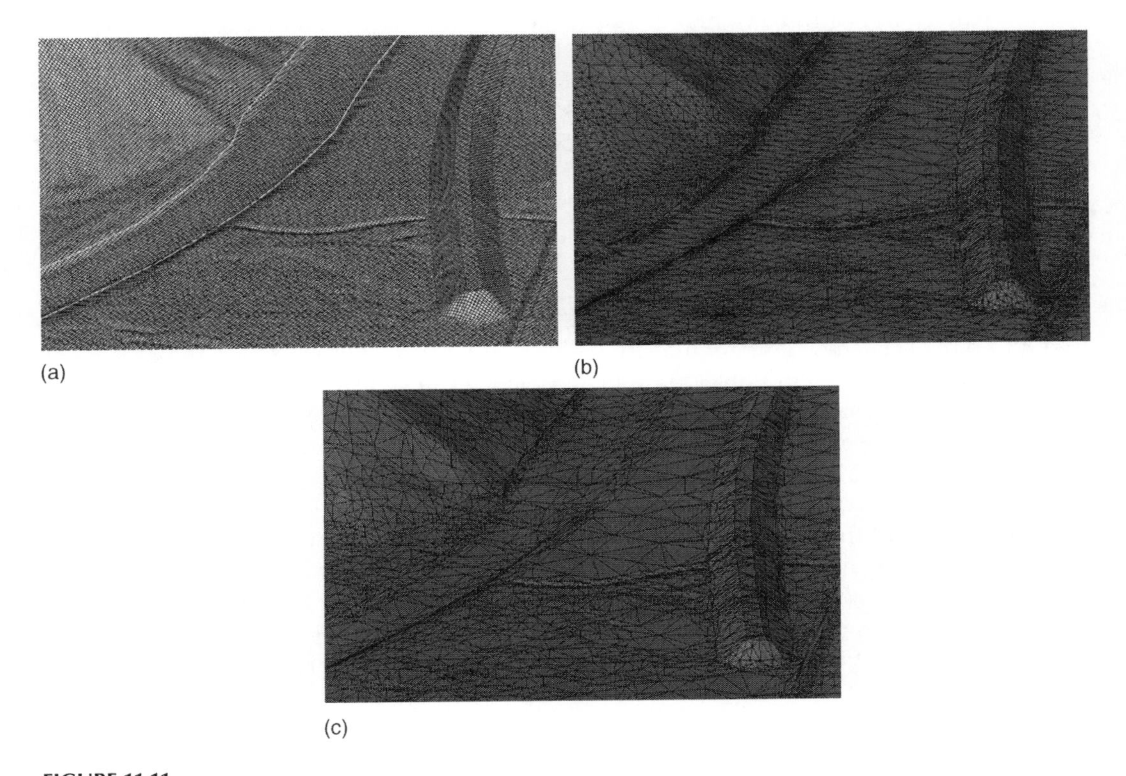

FIGURE 11.11
(a) Original hybrid DEM of Möllbrücke/Drau; base grid width: 2m; (b) Adaptive TIN using hierarchical division; (c) Adaptive TIN using irregular division. For (b) and (c) Δz_{max} = 0.25m and the L_∞-norm was applied as error measure in thinning.

11.4 Quality

The quality of a DTM derived from laser scanning has three major impact factors.

1. Quality of the original laser scanning point cloud
2. Quality of the filtering
3. Quality of the DTM interpolation, if applicable, including thinning and break line modeling

The first item is addressed in other chapters on data acquisition and calibration. The starting point is the quality of the trajectory, which is about 5–10 cm in planimetry and 7–15 cm in height, and 0.005° for roll and pitch and 0.008° in yaw (Skaloud, 2007). Rigorous models of strip adjustment with computation of the sensor system parameters as Inertial Measurement Unit (IMU) misalignment and Global Positioning System (GPS) antenna offset, together with the use of control patches, can lead to point clouds with a vertical accuracy of ±10 to 15 cm in height if acquisition is performed from platforms flying below 1000 m a.g.l. and over smooth targets. The accuracy of digital terrain models (item 3) may, however, be better, as the random measurement error component can be minimized within the interpolation. Furthermore, redundant point information (e.g., from overlapping laser scanner strips) improves the relative accuracy, i.e., the precision, of the DTM. Generally speaking, evaluating the accuracy of DTM generation from ALS data is difficult, because obtaining area-wise ground truth that has to be of higher precision than the laser scanning data itself is very expensive. Therefore, the discretization of the terrain surface by check points is much poorer, typically on small scale (only a few number of check areas over the total region) as well as on large scale (density of ground truth points in a check area) and stationarity of the errors has to be assumed. An approach for deriving local DTM accuracy measures in a general framework of error propagation is presented by Kraus et al. (2006). The accuracy of every DTM grid point is estimated from the data itself, by considering the original point density and distribution, terrain curvature, and the height accuracy of the original points.

11.4.1 Filtering Quality

In built-up regions, the quality of filtering (item 2) can be very high, because the distinction between ground points and object points is typically quite clear. At jump edges as found between street level and building roofs there is a clear divide. It is rather the question of the type of surface that the digital terrain model shall represent (e.g., at under passes), which accounts for filtering problems.

In areas covered by herbaceous vegetation, the quality of filtering cannot be separated from the quality of the original measurements. Laser ranging typically produces only one echo over low vegetation, i.e., for heights up to 50 cm, even if the canopy and the ground are separated surfaces. For grass and reed type of vegetation, the scattering volume may even be extended over larger vertical ranges, and one echo is recorded that represents ground and vegetation. In these cases, filtering cannot separate ground and off-terrain points, as the bulk of measurement results show no distinct ground-only echoes. Without additional assumptions, e.g., on vegetation height, the quality of digital terrain models depends on vegetation height and vegetation density within the footprint (Pfeifer et al., 2004; Hopkinson et al., 2004). For areas under forest cover, filtering quality has not been investigated extensively. Exceptions are the works of Hyyppä et al. (2001), treated later in the text, and the work of Kobler et al. (2007). One obvious problem is obtaining reference values and the general lack of alternative methods for obtaining area-wise ground height under forest cover. Again, the definition of the terrain surfaces becomes problematic, as tree roots, small gullies, etc. may be considered as terrain-forming or not.

Sithole and Vosselman (2004) have addressed the question for filtering quality by manually filtering a point cloud using the context knowledge and aerial imagery. The data was made available by EuroSDR (www.eurosdr.net). Eight test sites were chosen in order to represent urban and rural scenes, "steep slopes, dense vegetation, densely packed buildings with vegetation in between, large buildings (a railway station), multilevel buildings with courtyards, ramps, underpasses, tunnel entrances, a quarry (with breaklines), and data gaps" (Sithole and Vosselman, 2004). Researchers were asked to participate in the test, and eight groups responded. The participants were asked to filter the data sets with their algorithms and the classified ground points were compared to the manual reference by Sithole and Vosselman. Assuming that the manual ground point identification process is error-free, this measures exactly the quality of filtering, without influences of georeferencing or DTM interpolation. However, as some filters produce a DTM (i.e., a surface and not a classification of points), they worked with a 20 cm tolerance band and compared the DTMs to the manually filtered point cloud. First a qualitative assessment with respect to low and high outliers, very large and small objects (e.g., industrial complexes and cars), buildings on slopes, ramps, bridges, behavior at jump edges found in urban scenes, break lines, and high vegetation was performed. Most algorithms produce poor results at ramps, jump edges, and break lines. Then, the classification result was evaluated quantitatively. Considering all different scenes and all filter results, type I errors ranged from 0% to 60% (false rejection of ground points) and type II errors ranged from 0% to 20% (wrong acceptance of object points as ground). The conclusion of the study is that all filters work well in gently sloped terrain with small buildings and sparse vegetation. The most difficult scenes were complex cityscapes with multitier buildings, plazas, etc. and discontinuities in the bare-earth (Sithole and Vosselman, 2004). They further concluded that surface-based filters tend to provide better results.

11.4.2 DTM Quality

In the following, studies representing the final quality of the digital terrain model will be presented. The measure of accuracy is therefore including all aspects discussed so far.

A TopoSys Falcon I laser scanner was used to acquire data over an area in the Vienna city center covering 2.5 km^2 from an average flying height of 500 m a.g.l. using 19 strips, resulting in 10 points per m^2, although linearly distributed within each square meter. The DTM was determined by robust interpolation and checked manually for obvious filter errors. With over 800 terrestrially measured points, providing subcentimeter accuracy, the precision of data acquisition, filtering, and DTM generation was checked (Pfeifer et al., 2001). On average, this resulted in a r.m.s.e. of ±11 cm, but showed a strong dependency on surface type (see Figure 11.12). Densely forested park areas (area 5 in Figure 11.12) showed the highest r.m.s.e. with ±15 cm, for sparse trees it was reduced to ±11 cm (1), and reached ±9 cm in open park areas (2 and 3) as well as below parked cars (4). For open asphalt areas (6), a value of ±3 cm was obtained, which is far more precise than the navigation solution of the trajectory and can be explained by georeferencing control in the vicinity of these points. However, it shows the potential of laser scanning after elimination of all systematic errors. One should also keep in mind that random measurement errors in the range measurement are reduced because of qualified surface interpolation.

Ahokas et al. (2003) used TopoSys and TopEye laser scanning data from flying heights between 100 and 800 m a.g.l. with a point density varying from 1 to 8 points per m^2.

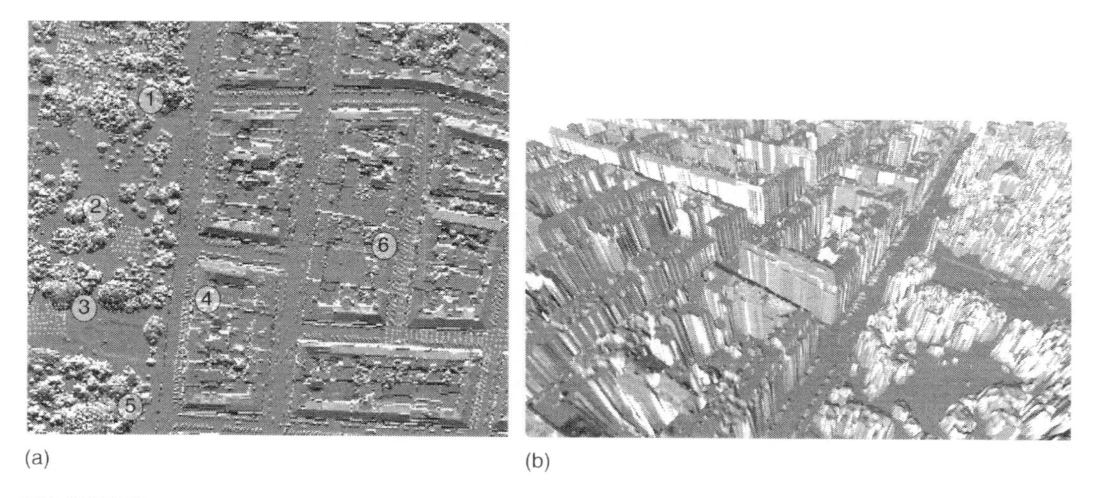

FIGURE 11.12
In (a), check points in a city area at different land cover classes are shown. In (b), a perspective view of the scene is shown. In Figure 11.7, the same area is shown.

Filtering was performed with a progressive densification method and the DTM, evaluated only at the check points, was calculated from the classified points by local cubic interpolation. For this comparison about 3500 check points were measured. Below forest cover, the standard deviation was between ±12 and ±18 cm (therefore excluding a systematic vertical bias). For grass the standard deviation was between ±6 and ±12 cm, and for asphalt between ±4 and ±10 cm. The systematic shifts encountered in the test areas (which are included in the r.m.s.e. reported for the previous study) reached up to 30 cm of absolute value.

References to other comparative studies are found in the cited literature and Pfeifer et al. (2004). Specifically for forest conditions, the outcome of different filter algorithms was compared to ground truth for the DTM within the HIGH-SCAN project (Hyyppä et al., 2001). They reported biases up to 40 cm and standard deviations of the heights below ±40 cm.

In brief, it can be said that under ideal circumstances with (1) proper georeferencing using additional ground control, over (2) hard flat targets, with (3) no need to remove vegetation points and (4) qualified noise reducing DTM interpolation techniques, an accuracy below ±1 dm can be achieved. For areas with low vegetation, a systematic upwards shift of the final DTM and an increase in the random error component occurs. While strongly depending on the vegetation type these values easily reach +1 and ±1 dm, respectively. Under forest cover and because of terrain surface definition insecurity, the DTM precision in forests is in the best case ±2 dm. Georeferencing shortcomings typically lead to an overall accuracy of ±3 dm. If planimetric errors in footprint geolocation are also considerd as adding to height errors, an additional loss of $\pm\sigma_P \tan(\gamma)$ is added. The term γ is the terrain inclination and σ_P is the planimetric accuracy, typically in the order of flying height/2000.

11.5 Conclusions

Extracting the ground from airborne laser scanning data is operational. However, the problems addressed by Sithol and Vosselman (2004) are not entirely solved, and new algorithms are being presented in order to overcome those problems. The usage of additional information from full-waveform echo attributes (Briese and Doneus, 2006) has shown a way to improve the filtering by including additional information and incorporating it into known filter approaches rather than inventing new filter algorithms. Terrain information is currently encoded quantitatively (stiffness or covariance function in the surface-based approaches), but explicit representation of knowledge on the terrain has not been investigated yet.

In commercial use, a high filtering quality is only achieved by human inspection, typically using shaded relief views of the original and the filtered data, and contour line plots. Stereoscopic viewing of the point cloud or the computed DTM can be an alternative giving more insight at the price of a longer procedure.

For terrain capturing, especially over forested areas and open areas, airborne laser scanning is currently the method of choice. The active, polar measurement method, together with the high precision, strongly supports its posterity. To widen the application field, calibration and strip adjustment methods will have to become standard, providing high-detail high-accuracy DTMs of unprecedented quality. Under optimal conditions and for well-defined surfaces, a precision of ±5 cm seems achievable in the near future for larger areas also.

In built-up areas featuring many edges and distinct texture, airborne laser scanning is "challenged" by digital aerial imagery. However, algorithms for data exploitation are lagging behind the technological development, and the rich information in laser scanning point clouds, augmented by sensor developments described in earlier chapters, is not fully exploited yet.

References

Ackermann, F., and Kraus, K. 2004. Reader commentary: Grid based digital terrain models. *Geoinformatics* 7(6), 28–31.

Ahokas, K. and Hyyppä, J. 2003. A quality assessment of airborne laser scanner data. IAPRS XXXIV 3/W13, Dresden, Germany.

Artuso, R., Bovet, S., and Streilein, A. 2003. Practical methods for the verification of countrywide produced terrain and surface models. IAPRS, XXXIV, 3/W13, Dresden, Germany.

Axelsson, P. 1999. Processing of laser scanner data—algorithms and applications. *ISPRS Journal of Photogrammetry and Remote Sensing* 54(2–3), 138–147.

Axelsson, P. 2000. DEM generation from laser scanner data using adaptive TIN models. IAPRS, XXXIII B4/1, Istanbul, Turkey.

Borokowski, A. 2004. Modellierung von Oberflächen mit Diskontinuitäten, Habilitation, Veröffentlichung der Deutschen Geodätischen Kommission (DGK), B and C 575.

Briese, C. 2004. Three-dimensional modeling of breaklines from airborne laser scanner data. IAPRS XXXV B/3, Istanbul, Turkey.

Brovelli, M., Cannata, M., and Longoni, U. 2003. Lidar data filtering and DTM interpolation within GRASS. *Transactions in GIS* 8(2), 155–174.

Brügelmann, R. 2000. Automatic break line detection from airborne laser scanner data. IAPRS XXXIII B3, Amsterdam.

Brügelmann, R., and Bollweg, A. 2004. Laser altimetry for river management. IAPRS XXXV/B2, Istanbul, Turkey.

Brzank, A., Lohmann, P., and Heipke, C. 2004. Automated extraction of pair wise structure lines using airborne laserscanner data in coastal areas. IAPRS XXXVI, 3/W19, Enschede, The Netherlands.

Doneus, M., and Briese, C. 2006. Full-waveform airborne laer scanning as a tool for archeological reconnaissance. International Conference on Remote Sensing in Archeology, Rome.

Elmqvist, M. 2001. Ground estimation of lasar radar data using active shape models. OEEPE workshop on airborne laser scanning and interferometric SAR for detailed digital elevation models, Stockholm.

El-Sheimy, N., Valeo, C., and Habib, A. 2005. *Digital Terrain Modeling*. Artech House, Boston, Mass.

Filin, S. 2005. Neighborhood systems for airborne laser scanner data. *Photogrammetric Engineering and Remote Sensing* 59(3), 743–755.

Filin, S., and Pfeifer, N. 2006. Segmentation of airborne laser scanning data using a slope adaptive neighborhood. *ISPRS Journal of Photogrammetry and Remote Sensing* 60, 71–80.

Grün, A., and Baltsavias, E. 1988. Geometrically constrained multiphoto matching. *Photogrammetric Engineering and Remote Sensing* 54(5), 633–641.

Haralick, R., and Shapiro, L. 1992. *Computer and Robot Vision*. Addison Wesley, Reading, Mass.

Heckbert, P., and Garland, M. 1997. Survey of polygonal surface simplification algorithms. Research report, School of computer science, Carnigie Mellon University, Pittsburgh, PA.

Hopkinson, C., Lim, K., Chasmer, L., Treitz, P., Creed, I., and Gynan, C. 2004. Wetland grass to plantation forest—Estimating vegetation height from the standard deviation of lidar frequency distributions. IAPRS XXXVI 8/W2, Freiburg, Germany.

Hyyppä, J. et al., 2001. HIGH-SCAN: The first European-wide attempt to derive single-tree information from laserscanner data. *Photogrammetric Journal of Finland* 17, 58–68.

Isenburg, M., Liu, Y., Shewchuk, J., and Snoeyink, J. 2006. Streaming computation of Delaunay Triangulations. *ACM Transactions on Graphics* 25(3), 1049–1056.

Jacobsen, K., and Lohmann, P. 2003. Segmented filtering of laser scanner DSMs. IAPRS, XXXIV 3/W13, Dresden, Germany.

Journel, A., and Huijbregts, C. 1978. *Mining Geostatistics*. Academic Press, London.

Kass, M., Witkin, A., and Terzopoulus, D. 1988. Snakes: Active contour models. *International Journal of Computer Vision* 1, 321–331.

Kilian, J., Haala, N., and Englich, M. 1996. Capture and evaluation of airborne laser scanning data. IAPRS XXXI 3, Vienna.

Kobler, A., Pfeifer, N., Ogrinc, P., Todorovski, L., Ostir, K., and Dzeroski, S. 2007. Repetitive interpolation: A robust algorithm for DTM generation from Aerial Laser Scanner Data in forested terrain. *Remote Sensing of Environment* 108, 9–23.

Koch, K. 1999. *Parameter Estimation and Hypothesis Testing in Linear Models*. 2nd ed., Springer, Berlin.

Köstli, A., and Sigle, M. 1986. The random access data structure of the DTM program SCOP. In: *International Archives of Photogrammetry and Remote Sensing*, Vol. XXVI, Comm. IV, Edinburgh, Scotland, pp. 42–45.

Kraus, K. 2000. Photogrammetrie, Band 3, Topographische Informationssysteme. Dümmler, Köln.

Kraus, K. 2007. *Photogrammetry*, 2nd ed. Walter de Gruyter, Berlin.

Kraus, K., and Mikhail, E. 1972. Linear least squares interpolation. *Photogrammetric Engineering* 38, 1016–1029.

Kraus, K., and Pfeifer, N. 1997. A new method for surface reconstruction from laser scanner data. IAPRS XXXII, 3/2W3, Haifa, Israel.

Kraus, K., and Pfeifer, N. 1998. Derivation of digital terrain models in wooded areas. *ISPRS Journal of Photogrammetry and Remote Sensing* 53(4), 193–203.

Kraus, K., and Pfeifer, N. 2001. Advanced DTM generation from lidar data. IAPRS XXXIV 3/W4, Annapolis, MD.

Kraus, K., Karel, W., Briese, C. and Mandlburger, G. 2006. Local accuracy measures for digital terrain models. *The Photogrammetric Record (Invited)* 21(116), 342–354.

Lohmann, P., Koch, A., and Schaeffer, M. 2000. Approaches to the filtering of laser scanner data. *International Archives of Photogrammetry and Remote Sensing* 33(B3/1), 534–541.

Mandlburger, G. 2006. Topographische Modelle für Anwendungen in Hydraulik und Hydrologie. PhD thesis, Vienna University of Technology, http://www.ipf.teewien.ac.at/phdtheses/diss_gm_06.pdf.

Maune, D. 2001. *Digital Elevation Model Technologies and Applications: The DEM Users Manual*. American Society of Photogrammetry and Remote Sensing, Bethesda, ML.

McGlone, J., Mikhail, E., and Bethel, J. 2004. *Manual of Photogrammetry*, 5th ed. American Society of Photogrammetry and Remote Sensing, Bethesda, MD.

Mikhail, E. 1976. Observations and Least Squares. IEP–A Dun–Donnelley Publisher, New York.

Nardinocci, C., Forlani, G., and Zingaretti, P. 2003. Classification and filtering of laser data. IAPRS XXXIV, 3/W13, Dresden, Germany.

Pfeifer, N. 2005. A subdivision algorithm for smooth 3D terrain models. *ISPRS Journal of Photogrammetry and Remote Sensing* 59(3), 115–127.

Pfeifer, N., Gorte, G., and Oude Elberink, S. 2004. Influences of vegetation on laser altimetry—analysis and correction approaches. IAPRS XXXVI 8/W2, Freiburg, Germany.

Pfeifer, N., Stadler, P., and Briese, C. 2001. Derivation of digital terrain models in the SCOP++ environment. OEEPE Workshop on Airborne Laserscanning and Interferometric SAR for Detailed Digital Elevation Models, Stockholm.

Schenk, T., and Cshato, B. 2007. Fusing imagery and 3D point clouds for reconstructing visible surfaces of urban scenes. *Proceedings of Urban Remote Sensing*, 2007, Paris.

Schiewe, J. 2001. Ein regionen-basiertes Verfahren zur Extraktion der Geländeoberfläche aus Digitalen Oberflächen-Modellen. *Photogrammetrie, Fernerkundung, Geoinformation* 2, 81–90.

Sithole, G. 2001. Filtering of laser altimetry data using a slope adaptive filter. IAPRS, XXXIV, 3/W4, Annapolis, MD.

Sithole, G. 2005. Segmentation and classification of airborne laser scanner data, Dissertation, TU Delft, ISBN 90 6132 292 8, Publications on Geodesy of the Netherlands Commission of Geodesy, Vol. 59.

Sithole, G., and Vosselman, G. 2004. Experimental comparison of filter algorithms for bare—Earth extraction from airborne laser scanning point clouds. *ISPRS Journal of Photogrammetry and Remote Sensing* 59, 85–101.

Skaloud, J. 2007. Reliability of Direct Georeferencing—Beyond the Achilles' Heel of Modern Airborne Mapping. Photogrammetric Week'07. Wichmann, Heidelberg.

Sohn, G., and Dowman, I. 2002. Terrain surface reconstruction by the use of tetrahedron model with the MDL criterion. IAPRS XXXIV 3A, 336–344, Graz, Austria.

Tovari, D., and Pfeifer, N. 2005. Segmentation based robust interpolation—a new approach to laser data filtering. IAPRS XXXVI, 3/W19, Enschede, The Netherlands.

von Hansen, N., and Vögtle, T. 1999. Extraktion der Geländeoberfläche aus flugzeuggetragenen Laserscanner-Aufnahmen. *Photogrammetrie, Fernerkundung, Geoinformation* 4, 229–236.

Vosselman, G. 2000. Slope based filtering of laser altimetry data. IAPRS XXXIII, B3/2, Amsterdam.

Zhang, K., Chen, S., Whitman, D., Shyu, M., Yan, J., and Zhang, C. 2003. A progressive morphological filter for removing nonground measurements from airborne LIDAR data. *IEEE Transactions on Geoscience and Remote Sensing* 41(4), 872–882.

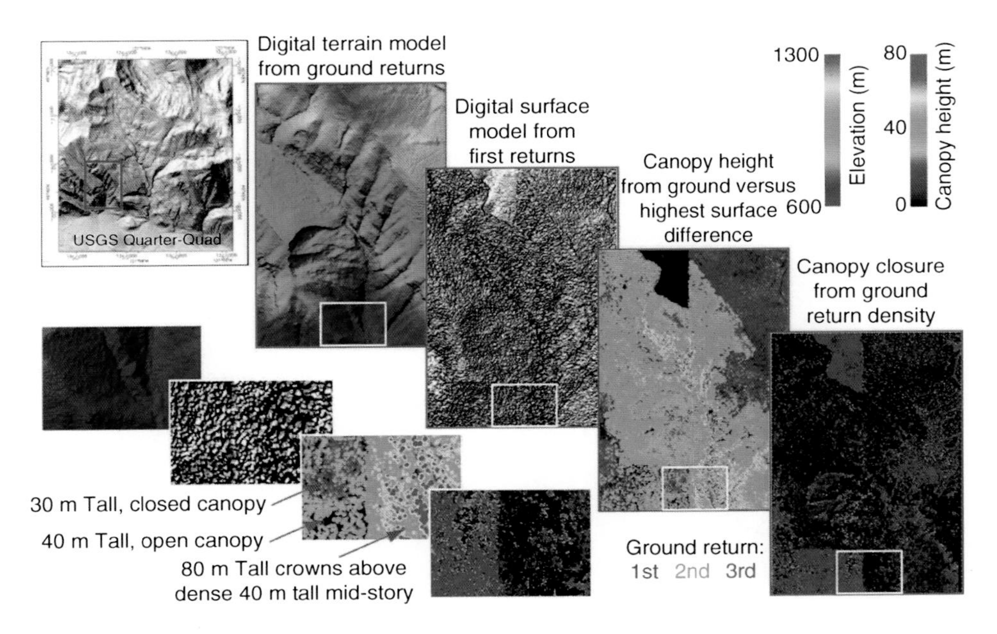

FIGURE 5.12
Discrete return point clouds and the DEMs generated from them provide highly detailed images of ground topography and forest canopy structure. Here commercial ALSM data acquired for the Puget Sound Lidar Consortium (From Haugerud, R., Harding, D.J., Johnson, S.Y., Harless, J.L., Weaver, C.S., and Sherrod, B.L., *GSA Today*, 13, 4–10, 2003) and gridded at six foot postings is used to depict several representations of topography and vegetation cover for the area indicated by the red rectangle in the upper left DTM shaded relief image corresponding to a USGS 7.5 in. quarter quadrangle. The area indicated by the white rectangles is enlarged in the insets to show image details.

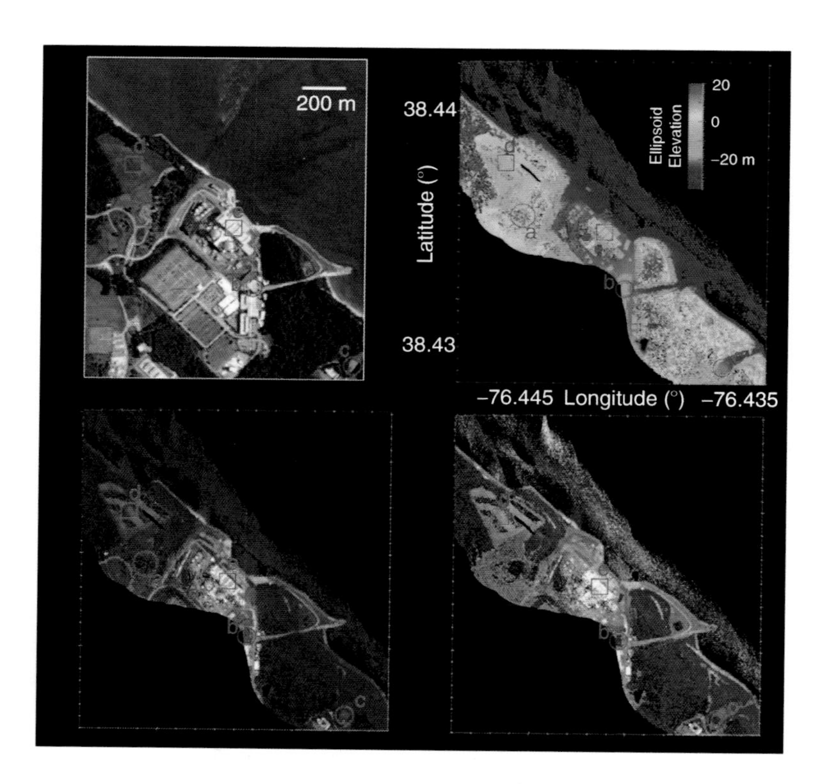

FIGURE 5.14
ATM mapping of the Chesapeake Bay western shoreline at Calvert Cliffs Nuclear Power Plant, Maryland acquired on September 17, 1997. Individual laser footprints from multiple flight passes are plotted showing elevation (upper right) and uncalibrated received 532nm laser energy (lower right). Also shown is the ATM measurement of uncalibrated reflected 532nm solar energy acquired between laser pulses (lower left) and a natural-color aerial photograph from Google Earth (upper left). Examples of intensity differences include areas where laser backscatter is lower, with respect to the surroundings, as compared to the reflected sunlight (circles: (a) parking lot, (b) building roof, (c) clay tennis court) and where the laser backscatter is higher (squares: (d) grass, (e) cooling tower roof). Also, intense specular reflections from offshore waves are observed in the laser backscatter image but not in the reflected sunlight image. (ATM images provided by Bill Krabill and Serdar Manizade, NASA Goddard Space Flight Center.)

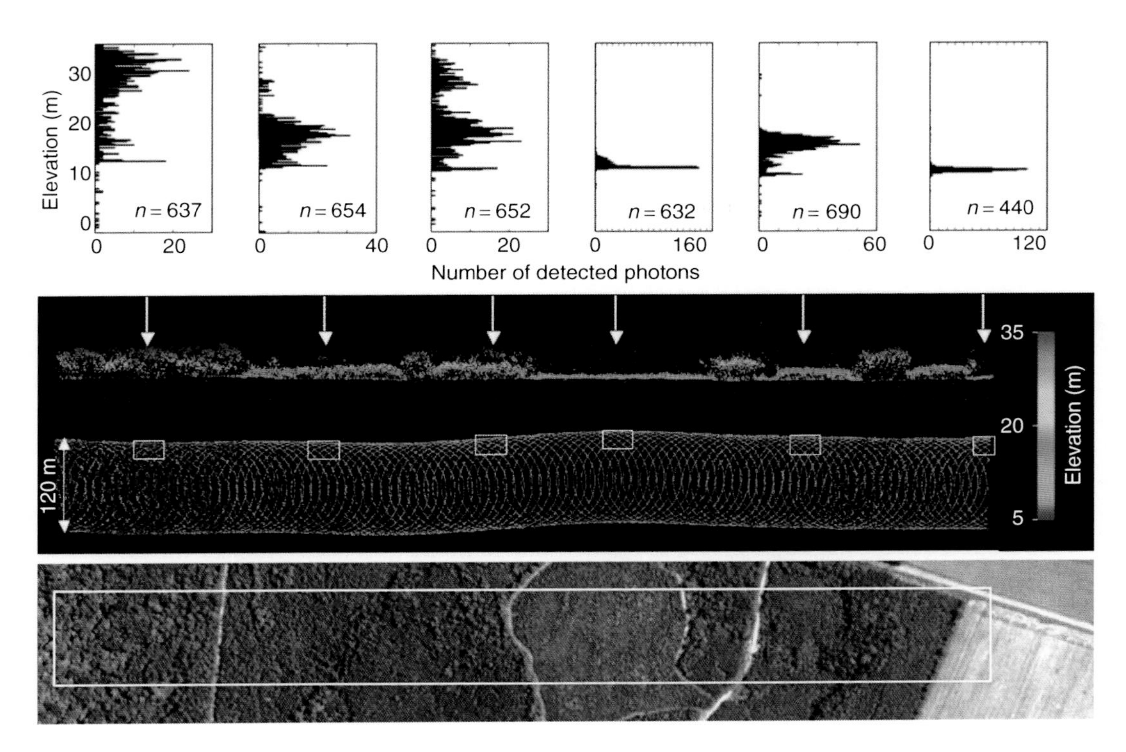

FIGURE 5.15
MMLA photon-counting, helical-scanning airborne data acquired in a 120 m wide swath across the Wicomico Demonstration Forest, MD Eastern Shore on September 12, 2002. Middle: single photon returns color-coded by elevation in map and cross-section views (ground = cyan, canopy = green through red); the cross-section portrays laser pulse returns and solar background noise (no noise filtering has been applied) from a ~20 m wide corridor along the top edge of the swath where returns are densest (~1 return per square meter). Top: height distributions (0.25 m bins) of single photon returns accumulated over ~20 × 30 m areas (white boxes) showing canopy vertical structure, distinct ground peaks, and low levels of solar background noise. Bottom: natural color aerial photograph showing approximate location of the MMLA swath (white rectangle). (Geolocated MMLA data was provided by Jan McGarry, NASA Goddard Space Flight Center.)

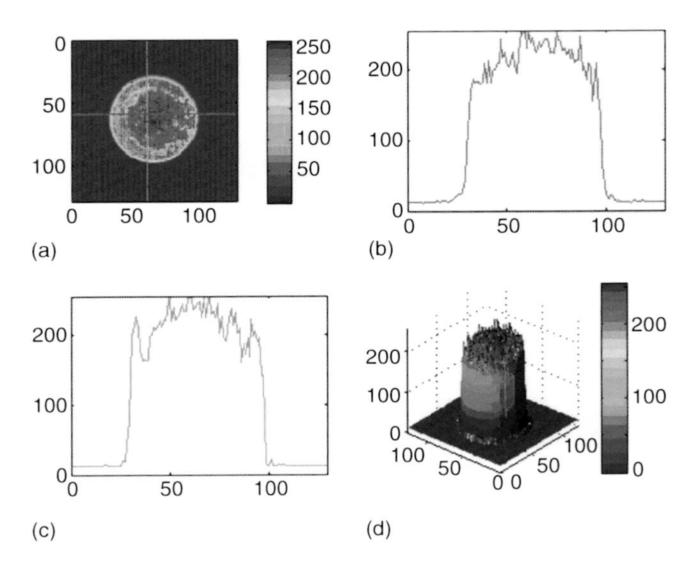

FIGURE 7.5
Measured cylindrical beam distribution (top-hat form). (a) 2D visualization, with the row (red) and column (green) of the maximum intensity indicated, (b) profile of the indicated row, (c) profile of the indicated column, and (d) 3D visualization.

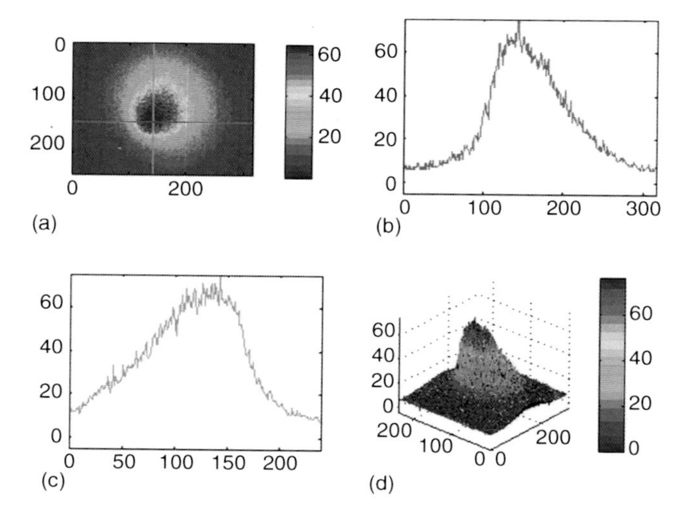

FIGURE 7.6
Measured Gaussian beam distribution. (a) 2D visualization, with the row (red) and column (green) of the maximum intensity indicated, (b) profile of the indicated row, (c) profile of the indicated column, and (d) 3D visualization.

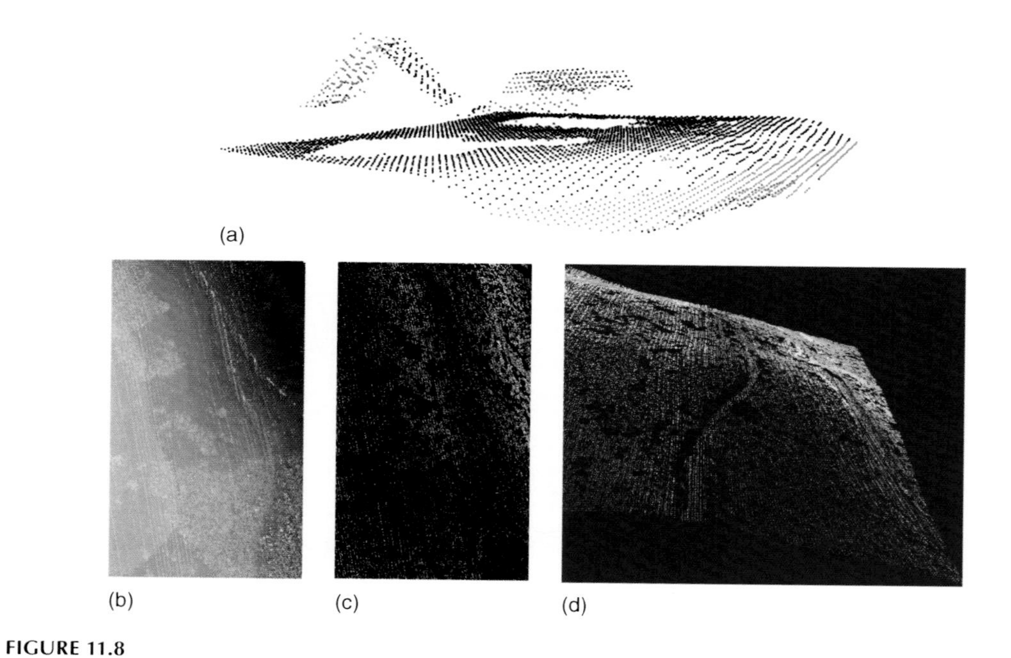

FIGURE 11.8
(a) Segmented point cloud of a small scene with two houses. Segments are distinguished by color. (b) Z-coded original point cloud of a forest and open area scene Figure 11.2b, white means higher). (c) Segmentation result, with points of the same segment are shown in the same color (only larger segments are displayed). (d) Perspective view of (c).

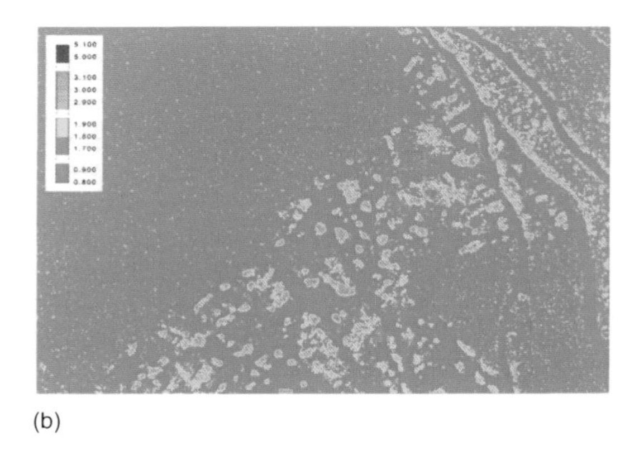

(b)

FIGURE 11.9
(b) Echo width (compare Figures 11.2 and 11.5b).

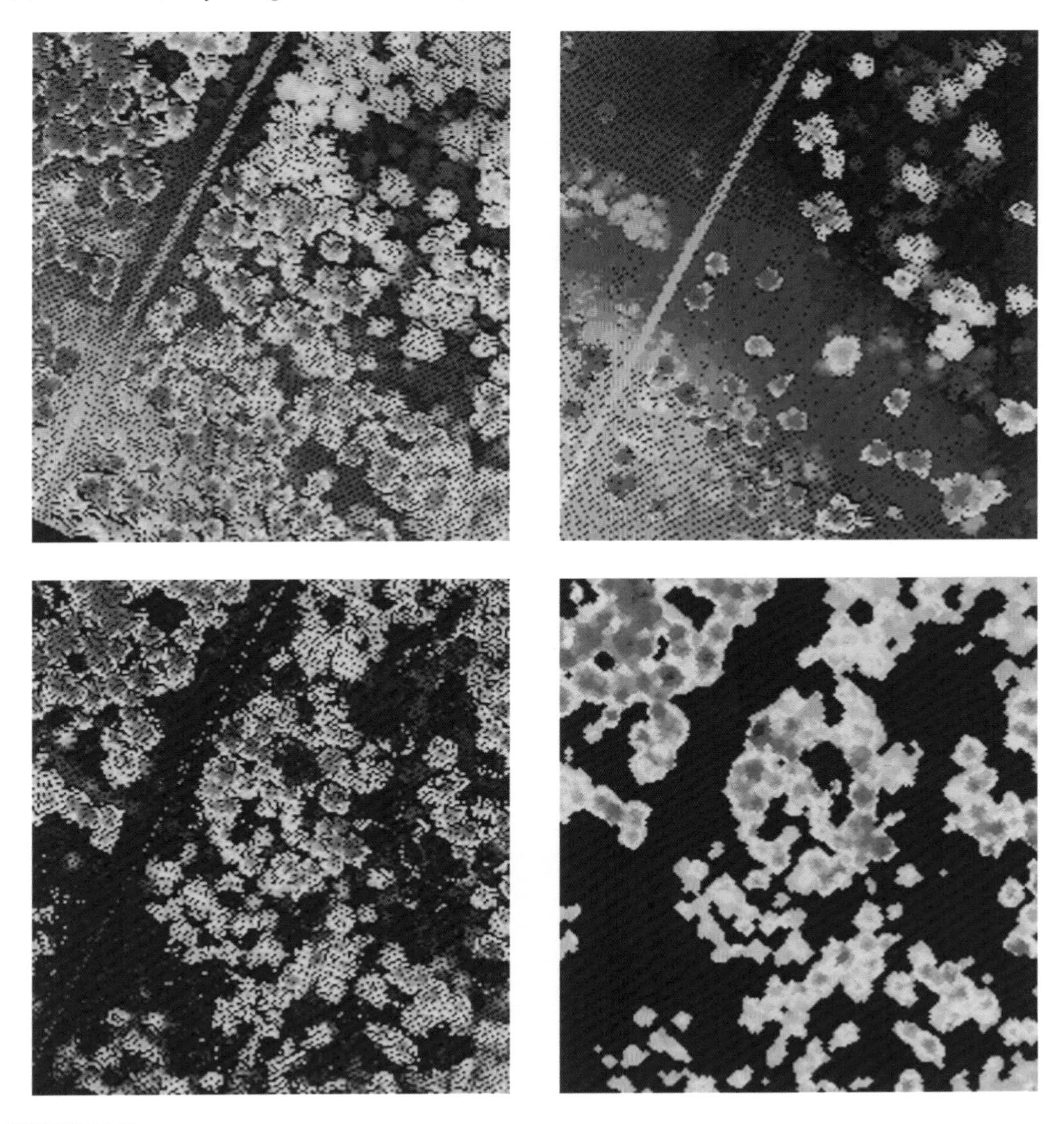

FIGURE 12.13
Change detection of harvested trees from two laser surveys, upleft: DSM from earlier year, upright: DSM from later year, lowleft: difference of DSMs, lowright: filtered difference image showing the harvested trees. Height was coded by colors.

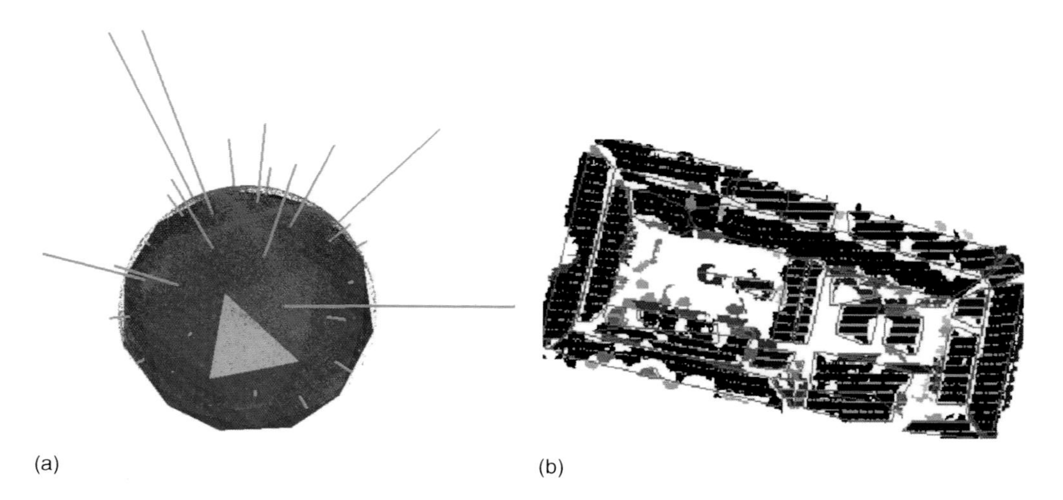

FIGURE 14.4
Aspects of the normal driven RANSAC algorithm. (a) Example of an EGI: Orientation histogram collected on a geodesic dome derived from the icosahedron. This is a discrete approximation of the EGI. The length of the green vectors attached to the center of a cell is proportional to the number of surface normals that fall within the range of directions spanned by that cell. Blue points are the projection of normals onto the Gaussian sphere. (b) Projection of lidar points (green points) and extracted facets (red polygons) onto focusing masks (coded in grey level depending on their surface).

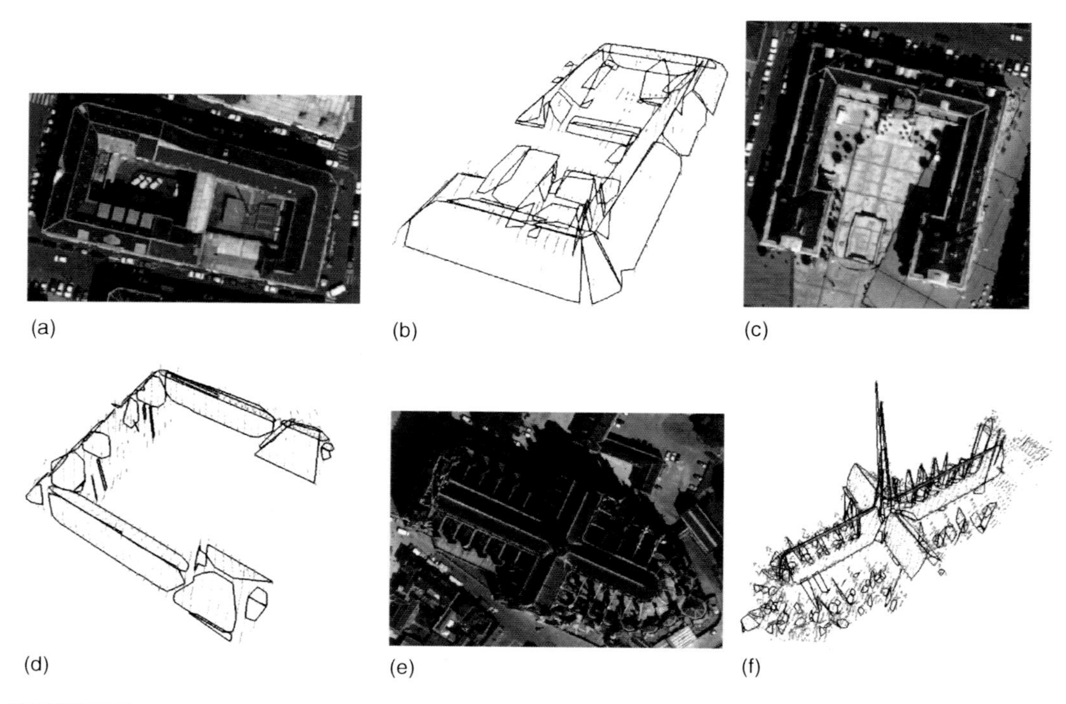

FIGURE 14.5
Right column (a, c, and e) is the result of 3D roof facet detection with normal-driven RANSAC. Left column (b, d, and f) is the corresponding aerial image.

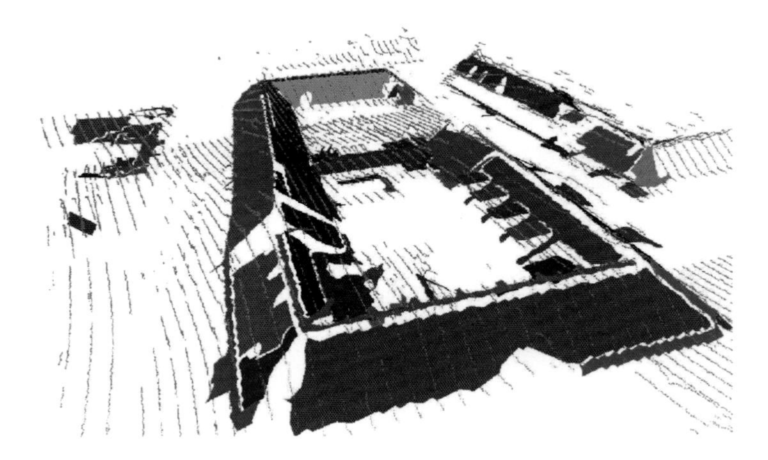

FIGURE 14.9
Extracted 3D facets of a complex building from the joint segmentation process lidar or image represented with lidar data (red points).

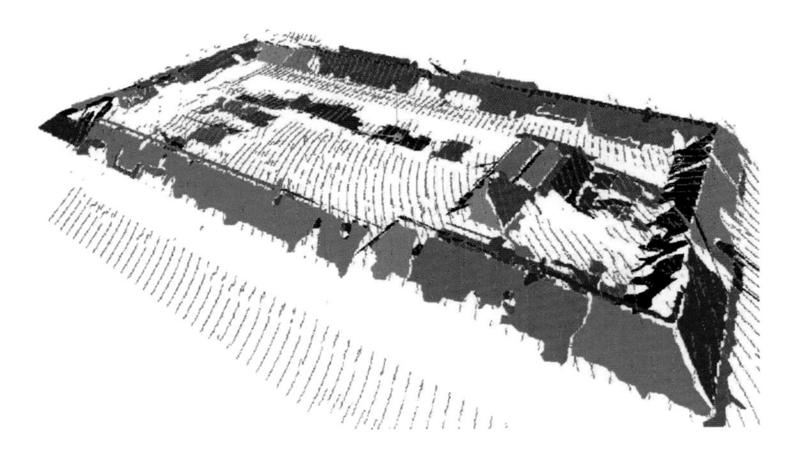

FIGURE 14.10
Extracted 3D facets of another building from the joint segmentation process lidar or image represented with lidar data (red points).

FIGURE 16.2
(a) Results of the per-pixel classification after postprocessing. Red: buildings, dark green: trees, light green: grass land, brown: bare soil. (b) Comparison of final results to reference data on a per-pixel level. Yellow: TP, red: FP, blue: FN. (Reprinted from Rottensteiner, F. et al., *ISPRS J. Photogram. Rem. Sens.*, 62(2), 135, 2007. With permission.)

FIGURE 16.4
DSM with grid width = 0.5 m (left), orthophoto (center), and planar segments (right) for two buildings. Upper row: width = 115 m; lower row: width = 60 m. Plane P_f will be eliminated later. (From Rottensteiner, F., et al., *Int. Arch. Photogram. Rem. Sens. Spatial Inform. Sci.*, *XXXVI*(3/W19), 221, 2005.)

12

Forest Inventory Using Small-Footprint Airborne LiDAR

Juha Hyyppä, Hannu Hyyppä, Xiaowei Yu, Harri Kaartinen, Antero Kukko, and Markus Holopainen

CONTENTS

12.1 Introduction

12.1.1 History of Profiling Measurements over Forests

The concept of producing forest stand profiles (i.e., height profiles) with high-precision instruments was demonstrated as early as 1939 (Hugershoff, 1939), and the concept was implemented with laser profilers around 1980 (Solodukhin et al., 1977; Nelson et al., 1984; Schreier et al., 1985; Aldred and Bonnor, 1985; Maclean and Krabill, 1986; Bernard et al., 1987; Currie et al., 1989). Since then, studies have used laser measurements, for example, for estimating the tree height, stem volume, and biomass (Nelson et al., 1984; Aldred and Bonnor, 1985; Schreier et al., 1985; Maclean and Krabill, 1986; Nelson et al., 1988; Currie et al., 1989; Nilsson, 1990). Nelson et al. (1984) proposed the use of the laser-derived stand profiles for the retrieval of stand characteristics. They also showed that the elements of the stand profile are linearly related to crown closure and may be used to assess tree height. Schreier et al. (1985) concluded that the near-infrared laser can produce terrain and vegetation canopy profiles, and examined laser intensity for vegetation discrimination. Nelson et al. (1988) demonstrated that the tree height, stem volume, and biomass can be predicted with reasonable accuracy using reference plots and averaging. Aldred and Bonnor (1985) presented laser-derived tree height estimates within 4.1 m of the field-measured stand heights at a 95% level of confidence. Currie et al. (1989) estimated the height of the flat-topped crowns with an accuracy of about 1 m. Nilsson (1990) proposed that the data of a laser mounted on a boomtruck correlates with volume changes, such as thinnings. In Hyyppä (1993) and Hyyppä and Hallikainen (1993), a radar-based profiling system and feasibility for forest measurements (tree height, basal area, and volume) were depicted. In order to obtain more information on the history of prior laser ranging measurements over forests, the reader is referred to Nelson et al. (1997); Nilsson (1996); Lim et al. (2003); and Holmgren (2003).

12.1.2 Background of Airborne Laser Scanning in Forestry

The first studies of small-footprint ALS for forests included the determination of terrain elevations (Kraus and Pfeifer, 1998; Vosselman, 2000), standwise mean height and volume estimation (Næsset, 1997a,b), individual-tree-based height determination and volume estimation (Hyyppä and Inkinen, 1999; Brandtberg, 1999; Ziegler et al., 2000, Hyyppä et al., 2001a), tree-species classification (Brandtberg et al., 2003; Holmgren and Persson, 2004), measurement of forest growth, and detection of harvested trees (Hyyppä et al., 2003b; Yu et al., 2003, 2004a). Laser-scanning experiences in Canadian, Finnish, Norwegian, and Swedish forestry can also be found in Wulder (2003); Hyyppä et al. (2003a); Næsset (2003) and Nilsson et al. (2003). A Scandinavian summary of laser scanning in forestry can be read in Næsset et al. (2004) and a summary of the methods used for forest inventory can be also read in Hyyppä et al. (2008). Since 2002 there have been annual meetings of the LiDAR-based forest mensuration society. Examples of such conferences include a workshop on three-dimensional analysis of forest structure and terrain using LiDAR technology in Victoria, an Australian workshop on airborne laser altimetry for forests and woodland inventory and monitoring in Brisbane in 2002, Scandlaser, 2003, in Umeå, Natscan, 2004, in Freiburg, Silvilaser, 2005, in Backsburg, Virginia, 3D Remote Sensing in Forestry in Vienna, 2006, Silvilaser, 2006, in Matsuyama, Japan, and Laser Scanning 2007, and Silvilaser, 2007, in Espoo. Articles on the development of methods can be found in the papers of these conferences. Additionally, there have been special issues on forestry and laser scanning in the *Canadian Journal of Remote Sensing* 2003, the *Scandinavian Journal of Forest Research* 2004,

FIGURE 12.1
Laser point cloud from sparse forest.

Photogrammetric Engineering and Remote Sensing, December 2006, and in the *International Journal of Remote Sensing* 2008. An example of the dense point cloud collected with small-footprint ALS is depicted in Figure 12.1.

In addition to small-footprint (0.2–2 m) ALS, several large footprint systems such as SLICER (Scanning LiDAR Imager of Canopies by Echo Recovery), RASCAL (Raster Scanning Airborne Laser Altimeter), and SLA 01/02 (Shuttle Laser Altimeter I and II) have been developed. Currently, since the small-footprint systems are commercially attractive we will focus our discussion on small-footprint ALS.

This chapter is divided into eight sections that intend to give a broad understanding of laser-based forest inventory. A summary of user requirements of forestry is given in Section 12.2. Section 12.3 introduces the LiDAR-scattering process in forests. Section 12.4 depicts canopy height retrieval and the corresponding accuracy obtained in height measurements—all forest inventories are based on accurate measurements of canopy height. Section 12.5 focuses on forest inventory approaches aimed at tree or stand parameter retrieval. Section 12.6 gives examples of change detection possibilities using multitemporal laser surveys. Section 12.7 is a summary of the ISPRS/EuroSDR tree extraction comparison describing mainly differences of individual-tree-based methods, which is a global comparison for tree parameter retrieval. Section 12.8 provides concluding remarks and future possibilities. This chapter is adapted using Hyyppä et al. (2008).

12.2 Users' Requirements from Forests

About 100 years ago, forest inventory was considered as the determination of the volume of logs, trees, and stands, and a calculation of the increment and yield. More recently, forest inventory has expanded to cover assessment of various issues including wildlife, recreation, watershed management, and other aspects of multiple-use forestry. However, a major emphasis of forest inventory still lies in obtaining information on the volume and growth of trees, forest plots, stands, and large areas. In the following, we will depict in more detail the major attributes of forests relevant from the point of view of laser scanning.

12.2.1 Individual Tree Attributes

Forests can be characterized by their attributes (parameters, features, variables). Basic attributes for a tree are height, diameter at breast height (dbh), upper diameter (diameter of the tree at the height of, for example, 6 m), height of crown base (height from the ground to the lowest green branch or to the lowest complete living whorl of branches), species, age, location, basal area (cross-sectional area defined by the dbh), volume, biomass, growth, and leaf area index. Some of them can be directly measured or calculated from these direct measurements, while others need to be estimated (predicted) through statistical or physical modeling.

Traditionally, individual tree attributes such as height, diameters at different heights along the stem, and crown diameter are measured in the field (Figures 12.2 and 12.3). The conventional strategy for collecting such data is a plot-wise field inventory, especially by measuring the diameters because diameter is convenient to measure and one of the directly measurable dimensions from which cross-sectional area of a tree, surface area, and volume can be computed. Various instruments and methods have been developed for measuring the dimension of a tree in the field (Husch et al., 1982; Päivinen et al., 1992; Gill et al., 2000; Korhonen et al., 2006), such as caliper, diameter tape, and optical devices for diameter measurements; level rod, pole, hypsometers for tree height measurements; and increment borer for diameter growth measurements. The method used in obtaining the measurements is largely determined by the accuracy required. Sometimes it is necessary to fell the tree to obtain more accurate measurements, for example, the only way to actually measure the stem volume is through destructive sampling of a tree. As a result, direct and indirect methods have been developed for the estimation of such variables. For example, volume estimation methods include graphical methods in which cross-sectional areas at different heights along the trunk have been plotted over height on paper and the area under the curve is equivalent to the volume, and the use of volume equations that estimate volume through a relationship with measurable parameters such as height and dbh. In practice, tree volume is estimated from dbh and possibly together with height and upper diameter for each tree species. The models for volume, especially based on diameter information, for individual trees exist in literature in each country.

Better measurements of a tree are required when the interest is in growth over time rather than size at a particular time. Estimates of height increment are usually

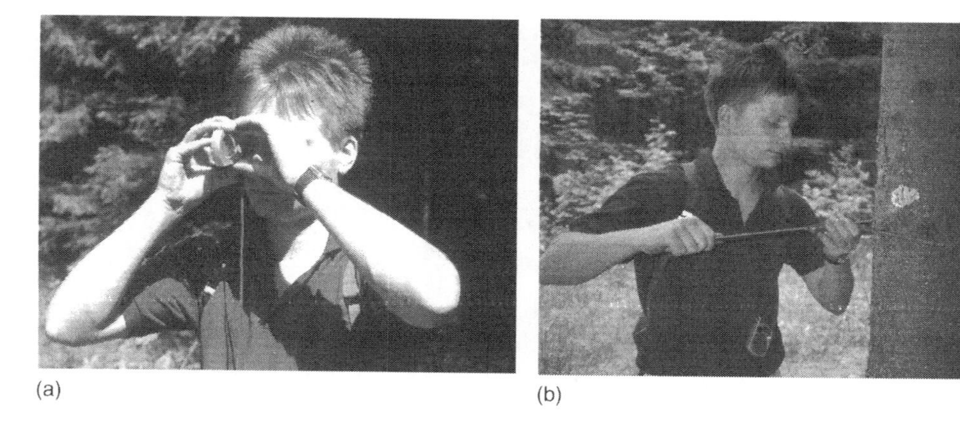

(a) (b)

FIGURE 12.2
Collecting individual tree information using hypsometer for height (a), and using caliper for diameter at breast height (b).

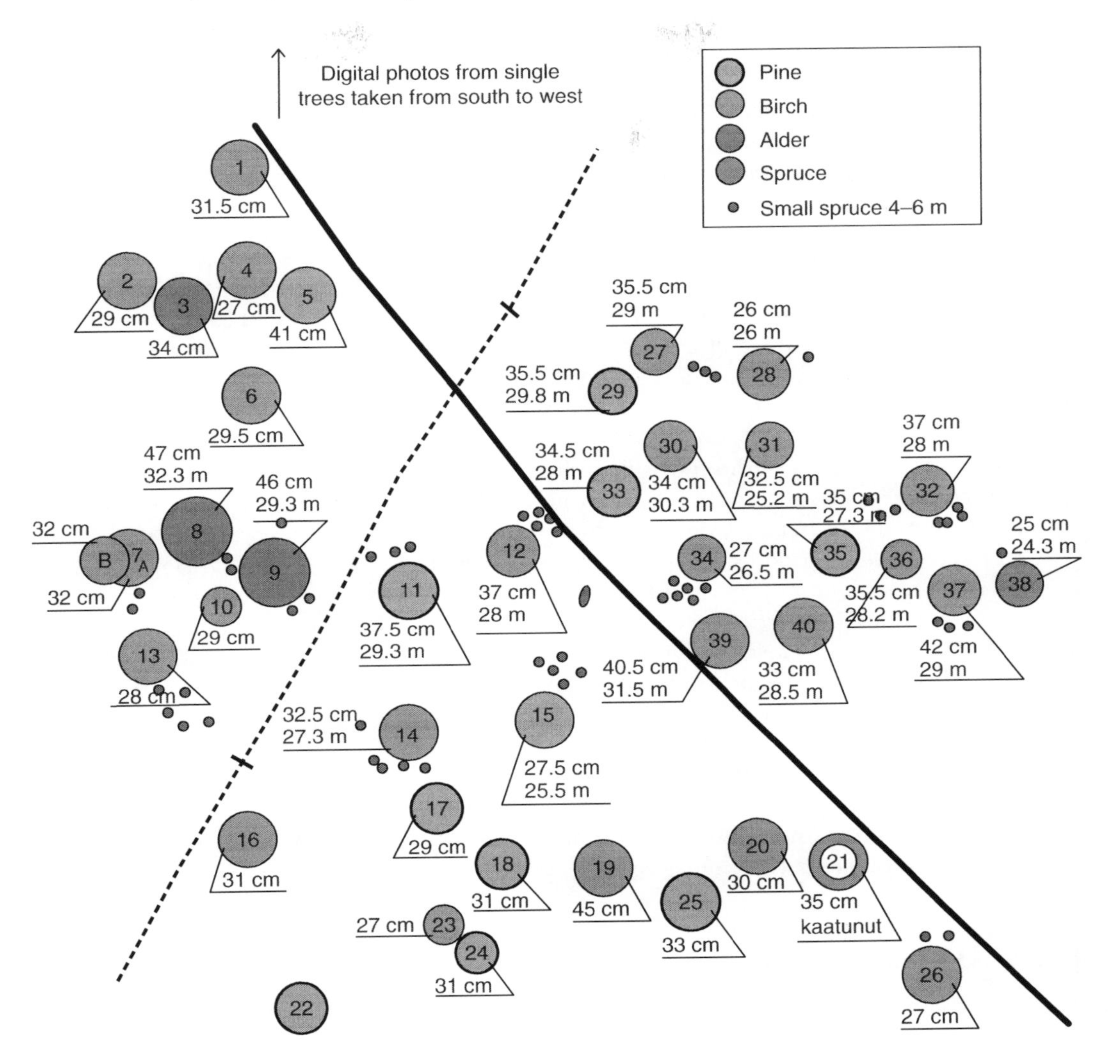

FIGURE 12.3
An example of the sample plot showing each individual tree and their characteristics.

satisfactory if height is measured by height sticks, but may be unsatisfactory if measured by a hypsometer (Husch et al., 1982). Estimates of diameter increment are much more reliable, particularly if the point of measurement on stems is marked permanently. The past growth of diameter can be obtained from increment borings or cross-section cuts (Poage and Tappeiner, 2002). The past height increment may be determined by stem analysis (Uzoh and Oliver, 2006). For species in which the internodal lengths on the stem indicate a year's growth, the past height growth may be determined by measuring internodal lengths (Husch et al., 1982). Growth is most commonly determined by repeated measurements at the beginning and at the end of a specified period, and by considering the difference (Husch et al., 1982; Uzoh and Oliver, 2006). In principal, only past growth can be measured but usually future growth is of interest and it has to be predicted by using the growth model (Hynynen, 1995; Hökkä and Groot, 1999; Hall and Bailey, 2001; Matala, 2005).

The accuracy of field measurements is usually high at a point level. Päivinen et al. (1992) reported that dbh could be measured with a standard deviation ranging from 2.3 mm to 4.6 mm and height with a standard deviation of 67 cm. The 5-year height increment of Scots pine and Norway spruce was measured with a standard deviation of 27 cm, and 20.5% in the estimate of volume increment for a 65-year-old Scots pine stand.

12.2.2 Stand Attributes

The common attributes used to describe stand and plots of even-aged forests of a single species include age, number of trees per hectare, mean diameter, basal area per hectare (sum of the cross-sectional areas per hectare), mean diameter, mean height (arithmetic mean height), dominant height (referring, e.g., to the mean height of the 100 trees per hectare with the largest dbh, dominant trees), Lorey's mean (each tree is weighted by its cross-sectional area) or weighted height, volume per hectare, mean form factor (coefficient to relate volume of trees using a product of basal area and Lorey's mean height), current annual increment per hectare, mean annual increment per hectare, and growth.

A more accurate method of estimating the volume of the plot is to sum up the volumes of individual trees using individual-tree models for each tree species and strata separately. In forest inventory, a practical method of measuring the basal area per hectare relative to the plot is to use a relascope, which is an instrument used in the forest inventory to discriminate between trees on the basis of whether or not the tree subtends an angle equal to or greater than that of the relascope when viewed from the sampling point (Philip, 1983). Volumes per hectare of even-aged forests of a single species can be predicted using the stand volume table. The commonest stand volume table is derived from a simple linear regression of volume per hectare on both the basal area per hectare and the height representative of the forest (mean height or dominant height). In order to get an estimate for a stand, several plots need to be measured.

12.2.3 Operative Compartment-Wise Inventory

There exist several types of operational forest inventory methods ranging from national and continent-wise forest inventory to compartment-wise forest inventory. In this presentation we concentrate on compartment-wise inventory due to its high commercial impact. Compartment-wise forest inventory is a widely used method in Finland, both in public and privately owned forests. The basic unit of forest inventories is a forest stand, which is used as the management-planning unit. The size of a forest stand is normally 0.5–3 ha. The forest stand is defined as a homogenous area according to relevant stand characteristics, e.g., site fertility, composition of tree species, and stand age. Forest inventory data are mostly collected with the aid of field surveys, which are both expensive and time-consuming. The compartments are typically measured separately by analyzing sample plots placed on the stands. From each plot, tree and stand attributes are measured. Finally, the standwise attributes describing the density and tree dimensions are derived from these plot measurements. The method is also sensitive to subjective measurement errors. Remote sensing is normally used for nothing more than delineation of compartment boundaries. The total costs of compartment-wise inventory in Finland were 17.9 €/ha in 2000, of which 7.9 €/ha, i.e., 45% of the costs, consisted of field measurements (Uuttera et al., 2002; Holopainen and Talvitie, 2006). For the total stem volume per hectare, basal area per hectare, and mean height, the required accuracy is roughly 15%.

In practice, the accuracy ranges from 10% to 30% depending on the heterogeneity of the forests (mixture of tree species, several strata, dense undervegetation, varying elevation). The stem volume for each tree species and each strata is obtained with significantly lower accuracy.

The forest inventory has been performed to most of the stands several times. Thus, information from previous inventories exists, and the minimum data requirements for standwise forest inventory are presently the total volume, basal area, and mean height for each tree species from the dominant tree storey. Other required attributes can then be derived from these data. In the near future, more accurate quantity and quality information of individual trees in the stand can be used as a base for felling and transporting round timber from the forests directly to the manufacturers according to the demand for raw material. One important benefit from improved accuracy of forest data is the ability to better plan the forest operations as well as the supply chain. As these activities constitute a significant part of the cost of raw material for the industry, it is of vital importance to control these costs effectively.

12.3 Laser Beam Interaction with Forest Canopies

The laser pulse hit on the forest canopy can be simple or complex (Figure 12.4). In the simplest case, a laser pulse may be scattered directly from the top of a very dense vegetation canopy resulting in a single return. Since a forest canopy is not a solid surface and there exist gaps in the canopy cover, the situation becomes more complex when a laser pulse hits a forest canopy passes through the top of the canopy and intercepts with different parts of the canopy such as the trunk, branches, and leaves before reaching the ground. This series of events may result in several returns being recorded for a single laser pulse, which is called multiple returns. In most cases, the first and last returns are recorded. The first

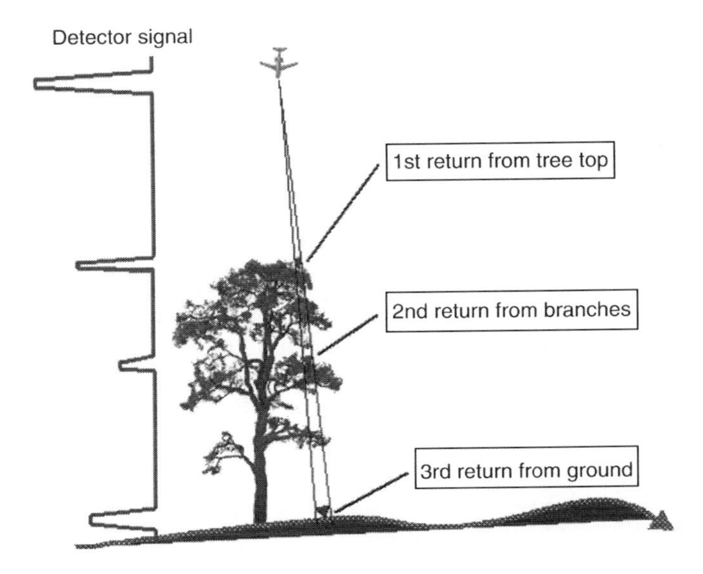

FIGURE 12.4
Interaction of laser pulse with forest canopy resulting multireturns. The amplitudes and echo widths are not to be considered as typical ones. Due to fading, absorption, and scattering, the amplitude and echo width can change dramatically between waveforms.

returns are mainly assumed to come from the top of the canopy and the last returns mainly from the ground, which is important for extracting the terrain surface. Multiple returns produce useful forest information regarding forest structure.

Trunks, branches, and leaves in dense vegetation tend to cause multiple scattering or absorption of the emitted laser energy so that fewer backscattered returns are reflected directly from ground (Harding et al., 2001; Hofton et al., 2002). This effect increases when the canopy closure, canopy depth, and structure complexity increase because the laser pulse is greatly obscured by the canopy. In practice, the laser system specification and configurations also play an important role in how the laser pulse interacts with forests. For example, it has been found that the small-footprint laser tends to penetrate the tree crown before reflecting a signal (Gaveau and Hill, 2003); ground returns decrease as the scanning angle increases (TopoSys, 1996); penetration rate is affected by the laser beam divergence (Aldred and Bonnor, 1985; Næsset, 2004); higher flight altitude alters the distribution of laser returns from the top and within the tree canopies (Næsset, 2004); and the distribution of laser returns through the canopy varies with the change in laser pulse repetition frequency (PRF) (Chasmer et al., 2006). Goodwin et al. (2006) used three different platform altitudes (1000, 2000, and 3000 m), two scan angles at 1000 m (10° and 15° half max. angle off nadir), and three footprint sizes (0.2, 0.4, and 0.6 m) in eucalyptus forests at three sites, which varied in vegetation structure and topography. They reported that higher platform altitudes record a lower proportion of the first and last return combinations that will further reduce the number of points available for forest structural assessment and development of digital elevation models; for discrete LiDAR data, increasing platform altitudes will record a lower frequency of returns per crown resulting in larger underestimates of the individual tree crown (ITC) area and volume. Furthermore, the sensitivity of the laser receiver, wavelength, laser power, and total backscattering energy from the treetops are also the factors that may influence the ability of laser pulses to penetrate and distribute laser returns from the forest canopy (Baltsavias, 1999).

One of the most crucial factors for an exact range measurement is the echo detection algorithm used (Wagner et al., 2004; Wagner, 2005). As the length of the laser pulse is longer than the accuracy needed (a few meters versus a few centimeters), a specific timing in the return pulse needs to be defined. In a nonwaveform ranging system, analogue detectors are used to derive discrete, time-stamped trigger pulses from the received signal in real time during the acquisition process. The timing event should not change when the level of signal varies, which is an important requirement in the design of analog detections as discussed by (Palojärvi, 2003). Unfortunately, in the case of commercial ALS systems, detailed information concerning the analog detection method is normally lacking, even though different detection methods may yield quite different range estimates. For full-waveform digitizing ALS systems, several algorithms can be used in the postprocessing stage (e.g., leading edge discriminator and threshold, center of gravity, maximum, zero crossing of the second derivative, and constant fraction). The most basic technique for pulse detection is to trigger a pulse whenever the rising edge of the signal exceeds a given threshold (leading edge discriminator).

12.4 Extraction of Canopy Height

Since laser scanning can provide 3D models of forest canopies, the basics for modern ALS-based forest measurements rely on the acquisition of the CHM (canopy height model),

DTM (digital terrain model) corresponding to the ground surface and DSM (digital surface model) corresponding to treetops. The errors in the DTM will result in errors in tree height and CHM.

12.4.1 Methods

The most appropriate technique to obtain a DSM relevant to treetops is to classify the highest reflections (i.e., by taking the highest point within a defined neighborhood) and interpolate missing points, e.g., by Delaunay triangulation. Then, the CHM is obtained by subtracting the DTM from the corresponding DSM. CHM is also called a normalized DSM (nDSM). The DSM is typically calculated by means of the first pulse echo and the DTM with the last pulse echo. In order to guarantee that there are no systematic errors between the first and last pulse data, calibration using flat, nonvegetated areas, such as roads, roofs, and sports grounds, should be performed, which is especially necessary when laser-scanning systems provide separate first and last echo recordings, e.g., see Figure 12.5 to notice small systematic shifts between the first and the last pulse data.

The processing of the point cloud data into canopy heights or normalized heights is an effective way of increasing the usability of the laser data. Laser canopy heights are simply obtained as the difference between the elevation values of laser hits and the estimated terrain elevation values at the corresponding location.

Concerning DTM generation, the reader is referred to previous chapters. For those reading mainly this chapter, a brief summary is given as a state of the art.

Scientists have developed various methods for obtaining DTMs from laser scanning point clouds (Kraus and Pfeifer, 1998; Pyysalo, 2000; Axelsson 1999, 2000, 2001; Elmqvist et al., 2001; Wack and Wimmer, 2002; Sithole, 2001; Vosselman and Maas, 2001). One of the first international comparisons of DTM methods was carried out within the European Commission-funded HIGH-SCAN project (1998–2001), in which three different DTM algorithms were compared in Finnish (test site Kalkkinen), Austrian (Hohentauern), and Swiss (Zumikon) forests (Hyyppä et al., 2001b). The more detailed comparison of the filtering techniques used for DTM extraction was later made within the ISPRS

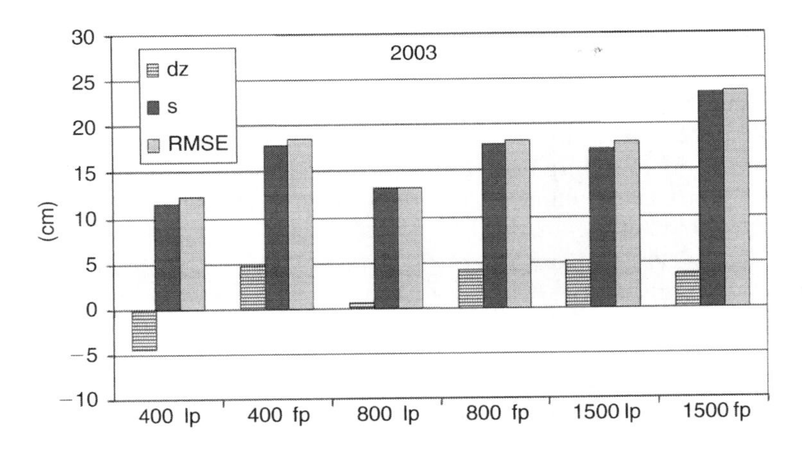

FIGURE 12.5
An example of DTM quality in boreal forests using different pulse densities and pulse modes for model derivation (lp refers to last pulse return, fp refers to first pulse return; 400 refers to 8–10 pulses per m^2, 800 refers to 4–5 pulses per m^2, and 1500 refers to 2–3 pulses per m^2 (from the named altitude)). dz refers to systematic error, s refers to random error, and RMSE to root mean squared error.

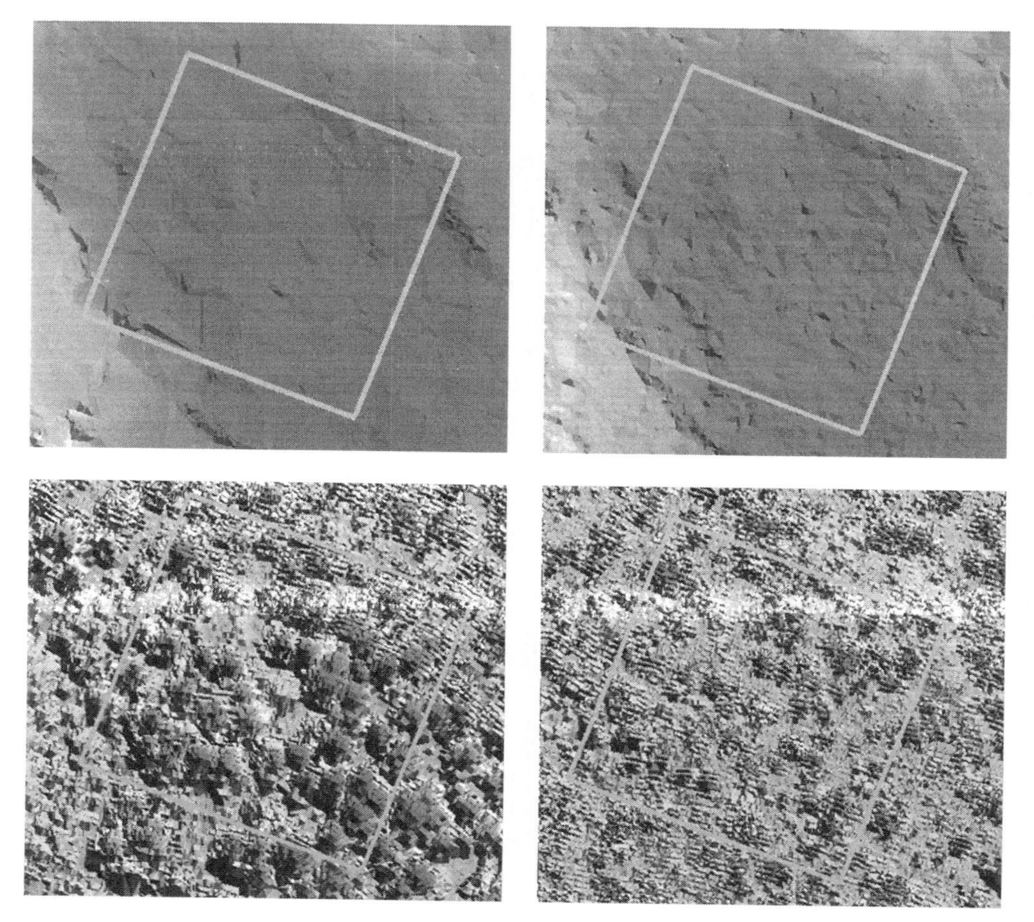

FIGURE 12.6
The use of first (left) and last pulse (right) returns in DTM (above) and canopy height model (below) derivation. Height coded by colors.

comparison of filters (Sithole and Vosselman, 2004). The variation of ALS-derived DTM quality with respect to date, forest cover, flight altitude, pulse mode, terrain slope, and within-plot variation in Finnish conditions was reported in (Hyyppä et al., 2005a). Commercial software that includes DTM generation are REALM, TerraScan, Geomatica LiDAR Engine, and SCOP++.

Figure 12.6 shows typical examples of DTM and CHM derived using the first and the last pulse mode. Obviously, the last pulse describes better ground elevation and the first pulse gives better description of the forest top.

12.4.2 DTM Quality in Forest Conditions

Kraus and Pfeifer (1998) reported an RMSE of 57 cm for DTM in wooded areas using ALTM 1020 and average point spacing of 3.1 m. Hyyppä et al. (2000) reported a random error of 22 cm for fluctuating forest terrain (variation for couple of tens of meters) using TopoSys-1 and nominal point density of 10 pts/m^2. During the European Commission-funded

HIGH-SCAN project (1998–2001), three different DTM algorithms were compared in Finnish (test site Kalkkinen, nominal point density of 10 pts/m^2), Austrian (Hohentauern, density 4–5 pts/m^2), and Swiss (Zumikon, density 4–5 pts/m^2) forests. The random errors obtained for DTM varied between 22 and 40 cm (Hyyppä et al., 2001b) using TopoSys-1. Ahokas et al. (2002) compared three algorithms in forested terrain in Finland and found random errors between 13 and 41 cm using TopoSys-1. Reutebuch et al. (2003) reported random errors of 14 cm for clearcut, 14 cm for heavily thinned forests, 18 cm for lightly thinned forests, and 29 cm for uncut forests using TopEye data with 4 pts/m^2. However, in dense forests, DTM errors of up to 10 to 20 m can occur (Takeda, 2004). Results described in a paper by Hyyppä et al. (2005a) can be partly used to optimize the laser flight parameters with respect to the desired quality, e.g., Figure 12.5. The paper analyzed the effects of the date, flight altitude, pulse mode, terrain slope, forest cover, and within-plot variation on the DTM accuracy in the boreal forest zone. Ahokas et al. (2005) proposed that the optimization of the scanning angle (i.e., field of view) is an important part of countrywide laser scanning, since significant savings can be obtained by increasing the scanning angle and flight altitude. The first results obtained with scanning angle analysis showed that the scanning angle had an effect on the precision, but other factors such as forest density dominate the process. Scanning angles up to 15° seem to be usable in the boreal forest zone. The effects of the scanning angle should be further studied. For comparison, the maximum field of view of Optech ALTM 3100EA is 50° and the corresponding value for Leica ALS50-II is 75°.

12.4.3 Canopy Height Quality

12.4.3.1 Factors Affecting Canopy Height

It was already demonstrated in the 1980s using small-footprint systems that the use of a laser leads to an underestimation of tree height. That was obvious with the use of profiling lasers, since it was expected that the laser beam would mainly hit the tree shoulder rather than the treetops (Nelson et al., 1988). Thus, detection of the uppermost portion of a forest canopy is expected to require a sufficient density of laser pulses to sample the treetops and a sufficient amount of reflecting material occupying each laser pulse footprint to cause a detectable return signal (Lefsky et al., 2002). If the ground elevation and the uppermost portion of a forest canopy are not detected, then the canopy height will be underestimated. Lefsky et al. (2002) also expected the sampling density to be the major issue while determining whether the canopy height with a small-footprint ALS is underestimated. Previously, tree-height underestimation has been reported for individual trees including both deciduous and coniferous trees (Hyyppä and Inkinen, 1999; Persson et al., 2002; Gaveau and Hill, 2003; Leckie et al., 2003; Yu et al., 2004b; Maltamo et al., 2004; Chasmer et al., 2006; Falkowski et al., 2006). As a summary of all these previous studies, it seems that the underestimation of tree height is affected by the density and coverage of laser pulses and beam; the algorithm used to obtain the CHM; the amount and height of undervegetation; the algorithm used to calculate the DTM; the sensitivity of the laser system and echo detection algorithms used in the signal processing as well as pulse penetration into the canopy; and the tree shape and tree species. Finding a universal correction factor for the underestimation is expected to be difficult, since the correction appears to be dependent on the sensor system, flight altitude, forest type, and the algorithm used. Gaveau and Hill (2003) used a terrestrial laser system and Rönnholm et al. (2004) used terrestrial image data to calibrate the underestimation.

12.4.3.2 Quality of Canopy Height Models Analyzed Using Individual Trees

Examples of reported tree underestimation values and the accuracy of individual tree assessments are given below including assessments where errors have not been calibrated or compensated for with reference data. For a comparison of mean tree height obtained in a forest inventory the reader is referred to Næsset et al. (2004).

Hyyppä and Inkinen (1999) reported individual tree height estimation with an RMSE of 0.98 m and a negative bias of 0.14 m (nominal point density about 10 pts/m^2), while Persson et al. (2002) reported an RMSE of 0.63 m and a negative bias of 1.13 m. Both forest sites mainly consisted of Norway spruce and Scots pine. Persson et al. (2002) explained their greater underestimation of the average tree height as resulting from a lower ALS sampling density (about 4 pulses per m^2). Næsset and Økland (2002) concluded that the estimation accuracy was significantly reduced by a lower sampling density. Gaveau and Hill (2003) reported a negative bias of 0.91 m for sample shrub canopies and 1.27 m for sample deciduous tree canopies. Leckie et al. (2003) presumed that some of the 1.3 m underestimation could be accounted for by the undergrowth. Yu et al. (2004a) reported a systematic underestimation of tree heights of 0.67 m for the laser acquisition carried out in 2000 and 0.54 m for another acquisition in 1998. The underestimation corresponded to 2–3 years' annual growth of those trees. Of that, the elevation model overestimation (due to undervegetation) was assumed to account for about 0.20 m. Maltamo et al. (2004) used 29 pines, the heights of which were measured with a tacheometer giving more precise field measurements than conventional methods, and found a 0.65 m underestimation of the height for single trees, including annual growth that was not compensated for in the plot measurements. They also found that the random error of 0.50 m for individual tree height measurements was better than that reported earlier (Hyyppä and Inkinen, 1999; Persson et al., 2002). In the studies of Rönnholm et al. (2004) and Brandtberg et al. (2003), it was shown that the tree height can be reliably estimated even under leaf-off conditions for deciduous trees.

12.4.3.3 Experimental Results of the Effect of Flight Altitude, Point Density, and Footprint Size on Canopy Height

Yu et al. (2004b) studied the effect of laser flight altitude on the tree height estimation at individual-tree level in a boreal forest area mainly consisting of Norway spruce, Scots pine, and silver and downy birch. The test area (0.5 × 2 km) was flown over at three altitudes (400, 800, and 1500 m) with a TopoSys II scanner (beam divergence 1 mrad) in spring 2003. A field inventory was performed on 33 sample plots (about 30 m × 30 m) in the test area during summer 2001. Evaluations of estimation errors due to flight altitudes, including beam size and point density, were carried out for different tree species. The results indicate that the accuracy of the tree height estimation decreases (from 0.76 to 1.16 m) with the increase in flight height (from 400 m to 1.5 km). The number of detectable trees also decreases. Point density had more influence on the tree height estimation than the footprint size; for more details the reader is referred to Yu et al. (2004b). Birch was less affected than coniferous trees by the change in the flight altitude in this study. Persson et al. (2002) reported that the estimates of tree height were not affected much by different beam diameters ranging from 0.26 to 2.08 m. With a larger beam diameter of 3.68 m acquired at a 76% higher altitude, the underestimations of tree heights were greater than with other beam diameters, which is probably due to the decreased point density. Nilsson (1996) did not find any significant effects of beam size on the height estimates over a pine-dominated test site. Aldred and Bonnor (1985) reported increased height estimates as the beam divergence increased, especially for deciduous trees. In the study conducted by Næsset (2004), it was concluded that the first pulse measurements of height are relatively stable regardless of flight altitude and beam size when the beam size varies in the 16–26 cm range. Goodwin et al. (2006) used three different platform altitudes (1000, 2000, and 3000 m), two scan angles at 1000 m (10° and 15° half max. angle off nadir), and three footprint sizes (0.2, 0.4, and 0.6 m) in eucalyptus forests at three sites, which varied in vegetation structure and topography, and observed no significant difference between the relative distribution of laser point returns, indicating that platform altitude and footprint size have no major influence on canopy height estimation.

The results seem to indicate that relatively good canopy height information can be collected with various parameter configurations. Point density is today expected to be the key parameter that affects the level of the inventory—this topic is discussed in detail in the next section. Additionally, the repeatability of canopy height distributions due to changes in survey parameters, sensors, and forest conditions is a topic that needs to be considered in the applied inventory approach. In the ISPRS Workshop on Laser Scanning 2007 and Silvilaser 2007, a panel discussion was carried out on the effects of sensor parameters on forest point clouds.

12.5 Main Feature Extraction Approaches

The two main feature extraction approaches for deriving forest information from laser scanner data have been those based on statistical canopy height distribution and individual-tree detection and possibly segmentation. These categories relate to the need for scale and accuracy of the forestry information and available point density. Both approaches use the CHM or the processed point clouds into canopy heights described in Section 12.4.1. In the distribution-based techniques, features and predictors are assessed from the laser-derived surface models and canopy height point clouds, which are directly used for forest parameter estimation, typically using regression, nonparametric, or discriminant analysis. The distribution-based techniques rely entirely on the reference data collected in the field. In the individual-tree-based approaches, the neighborhood information of point clouds and pixels of DSMs or CHMs are effectively used to derive physical features and measures, such as crown size, individual tree height, and location. The forest inventory data is then calculated or estimated using existing models and statistical techniques or a compilation of individual-tree information. Calibration of individual-tree models is done with reference data, but the amount of reference data is significantly lower than that of the distribution-based technique.

12.5.1 Extraction of Forest Variables by Canopy Height Distribution

Height percentiles of the distribution of canopy heights have been used as predictors in regression or nonparametric models for the estimation of the mean tree height, basal area, and volume (Næsset, 1997a,b; Lefsky et al., 1999; Magnussen et al., 1999; Means et al., 1999; Lim et al., 2002; Næsset and Økland, 2002; Næsset, 2002; Hopkinson et al., 2006; Maltamo et al., 2006b) (see Figures 12.7 and 12.8). In addition to the prediction of stand mean and sum characteristics, diameter distributions of a forest stand has also been

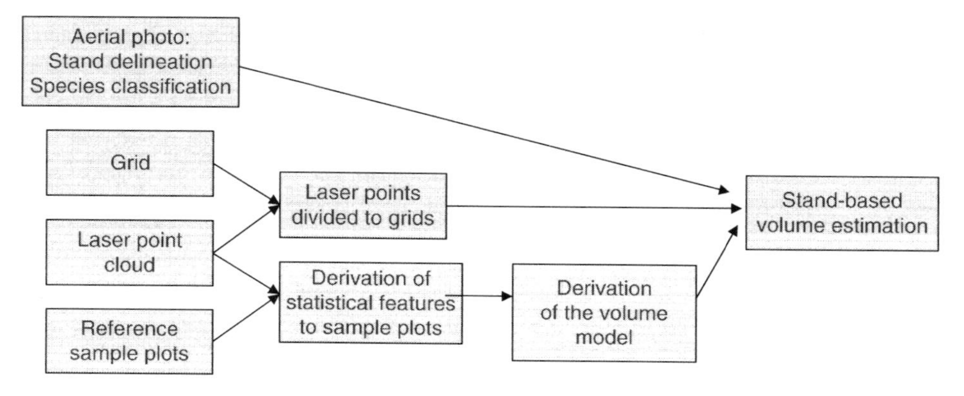

FIGURE 12.7
The basic principle of canopy height distribution-based techniques.

predicted by using the statistical canopy height distribution based approach (Gobakken and Næsset, 2005; Maltamo et al., 2006a). In Means et al. (1999), SLICER having large-footprints was used in the estimation of the tree height, basal area, and biomass in Douglas-fir dominated forests, with the tree height ranging from 7 to 52 m, with a coefficient of determination values of 0.95, 0.96, and 0.96, respectively. In addition to canopy height information, canopy reflection sum, canopy closure, and ground reflection sum were used. The canopy reflection sum is the sum of the portion of the waveform return reflected from the canopy. The ground reflection sum is the sum of the waveform return reflected from the ground multiplied by a factor correcting for the canopy attenuation. Canopy closure was approximated by dividing the canopy and ground reflection sums. In Næsset (2002), several forest attributes were estimated using canopy height and canopy density metrics and a two-stage procedure with field data. Canopy height metrics consisted of quantiles corresponding to the 0, 10, …, 90 percentiles of the first pulse laser canopy heights and corresponding statistics, whereas canopy density corresponded to the proportions of both first and last pulse laser hits above the 0, 10, …, 90 quantiles to the total number of pulses.

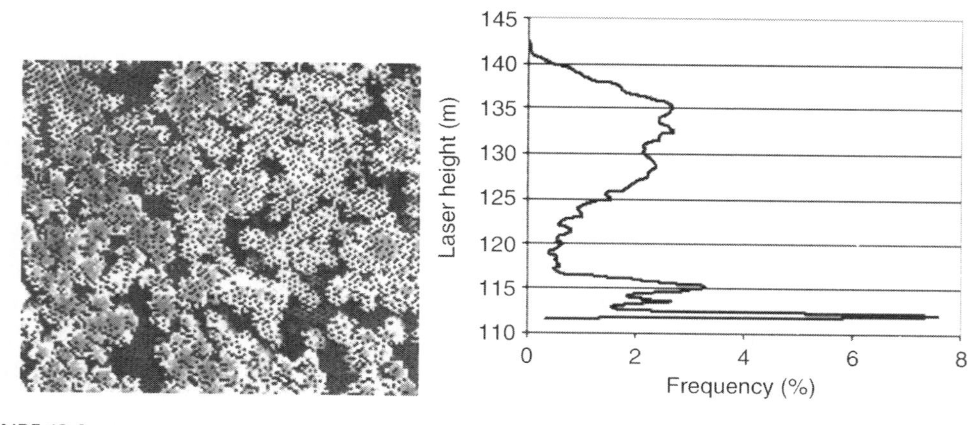

FIGURE 12.8
A DSM and corresponding elevation distribution of laser hits.

In Riano et al. (2003), several statistical parameters were defined for forest fire behavior modeling. Tree cover was calculated from the proportion of laser hits from the tree canopy divided by the total number of laser hits. Surface cover was defined as the proportion of laser hits from the surface and the total number of hits. Crown bulk density was obtained from the foliage biomass estimates. Crown volume was estimated as the crown area times the crown height after a correction for mean canopy cover.

In Holmgren and Persson (2004), a number of height and intensity-based variables were defined for tree species classification, for example, relative standard deviation of tree heights, the proportion of single returns and the proportion of first return, as well as the proportion of vegetation points (the number of returns that were located above the crown base height divided by the total number of returns from the segment), crown shape by fitting a parabolic surface to the laser point cloud, and mean intensity and standard deviation of both single and surface returns. They reported an overall tree species discrimination accuracy of 95% between Scots pine and Norway spruce. High classification accuracy was simply obtained by using the proportion of first returns and standard deviation of the intensity. There was also a strong correlation between the standard deviation of laser heights within a segment and the corresponding crown base height and also between the mean distance between the first and last return of a double return within a segment, and the corresponding crown length was found. In Hall et al. (2005), similarly, 39 metrics were derived from the LiDAR data.

Examples of the accuracy of techniques based on canopy height distribution can be found in Næsset et al. (2004). In Hollaus (2006), the distribution-based technique was tested for the alpine forests. It was stated that the multiple regression analyses led to different sets of independent variables if ALS data with different acquisition times or point densities were used for the calculations. Therefore, Hollaus (2006) recommended that for ALS datasets with different properties (e.g., point densities, acquisition times), separate regression models should be used. The proposed linear approach in Hollaus (2006) was based on the original work of Nelson et al. (1984, 1988) and adapted from canopy profiles to 3D canopy heights, showing that for ALS data with varying properties robust and reliable results of high accuracies (e.g., $R^2 = 0.87$, standard deviation of the stem volume residuals derived from a cross-validation is $90.0\,m^3/ha$) can be achieved. Because of the simplicity of this model, a physically explicit connection between the stem and the canopy volume is available.

12.5.2 Extraction of Individual-Tree-Based Information Using LiDAR

Recent developments in the computer analysis of very high spatial resolution images are leading towards the semiautomated production of forest inventories based on ITC information. The extraction of individual-tree-based information from remote sensing can be traditionally divided into finding tree locations, finding tree locations with crown size parametrization, or full crown delineation (Pouliot et al., 2002; Gougeon and Leckie, 2003). The methods used in laser scanning can utilize the methods already developed using high and very high resolution aerial imagery. Additionally, in laser scanning it is possible to improve the image-based approaches by utilizing the powerful ranging algorithms and knowledge-based approaches (e.g., we know the tree height and we can roughly estimate the size of the crown).

Tree locations can be obtained by detecting image or point cloud local maxima (Gougeon and Moore, 1989; Dralle and Rudemo, 1996; Hyyppä and Inkinen, 1999; Wulder et al., 2000; Hyyppä et al., 2001a). In laser scanning, the aerial image is replaced by the crown DSM, the CHM, or normalized point cloud. Provided that the filter size and image smoothing parameters are appropriate for the tree size and image resolution, the approach works relatively well with coniferous trees (Gougeon and Leckie, 2003). In Scandinavia, the filtering should be very modest due to narrow tall trees (Hyyppä, 2007). After the local maxima have been found, the edge of the crown can be found using region segmentation, edge detection, or local minima detection (Pinz, 1991; Uuttera et al., 1998; Hyyppä and Inkinen, 1999; Hyyppä et al., 2001a; Persson et al., 2002; Culvenor, 2002). Full crown delineation is also traditionally possible with techniques such as shade-valley-following (Gougeon, 1995), edge curvature analysis (Brandtberg and Walter, 1999), template matching (Pollock, 1994; Larsen and Rudemo, 1998), region growing (Erikson, 2003), and point cloud based reconstruction (Pyysalo and Hyyppä, 2002). In laser scanning, the individual tree approach typically provides tree counts, tree species, crown area, canopy closure, gap analysis, and volume or biomass estimation (Hyyppä and Inkinen, 1999; Hyyppä et al., 2001a; Gougeon and Leckie, 2003) (see Figure 12.9). In the following, the focus is on laser-based individual-tree-based solutions.

Hyyppä and Inkinen (1999); Friedlaender and Koch (2000); Ziegler et al. (2000) and Hyyppä et al. (2001a) demonstrated the individual-tree-based forest inventory using laser scanner tree finding with maxima of the CHM and segmentation for edge detection. They also presented the basic LiDAR-based ITC approach in which, from individual trees, location, tree height, crown diameter and species are derived using laser, possibly in combination with aerial image data, especially for tree species classification; then, other important

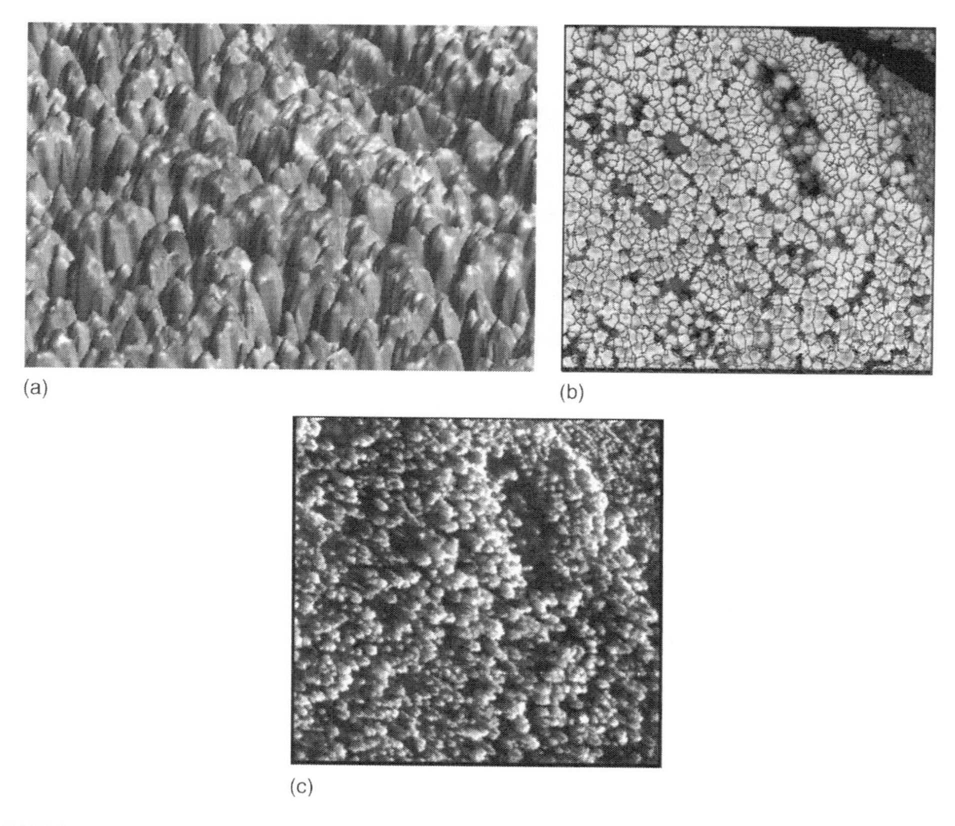

FIGURE 12.9
Canopy height model (a), segmented canopy height model (b), and corresponding aerial image (c) are shown.

variables such as stem diameter, basal area, and stem volume are derived using existing models. The methods were tested in Finnish, Austrian, and German coniferous forests and 40%–50% of the trees could be correctly segmented (Hyyppä et al., 2001b). Persson et al. (2002) could link 71% of the tree heights with the reference trees. Other attempts to use DSM or CHM image for ITC isolation or crown diameter estimation (or segmentation) have been reported by Brandtberg et al. (2003); Leckie et al. (2003); Straub (2003); Popescu et al. (2003); Tiede and Hoffman (2006), and Falkowski et al. (2006). Andersen et al. (2002) proposed to fit ellipsoid crown models in a Bayesian framework to the point cloud. Morsdorf et al. (2003) presented a two-stage procedure where tree locations were defined using the DSM and the local maxima, and crown delineation was performed using k-means clustering in the three-dimensional point cloud. Pitkänen et al. (2004) proposed three methods for individual-tree detection: smoothed CHM with the knowledge of the canopy height, elimination of candidate tree locations based on the predicted crown diameter and distance and valley depth between two locations studied (maximum elimination), and modified scale-space method used for blob detection. The maximum elimination method gave the best results of tree detection, however, with the cost of including several parameters to keep the number of false positives low. Sohlberg et al. (2006) presented methods for controlling the shape of the crown segments, and for residual adjustment of the CHM. The method was applied and validated in a Norway spruce dominated forest having heterogeneous structure. The number of trees detected varied with the social status of the trees, from 93% of the dominant trees to 19% of the suppressed trees.

A new approach to improve the quality of the individual-tree-based inventory was proposed by Villikka et al. (2007). They used tree-level laser height distribution characteristics of individual trees combined with conventional variables of individual tree recognition (height, dbh) for improving the prediction of individual tree stem volume. It is worth noting that approximate tree height and crown diameters were used in all of the constructed models, but the lower height quantiles and corresponding crown densities hold some additional statistical explanation for the tree characteristics.

Canopy height distribution approaches use the distribution of laser canopy heights to estimate stand heights. Individual-tree methods on the other hand are focused on determining the heights of individual trees. The expectation with individual-tree methods is that height can be determined with a low amount of the required project- or site-specific calibration. There has not been a careful comparison of canopy height distribution and individual-tree-based techniques; results obtained have been based on different types of reference datasets. Typically, distribution-based techniques have been calibrated with very large and accurate reference data and individual-tree-based approaches have not been calibrated at all. Thus, the study results obtainable are not comparable, even though a summary of the obtainable accuracy can be found in Næsset et al. (2004). A short summary of the general advantages and disadvantages of the techniques are given in Table 12.1. In the practical implementation of individual-tree-based techniques, calibration data either at tree level or plot level is recommended to calibrate systematic errors in the applied models and to improve the volume and diameter estimation.

12.5.3 Tree Cluster-Based Inventory

Depending on the density of the forest and the density of the point clouds, the discrimination of individual trees is a problem of varying complexity. In dense stands, individual-tree-based approach without any calibration leads to an underestimation of the number of tree stems. What happens if the individual-tree-based technique is applied directly to tree clusters?

TABLE 12.1

Comparison of Distribution and Individual-Tree-Based Methods

	Advantages	Disadvantages
Distribution-based methods	Easy to integrate with present forest inventory practices due to common reference plots	Requires extensive, accurate, representative, and expensive reference data
	Strong statistical approach used	Without a large amount of reference data, strong possibility of large errors in operational inventories
	Laser scanning data relatively inexpensive	
Individual-tree-based methods	Good physical correspondence (existing models) with volume estimation	More expensive laser data
	Low amount of reference data needed for calibration	More complex system to implement
	Allows precision forestry and increased amount of information on the forests	

Bortolot (2006) proposed a tree cluster approach that consists of calculating the percentage of grid pixels that are in tree clusters, percentage of cluster pixels that are core pixels, mean height of the cluster pixels, and standard deviation of the cluster pixels from the CHM. Core pixels were those pixels that were fully surrounded by the cluster pixels. The percentage of core pixels referred to cases where many trees are joined together. The percentage of core pixels and mean height metrics appeared to be the best for predicting density and biomass. Preliminary results showed better performance over the applied individual-tree-based technique.

Hyyppä et al. (2006) proposed a cluster-based technique using the individual-tree-based approach, low-density laser data, and by calculating the corresponding number of trees in the large segment based on statistics or by using the existing volume model to account for the large segment. The results indicated that individual trees volumes can be obtained, with random errors of about 30%, and the volume related to small tree clusters or segments, with a random error of 37%.

12.5.4 Tree Species—Synergy between Optical Imagery and ALS Data

The integration of laser scanning and aerial imagery can be based on simultaneous or separate data capturing. There is high synergy between the high-resolution optical imagery and laser scanner data for extracting forest information: laser data provides accurate height information, which is missing in nonstereo optical imagery, and also supports information on the crown shape and size depending on point density, whereas optical images provide more details about spatial geometry and color information usable for classification of tree species and health. In practice, tree species derivation is the major reason for the synergetic use needed between aerial image and laser scanner data.

12.5.4.1 Derivation of Tree Species Information Using LiDAR Data

Holmgren and Persson (2004) tested classification of species of Scots pine and Norway spruce using laser data (point spacing 0.4–0.5 m) at individual-tree level. The proportion of correctly classified trees on all the plots was 95%. Moffiet et al. (2005) suggested that the proportion of laser singular returns is an important predictor for the tree species classification for species such as poplar box and Cypress pine. While a clear distinction between these two species was not always visually obvious at the individual-tree level, due to some other extraneous sources of variation in the dataset, the observation was supported in general at the site level. Sites dominated by poplar box generally exhibited a lower proportion of singular returns compared to sites dominated by cypress pine. Brandtberg et al. (2003) used laser data under leaf-off conditions for the detection of individual trees and tree species classification using different indices. The results suggest a moderate to high degree of accuracy in deciduous species classification. Liang et al. (2007) used a simple technique, the difference of the first and last pulse return under leaf-off conditions, to discriminate between deciduous and coniferous trees at individual tree level. Classification accuracy of 90% was obtained. Reitberger et al. (2006) described an approach to tree species classification based on features derived by a waveform decomposition of full-waveform LiDAR data. Point distributions were computed in sample tree areas and compared with the numbers that result from a conventional signal detection. Unsupervised tree species classification was performed using special tree saliencies derived from the LiDAR points. The classification into two clusters (deciduous, coniferous) led to an overall accuracy of 80% in a leaf-on situation.

12.5.4.2 Derivation of Tree Species Information Using LiDAR and Optical Images

Conventionally, tree species information is extracted from high spatial resolution color-infrared aerial photographs (Brandtberg, 2002; Bohlin et al., 2006). Persson et al. (2006) derived tree species for trees through combining features of high resolution laser data (50 pts/m^2) with high resolution multispectral images (ground resolution 0.1 m). The tree species classification experiment was conducted in southern Sweden in a forest consisting of Norway spruce (*Picea abies*), Scots pine (*Pinus sylvestris*), and deciduous trees, mainly birch (*Betula* spp.). The results implied that by combining a laser-derived geometry and image-derived spectral features, the classification could be improved and accuracies of 95% were achieved. Packalén and Maltamo (2006) used a combination of laser scanner data and aerial images to predict species-specific plot level tree volumes. A nonparametric k-Most Similar Neighbor application was constructed by using characteristics of canopy height distribution approach of laser data and textural and spectral values of aerial images at plot level.

12.5.4.3 Other Synergetic Use of LiDAR and Optical Images

Gougeon et al. (2001) studied the synergy between aerial and LiDAR data and found that the LiDAR data, when used as a filter to the aerial data or on its own, made extremely obvious (and intuitive) distinction between the dominant and codominant level and the understorey level, or regeneration versus ground vegetation, thus permitting separate analyses. They also found that using a height-based threshold, the valley-following-based crown delineation algorithm is able to function (on aerial or LiDAR CHM) in wide-open and low-density areas, and valley-following-based crown delineation in the optical part of the spectrum is usually hampered in a direction perpendicular to the illumination angle (i.e., no shade between crowns). Similarly, delineation from a LiDAR-acquired CHM is hampered in the direction of the scan (off nadir); the synergistic effect of using the two datasets leads to more crowns being properly found. Leckie et al. (2003) tested a valley-following approach to individual-tree isolation of both digital frame camera imagery and a canopy height model created from high-density LiDAR data. The results indicate that optical data may be better at outlining crowns in denser situations and thus more weight should be given to optical data in such situations. LiDAR eliminated easily most of the commission errors that often occur in open stands, with optical imagery.

There are several possibilities of individual-tree-based retrieval methods using the combination of aerial images and laser data. It has been shown that a laser gives a more reliable tree height than, e.g., photogrammetry, since the ground surface is often obscured on aerial images and it is difficult to measure real tree heights. Alternative ways of obtaining the tree height for individual trees are (1) subtracting the old laser-derived DTM from the DSM of treetops obtained using stereophotogrammetry (as proposed by St-Onge et al. [2004]) and (2) interpolating a corresponding tree height from the low-density derived CHM. The tree-height information can then be joined with a properly segmented aerial image. In Hyyppä et al. (2005b), an aerial image was segmented and the tree height for each tree was obtained from a laser CHM. The accuracy of the inventory was comparable with that done with a high-density-laser, individual-tree-based technique, and there was a significant improvement over the performance compared with the fully aerial-image-based individual-tree-based approach, since in the latter there is no information of tree height without using stereo imagery.

Suarez et al. (2005) tested a segmentation method using a data fusion technique available from eCognition to identify individual trees using scale and homogeneity parameters from the image and the elevation values from the CHM. The segmentation process resulted in the aggregation of pixels sharing similar characteristics for reflectance and elevation. The object primitives were classified according to an empirical, rule-based system aiming to identify treetops. The purpose of this classification was to combine the segments into units representative of tree crowns. The classification was based on a fuzzy logic classification system where membership functions apply thresholds and weights for each data layer. Elevation in the CHM was weighted five times more than each layer in the visible bands in order to strengthen the importance of elevation compared with the three color bands.

12.5.5 Derivation of the Suppressed Tree Storey

The possibility of characterizing suppressed trees is a relatively new research area. The original point clouds instead of DSMs or CHMs need to be used for characterization. Since some of the laser pulses will penetrate under the dominant tree layer, it may be possible to analyze multilayered stands. For example, Zimble et al. (2003) showed that laser-derived tree height variances could be used to distinguish between single-story and multistory classes. Maltamo et al. (2005) examined the existence and number of suppressed trees by analyzing the height distributions of reflected laser pulses. The histogram thresholding method of Lloyd was applied to the height distribution of laser hits in order to separate different tree storeys. The number and sizes of suppressed trees were predicted with estimated regression models. The results showed that multilayered stand structures can be recognized and quantified using quantiles of laser scanner height distribution data. On the other hand, the accuracy of the results is dependent on the density of the dominant tree layer. Persson et al. (2005) reported the number of additional points that could be extracted using waveform signal ranged between 18% and 57%, depending on the type of vegetation. They proposed that additional points can give a better description of the vertical structure of vegetation and can possibly improve tree species classification as it was done by Reitberger et al. (2006). It is also expected that the waveform-based techniques of the small-footprint ALS will develop significantly in the coming years.

12.6 Forest Change

12.6.1 Methods and Quality of Forest Growth

Tree growth consists of the elongation and thickening of roots, stems, and branches (Husch et al., 1982); growth causes trees to change in weight, volume, and shape (form). Usually only the growth of the tree stem is considered by using the growth characteristics of the tree, diameter, height, basal area, and volume. In most cases, volume growth is the most interesting characteristic and it has to be derived from the change observed in other characteristics. In practice, height and diameter growth of individual trees are determined in the field from repeated measurements of permanent sample plots and from increment core measurements (e.g., boring) (Husch et al., 1982).

Laser data based methods for forest growth are relatively simple in principle. The height growth can be determined by several means: from the difference in the height of individual trees determined from repeated measurements (Yu et al., 2003) (see Figure 12.10), from the height difference of repeated DSMs (Hirata, 2005), from repeated height histograms (Næsset and Gobakken, 2005), or from the difference of the volumes of individual trees (Hyyppä, 2007). The changes in forests that affect the laser scanning response include the vertical and horizontal growth of crowns, the seasonal change of needle and leaf masses, the state of undergrowth and low vegetation, and the trees moving with the wind (especially for taller trees). Thus, the monitoring of growth using ALS can be relatively complicated in practice. The technique applied should be able to separate growth from other changes in the forest, especially those due to selective thinning or naturally fallen trees. The difference between DSMs is assumed to work in areas with wide and flat-topped crowns. In coniferous forests with narrow crowns, the planimetric displacement between two acquisitions can be substantial. Height histograms can be applied to point clouds corresponding to individual trees or plots or stands, but the information contents of histograms are corrupted if, e.g., thinning has occurred or the parameters of laser surveys are significantly different, thus change detection based on height histograms does not work in practice.

Yu et al. (2003, 2004a) demonstrated the application of laser data to forest growth at plot and stand level using an object-based tree-to-tree matching algorithm and statistical analysis. St-Onge and Vepakomma (2004) concluded that sensor-dependent effects such as echo triggering are probably the most difficult to control in multitemporal laser surveys for growth analysis purposes. Due to rapid technological developments, it is very likely that different sensors will be used, especially over long-term intervals that are needed in forest inventories (e.g., 10 year time interval). Næsset and Gobakken (2005) concluded that over a 2-year period, the prediction accuracy for plotwise and

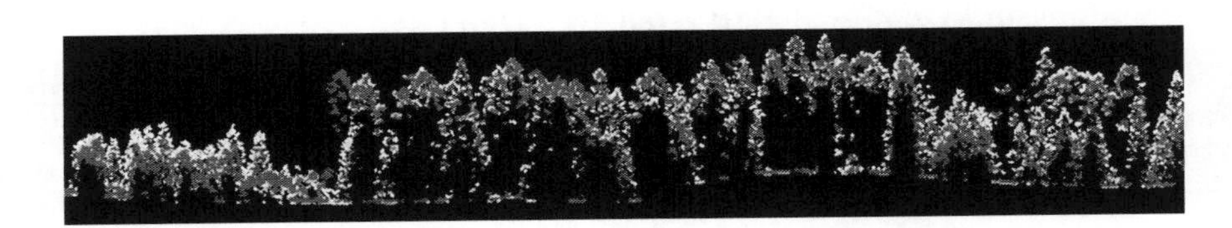

FIGURE 12.10
Profiles of 150 m sections. White, laser scanning of year 2003; dark gray, laser scanning of year 1998. Cutting tree was showed in 1998 data but not in 2003. Growth of young trees was clearly visible on the left part of the profiles.

standwise change in mean tree height, basal area, and volume was low when a point density of about 1 pt/m^2 and canopy height distribution technique were used. They also reported that certain height measurements, such as maximum height, seemed less suitable than many other height metrics because maximum height tends to be less stable—most probably due to low pulse density, narrow beam size, and relatively short growth period (2 years).

Yu et al. (2005) showed that height growth for individual trees can be measured with an accuracy better than 0.5 m using multitemporal laser surveys conducted in a boreal forest zone for a four-year time series and higher point density. In Yu et al. (2006), 82 sample trees were used to analyze the potential of measuring individual-tree growth of Scots pine in the boreal forest. Point clouds, having 10 pts/m^2 and illuminating 50% of the tree tops (i.e., the beams covering 50% of tree tops), were acquired in September 1998 and May 2003 with TopoSys 83 kHz LiDAR system. Three variables extracted from the point clouds representing each tree included the difference of the highest z value, difference between the DSMs of treetops, and difference of 85%, 90%, and 95% quantiles in the height histograms corresponding to a crown. An R^2 value of 0.68 and a standard deviation of 43 cm were derived with the best model. The results confirmed that it is possible to measure the growth of an individual tree with multitemporal laser surveys. They also demonstrated a better algorithm for tree-to-tree matching that is needed in operational individual-tree-based growth estimation in areas with narrow trees. The method is based on minimizing the sum of distances between treetops in an N-dimensional data space. The experiments showed that the location of trees (derived from laser data) and the height of the trees were together adequate to provide a reliable tree-to-tree matching. In future, the crown area should also be included in the matching as the fourth parameter.

In Yu et al. (2008), extended data set were used to estimate the tree mean height and volume growth at plot level in a boreal forest. Laser datasets were collected with a TopoSys laser scanner in 1998, 2000, and 2003 with a nominal point density of 10 points/m^2. Three techniques were used to predict the growth values based on individual-treetop differencing, DSM differencing, and canopy height distribution differencing. The regression models were developed for mean height growth and volume growth using single predictor derived from each method and using selected predictors from all methods. The best results were obtained for mean height growth (adjusted R^2 value of 0.86 and standard deviation of residual of 0.15 m) using the individual-treetop differencing method. The corresponding values for volume growth were 0.58 and 8.39 m^3 ha^{-1} (35.7%), respectively, using DSM differencing. The combined use of the three techniques yielded a better result for volume growth (adjusted R^2 = 0.75), but did not improve the estimation for mean height growth. In the treetop differencing methods, the most problematic part is to find pairs of treetops that represent the same tree (tree-to-tree matching) (Figure 12.11).

12.6.2 Methods and Quality of Harvested Tree Detection

Laser data can also be used for change detection. Yu et al. (2003, 2004a) also examined the applicability of airborne laser scanners in monitoring harvested trees (see Figures 12.12 and 12.13), using datasets with a point density of about 10 pts/m^2 over a 2-year period. The developed automatic method used for detecting harvested trees was based on image differencing. First, a difference image was calculated by subtracting the latter CHM (or DSM) from the former CHM (or DSM) after two datasets have been co-registered and sampled into raster images. The resulting difference image represented the pixel-wise changes between the two dates. Clustered high positive differences presented harvested

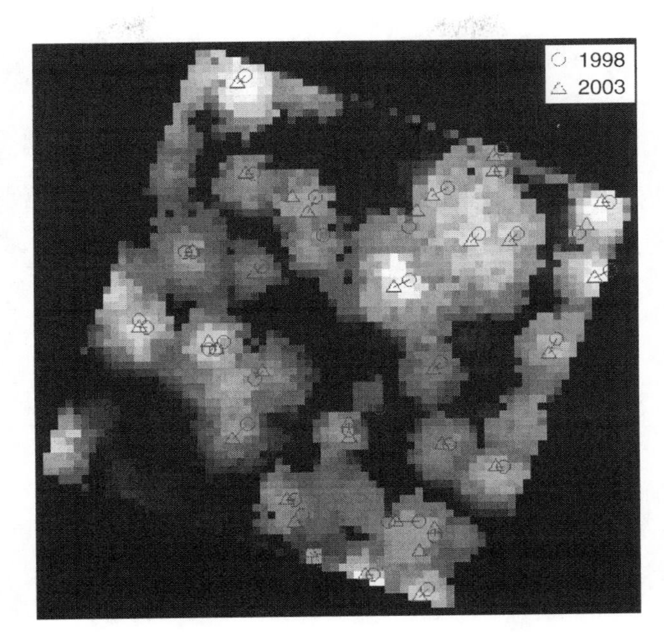

FIGURE 12.11
Result of tree matching using three variables (tree location-(x, y), tree height, H) for one coniferous tree dominated plot using laser scanner data acquired in 1998 and 2003. Background image is CHM of 2003. The location of the trees identified from the 1998 and 2003 data are marked with circles and triangles. Matched trees are linked by line.

trees. Most of the image values were close to zero (ground) or slightly below zero due to the tree growth. In order to identify harvested trees, a threshold was selected and applied to the different image for distinguishing no changes or small changes from big changes. A morphological opening was then performed to reduce noise-type fluctuation. The

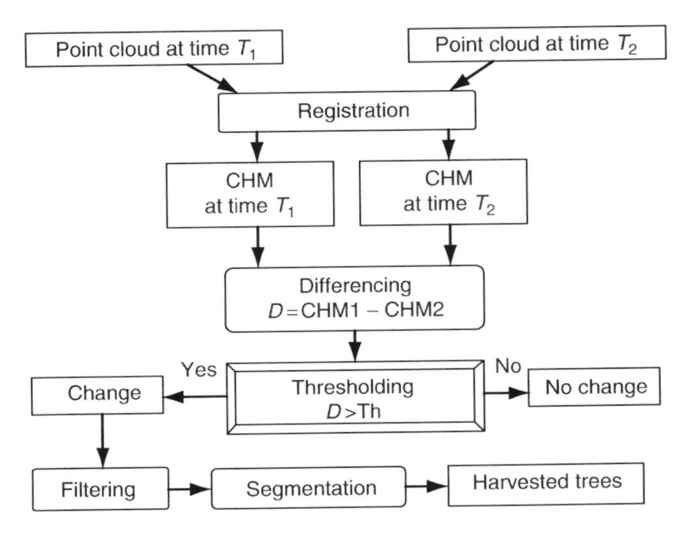

FIGURE 12.12
Flowchart for harvested tree detection.

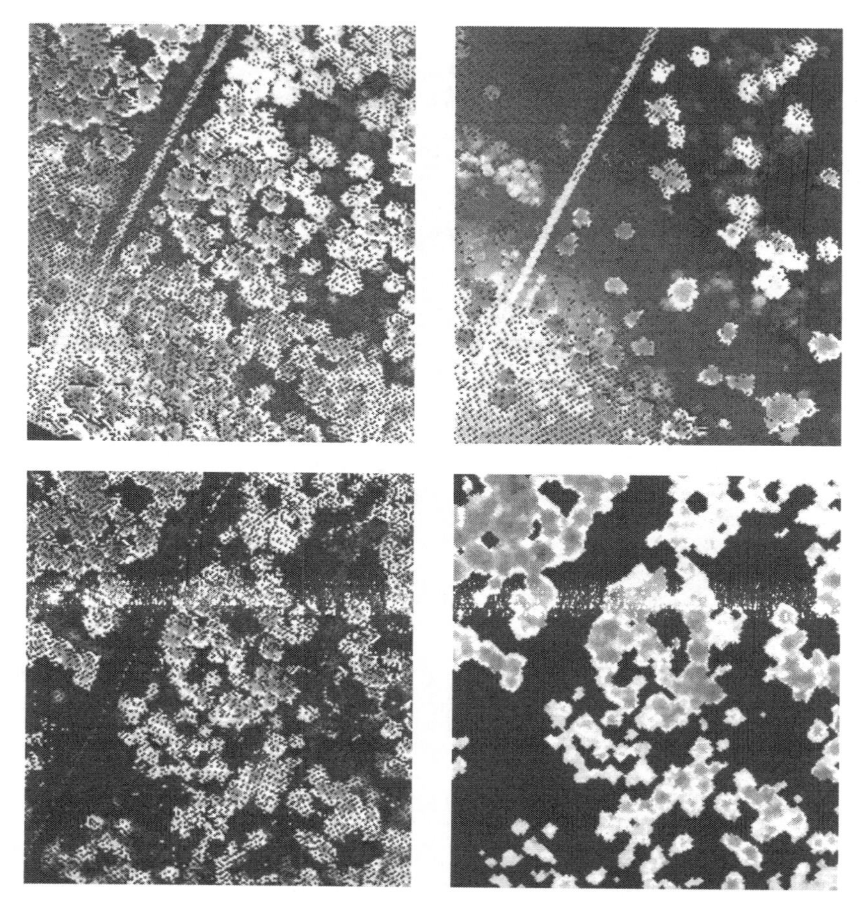

FIGURE 12.13 **(See color insert following page 334.)**
Change detection of harvested trees from two laser surveys, upleft: DSM from earlier year, upright: DSM from later year, lowleft: difference of DSMs, lowright: filtered difference image showing the harvested trees. Height was coded by colors.

location and the number of harvested trees was determined based on the segmentation of the resulting image. Out of 83 field-checked harvested trees, 61 were detected automatically and correctly. All the mature harvested trees were detected; it was mainly the smaller trees that were not detected, but were present.

12.7 EuroSDR/ISPRS Tree Extraction Comparison

The development is definitely towards individual-tree-based inventory, e.g., due to rapid development of ALS PRF (from few kilohertz to about 200 kHz in slightly more than 10 years) and saving due to better silvicultural treatments proposed with better data. In order to be able to compare various individual-tree-based solutions, an international comparison was established within the SilviLaser group and officially supported by

EuroSDR and ISPRS WG III. The objective of the joint EuroSDR and ISPRS Tree Extraction project was to evaluate the quality, accuracy, and feasibility of automatic or semi-automatic tree extraction methods based on high-density laser scanner data and digital image data. The sub-objectives included the following:

- How much variation is caused in tree extraction by the methods?
- How much does the point density of laser data affect tree extraction?
- How can the results be improved by integrating laser scanner data and aerial data?

The study site was located in Espoo close to the suburban area, about 15 km west of Helsinki. Because of the location, the study site was very diverse; the site was partly flat and partly steep terrain, with areas of mixed and more homogenous tree species at various stages of growth found in a small area. Main tree species were Scots pine, Norway spruce, and silver and downy birches. Laser scanner data with densities of 2, 4, and 8 pts/m^2 as well as Vexcel UltraCam-D images were provided to the participants. Digital images with GSD of 20 cm were given in CIR (3 channels, 8 bits/channel) and Color (4 channels, 16 bits/channel) formats. Camera calibration, image orientation, and ground control point information was also given together with image data. Laser data was collected with Optech ALTM 2033 having four repeated strips giving a point density of 2, 4 (two strips combined) and 8 (four strips combined) pts/m^2. Both first and last pulse data were given (first = first of many + only pulse; last = last of many + only pulse). A DTM with 0.5 m grid spacing obtained with the laser data using TerraScan was given to avoid the variability of DTM filtering techniques to affect the results of the test. Training data set of about 70 trees included species, location, dbh, and crown delineation given with 3–5 points.

Reference data was collected with a total station and terrestrial laser scanner Faro 880HE80 (max range 70 m, measurement rate 120 kHz, beam divergence 0.2 mrad, field of view 320° × 360°) measurements for six test plots. A ground control point (GCP) network was created at the test plots using RTK-GPS measurements on open areas and a total station was used to densify the GCP-network to tree-covered areas. Later these GCPs were used for the total station setup, and the total station was used to measure coordinates for the spherical reference targets, which were used for georeferencing of terrestrial laser scanner point clouds. The total station was also used for measuring the heights and crown outlines of selected trees for verification purposes. The results are based on 48 terrestrial laser scannings. The processing of point clouds into 3D models included georeferencing of point clouds, transferring point clouds to meshes, and editing. Characteristics of each tree, such as height, location, and species were interactively measured.

The participant list included Definiens AG, Germany; FOI, Sweden; Pacific Forestry Centre, Canada; University of Hannover, Germany; Joanneum Research, Austria; University of Joensuu/Finnish Forest Research Institute, National Ilan University, Taiwan; Texas A&M University; University of Zürich, Switzerland; Progea Consulting, Poland; and Universita' di Udine, Italy. Progea Consulting used only aerial images, Pacific Forestry Centre and Joanneum Research used aerial images and laser data, and other participants used only laser data.

The results included in this chapter are the quality of tree location, tree height, and percentage of detected trees. Only the methods based on laser scanning are included. The reference of detected trees was formed by taking those trees found by at least one partner.

Tree location accuracy was estimated by taking the distance measured from every reference tree to the nearest tree found in the extracted model. The maximum distance

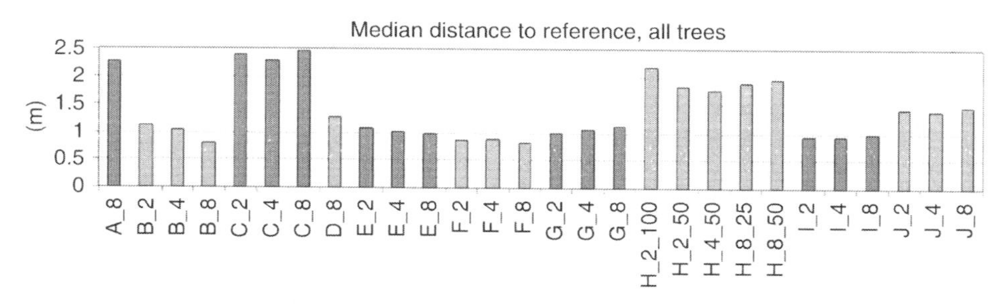

FIGURE 12.14
Tree location accuracy. The providers are marked with letters, the number shows the point density per m².

was set to 5 m. In all tree analysis, 357 reference trees were included of which 106 reference trees were more than 20 m in height. The results (Figures 12.14 through 12.17) clearly showed that the variability of tree location is small as a function of point density and it mainly changes as a function of the provider. Obviously, the calibration of the models has not been successful and several models assumed the trees to be significantly larger in width. With the best models for all the trees, the mean location error was less than 1 m and the difference with 2, 4, and 8 pts/m^2 was negligible. With trees over 20 m, accuracy of tree location of 0.5 m was obtained. Tree height quality analysis using selected 70 reference trees, (the reference height was known with an accuracy of 10 cm), showed again that the variability of the point density was negligible compared to method variability. With best models, RMSE of 50–80 cm was obtained for tree height. Even the 2 pts/m^2 seemed to be feasible for individual-tree detection. The percentage of the trees found by partners showed that the best algorithms found 90% of those trees that were found at least by one of the partners. There was again higher variation with the method used rather than point density.

The results of the test showed that the methods of individual-tree detection vary significantly and that the method itself is more significant for individual-tree-based inventories rather than the applied point density. The results confirm the early findings of Hyyppä et al. (2001b) in which dense point clouds were sparsified and individual-tree-based approach was implemented to laser data having point density of 1, 4, and more than 10 pulses per m^2. No significant variation with respect to the standard deviation of the stem volume estimates was obtained even though a noticeable increase in bias with lower pulse densities was observed.

Based on the final results, a final report will be published as EuroSDR reports.

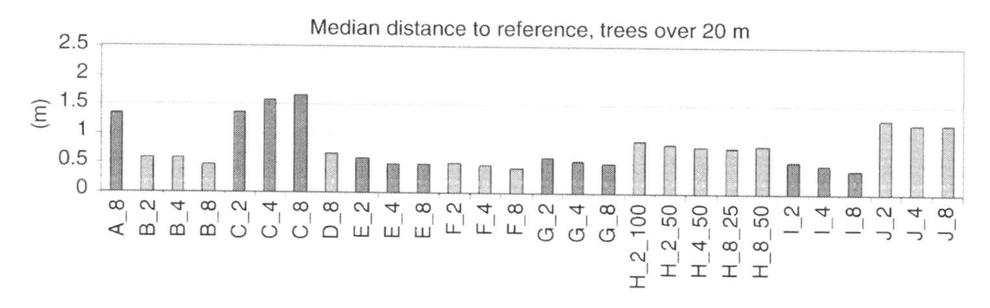

FIGURE 12.15
Tree location accuracy with trees over 20 m. The providers are marked with letters, the number shows the point density per m².

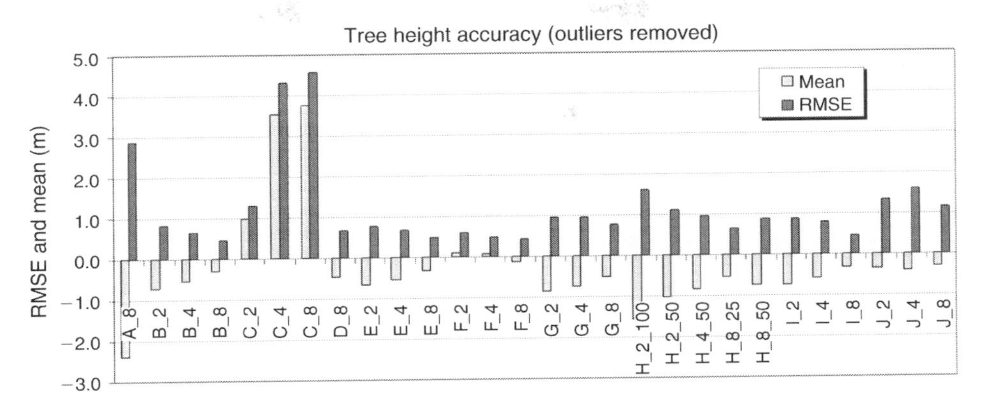

FIGURE 12.16
Tree height accuracy. The providers are marked with letters, the number shows the point density per m^2.

12.8 Outlook

The quality of the CHM has proved to be high. With relatively dense point clouds, the height values given for individual trees are about as accurate as conventional hypsometric measurements. The cause or causes of the underestimation of tree height needs to be further studied. This is not simple as a number of factors need to be isolated and considered including the amount and height of ground vegetation; the algorithm used to calculate the DTM; the power and sensitivity of the laser system; the signal processing and thresholding algorithms used to trigger the returns; and the quantity, geometry, and type of vegetation it takes to trigger a first return and pulse penetration into trees. This should be understood in general and for each sensor type and signal processing system. It is proposed that a cost-effective method of calibrating the underestimation of the laser-derived individual-tree height needs to be developed.

The methods for individual-tree isolation using laser scanner data are still under development and more empirical studies on the quality of the approaches are needed.

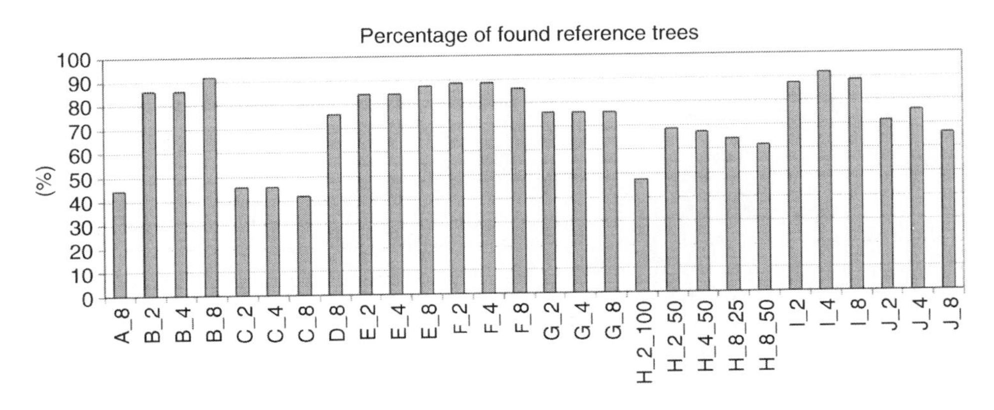

FIGURE 12.17
Percentage of found trees (relative to trees visible with laser scanning). The providers are marked with letters, the number shows the point density per m^2.

The first findings of the ISPRS/EuroSDR Tree Extraction comparison coordinated by the Finnish Geodetic Institute show that individual-tree-detection based on laser data seems to be more complicated than previously thought. On the basis of that experience, it is thought that it will be possible to improve the retrieval of individual-tree characteristics, especially the detection of the tree locations and segmentation of crown outlines. Preliminary results indicate that forest conditions play an important role in the determination of the accuracy of the methods, and thus, the accuracy obtained with each method could not be anticipated from literature surveys. The Finnish Geodetic Institute will continue the comparison and we are also providing companies possibilities of testing the quality of their retrieval systems using the experience gained in the ISPRS/EuroSDR test. The more accurate quantity and quality information of individual trees of the stands can be used as a base for felling and in transporting round timber from forests directly to the right manufacturing factories according to the demand for raw materials. One important benefit from the improved accuracy of forest data is the ability to better plan the forest operations as well as the supply chain. As these activities constitute a significant part of the cost of raw material for the industry, it is of vital importance to control these costs effectively.

There are several ways to improve growth estimates. In studies done by Yu et al. (2008), volume change was mainly predicted from height and height change information. The use of improved individual-tree-based volume techniques will also lead to an improvement in volume growth estimation. Seamless integration with optical data is important due to possible savings in data acquisition and increased automation in the processing. LiDAR-assisted individual-tree-based forest inventory is now starting its first commercial projects in Scandinavia. Several companies are establishing production chains based on individual trees.

Acknowledgments

We gratefully acknowledge the financial support from the Academy of Finland (Novel Aspects in Airborne Laser Scanning in Forests, 2007, Processing and use of 3D/4D information, 2004–2009). We acknowledge the assistance and comments from F. Gougeon, D. Leckie and M. Maltamo and the Finnish companies Blom Kartta Oy and UbiMap Oy.

References

Ahokas, E., Kaartinen, H., Matikainen, L., Hyyppä, J., and Hyyppä, H., 2002, Accuracy of high-pulse-rate laser scanners for digital target models. In observing our environment from space. New solutions for a new millennium, *Proceedings of the 21st EARSeL Symposium*, May 14–16, 2001, Paris: Balkema, pp. 175–178.

Ahokas, E., Yu, X., Oksanen, J., Hyyppä, J., Kaartinen, H., and Hyyppä, H., 2005, Optimization of the scanning angle for countrywide laser scanning. *Proceedings of ISPRS Workshop Laser scanning*, September 12–14, 2005, Enschede, The Netherlands, In *International Archives of Photogrammetry, Remote Sensing and Spatial Information Sciences*, XXXVI, Part 3/W19, 115–119.

Aldred, A.H. and Bonnor, G.M., 1985, Application of airborne lasers to forest surveys. Information Report PI-X-51, Canadian Forestry Service, Petawawa National Forestry Institute, 62 pp.

Andersen, H-E, Reutebuch, S., and Schreuder, G., 2002, Bayesian object recognition for the analysis of complex forest scenes in airborne laser scanner data. In *International Archives of the Photogrammetry, Remote Sensing and Spatial Information Sciences*, 34, Part 3A, 35–41.

Axelsson, P., 1999, Processing of laser scanner data—algorithms and applications. *ISPRS Journal of Photogrammetry and Remote Sensing*, 54, 138–147.

Axelsson, P., 2000, DEM generation from laser scanner data using adaptive TIN models. In *International Archives of Photogrammetry and Remote Sensing*, 33, Part B4, 110–117.

Axelsson, P., 2001, Ground estimation of laser data using adaptive TIN models. In *Proceedings of OEEPE Workshop on Airborne Laserscanning and Interferometric SAR for Detailed Digital Elevation Models*, Royal Institute of Technology, Stockholm, Sweden. Publication No. 40. CD-ROM. pp. 185–208.

Baltsavias, E.P., 1999. Airborne laser scanning: Basic relations and formulas. *ISPRS Journal of Photogrammetry and Remote Sensing*, 54(2–3), 199–214.

Bernard, R., Vidal-Madjar, D., Baudin, F., and Laurent, G., 1987, Nadir looking airborne radar and possible applications to forestry. *IEEE Transctions on Geoscience and Remote Sensing*, 21, pp. 297–309.

Bohlin, J., Olsson, H., Olofsson, K., and Wallerman, J., 2006, Tree species discrimination by aid of template matching applied to digital air photos. EARSeL SIG Forestry. In *International Workshop 3D Remote Sensing in Forestry Proceedings*, Vienna, February 14–15, 2006, pp. 199–203.

Bortolot, Z.J., 2006, Using tree clusters to derive forest properties from small footprint lidar data. *Photogrammetric Engineering & Remote Sensing*, 72(12): 1389–1397.

Brandtberg, T., 1999, Automatic individual tree-based analysis of high spatial resolution remotely sensed data. PhD thesis, Acta Universitatis Agriculturae Sueciae, Silvestria 118, Swedish University of Agricultural Sciences, Uppsala, Sweden. 47 pp.

Brandtberg, T., 2002, Individual tree-based species classification in high spatial resolution aerial images of forests using fuzzy sets. *Fuzzy Sets and Systems*, 132(3), 371–387.

Brandtberg, T. and Walter, F., 1999, An algorithm for delineation of individual tree crowns in high spatial resolution aerial images using curved edge segments at multiple scales. In *Proceedings of the International Forum on Automated Interpretation of High Spatial Resolution Digital Imagery for Forestry*. D.A. Hill and D.G. Leckie, (Eds.). February 10–12, 1998, Natural Resources Canada, Canadian Forest Service, Pacific Forestry Centre, Victoria, BC, pp. 41–54.

Brandtberg, T., Warner, T., Landenberger, R., and McGraw, J., 2003, Detection and analysis of individual leaf-off tree crowns in small footprint, high sampling density lidar data from the eastern deciduous forest in North America. *Remote Sensing of Environment*, 85, 290–303.

Chasmer, L., Hopkinson, C., and Treitz, P., 2006, Investigating laser pulse penetration through a conifer canopy by integrating airborne and terrestrial lidar. *Canadian Journal of Remote Sensing*, 32(2), 116–125.

Culvenor, D.S., 2002, TIDA: An algorithm for the delineation of tree crowns in high spatial resolution digital imagery of Australian native forest. PhD Thesis, Melbourne, Australia, University of Melbourne.

Currie, D., Shaw, V., and Bercha, F., 1989, Integration of laser rangefinder and multispectral video data for forest measurements. In *Proceedings of IGARSS'89 Conference*, July 4, 10–14, Vancouver, Canada, pp. 2382–2384.

Dralle, K. and Rudemo, M., 1996, Stem number estimation by kernel smoothing of aerial photos. *Canadian Journal of Forest Research*, 26, 1228–1236.

Elmqvist, M., Jungert, E., Lantz, F., Persson, Å., and Söderman, U., 2001, Terrain modelling and analysis using laser scanner data. In *International Archives of Photogrammetry and Remote Sensing*, 34, 3/W4, 219–226.

Erikson, M., 2003, Segmentation of individual tree crowns in colour aerial photographs using region growing supported by fuzzy rules. *Canadian Journal of Forest Research*, 33, 1557–1563.

Falkowski, M.J., Smith, A.M.S., Hudak, A.T., Gessler, P.E., Vierling, L.A., and Crookston, N.L., 2006, Automated estimation of individual conifer tree height and crown diameter via two-dimensional spatial wavelet analysis of lidar data. *Canadian Journal of Remote Sensing*, 32(2), 153–161.

Friedlaender, H. and Koch, B., 2000, First experience in the application of laserscanner data for the assessment of vertical and horizontal forest structures. In *International Archives of Photogrammetry and Remote Sensing*, XXXIII, Part B7, ISPRS Congress, Amsterdam, July 2000, 693–700.

Gaveau, D. and Hill, R., 2003, Quantifying canopy height underestimation by laser pulse penetration in small-footprint airborne laser scanning data. *Canadian Journal of Remote Sensing*, 29, 650–657.

Gill, S.J., Biging, G.S., and Murphy, E.C., 2000, Modeling conifer tree crown radius and estimating canopy cover. *Forest Ecology and Management*, 126(3), 405–416.

Gobakken, T. and Næsset, E., 2005, Weibull and percentile models for lidar-based estimation of basal area distribution. *Scandinavian Journal of Forest Research*, 20, 490–502.

Goodwin, N.R., Coops, N.C., and Culvenor, D.S., 2006, Assessment of forest structure with airborne LiDAR and the effects of platform altitude. *Remote Sensing of Environment*, 103(2), 140–152.

Gougeon, F.A., 1995, A crown-following approach to the automatic delineation of individual tree crowns in high spatial resolution aerial images. *Canadian Journal of Remote Sensing*, 21(3), 274–284.

Gougeon, F.A. and Moore, T., 1989, Classification individuelle des arbres à partir d'images à haute résolution spatiale. *Télédétection et gestion des resources*, 6, 185–196.

Gougeon, F.A., St-Onge, B.A., Wulder, M., and Leckie, D., 2001, Synergy of airborne laser altimetry and digital videography for individual tree crown delineation. In *Proceedings of the 23rd Canadian Symposium on Remote Sensing/10e Congrès de l'Association québécoise de télédétection* (CD-ROM), August 21–24, 200, Sainte-Foy, Québec, Canada, 1.

Gougeon, F.A. and Leckie, D., 2003, Forest information extraction from high spatial resolution images using an individual tree crown approach. Information report BC-X-396, Natural Resources Canada, Canadian Forest Service, Pacific Forestry Centre, 26 pp.

Hall, D.B. and Bailey, R.L., 2001, Modeling and prediction of forest growth variables based on multilevel nonlinear mixed models. *Forest Science*, 47(3), 311–321.

Hall, S.A., Burke, I.C., Box, D.O., Kaufmann, M.R., and Stoker, J.M., 2005, Estimating stand structure using discrete-return lidar: An example from low density, fire prone ponderosa pine forests. *Forest Ecology and Management*, 208(1–3), 189–209.

Harding, D.J., Lefsky, M.A., Parke, G.G., and Blair, J.B., 2001. Laser altimeter canopy height profiles. Methods and validation for closed canopy, broadleaf forests. *Remote Sensing of Environment*, 76(3), 283–297.

Hirata, Y., 2005, Relationship between crown and growth of individual tree derived from multitemporal airborne laser scanner data. Silvilaser 2005, September 29–October 1, 2005, Virginia Tech, Blacksburg, VA, Power-point Presentation.

Hofton, M.A., Rocchio, L., Blair, J.B., and Dubayah, R., 2002. Validation of vegetation canopy Lidar sub-canopy topography measurements for a dense tropical forest. *Journal of Geodynamics*, 34 (3–4), 491–502.

Hollaus, M., 2006, Large scale applications of airborne laser scanning for a complex mountainous environment. Dr. Ing. Thesis, TU Vienna, November 2006, 127 pp.

Holmgren, J., 2003, Estimation of forest variables using airborne laser scanning. PhD Thesis, Acta Universitatis Agriculturae Sueciae, Silvestria 278, Swedish University of Agricultural Sciences, Umeå, Sweden.

Holmgren, J. and Persson, Å., 2004, Identifying species of individual trees using airborne laser scanning. *Remote Sensing of Environment*, 90, 415–423.

Holopainen, M. and Talvitie, T., 2006, Effects of data acquisition accuracy on timing of stand harvests and expected net present value. *Silva Fennica*, 40(3), 531–543.

Hopkinson, C., Chasmer, L., Lim, K., Treitz, P., and Creed, I., 2006, Towards a universal lidar canopy height indicator. *Canadian Journal of Remote Sensing*, 32(2), 139–152.

Hugershoff, R., 1939, Die Bildmessung unde ihre forstlichen Anwendungen. *Der Deutsche Forstwirt*, 50(21), 612–615.

Husch, B., Miller, C.I., and Beers, T.W., 1982, *Forest Mensuration*. 2nd ed., New York: John Wiley & Sons.

Hynynen, J., 1995, Modelling tree growth for managed stands. Ph.d Thesis, The Finnish Forest Research Institute.

Hyyppä, H., Yu, X., Hyyppä, J., Kaartinen, H., Honkavaara, E., and Rönnholm, P., 2005a, Factors affecting the quality of DTM generation in forested areas. In *Proceedings of ISPRS Workshop Laser scanning 2005*, September 12–14, 2005, Enschede, The Netherlands, (Netherlands: GITC bv), XXXVI, Part 3/W19, 85–90.

Hyyppä, H., 2007, Personal communication based on experience of UbiForest software.

Hyyppä, J., 1993, Development and feasibility of airborne ranging radar for forest assessment. PhD Thesis, Helsinki University of Technology, Laboratory of Space Technology, 112 pp.

Hyyppä, J. and Hallikainen, M., 1993, A helicopter-borne eight-channel ranging scatterometer for remote sensing, Part II: Forest inventory, *IEEE Transactions on Geoscience and Remote Sensing.*, 31(1), 170–179.

Hyyppä, J. and Inkinen, M., 1999, Detecting and estimating attributes for single trees using laser scanner. *The Photogrammetric Journal of Finland*, 16, 27–42.

Hyyppä, J., Pyysalo, U., Hyyppä, H., Haggren, H., and Ruppert, G., 2000, Accuracy of laser scanning for DTM generation in forested areas. In *Proceedings of SPIE*, 4035, 119–130.

Hyyppä, J., Kelle, O., Lehikoinen, M., and Inkinen, M., 2001a, A segmentation-based method to retrieve stem volume estimates from 3-dimensional tree height models produced by laser scanner. *IEEE Transactions of Geoscience and Remote Sensing*, 39, 969–975.

Hyyppä, J., Schardt, M., Haggrén, H., Koch, B., Lohr, U., Scherrer, H.U., Paananen, R., Luukkonen, H., Ziegler, M., Hyyppä, H., Pyysalo, U., Friedländer, H., Uuttera, J., Wagner, S., Inkinen, M., Wimmer, A., Kukko, A., Ahokas, E., and Karjalainen, M., 2001b, HIGH-SCAN: The first European-wide attempt to derive single-tree information from laserscanner data. *The Photogrammetric Journal of Finland*, 17, 58–68.

Hyyppä, J., Hyyppä, H., Maltamo, M., Yu, X., Ahokas, E., and Pyysalo, U., 2003a, Laser scanning of forest resources—some of the Finnish experience. In *Proceedings of the ScandLaser Scientific Workshop on Airborne Laser Scanning of Forests*, September 3–4, 2003, Umeå, Sweden, pp. 53–59.

Hyyppä, J., Yu, X., Rönnholm, P., Kaartinen, H., and Hyyppä, H., 2003b, Factors affecting object-oriented forest growth estimates obtained using laser scanning. *The Photogrammetric Journal of Finland*, 18, 16–31.

Hyyppä, J., Mielonen, T., Hyyppä, H., Maltamo, M., Yu, X., Honkavaara, E., and Kaartinen, H., 2005b, Using individual tree crown approach for forest volume extraction with aerial images and laser point clouds. In *Proceedings of ISPRS Workshop Laser Scanning 2005*, September 12–14, 2005, Enschede, The Netherlands, (Netherlands: GITC bv), XXXVI, Part 3/W19, 144–149.

Hyyppä, J., Yu, X., Hyyppä, H., and Maltamo, M., 2006, Methods of airborne laser scanning for forest information extraction. EARSeL SIG Forestry. In *International Workshop 3D Remote Sensing in Forestry Proceedings*, Vienna, February 14–15, 2006, pp. 63–78.

Hyyppä, J., Hyyppä, H., Leckie, D., Gougeon, F., Yu, X., and Maltamo, M., 2008, Review of methods of small-footprint airborne laser scanning for extracting forest inventory data in boreal forests. *International Journal of Remote Sensing*, 29(5), 1339–1366.

Hökkä, H. and Groot, A., 1999, An individual-tree basal area growth model for black spruce in second-growth peatland stands. *Canadian Journal of Forest Research*, 29, 621–629.

Kraus, K. and Pfeifer, N., 1998, Determination of terrain models in wooded areas with airborne laser scanner data. *ISPRS Journal of Photogrammetry and Remote Sensing*, 53, 193–203.

Korhonen, L., Korhonen, K.T., Rautiainen, M., and Stenberg, P., 2006, Estimation of forest canopy cover: A comparison of field measurement techniques. *Silva Fennica*, 40(4), 577–588.

Larsen, M. and Rudemo, M., 1998, Optimizing templates for finding trees in aerial photographs. *Pattern Recognition Letters*, 19(12), 1153–1162.

Leckie, D., Gougeon, F., Hill, D., Quinn, R., Armstrong, L., and Shreenan, R., 2003, Combined high-density lidar and multispectral imagery for individual tree crown analysis. *Canadian Journal of Remote Sensing*, 29(5), 633–649.

Lefsky, M., Cohen, W., Acker, S., Parker, G., Spies, T., and Harding, D., 1999, Lidar remote sensing of the canopy structure and biophysical properties of Douglas-fir western hemlock forests. *Remote Sensing of Environment*, 70, 339–361.

Lefsky, M., Cohen, W., Parker, G., and Harding, D., 2002, Lidar remote sensing for ecosystem studies. *Bioscience*, 52, 19–30.

Liang, X., Hyyppä, J., and Matikainen, L., 2007, Deciduos-coniferous tree classification using difference between first and last pulse laser signatures. ISPRS Workshop on Laser Scanning 2007 and Silvilaser 2007, September 12–14, 2007, In *International Archives of Photogrammetry, Remote Sensing and Spatial Information Sciences*, XXXVI (3/W52), 253–257.

Lim, K., Treitz, P., Baldwin, K., Morrison, I., and Green, J., 2002, Lidar remote sensing of biophysical properties of northern tolerant hardwood forests. *Canadian Journal of Remote Sensing*, 29(5), 658–678.

Lim, K., Treitz, P., Wulder, M., St.Onge, B., and Flood, M., 2003, LIDAR remote sensing of forest structure. *Progress in Physical Geography*, 27(1), 88–106.

Maclean, G. and Krabill, W., 1986, Gross-merchantable timber volume estimation using an airborne lidar system. *Canadian Journal of Remote Sensing*, 12, 7–18.

Magnussen, S., Eggermont, P., and LaRiccia, V.N., 1999, Recovering tree heights from airborne laser scanner data. *Forest Science*, 45, 407–422.

Maltamo, M., Mustonen, K., Hyyppä, J., Pitkänen, J., and Yu, X., 2004, The accuracy of estimating individual tree variables with airborne laser scanning in boreal nature reserve. *Canadian Journal of Forest Research*, 34, 1791–1801.

Maltamo, M., Packalén, P., Yu, X., Eerikäinen, K., Hyyppä, J., and Pitkänen, J., 2005, Identifying and quantifying structural characteristics of heterogeneous boreal forests using laser scanner data. *Forest Ecology and Management*, 216, 41–50.

Maltamo, M., Eerikäinen, K., Packalen, P., and Hyyppä, J., 2006a, Estimation of stem volume using laser scanning-based canopy height metrics. *Forestry*, 79, 217–229.

Maltamo, M., Malinen, J., Packalen, P., Suvanto, A., and Kangas, J., 2006b, Nonparametric estimation of stem volume using airborne laser scanning, aerial photography, and stand-register data. *Canadian Journal of Forest Research*, 36, 426–436.

Matala, J., 2005, Impacts of climate change on forest growth: A modelling approach with application to management. Ph.D dissertation, Faculty of Forestry, University of Joensuu, Finland.

Means, J., Acker, S., Harding, D., Blair, J., Lefsky, M., Cohen, W., Harmon, M., and McKee, A., 1999, Use of large-footprint scanning airborne lidar to estimate forest stand characteristics in the western cascades of Oregon. *Remote Sensing of Environment*, 67, 298–308.

Moffiet, T., Mengersen, K., Witte, C., King, R., and Denham, R., 2005, Airborne laser scanning: Exploratory data analysis indicates potential variables for classification of individual trees or forest stands according to species. *ISPRS Journal of Photogrammetry and Remote Sensing*, 59, 289–309.

Morsdorf, F., Meier, E., Allgöwer, B., and Nüesch, D., 2003, Clustering in airborne laser scanning raw data for segmentation of single trees. In *International Archives of the Photogrammetry, Remote Sensing and Spatial Information Sciences*, 34, Part 3/W13, 27–33.

Næsset, E., 1997a, Determination of mean tree height of forest stands using airborne laser scanner data. *ISPRS Journal of Photogrammetry and Remote Sensing*, 52, 49–56.

Næsset, E., 1997b, Estimating timber volume of forest stands using airborne laser scanner data. *Remote Sensing of Environment*, 61, 246–253.

Næsset, E., 2002, Predicting forest stand characteristics with airborne scanning laser using a practical two-stage procedure and field data. *Remote Sensing of Environment*, 80, 88–99.

Næsset, E., 2003, Laser scanning of forest resources—the Norwegian experience. In *Proceedings of the ScandLaser Scientific Workshop on Airborne Laser Scanning of Forests*, September 3–4, Umeå, Sweden, pp. 35–42.

Næsset, E., 2004, Effects of different flying altitudes on biophysical stand properties estimated from canopy height and density measured with a small-footprint airborne scanning laser. *Remote Sensing of Environment*, 91(2), 243–255.

Næsset, E. and Økland, T., 2002, Estimating tree height and tree crown properties using airborne scanning laser in a boreal nature reserve. *Remote Sensing of Environment*, 79, 105–115.

Næsset, E., Gobakken, T., Holmgren, J., Hyyppä, H., Hyyppä, J., Maltamo, M., Nilsson, M., Olsson, H., Persson, Å., and Söderman, U., 2004, Laser scanning of forest resources: The Nordic experience. *Scandinavian Journal of Forest Research*, 19(6), 482–499.

Næsset, E. and Gobakken, T., 2005, Estimating forest growth using canopy metrics derived from airborne laser scanner data. *Remote Sensing of Environment*, 96(3–4), 453–465.

Nelson, R., Krabill, W., and Maclean, G., 1984, Determining forest canopy characteristics using airborne laser data. *Remote Sensing of Environment*, 15, 201–212.

Nelson, R., Krabill, W., and Tonelli, J., 1988, Estimating forest biomass and volume using airborne laser data. *Remote Sensing of Environment*, 24, 247–267.

Nelson, R., Oderwald, R., and Gregoire, T., 1997, Separating the ground and airborne laser sampling phases to estimate tropical forest basal area, volume, and biomass. *Remote Sensing of Environment*, 60, 311–326.

Nilsson, M., 1990, Forest inventory using an airborne LIDAR system. In *Proceedings from SNS/IUFRO Workshop in Umeå*, Remote Sensing Laboratory, Report 4, February, 26–28, 1990, (Umeå: Swedish University of Agricultural Sciences), pp. 133–139.

Nilsson, M., 1996, Estimation of tree heights and stand volume using an airborne lidar system. *Remote Sensing of Environment*, 56, 1–7.

Nilsson, M., Brandtberg, T., Hagner, O., Holmgren, J., Persson, Å., Steinvall, O., Sterner, H., Söderman, U., and Olsson, H., 2003, Laser scanning of forest resources—the Swedish experience. In *Proceedings of the ScandLaser Scientific Workshop on Airborne Laser Scanning of Forests*, September 3–4, 2003, Umeå, Sweden, pp. 43–52.

Packalén, P. and Maltamo, M., 2006, Predicting the plot volume by tree species using airborne laser scanning and aerial photographs. *Forest Science*, 52(6), 611–622.

Päivinen, R., Nousiainen, M., and Korhonen, K., 1992, Puutunnusten mittaamisen luotettavuus. English summary: Accuracy of certain tree measurements. *Folia Forestalia*, 787, 18 pp.

Palojärvi, P., 2003, Integrated electronic and optoelectronic circuits and devices for pulsed time-of-flight laser rangefinding. PhD thesis, Department of Electrical and Information Engineering and Infotech Oulu, University of Oulu, 54 pp.

Persson, Å., Holmgren, J., and Söderman, U., 2002, Detecting and measuring individual trees using an airborne laser scanner. *Photogrammetric Engineering & Remote Sensing*, 68(9), 925–932.

Persson, Å., Söderman, U., Töpel, J., and Ahlberg, S., 2005, Visualization and analysis of full-waveform airborne laser scanner data. In *Proceedings of ISPRS Workshop Laser Scanning 2005*, September 12–14, 2005, Enschede, The Netherlands, In *International Archives of Photogrammetry, Remote Sensing and Spatial Information Sciences*, XXXVI, Part 3/W19, 103–108.

Persson, Å., Holmgren, J., and Söderman, U., 2006, Identification of tree species of individual trees by combining very high resolution laser data with multi-spectral images. EARSeL SIG Forestry. In *International Workshop 3D Remote Sensing in Forestry Proceedings*, Vienna, February 14–15, 2006, pp. 91–96.

Philip, M., 1983, *Measuring Trees and Forests*, The Division of Forestry, University of Dar es Salaam, Great Britain, Aberbeen University Press, 338 pp.

Pinz, A.J., 1991, A computer vision system for the recognition of trees in aerial photographs. *Proceedings of Multisource Data Integration in Remote Sensing.* In *NASA Conference Publication*, University of Maryland, College Park, Maryland, 3099. pp. 111–124.

Pitkänen, J., Maltamo, M., Hyyppä, J., and Yu, X., 2004, Adaptive methods for individual tree detection on airborne laser based height model. *International Conference NATSCAN 'Laser-Scanners for Forest and Landscape Assessment—Instruments, Processing Methods and Applications'*, October 3–6, 2004, Freiburg, Germany, In *International Archives of Photogrammetry, Remote Sensing and Spatial Information Sciences*, XXXVI(8/W2), 187–191.

Poage, N.J. and Tappeiner, J.C., 2002, Long-term patterns of diameter and basal area growth of old-growth Douglas-fir trees in western Oregon. *Canadian Journal of Forest Research*, 32, 1232–1243.

Pollock, R.J., 1994, A model-based approach to automatically locating individual tree crowns in high-resolution images of forest canopies. In *Proceedings of First International Airborne Remote Sensing Conference and Exhibition*. September 11–15, 1994, Strasbourg, France, III, pp. 357–369.

Popescu, S., Wynne, R., and Nelson, R., 2003, Measuring individual tree crown diameter with lidar and assessing its influence on estimating forest volume and biomass. *Canadian Journal of Remote Sensing*, 29(5), 564–577.

Pouliot, D.A., King, D.J., Bell, F.W., and Pitt, D.G., 2002, Automated tree crown detection and delineation in high-resolution digital camera imagery of coniferous forest regeneration. *Remote Sensing of Environment*, 82, 322–334.

Pyysalo, U., 2000, Generation of elevation models in wooded areas from a three dimensional point cloud measured by laser scanning. MSc Thesis, Helsinki University of Technology, Espoo, Finland, 68 pp.

Pyysalo, U. and Hyyppä, H., 2002, Reconstructing tree crowns from laser scanner data for feature extraction. *International Society for Photogrammetry and Remote Sensing - ISPRS Commission III Symposium (PCV'02)*. September 9–13, 2002, Graz, Austria.

Reitberger, J., Krzystek, P., and Heurich, M., 2006, Full-waveform analysis of small footprint airborne laser scanning data in the Bavarian forest national park for tree species classification. In *International Workshop 3D Remote Sensing in Forestry Proceedings*. Vienna, February 14–15, 2006, pp. 218–227.

Reutebuch, S., McGaughey, R., Andersen, H., and Carson, W., 2003, Accuracy of a high-resolution lidar terrain model under a conifer forest canopy. *Canadian Journal of Remote Sensing*, 29, 527–535.

Riaño, D., Meier, E., Allgöwer, B., Chuvieco, E., and Ustin, S., 2003, Modeling airborne laser scanning data for the spatial generation of critical forest parameters in fire behavior modelling. *Remote Sensing of Environment*, 86, 177–186.

Rönnholm, P., Hyyppä, J., Hyyppä, H., Haggrén, H., Yu, X., Pyysalo, U., Pöntinen, P., and Kaartinen, H., 2004, Calibration of laser-derived tree height estimates by means of photogrammetric techniques, *Scandinavian Journal of Forest Research*, 19(6), 524–528.

Schreier, H., Lougheed, J., Tucker, C., and Leckie, D., 1985, Automated measurements of terrain reflection and height variations using airborne infrared laser system. *International Journal of Remote Sensing*, 6(1), 101–113.

Sithole, G., 2001, Filtering of laser altimetry data using a slope adaptive filter. In *International Archives of Photogrammetry and Remote Sensing*, 34–3/W4, 203–210.

Sithole, G. and Vosselman, G., 2004, Experimental comparison of filter algorithms for bare-Earth extraction from airborne laser scanning point clouds. *ISPRS Journal of Photogrammetry and Remote Sensing*, 59, 85–101.

Sohlberg, S., Næsset, E., and Bollandsas, O., 2006, Single tree segmentation using airborne laser scanner data in a structurally heterogeneous spruce forest. *Photogrammetric Engineering and Remote Sensing*, 72(12), 1369–1378.

Solodukhin, V., Zukov, A., and Mazugin, I., 1977, Possibilities of laser aerial photography for forest profiling, *Lesnoe Khozyaisto* (Forest Management), 10, 53–58. (in Russian).

St-Onge, B., Jumelet, J., Cobelli, M., and Véga, C., 2004, Measuring individual tree height using a combination of stereophotogrammetry and lidar. *Canadian Journal of Forest Research*, 34(10), 2122–2130.

St-Onge, B. and Vepakomma, U., 2004, Assessing forest gap dynamics and growth using multitemporal laser scanner data. International Conference NATSCAN 'Laser-Scanners for Forest and Landscape Assessment - Instruments, Processing Methods and Applications', October 3–6, 2004, Freiburg, Germany. In *International Archives of Photogrammetry, Remote Sensing and Spatial Information Sciences*, XXXVI(8/W2), 173–178.

Straub, B., 2003, A top-down operator for the automatic extraction of trees - concept and performance evaluation. In *Proceedings of the ISPRS Working Group III/3 Workshop '3-D Reconstruction from Airborne Laserscanner and InSAR Data'*, October 8–10, 2003, Dresden, Germany, pp. 34–39.

Suarez, J.C., Ontiveros, C., Smith, S., and Snape, S., 2005, Use of airborne LiDAR and aerial photography in the estimation of individual tree heights in forestry. *Computers and Geosciences*, 31, 253–262.

Takeda, H., 2004, Ground surface estimation in dense forest. *The International Archives of Photogrammetry, Remote Sensing and Spatial Information Sciences*, 35, Part B3, 1016–1023.

Tiede, D. and Hoffman, C., 2006, Process oriented object-based algorithms for single tree detection using laser scanning. In *International Workshop 3D Remote Sensing in Forestry Proceedings*, Vienna, February 14–15, 2006.

TopoSys, 1996, *Digital Elevation Models, Services and Products*, 10 pp.

Uuttera, J., Haara, A., Tokola, T., and Maltamo, M., 1998, Determination of the spatial distribution of trees from digital aerial photographs. *Forest Ecology and Management*, 110, 275–282.

Uuttera, J., Hiltunen, J., Rissanen, P., Anttila, P., and Hyvönen, P., 2002. Uudet kuvioittaisen arvioinnin menetelmät – Arvio soveltuvuudesta yksityismaiden metsäsuunnitteluun. *Metsätieteen aikakauskirja*, 3, 523–531. (In Finnish)

Uzoh, F.C.C. and Oliver, W.W., 2006, Individual tree height increment model for managed even-aged stands of ponderosa pine throughout the western United States using linear mixed effects models. *Forest Ecology and Management*, 221, 147–154.

Villikka, M., Maltamo, M., Packalén, P., Vehmas, M., and Hyyppä, J., 2007, Alternatives for predicting tree-level stem volume of Norway spruce using airborne laser scanner data. *The Photogrammetric Journal of Finland*, 20(2), 33–42.

Vosselman, G., 2000, Slope based filtering of laser altimetry data. *International Archives of Photogrammetry and Remote Sensing*, 33, B3/2, 935–942.

Vosselman, G. and Maas, H.-G., 2001, Adjustment and filtering of raw laser altimetry data. *Proceedings of OEEPE Workshop on Airborne Laserscanning and Interferometric SAR for Detailed Digital Elevation Models*, Royal Institute of Technology, Stockholm, Sweden. Paper 5, 11 pp.

Wack, R. and Wimmer, A., 2002, Digital terrain models from airborne laser scanner data – a grid based approach. In *International Archives of Photogrammetry and Remote Sensing*, 35, Part 3B, 293–296.

Wagner, W., 2005, Physical principles of airborne laser scanning. University course: Laser scanning, data acquisition and modeling, October 6–7, 2005.

Wagner, W., Ullrich, A., Melzer, T., Briese, C., and Kraus, K., 2004, From single-pulse to full-waveform airborne laser scanners: Potential and practical challenges. In *International Archives of Photogrammetry, Remote Sensing and Spatial Information Sciences* 2004, 35, Part B3, 201–206.

Wulder, M., Niemann, K.O., and Goodenough, D., 2000, Local maximum filtering for the extraction of tree locations and basal area from high spatial resolution imagery. *Remote Sensing of Environment*, 73, 103–114.

Wulder, M., 2003, The current status of laser scanning of forests in Canada and Australia. In *Proceedings of the ScandLaser Scientific Workshop on Airborne Laser Scanning of Forests*, September 3–4, 2003, Umeå, Sweden, pp. 21–33.

Yu, X., Hyyppä, J., Rönnholm, P., Kaartinen, H., Maltamo, M., and Hyyppä, H., 2003, Detection of harvested trees and estimation of forest growth using laser scanning, In *Proceedings of the Scandlaser Scientific Workshop on Airborne Laser Scanning of Forests*, September 3–4, 2003, Umeå, Sweden, final ed. Swedish University of Agricultural Sciences, pp. 115–124.

Yu, X., Hyyppä, J., Kaartinen, H., and Maltamo, M., 2004a, Automatic detection of harvested trees and determination of forest growth using airborne laser scanning. *Remote Sensing of Environment*, 90, 451–462.

Yu, X., Hyyppä, J., Hyyppä, H., and Maltamo, M., 2004b, Effects of flight altitude on tree height estimation using airborne laser scanning. International Conference NATSCAN 'Laser-Scanners for Forest and Landscape Assessment—Instruments, Processing Methods and Applications', October 3–6, 2004, Freiburg, Germany, In *International Archives of Photogrammetry, Remote Sensing and Spatial Information Sciences*, XXXVI(8/W2), 96–101.

Yu, X., Hyyppä, J., Kaartinen, H., Hyyppä, H., Maltamo, M., and Rönnholm, P., 2005, Measuring the growth of individual trees using multitemporal airborne laser scanning point clouds. In *Proceedings of ISPRS Workshop Laser Scanning 2005*, September 12–14, 2005, Enschede, Netherlands, (Netherlands: GITC bv), XXXVI Part 3/W19, 204–208.

Yu, X., Hyyppä, J., Kukko, A., Maltamo, M., and Kaartinen, H., 2006, Change detection techniques for canopy height growth measurements using airborne laser scanner data. *Photogrammetric Engineering and Remote Sensing*, 72(12), 1339–1348.

Yu, X., Hyyppä, J., Kaartinen, H., Maltamo, M., and Hyyppä, H., 2008, Obtaining plotwise mean height and volume growth in boreal forests using multitemporal laser surveys and various change detection techniques. *International Journal of Remote Sensing*, 29(5), 1367–1386.

Ziegler, M., Konrad, H., Hofrichter, J., Wimmer, A., Ruppert, G., Schardt, M., and Hyyppä, J., 2000, Assessment of forest attributes and single-tree segmentation by means of laser scanning. *Laser Radar Technology and Applications V*, 4035, 73–84.

Zimble, D.A., Evans, D.L., Carlson, G.C., Parker, R.C., Grado, S.C., and Gerard, P.D., 2003, Characterizing vertical forest structure using small-footprint airborne LiDAR. *Remote Sensing of Environment*, 87(2–3), 171–182.

13

Integration of LiDAR and Photogrammetric Data: Triangulation and Orthorectification

Ayman Habib

CONTENTS

13.1 Introduction

The steady evolution of mapping technology is leading to an increasing availability of multisensor geo-spatial datasets at a reasonable cost. For decades, analog frame cameras have been the traditional source of mapping data. The development of softcopy photogrammetric workstations, together with the improved performance of charge-coupled device (CCD) and complementary metal-oxide-semiconductor (CMOS) arrays, is stimulating the direct incorporation of digital imaging systems in mapping activities. However, current single-head digital frame cameras are incapable of providing imagery while simultaneously maintaining the geometric resolution and ground coverage of analog sensors. In spite of this deficiency, the low cost of medium-format digital frame imaging systems has led to their frequent adoption by the mapping community, especially when combined with light detection and ranging (LiDAR) systems. To offset the limitations of single-head digital frame cameras, multihead frame cameras and line cameras are being utilized in space-borne imaging systems and some airborne platforms. On another front, the improved performance of global navigation satellite system (GNSS)/inertial navigation system (INS) technology is having a positive impact in reducing the control requirements for photogrammetric triangulation. In addition to this contribution, the increased accuracy of direct geo-referencing systems is leading to the widespread implementation of LiDAR systems for 3-D data collection. The complementary nature of the spectral and spatial data acquired by imaging and LiDAR systems is motivating their integration for a better description of the object space. However, such integration is only possible after accurate co-registration of the collected data to a common reference frame. This chapter introduces algorithms for a multiprimitive and multisensor triangulation environment, which is geared towards taking advantage of the complementary characteristics of spatial data available from the above sensors. The triangulation procedure ensures the alignment of the involved data to a common reference frame, as defined by the control utilized. Such alignment leads to the straightforward production of orthophotos. The main advantages of orthophotos include the possibility of using them as an inexpensive alternative to base maps and the capability of easily integrating them with other forms of geo-spatial data such as surface models and existing geographic information systems (GIS) databases. In addition, orthophotos can be draped on top of LiDAR surface models to produce realistic visualization of urban environments. Differential rectification, which is the standard procedure for orthophoto generation, leads to serious artifacts (ghost images/double mapping problems) when dealing with large scale imagery over urban area. To eliminate these artifacts, true orthophoto generation techniques should be adopted. In this regard, this chapter outlines several methodologies for true orthophoto generation using LiDAR and photogrammetric data collected over urban areas. The devised methodologies are tested and proven to be efficient through experiments with real multisensor data.

13.2 Synergistic Characteristics of LiDAR and Photogrammetric Data

A diverse range of spatial data acquisition systems are now available onboard satellite, aerial, and terrestrial mapping platforms. The diversity starts with analog and digital frame cameras and continues, to include linear array scanners. In the past few years, imaging sensors witnessed vast development as a result of enormous advancement in digital technology. For example, the increasing sensor size and storage capacity of digital frame cameras led to their application in traditional and new mapping functions. However, due to technical limitations, current single-head digital frame cameras are not capable of simultaneously providing geometric resolution and ground coverage similar to those associated with analog frame cameras. To alleviate this limitation, multi-head frame cameras and push-broom scanners (line cameras) have been developed and used onboard satellite and aerial platforms. Rigorous modeling of the imaging process used by push-broom scanners is far more complicated than the modeling of frame sensors (Ebner et al., 1994; Ebner et al., 1996; Habib and Beshah, 1998; Lee et al., 2000; Habib et al., 2001; Lee and Habib, 2002; Lee et al., 2002; Poli, 2004; Toutin, 2004a,b). For example, in the absence of geo-referencing parameters, the narrow Angular Field of View (AFOV) associated with current push-broom scanners is causing instability in the triangulation procedure due to high correlations among the exterior orientation parameters (EOP). Therefore, there has been a tremendous body of research dealing with the derivation of alternative models for line cameras, to circumvent such instability (El-Manadili and Novak, 1996; Gupta and Hartley, 1997; Wang, 1999; Dowman and Dolloff, 2000; Ono et al., 2000; Tao and Hu, 2001; Hanley et al., 2002; Fraser et al., 2002; Fraser and Hanley, 2003; Habib et al., 2004; Tao et al.,

TABLE 13.1

Photogrammetric Weaknesses, Contrasted with LiDAR Strengths

LiDAR Pros	Photogrammetric Cons
Dense information along homogeneous surfaces	Almost no positional information along homogeneous surfaces
Day or night data collection	Day time data collection only
Direct acquisition of 3D coordinates	Complicated and sometimes unreliable matching procedures
The vertical accuracy is better than the planimetric accuracy	The vertical accuracy is worse than the planimetric accuracy

2004). In addition to line cameras, the low cost of medium-format digital frame cameras is leading to their frequent adoption by the mapping community, especially when combined with LiDAR systems. Apart from imaging systems, LiDAR scanning is rapidly taking its place in the mapping industry as a fast and cost-effective 3-D data acquisition technology. The increased accuracy and affordability of GNSS/INS systems are the main reasons behind the expanding role of LiDAR sensors in mapping activities.

Considering the characteristics of the acquired spatial and spectral data by imaging and LiDAR systems, one can argue that their integration will be beneficial for an accurate and better description of the object space. As an illustration of the complementary characteristics of imaging and LiDAR systems, Tables 13.1 and 13.2 list the advantages and disadvantages of each system, contrasted with the corresponding cons and the pros of the other system. As can be seen in these tables, the disadvantages of one system can be compensated for by the advantages of the other system (Baltsavias, 1999; Satale and Kulkarni, 2003). However, the synergic characteristics of both systems can be fully utilized only after ensuring that both datasets are geo-referenced relative to the same reference frame (Habib and Schenk, 1999; Chen et al., 2004).

Traditionally, photogrammetric geo-referencing has been either indirectly established with the help of Ground Control Points (GCP) or directly defined using GNSS/INS units onboard the imaging platform (Cramer et al., 2000; Wegmann et al., 2004). On the other hand, LiDAR geo-referencing is directly established through the GNSS/INS components of the LiDAR system. In this regard, this chapter presents some alternative methodologies for the utilization of LiDAR features as a source of control for photogrammetric geo-referencing. These methodologies have two main advantages. First, they ensure the co-alignment of the LiDAR and photogrammetric data to a common reference frame as defined by the GNSS/INS unit of the LiDAR system. Moreover, LiDAR features eliminate the need for GCP to establish the geo-referencing parameters for the photogrammetric data. The possibility of utilizing

TABLE 13.2

LiDAR Weaknesses, Contrasted with Photogrammetric Strengths

Photogrammetric Pros	LiDAR Cons
High redundancy	No inherent redundancy
Rich in semantic information	Positional information; difficult to derive semantic information
Dense positional information along object space break lines	Almost no positional information along break lines
The planimetric accuracy is better than the vertical accuracy	The planimetric accuracy is worse than the vertical accuracy

LiDAR data as a source of control for photogrammetric geo-referencing is furnished by the accuracy of the LiDAR point cloud (e.g., the horizontal accuracy is usually in the range of few decimeters while the vertical accuracy is in the subdecimeter range). LiDAR data can be used for image geo-referencing if and only if we can identify common features in both datasets. Therefore, the first objective of the developed methodologies is to select appropriate primitives (i.e., point, linear, or areal features). Afterwards, the mathematical models, which can be utilized in the triangulation procedure, to relate the LiDAR and photogrammetric primitives will be introduced. Another objective of the proposed methodologies is that they should be flexible enough to allow the incorporation of primitives identified in scenes captured by frame and line cameras. In other words, the developed methodologies should handle multisensor data, while using a wide range of primitives (i.e., they are multiprimitive and multisensor triangulation methodologies). The outcome of the multisensor triangulation contributes to a straightforward incorporation of photogrammetric and LiDAR data for orthophoto generation. However, to ensure the generation of accurate true orthophotos when dealing with large scale imagery over urban areas, one must make use of additional criteria to detect portions of the LiDAR surface that are invisible in the involved imagery. Therefore, the triangulation discussion will be followed by an investigation of potential true orthophoto generation methodologies that use LiDAR and imagery acquired by frame and line cameras.

The chapter will start by offering a brief discussion of LiDAR and photogrammetric principles. The discussion of photogrammetric principles will focus on the possibility of incorporating frame and line cameras in a single triangulation mechanism. Then, the selection of primitives for relating the LiDAR and photogrammetric data, as well as the respective mathematical models for their incorporation in the triangulation procedure, will be outlined. Afterwards, alternative methodologies for true orthophoto generation will be introduced. Finally, the feasibility and the performance of the developed multiprimitive and multisensor triangulation methodologies will be outlined in the experimental results section (in which real data is used), together with some concluding remarks.

13.3 LiDAR Principles

LiDAR has been conceived to directly and accurately capture digital surfaces. Although 30 years old, the commercial market for LiDAR has developed significantly only within the last decade (Faruque, 2003). The affordability, increased pulse frequency, and versatility of new LiDAR systems are causing an exponential profusion of these systems in the mapping industry. A LiDAR system is composed of two main units; these are the laser ranging and GNSS/INS units, shown in Figure 13.1. Positional information derived from LiDAR systems is based on calculating the range from the laser unit to the object space. As shown in Figure 13.1, the measured range is coupled with the position and orientation information, as determined by the onboard GNSS/INS unit, to directly determine the position of the laser footprint, through a vector summation process, using Equation 13.1 (El-Sheimy et al., 2005). As can be seen in Equation 13.1, there is no inherent redundancy in the computation of the captured LiDAR surface. Therefore, the overall accuracy depends on the accuracy and calibration quality of the different components comprising the LiDAR system.

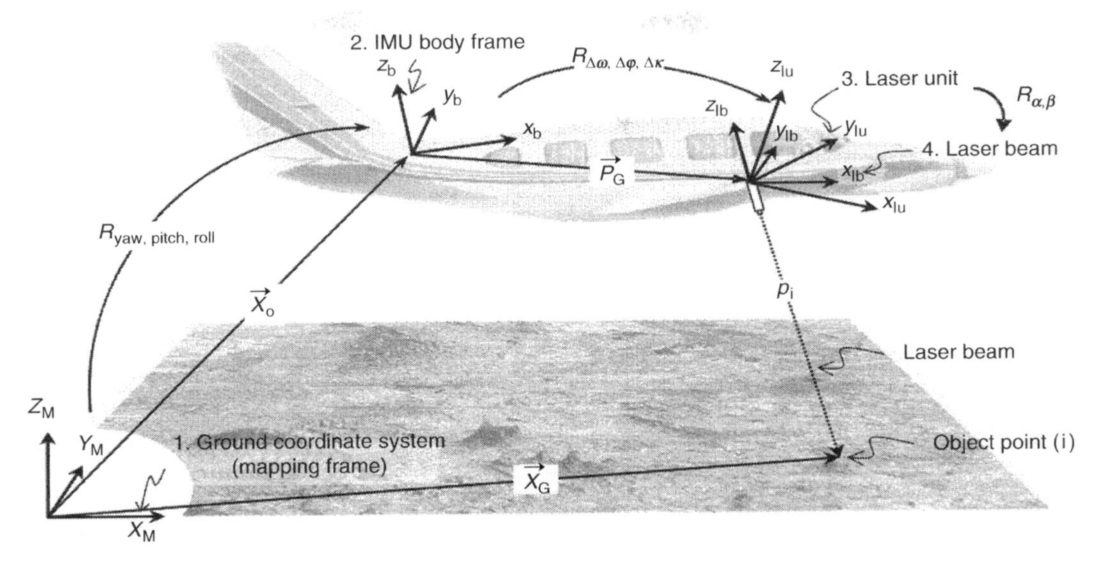

FIGURE 13.1
Coordinate systems and parameters involved in the direct geo-referencing of LiDAR systems.

$$\vec{X}_G = \vec{X}_o + R_{\text{yaw, pitch, roll}} \, R_{\Delta\omega,\, \Delta\phi,\, \Delta\kappa} \, \vec{P}_G + R_{\text{yaw, pitch, roll}} \, R_{\Delta\omega,\, \Delta\phi,\, \Delta\kappa} \, R_{\alpha,\, \beta} \begin{bmatrix} 0 \\ 0 \\ -\rho \end{bmatrix} \tag{13.1}$$

Besides the derived positional information, the intensity of the signal echo is also recorded by most LiDAR scanners, as shown in Figure 13.2. The intensity map can be used for object space segmentation and feature extraction. However, it is still difficult to derive semantic information regarding the captured surfaces (e.g., material and type of observed structures) from the intensity image (Wehr and Lohr, 1999; Baltsavias, 1999; Schenk, 1999). It is expected that the integration of photogrammetric and LiDAR data will improve the semantic content of the LiDAR data and will offer another check for the quality control of LiDAR surfaces.

FIGURE 13.2
Visualization of LiDAR coverage: Shaded-relief of range data (a) and intensity image (b).

13.4 Photogrammetric Principles

In this section, a brief overview of photogrammetric principles will be presented. Since photogrammetric operations using frame cameras are well established, the focus of this section will be on the utilization of line cameras in photogrammetric triangulation, while investigating the possibility of incorporating frame and line cameras in a single triangulation mechanism. The majority of imaging satellites make use of a line camera that has a single linear array in the focal plane, in contrast to the two-dimensional array of a frame camera, as shown in Figure 13.3. A single exposure of this linear array covers a narrow strip in the object space. Continuous coverage of contiguous areas on the ground is achieved by repetitive exposure of the linear array while observing a proper relationship between the sensor's integration time, ground sampling distance, and the velocity of the platform. In this regard, a distinction will be made between a scene and an image. An image is obtained through a single exposure of the light-sensitive elements in the focal plane. A scene, on the other hand, covers a two-dimensional area in the object space and might be composed of a single image or multiple images, depending on the nature of the camera utilized. Based on this distinction, a scene captured by a frame camera consists of a single image. A scene captured by a line camera, on the other hand, is composed of multiple images.

For frame and line cameras, the collinearity equations mathematically describe the constraint that the perspective center, the image point, and the corresponding object point are aligned along a straight line. For a line camera, the collinearity model can be described by Equation 13.2. It should be noted that the collinearity equations involve the image coordinates (x_i, y_i), which are equivalent to the scene coordinates (x_s, y_s) when dealing with a scene captured by a frame camera. For line cameras, however, the scene coordinates (x_s, y_s) need to be transformed into image coordinates. In this case, the x_s value is used to indicate the moment of exposure of the corresponding image. The y_s value, on the other hand, is directly related to the y_i image coordinate, as shown in Figure 13.4. It should be noted that the x_i image coordinate in Equation 13.2 is a constant that depends on the alignment of the linear array in the focal plane.

$$
\begin{aligned}
x_i &= x_p - c\frac{r_{11}^t(X_G - X_O^t) + r_{21}^t(Y_G - Y_O^t) + r_{31}^t(Z_G - Z_O^t)}{r_{13}^t(X_G - X_O^t) + r_{23}^t(Y_G - Y_O^t) + r_{33}^t(Z_G - Z_O^t)} \\
y_i &= y_p - c\frac{r_{12}^t(X_G - X_O^t) + r_{22}^t(Y_G - Y_O^t) + r_{32}^t(Z_G - Z_O^t)}{r_{13}^t(X_G - X_O^t) + r_{23}^t(Y_G - Y_O^t) + r_{33}^t(Z_G - Z_O^t)}
\end{aligned}
\tag{13.2}
$$

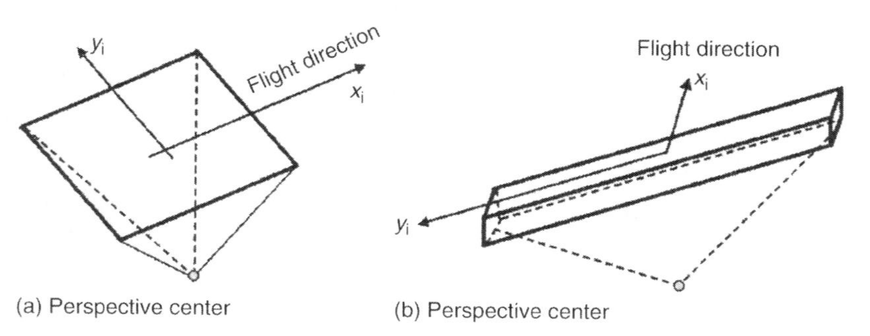

FIGURE 13.3
Imaging sensors for frame (a) and line (b) cameras.

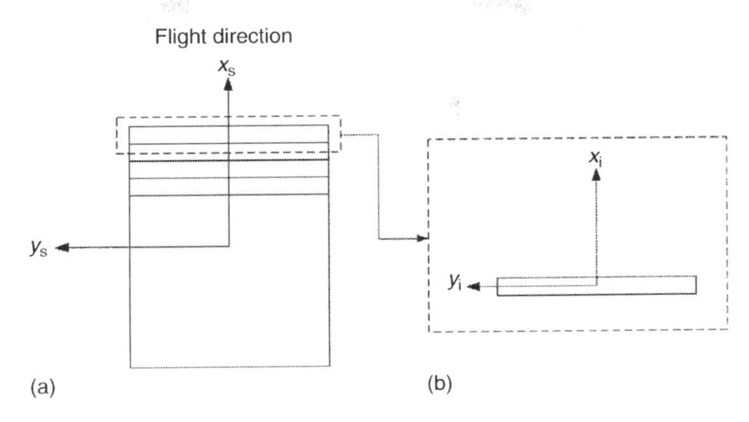

FIGURE 13.4
Scene (a) and image (b) coordinate systems for a line camera.

where
- (X_G, Y_G, Z_G) are the ground coordinates of the object point
- (X_O^t, Y_O^t, Z_O^t) are the ground coordinates of the perspective center at the moment of exposure
- r_{11}^t to r_{33}^t are the elements of the rotation matrix at the moment of exposure
- (x_i, y_i) are the image coordinates of the point in question
- (x_p, y_p, c) are the Interior Orientation Parameters (IOP) of the imaging sensor

Another difference between the collinearity equations for frame and line cameras is the multiple exposures associated with a line camera scene, in contrast to the single exposure for an image captured by a frame camera. Because of the multiple exposures, the EOP associated with a line camera scene are time dependent and vary depending on the image considered within the scene. For practical reasons, the bundle adjustment of scenes captured by line cameras does not consider all the EOP involved since this would lead to an excessive number of parameters to be solved, requiring an extensive amount of control. To avoid such a problem, bundle adjustments of scenes captured by line cameras make use of one of two parameter reduction approaches (Ebner et al., 1994, 1996). In the first approach, the system's trajectory and attitude are assumed to follow a polynomial trend, with time as the independent variable. Thus, the number of unknown EOP is reduced to the number of coefficients involved in the assumed polynomials. Available knowledge of the trajectory of the imaging system can be used as prior information regarding the polynomial coefficients, leading to what is known as the physical sensor model (Toutin, 2004a,b). Another approach to reducing the number of EOP involved is based on the use of orientation images that are equally spaced along the system's trajectory. The EOP at any given time are modeled as a weighted average of the EOP associated with the neighboring orientation images. In this way, the number of EOP for a given scene is reduced to the number of parameters associated with the orientation images involved.

It should be noted that the imaging geometry associated with line cameras (including the reduction methodology used for the involved EOP) is more general than that of frame cameras. In other words, the imaging geometry of a frame camera can be derived as a special case of that for a line camera. For example, an image captured by a frame camera can be considered to be a special case of a scene captured by a line camera, in which the trajectory and attitude are represented by a zero-order polynomial. Alternatively, when working with orientation images, a frame image can be considered to be a line camera scene with one orientation image. The general nature of the imaging geometry of line cameras lends itself to straightforward development of multisensor triangulation procedures capable of incorporating frame and line cameras (Habib and Beshah, 1998; Lee et al., 2002).

Having discussed LiDAR and photogrammetric principles, the focus will now be shifted towards algorithm development for the direct incorporation of LiDAR data in photogrammetric triangulation. To address this, the Section 13.5 will deal with the selection, representation, and extraction of appropriate primitives from LiDAR and photogrammetric data. This discussion will be followed by the derivation of the necessary mathematical models used to incorporate these primitives in the photogrammetric triangulation.

13.5 Triangulation Primitives

A triangulation process relies on the identification of common primitives to relate the involved datasets to the reference frame defined by the control information. Traditionally, photogrammetric triangulation has been based on point primitives. When considering photogrammetric and LiDAR data, relating a LiDAR footprint to the corresponding point in the imagery is almost impossible, as shown in Figure 13.5. Therefore, point primitives are not appropriate for the task at hand. Linear and areal features are other potential primitives that can be more suitable for relating LiDAR and photogrammetric data. Linear features can be manually identified in the imagery, while conjugate LiDAR lines can be extracted through planar patch segmentation and intersection. Alternatively, LiDAR lines can be directly identified in the intensity images produced by most of today's LiDAR systems. It should be noted that linear features extracted through planar patch segmentation and intersection are more accurate than features extracted from intensity images. The lower quality of features extracted from intensity images is caused by the interpolation process utilized to convert the irregular LiDAR footprints to a raster grid (Ghanma, 2006).

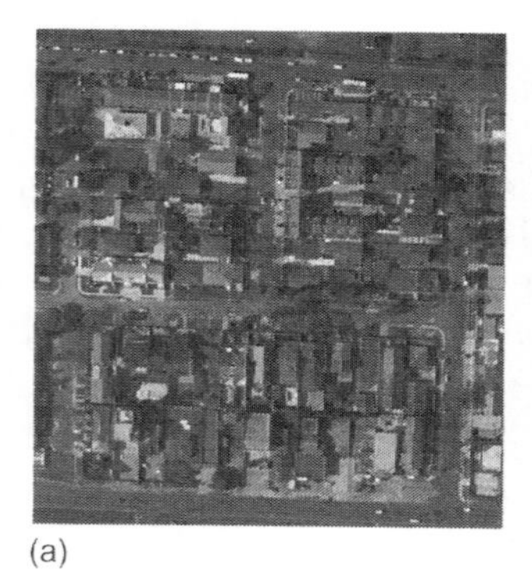
(a)

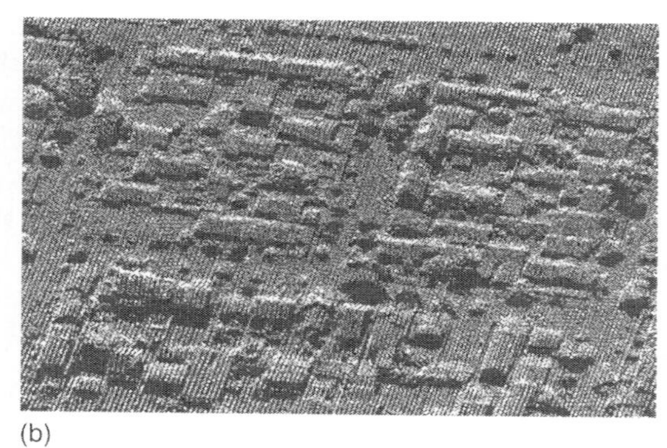
(b)

FIGURE 13.5
Imagery (a) and LiDAR (b) coverage of an urban area.

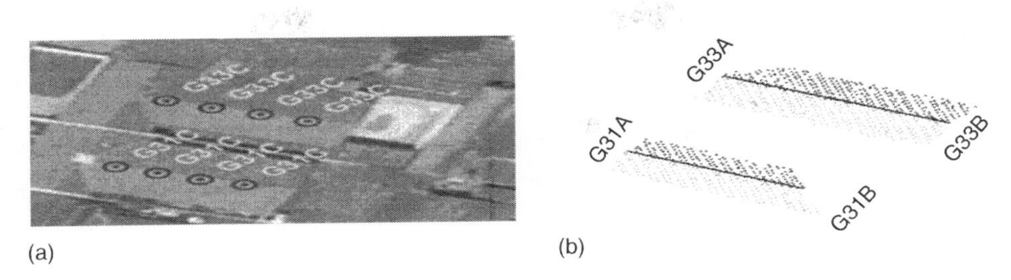

FIGURE 13.6
Image space linear features defined by sequences of intermediate points (a), while corresponding LiDAR lines are defined by their end points (b).

Instead of using linear features, areal primitives in photogrammetric datasets can be defined by their boundaries, which can be identified in the imagery. Such primitives include, for example, rooftops, lakes, and other homogeneous regions. In LiDAR datasets, areal regions can be derived through planar patch segmentation techniques.

Another issue that is related to the selection of primitives is their representation in both photogrammetric and LiDAR data. Image space lines can be represented by a sequence of image points along the feature, which can be manually identified, as shown in Figure 13.6a. This is an appealing representation since it can handle image space linear features in the presence of distortions, which cause deviations from straightness in the image space. Furthermore, such a representation allows the inclusion of linear features in scenes captured by line cameras, since perturbations in the flight trajectory also lead to deviations from straightness in image space linear features corresponding to object space straight lines (Habib et al., 2002). It should be noted that the intermediate points selected along corresponding line segments in overlapping scenes need not be conjugate. As for the LiDAR data, object space lines can be represented by their end points, as shown in Figure 13.6b. The points defining the LiDAR lines need not be visible in the imagery.

When using areal primitives, photogrammetric planar patches can be represented by three points (e.g., corner points, as in Figure 13.7a) along their boundaries. These points should be identified in all overlapping imagery. Similarly to the representation of linear features, this representation is valid for scenes captured by both frame and line cameras. LiDAR patches, on the other hand, can be represented by the footprints defining the patches, as shown in Figure 13.7b. These points can be derived directly through planar patch segmentation techniques. Having settled the issues of primitive selection and representation, the Section 13.6 will focus on explaining the proposed mathematical models to relate the photogrammetric and LiDAR primitives within the triangulation procedure.

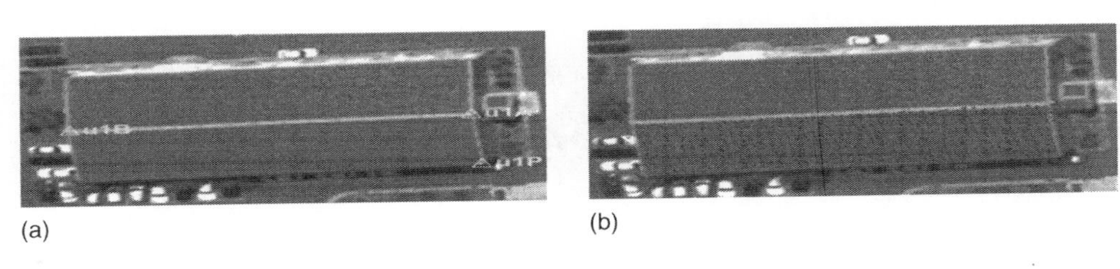

FIGURE 13.7
Image space planar features are represented by three points (a), while LiDAR patches are defined by the points comprising the patches (b).

13.6 Mathematical Models

13.6.1 Utilizing Straight Linear Primitives: Coplanarity-Based Constraint

This section will focus on deriving the mathematical constraint used to relate LiDAR and photogrammetric lines, which are represented by the end points in the object space and a sequence of intermediate points in the image space, respectively. From this perspective, the photogrammetric datasets will be aligned to the LiDAR reference frame through the direct incorporation of LiDAR lines as the source of control. The photogrammetric and LiDAR measurements along corresponding linear features can be related to one another through the coplanarity constraint of Equation 13.3. This constraint indicates that the vector from the perspective center to any intermediate point along the image line is contained within the plane defined by the perspective center of that image and the two points defining the LiDAR line. In other words, for a given intermediate point, k″ the points $\{(X_1, Y_1, Z_1), (X_2, Y_2, Z_2), (X_o'', Y_o'', Z_o''), \text{and } (x_{k''}, y_{k''}, 0)\}$ are coplanar, as illustrated in Figure 13.8.

$$(\vec{V}_1 \times \vec{V}_2) \cdot \vec{V}_3 = 0 \tag{13.3}$$

where

$\vec{V}_1$ is the vector connecting the perspective center to the first end point along the LiDAR line

$\vec{V}_2$ is the vector connecting the perspective center to the second end point along the LiDAR line

$\vec{V}_3$ is the vector connecting the perspective center to an intermediate point along the corresponding image line

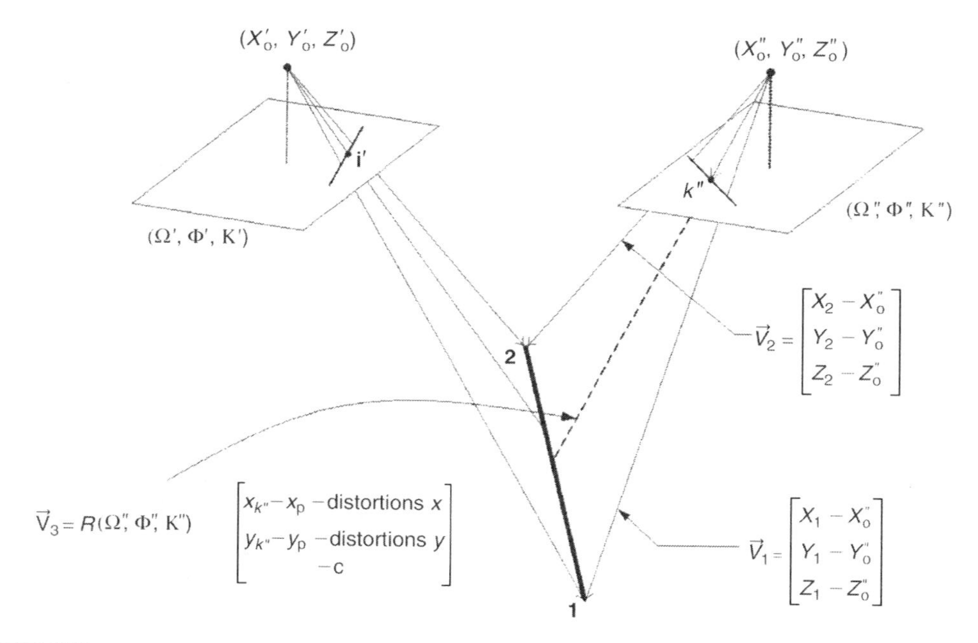

FIGURE 13.8
Perspective transformation between image and LiDAR straight lines, and the coplanarity constraint for intermediate points along the line.

It should be noted that the above constraint can be introduced for all the intermediate points along the image space linear feature. Moreover, the coplanarity constraint is valid for both frame and line cameras. For scenes captured by line cameras, the EOP involved should correspond to the image associated with the intermediate point under consideration. For frame cameras with known IOP, a maximum of two independent constraints can be defined for a given image. However, in self-calibration procedures, additional constraints aid in the recovery of the IOP, since the distortion pattern changes from one intermediate point to the next along the image space linear feature. The coplanarity constraint also aids in the recovery of the EOP associated with line cameras. This contribution is attributed to the fact that the system's trajectory affects the shape of the linear features in the image space.

For an image block, a minimum of two noncoplanar line segments are needed to establish the datum of the reconstructed object space, namely, the scale, rotation, and shift components. This requirement assumes that a model can be derived from the image block, and is explained by the fact that a single line defines two shift components across the line, as well as two rotation angles (as defined by the line heading and pitch). Having another noncoplanar line helps in the estimation of the remaining shift and rotation components, as well as the scale factor.

13.6.2 Utilizing Straight Linear Primitives: Collinearity-Based Constraint

Instead of the above approach, existing bundle adjustment procedures, which are based on the collinearity equations, can be used to incorporate linear features after the modification of the variance–covariance matrices associated with the line measurements in either the object space or the image space. Before getting into the implementation details of this approach, we will start by outlining the method of representing the linear features. Similarly to the previous approach, derived object space lines from the LiDAR data are represented by two points. Image space linear features, on the other hand, are represented by two points that are not conjugate to those defining the line in the object space, as illustrated in Figure 13.9. However, the image and object space points are assigned the same

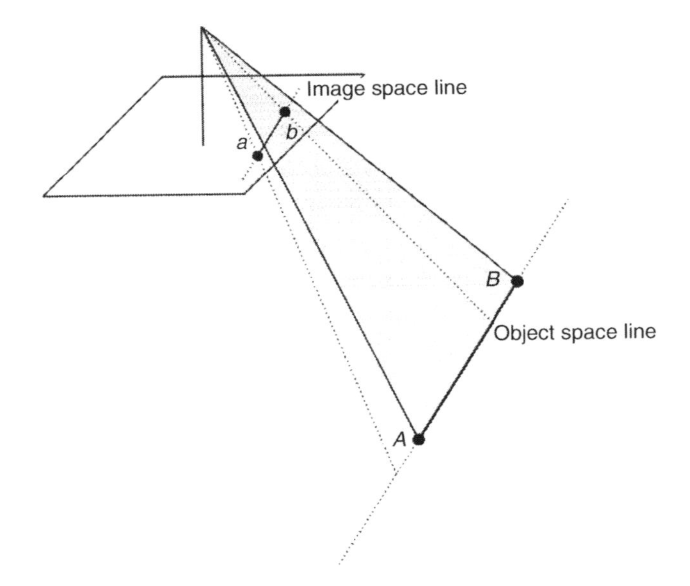

FIGURE 13.9
Representation of image and object lines for the collinearity-based incorporation of linear features in a bundle adjustment.

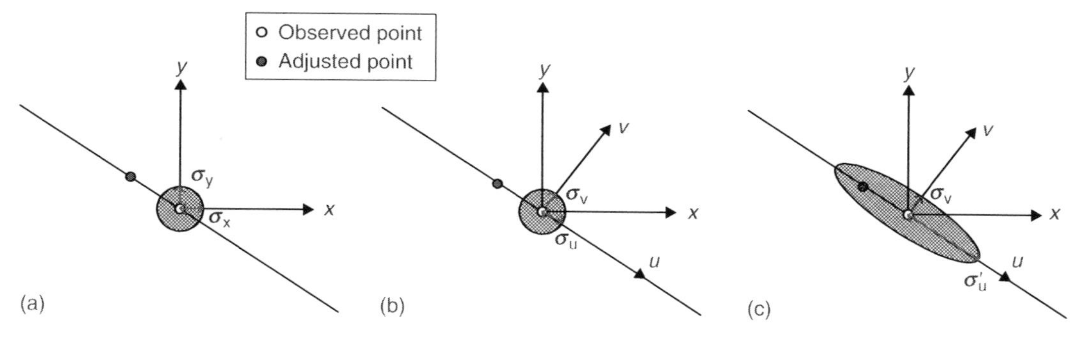

FIGURE 13.10
Steps involved in the expansion of the error ellipse associated with an end point of an image line. (a) Error ellipse in the (x,y) coordinate system, (b) error ellipse in the (u,v) coordinate system, and (c) error ellipse in the (u,v) coordinate system after expanding the error along the line orientation.

identification code. The fact that these points are not conjugate is tolerated by manipulating their variance–covariance matrices in either the image space or the object space. In other words, the variance–covariance matrix of a given point is expanded along the linear feature to allow for its movement along the direction of the line. For example, Equations 13.4 through 13.8 are used in the expansion of the error ellipse associated with an end point in the image space (refer to Figure 13.10 for a conceptual illustration). First, we need to define the xy- and uv-coordinate systems. The xy-coordinate system is the image coordinate system, shown in Figure 13.10a. The uv-coordinate system, on the other hand, is a coordinate system with the u-axis aligned along the image line, as illustrated in Figure 13.10b. The relationship between these coordinate systems is defined by a rotation matrix, in Equation 13.4, which, in turn, is defined by the image line orientation. Using the law of error propagation, the variance–covariance matrix of a line end point in the uv-coordinate system can be derived from its variance–covariance matrix in the xy-coordinate system, Equation 13.5, according to Equation 13.6. To allow for the movement of the point along the line, its variance along the u-axis is replaced by a large value (simply by multiplying the variance by a constant m), as in Equation 13.7 and shown in Figure 13.10c. Finally, the modified variance–covariance matrix in the xy-image coordinate system is calculated through the utilization of the inverse rotation matrix, as in Equation 13.8. The same procedure can be alternatively used to expand the variance–covariance matrices of the end points of the LiDAR lines.

$$\begin{bmatrix} u \\ v \end{bmatrix} = R \begin{bmatrix} x \\ y \end{bmatrix} \tag{13.4}$$

$$\sum\nolimits_{xy} = \begin{bmatrix} \sigma_x^2 & \sigma_{xy} \\ \sigma_{yx} & \sigma_y^2 \end{bmatrix} \tag{13.5}$$

$$\sum\nolimits_{uv} = R \sum\nolimits_{xy} R^{\mathrm{T}} = \begin{bmatrix} \sigma_u^2 & \sigma_{uv} \\ \sigma_{uv} & \sigma_v^2 \end{bmatrix} \tag{13.6}$$

$$\sum_{uv}' = \begin{bmatrix} m \times \sigma_u^2 & \sigma_{uv} \\ \sigma_{vu} & \sigma_v^2 \end{bmatrix} \tag{13.7}$$

$$\sum_{xy}' = R^{\mathrm{T}} \sum_{uv}' R \tag{13.8}$$

The decision of whether to expand the variance–covariance matrix in the image space or the object space depends on the procedure under consideration (e.g., single photo resection, bundle adjustment of an image block using control lines, or bundle adjustment of an image block using tie lines), as explained below.

1. Single photo resection using control lines (Figure 13.11a): In this approach, the image line is represented by two end points with their variance–covariance matrices defined by the expected image coordinate measurement accuracy. The variance–covariance matrices of the end points of the control line, on the other hand, are expanded to compensate for the fact that the image and the object end points are not conjugate. It should be noted that this approach can also be used in single photo resection of a scene captured by a line camera.
2. Single photo resection using control lines (Figure 13.11b): In this approach, the object line is represented by its end points, whose variance–covariance matrices are defined by the expected accuracy of the procedure used to define the points. The variance–covariance matrices of the image line end points, on the other hand, are expanded along the direction of the line. One should note that this approach is not appropriate for scenes captured by line cameras, since the image line orientation cannot be rigorously defined at a given point, due to perturbations in the flight trajectory.
3. Bundle adjustment of an image block using control lines (Figure 13.11c): In this approach, the object line is represented by its end points, with their variance–covariance matrices defined by the procedure used to provide the control lines. The variance–covariance matrices of the image lines, on the other hand, are expanded to compensate for the fact that the end points of the image lines are not conjugate to those defining the object line. It should be noted that this approach is not appropriate for scenes captured by line cameras since the image line orientation cannot be rigorously defined.
4. Bundle adjustment of an image block using control lines (Figure 13.11d): In this case, the image lines in overlapping images are represented by nonconjugate end points whose variance–covariance matrices are defined by the expected image coordinate measurement accuracy. To compensate for the fact that these points are not conjugate, the selected end points are assigned different identification codes. The object line, on the other hand, is defined by a list of points whose variance–covariance matrices are expanded along the line. The number of object points used depends on the number of selected points in the image space. It should be noted that this approach can be used for scenes captured by frame or line cameras since it does not require the expansion of the variance–covariance matrices in the image space.
5. Bundle adjustment of an image block using tie lines (Figure 13.11e): In this approach, all variance–covariance matrices associated with the end points of the image lines are expanded, except for two points, which are used to define the object line. Since this approach requires an expansion of the error ellipses in

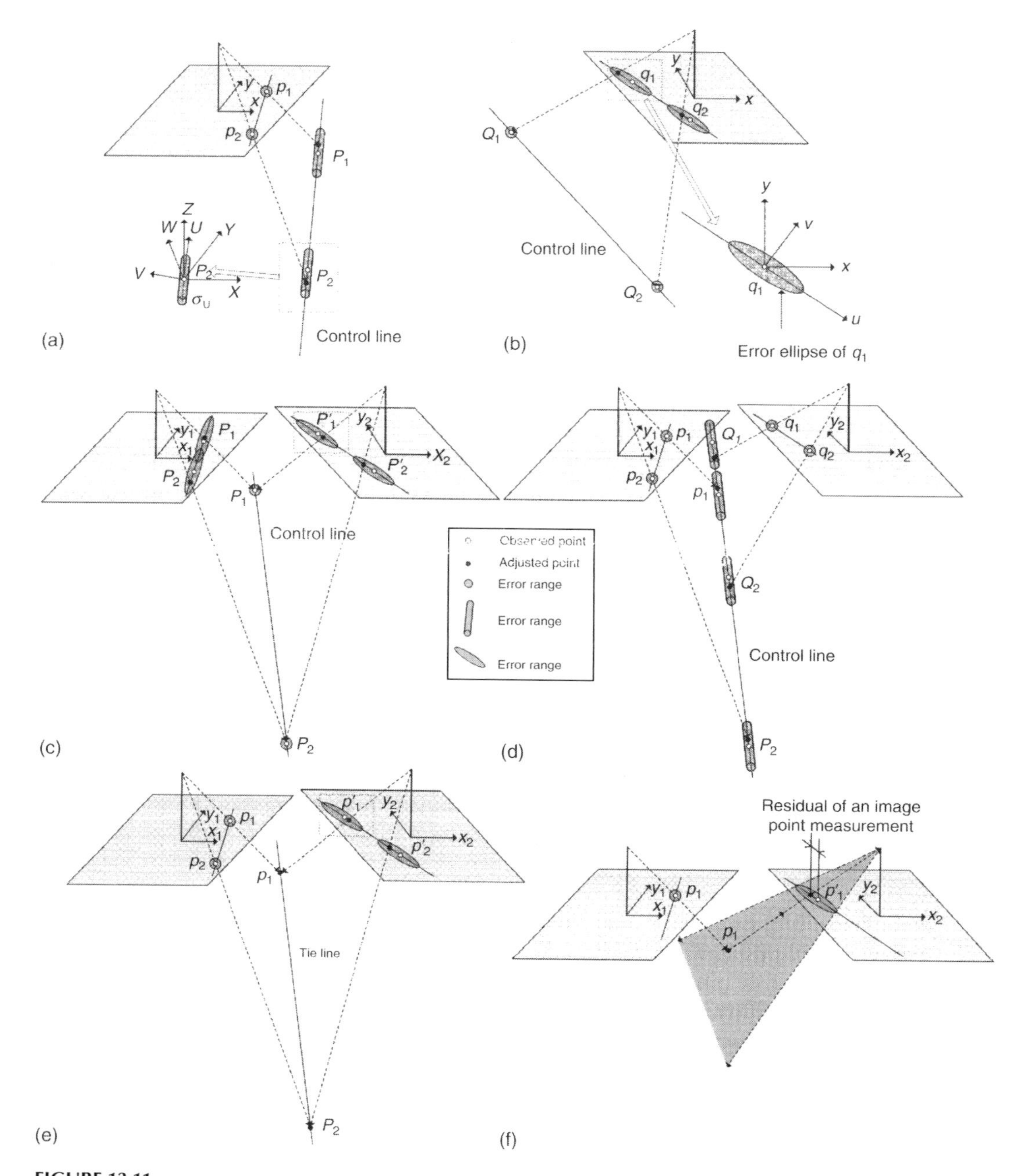

FIGURE 13.11
(a–f) Variance expansion options for line end points, for their incorporation in a collinearity-based bundle adjustment procedure.

the image space, it is not appropriate for scenes captured by line cameras, since the image line orientation cannot be rigorously defined due to perturbations in the flight trajectory. Figure 13.11f illustrates the conceptual basis of the derivation of the object point through the proposed variance–covariance manipulation. An image point used to define the object line, and its error ellipse, which is established by the image coordinate measurement accuracy, together with the corresponding perspective center, define an infinite light ray. An image point and its expanded error ellipse, on the other hand, together with the corresponding perspective center, define a plane. The intersection of the aforementioned light ray and the plane is used to estimate the ground coordinates of the point in question.

13.6.3 Utilizing Planar Patches: Coplanarity-Based Constraint

This section will focus on deriving the mathematical constraint used to relate LiDAR and photogrammetric patches, which are represented by a set of points in the object space and three points in the image space, respectively. As an example, let us consider a surface patch that is represented by two sets of points, namely the photogrammetric set $S_{PH} = \{A, B, C\}$ and the LiDAR set $S_L = \{(X_P, Y_P, Z_P), P = 1 \text{ to } n\}$, as shown in Figures 13.12 and 13.14a. Since the LiDAR points are randomly distributed, no point-to-point correspondence can be assumed between the datasets. For the photogrammetric points, the image and object space coordinates are related to one another through the collinearity equations. LiDAR points belonging to a certain planar patch, on the other hand, should coincide with the photogrammetric patch representing the same object space surface, as illustrated in Figure 13.12. The coplanarity of the LiDAR and photogrammetric points can be mathematically expressed through the determinant constraint in Equation 13.9.

$$V = \begin{vmatrix} X_P & Y_P & Z_P & 1 \\ X_A & Y_A & Z_A & 1 \\ X_B & Y_B & Z_B & 1 \\ X_C & Y_C & Z_C & 1 \end{vmatrix} = \begin{vmatrix} X_P - X_A & Y_P - Y_A & Z_P - Z_A \\ X_B - X_A & Y_B - Y_A & Z_B - Z_A \\ X_C - X_A & Y_C - Y_A & Z_C - Z_A \end{vmatrix} = 0 \tag{13.9}$$

The above constraint is used as the mathematical model to incorporate LiDAR points into the photogrammetric triangulation. In physical terms, this constraint describes the fact that the normal distance between any LiDAR point and the corresponding

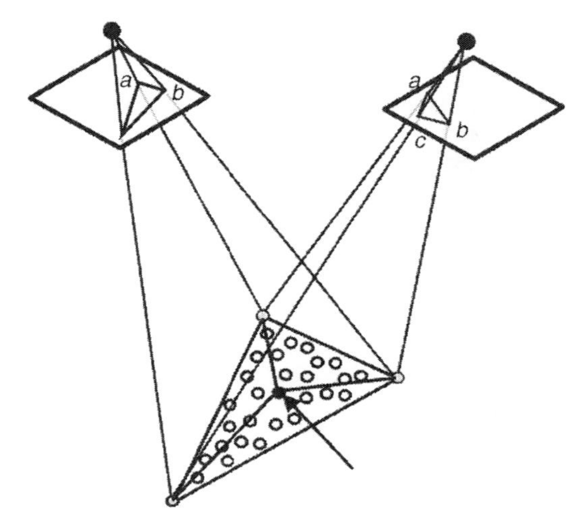

FIGURE 13.12
Coplanarity of photogrammetric and LiDAR patches.

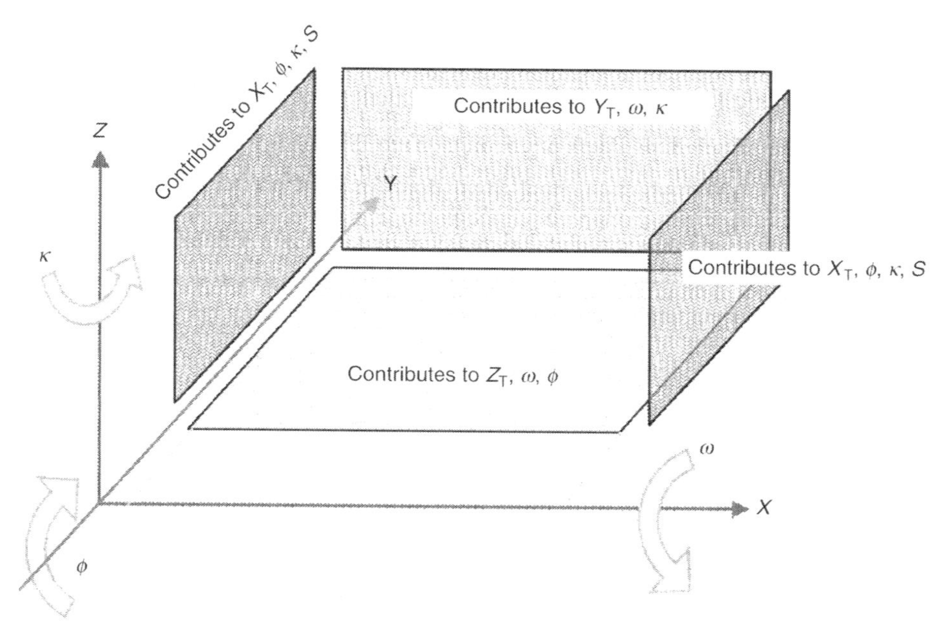

FIGURE 13.13
Optimal configuration for establishing the datum using planar patches as the source of control.

photogrammetric surface should be zero, or the volume of the tetrahedron composed of the four points should be zero. This constraint can be applied to all LiDAR points comprising a surface patch.

It should be noted that the above constraint is valid for both frame and line cameras. Another advantage of this approach is the possibility of using such a constraint for LiDAR system calibration. If the system model is available with explicit values of the systematic error terms, the raw LiDAR points (X_P, Y_P, Z_P) in Equation 13.9 can be replaced with the LiDAR measurements, as in Equation 13.1. In this case, the systematic error terms can be solved for in the bundle adjustment procedure, and photogrammetric planar patches serve as the source of control for the LiDAR calibration. Therefore, additional control information is needed to establish the geo-referencing parameters for the photogrammetric data, which, in turn, serve as the control for LiDAR calibration.

To be sufficient as the only source of control, LiDAR patches should be able to provide all the datum parameters; three translations (X_T, Y_T, Z_T), three rotations (ω, ϕ, κ), and one scale (S) parameter. By inspecting Figure 13.13, it is evident that a patch normal to one of the axes will provide the shift in the direction of that axis, as well as the rotation angles across the other axes. Therefore, three nonparallel patches are sufficient to determine the position and orientation components of the datum. To enable scale determination, the three planar patches should not intersect at a single point (e.g., they should not be facets of a pyramid). Alternatively, the scale can be determined by incorporating a fourth plane, as shown in Figure 13.13.

13.6.4 Utilizing Planar Patches: Collinearity-Based Constraint

In this approach, existing bundle adjustment procedures, which are based on the collinearity equations, can be used to incorporate areal control features after the manipulation of the variance–covariance matrices defining the control patches. Similarly to the previous approach, the planar patch in the image space is defined by a set of vertices that can be

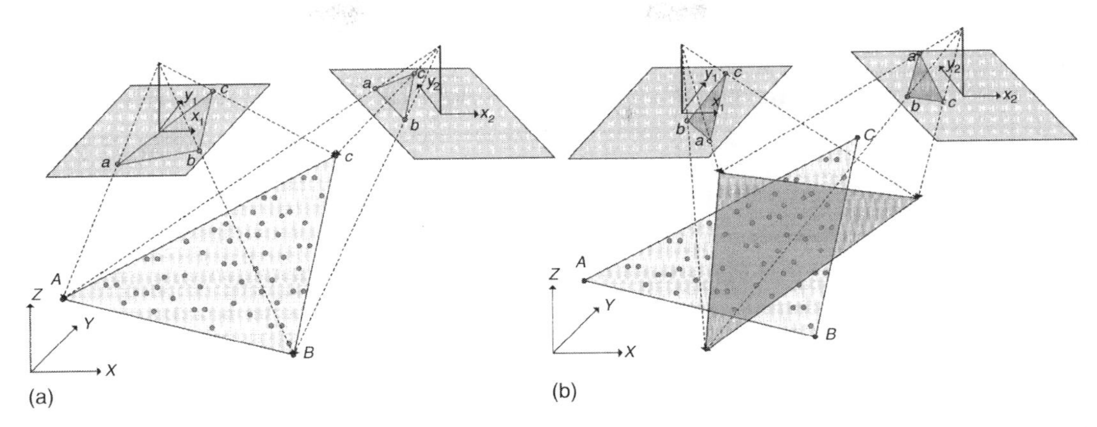

FIGURE 13.14
Coplanarity-based (a) and collinearity-based (b) incorporation of LiDAR patches in photogrammetric triangulation.

identified in overlapping imagery. The control patch, on the other hand, is defined by an equivalent set of object points (i.e., there is an identical number of vertices defining the image and object patches). However, the object points need not be conjugate to those identified in the image space, as illustrated in Figure 13.14b. To compensate for the fact that these sets of points are not conjugate, the variance–covariance matrices of the object points are expanded along the plane direction (Equations 13.10 through 13.14, Figure 13.15). In these equations, the *XYZ*-coordinate system defines the ground reference frame, and the *UVW*-coordinate system is defined by the orientation of the control patch in the object space. Specifically, the *UV*-axes are defined to be parallel to the plane through the patch (i.e., the *W*-axis is aligned along the normal to the patch). The expansion of the error ellipsoid

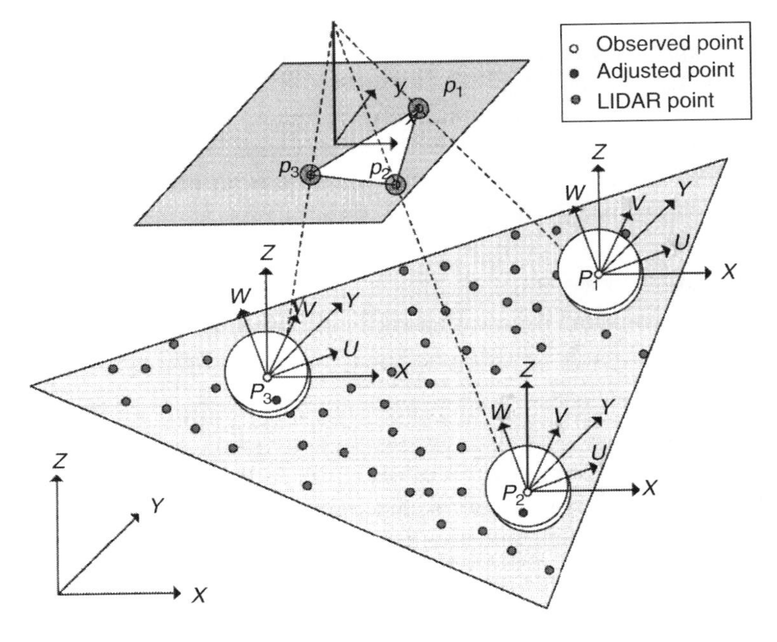

FIGURE 13.15
Conceptual basis of the collinearity-based incorporation of areal control patches in existing bundle adjustment procedures.

along the plane is facilitated by multiplying the variances along the U and V directions by large numbers (e.g., m_U and m_V in Equation 13.13). It should be noted that this approach is suitable for scenes captured by both frame and line cameras.

$$\begin{bmatrix} U \\ V \\ W \end{bmatrix} = R \begin{bmatrix} X \\ Y \\ Z \end{bmatrix} \quad (13.10)$$

$$\sum_{XYZ} = \begin{bmatrix} \sigma_X^2 & \sigma_{XY} & \sigma_{XZ} \\ \sigma_{YX} & \sigma_Y^2 & \sigma_{YZ} \\ \sigma_{ZX} & \sigma_{ZY} & \sigma_Z^2 \end{bmatrix} \quad (13.11)$$

$$\sum_{UVW} = R \sum_{XYZ} R^T = \begin{bmatrix} \sigma_U^2 & \sigma_{UV} & \sigma_{UW} \\ \sigma_{VU} & \sigma_V^2 & \sigma_{VW} \\ \sigma_{WU} & \sigma_{WV} & \sigma_W^2 \end{bmatrix} \quad (13.12)$$

$$\sum_{UVW}' = \begin{bmatrix} m_U \times \sigma_U^2 & \sigma_{UV} & \sigma_{UW} \\ \sigma_{VU} & m_V \times \sigma_V^2 & \sigma_{VW} \\ \sigma_{WU} & \sigma_{WV} & \sigma_W^2 \end{bmatrix} \quad (13.13)$$

$$\sum_{XYZ}' = R^T \sum_{UVW}' R \quad (13.14)$$

In summary, the previous Sections 13.6.1–13.6.4 introduced various approaches for performing multisensor and multiprimitive triangulation. These methodologies are intended to incorporate photogrammetric datasets acquired by frame and line cameras, in addition to LiDAR derived features, in a single triangulation mechanism. Having discussed multisensor triangulation for the alignment of the LiDAR and photogrammetric data to a common reference frame, Section 13.7 will address the utilization of this data for orthophoto generation.

13.7 LiDAR and Photogrammetric Data for Orthophoto Generation

Orthophoto production focuses on the elimination of the sensor tilt and terrain relief effects from captured perspective imagery. Uniform scale and the absence of relief displacement make orthophotos an important component of GIS databases, in which the user can directly determine geographic locations, measure distances, compute areas, and derive other useful information about the area in question. Recently, with the increasing availability of overlapping LiDAR and photogrammetric data, which is acquired from aerial and satellite imaging platforms, there has been a persistent need for an orthophoto generation methodology that is capable of dealing with the imagery acquired from such systems. Moreover, the increasing resolution of available data mandates the development of true orthophotos, in which occluded portions of the LiDAR surface are reliably detected in the involved imagery.

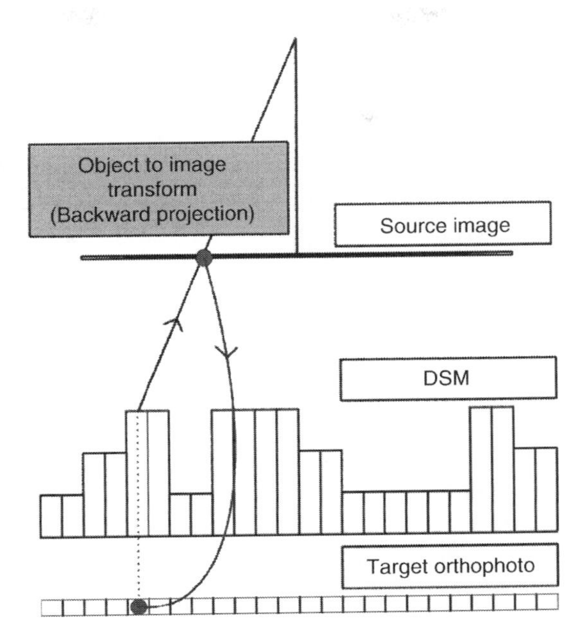

FIGURE 13.16
Orthophoto generation through differential rectification.

Differential rectification has traditionally been used for orthophoto generation, and is illustrated in Figure 13.16. The differential rectification procedure starts by projecting the digital surface model (DSM) cells (as derived from the LiDAR data) onto the image space (Novak, 1992). This is followed by resampling the gray values at the projected points from neighboring pixels in the image plane. Finally, the resampled gray values are assigned to the corresponding cells in the orthophoto plane. It should be noted that the projection from the DSM cells to the image plane is a straightforward process when working with images acquired by frame cameras (i.e., the projection involves solving two collinearity equations in two unknowns, which are the image coordinates). However, the object-to-image projection requires an iterative process when working with scenes captured by line cameras, as shown in Figure 13.17. The iterative procedure is essential to estimate the location of the exposure station that captured the object point in question (Habib et al., 2006).

The performance of differential rectification procedures has been quite acceptable when dealing with medium resolution imagery over relatively smooth terrain. However, when dealing with high resolution imagery over urban areas, differential rectification produces artifacts in the form of ghost images (double-mapped areas) at the vicinity of abrupt surface changes (e.g., at building boundaries and steep cliffs), as shown in Figure 13.18. The double-mapped areas arise because the gray values of occluded areas are erroneously imported from the perspective imagery. The effects of these artifacts can be mitigated by true orthophoto generation methodologies. The term "true orthophoto" is generally used for orthophotos in which surface elements that are not included in the digital terrain model (DTM) are rectified to the orthogonal projection. These elements are usually buildings and bridges (Amhar et al., 1998). Kuzmin et al. (2004) proposed a polygon-based approach for the detection of these obscured areas in order to generate true orthophotos. In this method, conventional digital differential rectification is first applied, then hidden areas are detected through

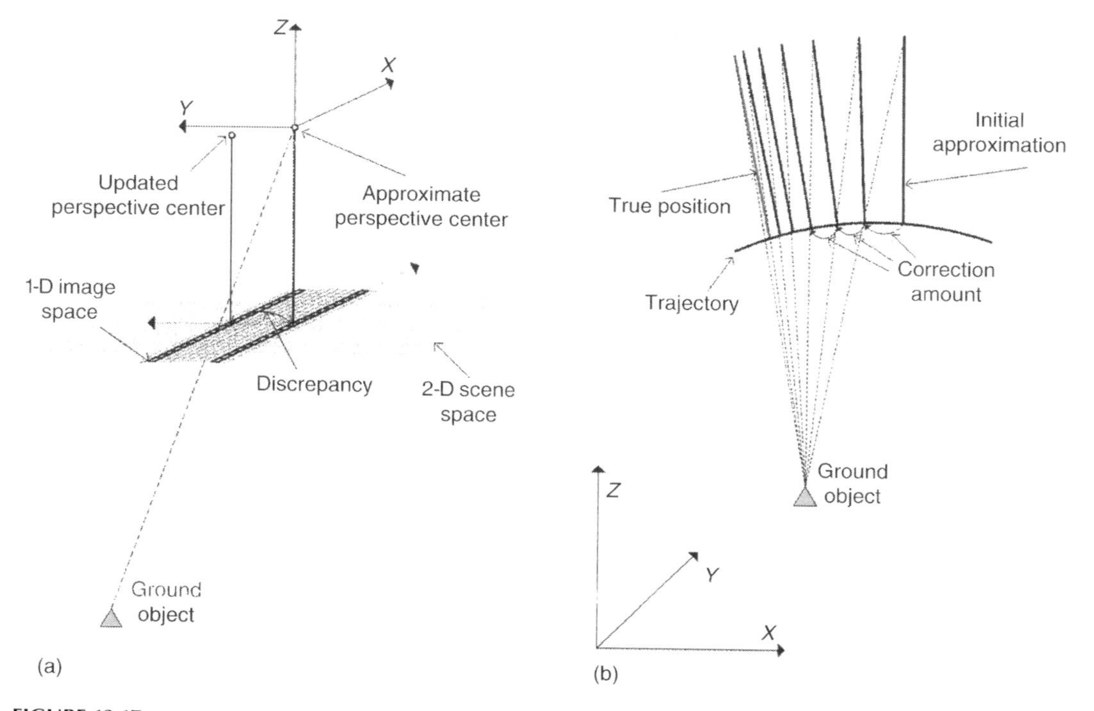

FIGURE 13.17
Iterative procedure for determining the exposure station that captured a given object point (b), starting from an approximate location (a).

the use of polygonal surfaces generated from a Digital Building Model (DBM). With the exception of the methodology proposed by Kuzmin et al., the majority of existing true orthophoto generation techniques are based on the z-buffer method, which is

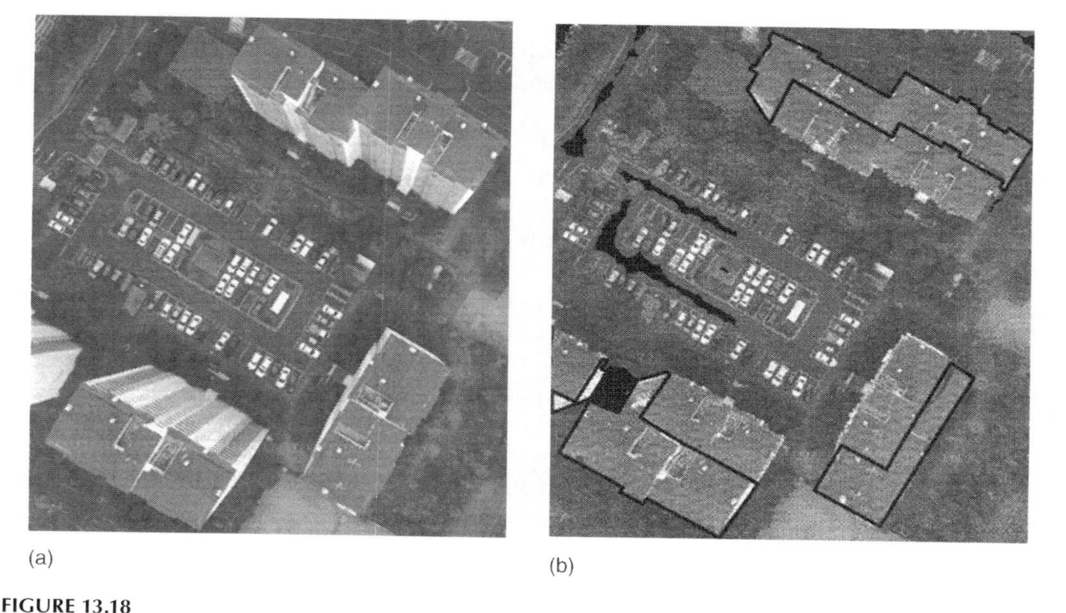

FIGURE 13.18
Double-mapped areas where relief displacement in the perspective image (a) causes duplication of the gray values projected onto the orthophoto plane (b).

briefly explained in Section 13.7.1 (Catmull, 1974; Amhar et al., 1998; Rau et al., 2000, 2002; Sheng et al., 2003).

13.7.1 z-Buffer Method

The *z*-buffer algorithm proposed by Amhar et al. (1998) has mainly been used for true orthophoto generation. As can be seen in Figure 13.19a, double-mapped areas arise when two object space points (e.g., *A* and *D*, *B* and *E*, or *C* and *F*) are competing for the same image location (e.g., *d*, *e*, or *f*, respectively). The *z*-buffer method resolves the ambiguity regarding which object point should be assigned to a certain image location by considering the distances between the perspective center and the object points in question. Among the competing object points, the point closest to the perspective center of an image is considered to be visible, while the other points are considered to be invisible in that image. It should be noted that the *z*-buffer method can be applied to scenes captured by both frame and line cameras.

The *z*-buffer has several limitations, which include its sensitivity to the sampling interval of the digital surface model (DSM), as it is related to the ground sampling distance (GSD) of the imaging sensor. Another significant drawback of the *z*-buffer methodology is the false visibility associated with narrow vertical structures. This problem is commonly known in the photogrammetric literature as the M-portion problem, which can be resolved by introducing additional artificial points (pseudo-ground points) along building facades (Rau et al., 2000, 2002). Pseudo-ground points should be carefully introduced along the building facades to avoid false visibility while optimizing the computational speed, as illustrated in Figure 13.19b. If the pseudo-ground points are too dense, the true orthophoto generation process will be time intensive. On the other hand, pseudo-ground points that are too sparse can cause false visibilities. As shown in Figure 13.19b, the spacing between two successive pseudo-ground points should be small enough to handle facades located at image boundaries (i.e., the relative displacement between their projections onto the image

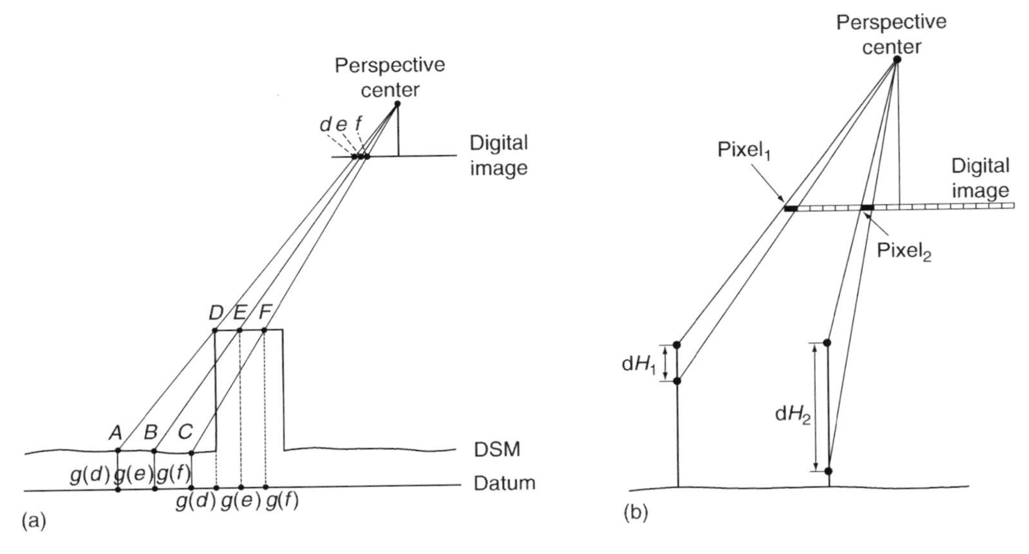

FIGURE 13.19
(a) True orthophoto generation using the *z*-buffer method, and (b) optimal spatial interval for the pseudo-ground points.

space should not exceed one pixel). The separation between successive pseudo-ground points should increase as one moves towards the nadir point, where no pseudo-ground points are needed. For more detailed analysis of the limitations of the z-buffer method, interested readers can refer to Habib et al. (2007).

13.7.2 Angle-Based Method

The top and bottom of a vertical structure are projected onto the same location in the absence of relief displacement. However, in a perspective projection, the top and bottom of that structure are projected as two points. These points are not at the same location; they are spatially separated by the relief displacement. This displacement takes place along a radial direction from the image space nadir point (Mikhail, 2001); relief displacement is the cause of occlusions in perspective imagery. The presence of occlusions can be discerned by sequentially checking the off-nadir angles to the lines of sight connecting the perspective center to the DSM points, along a radial direction starting with the object space nadir point, as shown in Figure 13.20. More specifically, as one moves away from the nadir point, the angle between the projection ray and the nadir line should increase. Whenever there is a sudden decrease in the off-nadir angle, invisibility has been detected. This invisibility persists until the off-nadir angle exceeds that of the last visible point, as illustrated in Figure 13.20. This approach for true orthophoto generation is referred to as the angle-based method. The angle-based methodology for true orthophoto generation can be applied, with minor variations, to scenes captured by both frame and line cameras. For scenes captured by frame cameras, the angle-based method inspects the off-nadir angles starting from the nadir point and moving towards the point in question. For a line camera scene, on the other hand, we have multiple exposure stations, which, in turn, lead to multiple nadir points. Therefore, for scenes captured by line cameras, the angle-based method starts from the object point and moves inward towards the nadir point of the image associated with the object point

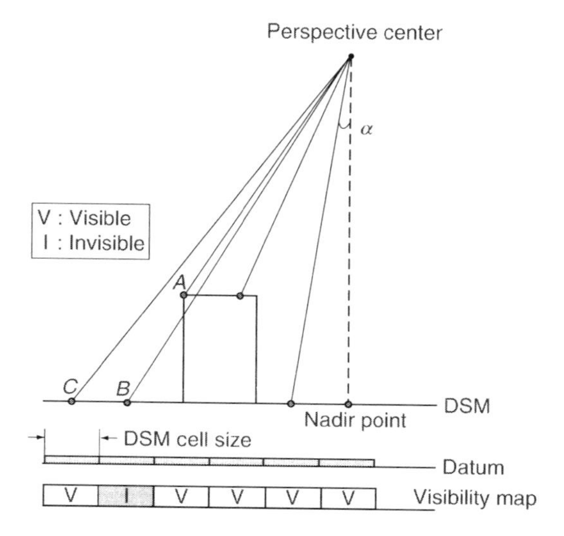

FIGURE 13.20
True orthophoto generation using the angle-based method.

under consideration. For more details regarding the angle-based method of orthophoto generation, interested readers can refer to Habib et al. (2006, 2007).

13.8 Case Study

To validate the feasibility and applicability of the above methodologies, multisensor datasets will be analyzed. The conducted experiments will involve three types of sensors, namely, a digital frame camera equipped with a global positioning system (GPS)/INS unit, a satellite-based line camera, and a LiDAR system. These experiments will focus on investigating the following issues:

- Validity of using the line-based geo-referencing procedure for scenes captured by frame and line cameras.
- Validity of using the patch-based geo-referencing procedure for scenes captured by frame and line cameras.
- Impact of integrating satellite scenes, aerial scenes, LiDAR data, and GPS exposure positions in a unified bundle adjustment procedure.
- Qualitative investigation of the methodologies presented for true orthophoto generation.

The first dataset is composed of three blocks, each consisting of six digital frame images, captured in April 2005 by the Applanix Digital Sensor System (DSS) over the city of Daejeon in South Korea from an altitude of 1500 m. The DSS camera has 16 megapixels (9 μm pixel size) and a 55 mm focal length. The position and the attitude of the DSS camera were tracked using the onboard global positioning system (GPS)/inertial navigation system (INS) unit. However, only the position information was available for this test. The second dataset consists of an IKONOS stereo-pair, which was captured in November 2001, over the same area. It should be noted that these scenes are raw imagery that

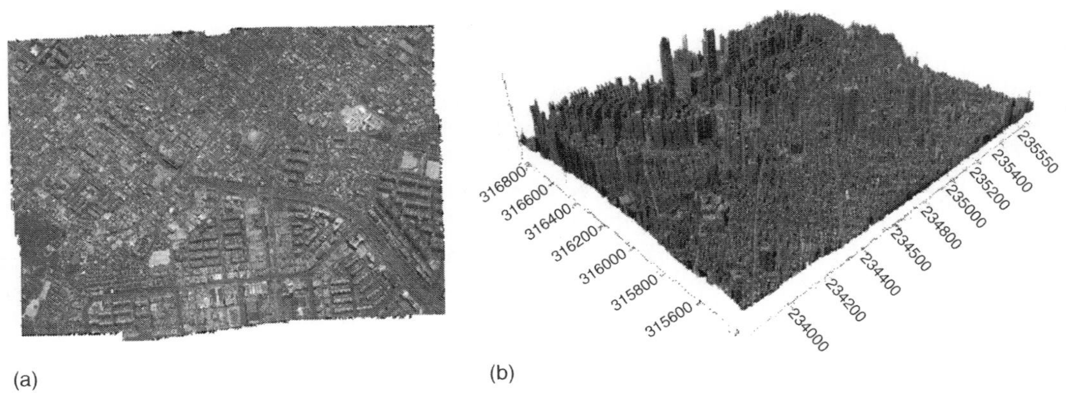

FIGURE 13.21
DSS middle image block (a) and the corresponding LiDAR cloud (b); the circles in (a) indicate the locations of linear and areal primitives extracted from the LiDAR data.

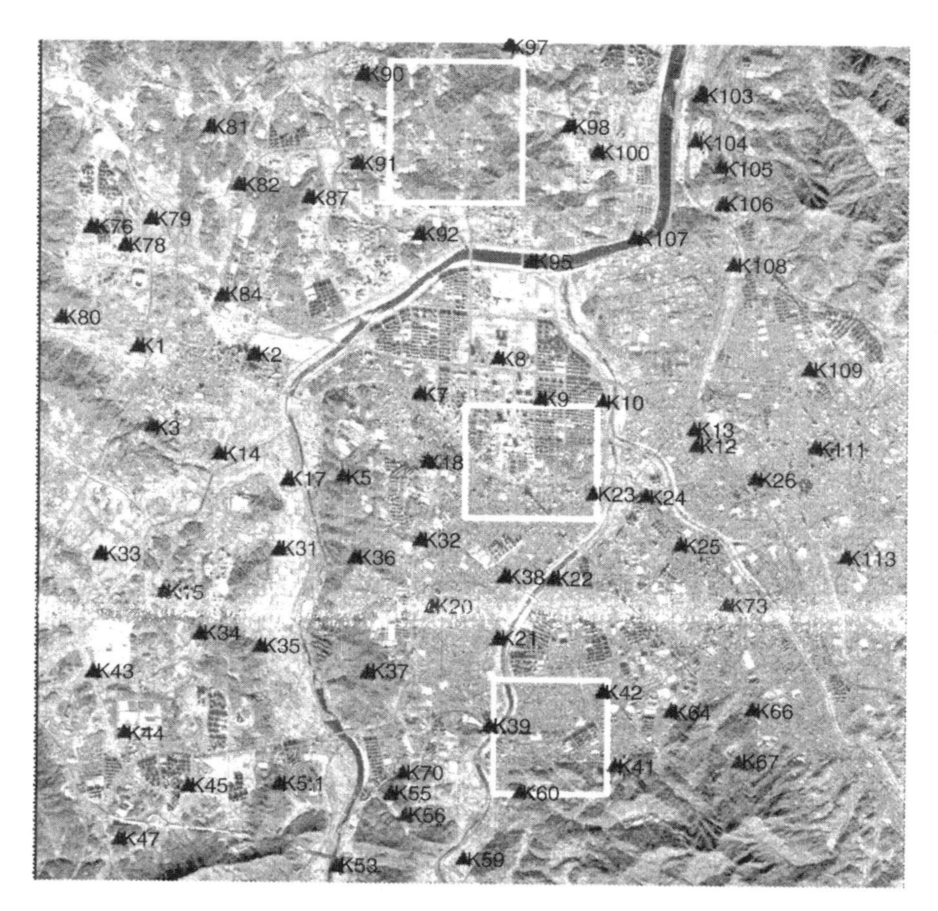

FIGURE 13.22
IKONOS scene coverage, showing the three patches covered by the DSS imagery and LiDAR data, together with the distribution of the GCP.

did not go through any geometric correction and are provided for research purposes. Finally, a multistrip LiDAR coverage, corresponding to the DSS coverage, was collected using the OPTECH ALTM 2050 system with an average point density of 2.67 point/m^2 from an altitude of 975 m. An example of one of the DSS image blocks and a visualization of the corresponding LiDAR coverage can be seen in Figure 13.21. Figure 13.22 shows the IKONOS coverage and the locations of the DSS image blocks.

To extract the LiDAR control, a total of 139 planar patches and 138 linear features were manually identified through planar patch segmentation and intersection, respectively. Figure 13.21 shows the locations of the features extracted from the middle LiDAR point cloud, within the IKONOS scenes. The corresponding linear and areal features were digitized in the DSS and IKONOS scenes. To evaluate the performance of the different geo-referencing techniques, a set of 70 GCP was also acquired; refer to Figure 13.22 for the distribution of these points. The performance of the point-, line-, patch-based, and GPS-assisted geo-referencing techniques was evaluated using Root Mean Square Error (RMSE) analysis. In the experiments, a portion of the available GCPs was used as control in the bundle adjustment, while the rest was used as check points. It is worth mentioning that none of the control points are visible in any of the available DSS imagery for this test.

TABLE 13.3

RMSE (m) for the Checkpoints in MultiSensor and MultiPrimitive Triangulation

	IKONOS Only	IKONOS + 18 DSS Frame Images					
				Control Points Plus			
				Control Lines		Control Patches	
Number of GCPs	Control Points Only	Control Points Only	DSS GPS	138	45	139	45
0	N/A	N/A	3.07	3.07	3.09	5.41	5.86
1	N/A	N/A	3.37	3.01	3.08	5.39	6.40
2	N/A	N/A	3.08	3.13	3.20	4.82	5.25
3	N/A	21.32	2.86	2.86	2.85	2.93	3.09
4	N/A	19.96	2.76	2.71	2.75	2.64	3.05
5	N/A	4.34	2.73	2.72	2.74	2.63	2.72
6	3.67	3.35	2.70	2.69	2.74	2.61	2.71
7	3.94	3.04	2.62	2.70	2.73	2.55	2.56
8	3.59	3.42	2.58	2.57	2.53	2.45	2.72
9	4.08	2.54	2.51	2.55	2.52	2.44	2.51
10	3.07	2.50	2.50	2.56	2.53	2.43	2.49
15	3.15	2.41	2.46	2.46	2.43	2.40	2.44
40	2.01	2.09	2.11	2.11	2.10	2.03	2.05

To investigate the performance of the various geo-referencing methodologies, we conducted the following experiments (the total RMSE values resulting from these experiments are reported in Table 13.3):

- Photogrammetric triangulation of the IKONOS scenes, while varying the number of GCP utilized (second column in Table 13.3).
- Photogrammetric triangulation of the IKONOS and DSS scenes, while varying the number of GCP utilized (third column in Table 13.3).
- Photogrammetric triangulation of IKONOS and DSS scenes, while considering the GPS observations associated with the DSS exposures and varying the number of GCP utilized (fourth column in Table 13.3).
- Photogrammetric triangulation of IKONOS and DSS scenes, while varying the number of LiDAR lines (45 and 138 lines) and changing the number of GCP (fifth and sixth columns in Table 13.3).
- Photogrammetric triangulation of IKONOS and DSS scenes, while varying the number of LiDAR patches (45 and 139 patches) and changing the number of GCP (seventh and eighth columns in Table 13.3).

In Table 13.3, N/A means that no solution was attainable (i.e., the control provided was not sufficient to establish the datum for the triangulation procedure). The reported RMSE values in the last four columns of Table 13.3 are based on the coplanarity-based approaches for incorporating linear and areal features. The utilization of the collinearity-based approaches for incorporating linear and areal features resulted in almost identical results. It is worth mentioning that the collinearity-based incorporation of linear and areal features can be used in existing point-based triangulation packages. In such a case, the input files should be modified by expanding the variance–covariance matrices of the utilized points along the linear/areal features. For more clarity, the results in Table 13.3

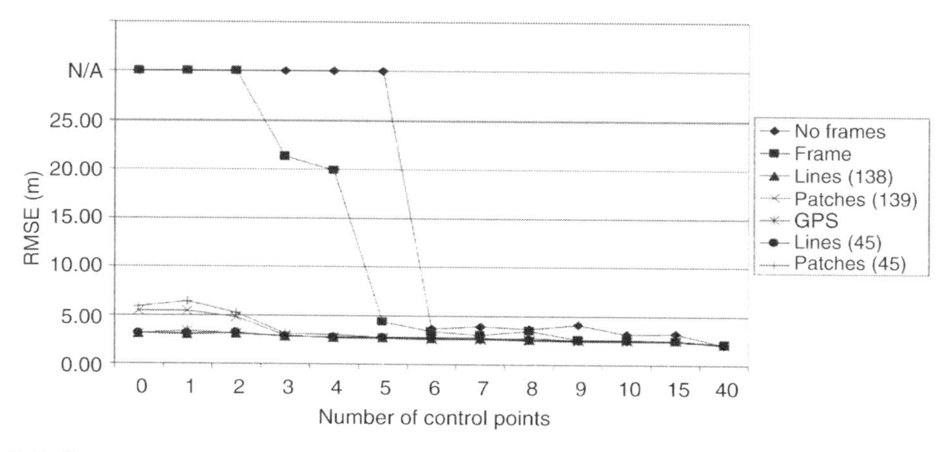

FIGURE 13.23
Check point analyses from the bundle adjustments involving IKONOS and DSS imagery as well as LiDAR features, for various control configurations.

are visually aggregated in Figure 13.23, in which the Total RMSE values are plotted against the number of control points, for the different geo-referencing techniques. By examining Table 13.3 and Figure 13.23, one can make the following remarks:

- Utilizing points as the only source of control for the triangulation of the stereo IKONOS scenes required a minimum of six GCPs.
- Including DSS imagery in the triangulation of the IKONOS scenes reduced the control requirement for convergence to three GCPs. Moreover, the incorporation of the GPS observations at the DSS exposure stations enabled convergence without the need for any GCP. Therefore, it is clear that incorporating satellite scenes with a few frame images enables photogrammetric reconstruction with a reduced GCP requirement.
- LiDAR linear features are sufficient to geo-reference the IKONOS and DSS scenes without the need for an additional source of control. As can be seen in the fifth and sixth columns of Table 13.3, incorporating additional control points in the triangulation procedure did not significantly improve the reconstruction outcome. Moreover, by comparing the fifth and sixth columns, one can see that increasing the number of linear features from 45 to 138 does not significantly improve the quality of the triangulation outcome.
- LiDAR patches are sufficient to geo-reference the IKONOS and DSS scenes without the need for an additional source of control. However, as it can be seen in the seventh and eighth columns of Table 13.3, incorporating a few control points had a significant impact in improving the results (through the use of 3 GCPs and 139 control patches, the total RMSE was reduced from 5.41 to 2.93 m). Incorporating additional control points (i.e., beyond three GCPs) did not have a significant impact. The improvement in the reconstruction outcome achieved by using a few GCPs can be attributed to the fact that the majority of the utilized patches were horizontal or with mild slopes, as they represented building roofs. Therefore, the estimation of the model shifts in the X and Y directions was not accurate enough. Incorporating vertical or steep patches could have solved this problem; however, such patches were not available in the dataset provided. Moreover, by comparing the seventh and eighths

columns, it can be seen that increasing the number of control patches from 45 to 139 did not significantly improve the quality of the triangulation outcome.

- By comparing the different geo-referencing techniques, it can be seen that the patch-based, line-based, and GPS-assisted geo-referencing techniques produced better outcomes than the point-based geo-referencing technique. Such an improvement attests to the advantage of adopting multisensor and multiprimitive triangulation procedures.

In an additional experiment, we utilized the EOP derived from the multisensor triangulation of frame and line camera scenes together with the LiDAR surface to generate orthophotos. Figure 13.24 shows sample patches, in which the IKONOS and DSS orthophotos are laid side by side. As can be seen in Figure 13.24a, the generated orthophotos are quite compatible, as indicated by the smooth continuity of the observed features between the DSS and IKONOS orthophotos. Figure 13.24b reveals changes in the object space between the moments of capture of the IKONOS and DSS imagery. Therefore, it is evident that the multisensor triangulation of imagery from frame and line cameras improves the quality of the derived object space, while offering an environment for the accurate geo-referencing of temporal imagery. Following the geo-referencing, the involved imagery can be analyzed for change detection applications using the derived and properly geo-referenced orthophotos.

Figures 13.18, 13.25, and 13.26 illustrate the comparative performance of the differential rectification, *z*-buffer, and angle-based methods of orthophoto generation using frame and line cameras, respectively. As can be seen in these figures, differential rectification produces ghost images (refer to the highlighted areas in Figure 13.18b for frame cameras and Figure 13.26b for line cameras). However, ghost images are not present in the true orthophotos generated by the *z*-buffer and angle-based methods (refer to the highlighted areas in Figure 13.25a and b for frame cameras and Figure 13.26c and d for line cameras). However, visual inspection of Figure 13.25a and b, and Figure 13.26c and d reveals the superior performance of the angle-based method when compared to that of the *z*-buffer method.

(a)

(b)

FIGURE 13.24
Change detection between DSS (left) and IKONOS (right) orthophotos. A smooth transition between the two orthophotos can be observed in (a), while discontinuities are observed in (b) due to changes in the object space.

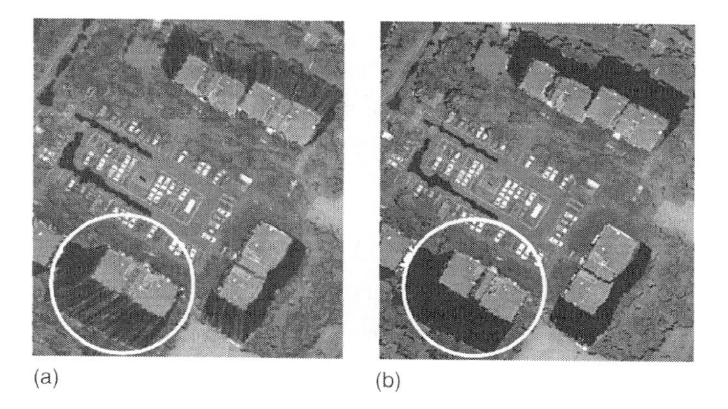

FIGURE 13.25
True orthophotos, corresponding to the illustrated image in Figure 13.18a, generated using *z*-buffer (a), and angle-based (b) techniques.

FIGURE 13.26
A scene captured by a line camera (a), and orthophotos generated using the differential rectification (b), *z*-buffer (c), and angle-based (d) techniques.

FIGURE 13.27
Photo-realistic 3D model produced by rendering a true orthophoto onto LiDAR data.

13.9 Concluding Remarks

The continuous advancement in mapping technology demands the development of commensurate processing methodologies to take advantage of the synergistic characteristics of available geo-spatial data. In this regard, it is quite evident that integrating LiDAR and photogrammetric data is essential to ensure the accurate and better description of the object space. This chapter presented different methodologies for aligning LiDAR and photogrammetric data relative to a common reference frame using manually extracted linear and areal primitives. The developed methodologies are suited to the characteristics of these datasets. Moreover, the introduced methodologies are general enough to be directly applied to scenes captured by both line and frame cameras. The experimental results have shown that the utilization of LiDAR derived primitives as the source of control for photogrammetric geo-referencing yields slightly better results when compared with point-based geo-referencing techniques. Moreover, it has been shown that the incorporation of sparse frame imagery, together with satellite scenes, improves the results by taking advantage of the geometric strength of frame cameras to improve the inherently weak geometry of line cameras onboard imaging satellites. Therefore, the combination of aerial and satellite scenes improves the coverage extent as well as the geometric quality of the derived object space. The incorporation of LiDAR data, aerial images, and satellite scenes in a single triangulation procedure also ensures the co-registration of these datasets relative to common reference frame, which is valuable for orthophoto generation and change detection applications. Future work should focus on the automated extraction of the triangulation primitives from the image and LiDAR data. In addition to the multiprimitive triangulation of multisensor data, this chapter outlined several methodologies for orthophoto generation, to take advantage of the synergistic characteristics of photogrammetric and LiDAR data. With the increasing resolution of spatial data, true orthophoto generation techniques should be adopted to ensure the absence of double-mapped areas (ghost images). Such artifacts are expected even when working with satellite scenes. The true orthophotos generated, after being draped on top of the LiDAR DSM, can be used to produce photo-realistic 3D models, as in Figure 13.27.

Acknowledgments

This research work has been conducted under the auspices and financial support of the GEOIDE Research Network. The author would like to thank Mr. Paul Mrstik, Terrapoint

Canada Inc. for the valuable technical support and providing the LiDAR and image datasets, which was utilized in the experimental results section.

References

Amhar, F., et al., 1998. The generation of true orthophotos using a 3D building model in conjunction with a conventional DTM, *International Archive Photogrammetry and Remote Sensing*, 32: 16–22.

Baltsavias, E., 1999. A comparison between photogrammetry and laser scanning, *ISPRS Journal of Photogrammetry and Remote Sensing*, 54(1): 83–94.

Catmull, E., A subdivision algorithm for computer display of curved surfaces, PhD dissertation, Department of Computer Science, University of Utah, Salt lake city, UT, 1974.

Chen, L., Teo, T., Shao, Y., Lai, Y., and Rau, J., 2004. Fusion of LiDAR data and optical imagery for building modeling, *International Archives of Photogrammetry and Remote Sensing*, 35(B4): 732–737.

Cramer, M., Stallmann, D., and Haala, N., 2000. Direct georeferencing using GPS/Inertial exterior orientations for photogrammetric applications, *International Archives of Photogrammetry and Remote Sensing*, 33(B3): 198–205.

Dowman, I. and Dolloff, J., 2000. An evaluation of rational functions for photogrammetric restitutions, *International Archives of Photogrammetry and Remote Sensing*, 33(B3): 254–266.

Ebner, H., Kornus, W., and Ohlhof, T., 1994. A simulation study on point determination for the MOMS-02/d2 space project using an extended functional model, *Geo-Information Systems*, 7(1): 11–16.

Ebner, H., Ohlhof, T., and Putz, E., 1996. Orientation of MOMS-02/D2 and MOMS-2P Imagery, *International Archives of Photogrammetry and Remote Sensing*, 31(B3): 158–164.

El-Manadili, Y. and Novak, K., 1996. Precision rectification of SPOT imagery using the direct linear transformation model, *Photogrammetric Engineering and Remote Sensing*, 62(1): 67–72.

El-Sheimy, N., Valeo, C., and Habib, A., 2005. *Digital Terrain Modelling: Acquisition, Manipulation and Applications*, Artech House, Inc., Norwood, MA.

Faruque, F., 2003. LiDAR image processing creates useful imagery. *ArcUser Magazine*, January–March 2003, 6(1).

Fraser, C. and Hanley, H., 2003. Bias compensation in rational functions for IKONOS satellite imagery, *Photogrammetric Engineering and Remote Sensing*, 69(1): 53–57.

Fraser, C., Hanley, H., and Yamakawa, T., 2002. High-precision geopositioning from IKONOS satellite imagery, *Proceeding of ACSM-ASPRS 2002*, Washington, DC, unpaginated CD-ROM.

Gupta, R. and Hartley, R., 1997. Linear push-broom cameras, *IEEE Transactions on Pattern Analysis and Machine Intelligence*, 19(9): 963–975.

Ghanma, M., 2006. Integration of photogrammetry and LiDAR. PhD thesis, Department of Geomatics Engineering, University of Calgary, April 2006, UCGE Report 20241, 141 pp.

Habib, A. and Beshah, B., 1998. Multi sensor aerial triangulation, *Proceeding of ISPRS Commission III Symposium*, 6–10 July 1998, Columbus, OH, pp. 37–41.

Habib, A. and Schenk, T., 1999. A new approach for matching surfaces from laser scanners and optical sensors, *International Archives of Photogrammetry and Remote Sensing*, 32(3W14): 55–61.

Habib, A., Lee, Y., and Morgan, M., 2001. Bundle adjustment with self-calibration of line cameras using straight lines, *Joint Workshop of ISPRS WG I/2, I/5 and IV/7: High Resolution Mapping from Space 2001*, 19–21 September 2001, University of Hanover, Hanover, Germany, unpaginated CD-ROM.

Habib, A., Morgan, M., and Lee, Y., 2002. Bundle adjustment with self-calibration using straight lines, *Photogrammetric Record*, 17(100): 635–650.

Habib, A., Kim, E., Morgan, M., and Couloigner, I., 2004. DEM Generation from high resolution satellite imagery using parallel projection model, *Proceeding of the XXth ISPRS Congress*, Commission 1, TS: HRS DEM Generation from SPOT-5 HRS Data, 12–23 July 2004, Istanbul, Turkey, pp. 393–398.

Habib, A., Bang, K., Kim, C., and Shin, S., 2006. True orthophoto generation from high resolution satellite imagery. *Innovations in 3D Geo Information Systems*, Lecture Notes in Geo Information and

Cartography, pp. 641–656. Editors: Abdul-Rahman, A., Zlatanova, S., and Coors, V., Springer, 2006, Berlin, Heidelberg.

Habib, A., Kim, E., and Kim, C., 2007. New methodologies for true orthophoto generation. *Photogrammetric Engineering and Remote Sensing; 2007,* 73(1): 25–36.

Hanley, H., Yamakawa, T., and Fraser, C., 2002. Sensor orientation for high resolution satellite imagery, Pecora 15/Land Satellite Information IV/ISPRS Commission I/FIEOS, 10–15 November 2002, Denver, CO, unpaginated CD-ROM.

Kuzmin, P., Korytnik, A., and Long, O., 2004. Polygon-based true orthophoto generation, *XXth ISPRS Congress*, 12–23 July, Istanbul, pp. 529–531.

Lee, Y. and Habib, A., 2002. Pose estimation of line cameras using linear features. *Proceedings of ISPRS Commission III Symposium Photogrammetric Computer Vision*, Graz, Austria, September 9–13, 2002.

Lee, C., Thesis, H., Bethel, J., and Mikhail, E., 2000. Rigorous mathematical modeling of airborne pushbroom imaging systems, *Photogrammetric Engineering and Remote Sensing*, 66(4): 385–392.

Lee, Y., Habib, A., and Kim, K., 2002. A study on aerial triangulation from multi-sensor imagery. *Proceedings of the International Symposium on Remote Sensing (ISRS) 2002*, Sokcho, Korea, (October 30 to November 1, 2002).

Mikhail, E., Bethel, J., and McGlone, J., 2001. *Introduction to Modern Photogrammetry*, John Wiley & Sons, New York.

Novak, K., 1992. Rectification of digital imagery, *Photogrammetric Engineering and Remote Sensing*, 58(3): 339–344.

Ono, T., Hattori, S., Hasegawa, H., and Akamatsu, S., 2000. Digital mapping using high resolution satellite imagery based on 2-D affine projection model, *International Archives of Photogrammetry and Remote Sensing*, 33(B3): 672–677.

Poli, D., 2004. Orientation of satellite and airborne imagery from multi-line pushbroom sensors with a rigorous sensor model, *International Archives of Photogrammetry and Remote Sensing*, Istanbul, Turkey, 34(B1): 130–135.

Rau, J., Chen, N., and Chen, L., 2000. Hidden compensation and shadow enhancement for true orthophoto Generation, *Proceedings of Asian Conference on Remote Sensing*, 4–8 December 2000, Taipei, CD-ROM.

Rau, J., Chen, N., and Chen, L., 2002. True orthophoto generation of built-up areas using multi-view images, *Photogrammetric Engineering and Remote Sensing*, 68(6): 581–588.

Satale, D. and Kulkarni, M., 2003. LiDAR in mapping. Map India Conference GISdevelopment.net. Available at http://www.gisdevelopment.net/technology/gis/mi03129.htm (last accessed March 28, 2006).

Schenk T., 1999. Photogrammetry and laser altimetry. 32, Part 3-W14, *Proceedings of ISPRS Workshop Mapping Surface Structure and Topography by Airborne and Spaceborn Lasers*, La Jolla, CA, 9–11 November 1999, pp. 3–12.

Sheng, Y., Gong, P., and Biging, G., 2003. True orthoimage production for forested areas from large-scale aerial photographs, *Photogrammetric Engineering and Remote Sensing*, 69(3): 259–266.

Tao, V. and Hu, Y., 2001. A comprehensive study of rational function model for photogrammetric processing, *Photogrammetric Engineering and Remote Sensing*, 67(12): 1347–1357.

Tao, V., Hu, Y., and Jiang, W., 2004. Photogrammetric exploitation of IKONOS imagery for mapping applications, *International Journal of Remote Sensing*, 25(14): 2833–2853.

Toutin, T., 2004a. DTM generation from IKONOS in-track stereo images using a 3D physical model, *Photogrammetric Engineering and Remote Sensing*, 70(6): 695–702.

Toutin, T., 2004b. DSM generation and evaluation from quickbird stereo images with 3D physical modeling and elevation accuracy evaluation, *International Journal of Remote Sensing*, 25(22): 5181–5192.

Wang, Y., 1999. Automated triangulation of linear scanner imagery, *Joint Workshop of ISPRS WG I/1, I/3 and IV/4 on Sensors and Mapping from Space*, 27–30 September 1999, Hanover, Germany, unpaginated CD-ROM.

Wegmann, H., Heipke, C., and Jacobsen, K., 2004. Direct sensor orientation based on GPS network solutions, *International Archives of Photogrammetry and Remote Sensing*, 35(B1): 153–158.

Wehr, A. and Lohr, U., 1999. Airborne laser scanning—an introduction and overview, *ISPRS Journal of Photogrammetry and Remote Sensing*, 54(2–3): 68–82.

14

Feature Extraction from LiDAR Data in Urban Areas

Frédéric Bretar

CONTENTS

14.1 Introduction

14.1.1 Background

Automatic mapping of urban areas from aerial images is a challenging task for scientists and surveyors because of the complexity of urban scenes. From a photogrammetric point of view, aerial images can be used to produce 3D points provided that their acquisition have been performed in a (multi-) stereoscopic context [12]. Altitudes are processed using automatic correlation algorithms to generate Digital Surface Models (DSM) [1,14].

For the last decade, LiDAR data have become an alternative data source for generating a tridimensional representation of landscapes. Basically, LiDAR data are acquired by strips

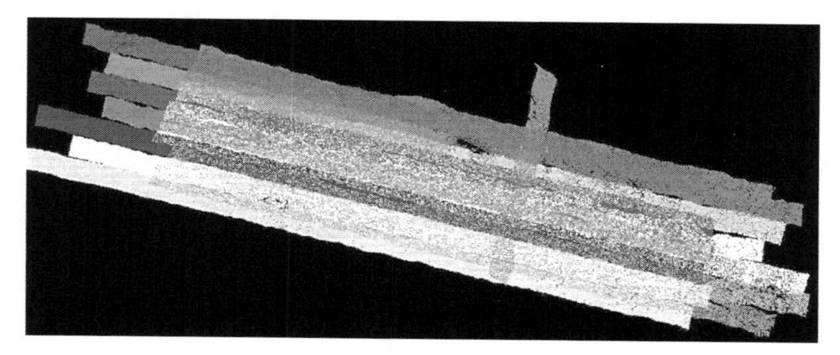

FIGURE 14.1
Example of a global survey over a city area, Amiens, France. Each strip is colored and has a width of ~170 m.

and finally gathered as 3D point clouds (Figure 14.1). Raw LiDAR data (3D points) are not expressed in an easy topology since there is no relationship between neighboring points, unlike the intrinsic 4/8-connexity of image-based representations. The access to the data is obtained using Kd-tree structures or graphs. Moreover, airborne LiDAR systems do not provide textural information that can be easily exploited.*

Whereas the spatial distribution of a LiDAR point cloud is not uniform, punctual 3D information is an accurate representation of the topography. But a higher description level of landscapes is often wished to fill geographic information system (GIS) databases (vector-based features, higher order semantic). Beyond the production of digital surface or terrain models, the understanding of airborne LiDAR scenes depends on our capability of finding specific shapes to derive a higher order cartography (vector-based) than 3D points.

14.1.2 Line Extraction

In the context of building reconstruction, lines and planar surfaces are particularly of interest. The extraction of lines from a LiDAR point cloud is a difficult task since LiDAR points are randomly distributed over surfaces: Depending on the point density, building edges are approximately delineated. Points considered as edge points may not belong to the true building edge. Nevertheless the extraction of 2D or 3D lines is considered in different problems such as the reconstruction of building footpoints [19] or the registration of LiDAR data with regard to images. In case of a joint registration of LiDAR and images, we need to detect invariant features in both data sets before estimating a transform between reference frames [15]. On the one hand, 3D segments can be generated by intersecting preprocessed 3D planar surfaces considered as adjacent ones [6]. On the other hand, 3D segments can be extracted directly from the point cloud [4]. The idea of such approaches (which is very sensitive to parameters) is to classify each LiDAR point depending on its local geometric environment by computing 3D moments. The eigenvalues and the eigenvectors of a covariance matrix deliver geometric characteristics of the distribution of each LiDAR point. In case of an ideal line, two eigenvalues are null while the third one is greater than zero. In case of a plane, two eigenvalues are similar to a nonzero value while the third one is zero. In case of an edge line, one eigenvalue is zero while there is a significant difference between the two other ones.

Building footprints are fundamental information for mapping applications. Different steps are involved in the process:

* Intensity values are not considered here since researches are not conclusive regarding their usefulness.

FIGURE 14.2
Extracted contours.

1. Generating a DSM
2. Computing the DSM gradient (Canny-Deriche filter)
3. Processing an hysteresis thresholding to identify high gradient values that are meant to describe roof gutters
4. Chaining segments

This simple flowchart provides initial solutions for further developments (Figure 14.2). Indeed, the interpolation influence can be significant on the surface quality, particularly when working on urban areas. Moreover, connecting segments are still a challenging problem in the image-processing community.

14.1.3 Surface Extraction: Hough Transform

This chapter focuses on the extraction of 3D planar primitives. It is critical to extract high quality of 3D planar surfaces since they serve as important primitives to reconstruct 3D building models.

A classical methodology for extracting geometrical entities is the Hough transform. A plane $\mathcal{P}$ that includes a given LiDAR point (x, y, z) can be written in spherical coordinates as

$$\begin{aligned} \mathcal{P} : \left[0,\pi\right[\times\left[0,\pi\right[&\rightarrow \mathbb{R} \\ (\theta,\phi) &\rightarrow \rho = \cos\theta\cos\phi x + \sin\theta\cos\phi y + \sin\phi z \end{aligned} \tag{14.1}$$

Therefore, each LiDAR point (x, y, z) is represented as a complex surface in the Hough space. This surface is represented in a discretized volume (accumulator) in the (ρ, θ, ϕ)-directions. Looking for 3D planar surfaces in the object space is equivalent to looking for 3D points in the Hough space (ρ, θ, ϕ). They are the local maxima of the accumulator volume. Depending on the discretization step, calculations can become time consuming. Within the building reconstruction context, the parameter space should be limited by some constraints that are often used, such as

1. Plane normal vector is perpendicular to at least one of the directions of the building facades, if they are available
2. Roof slopes are beneath 45° for instance
3. ρ is limited to an embedded volume delimited by the DSM

In a cadastral map, roofs may have a complex shape. The extraction of relevant planes within the accumulator is not an obvious process. First, a plane may include two disconnected sets of LiDAR points within a region. Vosselman proposes to avoid this problem by comparing the generated triangulated irregular network (TIN) from assigned points to a global TIN: only the connected piece with minimal surface is considered to be a planar face [17]. Selected points are then removed from the initial point set and the process is repeated. Another preprocessing step could be to refine the designed cadastral parcel in light of the DSM. Jibrini has proposed to perform a 3D segmentation of the accumulator volume by watershed [10], but prefers searching plane hypothesis satisfying building model hypothesis [11]. With no a priori knowledge on the building shape, this methodology proposes plane hypothesis that are all intersected before searching for the best polyhedra describing the roof surface [16].

The aim of this chapter is to present two approaches for detecting 3D roof facets; the first one is based on the analysis of the 3D point cloud while the second one integrates aerial images. We would like to show that searching for planar primitives in a LiDAR point cloud has limitations with regard to a joint use of aerial images and LiDAR data. Two approaches will be detailed and discussed: a robust RANSAC*-based approach and eventually an hybrid image segmentation approach.

14.2 RANSAC-Based Approach for Plane Extraction

14.2.1 Background

The RANSAC algorithm introduced by Fischler and Bolles is a general robust approach to estimate model parameters [3]. Instead of using the largest amount of data to obtain an initial solution and then attempting to eliminate the invalid data points, RANSAC uses the smallest feasible data set and enlarges this set with consistent data when possible. With applications to roof facet detection, classical RANSAC would be formulated as follows (cf. Algorithm 14.1): (i) randomly select a set of $N \in \mathbb{N}$ planar surfaces $\mathcal{P}$ within a LiDAR point cloud S and keep a count of the number of points (also called supports) of Euclidean distances from the associated planes $\mathcal{P}$ less than a critical distance d_{cr}. (ii). A least square estimation of the final plane ($\mathcal{P}_{\text{final}}$) is performed with the set of supports ($\mathcal{M}_k$) of maximum cardinal. (iii) The set $\mathcal{M}_k$ is then removed from the initial point cloud S. The algorithm runs until card(S) < 3, where card(S) is the cardinal of set S.

Theoretically, all triplets of S have to be tried to ensure that the best plane was drawn at each iteration. N_{th} is therefore defined as

$$N_{th} = \frac{\text{Card}(S)!}{3!(\text{card}(S)-3)!} \tag{14.2}$$

* Random sample consensus.

Nevertheless, this approach may be extremely time consuming. Most often it is not worthy to try all possible draws [8]. In other words, for a given probability t of drawing a correct plane $\mathcal{P}$ (that is three points without outlier), we would like to maximize the probability w that any selected point is an inlier (w^3 for 3 points). We can derive a relationship between t, w, and N:

$$(1-t)=(1-w^3)^N \Leftrightarrow N=\frac{\log(1-t)}{\log(1-w^3)} \tag{14.3}$$

The number of draws N can therefore be calculated directly from the knowledge of t and w. If t is generally kept constant to 0.99, w has to be estimated with a priori knowledge. The general idea of our approach is to improve the efficiency of a classical RANSAC approach by focusing the drawing of triplets on presegmented areas. In our context, main plane directions correspond to roof facet orientations. As a result, focusing the consensus onto regions sharing the same normal orientation will constrain the probability w to follow specific statistical rules as developed in Section 14.2.3.2.

Algorithm 14.1: Classical RANSAC for detecting roof facets

begin
  repeat
    # Selection of the sets of supports
    while $n \leq N$ **do**
      Randomly select a plane $\mathcal{P}$(3 points)
      # Selecting points within a critical distance of the plane
      $\mathcal{M}_n=\left\{m \in \mathcal{S} / \|m-\mathcal{P}(m)\|^2 \leq d_{cr}\right\}$
      $++n$
    # Select of the set of highest cardinal
    $\exists k \leq N / \forall n \leq N, \text{card}(\mathcal{M}_k) > \text{card}(\mathcal{M}_n)$
    # Estimation of the final plane over all planes
    $\mathcal{P}_{final}=\arg\min_{\mathcal{P}'} \sum_{m \in \mathcal{M}_k} \left\|m-\mathcal{P}'(m)\right\|^2$
    # Removing previous supports from the main point cloud
    $\mathcal{S} \leftarrow \mathcal{S} \backslash \mathcal{M}_k$
  until $card(\mathcal{S})<3$
end

14.2.2 Surface Normal Vector Clustering Based on Gaussian Sphere Analysis

Computing local surface properties in a 3D scene may help the automatic shape recognition step, providing relevant feature descriptions for segmentation purpose. There are several methods for obtaining local surface properties from range data depending on the 3D scene. For terrestrial applications when curved objects are present, quadric or superquadric fitting becomes appropriate [13].

Here, we propose a planar segmentation of the LiDAR point cloud by analyzing the Gaussian sphere of the scene. Normal vectors are calculated for each LiDAR point by extracting a spherical neighboring. A plane is estimated using a robust regression of *M*-estimators' family with norm $L_{1.2}$ [18]. Unlike the standard least-square method that tries to minimize $\sum_i r_i^2$ where r_i is the *i*th residual, *M*-estimators try to reduce the effect of outliers by substituting the squared residuals r_i^2 for another function of the residuals, thus yielding to minimize $\sum_i \rho(r_i)$ where ρ is a symmetric, positive function with a unique maximum at zero, and is chosen to be less increasing than square. This optimization is implemented as an iterative reweighted least power algorithm. In a robust cost model, nothing special needs to be done with outliers. They are just normal measurements that happen to be down-weighted owing to their large deviation.

The mass density of the normal vectors on the Gaussian sphere is described by an Extended Gaussian Image (EGI) [9]. First, the Gaussian sphere can be approximated by a tessellation of the sphere based on regular polyhedrons. Such tessellation is computed from a geodesic dome based on the icosahedron divisions (Figure 14.4a). We define $\mathbf{\Phi}$ as a face of the Gaussian sphere. The EGI can be computed locally by counting the number of surface normals that belong to each cell. The values in the cells can be thought of as a histogram of the orientations. The angular spread (related to the number of faces) depends on the error we tolerate for the coherence of the normal vectors in a cell. As a result, within a specific cell, normals n_x, n_y, n_z will be distributed following a certain density of probability $p(n_x, n_y, n_z)$, which will be analyzed in Section 14.2.3.

Algorithm 14.2 summarizes the EGI generation, where $\mathcal{N}^\circ$ is the set of normal vectors of the scene, $\mathbf{\Phi}$ is an EGI's face, and $[\mathbf{\Phi}] \in \mathbb{R}^3$ is the face's normal vector (Figure 14.3a).

Algorithm 14.2: Computation of an **Extended Gaussian Image**

Data: $\mathcal{N}^\circ$

Result:EGI

begin

  # Select each normal vector

  foreach $\vec{n} \in \mathcal{N}^\circ$ **do**

    # Attribution of each normal vector to an EGI's face

    $[\mathbf{\Phi}] = Argmin_{\mathbf{\Phi} \in \mathrm{EGI}} \left\| \vec{n} - \mathbf{\Phi} \right\|^2$

    # Increment of the 3D histogram of orientations

    ++EGI[Φ]

end

A Gaussian sphere only describes facet orientations, and not their locations. In order to separate two parallel roof facets, an image-based representation is used. Normal vectors belonging to each EGI's face $\mathbf{\Phi}$ are resampled in a 3-channel image, each of them representing a component of the normal vector. There are as much images as EGI faces. Once the image background is removed (binary segmentation), a labeling algorithm is applied to extract the contours of coplanar regions. Connected areas are finally extracted and a set of masks containing single planar regions is generated. Several masks are therefore associated to a single EGI's face $\mathbf{\Phi}$. They are denoted $C_{\mathbf{\Phi}}$.

The set of all $C_{\mathbf{\Phi}}$ describes the coarse delineation of roof facets. We prefer to under-sample the EGI so that each focusing mask should contain more LiDAR points than the real facet

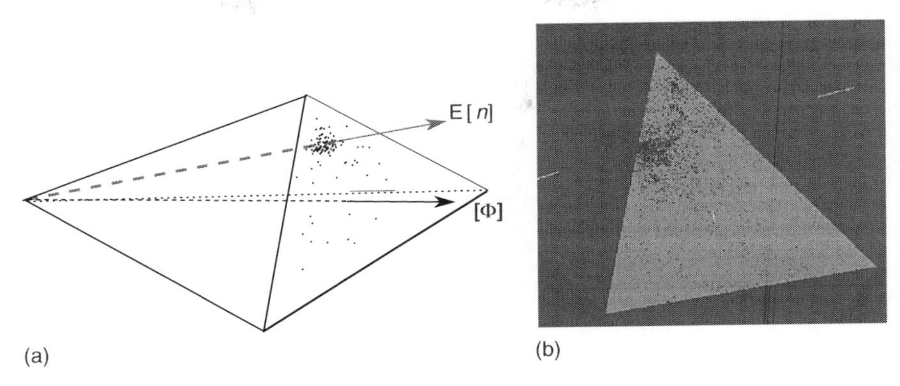

FIGURE 14.3
Generation of an EGI. (a) Details of an EGI's face [Φ]. E[n] is the mean of the orientation vectors. (b) Details of a Gaussian sphere (dark points).

does. RANSAC will therefore be applied sequentially on these sets of masks to extract 3D planar surfaces. This focusing step is then used to estimate two main parameters of RANSAC: the number of draws N to be done and the critical distance d_{cr}.

14.2.3 Parameter Estimation

The methodology developed in this section aims at automatically estimating the parameters of the RANSAC algorithm, which are the critical distance and the number of draws. The first one depends on the S/N ratio of the data while the second one depends on the statistical distribution of points on each face of the Gaussian sphere.

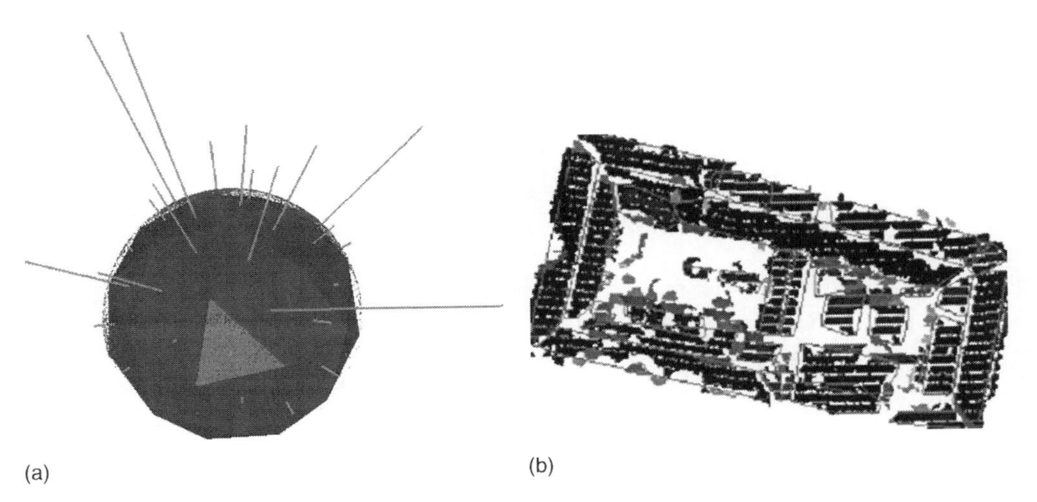

FIGURE 14.4 **(See color insert following page 334.)**
Aspects of the normal driven RANSAC algorithm. (a) Example of an EGI: Orientation histogram collected on a geodesic dome derived from the icosahedron. This is a discrete approximation of the EGI. The length of the green vectors attached to the center of a cell is proportional to the number of surface normals that fall within the range of directions spanned by that cell. Blue points are the projection of normals onto the Gaussian sphere. (b) Projection of LiDAR points (green points) and extracted facets (red polygons) onto focusing masks (coded in grey level depending on their surface).

14.2.3.1 Critical Distance

We noticed in Section 14.2.1 that supports were considered in a set $\mathcal{M}_n$ only if their distances to the associated random plane are less than a critical distance d_{cr}. This distance may be seen as the standard deviation of the supports with regard to the 3D plane. d_{cr} is therefore defined for each focusing mask C_Φ as proportional to the square root of the final residuals of a least square fitted plane estimated from the entire LiDAR points $m \in C_\Phi$. If $\{m'\}$ is the set of orthogonal projection of LiDAR points $\{m\}$ onto the fitted plane, then

$$d_{cr} = \sqrt{\sum_{m \in C_\Phi} \|m - m'\|^2} \tag{14.4}$$

14.2.3.2 Number of Draws

Within a focusing mask C_Φ, normal vectors $\vec{n} \in \mathbb{R}^3$, considered as a random variable, have an empirical probability density function $pc_{-\Phi}(\vec{n})$. We define $w_{C\Phi}$ as the probability for a LiDAR point belonging to C_Φ to be a support of the final plane (inlier). Picking up a point whose orientation is close to the mathematical expectation $\mathbf{E}[\vec{n}]$ is the standard deviation of the distribution, we have:

$$w_{C_\Phi} = \int_{\mathbf{E}[\vec{n}]-\sigma}^{\mathbf{E}[\vec{n}]+\sigma} pc_{-\Phi}(\vec{n})d\vec{n} \tag{14.5}$$

Practically, we consider the component of $\vec{n}$ as 3 independent random variables and

$$w_{C_\Phi} = \int_{\mathbf{E}(n_x)-\sigma_{n_x}}^{\mathbf{E}(n_x)+\sigma_{n_x}} p_x(n_x)dn_x \cdot \int_{\mathbf{E}(n_y)-\sigma_{n_y}}^{\mathbf{E}(n_y)+\sigma_{n_y}} p_y(n_y)dn_y \cdot \int_{\mathbf{E}(n_z)-\sigma_{n_z}}^{\mathbf{E}(n_z)+\sigma_{n_z}} p_z(n_z)dn_z \tag{14.6}$$

The probability density function of each component is explicitly calculated as the derivative of the empirical cumulative distribution function that is described, for the x component as

$$F_{K_x}(x) = \begin{cases} 0 & \text{if } x < \mathbf{inf}\ n_i \\ \dfrac{n_i}{K_x} & \text{if } n_i < x \leq n_{i+1} \\ 1 & \text{if } x > \mathbf{sup}\ n_i \end{cases} \tag{14.7}$$

where

n_i is the proportion of values less than x
K_x the number of realizations of the random variable n_x

Finally, an optimal number of draws is calculated for each focusing mask C_Φ using Equation 14.3 and the previous calculation of $w_{C\Phi}$ and keeping $t = 0.99$.

14.2.4 Results

Figure 14.5 show some results of the extraction method. The normal-driven RANSAC approach enhances the classical RANSAC by parsing the initial 3D scene in pseudo-homogeneous

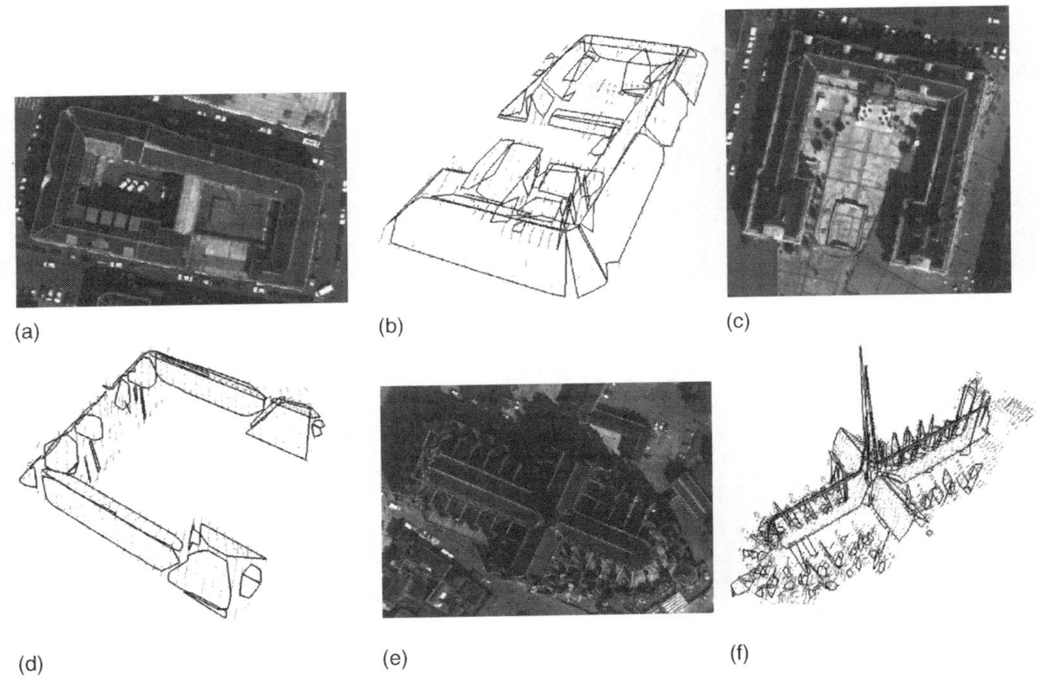

FIGURE 14.5 (See color insert following page 334.)
Right column (a, c, and e) is the result of 3D roof facet detection with normal-driven RANSAC. Left column (b, d, and f) is the corresponding aerial image.

planar regions. This strategy reduces the number of draws in each focusing areas and makes RANSAC valuable for large LiDAR data sets. We can notice that the more accurate the normal vector map (in terms of homogeneity intra-regions), the less the number of draws.

The approximation of the Gaussian sphere by a regular polyhedron, that is its discretization with a fixed step, has advantages with regard to other methods like, for instance, the *K*-means one. The main one is no doubt to manage with the extension of each cell. It represents the tolerance granted around one direction for gathering normal vectors. A rough discretization provides large areas of similar normal vectors. Indeed, this clustering stage is only considered as a focusing stage. Therefore, there is no need to have accurate boundaries since the final facet estimation is performed directly onto the point cloud. As a result, a good parameterization includes a not-so-fine Gaussian sphere approximation to provide large focusing areas. Analyzing the distribution of the 3D points within these focused areas will provide the best robust plane.

This methodology provides patches of planar surfaces where borders are the convex envelop of LiDAR points. These borders have no physical reality, but these primitives bring strong geometric constraints to advanced reconstruction strategies [16].

14.3 Joint Use of Image and LiDAR Data for Roof Facet Extraction

Three-dimensional primitives are initial features for further reconstruction strategies. Integrating aerial images with LiDAR data in a primitive detection process may highly enhance the resulting facets. First, the image geometry integrates the continuity of described objects: intensity changes generally corresponds to object ruptures. Most of the time, roof edges are visible in aerial images. Then, images describe sets of radiometric elements (pixels) that have a strong correlation between each other when describing the same surface (radiometry similarity). Having investigated the potential of extracting planar entities solely from 3D LiDAR data, we present in the following sections, an attempt to integrate image radiometry, LiDAR geometry, and LiDAR semantic in an image segmentation framework for detecting 3D roof facets.

14.3.1 Background

Segmenting an image $\mathcal{I}$ consists of determining a partition $\Delta_N(\mathcal{I})$ of N regions $R_{i\in[1,N]}$. A region is a connected set of pixels that satisfies certain predefined homogeneity criteria satisfying:

$$\Delta_N(\mathcal{I}) = \bigcup_{i\in[1,N]} R_i,\ R_i \bigcap_{i\neq j} R_j = \phi, \forall i, R_i \text{ is connected}$$

The segmentation problem may be considered under various points of views seeing that a unique and reliable partition does not exist. Beyond classical region-growing algorithms, approaches based on a hierarchical representation of the scene retained our attention. These methodologies open the field of multiscale descriptions of images [5]. Here, we are interested in obtaining an image partition wherein roof facets are clearly delineated and understandable as unique entities.

A data structure for representing an image partition is the Region Adjacency Graph (RAG). The RAG of an N-partition (a N-RAG) is defined as an undirected graph $G = (V, E)$ where V (card(V) = N) is the set of nodes related to an image region and E the set of edges related to adjacency relationships between two neighboring regions. Each edge is weighted by a cost function (or energy) that scores the dissimilarity between two adjacent regions. The general idea of a hierarchical ascendant segmentation is to merge sequentially the most similar pair of regions (or the one that minimizes the cost function) until a single region remains. The fusion of two regions (or the contraction of the RAG minimal edge) creates a node in the hierarchy and two father–child relationships in the case of a binary tree. The root of the tree corresponds to image $\mathcal{H}$ and the leaves to the image initial segmentation. A hierarchy $\mathcal{H}$ can be considered as a union of partition sets. Figure 14.6 sketches the generation process of the hierarchy from an initial partition of the image.

A cut of $\mathcal{H}$ is a set of nodes that intersects all branches of $\mathcal{H}$ once and only once. A cut is not necessarily horizontal and provides a partition of $\mathcal{I}$ (Figure 14.7). Both the hierarchy and the cut depend on the segmentation context, and their definitions will vary depending on the applications.

14.3.2 Theory

The shape of the hierarchy (therefore the region-merging order) constrains the existence of an eligible partition. In other words, initial regions that theoretically belong to a roof facet must be mutually merged until a node in the hierarchy appears as a roof facet entity. If it appears that subregions of a facet merge with adjacent regions that do no belong to their supporting facet, the embedded geometry is broken.

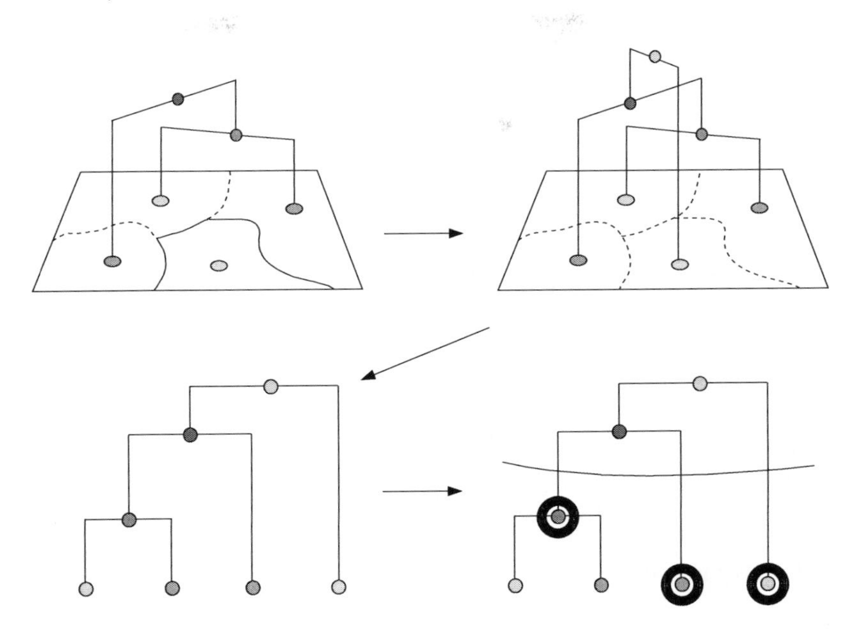

FIGURE 14.6
Construction of a hierarchy based on a RAG (top left).

14.3.2.1 Cost Function

The shape of the hierarchy, as well as the existence of the adequate eligible *cut*, which is a partition of roof facet regions, depends on the region-merging order, and therefore on the definition of the cost function ε. We can write ε as a sum of two terms ε_r, ε_l, and ε_s, respectively, related to the image radiometry, to the LiDAR geometry, and to the semantic extracted from LiDAR data.

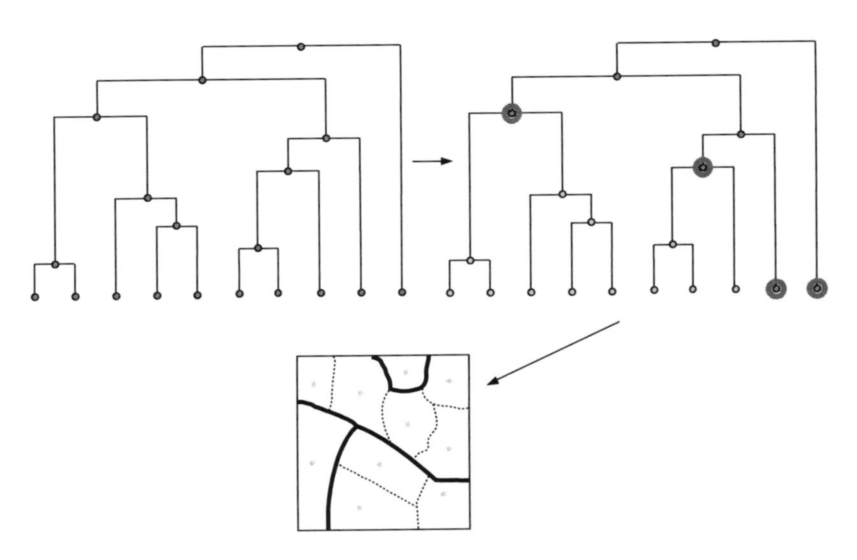

FIGURE 14.7
Sketch of a cut in a hierarchy (dendrogram). Red circles are the selected cut nodes and correspond to the presented image partition.

ε_r is defined to minimize the loss of information when describing the image from n to $n-1$ regions. We retained the cost function given by Haris [7] for merging two neighboring regions R_i and R_j:

$$\varepsilon_r(R_i, R_j) = \frac{\|R_i\|_r \|R_j\|_r}{\|R_i\|_r + \|R_j\|_r}(\mu_i - \mu_j)^2 \tag{14.8}$$

where
$\|\cdot\|_r$ is the number of pixels in each region
$\mu = \frac{1}{\|R\|_r}\sum_k \mathrm{I}(k)$ is the average value of the radiometries at image sites k of the region

In our context, ε_l is defined to take advantage of both the accuracy and the regularity of LiDAR measurements onto roof surfaces to make appear in the hierarchy nodes corresponding to roof facet entities. It is therefore expected that all regions included in a single roof facet all together merge before one of them merge with the neighboring facet. Higher levels of the hierarchy are not of interest here. The adequation of LiDAR points to lie on a roof facet is measured by estimating a plane onto those included in $R_i \cup R_j$. A nonrobust least square estimator is applied specifically to neighboring regions such that they do not merge when the estimated plane is corrupted by noncoplanar points. Such is the case when attempting to merge two regions apart from the roof top before other couples of regions belonging to the same roof facet with possible significant radiometric dissimilarities. If $\|N_i\|_l$ (resp. $\|N_j\|_l$) is the number of LiDAR points in region R_i (resp. R_j) and r_p the residuals of a LiDAR point to the estimated plane, ρ_l^2 is the average square distance of LiDAR points to the estimated plane with

$$\rho_l^2 = \frac{1}{\|N_i\|_l + \|N_j\|_l} \sum_{p=1}^{\|N_i\|_l + \|N_j\|_l} r_p^2$$

If we consider a similar weighting factor as for ε_r depending on the number of LiDAR points $\|.\|_l$ in image regions, ε_l is expressed as

$$\varepsilon_l(R_i, R_j) = \frac{\|N_i\|_l \|N_j\|_l}{\|N_i\|_l + \|N_j\|_l}\rho_l^2$$

3D LiDAR points that have been previously processed to extract a binary semantics: ground and off-ground points. Theoretically, the off-ground class includes building and vegetation. However, we will only consider the segmentation algorithm that focuses on buildings. The process is performed with a high level of relevancy over urban areas owing to the sharp slope breaking onto building edges [2]. An image region will be classified as ground if it contains at least one projected LiDAR ground point. Otherwise, the region is considered to be a built-up area. This binary semantics provides a reliable ground mask that can be integrated into the initial segmentation. Two regions of different classes are kept disjoint until the highest levels of the hierarchy. The methodology is summarized in Figure 14.8. Finally, we can write

$$\varepsilon_s(R_i, R_j) = \begin{cases} \infty & \text{if } R_i \text{ or } R_j \text{ is a } \textbf{ground} \text{ region} \\ 0 & \text{otherwise} \end{cases}$$

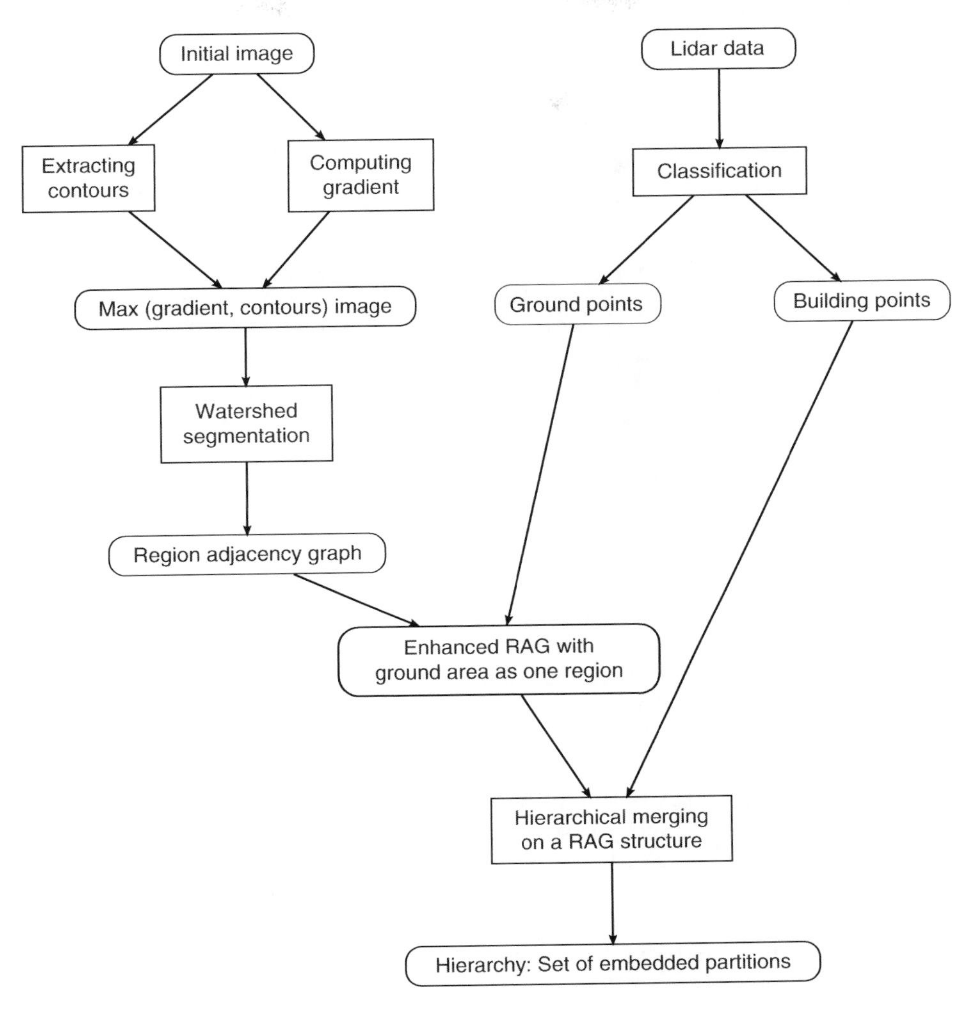

FIGURE 14.8
Flow chart of the joint segmentation methodology.

14.3.2.2 *Optimal Eligible Cut*

A roof facet is defined as a 3D planar polygon whose average square distance to LiDAR support points (ρ_1^2) is less than a threshold, *s*. The final partition is obtained by recursively exploring the binary tree structure from its root comparing ρ_1^2 of each node to *s*.

14.3.3 Results

Planar 3D primitives extracted by the joint segmentation process LiDAR or image are presented in Figures 14.9 and 14.10. For these applications, an orthorectified image has been used in order to avoid the facade detection problem. Red points are LiDAR points belonging to a single strip. We clearly distinguish the optic fiber LiDAR device used for processing the planar primitives. We used a threshold, *s*, of 0.5 m in the first figure and 0.7 m in the second case. This reconstruction considers LiDAR points belonging to an image region larger than 30 pixels and whose orientation is greater than 30° from the zenith direction. The presented

FIGURE 14.9 (See color insert following page 334.)
Extracted 3D facets of a complex building from the joint segmentation process LiDAR or image represented with LiDAR data (red points).

3D scenes give a realistic representation of the buildings wherein hyperstructures such as dormer windows are particularly visible. In order to enforce the region borders to lie on real discontinuities, we applied a contour detection algorithm (hysteresis thresholding) on the gradient image. The gradient was computed with a Canny-Deriche operator. The watershed algorithm is finally applied on a combination of both images (maximum of gradient and contour images).

When comparing Figure 14.9 with Figure 14.5b, we can notice a better detection of planar primitives as well as a fine description of plane orientation. The introduction of a semantic term in the energy produced a sharp separation between points belonging to the ground and the others, and can be observed in Figure 14.11b where regions belong only to build-up areas. The delineation of planar entities is also improved with regard to the ones extracted directly from the point cloud without image information. Figure 14.11 is a visual comparison between both methodologies developed in this chapter. The patch effect is removed in

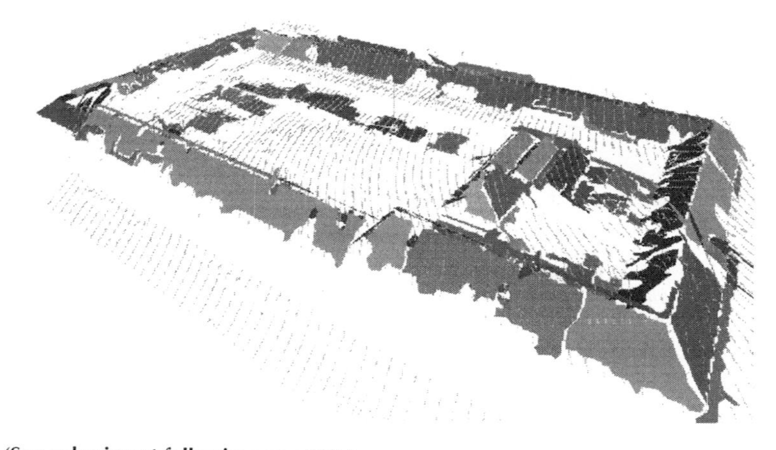

FIGURE 14.10 (See color insert following page 334.)
Extracted 3D facets of another building from the joint segmentation process LiDAR or image represented with LiDAR data (red points).

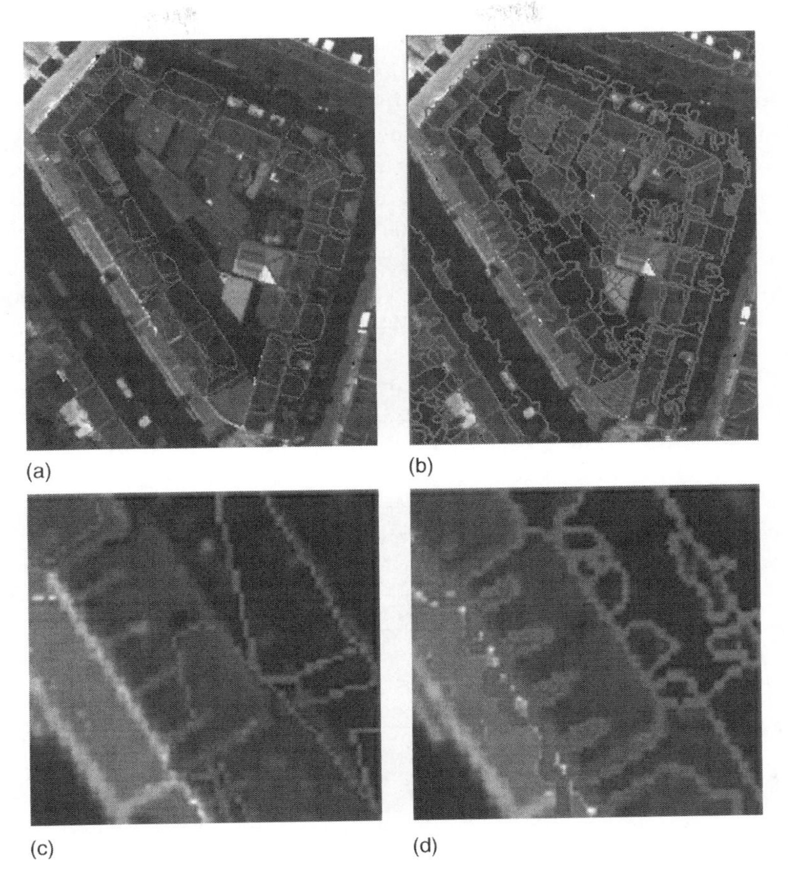

FIGURE 14.11
Comparison of 3D facets extracted from the normal-driven RANSAC algorithm to the hierarchical segmentation process. (a) 3D facets extracted from normal-driven RANSAC (red) and projected onto an ortho-image. (b) Result of the joint segmentation process. Red polygons are eligible roof facets. (c) Zoom in image (a) of a particular facet extracted from the normal-driven RANSAC algorithm. (d) Zoom in image (b) of the same facet calculated from the joint segmentation process.

Figure 14.11d and region frontiers are coherent with regard to the image-based building edges. One can remark on some remaining small regions near the building edges that can be attributed either to radiometric artefacts introduced during the orthorectification process (resampling) or to building shadows whose radiometry is similar to the roof's one.

The actual implementation of the methodology depends on parameter, s, which is set a priori. The threshold theoretically varies with each building, and even with each roof facet depending on its regularity (dormers, chimneys, antennas, etc.).

14.4 Conclusion

We have presented in this chapter two methodologies for extracting 3D planar primitives in a building reconstruction context: the first one is based on the well-known RANSAC paradigm, and the second one investigates the potentialities of using LiDAR data and aerial images jointly. The algorithm developed in Section 14.2 shows that the recognition of planar surfaces in a point cloud is a challenging task. Firstly, the point density may be a limiting factor for the building description: the higher the point density, the better the roof facet description. However, the orientation of planar surfaces is of quality due to the robustness of the LiDAR measurements. Secondly, the topology is not well adapted to generate fine edges since we cannot predict a LiDAR impact to hit a particular area. As a result, the integration of aerial images may overcome this problem. The radiometric correlation between two neighboring pixels is high. It ensures to retrieve the continuity of particular linear structures, and tends to describe a single facet with neighboring radiometries. The hierarchical segmentation framework allows to integrate as a unique energy function, information from LiDAR data and from aerial images. Even if this approach need improvements, results are qualitatively better than the RANSAC approach.

Photogrammetry and airborne LiDAR are not antagonist techniques. On the contrary, a relevant combination of images and LiDAR data is the future research direction for improving performances of feature-extraction algorithms. The use of full-waveform LiDAR data may also be relevant in urban areas beyond the densificatiorn of the point cloud. Among many dependencies, waveforms vary depending on the roof material and on its geometry. Future researches would consist of establishing neighboring relationships between waveforms by extracting morphological characteristics.

References

1. C. Baillard and O. Dissard. A stereo matching algorithm for urban digital elevation models. *Photogrammetric Engineering and Remote Sensing*, 66(9):1119–1128, Sep. 2000.
2. F. Bretar, M. Chesnier, M. Pierrot-Deseilligny, and M. Roux. Terrain modeling and airborne laser data classification using multiple pass filtering. In: *Proceedings of the XXth ISPRS Congress*, Volume XXXV Part B of *The International Archives of the Photogrammetry, Remote Sensing and Spatial Information Sciences*, International Society for Photogrammetry and Remote Sensing (ISPRS), Istanbul, Turkey, Jul 2004, pp. 314–319.
3. M.A. Fischler and R.C. Bolles. Random sample consensus: A paradigm for model fitting with applications to image analysis and automated cartography. *Graphics and Image Processing*, 24(6):381–395, Jun. 1981.
4. H. Gross and U. Thoennessen. Extraction of lines from laser point cloud. In: *Proceedings of the ISPRS Commission III Symposium on Photogrammetric and Computer Vision*, Volume XXXVI in *The International Archives of the Photogrammetry, Remote Sensing and Spatial Information Sciences*, ISPRS, Bonn, Germany, 2006, pp. 86–91.
5. L. Guigues, J.-P. Cocquerez, and H. Le Men. Scale sets image analysis. *International Journal of Computer Vision*, 68(3):289–317, Jul. 2006.
6. A. Habib, M. Ghanma, M. Morgan, and R. Al-Ruzouq. Photogrammetric and lidar data registration using linear features. *Photogrammetric Engineering and Remote Sensing*, 71(6):699–707, Jun. 2005.
7. K. Haris, S. Efstratiadis, N. Maglaveras, and A.K. Katsaggelos. Hybrid image segmentation using watersheds and fast region merging. *IEEE Transactions on Image Processing*, 7(12):1684–1689, Dec. 1998.
8. R. Hartley and A. Zisserman. *Multiple View Geometry in Computer Vision*. Cambridge University Press, U.K. 2002.
9. B.K.P. Horn. Extended Gaussian images. *PIEEE*, 72(12):1656–1678, Dec. 1984.

10. H. Jibrini. Reconstruction automatique de batiments en modeles polyhedriques 3-D a partir de donnees cadastrales vectorisees 2-D et un couple d'images aeriennes a haute resolution. PhD thesis, Ecole Nationale Superieure des Telecommunications de Paris, 2002.
11. H. Jibrini, M. Pierrot-Deseilligny, N. Paparoditis, and H. Maitre. Automatic building reconstruction from very high resolution aerial stereopairs using cadastral ground plans. In: *Proceedings of the XIXth ISPRS Congress, The International Archives of the Photogrammetry, Remote Sensing and Spatial Information Sciences*, ISPRS, Amsterdam, The Netherlands, Jul. 2000.
12. M. Kasser and Y. Egels. *Digital Photogrammetry*. Taylor & Francis, 2002.
13. A. McIvor and R.J. Valkenburg. A comparison of local surface geometry estimation methods. In: *MVA IAPR Conference on Machine Vision Applications*, Vol. 10, 1997, pp. 17–26.
14. M. Pierrot-Deseilligny and N. Paparoditis. A multiresolution and optimization-based image matching approach: An application to surface reconstruction from spot5-hrs stereo imagery. In: *Proceedings of the ISPRS Conference Topographic Mapping from Space (with Special Emphasis on Small Satellites)*, ISPRS, Ankara, Turkey, Feb. 2006.
15. T. Schenk and B. Csatho. Fusion of lidar data and aerial imagery for a more complete surface description. In: R. Kalliany and F. Leberl, Eds., *Proceedings of the ISPRS Commission III Symposium on Photogrammetric and Computer Vision*, Volume XXXIV of *The International Archives of the Photogrammetry, Remote Sensing and Spatial Information Sciences*, Institute for Computer Graphics and Vision, Graz University of Technology, Graz, Austria, 2002, pp. 310–317.
16. F. Taillandier. Automatic building reconstruction from cadastral maps and aerial images. In: U. Stilla, F. Rottensteiner, and S. Hinz, Eds., *Proceedings of the ISPRS Workshop CMRT 2005: Object Extraction for 3D City Models, Road Databases and Traffic Monitoring—Concepts, Algorithms and Evaluation*, Vienna, Austria, Aug. 2005, pp. 105–110.
17. G. Vosselman and S. Dijkman. 3d building reconstruction from point cloud and ground plans. In: *Proceedings of the ISPRS Workshop on Land Surface Mapping and Characterization Using Laser Altimetry*, Volume XXXIV of *International Archives of Photogrammetry and Remote Sensing*, Annapolis, MD, Oct. 2001, pp. 37–43.
18. G. Xu and Z. Zhang. *Epipolar Geometry in Stereo, Motion and Object Recognition*. Kluwer Academic, Dordrecht, The Netherlands, 1996.
19. K. Zhang, J. Yan, and S.C. Chen. Automatic construction of building footprints from airborne lidar data. *IEEE GRS*, 2006.

15

Building Extraction from LiDAR Point Clouds Based on Clustering Techniques

Jie Shan and Aparajithan Sampath

CONTENTS

15.1 Introduction

Modeling buildings has been of interest among a wide range of research communities, including computer vision, computer graphics, geoinformatics, urban science, and telecommunication. Recently, this interest is rapidly growing due to the increasing popularity and availability of the light ranging and detection (LiDAR) technique, which directly provides 3D dense point representations (the so-called point clouds) of the terrain surface. LiDAR point clouds collected over urban areas consist of points reflected from man-made objects like buildings, cars, etc. and from natural surfaces such as bare earth terrain, trees, etc. The points from these different reflective surfaces must first be separated from each other so that they can be modeled. As for building modeling, a well-accepted procedure is to determine bare earth (also called filtering) as the first step, followed by building extraction as the second step. The basic idea for bare earth determination is to apply a filter to remove the nonground points in the raw LiDAR point clouds, often in a recursive manner, such that the ground points can be separated. For a detailed description on different filters, the reader is directed to Kilian et al. (1996), Kraus and Pfeifer (1998), Vosselman (2000), Sithole and Vosselman (2004), Shan and Sampath (2005), Zhang and Whitman (2005).

Building extraction usually starts from the nonground dataset obtained through the filtering process. A number of methods have been developed for this purpose. In terms of the data involved and utilized, some methods use LiDAR data alone, while others use them along with additional data, such as images or maps. Building extraction methods can also be characterized as model-driven or data-driven with respect to the underlying assumptions made to the buildings. Finally, the methods applied to tackle the building extraction task can be distinguished as geometric or statistical ones based on the dominant principles applied in this task. It should be noted that the above characterization is mostly for the convenience of discussions and review. Many building extraction practices may actually use a combination of different techniques. A brief review of these methods is presented below.

Auxiliary data have been used along with LiDAR data. Haala and Brenner (1999) presented an approach to derive 3D building geometry using the LiDAR surface model, 2D (building) ground plans, and terrestrial photographs. Building outlines were extracted from 2D ground plans, while the LiDAR data were used to build the 3D views. Vosselman and Dijkman (2001) used 3D Hough transform to detect planar faces from LiDAR point clouds. The 2D ground plans were then exploited to determine the intersection of roof faces and height jump edges as a critical step towards building reconstruction. Similar studies were reported by Overby et al. (2004), Hofman et al. (2002), and Schwalbe et al. (2005). The possible incompleteness, limited accuracy of the available 2D ground plans, and especially the fact that roof details may not be available in 2D databases pose limitations to such methods. Vögtle and Steinle (2000) used color images to exclude vehicles, trees, and vegetations from the initial buildings detected from LiDAR points. Schenk and Csathó (2002) detected roof breaklines from panchromatic images based on texture discontinuity and utilized them to refine the results from LiDAR data. Rottensteiner et al. (2005) used multispectral images along with LiDAR data for building reconstruction. Initially the building pixels were detected by a classification technique based on surface model and ground model obtained from image matching, (normalized difference vegetation index), texture strength, texture direction, and the elevation difference between the first and the last laser pluses. In the subsequent steps, the roofs segmented from LiDAR data were improved and precisely located by using edges detected from the imagery. A similar study was reported by Sohn and Dowman (2007), where IKONOS pan-sharpened multispectral images were used for building candidate detection. Although imagery is useful in building extraction from LiDAR data, there are some limitations, such as not all surface discontinuities are detectable due to weak contrast, occlusion, incomplete breaklines, and false alarm breaklines on the imagery. Moreover, other possible limitations include misclassification due to similar spectral reflectance, radiometric effects, and the large number of object classes in the classification procedure.

Once specific assumptions are made on the building models, they can be determined by the so-called model-driven approach. Mass and Vosselman (1999) utilized invariant moments to determine the roof parameters of regular building types using the raw laser scanning data. They worked on simple roof models, such as the gable roof and its variations, and determined the parameter values for such models. Studies have shown that some complex buildings can be divided into smaller parts so that they can be replicated as a series of simpler models. The difficulty lies in the fact that invariants may not exist for complex buildings. Recently, Tarsha-Kurdi et al. (2007b) compared the properties of the model- and data-driven approaches. It is concluded that the model-driven approach can rapidly create building models with appealing appearance, while the data-driven approach is able to consider faithful details pertinent to the resolution of the input data.

If buildings to be extracted are hard to parameterize or the data are quite noisy, the data-driven approach is a reasonable option, which actually constitutes the majority of the existing studies. In this approach, only generic assumptions such as planar roof are made about the buildings. Therefore, finding the constituent planes or other primitives such as lines and vertices becomes the key issue. Rottensteiner and Briese (2003) generated a digital surface model and determined a few homogenous pixels, i.e., pixels that are most likely to be planar. Connected homogenous pixels were used as initial seeds to generate planar regions. Planar facets were then used to determine lines of intersection of planes and step edges. This method was improved in (Rottensteiner et al., 2005) by introducing statistical tests to minimize its dependence on the thresholds at all stages, including detection and classification of roof planes, breaklines, and step edges. Forlani et al. (2003) used a similar approach, where building "pixels" were segmented and for each "pixel" a gradient slope was calculated. The connected building "pixels" with similar gradient slopes were then clustered. Peternell and Steiner (2004) separated the LiDAR data into a number of grids. For each grid and its eight neighbors, the surface normals were determined and the grids with similar surface normals were linked using a connected component analysis. Alharthy and Bethel (2004) used a technique of moving windows, which is similar to the above technique. They determined the slope in X, slope in Y, and the Z interception for points in each grid. Then a region growing approach was used to extract planar segments. Tarsha-Kurdi et al. (2007a) used Hough transform in combination with the RANSAC (random sample consensus) approach for robust roof plane determination. Brenner (2005) summarized various building reconstruction techniques using image, LiDAR, and map data that have been suggested by different authors. Rabbani et al. (2007) reviewed the techniques to extract different surfaces, including planar, cylindrical, and spherical surfaces from LiDAR point clouds.

Statistical approaches dominantly use classification or clustering techniques to seek building constituents such as planar segments in the LiDAR data. Nizar et al. (2006) demonstrated an approach to building reconstruction similar to (Sampath and Shan, 2006). They defined feature vectors for each LiDAR point based on a tangential plane (three measures of a normal vector) and a height difference measure. These four measures were then used to cluster points into surface classes and separate similar surface classes based on their spatial proximity. Despite the similarities, Sampath and Shan (2006) did not agree that all the LiDAR points in the building subset will have planar characteristics even under the assumption of planar roofs. Fitting a plane at roof edges, roof ridges, or trees would lead to errors in clustering. Therefore, they removed such ambiguous points (nonplanar points) first, which led to a better performance in clustering (Sampath and Shan, 2006). As a matter of fact, once the LiDAR technology started to become popular, it was realized that breaklines are necessary for building extraction from LiDAR data and yet difficult to extract. Significant efforts were made towards this objective, including fusing imagery with LiDAR data as addressed above. Schenk and Csathó (2002, 2007) pointed out that detecting breaklines as the first step in building extraction intrinsically follows the human perception process and principle, which states that breaklines are essential for detecting and recognizing an object.

In this chapter, building extraction is treated as a machine learning or data mining problem. The general objective is to find spatial patterns in the data, which in this case are the points lying on the same roof segment. The generic assumption about the object is that the building roofs are planar in nature. Based on this assumption, all the planar segments that make up a building roof will be separated. We accomplish this task in a step-by-step manner. The first step is to separate the planar parts of the dataset from the nonplanar parts, such as trees, roof ridges, and roof edges. This separation is accomplished through the principal component analysis (PCA) technique by employing its capability and property in reducing the dimensionality of a high-dimensional dataset. In the second step, the separated planar points are clustered into planes with the same direction through a potential-based *k*-means clustering algorithm. The local surface normals or eigen vectors resulting from PCA in the first step are used as feature vectors for clustering. The joint use of the potential calculation and the *k*-means clustering allows us to autonomously determine the number of directional planes for a building and their normal directions. The third step separates parallel planes with distinct vertical offset, and coplanar segments that have the same mathematical expression. In the final step, a set of mutually adjacent planar segments is found and their intersection is determined as a roof vertex. The connection of all such determined vertices based on the adjacency of the planar segments will ultimately lead to the reconstructed building models.

The rest of the chapter is structured as follows. It will first present a brief introduction to the clustering technique, including the related concepts and terms, and point out the difficulties when this technique is applied to LiDAR data. The next section will describe the breakline detection, i.e., the separation of planar points with nonplanar points, using the PCA technique. The section on clustering mainly applies the *k*-means algorithm to separate planar points into individual directional planes, with the discussion focused on the determination of the number of directional planes and their directions. The clustering results are evaluated in a separate section based on their statistics and visual comparison with reference to the LiDAR data and higher resolution images. The building reconstruction section presents an approach that uses mutually adjacent planar segments to define a building vertex and connects all vertices to form the final building model. Our experience and conclusions are summarized in the last section.

15.2 Clustering Techniques

Data clustering is a widely used technique in pattern recognition, remote sensing, and data mining. Through this technique, a heterogeneous dataset with different properties will be partitioned into a number of subsets such that each subset should share similar properties. Obviously, to achieve this objective, we must first define the "properties" and measure the "similarity," which are two fundamental and difficult issues in clustering. In general, the data properties are often defined by the selection of a feature space, while the similarity is measured by distance.

15.2.1 Data Space and Feature Space

Data can be represented in different spaces or coordinate systems. The space in which the data are initially measured or represented during acquisition is often called data space. This can be the pixel values (r, g, b, etc.) of images or locations (X, Y, Z coordinates) of the LiDAR points. Feature space is used to characterize certain properties of the data and may be different from the data space. Depending on the type of pattern that is being sought, the feature space can be formed differently. For instance, in image classification, the feature space is the pixel values of the entire or part of the multispectral or hyperspectral bands. Such selection of feature space would allow us to differentiate the physical properties of the pixels with respect to reality. For LiDAR data clustering, the original data space often needs to be converted to a feature space because the initial data space (metric space) only contains information about LiDAR data locations. For a gable roof building with two intersected planar segments, if we are to cluster the LiDAR points on one plane, we would not directly use their coordinates since they only represent locations. Instead, if the feature space consists of the direction cosines (i.e., surface normal) defined at the vicinity of a LiDAR point, clustering techniques can easily separate the two planes. Choosing the relevant properties to represent the data and form the corresponding feature space is called feature selection in pattern recognition (Duda et al., 2001). The selected feature properties will form a feature vector $\mathbf{F} = [F^1, F^2, \ldots, F^m]^T$, where m is the number of selected features F^u ($u = 1,2,\ldots,m$). A cluster is formed by one or more adjacent data points in the feature space. The subsequent clustering computations will occur in the feature space by working with the feature vectors.

15.2.2 Similarity Measures

Distance is used to measure the similarity of features such that the data points (in feature space) with significant distance can be separated. The type of distance measures used can influence the shape and the size of the clusters. Some of the popular distance measures between two feature vectors $\mathbf{F}_i$ and $\mathbf{F}_j$ are listed below

Euclidean distance

$$d(\mathbf{F}_i, \mathbf{F}_j) = \sqrt{(\mathbf{F}_i - \mathbf{F}_j)^T (\mathbf{F}_i - \mathbf{F}_j)} \tag{15.1}$$

Manhattan distance (also called taxicab norm or 1-norm)

$$d_1(\mathbf{F}_i, \mathbf{F}_j) = \sum_{u=1}^{m} \left| F_i^u - F_j^u \right| \tag{15.2}$$

Mahalanobis distance

$$d_M(\mathbf{F}_i, \mathbf{F}_j) = \sqrt{(\mathbf{F}_i - \mathbf{F}_j)^T \Sigma^{-1} (\mathbf{F}_i - \mathbf{F}_j)} \tag{15.3}$$

where $\mathbf{\Sigma}$ is the covariance matrix of the feature vector $\mathbf{F}$. The Mahalanobis distance corrects the scale differences and the correlations in the constituent elements in the feature vector. In addition, the angle between two vectors can also be used as a distance measure when clustering high-dimensional data.

$$d_A(\mathbf{F}_i, \mathbf{F}_j) = \frac{\mathbf{F}_i \bullet \mathbf{F}_j}{\|\mathbf{F}_i\| \cdot \|\mathbf{F}_j\|} \tag{15.4}$$

where $\bullet$ is the inner product of two vectors.

15.2.3 Clustering Methods

Several types of clustering methods are available with each having its own advantages and disadvantages (Duda et al., 2001; Berkhin, 2002; Jain et al., 1999). Hierarchical methods build a cluster hierarchy or a tree of clusters known as a dendrogram. Every cluster node contains child clusters; sibling clusters partition the nodes covered by their common parent. Hierarchical algorithms can be agglomerative or divisive. Agglomerative algorithms are a bottom-up method that begins with the assumption that each data point is a separate cluster and then merges some data points into one cluster based on a distance criterion. The process continues until a stopping criterion, such as the number of clusters *k*, is achieved. The divisive algorithms are a top-down method, which starts with a single cluster containing all data points and then successively splits the resulting clusters until the predefined number of clusters *k* is reached.

Partitioning methods iteratively relocate the data points among various clusters until they achieve the greatest distinction. The popular *k*-means algorithm, for instance, belongs to this category, and will produce exactly *k* different clusters of greatest possible distinction (Duda et al., 2001). The *k*-means algorithm is explained in detail later in the next section. The partitions may be formed using an objective function that leads to what are called decision boundaries. As an extension of the traditional partitioning method, the model-based methods assume that the data are generated from a mixture of probability distributions with each component corresponding to a different cluster. The goal of the clustering algorithm is then to maximize the overall probability or likelihood of the data (Duda et al., 2001; Jain et al., 1999).

Density-based methods mostly work in the original data space (Berkhin, 2002). The nature of the problem is similar to image segmentation where the data are separated into different spatial regions. They are particularly useful when the shape and the size of the clusters are random. These methods calculate the density at each data point and use a connectivity criterion to categorize all connected points that have similar density. We will use the concept of density-based clustering for the segmentation of coplanar roof segments in the original data space.

15.2.4 Number of Clusters

Many clustering algorithms require an initial estimate on the number of clusters in the dataset. This is used either as the stopping condition (hierarchical methods) or initial condition (*k*-means method). Estimating the number of clusters in the dataset is one of the trickiest aspects in clustering. The elbow criterion (Duda et al., 2001) is a popular method to determine what number of clusters should be chosen in a clustering method. Since the initial assignment of cluster seeds affects the final clustering performance, it is necessary to run the cluster analysis multiple times. The elbow criterion states that one should choose the number of clusters such that adding another cluster will not yield significant variation in the similarity or likelihood measures of the clusters. More precisely, in a graph showing the similarity of clusters (vertical direction) against the number of clusters (horizontal direction), the first few clusters often correspond to high variation in the similarity and at some point the variation becomes small and stable, yielding an angle or elbow in the plot (see Figure 15.4). This elbow point suggests the number of clusters to be chosen.

15.2.5 Noise and Ambiguities

Data to be clustered are not free of noise and may contain outliers that do not comply with the underlying assumptions about the data or the objects we are seeking. In laser ranging measurements, noise may be caused by atmospheric conditions or multiple reflections. Trees may also remain in the nonground dataset as a result of imperfect filtering. Such data points should be regarded as noise since they do not belong to the set of objects (buildings) to be detected. Ambiguity occurs when some data points belong to more than one cluster. This is the case for LiDAR points at roof ridges or roof edges, where they are located at the intersection of two roof planes, or of one roof plane and a vertical wall.

Noisy and ambiguous data affect the clustering results. Because of their presence in computation, feature vectors such as surface normals will not be calculated accurately. Besides, due to the limitation of LiDAR data resolution, feature vectors of small roof components such as chimneys and attachments may not be reliably estimated either. The participation of inaccurate or unreliable feature vectors in clustering will in turn cause confusion and difficulty to the clustering algorithm. This is illustrated in Figure 15.1, which shows a point cloud over a building and its neighborhood (a) and its normal vectors (b). Figure 15.1c and d display the same plots, but after the removal of noisy (trees) and ambiguous (roof ridges and edges) points from the feature vector calculation. It is clear that Figure 15.1a and b are much noisier than Figure 15.1c and d, and hard to differentiate the clusters. Therefore, as a general guideline, noisy and ambiguous points should be excluded from the clustering as early as possible, even from the very first step of feature selection to ensure satisfactory clustering performance. This again justifies the breakline detection as the first step in building extraction under the scope of clustering techniques.

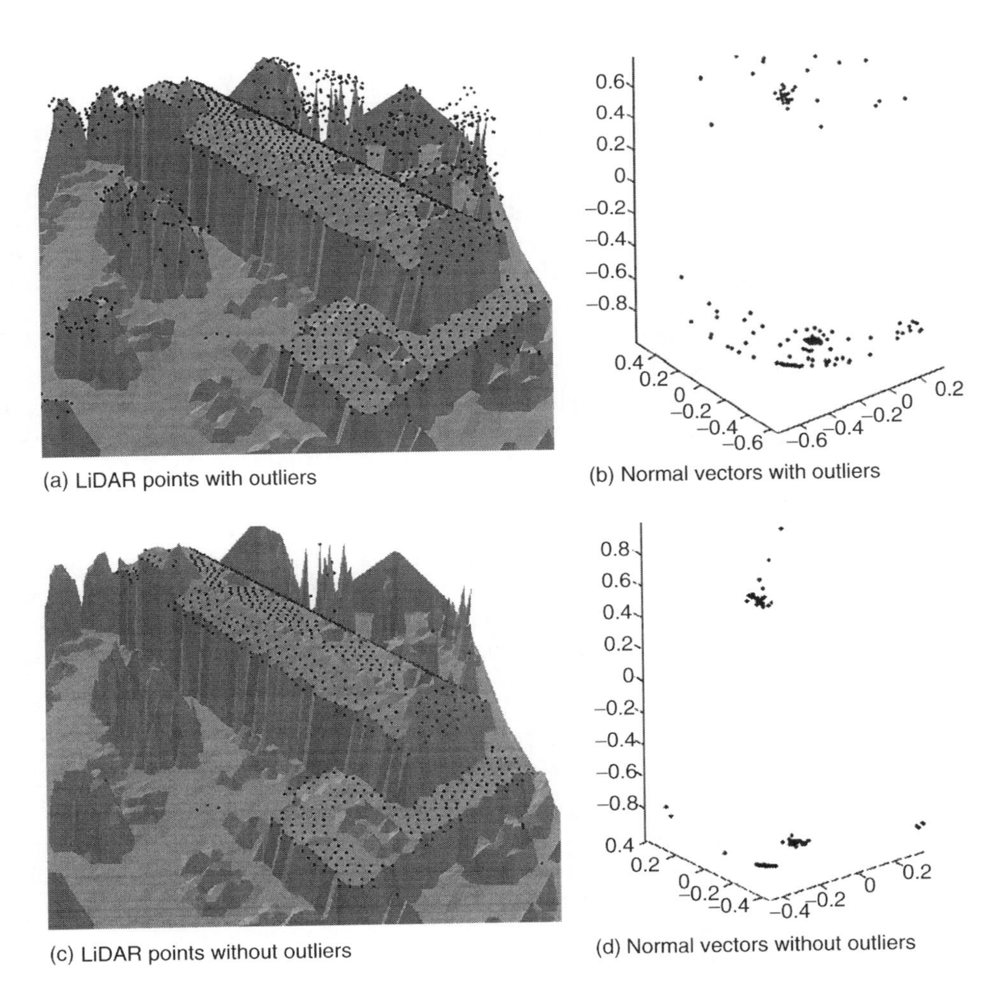

FIGURE 15.1
Effects of outliers (noisy and ambiguous points) on clustering. (a) The point cloud of a building and its neighborhood and (b) its surface direction cosines, (c) the point cloud after removing outliers and (d) its surface direction cosines.

15.3 Breakline Detection

Based on the above principles, we devise the following building extraction strategy. Our underlying assumption is that the building is piecewise planar or polyhedral and our ultimate objective is to determine each planar roof segment. The first step is to separate outliers (noisy and ambiguous points) from the LiDAR data and exclude them from clustering. As pointed out earlier, noisy points may include trees close to, or hanging over a building. Examples of ambiguous points are LiDAR points at the intersection of roof planes or a step-edge that is elevated from its surroundings. All such outliers in this study are regarded as breakline points or, more technically, nonplanar points. This term is chosen to reflect the fact that the proposed strategy should still be followed even when filtering is perfect and no trees remain in the dataset. Besides, we use this term to emphasize the fact that breakline detection actually is a broader and profound concept in visual perception and computer vision. An object is best characterized by places with sudden changes in surface orientation, which correspond to roof edges and ridges for a building. Therefore, breakline detection is also a necessary step for building extraction from visual perception point of view.

Under a certain homogeneity measure, the breaklines would be the locations where the data are least homogenous. The homogeneity measure might vary among the sensors with which the data are collected. Edge detectors for imagery, such as Canny and Laplacian of Gaussian, employ the intensity change at a particular pixel. However, the fact is that LiDAR data are geometric, both discrete and irregular, and at much lower resolution than imagery. It is unreasonable to expect any LiDAR returns exactly from building breaklines. Therefore, one can only extract a small strip of LiDAR points that surround the actual breakline as its approximation, which needs further refinement based on certain assumptions about the building (Brugelmann, 2000; Briese, 2004).

This study applies the linear space theory to separate breaklines (nonplanar points) from planar points. An n -dimensional linear space can be defined by n mutually orthogonal basis vectors. Any point in this space can be represented as a linear combination of the basis vectors. In our problem, if a group of LiDAR points can be represented by only two other than three basis vectors (i.e., the third coefficient in the linear combination is zero), then this group of points lies on a plane. By the same logic, for breakline points they need all three basis vectors. The PCA is applied to determine the necessary basis vectors. It is based on the eigenvalues and eigenvectors of the covariance matrix of a (small) group of LiDAR points in a neighborhood (Fransens et al., 2003). Let $\mathbf{P}$ be the point group with points $\mathbf{X}_i$ $(i = 1,\ldots, p)$ in $\mathcal{R}^3$ and $\bar{\mathbf{X}}$ be their mean vector, the covariance matrix Σ_{XX} is calculated as

$$\Sigma_{XX} = \sum_{i=1}^{p}(\mathbf{X}_i - \bar{\mathbf{X}})(\mathbf{X}_i - \bar{\mathbf{X}})^{T} \tag{15.5}$$

For this covariance matrix Σ_{XX}, we determine its eigenvalues λ_1, λ_2, λ_3 and their corresponding eigenvectors Λ_1, Λ_2, Λ_3. The eigenvalues are sorted such that $\lambda_1 \leq \lambda_2 \leq \lambda_3$. If the LiDAR point set $\mathbf{P}$ consists of points that lie on a plane, the minimum eigenvalue λ_1 should be very close to zero and the corresponding eigenvector Λ_1 should be in the direction perpendicular to the plane, i.e., the normal of the plane. On the other hand, if λ_1 is not negligible, it can be concluded that the point set $\mathbf{P}$ does not lie on a plane. For buildings consisting of only planar roofs, such a point set $\mathbf{P}$ actually lies at the intersection of two planes, i.e., breaklines. In the implementation of the above principle, the LiDAR data are divided into small nonoverlap

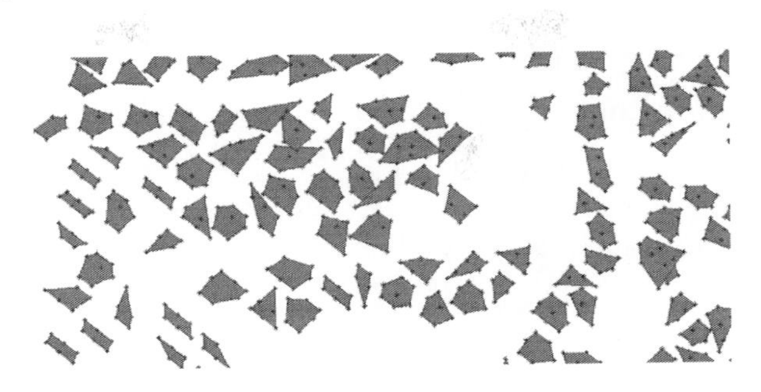

FIGURE 15.2
Patch distribution of a subset of LiDAR points.

groups called patches. A patch is formed by selecting a single LiDAR point and then gathering its neighboring points in all directions. Ideally, each patch should consist of six to eight points. Mathematically, any three points can be shown to lie on a plane. To judge whether a group of points are planar or not, we need at least four points. To balance the reliability for surface normal determination and the detectability for small roof objects, we recommend six to eight points as the size of a patch. This suggestion can be compared to the four and eight neighborhoods used in image processing and computer vision. Figure 15.2 below shows the distribution of patches formed by a subset of LiDAR points.

For each patch, the PCA is carried out. We define the normalized eigenvalue $\bar{\lambda}$ as

$$\bar{\lambda} = \frac{\lambda_1}{(\lambda_1 + \lambda_2 + \lambda_3)} \tag{15.6}$$

Patches with the normalized eigenvalue $\bar{\lambda} < \bar{\lambda}_T$, where $\bar{\lambda}_T$ is a given threshold, are regarded as being formed by planar points, otherwise they will be detected as breakline points. The white crosses in Figure 15.3 show the detected breakline points, whereas the black dots are planar points.

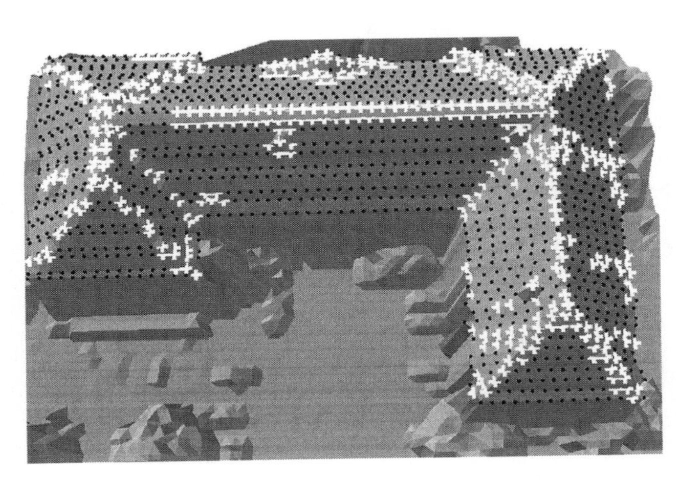

FIGURE 15.3
Detected building breakline points (white crosses) and planar points (black dots).

15.4 Roof Segmentation

Once the breakline points are detected, we are left with a dataset that should consist of only planar roof points. The next task is to find which points belong to which specific roof segment. This is treated as a clustering problem.

15.4.1 Feature Vector

As addressed above, selecting the feature space or feature vector is the first step in clustering. During the above breakline detection process, the eigenvector corresponding to the smallest eigenvalue gives the normal vector $\mathbf{N} = (N_X, N_Y, N_Z)^T$ (also called surface normal or direction cosines) to a planar patch. Such normal vector (**N**) uniquely determines the direction of a roof plane and thus is selected as the feature vector **F** for clustering.

15.4.2 *k*-Means Clustering

As one of the most common and simple algorithms, the *k*-means algorithm is used for clustering in this study. To determine *k* number of clusters in a dataset, it follows the following steps:

1. Initially *k* cluster centers are chosen randomly among the data points.
2. Each data point (in feature space) is assigned to its nearest cluster center.
3. Cluster centers are recomputed using the current clusters.
4. Go to step 2 if the convergence criterion is not met, which is usually a threshold in the average of the squared distances of each data point to its cluster center.

Mathematically, the *k*-means algorithm can be described as a clustering scheme that partitions the data points into *k* clusters, such that the following objective function is minimized:

$$\Phi = \sum_{j=1}^{k} \sum_{i=1}^{m} d(\mathbf{F}_i^{(j)} - \bar{\mathbf{F}}_j) \tag{15.7}$$

where

$\mathbf{F}_i^{(j)}$ is the *i*th feature vector assigned to the *j*th cluster
$\bar{\mathbf{F}}_j$ is the mean feature vector of cluster *j*
d is the distance function defined in Equations 15.1 through 15.5
k is the number of clusters
m is the total number of data points

For a detailed discussion on the *k*-means clustering algorithm, please refer to Duda et al. (2001), Jain et al. (1999), and Tibshirani et al. (2001).

The main disadvantage of the *k*-means algorithm is that the number of clusters *k* should be given. In our case, the number of clusters corresponds to the number of directional roof planes in a building and is therefore usually not known before clustering. Another disadvantage of the *k*-means method is that the initial choice of cluster centers is critical to the convergence of the objective function to the global other than a local minimum. Therefore, such shortcomings need to be properly considered in our implementation.

15.4.3 Cluster Centers and the Number of Clusters

This section introduces a potential-based clustering approach and applies the elbow joint method to determine the number of clusters. The potential-based clustering concept was proposed by Yager and Filev (1994). The feature space is first divided into grids and the potential (number of data points per grid, equivalent to density) of each grid is computed based on the distance from the grid center to the data points. A grid with many data points nearby will have a high potential value. The grid with the highest potential value is chosen as the first cluster center. Once the first cluster center (grid) is chosen, the potential of the nearby grids is reduced based on their distance from the cluster center. This is to make sure that two grids that are close together do not become different clusters. The next cluster center is then chosen from the remaining grids with the highest potential. Chiu (1994) further developed this concept by using the actual data points instead of the grids as the cluster centers. Chiu's study calculated the potential for each point $\mathbf{F}_i$ as

$$p_i = \sum_{j=1}^{n} \exp\left\{-\frac{4}{r_a^2}\left\|\mathbf{F}_i - \mathbf{F}_j\right\|^2\right\} \quad (15.8)$$

where

n is the number of data points

r_a is a positive constant

As is shown in the above equation, the potential of a data point is dependent on its distances to all other data points. A data point with many neighboring data points will have a high potential value. The first cluster center is chosen as the data point that has the greatest potential (i.e., the maximum value of p_i for all i). To reduce the possibility of the next cluster center being close to the first cluster center, the potential for every data point is then reduced as a function of its distance to the first cluster center.

$$p_{i\_\text{new}} = p_{i\_\text{old}} - p_1^*.\exp\left\{-\frac{4}{r_b^2}\left\|\mathbf{F}_i - \mathbf{F}_1^*\right\|^2\right\} \quad (15.9)$$

where

p_{i_new} and p_{i_old} are respectively the new and old potentials for data point i

$\mathbf{F}_1^*$ and p_1^* are respectively the selected first cluster center and its potential

Chiu (1994) recommended a ratio of 1.5 between r_b and r_a, such that the data points near the first cluster center will have greatly reduced potentials, and therefore will unlikely be selected as the next cluster center. Once the potentials of all the data points are updated based on Equation 15.9, the data point with the current largest potential will be selected as the second cluster center. The above process of acquiring new clusters stops when the cumulative potential becomes too small.

It is clear that the value of the radius r_a is critical to the clustering results. A smaller r_a will yield a higher number of clusters and vice versa. Since we are operating in feature space, it is difficult to design a reasonable threshold for r_a. To overcome this problem, the method described by Chiu (1994) is iteratively implemented. Starting from a smaller radius and increasing it gradually, fewer and fewer number of cluster centers will be obtained. The cluster centers generated for each r_a are used as the input to the k-means clustering algorithm. As the result of clustering, a likelihood estimate for each cluster will be produced. It measures the compactness of the clustering and can be used to determine the optimal

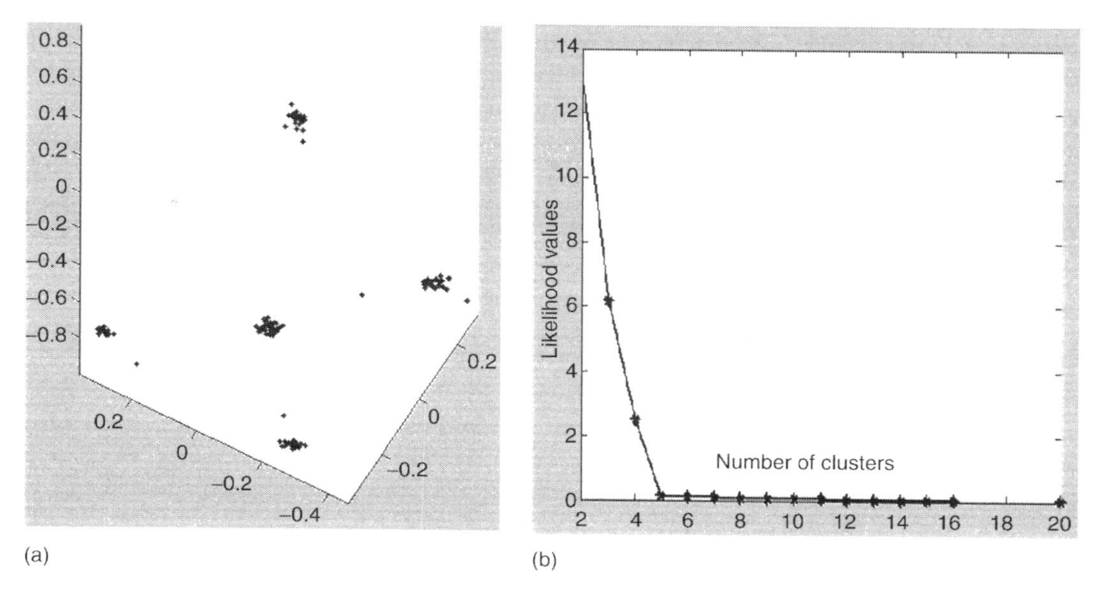

FIGURE 15.4
Plots of the direction cosines of (a) planar patches and (b) the elbow joint graph.

value of r_a, and hence the number of clusters. Figure 15.4 plots the likelihood estimates with respect to the number of clusters. We define the likelihood estimates, alternatively termed as similarity measure, as the average distance from the cluster center to data points. Considering a case of k clusters, the mean distance of data points from their respective cluster centers is calculated as

$$\bar{d}_i = \sum_{j=1}^{n_i} \frac{d_{ij}}{n_i} \tag{15.10}$$

where
d_{ij} is the distance of a data point j in cluster i to its respective cluster center
n_i is the number of data points in the i th cluster

The likelihood value is then defined as

$$\bar{d} = \sum_{i=1}^{k} \frac{\bar{d}_i}{k} \tag{15.11}$$

As is shown in Figure 15.4, the likelihood falls sharply onto a place, after which the decrease becomes stable. This effect is likened to an elbow and the elbow joint is considered to be a good estimate on the number of clusters in the dataset. A further study of the phenomenon can be found in Duda et al. (2001) and Tibshirani et al. (2001).

At the end of this process we know (a) the number of clusters, (b) the cluster centers, and (c) the constituent planar patches for each cluster. In data space, they respectively correspond to (a) the number of distinct directions on the roof, (b) their distinct normal vectors, and (c) the LiDAR point patches, and thus the LiDAR points that form the directional roof planes.

15.4.4 Separation of Parallel and Coplanar Planes

The above clustering process returns the planes with the same normal vectors. However, our task is still incomplete, because the planes may have roof segments that are parallel to each other or roof segments that are mathematically the same but are separated spatially. Therefore, further separation is needed to separate parallel planes and coplanar planes in the above clustering results.

To separate parallel planes, for each patch in the same cluster, its centroid (X_C, Y_C, Z_C) can be calculated by taking the average of the LiDAR points in the patch. The distance from the coordinate origin to the directional plane that contains this patch can then be computed as

$$\rho = \frac{(N_X X_C + N_Y Y_C + N_Z Z_C)}{\sqrt{N_X^2 + N_Y^2 + N_Z^2}} \tag{15.12}$$

where N_X, N_Y, and N_z are the normal vectors of the plane.

If a given cluster has parallel planes, the values of ρ will vary significantly among the patches. Parallel planes can thus be separated based on these different ρ values.

At this stage, all parameters that define a mathematical plane on the roof are determined. The initially determined breakline points are checked one by one against all planes by substituting their coordinates into the determined planar equations. If the discrepancy between a breakline point and a plane is within a tolerance, it is assigned to the plane that yields the minimum discrepancy. For breakline points that do not belong to any plane, they will be treated as tree points and excluded from further processing. This way we are able to bring back all roof points that we initially excluded for clustering and separate tree points from roof points.

Building roofs may have two or more planar segments that are mathematically the same, but spatially separated. Such coplanar segments can be separated in the conventional data space based on the concept of density clustering and connectivity analysis. In Figure 15.5, points A and B are directly density connected, and point C is density connected to points A and B. However, point D is neither density connected nor density reachable to any of the points A, B, or C. Therefore, point D lies in a separate cluster from A, B, and C. This concept depends on the neighborhood defined by its radius R. For a further reading in this topic, readers are referred to (Berkhin, 2002). In our case, we have clusters of points that lie on the same plane, but are spatially separated. Using an appropriate value of about two times the LiDAR data spacing for R, these clusters can be easily separated. It should be noted that the isolated planar trees possibly left from the previous steps can also be separated from buildings in this step if their size is significantly small.

Figure 15.6 illustrates the series of steps that have been outlined above with an example building. The dataset has a ground spacing about 1.0 m over the Purdue campus at West

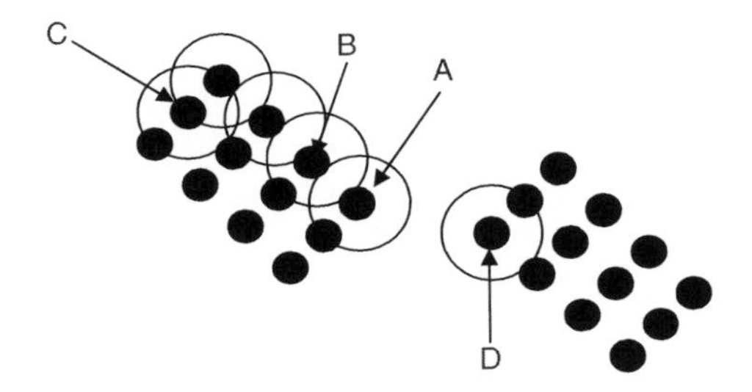

FIGURE 15.5
Separation of coplanar roof segments with density-based clustering.

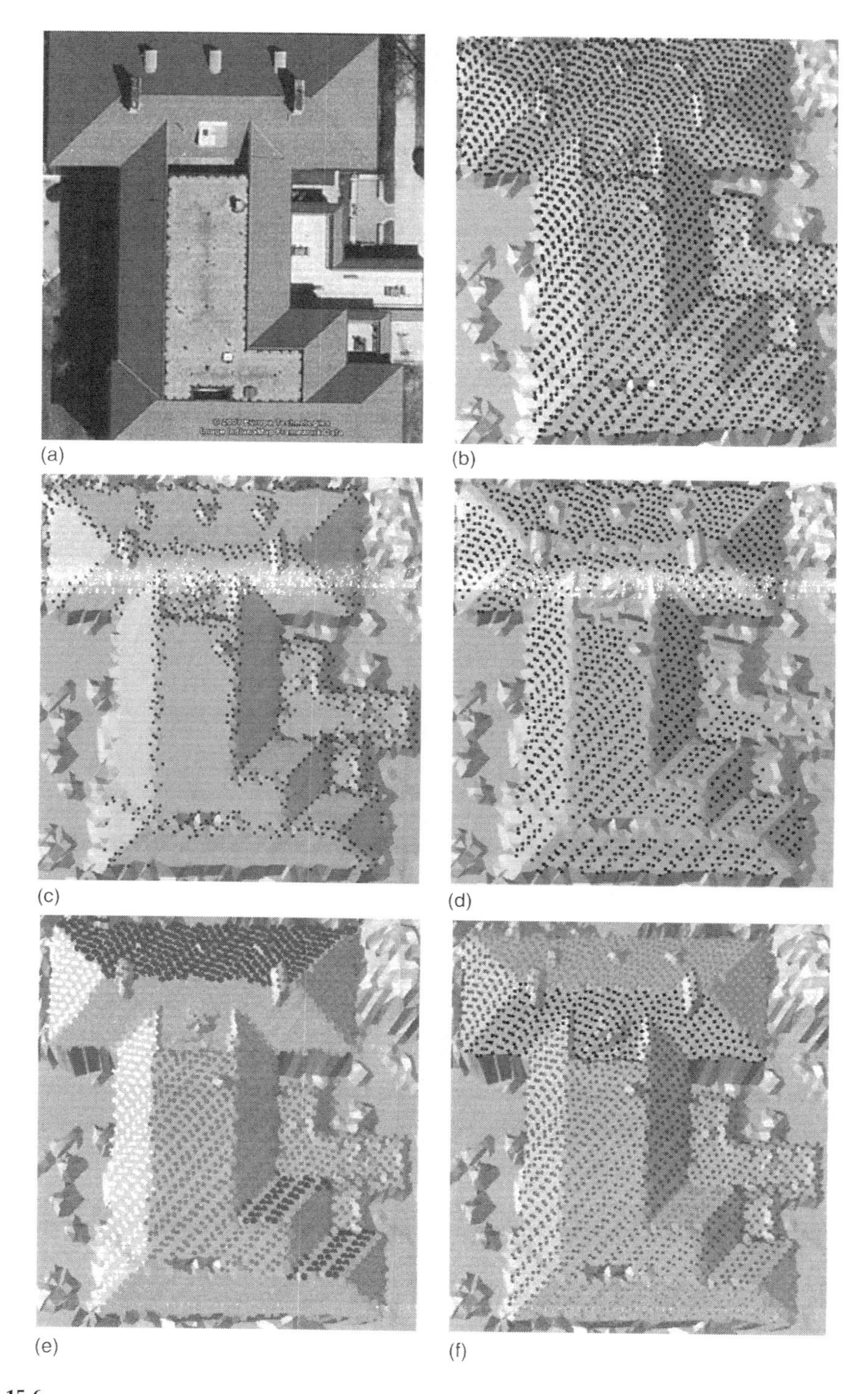

FIGURE 15.6
Roof LiDAR points clustering and segmentation. (a) Image (from Google), (b) LiDAR points, (c) breakline points, (d) planar points, (e) clustered points, and (f) planar segments.

Lafayette, Indiana. From left to right in the top row of the figure, they are the building aerial image (from Google Earth, a), and LiDAR points atop the triangulated irregular network (TIN) model (b). The second row displays the separated breakline points (c) and the planar points (d). It is clear that almost all breakline points are detected correctly and the performance is satisfactory. Returns from nonplanar points, such as the small roof attachments, are also well detected as breakline points. However, a closer look suggests that the separation is not always perfect as there exists over-removal of planar points. This usually should not be a concern because of the large number of available LiDAR points on the roof. Moreover, the mistakenly detected breakline points will be back-projected to the planar roofs after clustering. The opposite situation, "under-removal," occurs when real breakline points are not correctly detected. This takes place at relatively flat or smoothed roof ridges or edges where breaklines are wide and form a small flat region. Both mistakes may potentially cause inaccurate determination of roof plane boundaries in the subsequent steps, particularly for the smaller roof planes. Finally, we have five clustered directional planes (e) shown with different colors. After separating parallel planes based on their distance from the coordinate origin, and coplanar segments based on density connectivity, we obtain the segmented results, where the LiDAR returns from the roof are segmented into 14 individual planar segments as color coded in (f).

15.5 Discussions and Evaluation

The section will discuss the experience gained from the implementation of the above outlined clustering approach and present a framework to evaluate the clustering results.

15.5.1 Discussions

The first discussion is about the selection of the patch size. Two related factors should be considered in forming the patches: the number of points in a patch and the total number of patches for clustering. In general, the patch size should be smaller than the smallest roof entity for its detection and be large enough so that there will be enough LiDAR points on each patch to define a plane. Patches with three to four points are more likely collinear or coplanar, susceptible to data noise, and may not be a good representative of the planar segment that it belongs to. Our study shows it is appropriate to form a patch with six to eight points.

On the other hand, the number of patches plays an important role in the clustering process. In general, it is preferable to have a large number of small patches rather than a small number of large patches. A large patch and a small patch carry equal weight in the clustering process, and having a large number of small patches means we have more data points in the feature space for clustering. This helps in obtaining robust clustering outcome.

The next issue is the selection of the threshold $\bar{\lambda}_T$ for the normalized eigenvalue $\bar{\lambda}$. For an ideal planar surface, the smallest eigenvalue and hence the normalized eigenvalue $\bar{\lambda}$ will always be zero. However, in reality, the $\bar{\lambda}$ for a planar surface is usually finitely small but zero. Therefore, the test for breaklines needs a threshold $\bar{\lambda}_T$ below which the LiDAR point patch shall be considered as planar. A value that is too close to zero may result in a planar region being classified as breaklines, and a larger threshold will result in the opposite.

Our experiments with different buildings found that $\bar{\lambda}_T = 0.005$ gives good and satisfactory results. It is appropriate and can balance the two types of errors, namely, breakline points being clustered as planar (under-removal) and planar points being categorized as breaklines (over-removal). A slight bias towards the second type of error, i.e., a smaller threshold, does not significantly affect the final results since the high density of LiDAR data usually provides sufficient redundancy to determine a plane and the initially removed breakline points will ultimately be mapped to their corresponding planes after clustering.

The next discussion is about the planar patches clustering. The left graph in Figure 15.4 shows a plot of normal vector values for all of the planar patches in clustering. In the plot, each dot represents a planar patch and its coordinates are determined by the patch normal vector $\mathbf{N} = (N_X, N_Y, N_Z)^T$. As shown, the dots are located in distinct clusters, indicating that there are several planes facing different directions in the building. Note, some dots seem to be slightly away from the cluster centers, which are most likely formed by LiDAR points at the breaklines or other nonplanar features that have not been completely removed in the breakline detection step. The existence of some isolated dots rather away from the cluster centers imply that they are likely trees remaining from the breakline detection step. The Chiu's potential-based clustering approach can reduce the effect of such off-center outliers that are mixed in the data points, and correctly determine the number and location of these clusters. In the elbow-joint graph of Figure 15.3 (right) that plots the likelihood values for different number of clusters, we choose the number of clusters as five at the place where the distinct change occurs in the elbow-joint graph. In this manner, the clustering process is able to determine the number of distinct planes with different normals, the constituent patches, and LiDAR points for each directional plane.

15.5.2 Evaluation

To evaluate the clustering results, three tables are generated along the steps of the above process to assess its performance. They respectively address the quality of patch-to-direction clustering, direction-to-plane clustering, and plane-to-segment clustering. The following tables are referred to the example building shown in Figure 15.6.

The first evaluation is about the quality of patch-to-direction clustering. Table 15.1 lists the results and statistics for the clustering of the planar patches based on their normal vectors. The ClustNo is the sequential cluster number, while the NumPat is the resultant number of patches for a cluster of directional planes with the similar normal. In total there are five directional planes (corresponding to five colors in Figure 15.6e) found through this clustering process. The table also collects the within-cluster minimum, maximum, mean, and standard deviation of the differences (in degrees) for the patches with reference to their respective cluster center. The last row is the same statistics with reference to the center of all clusters. It stands for the overall directional variation among all roof planes. The minimum between-cluster distance shows that the minimum angle between two planar directions in this building is about 26°. The relatively small values in the mean, minimum, and standard deviation within each cluster suggest that the clustering results are compact and that all the patches within a single cluster face the same direction with a variation of several (up to 9.8) degrees. However, the large maximum differences (up to 60.6°) from the cluster centers indicate that some patches in a cluster include LiDAR points at nonplanar regions such as trees, breaklines, chimneys, pipes, etc. on the roof.

The second evaluation is to check the coherence of LiDAR points to their corresponding (mathematical) plane. Once the parallel planes are separated, the distance d for each

TABLE 15.1

Statistics (in Degrees) of Patch-to-Direction Clustering

	ClustNo	NumPat	Min	Max	Mean	StdDev
Within cluster	1	50	0.297	33.399	2.686	4.960
	2	46	0.143	18.391	2.145	2.802
	3	53	0.137	11.473	1.932	1.857
	4	36	1.117	60.609	3.849	9.814
	5	42	0.235	7.040	2.115	1.657
Between the five clustered planes			25.796	152.941	77.138	56.361

Notes: ClustNo, sequential cluster number; NumPat, the number of patches; min, max, and mean, the minimum, maximum, and mean angular difference (in degrees) from the mean of one cluster (for within cluster) or entire clusters (for between clusters); StdDev, standard deviation (in degrees) of the angular difference from the mean of one cluster or entire clusters.

LiDAR point to its corresponding plane can be calculated. The statistics (min, max, mean, and standard deviation) of such distances of all points in a plane are listed in Table 15.2, which provides a measure on the closeness of the points to their corresponding plane. Each row in the table is a plane sequentially numbered under the column PlnNo. The third column NumPnt represents the number of points in the plane. N_X, N_Y, N_Z and ρ are the four plane parameters. Smaller plane segments (typically with six to eight points) are discarded in this table since they are likely to lie on some small parts of the roof, which are hard to extract and reconstruct due to the resolution limit of the LiDAR data. Table 15.2 shows the maximum distance magnitude for a LiDAR point off its detected plane in the entire building is 2.2 m, and most distances are less than 0.4 m. The standard

TABLE 15.2

Plane Parameters and Statistics (in Meters) for Direction-to-Plane Clustering

ClustNo	PlnNo	NumPnt	N_X	N_Y	N_Z	ρ	Min	Max	StdDev
1	1	390	0.343	0.278	0.897	−217.66	−1.70	2.15	0.27
1	2	39	0.355	0.279	0.892	−200.49	−0.14	0.38	0.08
1	3	35	0.349	0.291	0.891	−196.85	−0.08	0.14	0.04
2	4	171	−0.338	−0.276	0.899	172.55	−0.12	0.53	0.10
2	5	295	−0.336	−0.280	0.898	−155.37	−1.03	0.40	0.13
3	6	347	0.002	0.004	0.999	−206.12	−1.24	0.39	0.09
3	7	124	−0.015	0.001	0.999	−194.74	−0.93	0.38	0.21
3	8	32	−0.057	0.034	0.999	−197.52	0.70	0.50	0.31
4	9	165	0.134	−0.161	0.978	−201.79	−1.54	1.52	0.71
4	10	142	0.289	−0.342	0.893	−182.04	−0.63	1.02	0.11
5	11	98	−0.281	0.336	0.898	−196.50	−0.06	0.18	0.04
5	12	213	−0.270	0.343	0.899	−192.96	−0.33	0.29	0.10

Notes: ClustNo, sequential cluster number; PlnNo, sequential plane number; NumPnt, the number of LiDAR points in a plane; N_X, N_Y, N_Z, and ρ: plane parameters of directional consines and intercept; Min, Max, and Std.Dev: minimum, maximum, and standard deviation (in meters) of distances of LiDAR points to their corresponding plane.

TABLE 15.3

Statistics (in Meters) of Plane-to-Segment Clustering

ClustNo	PlnNo	SgmtNo	NumPnt	Min	Max	Mean	StdDev
1	1	1	390	0.96	2.19	1.27	0.27
1	2	2	39	0.52	2.17	1.35	0.38
1	3	3	35	0.54	1.51	1.09	0.22
2	4	4	171	0.80	2.12	1.26	0.29
2	5	5	295	0.52	1.64	1.16	0.25
3	6	6	347	0.75	1.96	1.29	0.30
3	7	7	124	0.76	1.78	1.23	0.23
3	8	8	32	0.82	2.00	1.37	0.33
4	9	9	32	0.83	1.82	1.25	0.29
4	9	10	89	0.47	1.53	1.11	0.28
4	9	11	44	0.50	1.84	1.18	0.31
4	10	12	142	0.57	1.66	1.14	0.27
5	11	13	98	0.83	1.76	1.23	0.20
5	12	14	213	0.89	1.57	1.21	0.17

Notes: ClustNo, sequential cluster number; PlnNo, sequential plane number; SgmtNo, sequential segment number; NumPnt, the number of LiDAR points in a segment; Min, Max, and StdDev, minimum, maximum, and standard deviation (in meters) of the triangle lengths in the triangulation of LiDAR points.

deviation (0.10 m in average) suggests that there is no significant bias in the direction-to-plane clustering process.

The final evaluation occurs after the individual coplanar segments in a plane are spatially separated. To evaluate the quality of this final separation, the points of each segment are triangulated and the edge lengths of each triangle are recorded. Table 15.3 shows the statistics of the triangle edge lengths for all roof segments. The table shows the compactness of the resultant roof segments. It is shown that the average edge length in a roof segment is very close to the LiDAR point density (1.0 m), and the small standard deviation of the lengths suggests that the triangles are mostly equilateral, which reflects the distribution of LiDAR points on a building roof. This shows that the final segmentation is density-compact and there is no apparent outlier in the results.

15.6 Building Reconstruction

The purpose of reconstruction is to represent a building with as few point vertices as possible, i.e., to determine the topological primitives and their connectivity for the building.

The topologic primitives of a building consist of vertices, breaklines, and segments, which are all in 3D. It is important to point out that these primitives are not fully independent and can be formed completely or in part from each other. For example, a breakline can be regarded as the intersection of two planar segments or a connection of two vertices. Similarly, a segment can be formed by an ordered connection of lines or vertices. The type of topologic formulation scope used is critical to achieve effective building reconstruction. From the previous clustering step, we have the equations for each segment in a building roof and the LiDAR points that belong to a segment. For building reconstruction purpose, one option is to determine a breakline as an intersection of two adjacent segments; however, this would not directly lead to its two topologic vertices, which ultimately determine the building shape. Unlike the above approach, this study treats a vertex as an intersection of three or more planar segments that are mutually adjacent. The algorithm is described in detail as follows.

First of all, the adjacency of the planar segments in a building needs to be found out. This is not difficult since we already know the constituent LiDAR points associated with each roof segment after the previous steps. We define the distance of two planar roof segments as the minimum distance of all possible point combinations between the two segments

$$d(P,Q) = \min(d(p_i, q_j)) \quad \forall p_i \in P, \quad \forall q_i \in Q \tag{15.13}$$

where

$d(P, Q)$ is the distance between the two planar segments P and Q

$d(p_i, q_j)$ stands for the distance between any two points p_i and q_j, respectively, in the segments P and Q

Under this definition, any segments adjacent to a given segment P are found if their distances to P are below a given threshold, which is taken approximately as two times the LiDAR point spacing. As the result of such neighborhood analysis, an adjacency matrix is formed for the planar segments of a building. Based on this matrix, the mutual adjacency of segments can be obtained. Take the building in Figure 15.6 as an example, which is reproduced in its vector form in Figure 15.7. Its adjacency matrix is formed in Table 15.4. To find all the mutually intersecting segments (i.e., all the segments intersect at one vertex), take any two intersection segments, e.g., 1 and 4 {1,4} from the first row, which will determine a breakline. The next step needs to check all segments that intersect with both segments 1 and 4. Thus, segment 10 and 13 are selected and the existing intersection segment set {1,4} is expanded to {1,4,10} and {1,4,13}. Repeat this process until every segment that intersects with all segments in {1,4,10} or {1,4,13} is reached. The resultant mutual intersection set that has at least three segments, e.g., {1,4,10}, will then be used to

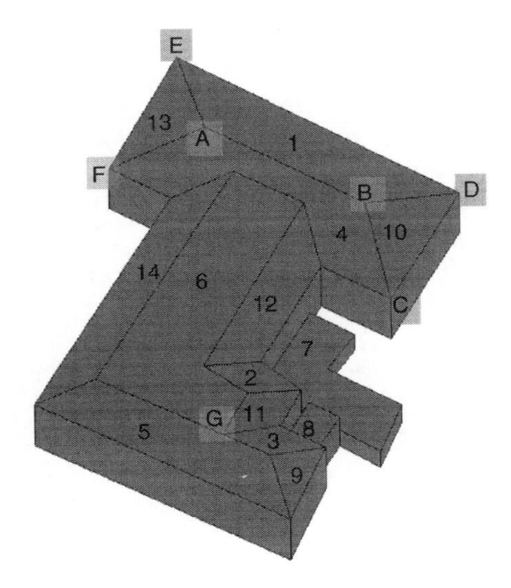

FIGURE 15.7
Planar segment adjacency and vertex determination for building reconstruction.

TABLE 15.4

Adjacency Matrix of Planar Segments in a Building (Figure 15.7)

SgmtNo	1	2	3	4	5	6	7	8	9	10	11	12	13	14
1				*						*			*	
2						*					*	*		
3					*	*			*		*			
4	*					*				*		*	*	*
5			*			*			*		*			*
6		*	*	*	*						*	*		*
7														
8														
9			*		*									
10	*			*										
11		*	*		*	*								
12		*		*		*								
13	*			*										
14				*	*	*								

Note: * Stands for the two segments represented in the corresponding column and row are adjacent.

determine the vertex A in this example. Similarly, the mutual intersection plane set {1,4,13} will determine the vertex B. The coordinates of A or B are determined by a system of planar equations of all involved segments.

Generally, when n mutually intersecting segments are used to determine a vertex, we call it an n -adjacency vertex. Singularities occur when less than three mutual intersection segments can be found, i.e., $n \leq 2$. When two segments intersect, they can only determine a breakline other than a vertex. Boundary conditions or constraints must be introduced to resolve this singularity. For each of the two involved segments, one vertical wall is introduced as an additional plane or segment, which will join the roof segments to determine the vertex. For example, the vertex C in Figure 15.7 is determined by the segment 4, 10, and two vertical planes respectively passing their external boundaries. The readers are referred to Sampath and Shan (2007) for the methodology to determine the building boundary in a point cloud.

Another singular situation is that no adjacent plane ($n = 0$) is found for a given plane (segment 7 and 8 in Figure 15.7). This occurs when buildings have one or more flat roof segments. Such roof segments can be reconstructed by determining their boundary and regularizing them (Sampath and Shan, 2007). It should be noted that the use of vertical walls as a boundary condition is equivalent to the use of roof segment boundary delineated from the LiDAR points. As shown in Figure 15.7, every constituent line in the isolated roof segments can form a vertical wall downwards to the ground or at a lower level to reconstruct the building.

Finally, the building is reconstructed based on the determined vertices and the above adjacency matrix. A roof edge is formed by connecting two ending vertices where two constituent planes intersect. Roof plane boundary can be represented by a sequential connection of the vertices. Figure 15.8 illustrates some reconstructed buildings along with their images and LiDAR data.

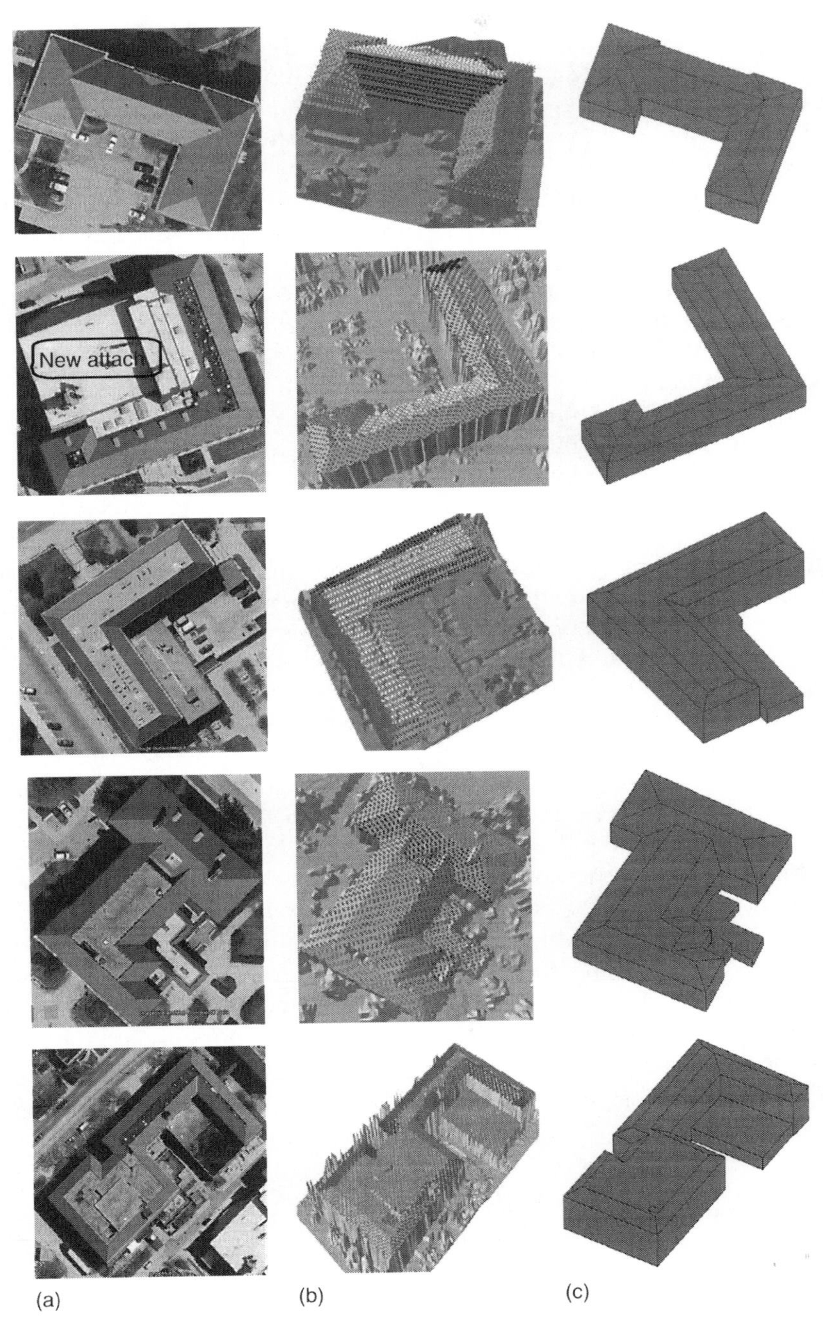

FIGURE 15.8
(a) Building images, (b) segmented LiDAR points, and (c) reconstructed buildings.

15.7 Conclusions

This study does not assume any specific building model. Instead, the building model is determined from the LiDAR data. The described technique is designed to work on raw point clouds, and does not require the point clouds be processed to a regular format. Besides, it is also shown that the method is robust to the existence of outliers, such as trees, breaklines, and other nonplanar objects.

Building extraction from LiDAR point clouds can be treated as a clustering problem. Reliable feature selection is the key to the success of this process. In the techniques presented in this chapter, instead of assuming the absence of trees, reflections from vertical walls, chimney, etc, we present an automated breakline detection process to remove such outlier points before clustering. It is shown that the performance of the clustering process is greatly affected. Implementing breakline detection as the first step raises the reliability and reduces the ambiguity of the clustering operation. The patch-based eignevalue analysis can yield satisfactory separation of planar points from nonplanar outlier points. It is shown that a patch of six to eight points is suitable for this propose.

The subsequent roof segmentation follows three sequential steps: clustering LiDAR points to directional planes, separating parallel planes, and segmenting coplanar roof segments. The iterative use of the potential-based k-means clustering method demonstrates satisfactory and promising results for moderate and complex buildings. The subsequent separations of parallel planes and coplanar roof segments are straightforward based on distance and point density without much difficulty. It is also shown that the compactness or coherences of the separated directional planes, parallel planes, and coplanar segments can be respectively assessed by the proposed quality matrices. The study shows through the above steps the building planar segments can be ultimately determined up to the resolution of the LiDAR points.

To reconstruct the final 3D vector building models, the proposed solution uses the roof breaklines and vertices determined by the intersection of two or more roof planar segments. The estimates of the planar equations during the clustering process are directly used in the building reconstruction process, which is treated as a computational geometry problem. It is shown that such a solution is a general approach, which can consider different types of buildings by introducing vertical walls as boundary conditions to assure a unique solution. The solution directly gives all the elements required for the complete topological description of a building. Tests with a number of buildings demonstrate that the proposed segmentation and reconstruction approaches not only are robust to outliers in the building data sets and to the complexities of building roofs but can also provide quantitative quality measures in each step for effective quality control and evaluation.

References

1. Alharthy, A. and Bethel, J., 2004. Detailed building reconstruction from airborne laser data using a moving surface method. *International Archives of the Photogrammetry and Remote Sensing*, Vol. XXXV-B3, pp. 213–218.
2. Berkhin, P., 2002. Survey of Clustering Data Mining Techniques. Available at http://citeseer.ist.psu.edu/berkhin02survey.html.
3. Brenner, C., 2005. Building reconstruction from images and laser scanning. *International Journal of Applied Earth Observation and Geoinformation*, 6, 187–198.

4. Briese, C., 2004. Three-Dimensional Modelling of Breaklines from Airborne Laser Scanner Data, XXth ISPRS Congress: Geo-imagery bridging continents. *International Archives of the Photogrammetry and Remote Sensing*, Istanbul, Turkey.
5. Brugelmann, R., 2000. Automatic breakline detection from sairborne laser range data. *International Archives of the Photogrammetry and Remote Sensing*, Vol. XXXIII, Part B3, Amsterdam, The Nether-lands, pp. 109–116.
6. Chiu, S., 1994. Fuzzy model identification based on cluster estimation. *Journal of Intelligent and Fuzzy Systems*, 2(3), September, pp. 267–278.
7. Duda, R.O., Hart, P.E., and Stork, D.G., 2001. *Pattern Classification* (2nd ed.), John Wiley & Sons.
8. Forlani, G., Nardinocchi, C., Scaioni, M., and Zingaretti, P., 2003. Building reconstruction and visualization from LIDAR data, *International Archives of the Photogrammetry, Remote Sensing and Spatial Information Sciences*, Vol. XXXIV, Part 5/W12. Available at http://www.commission5.isprs.org/wg4/workshop_ancona/proceedings/32.pdf.
9. Fransens, J., Bekaert, P., and van Reeth, F., 2003. A hierarchical PCA-based piecewise planar surface reconstruction algorithm. In *Proceeding of the Eighth SIAM Conference on Geometric Design and Computing*, pp. 203–214, Nashboro Press, Brentwood, TN, November.
10. Haala, N. and Brenner, C., 1999. Virtual city models from laser altimeter and 2D map data. *Photogrammetric Engineering and Remote Sensing*, 65(7), 787–795.
11. Hofmann, A.D., Maas, H.-G., and Streilein, A., 2002. Knowledge-based building detection based on laser scanner data and topographic map information. Available at http://www.isprs.org/commission3/proceedings02/papers/paper025.pdf.
12. Jain, A.K., Murty, M.N., and Flynn, P.J., 1999. Data clustering: A review. *ACM Computing Surveys*, 31(3), September, 264–323.
13. Kilian, J., Haala, N., and English, M., 1996. Capture and evaluation of airborne laser scanner data, *International Archives of Photogrammetry and Remote Sensing*, Vol. XXXI, Part B3, Vienna, pp. 383–388.
14. Kraus, K. and Pfeifer, N., 1998. Determination of terrain models in wooded areas with aerial laser scanner data. *ISPRS Journal of Photogrammetry and Remote Sensing*, 53(4), 193–203.
15. Maas, H.-G. and Vosselman, G., 1999. Two algorithms for extracting building models from raw laser altimetry data. *ISPRS Journal of Photogrammetry and Remote Sensing*, 54, 153–163.
16. Nizar, A., Filin S., and Doytsher, Y., 2006. Reconstruction of Buildings from Airborne Laser Scanning Data, ASPRS Annual Conference, Nevada.
17. Overby, J., Bodum, L., Kjems, E., and Iisoe, P.M., 2004. Automatic 3D building reconstruction from airborne laser scanning and cadastral data using Hough transform. *International Archives of Photogrammetry and Remote Sensing*, Vol. 35, Part. B3, pp. 296–301. Available at http://www.isprs.org/istanbul2004/comm3/papers/284.pdf.
18. Peternell, M. and Steiner, T., 2004. Reconstruction of piecewise planar objects from point clouds. *Computer-Aided Design*, 334(36), 333–342.
19. Rabbani, T., Dijkman, S., Heuvel van den, F.A., and Vosselman, G., 2007. An integrated approach for modelling and global registration of point clouds. *ISPRS Journal of Photogrammetry and Remote Sensing*, 61(6), 355–370.
20. Rottensteiner, F. and Briese, C., 2003. Automatic generation of building models from lidar data and the integration of aerial images. ISPRS, XXXIV, Dresden. Available at http://www.isprs.org/commission3/wg3/workshop_laserscanning/papers/Rottensteiner_ALSDD2003.pdf.
21. Rottensteiner, F., Trinder, J., Clode, S., and Kubik, K., 2005. Using the Dempster–Shafer method for the fusion of LIDAR data and multi-spectral images for building detection. *Information Fusion*, 6(4), December, 283–300.
22. Sampath, A. and Shan, J., 2006. Clustering based planar roof extraction from lidar data. American Society for Photogrammetry and Remote Sensing Annual Conference, Reno, Nevada, May 1–6.
23. Sampath, A. and Shan, J., 2007. Building boundary tracing and regularization from airborne lidar point clouds. *Photogrammetric Engineering and Remote Sensing*, 73(7), 805–812.
24. Schwalbe, E., Maas, H.-G., and Seidel, F., 2005. 3D building model generation from airborne laser scanner data using 2D GIS data and orthogonal point cloud projections. *International Archives of Photogrammetry and Remote Sensing*, Vol. 36, Part 3/W19, 209–214. Available at http://www.commission3.isprs.org/laserscanning2005/papers/209.pdf.

25. Schenk, T. and Csathó, B., 2002. Fusion of LIDAR Data and Aerial Imagery for a More Complete Surface Description, Photogrammetric Computer Vision, ISPRS Commission III Symposium, September 9–13, Graz, Austria. Available at http://www.isprs.org/commission3/proceedings02/papers/paper179.pdf.
26. Schenk, T. and Csathó, B., 2007. Fusing Imagery and 3D Point Clouds for Reconstructing Visible Surfaces of Urban Scenes, IEEE GRSS/ISPRS Joint Workshop on Remote Sensing and Data Fusion over Urban Areas.
27. Shan, J. and Sampath, A., 2005. Urban DEM generation from raw lidar data: A labelling algorithm and its performance. *Photogrammetric Engineering and Remote Sensing*, 71(2), 217–226.
28. Sohn, G. and Dowman, I., 2007. Data fusion of high-resolution satellite imagery and LiDAR data for automatic building extraction. *ISPRS Journal of Photogrammetry and Remote Sensing*, 62(1), May, 43–63.
29. Sithole, G. and Vosselman, G., 2004. Experimental comparison of filter algorithms for bare-Earth extraction from airborne laser scanning point clouds. *ISPRS Journal of Photogrammetry and Remote Sensing*, 59(1–2), August, 85–101.
30. Tarsha-Kurdi, F., Landes, T., and Grussenmeyer, P., 2007a. Hough-Transform and Extended RANSAC Algorithms for Automatic Detection of 3d Building Roof Planes From Lidar Data, *International Archives of Photogrammetry and Remote Sensing*, Vol. XXXVI, Part 3/W52. Available at http://www.commission3.isprs.org/laser07/final_papers/Tarsha-Kurdi_2007.pdf.
31. Tarsha-Kurdi, F., Landes, T., Grussenmeyer, P., and Koehl, M., 2007b. Model-driven and data-driven approaches using LIDAR data: Analysis and comparison, Photogrammetric Image Analysis. Available at http://www.ipk.bv.tum.de/isprs/pia07/pia07_prg.html.
32. Tibshirani, R., Walther, G., and Hastie, T., 2001. Estimating the number of clusters in a data set via the gap statistic. *Journal of the Royal Statistical Society: Series B (Statistical Methodology)*, 63, 411–423.
33. Vögtle, T. and Steinle, E., 2000. 3D modelling of buildings using laser scanning and spectral information. *International Archives of Photogrammetry and Remote Sensing*, Vol. XXXIII, Part B3/2, 927–934.
34. Vögtle, T. and Steinle, E., 2003. On the quality of object classification and automated building modeling based on laser scanning data. *International Archives of the Photogrammetry, Remote Sensing and Spatial Information Sciences*, Vol. XXXIV/3W13, pp. 149–155.
35. Vosselman, G., 2000. Slope based filtering of laser altimetry data. *International Archives of Photogrammetry and Remote Sensing*, Vol. 33, Part B3/2, Amsterdam, the Netherlands, pp. 935–942.
36. Vosselman, G. and Dijkman, S., 2001. 3D building model reconstruction from point clouds and ground plans. *International Archives of Photogrammetry and Remote Sensing*, Vol. 34(3W4), pp. 37–43.
37. Vosselman, G., Kessels, P., and Gorte, B., 2005. The utilisation of airborne laser scanning for mapping. *International Journal of Applied Earth Observation and Geoinformation*, 6(3–4), 177–186.
38. Yager, R. and Filev, D., 1994. Approximate clustering via the mountain method. *IEEE Transactions on Systems, Man and Cybernetics*, 24(8), August.
39. Zhang, K. and Whitman, D., 2005. Comparison of three algorithms for filtering airborne LiDAR data. *Photogrammetric Engineering and Remote Sensing*, 71(3), 313–324.

16

Building and Road Extraction by LiDAR and Imagery

Franz Rottensteiner and Simon Clode

CONTENTS

16.1 Introduction

LiDAR sensors deliver a 3D point cloud with the intensity of the returned signal. Sometimes multiple pulses or even the full waveform of the returned signal is provided [1]. In any case, just as with aerial or satellite imagery, a considerable amount of processing is required in order to extract semantically meaningful information from the LiDAR point cloud. Whereas a human being can easily recognize and consequently determine the shape of topographic objects in 3D point clouds or digital imagery, this is not a straightforward task that can be performed by a computer. Mapping of topographic objects, especially man-made structures such as buildings and roads, has always been one of the most important tasks in photogrammetry and remote sensing. Due to the increasing demand for spatial data as they are provided by Geographical Information Systems (GIS) or 3D City Models, the automation of that task has gained considerable importance in photogrammetric research.

Automated object extraction as it is understood in this chapter aims at the automatic recognition of topographic objects and their geometric reconstruction by vectors that can easily be used in GIS or computer-aided design (CAD) systems. Thus, it is usually carried out in two steps:

1. Object detection or object recognition is essentially a classification of the input data using some model of the appearance of the object in the sensor data. It results in detected objects with coarse boundaries in object space.
2. Object reconstruction or object vectorization is carried out in the previously detected object regions in order to reconstruct the shape of the detected objects by vectors.

This chapter will focus on the extraction of buildings and roads from LiDAR data, and in the case of buildings also from digital aerial imagery. In both cases the strategy of separating object detection from object reconstruction will be applied. In general, both buildings and roads consist of smooth surfaces with specific reflectance properties that are bounded by smooth curves. For buildings, the object surfaces are in most cases planes that are limited by polygons. Road surfaces are also smooth, though not necessarily planar, and their boundaries can be modeled by a concatenation of low-order polynomial curves, e.g., by splines or, less accurately, polygons.

Compared to aerial imagery, LiDAR has both advantages and disadvantages with respect to automatic object extraction [2]. First, LiDAR directly provides 3D points, whereas with aerial imagery, matching is required to obtain the third dimension. Being an active sensing technique, LiDAR does not have any problems with cast shadows, and occlusions are less problematic because LiDAR sensors typically have a smaller opening angle than aerial cameras. Due to its explicitly 3D nature, LiDAR data give better access to deriving geometrical properties of surfaces such as surface roughness, which is a distinguishing feature of man-made objects. However, LiDAR only provides limited information about the reflectance properties of the object surfaces, whereas modern aerial cameras deliver multispectral images. The boundaries of objects are usually better defined in aerial imagery than they are in LiDAR data especially if they correspond to height discontinuities, and finally, the resolution of LiDAR data, though astonishingly high, does not yet match the resolution of aerial imagery. Given the pros and cons of LiDAR and aerial imagery, it has been suggested to combine these data to improve the degree of automation and the robustness of automatic object extraction [2].

16.2 Building Extraction from LiDAR Data

In this section, a method for building extraction from LiDAR data, which follows the two-step procedure of detection followed by reconstruction, will be described. In the classification stage, multispectral imagery will be considered as an optional input source. Building reconstruction will be based on LiDAR data alone.

16.2.1 Building Detection by Dempster–Shafer Fusion of LiDAR Data and Multispectral Imagery

16.2.1.1 Cues for Building Detection

LiDAR data essentially deliver a Digital Surface Model (DSM) representing the surface from which the laser pulse is reflected, i.e., trees, terrain surface, buildings, etc. In the case of undulating terrain, the elevations reflect both the terrain heights and the height differences between points on elevated objects and the terrain. A normalized DSM directly reflects the heights of objects relative to the terrain [3]. In order to generate a normalized DSM, a Digital Terrain Model (DTM) has to be generated first, and its heights have to be subtracted from the DSM heights.

A DSM also provides information about surface roughness via an analysis of the second derivatives of the DSM. The actual measures used for surface roughness are far from unique and comprise the output of a Laplace filter applied to the DSM [4,5], local curvature [5], or the local variance of the surface normal vectors [6]. The representation of surface roughness used in this chapter is based on applying the concept of polymorphic feature extraction [7] to the first derivatives of the DSM and, thus, on an analysis of the second derivatives of the DSM, as shall be outlined here. Assuming the DSM to be represented by a height grid $z(x,y)$, a matrix $\mathbf{N}$ can be computed from the second derivatives of $z(x,y)$:

$$\mathbf{N} = \frac{1}{\overline{\sigma}_x^2} \cdot \mathbf{L} * \begin{pmatrix} \left(\frac{\partial^2 z}{\partial x^2}\right)^2 & \left(\frac{\partial^2 z}{\partial x^2}\right) \cdot \left(\frac{\partial^2 z}{\partial x \partial y}\right) \\ \left(\frac{\partial^2 z}{\partial x^2}\right) \cdot \left(\frac{\partial^2 z}{\partial x \partial y}\right) & \left(\frac{\partial^2 z}{\partial x \partial y}\right)^2 \end{pmatrix} + \frac{1}{\overline{\sigma}_y^2} \cdot \mathbf{L} * \begin{pmatrix} \left(\frac{\partial^2 z}{\partial y \partial x}\right)^2 & \left(\frac{\partial^2 z}{\partial y \partial x}\right) \cdot \left(\frac{\partial^2 z}{\partial y^2}\right) \\ \left(\frac{\partial^2 z}{\partial y \partial x}\right) \cdot \left(\frac{\partial^2 z}{\partial y^2}\right) & \left(\frac{\partial^2 z}{\partial y^2}\right)^2 \end{pmatrix} \quad (16.1)$$

In Equation 16.1, $\mathbf{L}$ is a lowpass filter by which the matrix elements are convolved, e.g., a binomial filter kernel of size $n \times n$; $\overline{\sigma}_x^2$ and $\overline{\sigma}_y^2$ are the variances of the smoothed matrix elements. They can be derived from an estimate of the variance σ_z^2 of the DSM heights by error propagation. Denoting the eigenvalues of $\mathbf{N}^{-1}$ by λ_1 and λ_2, a measure R of the strength of surface roughness and a measure D for the directedness of surface roughness can be defined:

$$R = \mathrm{tr}(\mathbf{N}) = \mathbf{L} * \left\{ \frac{1}{\overline{\sigma}_x^2} \cdot \left[\left(\frac{\partial^2 z}{\partial x^2}\right)^2 + \left(\frac{\partial^2 z}{\partial x \partial y}\right)^2 \right] + \frac{1}{\overline{\sigma}_y^2} \cdot \left[\left(\frac{\partial^2 z}{\partial y \partial x}\right)^2 + \left(\frac{\partial^2 z}{\partial y^2}\right)^2 \right] \right\} \quad (16.2)$$

$$D = 1 - \left(\frac{\lambda_1 - \lambda_2}{\lambda_1 + \lambda_2}\right)^2 = \frac{4 \cdot \det(\mathbf{N})}{\mathrm{tr}^2(\mathbf{N})} \quad (16.3)$$

R can also be interpreted as a measure of local coplanarity. The measures R and D are the basis for a three-way classification of the DSM pixels using two thresholds R_{min} and D_{min}.

If $R < R_{min}$, a pixel is classified as homogeneous. The nonhomogeneous pixels can be further classified according to D. If $D < D_{min}$, the pixel is classified as a line pixel, otherwise it is a point pixel. Line pixels correspond to surface discontinuities and surface intersections, whereas point pixels typically occur at building corners and with trees.

The height differences between the first and last pulse data have also been used to improve the results of building detection. They can be used as an indicator of vegetation, but large differences also occur at the edges of buildings.

The intensity of the returned laser beam provides information about the surface reflectance properties in the infrared part of the electromagnetic spectrum. The advantage of LiDAR intensities is that they are less affected by natural illumination conditions and cast shadows, since LiDAR is an active remote sensing technique. However, the intensities are often very noisy [8] because the diameter of the laser beam's footprint is usually much smaller than the sampling distance.

As an additional data source, multispectral images are often used due to their spectral content. As the most challenging task in building detection from LiDAR data is the separation of buildings and trees, color infrared imagery [9] or, alternatively, the normalized difference vegetation index (NDVI) derived from the near-infrared and the red portions of the spectrum [10] can be applied for their potential in discriminating vegetation.

In addition to the local characteristics described up to now, shape parameters can be evaluated to eliminate building candidate regions being either oddly shaped, too large, or too small [3,11].

16.2.1.2 Classification Techniques for Building Detection

Any building detection algorithm will use more than one of the cues described in Section 16.2.1.1. In this section, an overview on methods for combining these cues (often referred to as features) in the classification process will be given. The classification can be applied either to each pixel of the DSM (if the DSM is represented by a grid), to each LiDAR point, or to each building candidate region.

Rule-based classification is based on expert knowledge about the appearance of certain object classes in the data that are used to define rules by which the classes can be separated. These rules often involve thresholding operations. Rule-based classifications are easily implemented, but selecting the thresholds properly is often critical. In addition, the hierarchical structure of many rule-based approaches, where first a subset of the cues is selected to make an initial classification and then the other cues are used to resolve any ambiguities [12], makes it impossible to recover from previous errors in the classification process.

Fuzzy logic can be used to model vague knowledge about class assignment in order to avoid hard thresholds as in rule-based algorithms [13]. This requires the definition of membership functions for all shape cues and all classes, and their parameters have to be determined in a training phase. In a second step, these membership values are combined to obtain a final decision [5].

Unsupervised classification algorithms such as the ISODATA [9] or the K-means algorithm [10] aim at the detection of distinct clusters in feature space that correspond to objects having similar properties, without assigning these clusters to semantic classes such as building or tree. This assignment has to be done in a separate classification and is sometimes carried out interactively [9].

Probabilistic reasoning aims at assigning an object, s, to a class, C, of a given set of classes, θ, given the feature vector $\mathbf{x}_s$ of s. The optimum class C_{opt} is chosen as the class maximizing the a posteriori conditional probability $P(C_i \mid \mathbf{x}_s)$ of C_i given the data vector $\mathbf{x}_s$. These conditional probabilities are computed using the theorem of Bayes [14]:

$$P(C_i \mid \mathbf{x}_s) = \frac{P(\mathbf{x}_s \mid C_i) \cdot P(C_j)}{\sum_j P(\mathbf{x}_s \mid C_j) \cdot P(C_j)} \tag{16.4}$$

In Equation 16.4, $P(\mathbf{x}_s \mid C_i)$ is the a priori probability of data vector $\mathbf{x}_s$ under the assumption of a class C_i and $P(C_i)$ is the a priori probability of classifying an object as belonging to class C_i. Initially, the prior $P(C_i)$ is often assumed to be equal for all classes, and then recomputed from the relative numbers of objects classified as belonging to class C_i in the first iteration. The probabilities $P(\mathbf{x}_s \mid C_i)$ are often modeled by multivariate Gaussian distributions. If probabilistic reasoning is applied in an unsupervised way, the parameters of all these distributions (central value and covariance matrix) must be determined by some method, e.g., by heuristic assumptions about the average building height [6]. Such assumptions are unrealistic for built-up areas. For instance, the distribution of the DSM heights of the building roofs is a mixture of several normal distributions, each corresponding to a specific building type, rather than a simple normal distribution; thus, there are several clusters in feature space corresponding to buildings. An alternative is supervised classification, in which the parameters of the Gaussians are determined from the distribution of the data in training areas that have to be identified interactively.

If no a priori information about the distribution of the classes is used, i.e., if $P(C_i)$ is assumed to be equal for all classes, maximizing $P(C_i \mid \mathbf{x}_s)$ is equivalent to maximizing $P(\mathbf{x}_s \mid C_i)$. This is the approach used in the Maximum Likelihood classification, which has been applied to building detection in a supervised way [4].

The Dempster–Shafer theory of evidence was introduced as an expansion of the probabilistic approach that can also handle imprecise and incomplete information as well as conflict within the data [12,15]. Again, we consider a classification problem where the input data are to be classified into N classes $C_j \in \theta$, where θ is the set of all classes. Denoting power set of θ by 2^θ, a probability mass $m(A)$ is assigned to every class $A \in 2^\theta$ by a sensor (a classification cue) such that $m(\emptyset) = 0$, $0 \leq m(A) \leq 1$, and $\Sigma\, m(A) = 1$, where the sum is to be taken over all $A \in 2^\theta$ and $\emptyset$ denotes the empty set. Imprecision of knowledge can be handled by assigning a nonzero probability mass to the union of two or more classes C_j. The support Sup(A) of a class $A \in 2^\theta$ is the sum of all masses assigned to that class. The plausibility Pls(A) sums up all probability masses not assigned to the complementary hypothesis A' of A with $A \cap A' = \emptyset$ and $A \cup A' = \theta$. Sup(A') is called dubiety. It represents the degree to which the evidence contradicts a proposition. If p sensors are available, probability masses $m_i(B_j)$ have to be defined for all these sensors i with $1 \leq i \leq p$ and $B_j \in 2^\theta$. A combined probability mass can be computed for each class $A \in 2^\theta$:

$$m(A) = \frac{\sum_{B_1 \cap B_2 \cap \cdots \cap B_p = A} \left[\prod_{1 \leq i \leq p} m_i(B_j) \right]}{1 - \sum_{B_1 \cap B_2 \cap \cdots \cap B_p = \emptyset} \left[\prod_{1 \leq i \leq p} m_i(B_j) \right]} \tag{16.5}$$

The sum in the denominator of Equation 16.5 is a measure of the conflict in the evidence. Once the combined probability masses $m(A)$ have been determined, both Sup(A) and Pls(A) can be computed. The accepted hypothesis $C_{\text{opt}} \in \theta$ is determined according to a decision rule, e.g., as the class of maximum plausibility or the class of maximum support.

16.2.1.3 Workflow for Building Detection

The input to building detection is given by three data sets that have to be generated from the raw data by preprocessing. The DSM corresponding to the last pulse data is a regular height grid interpolated by linear prediction using a straight line as the covariance function, thus almost without filtering [11]. The first pulse data are also sampled into a regular grid, and by computing the height differences of the first and the last pulse DSMs, a grid ΔH_{FL} of the height differences between the first and the last pulses is obtained. The NDVI is computed from the near-infrared and the red bands of the multispectral images [10]. The image data must be geocoded so that the data are already aligned for the subsequent processes.

The work flow for building detection consists of three stages. The first stage is DTM generation in order to obtain a normalized DSM. The second stage of building detection is the detection of building candidate regions. An initial classification is carried out on a per-pixel level. Each DSM pixel is classified according to whether it is a building candidate pixel or not. Connected components of building pixels then become building candidate regions. The third stage is the evaluation of the building candidate regions. For the initial regions, average parameters describing the DSM heights, the spectral characteristics, surface roughness, and the region size are also evaluated to separate buildings from trees [10,11].

16.2.1.4 DTM Generation by Hierarchic Morphologic Filtering and Elimination of Large Buildings

DTM generation and building detection are closely interrelated. For computing a DTM, the LiDAR points on the tops of buildings and trees have to be eliminated, and thus information about the positions of such objects is required, whereas on the other hand, a DTM is required if buildings or trees are to be detected. This is why a hierarchical or coarse-to-fine strategy is applied for DTM generation. First, morphological gray-scale opening is applied to the DSM to generate a first approximation for the DTM. The size of the structural element has to be larger than the extent of the largest building in the data set. In the case that large buildings are present in the data set, this approximation will be too coarse. This would lead to small hills being classified as buildings in the later processing stages; however, if too small a structural element were used for morphological filtering, large building structures would be classified as terrain. This is why after morphological opening a rule-based algorithm [11] is used to detect large buildings in the data. The positions of large buildings are used in the next iteration of DTM generation, when a smaller structural element is used for morphological opening, to eliminate large buildings. DTM heights computed in the previous iteration are substituted for the results of the morphological filter. The process is finished when the minimum size for the structural element is reached [18]. The rule-based algorithm for building detection consists of a sequence of thresholding operations. It is only used to eliminate large buildings in DTM generation, when it is simple to select some of the thresholds because large buildings are usually characterized by large roof planes.

16.2.1.5 Detection of Building Candidate Regions by Dempster–Shafer Fusion

In this process, each pixel of the DSM is classified into one of four classes: buildings (B), trees (T), grass land (G), and bare soil (S). This is achieved by the Dempster–Shafer fusion of five sensors. In our model for the distribution of the evidence from each sensor to the four classes, we assume in general that each sensor, i, can separate two complementary subsets

U_{Ci} and U'_{Ci} of θ. The probability mass $P_i(x_i)$ assigned to U_{Ci} by the sensor i depending on the sensor output x_i is assumed to be equal to a constant P_1 for $x_i < x_1$. For $x_i > x_2$, it is assumed to be equal to another constant P_2, with $0 \leq P_1 < P_2 \leq 1$. Between x_1 and x_2, the probability mass is assumed to be a cubic parabola with horizontal tangents at $x_i = x_1$ and $x_i = x_2$, yielding a smooth transition between the probability levels P_1 and P_2:

$$P_i(x_i) = P_1 + (P_2 - P_1)\left[3\left(\frac{x_i - x_1}{x_2 - x_1}\right)^2 - 2\left(\frac{x_i - x_1}{x_2 - x_1}\right)^3\right] \tag{16.6}$$

The probability mass $[1-P_i(x_i)]$ will be assigned to class U'_{Ci}. The combined probability masses are computed for each pixel, and the pixel is assigned to the class of maximum support. In the following sections, the application of this model will be described for each of the five sensors.

16.2.1.5.1 *Height Differences ΔH between DSM and DTM*

ΔH distinguishes elevated objects from others. We assign a probability mass $P_{\Delta H} = P_{\Delta H}(\Delta H)$ according to the model described above to $B \cup T$, and $(1 - P_{\Delta H})$ to $G \cup S$. The last pulse DSM should be used to optimize the classification accuracy for buildings. An evaluation has shown that the quality of the classification results largely depends on the selection of $x_C = (x_1 + x_2)/2$, which has to be larger than the minimum building height in the scene. The values $(P_1, P_2) = (5\%, 95\%)$ and $(x_1, x_2) = (0m, 4m)$ have been shown to yield good results for different data sets [16].

16.2.1.5.2 *Strength R of Surface Roughness*

Surface roughness strength R as defined by Equations 16.1 and 16.2 is large in areas of great variations of the surface normal vectors, which is typical for trees. The absolute values of R will vary with the scene, so that it is impossible to find values for the parameters of the model described by Equation 16.6 that are generally applicable. The situation can be improved by a reparameterization of R. Rather than using R, we characterize surface roughness by the percentage $R_P(R)$ of pixels for which the surface roughness is smaller than R. $R_P(R)$ is limited to the interval $[0\%, \ldots, 100\%]$. We assign a probability mass $P_R = P_R(R_P)$ to class T, and $(1 - P_R)$ to $B \cup G \cup S$, neglecting that large values of R might also occur at the borders of buildings and at step edges of the terrain.

Assuming that the trees correspond to the areas of maximum surface roughness, an estimate for the percentage of the scene covered by trees can be used to derive the values of the parameters of the model in Equation 16.6. Denoting the percentage of trees by T_P, x_1 can be determined so that $P_R(100\% - T_p) = 50\%$. Using $(P_1, P_2) = (5\%, 95\%)$ and $x_2 = 100\%$ yields:

$$x_1 = 100\% - 2T_P \tag{16.7}$$

An evaluation of the method has shown that, depending on the dominant tree species and the time of the year the LiDAR data were captured, the model sometimes fits better to the first pulse DSM than to the last pulse DSM (especially with deciduous trees before the vegetation period), and sometimes vice versa [16].

16.2.1.5.3 *Directedness D of Surface Roughness*

The directedness D of surface roughness as defined by Equations 16.1 and 16.3 is another indicator for trees, but only if R differs significantly from 0; otherwise, D is dominated by noise. A probability mass $P_D = P_D(R, D)$ is assigned to class T, and $(1 - P_D)$ to $B \cup G \cup S$. In order to decide whether R is significant or not, it is compared to a threshold R_{min} that is determined so that T_P percent of the data have $R > R_{min}$. We select $(P_1, P_2) = (10\%, 70\%)$ and $(x_1, x_2) = (0, 1)$.

16.2.1.5.4 *Height Differences DH_{FL} between First and Last Pulse*

ΔH_{FL} is large in areas covered by trees. A probability mass $P_{FL} = P_{FL}(\Delta H_{FL})$ is assigned to class T. As a small value of ΔH_{FL} does not necessarily mean that there are no trees, the probability mass $(1 - P_{FL})$ is assigned to θ and not to $B \cup G \cup S$. The values $(P_1, P_2) = (5\%, 95\%)$ and $(x_1, x_2) = (0m, 4m)$ have been shown to be applicable for different scenes [16].

16.2.1.5.5 *NDVI*

The NDVI is an indicator of vegetation, thus, for classes T and G. The general model described above has to be improved to compensate for systematic classification errors that would be caused by the uncertainty of the NDVI in shadow areas. This can be achieved by modulating the probability masses depending on the standard deviation σ_{NDVI} of the NDVI. The NDVI is defined as the ratio between the difference and the sum of the infrared band IR and the red band Rd. Its standard deviation σ_{NDVI} can be computed from

$$\sigma_{NDVI} = \frac{2\sqrt{\mathrm{Rd}^2\sigma_{IR}^2 + \mathrm{IR}^2\sigma_{Rd}^2}}{(\mathrm{IR} + \mathrm{Rd})^2} \tag{16.8}$$

In Equation 16.8, σ_{Rd} and σ_{IR} are the standard deviations of Rd and IR. They are determined by analyzing the first derivatives of IR and Rd [7]. For $\|\sigma_{NDVI}\| \geq 0.25$, a probability mass of 1.0 is assigned to θ, i.e., the NDVI will not contribute to the classification. If $\|\sigma_{NDVI}\| < 0.25$, $P_N^0 = 2\|\sigma_{NDVI}\|$ is assigned to θ. Using $P_N^0 = P_N$ (NDVI) according to Equation 16.6, $P_N = (1 - 2\|\sigma_{NDVI}\|) \cdot P_N^0$ is assigned to class $T \cup G$ and $P_N = (1 - 2\|\sigma_{NDVI}\|) \cdot (1 - P_N^0)$ to $B \cup S$.

We choose $(P_1, P_2) = (10\%, 90\%)$. As the NDVI depends on the lighting conditions, the spectral characteristics of the sensor, and the predominant type of vegetation, the parameters x_1 and x_2 have to be determined in an exploration phase. In the example given in Section 16.2.1.7, we chose $(x_1, x_2) = (-0.3, 0.3)$.

16.2.1.5.6 *Postclassification*

A rule-based technique for postclassification is applied in order to eliminate single-building pixels and to compensate for classification errors at the building outlines. This classification takes into consideration the conflict in the data and the classification results in a local neighborhood of each pixel [16]. After postprocessing, a binary building image is generated. In this building image, small elongated areas of building pixels are eliminated by binary morphologic opening. After that, a building label image is created by a connected component analysis.

16.2.1.6 Evaluation of Building Candidate Regions

A second classification based on the Dempster–Shafer theory is applied to the initial building regions. The average height differences ΔH_a between the DSM and the DTM and the average NDVI (NDVI_a) are used in the same way as ΔH and NDVI in the initial classification. The percentage P of pixels classified as point-like in polymorphic feature extraction is an indicator for trees. Thus, a probability mass $P_P = P_P(P)$ is assigned to class T, and $(1 - P_P)$ to $B \cup G \cup S$.

The mathematical model described by Equation 16.6 is also used for computing the probability masses for ΔH_a, P, and $NDVI_a$. For $NDVI_a$ we choose the same model as for the NDVI. For the remaining sensors, we select $(P_1, P_2) = (5\%, 95\%)$. For ΔH_a, the values for $(x_1, x_2) = (1m, 3m)$ are chosen to be a bit tighter than for ΔH. The parameters for P depend on the typical size of a roof plane in relation to the LiDAR resolution and have to be determined in a training phase. The combined probability masses are evaluated for each initial building region, and any region assigned to another class than building is eliminated.

16.2.1.7 Results and Discussion

In this section, results will be presented for a test data set captured over Fairfield (Australia) using an Optech ALTM 3025 laser scanner. The data set covers an area of $2 \times 2\,\text{km}^2$. Both the first and the last echoes of the laser beam were recorded with an average point separation of about 1.2 m. DSMs with a grid width $\Delta = 1$ m were generated for both the first and the last pulse data. An RGB orthophoto with a resolution of 0.15 m was also available. For lack of an infrared band in the orthophoto, a pseudo-NDVI-image was generated at a resolution of 1 m, using the red band from the orthophoto and substituting the LiDAR intensity values for the infrared band. Reference data were captured by digitizing buildings in the orthophoto, resulting in 2424 building polygons. As the orthophoto and the LiDAR data correspond to different epochs, 49 buildings that were only available in one data set had to be excluded. Figure 16.1 shows the DSM and the NDVI image of the Fairfield data set.

The results of the initial Dempster–Shafer classification after postprocessing are shown in Figure 16.2a. Step edges at the building boundaries are often classified as trees, an effect that is reduced but not completely eliminated by postclassification. After region-based classification, this resulted in 2057 building regions.

For an evaluation of automatic feature extraction using a reference data set, two numbers of interest are the completeness and the correctness of the results [17]:

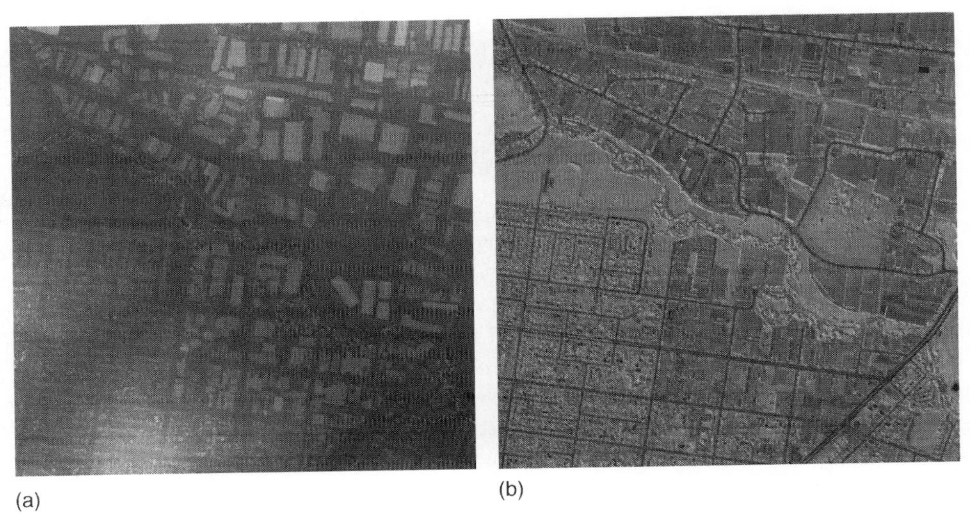

FIGURE 16.1
DSM (a) and NDVI image (b) of the Fairfield data set. (Reprinted from Rottensteiner, F. et al., *Inform. Fusion*, 6(4), 283, 2005. With permission.)

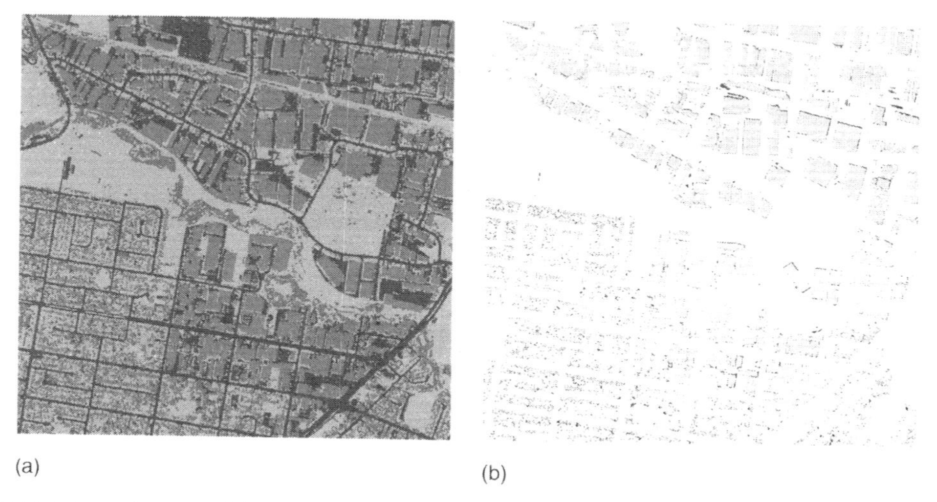

FIGURE 16.2 (See color insert following page 334.)
(a) Results of the per-pixel classification after postprocessing. Red: buildings, dark green: trees, light green: grass land, brown: bare soil. (b) Comparison of final results to reference data on a per-pixel level. Yellow: TP, red: FP, blue: FN. (Reprinted from Rottensteiner, F. et al., *ISPRS J. Photogram. Rem. Sens.*, 62(2), 135, 2007. With permission.)

$$\text{Completeness} = \frac{\text{TP}}{\text{TP} + \text{FN}} \tag{16.9}$$

$$\text{Correctness} = \frac{\text{TP}}{\text{TP} + \text{FP}} \tag{16.10}$$

where (Equations 16.9 and 16.10)
- TP denotes the number of true positives, i.e., the number of entities found to be available in both data sets
- FN is the number of false negatives, i.e., the number of entities in the reference data set that were not detected automatically
- FP is the number of false positives, i.e., the number of entities that were detected, but do not correspond to an entity in the reference data set

The methodology for evaluation is based on a comparison of two label images: the output of the building detection algorithm and the reference label image generated by rasterizing the reference polygons. Completeness and correctness are determined both on a per-pixel level and on a per-building level. Note that in the latter case, it is not straightforward to determine TP, FN, and FP because of problems related to handling multiple overlaps [18].

Figure 16.2b shows a comparison between the final classification results and the reference data. On a per-pixel level, completeness was 87%. The missed buildings were mostly small residential buildings having roofs consisting of many small faces, or they were too small to be detected given the resolution of the LiDAR data. Correctness was somewhat better than completeness at 91%. False positives mostly occur at bridges, at small terrain structures not covered by vegetation, and at container parks. These numbers have to be interpreted with caution because they are affected by errors in the reference data that can be up to 5 m.

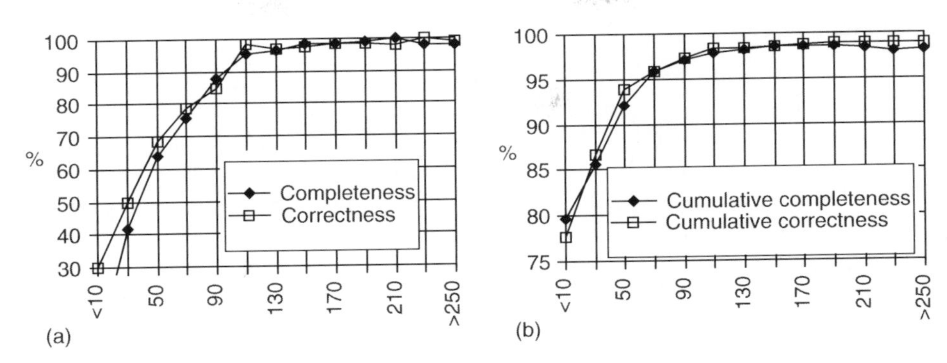

FIGURE 16.3
Completeness and correctness of the detection results as a function of the building size (m^2). (a) Completeness and correctness for buildings of the size given in the abscissa. (b) Cumulative completeness and correctness for all buildings larger than the size given in the abscissa. (Reprinted from Rottensteiner, F. et al., *ISPRS J. Photogram. Rem. Sens.*, 62(2), 135, 2007. With permission.)

Figure 16.3 shows how the completeness and correctness of the detection results decrease with decreasing area covered by the buildings. Figure 16.3a, completeness and correctness are given for buildings of the size shown in the abscissa. Buildings larger than 110 m^2 can be detected reliably, with both completeness and correctness being larger than 95%. The algorithm could detect more than 80% of the buildings with an area between 80 and 110 m^2, and the majority of buildings with an area between 40 and 80 m^2 could still be detected. Buildings smaller than 30 m^2 were not detectable. The cumulative completeness and correctness graphs in Figure 16.3a show that 95% of all buildings larger than 70 m^2 could be detected, whereas 95% of all detected buildings larger than 70 m^2 were correct.

The results described so far were achieved using the standard parameter settings as explained above. The most important control parameters are the values (x_1, x_2) used for the model of the probability masses for ΔH, ΔH_{FL}, and the NDVI in the initial classification, and the estimate T_p for the percentage of trees in the scene, which is essential to derive the parameters for the probability masses of the surface roughness parameters. An assessment of the sensitivity of the results to the parameter settings showed that the quality of the results mainly depends on the centre point $x_C = (x_1 + x_2)/2$ of the model for the probability masses [16]. For ΔH, it has to correspond to the minimum building height expected in the scene. Changes of x_C of 0.5 m can deteriorate the results for smaller buildings, but the standard parameter settings are well-chosen and will usually not need adaptation. The results are not very sensitive with respect to changes in the model for ΔH_{FL}. With respect to the NDVI, changes of x_C of 10% reduce the completeness results for buildings smaller than 150 m^2 by 10%–20%, without improving correctness accordingly. It was also found that the percentage T_p of trees should be known with an accuracy of 5% in order to achieve optimal results for smaller buildings. Thus, the parameters for the probability masses have to be known relatively precisely if excellent results are to be achieved for buildings smaller than 150 m^2; otherwise, good results can still be achieved, but the full potential of the method for detecting small buildings will not be exploited [16].

The limitations of building detection based on the sensor resolution were also investigated. In order to achieve good results for residential buildings covering an area of about 100–150 m^2, the sensor resolution must be at least 1.5 m. At a resolution of 3 m, only large structures can be detected reliably [16]. An investigation into the contributions of the individual cues showed that the main contribution of the NDVI is to increase the correctness by up to 20% for small-to medium-sized buildings. First pulse data also help, though to a lesser degree [16].

16.2.2 Building Reconstruction from LiDAR Data

The goal of automated building reconstruction is the generation of CAD models of the buildings. For visualization purposes, the most common way of modeling buildings is by boundary representation (B-rep) [19], i.e., the representation of a building by its bounding faces, edges, vertices, and their mutual topological relations. For building reconstruction it is assumed that the location of buildings is known at least in an approximate way. This can be the result of automatic building detection, as described in Section 16.2.1. As an alternative, building locations can be provided by an existing 2D GIS data base [20]. This has the advantage of giving relatively precise positions of the building outlines, which are not very well defined in LiDAR data. On the other hand, the up-to-date-ness of existing GIS data can become a problem.

Two strategies can be applied for the geometrical reconstruction of buildings. Firstly, parametric primitives can be instantiated and fit to the data, e.g., in rectangular regions derived by splitting a 2D ground plan [20]. This corresponds to a top-down or model-based approach. It has the advantage that the resulting building models are visually appealing because the primitives usually are characterized by regular shapes. Regularization of the resulting building models thus is an implicit part of the reconstruction process. On the other hand, this can result in an over-regularization: buildings having nonrectangular footprints are usually not accurately reconstructed in that way. Secondly, polyhedral models can be generated by extracting planar segments in a DSM and grouping these planes [20–22]. This corresponds to a data-driven or bottom-up strategy. As building outlines are difficult to locate precisely in LiDAR data, ground plans are also often used [20]. Polyhedral models are more generally applicable than parametric primitives, and a higher level of detail can be achieved if they are used. However, polyhedral models are sometimes not as visually appealing as models generated by primitives, because geometric regularities are not an implicit part of the model, so that regularization has to be applied in a separate processing stage.

In this section, the second strategy will be followed. The locations of buildings are assumed to be known with an accuracy of 1–3 m. Further, we assume the LiDAR data are sampled into a DSM in the form of a regular grid of width Δ by linear prediction [11]. The work flow for the geometric reconstruction of buildings as presented in this section consists of three steps:

1. Detection of roof planes based on a segmentation of the DSM to find planes that are expanded by region growing.
2. Grouping of roof planes and roof plane delineation: Coplanar roof segments are merged, and hypotheses for intersection lines and step edges are created based on an analysis of the neighborhood relations.
3. Consistent estimation of the building parameters using all available sensor data. This includes model regularization by introducing hypotheses about geometric constraints into the estimation process.

In the course of the reconstruction process, many decisions have to be taken, e.g., with respect to the actual shape of the roof polygons or to the mutual configuration of roof planes. Traditionally, such decisions are based on comparing geometric entities, e.g., distances, to thresholds. In order to avoid threshold selection as far as possible, decisions can be taken based on hypothesis tests, e.g., about the incidence of two geometric entities. Thus, the selection of thresholds is replaced by the selection of a significance level for the hypothesis test. This requires rigorous modeling of the uncertainty of the geometrical entities involved. The concept of uncertain projective geometry [23] is well-suited for that purpose and will be applied in all stages of the workflow, as will be presented in the subsequent sections.

16.2.2.1 Detection of Roof Planes in LiDAR DSMs

The problem of detecting planar segments in DSMs has been tackled in two different ways: clustering and region growing. In the first case, local planes are estimated for each LiDAR point (or each DSM pixel) from the points in a small neighborhood. Each plane corresponds to a point in a space whose axes are the plane parameters. Clustering techniques try to detect planes by finding clusters of such points in the parameter space [24]. In the second case, seed regions for planes are determined based on some measure of local coplanarity. For each seed region, the parameters of a plane passing through all points of the seed region are estimated, and the seed regions are expanded by neighboring points found to be situated in this plane [25]. In this section, a method for roof plane detection in DSM grids based on the second principle is described.

In Section 16.2.1.1, a measure R for surface roughness was introduced. This measure can be used to detect seed regions for planar segment in the DSM grid: pixels having a low value of R are surrounded by coplanar points. A straightforward method for roof plane detection could start by generating a binary image of homogeneous pixels (pixels with $R < R_{min}$) inside the coarse building outlines. Assuming connected components of homogeneous pixels to be seed regions, a plane can be fitted to all DSM pixels belonging to such a seed region, and seed regions that are actual planar (i.e., having an r.m.s.e of planar fit smaller than a threshold, e.g., ± 0.15 m) are grown. In each of the iterations of region growing, all pixels being adjacent to any seed region are tested according to whether they are inside the adjusting plane of that seed region, and if they are, they are added to that planar region. This test can be a hypothesis test for the incidence of a point and a plane [23,26]. The process is terminated when no more pixels can be added to any plane.

Unfortunately, this naïve approach does not work very well, because it is impossible to find an appropriate value for the threshold R_{min}: If a very low value is chosen, some planes will not be detected, whereas if it is too high, planes that should be separated will be merged. As a consequence, it is advisable to use the method outlined above in an iterative way, using different values for R_{min} in this process. In the first iteration, only the α% most homogeneous pixels can be used to determine the seed regions. Using for instance $\alpha = 5\%$ or $\alpha = 10\%$ results in very small seed regions and an over-segmentation of the DSM, whereas smaller roof planes might not yet be detected because these thresholds are very tight. The region-growing procedure is repeated several times, each time using a higher value for α, and not considering pixels that have already been assigned to a roof plane. In each iteration, less and less significantly coplanar seed regions are used. The iteration process is completed when a threshold α_{max} for α is reached. The value for this threshold has to take into account that pixels at the transition between neighboring roof planes and pixels at step edges, especially at the building outlines, will usually not be classified as homogenous. A typical value is $\alpha_{max} = 70\%$.

As stated above, this iterative scheme will result in an over-segmentation. Thus, neighboring planar segments have to be checked according to whether they are coplanar or not, and coplanar segments have to be merged. This can be achieved by an F-test comparing the r.m.s.e. of planar fit achieved using two models (identical vs. separate planes) [26]. Figure 16.4 shows two examples for the results of planar segmentation after merging of coplanar segments.

FIGURE 16.4 (See color insert following page 334.)
DSM with grid width = 0.5 m (left), orthophoto (center), and planar segments (right) for two buildings. Upper row: width = 115 m; lower row: width = 60 m. Plane P_f will be eliminated later. (From Rottensteiner, F., et al., *Int. Arch. Photogram. Rem. Sens. Spatial Inform. Sci.*, *XXXVI*(3/W19), 221, 2005.)

16.2.2.2 Grouping of Roof Planes and Roof Plane Delineation

16.2.2.2.1 Initialization of the Roof Boundary Polygons

Once the roof planes have been detected, their boundary polygons are determined. First, a Voronoi diagram of the planar segments is determined by chamfering [27]. The boundaries of the planes in the Voronoi diagram deliver approximate values for the boundary polygons of these planes. By using the Voronoi diagram for approximations, segmentation problems such as gaps between neighboring planes or incomplete planes due to occluding trees are overcome. The heights of the vertices of the boundary polygons are derived from the roof plane parameters. Each boundary polygon is split into an ordered set of polygon segments so that each segment separates exactly two neighboring planes. Then, each polygon segment is classified according to whether it corresponds to a step edge, to an intersection line, or, in the case of dormers, to both. This classification has to take into account the uncertainty of both the two neighboring planes and the approximate positions of the vertices of the initial polygon segments.

All vertices of the approximate boundary polygon segment are tested for incidence with the intersection line of the two neighboring planes. If all the vertices of the polygon segment are found to be incident with the intersection line, the segment is classified as an intersection. If the polygon segment is an outer boundary or if no vertex is found to be on the intersection, the polygon segment is classified as a step edge. If some vertices are determined to be on the intersection and others are not, the polygon segment will be split up into several new segments, each having a different classification. Of these new segments, very short ones and intersection segments whose average distance from the approximate polygon is larger than the segment length are discarded. For polygon segments classified as intersections, the intersection line might cut off considerable portions of the two roof

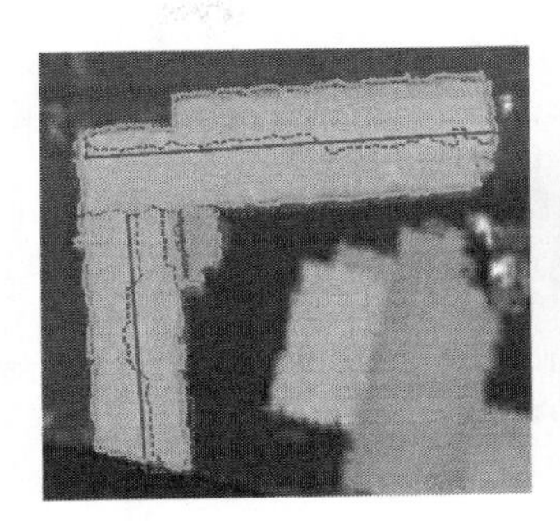
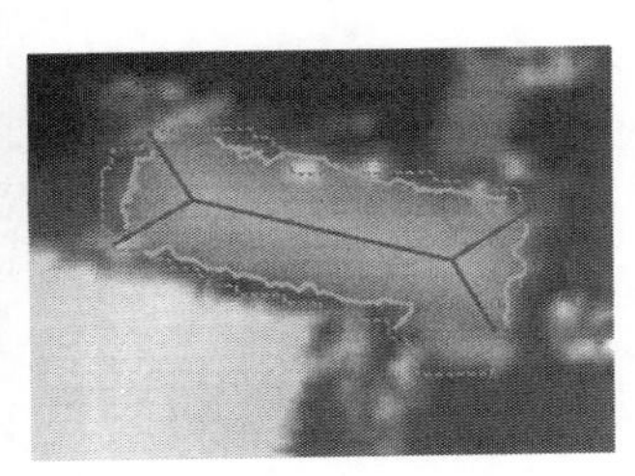

FIGURE 16.5
Classification of the roof boundary polygons and approximate step edges. Dashed lines: approximate boundary polygons. Light solid lines: original step edges. Dark solid lines: intersections. (From Rottensteiner, F. et al., *Int. Arch. Photogram. Rem. Sens. Spatial Inform. Sci.*, XXXVI(3/W19), 221, 2005.)

planes if the planes are nearly horizontal (e.g., the building in upper part of Figure 16.4). Thus, another hypothesis test checks whether the cut-off parts of the roof planes are coplanar with the neighbor. Only in this case the classification as an intersection will be accepted, otherwise, a step edge will be assumed. Figure 16.5 shows the results of the classification of the polygon segments for the two buildings in Figure 16.4.

16.2.2.2.2 Refinement of Roof Polygons

If two planes intersect, the part of the roof boundary polygons separating these two planes is replaced by the intersection line. Step edges have to be positioned more precisely in the LiDAR data. The original polygon segments at step edges are sampled at the DSM grid width. For each vertex **P** of such a polygon segment, a profile of the DSM passing through **P** is analyzed to determine a step edge point. The profile is orthogonal to the approximate step edge, and it is ordered so that it starts from the interior of the roof plane (Figure 16.6).

At outer boundaries, the analysis starts with a search of the first point $\mathbf{P}_1$ in the profile found not to be incident with the roof plane [26]. If no such point is found, the profile is supposed not to contain a step edge. Otherwise, the point $\mathbf{P}_{max}$ of maximum height difference is searched for in the profile, starting from $\mathbf{P}_1$. Since the terrain has to be lower than the roof, the search for $\mathbf{P}_{max}$ is stopped if the height difference between neighboring points in the profile becomes positive, which happens if the roof boundary is occluded by a high tree. In order to eliminate points on low vegetation next to a roof, $\mathbf{P}_{max}$ is discarded if it is

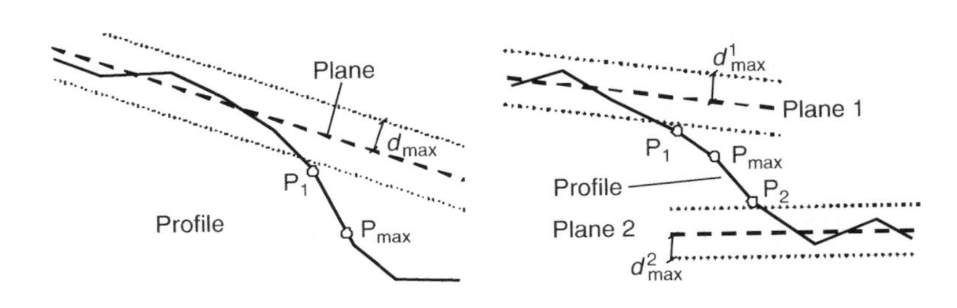

FIGURE 16.6
(a) Step edge detection at building outlines. $\mathbf{P}_1$: the first point on the profile outside the tolerance band of width d_{max}. $\mathbf{P}_{max}$: the step edge point determined as the point of maximum height gradient. (b) Step edge between two planes. In this case, $\mathbf{P}_1$ and $\mathbf{P}_2$ are the first points outside the tolerance bands of the two planes. (From Rottensteiner, F. et al., *Int. Arch. Photogram. Rem. Sens. Spatial Inform. Sci.*, XXXVI(3/W19), 221, 2005.)

above the roof plane. If no valid step edge point $\mathbf{P}_{max}$ is found, the step edge is assumed to be a straight line between the two closest step edge points visible to the sensor, which is what a human operator would do in such a situation.

For step edges separating two roof planes, the first points $\mathbf{P}_1$ and $\mathbf{P}_2$ of the profile that are not incident with the two planes are determined. If the order of $\mathbf{P}_1$ is found to be "behind" $\mathbf{P}_2$ in the profile, no step edge point can be determined; otherwise the position $\mathbf{P}_{max}$ of the step edge is determined as the position of the maximum height difference between $\mathbf{P}_1$ and $\mathbf{P}_2$.

The detected step edge polygons are quite noisy (green lines in Figure 16.5) and need to be generalized. First, step edge points having a distance larger than two times the original LiDAR point distance from both their predecessor and successor are eliminated as outliers. The remaining step edge polygon is approximated by longer polygon segments using a simple recursive splitting algorithm. Polygon segments containing less than three step edge points are discarded and in an iterative procedure, neighboring polygon segments are merged if they are found to be incident based on a statistical test [26]. In a second merging step we search for very short polygon segments. If both neighbors of such a segment are found to be incident, the short segment is eliminated and its neighbors are merged. Finally, the vertices of the generalized step edge polygon are determined by intersecting neighboring polygon segments; only if these segments are nearly parallel, their end points are connected by a new polygon edge.

16.2.2.2.3 Improving the Planar Segmentation

Replacing the approximate roof boundary polygons by the intersection lines and by the generalized step edges can affect the planar segmentation and also the neighborhood relations. Hence, after generating the roof polygons, all pixels inside a roof polygon assigned to another roof plane and all pixels assigned to a roof plane that are outside the roof polygon are eliminated from their respective planes. This is followed by a new iteration of region growing, where first the regions are only allowed to grow within their polygons. Spurious segments such as the plane $\mathbf{P}_f$ in Figure 16.4 might be eliminated in this process. Having improved the segmentation, the boundary classification and step edge detection are repeated. Figure 16.7 shows the final positions of all roof polygon segments for the buildings in Figure 16.4.

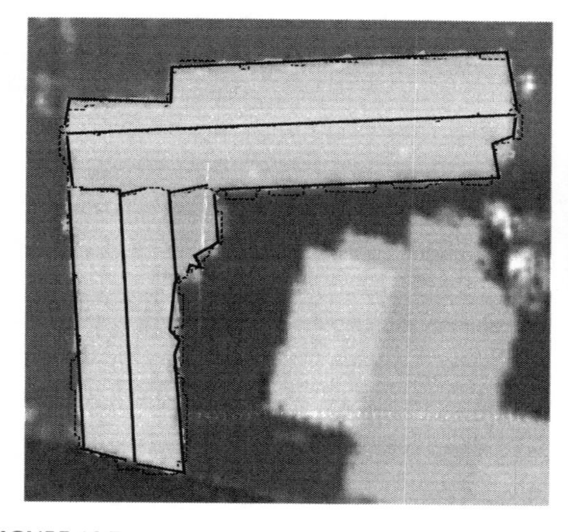
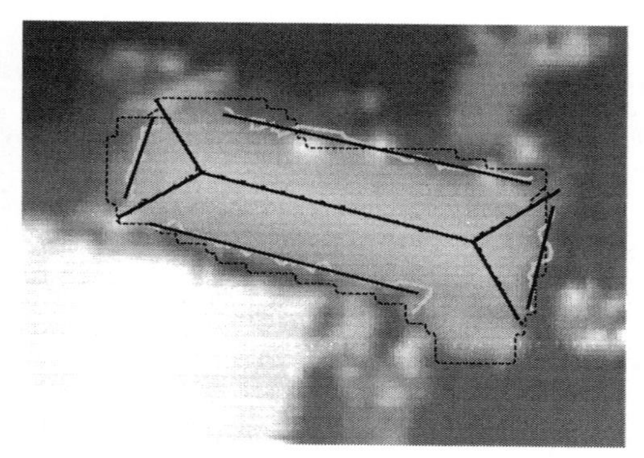

FIGURE 16.7
Approximate roof polygons (dashed lines), approximate step edges (light solid lines), and generalized roof polygon segments (dark solid lines) after improving the planar segmentation. (From Rottensteiner, F. et al., *Int. Arch. Photogram. Rem. Sens. Spatial Inform. Sci.*, XXXVI(3/W19), 221, 2005.)

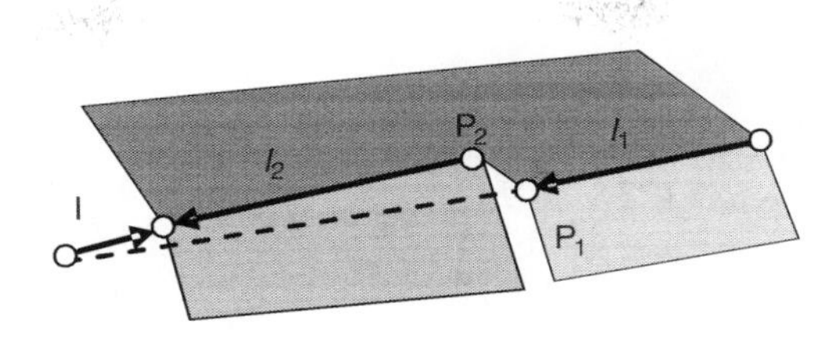

FIGURE 16.8
Consistency check: a new step edge is inserted between l_1 and l_2. (From Rottensteiner, F. et al., *Int. Arch. Photogram. Rem. Sens. Spatial Inform. Sci.*, XXXVI(3/W19), 221, 2005.)

16.2.2.2.4 *Combination of Roof Polygon Segments*

Until now, all segments of all roof boundary polygons were handled individually. Before combining these segments, their consistency is checked. If within a roof boundary there are two consecutive segments l_1 and l_2 classified as intersections, l_1 and l_2 must intersect. If replacing the segment endpoints $\mathbf{P}_1$ and $\mathbf{P}_2$ by the intersection point $\mathbf{I}$ changes the direction of one of the segments, a new step edge has to be inserted between $\mathbf{P}_1$ and $\mathbf{P}_2$ (Figure 16.8). A second consistency test checks whether for each roof polygon segment separating two planes, there is a matching opposite segment of the same type belonging to the neighboring plane. If this is not the case, it has to be inserted [26].

Now all the polygon segments have to be combined. This involves an adjustment of the vertices at the transitions between consecutive polygon segments. First, all polygon segments intersecting at one planimetric position have to be found. Figure 16.9 shows an example involving three roof planes. There are altogether three polygon segments. One of them is the intersection line between two roof planes, and the other two are step edges. There are two intersection points $\mathbf{P}_1$ and $\mathbf{P}_2$ having the same planimetric position but different heights. For adjustment, we consider all the planes in the vicinity of the vertices $\mathbf{P}_1$ and $\mathbf{P}_2$, i.e., the roof planes (ε_1, ε_2, ε_3) and the walls corresponding to the step edge segments (ε_{13}, ε_{23}). For each plane, we observe the distance between the plane and the point $\mathbf{P}_i = (X,Y,Z_i)^T$ to be 0. The weights of such an observation are determined from the standard deviation σ_i of the distances between the approximate position $\mathbf{P}_i^0 = (X^0, Y^0, Z_i^0)^T$ and the respective planes. The observation equations for a roof plane, u, giving support to height Z_i and for a wall, w, look as follows:

$$\begin{aligned} 0+v_{ui} &= A_u(X^0+\delta X)+B_u(Y^0+\delta Y)+C_u(Z_i^0+\delta Z_i)+W_u; \\ 0+v_w &= A_w(X^0+\delta X)+B_w(Y^0+\delta Y)+W_w; \end{aligned} \tag{16.11}$$

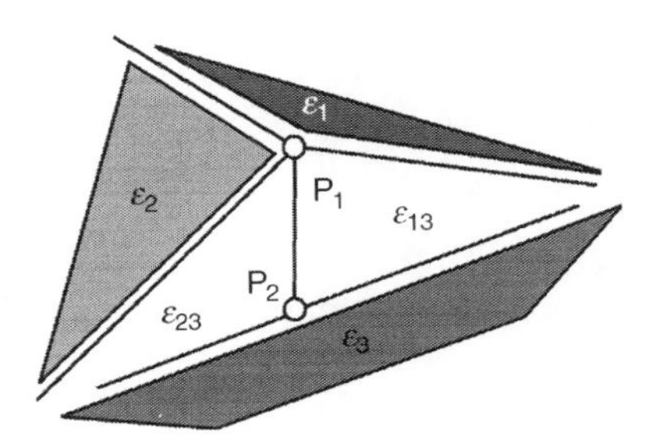

FIGURE 16.9
Vertex adjustment. (From Rottensteiner, F. et al., *Int. Arch. Photogram. Rem. Sens. Spatial Inform. Sci.*, XXXVI (3/W19), 221, 2005.)

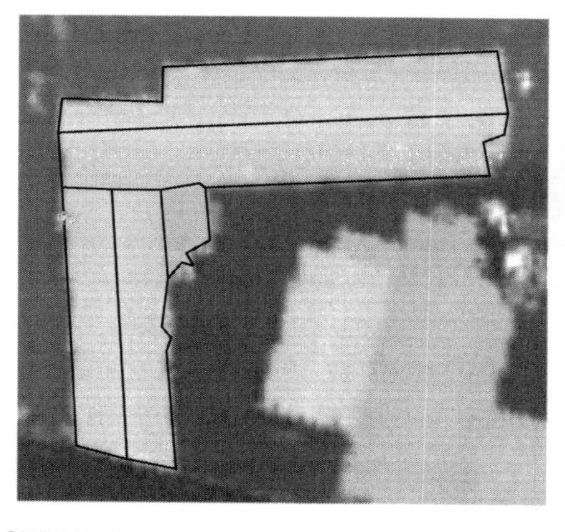

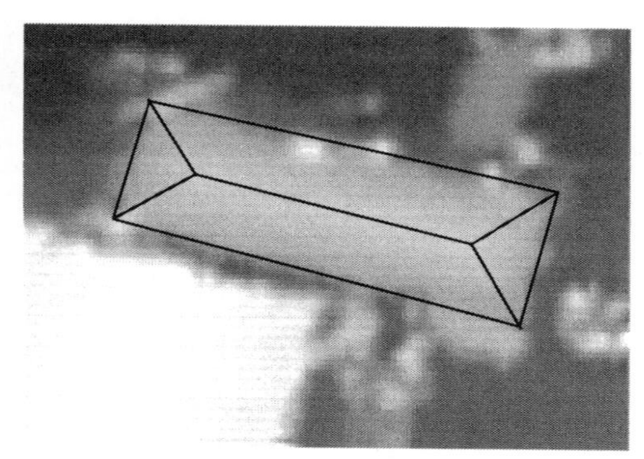

FIGURE 16.10
Combined roof polygons after local adjustment of vertices. (From Rottensteiner, F. et al., *Int. Arch. Photogram. Rem. Sens. Spatial Inform. Sci.*, XXXVI(3/W19), 221, 2005.)

where

v denotes the correction of the observation
(A_u, B_u, C_u, W_u) and (A_w, B_w, W_w) are the plane parameters
$(\delta X, \delta Y, \delta Z_i)$ are the corrections to the coordinates of the vertices that are to be determined in the adjustment

Equation 16.11 is used in an iterative least squares adjustment. This model assumes that all walls intersect in one vertical line. Due to errors in step edge extraction, small step edge segments might have been missed and the extracted step edge segment might pass by the intersection point at (X,Y). To find such segments, we compute the normalized corrections $v_w^n = v_w/\sigma_w$ of the wall observations after each iteration and exclude the wall with a maximum value of v_w^n if $v_w^n > 3.5$. For all excluded walls, a new step edge segment is introduced between the original end point of the step edge and the adjusted position of the vertex. Figure 16.10 shows the resulting roof boundaries for the buildings in Figure 16.4. From these combined roof polygons, a topologically consistent building model in B-rep is generated.

16.2.2.3 Consistent Estimation of Building Parameters and Model Regularization

Besides resulting in a more regular visual appearance, geometric regularities help to improve the geometric accuracy of the models if the sensor geometry is weak. The geometric constraints are not an implicit part of the building model, but rather are added as additional information to the estimation of the building parameters and thus only have to be considered where enough evidence is found in the data. In parameter estimation, geometric regularities can be considered in the adjustment by constraint equations. This strategy will result in models precisely fulfilling these hard constraints [28]. The alternative is to add soft constraints, i.e., direct observations for entities describing a geometric regularity, to the adjustment of the sensor-based observations. In this case, the constraints will not be fulfilled exactly, but there will be residuals to the observations. Using the second strategy, robust estimation techniques can be applied to the soft constraints to determine whether a hypothesis about a regularity fits the sensor data or not [29].

The result of the previous building reconstruction steps is a polyhedral building model in B-rep. The faces of the model are labeled as being a roof face, a wall, or the floor.

The model parameters are the coordinates of the model vertices and the plane parameters of the model faces. The topology of the model and some meaningful initial values for its parameters are known. The coarse model has to be analyzed for geometric regularities, and the model parameters have to be estimated. In the subsequent sections, the observations and the adjustment model used for parameter estimation will be described.

16.2.2.3.1 *Observations Representing Model Topology*

Parameter estimation is based on a mapping between the B-rep of the polyhedral model and a system of GESTALT (shape) observations representing the model topology in adjustment. GESTALT observations are observations of a point **P** being situated on a polynomial surface [30] that is parameterized in an observation coordinate system (u, v, w) related to the object coordinate system by a shift $\mathbf{P}_0$ and three rotations $\Theta = (\omega, \phi, \kappa)^T$. The observation is **P**'s distance from the surface, which has to be 0. Using $(u_R, v_R, w_R)^T = \mathbf{R}^T(\Theta) \cdot (\mathbf{P} - \mathbf{P}_0)$, with $\mathbf{R}^T(\Theta)$ being a transposed rotational matrix parameterized by Θ, and modeling walls to be strictly vertical, there are three possible formulations of GESTALT observation equations:

$$
\begin{aligned}
r_u &= \frac{m_u u_R + a_{00} + a_{01} m_v v_R}{\sqrt{1 + a_{01}^2}} \\
r_v &= \frac{m_v v_R + b_{00} + b_{10} m_u u_R}{\sqrt{1 + b_{10}^2}} \\
r_w &= \frac{m_w w_R + c_{00} + c_{10}(m_u u_R) + c_{01}(m_v v_R)}{\sqrt{1 + c_{10}^2 + c_{01}^2}}
\end{aligned}
\tag{16.12}
$$

In Equation 16.12, r_i are the corrections of the fictitious observations of coordinate i and $m_i \in \{-1, 1\}$ are mirror coefficients. With GESTALT observations, one is free to decide which of the parameters in Equation 16.12 are to be determined in the adjustment and how to parameterize a surface. Different GESTALTs can refer to identical transformation or surface parameters, which will be used to handle geometric regularities. Here, the rotations are 0 and constant. $\mathbf{P}_0$ is a point situated inside the building and constant. For each face of the B-rep of the building model, a set of GESTALT observations is defined, taking one of the first two Equations 16.12 for walls and the third one for roofs. The unknowns to be determined are the coordinates of each vertex **P** and the plane parameters (a_{jk}, b_{ik}, c_{ij}). As each vertex is neighbored by at least three faces, the vertex coordinates are determined by these GESTALT observations and need not be observed directly in the sensor data. These observations link the vertex coordinates to the plane parameters and thus represent the building topology in the adjustment.

16.2.2.3.2 *Observations Representing Geometric Regularities*

Geometric regularities are considered by additional GESTALT equations, taking advantage of specific definitions of the observation coordinate system and specific parameterizations of the planes. In our current implementation, geometric regularities can occur between two planes of the model that intersect in an edge or between two vertices connected by an edge. The observation coordinate system is centered in one vertex $\mathbf{P}_1$ of that edge and the w-axis is vertical, thus $\omega = \phi = 0 =$ constant. Four types of geometric regularities are considered (Figure 16.11). The first type, a horizontal roof edge, involves the edge's end points: Its two vertices $\mathbf{P}_1$ and $\mathbf{P}_2$ must have identical heights. The two points are declared to be in a horizontal plane ε_h that is identical to the (u,v)—plane of the observation coordinate system. One observation is inserted for $\mathbf{P}_2$: $r_w = w_R = Z_2 - Z_1$.

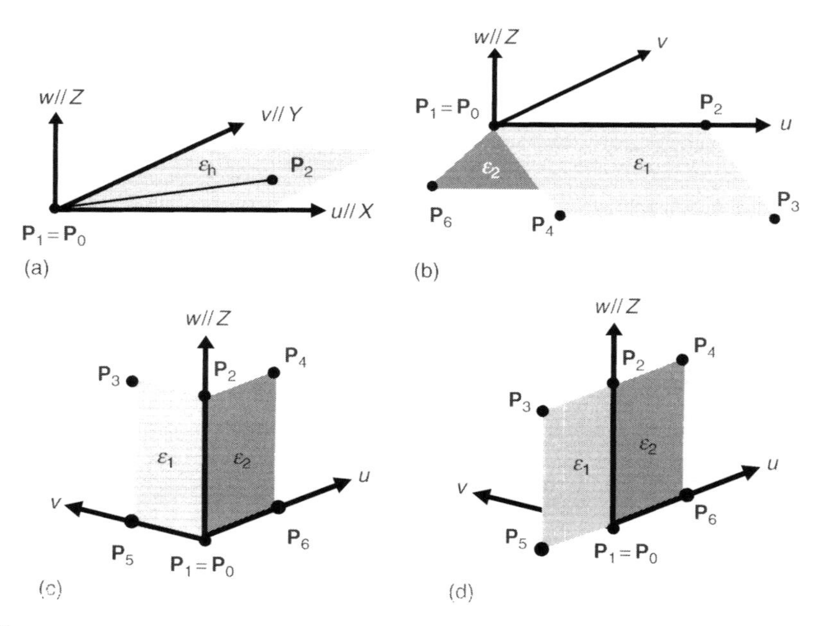

FIGURE 16.11
Geometric regularities: (a) horizontal edge; (b) symmetric roof planes intersecting in a horizontal edge; (c) perpendicular walls; and (d) identical walls. (From Rottensteiner, F. et al., *Int. Arch. Photogram. Rem. Sens. Spatial Inform. Sci.*, XXXVI(3/W19), 221, 2005.)

The other cases involve the two neighboring planes of an edge. One of the axes of the observation coordinate system is defined to be the intersection of these two planes ε_1 and ε_2. There is one additional unknown rotational angle κ describing the direction of the u-axis. For each vertex $\mathbf{P}_i$ of the planes, GESTALT observations are added for ε_1 or ε_2. For the edge's second vertex $\mathbf{P}_2$ two observations (one per plane) are added. The GESTALT observations for ε_1 and ε_2 are parameterized in a specific way.

In the first case, the edge is the horizontal intersection of two symmetric roof planes ε_1 and ε_2. There is only one tilt parameter c_{01}^1. Symmetry is enforced by selecting $m_v = -1$ for ε_2:

$$\varepsilon_1 : r_w = \frac{w_R + c_{01}^1 \cdot v_R}{\sqrt{1+(c_{01}^1)^2}}; \qquad \varepsilon_2 : r_w = \frac{w_R - c_{01}^1 \cdot v_R}{\sqrt{1+(c_{01}^1)^2}} \tag{16.13}$$

In the second case, the edge is the intersection of two perpendicular walls: $\varepsilon_1 : r_u = u_R$, $\varepsilon_2 : r_v = v_R$. There is no additional surface parameter to be determined. In the third case, two walls are identical and the edge does not really exist in the object: $\varepsilon_1 : r_v = v_R$, $\varepsilon_2 : r_v = v_R$. There is no additional surface parameter. $\mathbf{P}_1$ or $\mathbf{P}_2$ might become undetermined, so that direct observations for the coordinates of these vertices have to be generated.

The GESTALT observations corresponding to the geometrical constraints can be subject to robust estimation for gross error detection. If the sensor observations contradict the constraints, the respective GESTALT observations should receive large residuals, which can be used to modulate the weights in an iterative robust estimation procedure [30]. Thus, wrong hypotheses about geometric regularities can be eliminated. Whether or not a hypothesis about a constraint is introduced is decided by analyzing the coarse model, comparing geometric parameters such as height differences or angles to thresholds. For instance, if two neighboring walls differ from 90° by less than a threshold ε_α, a constraint about perpendicular walls can be inserted.

16.2.2.3.3 *Sensor Observations and Observations Linking the Sensor Data to the Model*

The observations described so far link the plane parameters to the vertices or to the parameters of other planes. In order to determine the surface parameters, observations derived from the LiDAR data are necessary. LiDAR points give support to the determination of the roof plane parameters. As a LiDAR point is not a part of the model, its object coordinates have to be determined as additional unknowns. Each LiDAR point gives four observations, namely, its three coordinates and one GESTALT observation for the roof plane the point is assigned to. In order to determine the parameters of the wall planes, the vertices of the original step edge polygons (cf. Section 16.2.2.2.2) are used as observations in the same way as the original LiDAR points are used to determine the roof plane parameters.

16.2.2.3.4 *Overall Adjustment*

In an overall adjustment process using all the observations discussed in the previous sections, the weights of the observations are determined from their a priori standard deviations. Robust estimation is carried out by iteratively reweighting the observations depending on their normalized residuals in the previous adjustment [30]. The reweighting scheme is only applied to the LiDAR and step edge point observations and to the observations modeling geometric constraints, in order to eliminate gross observation errors and wrong hypotheses about geometric regularities. The surface parameters and the vertex coordinates determined in the adjustment are used to derive the final building model [29].

16.2.2.4 Results and Discussion

The method for building reconstruction described in this section was tested using the Fairfield data set (cf. Section 16.2.1.7). Reference data were generated in a semi-automatic working environment using digital aerial images and LiDAR data. The precision of the building vertices was ±17 cm in X and Y and ±5 cm in Z [29]. From the LiDAR data, a DSM with a grid width of $\Delta = 0.5$ m was generated. Eight buildings of different size and complexity chosen to highlight the method's potential to handle buildings of both regular and irregular shapes were reconstructed, and the reconstruction results were compared to reference data.

The roof polygons before adjustment are shown in Figure 16.12. In general the models look quite good except for building 8, which is partly occluded by trees. There is some noise in the outlines of buildings 1 and 2. Buildings 4, 6, 7, and 8 and the main part of building 3 should have a rectangular footprint, which is not preserved in the initial models. The initial models, the original LiDAR points, and the step edge points provide the input for the overall adjustment. Hypotheses about geometric regularities were introduced just on the basis of a comparison of angles and height differences to thresholds. Robust estimation was applied to the soft constraints and to the LiDAR and step edge points.

Figure 16.13 shows the final results of building reconstruction and a comparison to the reference data. Compared to Figure 16.12, the building models appear to be more regular. For buildings 1–6, the number of extracted roof planes was correct. The intersection lines are very accurate, and step edges are in general also determined quite well. Some small roof structures are generalized, e.g., the outline of the smallest roof plane of building 1 or of roof plane a of building 2. The step edge between planes a and b of building 2 was not determined very precisely because that step edge was poorly defined. Roof plane a was horizontal, its western vertex being higher and its eastern vertex lower than roof plane b, with a maximum height difference of only 0.3 m. Building 7 was reconstructed as being flat. The intersection of the two roof planes is only 0.15 m lower than the eaves, which is the reason why the two planes were merged. Building 8 was also reconstructed as a flat roof.

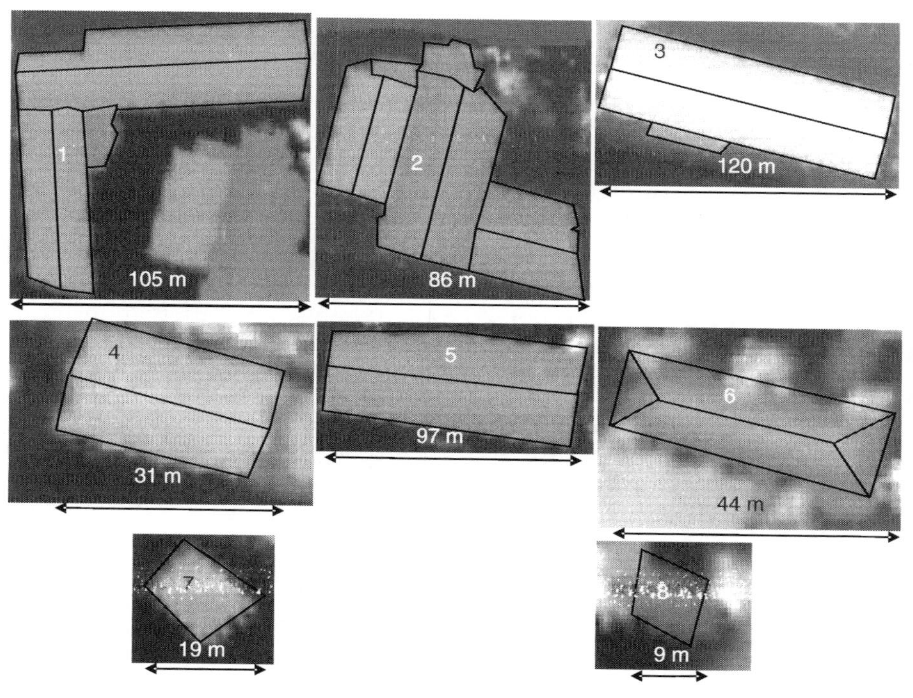

FIGURE 16.12
Initial roof boundary polygons for the eight buildings superimposed to the DSM. The buildings are shown in different scales, according to the extents shown in the figure. (From Rottensteiner, F., *Int. Arch. Photogram. Rem. Sens. Spatial Inform. Sci.*, XXXVI(3), 13, 2006.)

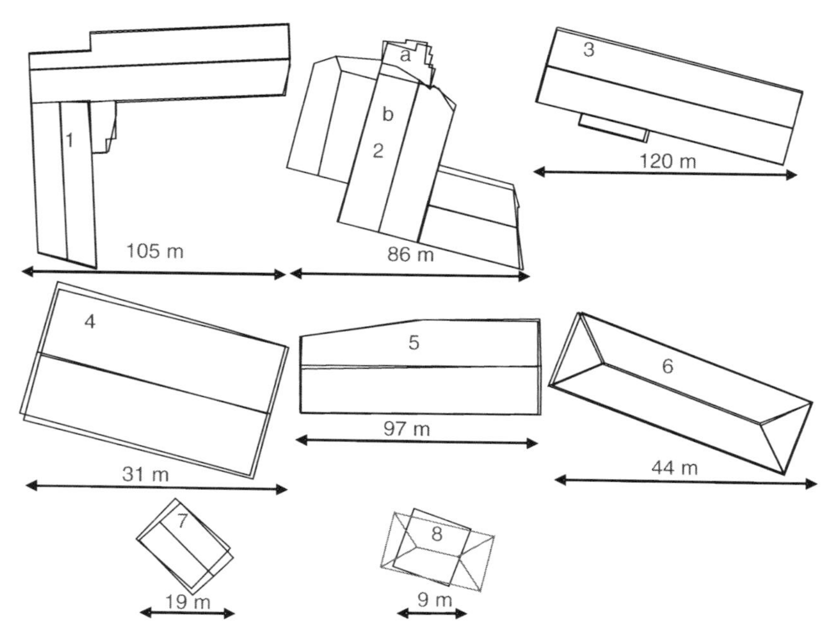

FIGURE 16.13
Final roof boundary polygons and reference data. A part of building 2 is missing in the reference data since it does not exist in the aerial images used to create the reference data. (From Rottensteiner, F., *Int. Arch. Photogram. Rem. Sens. Spatial Inform. Sci.*, XXXVI(3), 13, 2006.)

TABLE 16.1

Evaluation of the Geometric Accuracy of the Roof Vertices after Adjustment with Regularization

B	P	RMS_{XY} (m)	RMS_Z (m)	Δ_{XY} (m)	Δ_Z (m)
1	5	0.76	0.12	0.24	0.01
2	5	2.27	0.20	0.00	−0.02
3	3	0.82	0.10	0.07	0.16
4	2	0.60	0.02	0.13	0.03
5	2	1.31	0.08	−0.08	−0.02
6	4	0.48	0.09	0.36	0.17
7	2	1.43	0.14	0.44	0.03
8	0	2.74	—	−0.02	—

Source: From Rottensteiner, F., *Int. Arch. Photogram. Rem. Sens. Spatial Inform. Sci.*, XXXVI(3), 13, 2006.
Notes: *B*, building; *P*, number of planes; RMS_{XY}, RMS_Z, combined RMS values in planimetry/height; and Δ_{XY}, Δ_Z, improvement of RMS_{XY}/RMS_Z caused by regularization.

It was the smallest building in the sample with only a few LiDAR points on the roof planes, and both ends occluded by trees. The outlines at the occluded ends are not very well detected either. Apart from the visual inspection of the building models, a numerical evaluation of these results was carried out. RMS values of the coordinate differences of corresponding vertices in the reconstruction results and the reference data were computed for each roof plane. For buildings 7 and 8 only the outlines were evaluated.

Table 16.1 gives combined RMS values for all the test buildings. The large value for RMS_{XY} for building 2 of ±2.27 m is caused by the erroneous step edge; the combined value without that edge would be ±1.43 m. For most buildings, RMS_{XY} is better than the average point distance across the flight direction. Apart from problems with low step edges, errors occurred at the outlines of some of the larger building due to occlusions: as the test area was at the edge of the swath, the positions of the step edges were very uncertain. The height accuracy is good, with the largest value of ±0.20 m occurring at building 2, again at the problematic step edge. Table 16.1 also gives the impact of the overall adjustment to the RMS values. With building 5, the RMS values get worse by a small value after adjustment, but in most cases the RMS values are improved by the overall adjustment. The improvement can be up to 45% (building 6).

16.3 Road Extraction from LiDAR Data

Compared with other remote sensing data sources, extraction of roads from LiDAR data is in its infancy [31–34]. Often, LiDAR data are only used to support road extraction from imaging sensors [32,34]. In this section, we will describe a method for road extraction from LiDAR data alone that follows the two-step procedure of detection followed by reconstruction.

16.3.1 Road Detection

For road detection, similar cues and similar classification principles as for building detection can be used (cf. Sections 16.2.1.1 and 16.2.1.2). Again, the normalized DSM provides information about the heights of points above the terrain. However, unlike buildings, road points are situated on the terrain. The LiDAR intensities are better suited for road detection than for building detection, because roads have very specific reflectance properties in the wavelength of the LiDAR pulse. The effect of the fact that LiDAR intensities are very noisy is minimized by the typically uniform and consistent nature of road material along a section of a road. Topological properties can be used for classification, since there are usually no isolated road points. Finally, shape parameters are very helpful for classification: in data having a resolution of 1 m or better, roads appear as relatively thick lines.

The classification algorithm presented in this section consists of three stages. The first stage is DTM generation in order to obtain a normalized DSM. The algorithm used for DTM generation in building detection (cf. Section 16.2.1.4) can also be used for road detection. Once the normalized DSM has been determined, the detection process continues with a rule-based classification of road points. The original LiDAR points are classified according to whether or not they are situated on a road. Finally, a binary road image is generated from the classification results, and postclassification is carried out to correct some errors in the original classification.

16.3.1.1 Rule-Based Classification of Road Points

A rule-based classification technique is used to classify the LiDAR points into the classes road or nonroad. Let any LiDAR point be $\mathbf{P}_k$ described by its 3D coordinates x_k, y_k, and z_k, and the intensity of the last pulse strike, i_k, thus $\mathbf{P}_k = (x_k, y_k, z_k, i_k)^T$. Further, let $S = \{\mathbf{P}_1, \mathbf{P}_2, \mathbf{P}_3, \ldots, \mathbf{P}_N\}$ represent the set of all laser points collected, where $\{\mathbf{P}_1, \mathbf{P}_2, \mathbf{P}_3, \ldots, \mathbf{P}_N\}$ are the individual LiDAR points. Roads lie on or near the DTM. This is true except for elevated roads, bridges, and tunnels. Thus, all LiDAR points outside a given tolerance of the DTM can be disregarded, and subset S_1 of LiDAR points that are potential road points can be created:

$$S_1 = \{\mathbf{P}_k \in S : |z_k - \text{DTM}(x_k, y_k)| < \Delta h_{\max}\} \tag{16.14}$$

where

DTM (x_k, y_k) is the height value of the DTM at location (x_k, y_k)

$\Delta h_{\max}$ is the maximum allowable difference between z_k and the DTM

The classification procedure then filters LiDAR points based on their intensity values. In LiDAR intensity images, roads are clearly visible as dark thick connected lines. The algorithm requires some training in order to determine the reflectance properties of the road material to be detected. LiDAR points that have last pulse intensity values that appropriately represent the reflectance properties of the road material form a new subset S_2:

$$S_2 = \{\mathbf{P}_k \in S_1 : i_{\min} < i_k < i_{\max}\} \tag{16.15}$$

The values $i_{\min}$ and $i_{\max}$ are the minimum and maximum acceptable LiDAR intensities at any point $\mathbf{P}_k$. The result of Equation 16.15 is a set of LiDAR points (S_2) that were reflected from the road along with some other false positive (nonroad) detections. If more than one type of road material is to be detected in the surveyed region, different subsets S'_2 can be created according to the individual reflectance properties of the different materials being detected. A combined and complete set S_2 can be created by considering the union of the different subsets S'_2.

Roads are depicted as a continuous network of points that form thick lines. Due to this continuous nature of a road network, LiDAR points that have struck the middle of the road are expected to be surrounded by other points that have struck the road. The percentage of road points within a local neighborhood (e.g., defined by a circle of radius d around any point $\mathbf{P}_k$) will be called the "local point density." For points in S_2 that are situated in the middle of a road, the local point density should be close to 100%. For any point that lies on the edge of a road it can be considered to be close to 50%, and 25% for a LiDAR point in the corner of a sharp 90° bend. By testing all points against a chosen minimum local point density ρ_{min}, a new subset of points, S_3, is described as per Equation 16.16:

$$S_3 = \left\{ \mathbf{P}_k \in S_2 : \frac{\left| \left\{ \mathbf{P}_j \in S_2 : \left\| \mathbf{P}_k - \mathbf{P}_j \right\|_2 < d \right\} \right|}{\left| \left\{ \mathbf{P}_j \in S : \left\| \mathbf{P}_k - \mathbf{P}_j \right\|_2 < d \right\} \right|} > \rho_{min} \right. \tag{16.16}$$

where

d is the maximum distance from $\mathbf{P}_k$ or the radius of the local neighborhood

$|\{\ldots\}|$ denotes the number of points $\mathbf{P}_j$ in the respective set

$\|\mathbf{P}_k - \mathbf{P}_j\|_2$ is the Euclidean distance from $\mathbf{P}_j$ to $\mathbf{P}_k$

An upper bound should be placed on the possible values of d so that it is any value less than or equal to half of the expected maximum road width [33].

16.3.1.2 Generation of a Binary Road Image

A binary image, $F(x, y)$, is now created from the final subset S_3 with a pixel size Δ loosely corresponding to the original average LiDAR point density. The pixel values $f(x, y)$ of the binary image are determined according to whether a point $\mathbf{P}_k \in S_3$ exists inside the area represented by the pixel at position (x, y) or not

$$f(x,y) = \begin{cases} 1 & \text{if } \exists\, \mathbf{P}_k \in S_3 : \left(x - \dfrac{\Delta}{2} < x_k \le x + \dfrac{\Delta}{2} \right) \wedge \left(y - \dfrac{\Delta}{2} < y_k \le y + \dfrac{\Delta}{2} \right) \\ 0 & \text{otherwise} \end{cases} \tag{16.17}$$

The pixels in that binary image characterized by $f(x, y) = 1$ represent roads. Many small gaps exist between these road pixels. This is caused by reflections from other objects such as vehicles and overhanging trees and by the fact that in some image regions, the pixel size will be smaller than the LiDAR point density. These gaps are removed using a two-step approach based on morphologic filtering. First, a morphological closing with a small structural element is initially performed to connect neighboring road pixels. The Not-Road image (i.e., $1 - f(x, y)$) is then labeled using a connected component analysis, and the values of all pixels corresponding to Not-Road segments with a small area are switched to 1 in the binary image $f(x, y)$, which results in a road image that contains all public roads, private roads, car parks, and some noise. In a second stage of processing, another label image is created from this binary image in order to identify individual continuously connected road segments. This time, small road segments are erased in the binary image of road pixels, thus removing most of the noise present. This ensures that our detected roads are continuous in nature.

Car parks are not considered to be roads. However, since car parks and roads have similar surface and reflectance properties, it is difficult to detect and eliminate car parks. By defining a maximum acceptable road width prior to processing, very wide unconnected car parks can be removed from the binary image. As roads form a network of long, thin connected objects, the area ratio of each individual road segment and the corresponding minimum bounding rectangle (MBR) will decrease as the length of the smallest side in the MBR increases. Large isolated blobs can be detected in the image using this ratio, thus allowing the removal of any unconnected car parks from the final binary image of road pixels.

16.3.1.3 Results and Discussion

Again, the Fairfield data set (cf. Section 16.2.1.7) was used for testing purposes. A reference data set was created by digitizing roads interactively from the digital orthophoto of the area. The guideline used during digitizing was that public roads were to be classified as roads, but car parks and private roads (driveways and roads leading to car parks) were not.

Figure 16.14 shows the results of the classification. The classification results from the workflow described in the previous section are displayed as a binary image of road pixels in Figure 16.14a (as a negative). The spatial distribution of the TP, FP, and FN pixels (cf. Section 16.2.1.7, especially Equations 16.9 and 16.10) are displayed in Figure 16.14b in yellow, red, and blue, respectively. A perusal of the spatial distribution reveals that the majority of the FP detections correspond to car parks while the majority of FN detections have occurred at the ends of detected roads or on the edge of the image. These road components are disconnected from the road network and exist in small sections due to overhanging trees and were removed when small, disconnected components were removed.

A per-pixel evaluation of these results resulted in a completeness of 88% and a correctness of 67% (cf. Equations 16.9 and 16.10). There were several problems encountered during the classification phase of the algorithm. One of the problems was the detection of elevated roads and bridges (cf. the circle in Figure 16.14a). Our road network model assumed that roads lie on the DTM, an assumption that is not true for bridges. Another problem is car parks. Due to the industrial nature of sections of the test data, there are quite a few large car parks connected to the road network, and the classification method did not succeed in eliminating most of them; the majority of these car parks still remain. This difference is reflected in the correctness value, and better results are expected in nonindustrial areas.

16.3.2 Road Vectorization

In LiDAR data of a resolution better than or equal to 1 m, roads appear as two-dimensional areas rather than one-dimensional lines; they have width as well as length. The centerline

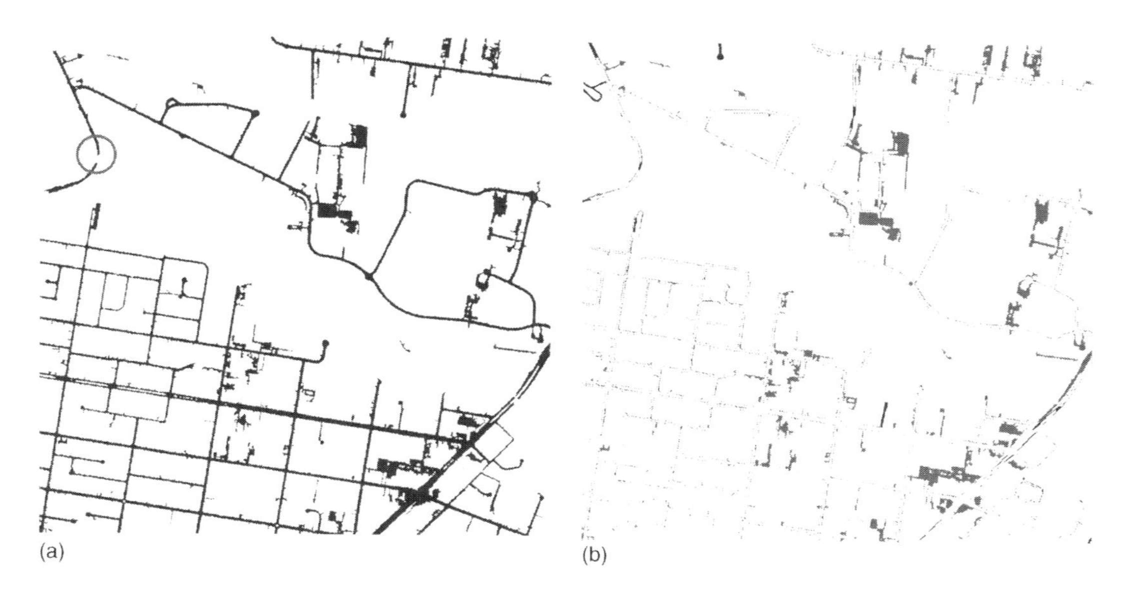

FIGURE 16.14
(a) Results of road network classification. The circle highlights a problem that occurs with bridges. (b) Spatial distribution of the classified pixels. TP pixels are not shaded, FP appear dark gray, and FN appear black. (From Clode, S. et al., *Photogram. Eng. Rem. Sens.*, 73(5), 517, 2007. With permission.)

can no longer be observed directly, but has to be derived by other methods, for instance from the road boundaries. Note that, whereas the classification method described in Section 16.3.1 uniquely exploits the properties of LiDAR data, the method of road vectorization described in this section is based on a binary road image and could be applied to such a road image derived by any classification method. Unlike other methods [34], it is not restricted to roads that form a grid-like pattern. It does overcome the problem of common line detection methods such as the Hough transform that detects the diagonal of a straight road segment rather than the centerline. Other road parameters, such as the width and direction, are important information that can also be extracted from high-resolution data, although many of the methods described so far do not extract some or all of this information directly [35]. The road vectorization method described in this section is capable of delivering these road parameters.

The method is based on the convolution of the binary road image with a Phase Coded Disk (PCD). The PCD is a complex kernel that uses phase to code for the angle of the line. By convolving the original image with a PCD, the centerline, direction, and width can be accurately extracted at any point along the detected centerline. The PCD is defined by Equation 16.18 [33]:

$$O_{\text{PCD}} = e^{j2\tan^{-1}(b/a)} = e^{j2\vartheta} \tag{16.18}$$

The variables a and b are x and y coordinates relative to the center of the PCD. Further, $a^2+b^2 \le r^2$, $\vartheta = \tan^{-1}(b/a)$, $j^2 = -1$, and r is the radius of the disk. The constant 2 in the exponent has been introduced into the definition of the PCD in order to ensure that pixels that are diametrically opposite in their direction from the center of the kernel during the convolution process indicate the same direction after convolution and do not cancel out. The convolution of the PCD with the binary image takes the form:

$$Q(x,y) = F(x,y) \otimes O_{\text{PCD}} \tag{16.19}$$

where

$Q(x, y)$ is the resultant image
$\otimes$ is the convolution operation
O_{PCD} is the PCD
$F(x, y)$ is the binary road image

The result of the convolution defined in Equation 16.19 yields a magnitude image M and a phase image Φ that are defined by Equations 16.20 and 16.21, respectively:

$$M = \left|F(x,y) \otimes O_{\text{PCD}}\right| \tag{16.20}$$

$$\Phi = \frac{1}{2}\arg(F(x,y) \otimes O_{\text{PCD}}) \tag{16.21}$$

The magnitude M and phase Φ images can be used to determine the desired parameters of a road. The result $q(x, y)$ of the convolution at any position (x, y) is given by the complex integral over the entire disk:

$$q(x,y) = \int_{-\pi}^{\pi} \int_{0}^{r} f(x,y)e^{j2\vartheta} \; u\,\mathrm{d}u\,\mathrm{d}\vartheta \tag{16.22}$$

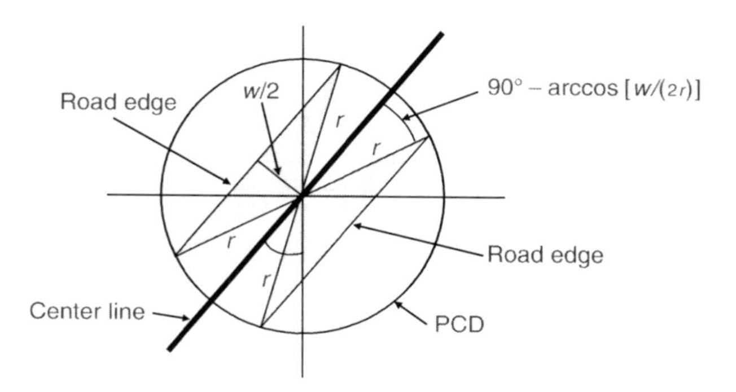

FIGURE 16.15
Extracting line information from the PCD. The thick line represents the road centre line, passing through the center of the PCD. The road orientation is ϕ. The integral over the road pixels in Equation 16.22 can be separated into two parts: the two wedges of opening angle 180° – 2 arccos ($w/2r$), where the integral is to be taken over the full radius r of the PCD, and the grey triangles, where the integral is limited by the road edges. (From Clode, S. et al., *Photogram. Eng. Rem. Sens.*, 73(5), 517, 2007. With permission.)

In Equation 16.22, the variable u is a substitute variable that has been introduced to represent the radius of the PCD as r is in the limits of the integral and the function $f(x, y)$ is understood to be translated to appropriate polar coordinates. Remembering that the road image is binary, the result of Equation 16.22 is identical to the integral over only the road area covered by the disk (i.e., all areas where $f(x, y) = 1$). Thus, for the model of a straight line segment of width w and orientation ϕ passing through the area covered by the PCD, the integral in Equation 16.22 can be solved analytically by changing the limits of the inner most integral so that the area defined is only the road contained within the disk. The integral consists of two parts: some areas can be integrated over the full radius (r) of the PCD, whereas others can only be integrated to a distance of $w/2 \cos(\vartheta)$ (Figure 16.15). The variable R is used to represent the limits of the integral at all different angles of ϑ over the road section contained within the PCD [33]:

$$R = \min\left(\frac{w}{2\cos(\vartheta)}, r\right) \tag{16.23}$$

Further, we substitute $\vartheta = \phi + \alpha$, thus relating the directional angle to the road direction ϕ. The integral over the road can be described as the integral over the disk where $f(x, y) = 1$:

$$q(x,y) = \int_{-\pi}^{\pi} \int_{0}^{R} e^{j2(\phi+\alpha)} u \, du \, d\alpha \tag{16.24}$$

Roads correspond to ridges in the magnitude image M of the convolution, with the relative maximum of M corresponding to the road centerline. The radius r of the PCD must be larger than the maximum road width to be detected. It can be shown [33] that the phase of $q(x,y)$ in Equation 16.24 only depends on the road orientation ϕ, and thus ϕ is related to the phase Φ of the complex convolution (Equation 16.21) by

$$\phi = \frac{\Phi}{2} \tag{16.25}$$

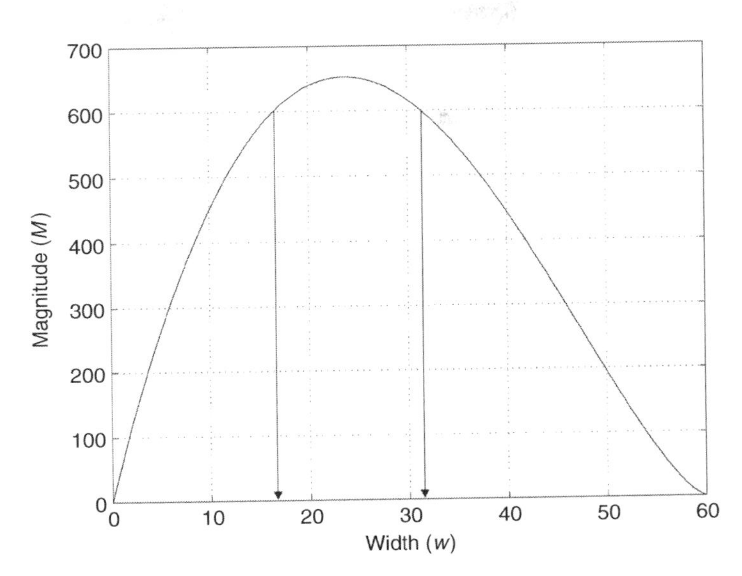

FIGURE 16.16
Graphical representation of the relationship between the magnitude of the convolution and the width of the road for a PCD of radius 30. (From Clode, S. et al., *Photogram. Eng. Rem. Sens.*, 73(5), 517, 2007. With permission.)

Further, evaluating the integral in Equation 16.24 results in an equation relating the magnitude M to the road width w and the PCD radius r [33]:

$$M = \left| w^2 \cos^{-1}\left(\frac{w}{2r}\right) - 2w\sqrt{r^2 - \frac{w^2}{4}} \right| \tag{16.26}$$

A graph can be generated for a PCD of a fixed radius thus enabling the width of a road to be determined at any point. Figure 16.16 illustrates the relationship between width and magnitude. If the road width w is to be determined from the magnitude M, ambiguity resolution is required. For instance, a magnitude of 600 could imply a width of 17 or 32. A constraint placed on r will resolve this ambiguity as well as avoid saturation problems that will occur when the road width is greater than the kernel radius.

16.3.2.1 Road Tracking and the Generation of a Road Network

The final road network consists of a set of road segments, each of them represented by three polylines, namely, the centerline and two road edges. Vectorization consists of three steps. First, the road centerline of each road segment is extracted. Second, the extracted road centerlines are joined or intersected with other centerlines of the neighboring segments to create a continuous network of road centerlines. An exact intersection is created by splitting the crossing road centerlines at the intersection point. Where a *T*-junction is encountered, the road segment at the base of the *T* is extended straight ahead until the centerlines of both roads intersect. A list of all crossing points and *T*-junctions is kept in conjunction with the connecting road segments in order to maintain the topology of the road network. Third, the polylines representing the road edges are constructed as being parallel to the centerline at a distance of half the road width. The polygon formed by the two edge polylines defines the road segment. The set of all road segments and their topology describes the continuous road network.

16.3.2.1.1 Vectorizing the Road Centerlines

The ridge of the magnitude image is traced in order to extract the road centerlines. Tracing is achieved by initially masking the magnitude image with the binary road mask in order to limit the search space and remove noisy edge areas. The maximum of the magnitude image is found and the corresponding line direction is read from the phase image. The tracing algorithm moves along the line pixel by pixel ensuring that the point is still a maximum against its neighbors until the line ends at a pixel of zero magnitude. A polygon is created that consists of a series of centerline segments. Points to the side of each centerline segment within the calculated road width are zeroed as the centerline is traced, ensuring a similar path is not retraced. Once the road segment is completed the process is repeated from the original maximum but with the diametrically opposed line direction. Then, the maximum of the remaining untraced magnitude pixels is found and tracing is recommenced from this pixel. The process is repeated until all relevant pixels have been traced. Thus, one centerline segment after the other is extracted until the masked magnitude image is completely blank.

16.3.2.1.2 Connecting Road Segments and Determining Junctions

Due to noise in the magnitude image, the tracing of some road centerlines is terminated prematurely by the tracing algorithm. In such cases, the tracing algorithm will extract another centerline that will terminate close to the original prematurely terminated centerline end. Connection of road centerline ends is performed by concatenating road segment chains that have ends that are both close and pointing to each other. The segments are close if they are within one road width of the centerline end. A check is made that the midpoint of the link actually lies on a ridge in the original magnitude image ensuring that dead end streets do not erroneously get connected. Once two road segments have been connected the process is repeated until no more concatenation can occur. During concatenation, all road crossings are found by identifying crossing road segments. These segments are split at the intersection and the road crossing is created, leaving only the determination of *T*-junctions. This requires that the end of each centerline be checked against all other centerlines. In a manner similar to the way individual centerlines were concatenated to form longer centerlines, the closest point on any other centerline is found for each centerline end point. If the direction to the closest other centerline position matches the current centerline direction, the road segment is extended if the midpoint corresponds to a road in the binary image.

16.3.2.1.3 Vectorizing the Road Edges

The edges of the road are also represented as polygons that are created by calculating the width of the road and the orientation of the road ϕ. The road width is then smoothed by applying a low pass filter to the widths of each road segment. At each point along the centerline of each road segment, two new road edge points are created based on the road orientation ϕ plus or minus 90° and half the smoothed road width. Two edge polylines are defined for each road segment thus defining the road segment polygon. At the end of each centerline, the intersections of the accompanying road edges are calculated. The edges within the intersection are kept to ensure that the connection between the centerline and the edges within a road segment are kept intact. To complete the visualization at the intersections, blanked road edge ends on the same side of the centerline are then joined.

16.3.2.2 Results and Discussion

Figure 16.17 shows the result of the convolution of the binary road image for Fairfield (Figure 16.14) with the PCD. Figure 16.17a displays the magnitude image M with the highest values displayed as white and the lowest as black. The centerlines of the road image are represented by white ridges running through the magnitude image. Figure 16.17b displays the resultant phase image Φ, which is related to the road direction as described above.

Vectorization of the convolution results was performed by the tracing algorithm as previously described. The results of the vectorization process can be seen in Figure 16.18. The centerline vectors appear visually to be a good approximation of the road network. In areas where there are very few car parks the vectors are smooth, predictable, and continuous. The results show that roads that have successfully been classified as dual carriageways have also been vectorized as dual carriageways in these areas. The curvilinear road that runs diagonally from the northwest corner to the eastern edge of the image is quite smooth and characterizes the classified image well. The grid pattern in the southwestern corner is very good. The behavior of the tracing algorithm is predominately as expected, although some anomalies could be seen. In the southwest corner of Figure 16.18 there are two intersections that are represented as a set of two 90° bends. In these instances, the ridge in the magnitude image was more dominant towards the entering road rather than the continuing road. Another problem also seen in the southwest corner of the image is the tracing of roundabouts. There are several regions that appear to be quite noisy, in particular, the area in the southeast of the image. This apparent noise is due to the presence of many car parks in the area. As far as the bridge highlighted Figure 16.14 is concerned, Figure 16.17 shows that it still yielded a continuous ridge, albeit reduced in magnitude, so that the tracing algorithm correctly vectorized this area, and information that was lost in the initial classification has been recovered. The edges of the roads are also displayed in Figure 16.18. In areas of relatively low noise, the road edges have been calculated consistently. Smooth edges are obtained along areas of clearly defined roads. In areas where

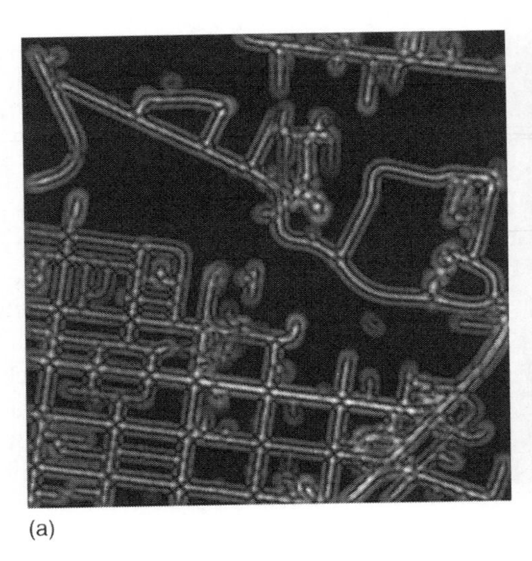
(a)

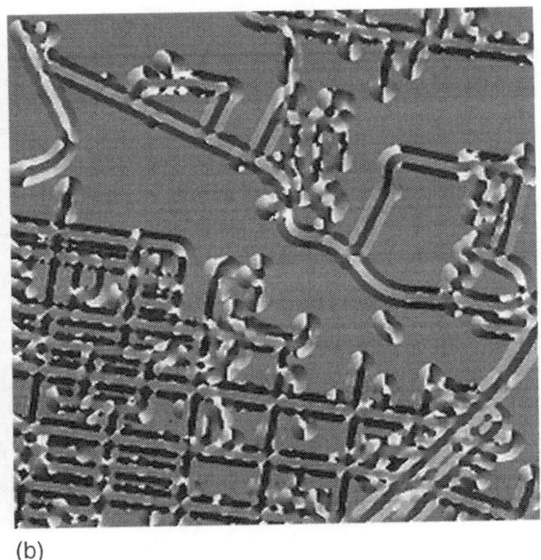
(b)

FIGURE 16.17
Results of the convolution with the PCD. (a) Magnitude M with the highest values displayed as white and the lowest as black. (b) Phase image Φ. (From Clode, S. et al., *Photogram. Eng. Rem. Sens.*, 73(5), 517, 2007. With permission.)

FIGURE 16.18
(a) Vectorized road centerlines and (b) vectorized road edges as calculated from the road centerline and the detected road width. (From Clode, S. et al., *Photogram. Eng. Rem. Sens.*, 73(5), 517, 2007. With permission.)

roads are less clearly defined, e.g., with car parks or road junctions, the resultant magnitude is less representative of the road due to a saturation effect within the PCD.

In order to quantify the vectorization results, evaluation methods described in the literature [36] were adopted. First, the reference centerlines were estimated from the reference road image. After that, both extracted and reference data are compared in vector format where the first step is to match the two data sets while the second is to calculate the quality measures. We use the buffer width for the simple matching technique described in the literature [36]. Completeness describes how well reference data is explained by the extracted data and is determined by the ratio of the length of the matched reference data to the total length of the reference data. For the results shown in Figure 16.18, completeness was 84%. Correctness represents the percentage of correctly extracted road data and is determined by the ratio of the length of the matched extraction to the total length of the matched extraction. Correctness was 75% in our test. The centerline RMS of all points expresses the quadratic mean distance between the matched extracted and the matched reference points. It was determined to be ±1.70 m. The centerline RMS of all road segments considers the RMS values for each individual road segment and expresses the quadratic mean distance of these results, and it was ±1.56 m. The width RMS of all points and road segments are calculated in a similar manner to the centerline quality values but consider the difference in extracted and reference widths as opposed to the centerlines. The value for all points was ±1.66 and ±1.48 m for the segments. Detected intersections were manually classified against the reference data into TP, FP, and FN detections. Topological completeness and correctness are then calculated using Equations 16.9 and 16.10. The algorithm achieved a topological completeness of 87% and a topological correctness of 73%. The final evaluation of the vectorized road network is promising, and the results resemble the classification results. This suggests that the vectorization process was performed well, and that improvements need to be made to the classification algorithm in order to achieve better final quality values. The topological evaluation of the network extraction suggests that improvement in the current network model can be made. The removal of car parks from the classified road image would greatly improve the classification results. The application of several PCDs with different radii within a hierarchical framework during the vectorization stage may improve the vectorization of the road edges and allow a more robust determination of the road width.

References

1. Kraus, K. et al., Gaussian decomposition and calibration of a novel small-footprint full-waveform digitising airborne laser scanner, *ISPRS Journal of Photogrammetry and Remote Sensing*, 60(2), 100, 2006.
2. Schenk, T. and Csatho, B., Fusion of LIDAR data and aerial imagery for a more complete surface description, *International Archives of the Photogrammetry, Remote Sensing and Spatial Information Sciences*, XXXIV(3A), 310, 2002.
3. Weidner, U. and Förstner, W., Towards automatic building reconstruction from high resolution digital elevation models, *ISPRS Journal of Photogrammetry and Remote Sensing*, 50(4), 38, 1995.
4. Maas, H.-G., Fast determination of parametric house models from dense airborne laserscanner data, *International Archives of Photogrammetry and Remote Sensing*, XXXII(2W1), 1, 1999.
5. Vögtle, T. and Steinle, E., On the quality of object classification and automated building modeling based on laserscanning data, *International Archives of the Photogrammetry, Remote Sensing and Spatial Information Sciences*, XXXIV(3W13), 149, 2003.
6. Brunn, A. and Weidner, U., Extracting buildings from digital surface models, *International Archives of Photogrammetry and Remote Sensing*, XXXII(3–4W2), 27, 1997.
7. Förstner, W., A framework for low level feature extraction, In *Computer Vision—ECCV '94. Proceedings of the Third European Conference on Computer Vision*, Stockholm, Sweden. Eklundh, J.O., Ed., Lecture Notes in Computer Science, Springer Verlag Berlin/Heidelberg, 1994, Vol. II, p. 383.
8. Vosselman, G., On the estimation of planimetric offsets in laser altimetry data, *International Archives of the Photogrammetry, Remote Sensing and Spatial Information Sciences*, XXXIV(3A), 375, 2002.
9. Haala, N. and Brenner, C., Extraction of buildings and trees in urban environments, *ISPRS Journal of Photogrammetry and Remote Sensing*, 54, 130, 1999.
10. Lu, Y.H., Trinder, J.C., and Kubik, K., Automatic building detection using the Dempster-Shafer algorithm, *Photogrammetric Engineering and Remote Sensing*, 72(4), 395, 2006.
11. Rottensteiner, F. and Briese, C., A new method for building extraction in urban areas from high-resolution LIDAR data, *International Archives of the Photogrammetry, Remote Sensing and Spatial Information Sciences*, XXXIV(3A), 295, 2002.
12. Lee, T., Richards, J.A., and Swain, P.H., Probabilistic and evidential approaches for multisource data analysis, *IEEE Transactions RS and GE*, 25(3), 283, 1987.
13. Binaghi, E. et al., Approximate reasoning and multistrategy learning for multisource remote sensing data interpretation, in *Information Processing for Remote Sensing*, Chen, C.H., Ed., World Scientific Publishing, Singapore, 1999, p. 397.
14. Gorte, B., Supervised image classification, in *Spatial Statistics for Remote Sensing*, Stein, A., van der Meer, F., Gorte, B., Eds., Kluwer Academic, Dordrecht, The Netherlands, 1999, p. 153.
15. Klein, L., *Sensor and Data Fusion, Concepts and Applications*, SPIE Optical Engineering Press, Bellingham, WA, 1999.
16. Rottensteiner, F. et al., Building detection by fusion of airborne laserscanner data and multispectral images: Performance evaluation and sensitivity analysis, *ISPRS Journal of Photogrammetry and Remote Sensing*, 62(2), 135, 2007.
17. Heipke, C. et al., Evaluation of automatic road extraction, *International Archives of Photogrammetry and Remote Sensing*, XXXII(3/2W3), 56, 1997.
18. Rottensteiner, F. et al., Using the Dempster Shafer method for the fusion of LIDAR data and multi-spectral images for building detection, *Information Fusion*, 6(4), 283, 2005.

19. Mäntylä, M., *An Introduction to Solid Modeling. Principles of Computer Science*, Computer Science Press, Rockville, MA, 1988.
20. Vosselman, G. and Dijkman, S., 3D building model reconstruction from point clouds and ground plans, *International Archives of the Photogrammetry, Remote Sensing and Spatial Information Sciences*, XXXIV(3W4), 37, 2001.
21. Alharty, A. and Bethel, J., Detailed building reconstruction from airborne laser data using a moving surface method, *International Archives of the Photogrammetry, Remote Sensing and Spatial Information Sciences*, XXXV(B3), 213, 2004.
22. Vosselman, G., Building reconstruction using planar faces in very high density height data, *International Archives of Photogrammetry and Remote Sensing*, XXXII(3–2W5), 87, 1999.
23. Heuel, S., *Uncertain Projective Geometry. Statistical Reasoning for Polyhedral Object Reconstruction*. Springer-Verlag, Berlin Heidelberg, Germany, 2004.
24. Hofmann, A., Maas, H.-G., and Streilein, A., Extraction of road geometry parameters from laser scanning and existing databases, *International Archives of Photogrammetry, Remote Sensing and Spatial Information Sciences*, XXXIV(3/W13), 112, 2003.
25. Rottensteiner, F., Automatic generation of high-quality building models from lidar data, *IEEE Computer Graphics and Applications*, 23(6), 42, 2003.
26. Rottensteiner, F. et al., Automated delineation of roof planes in lidar data, *International Archives of the Photogrammetry, Remote Sensing and Spatial Information Sciences*, XXXVI(3/W19), 221, 2005.
27. Ameri, B., Automatic recognition and 3D reconstruction of buildings from digital imagery, PhD thesis, German Geodetic Commission (DGK) C 526, Institue for Photogrammetry, Stuttgart University, 2000.
28. Brenner, C., Constraints for modelling complex objects. *International Archives of the Photogrammetry, Remote Sensing and Spatial Information Sciences*, XXXVI(3/W24), 49, 2005.
29. Rottensteiner, F., Consistent estimation of building parameters considering goemetric regularities by soft constraints, *International Archives of the Photogrammetry, Remote Sensing and Spatial Information Sciences*, XXXVI(3), 13, 2006.
30. Kager, H., Adjustment of algebraic surfaces by least squared distances, *International Archives of Photogrammetry and Remote Sensing*, XXXIII(B3), 472, 2000.
31. Alharthy, A. and Bethel, J., Automated road extraction from LIDAR data, in *Proceedings of the ASPRS Annual Conference*, Anchorage, Alaska, 2003, unpaginated CD-ROM.
32. Hu, X., Tao, C.V., and Hu, Y., Automatic road extraction from dense urban area by integrated processing of high resolution imagery and LIDAR data, *International Archives of Photogrammetry, Remote Sensing and Spatial Information Science*, XXXV(B3), 288, 2004.
33. Clode, S. et al., Detection and vectorization of roads from lidar data, *Photogrammetric Engineering and Remote Sensing*, 73(5), 517, 2007.
34. Zhu, P. et al., Extraction of city roads through shadow path reconstruction using laser data, *Photogrammetric Engineering and Remote Sensing*, 70(12), 1433, 2004.
35. Auclair-Fortier, M.-F. et al., Survey of work on road extraction in aerial and satellite images, *Canadian Journal of Remote Sensing*, 27(1), 76, 2001.
36. Wiedemann, C., External evaluation of road networks, *International Archives of Photogrammetry and Remote Sensing*, XXXIV(3/W8), 93, 2003.

17

A Data-Driven Method for Modeling 3D Building Objects Using a Binary Space Partitioning Tree

Gunho Sohn, Xianfeng Huang, and Vincent Tao

CONTENTS

17.1 Introduction

Urban regions, which cover only approximately 0.2% of the Earth's land surface, contain approximately one half of the human population and are expected to grow to 60% of the total by 2025 (World Resources Institute, 1996). Considering this rapid urbanization, accurate mapping and seamless monitoring of three-dimensional models of urban infrastructures are becoming ever more important for supporting human welfare-related applications. It is expected that widely available 3D city models will result in dramatic benefits and user demand for a range of existing and new applications, including entertainment, tourism, real estate, planning, marketing, military, web-based mapping, engineering, decision-support, emergency services, environmental services, etc. (Ameri, 2000). Recent advancements in web mapping services (e.g., Microsoft Virtual Earth, Google Earth, NASA World Wind, etc.) show great potential and demand for a 3D cityscape, which can provide an enriching, interactive, and visually appealing user experience that is in tune with the users' natural visual-based thinking and imagination.

Despite the enormous emerging demand for 3D building models, coverage is still limited primarily to the downtown core of only a very small number of large cities. This stems largely from the inefficiency of existing systems in achieving the required rapid, accurate extraction of realistic models for large urban areas. Thus, a new technique and strategy to greatly increase the efficiency for the generation of 3D city models is urgently required. Although in the last few decades considerable research effort has been directed toward achieving complete automation of 3D building reconstruction, this still remains a daunting challenge.

Thus far, we understand that a general solution to the building reconstruction system entails the collection of modeling cues (e.g., lines, corners, and planes), which represent the major components of a building structure. By correctly grouping these cues, one can create geometric topology among adjacent cues, thereby describing realistic building shapes. A significant bottleneck hindering the automation of the building reconstruction process stems from the fact that extraction of modeling cues is often missed, or fragmented, or geometrically distorted, by noise inherited from imaging sensors and object complexity. For solving this problem, 3D reconstruction systems have performed most effectively by constraining the knowledge of building geometry, either explicitly (model-driven reconstruction) or implicitly (data-driven reconstruction), in order to recover incomplete modeling cues. The former approach prespecifies particular types of building models so that geometric and topological relations across modeling cues are provided. By fitting the model to observed data, the model parameters are then determined. The latter method aims to group the cues by using a set of heuristic rules on proximity and symmetry between the cues, rather than heavily relying on a priori knowledge of the generic building models.

Many buildings in modern cities exhibit a very high degree of shape complexity, being composed of a mixture of various building primitives, with multiple stories and numerous roof superstructures (chimneys, air vents, water tanks, and roof fences, etc.). Under these circumstances, it becomes problematic to accommodate high variations in building shapes using predefined models. On the other hand, strong constraints in analyzing topological relations and grouping adjacent modeling cues, which are commonly used in data-driven reconstruction, may degrade the robustness of the reconstruction system if the fragmentary level of cue extraction becomes higher.

We understand that the building reconstruction system is highly influenced by the imaging sensor used. Many factors characterize the primary information source: the ground spacing resolution, S/N ratio, georeferencing accuracy, digital quantization level, scanning patterns, surveying quality, etc. These limit a priori knowledge that can be utilized for reconstructing buildings, and thus significantly affect the system design for developing both elementary algorithms and the entire strategy. Over the past few years, the ever increasing numbers of new remote sensors have been introduced for the purpose of providing primary information to enable 3D cityscape modeling. Chief amongst newcomers is the topographic airborne LiDAR (Light Detection And Ranging), which can collect elevation data at a vertical accuracy of 15 cm, and at a rate of higher than 100 KHz. In addition, the outgoing laser beams can penetrate foliage and the reflected laser energy can be digitized in discrete or continuous waveforms, together with reflectivity information of illuminated objects. The unique characteristics of LiDAR surveying provide great potential through complementarity with optical sensors, for automating the sophisticated tasks involved in 3D cityscape modeling. However, even with the use of LiDAR data, the technology for automatically reconstructing urban features is thus far poorly understood. This is mainly because it is not expected that one can obtain a complete set of modeling cues from LiDAR data; and regardless of the sensor type used, there is a serious lack of generalized methodology that is focused on object representation that is continually adaptive to intraclass variations (i.e., within the same building class), but presents a maximal contrast in distinguishing between interclasses (i.e., building vs. nonbuilding classes).

This chapter presents a unique framework of feature grouping for modeling 3D buildings by employing a Binary Space Partitioning Tree (BSP-Tree). This is mainly based on research works conducted by Sohn and Dowman (2007) and Sohn et al. (2008). Following the philosophy of data-driven reconstruction, the methods show an autonomous integration of linear features and area-features, which correspond to the saliencies of building shapes. Thereby the system achieves the modeling of two different types of building models, either prismatic or polyhedral models, subject to the point density acquired from topographic airborne LiDAR data. The chapter begins by discussing the background knowledge pertaining to building reconstruction, a general hierarchy of the building reconstruction process is then discussed, different types of building object representations are outlined, and finally a literature review is presented. Next, a building reconstruction system, based on the integration of airborne LiDAR data with high-resolution satellite data is presented for creating prismatic building models. This is followed by a description of a system for modeling polyhedral buildings using only airborne LiDAR data. An experimental result is presented for each method in order to aid in the assessment and discussion of system performance.

17.2 Background

17.2.1 General Hierarchy of Building Reconstruction Process

Automatic detection and modeling of building objects has been a major research topic in computer vision and photogrammetry communities for many years. The process of building reconstruction varies considerably, depending on the supporting models and data sources used. A wide range of approaches has been published in the literature (Ameri, 2000). In general, automatic building reconstruction can be achieved through a particular sequence of processing stages, namely, detection, extraction, and reconstruction. These tasks are defined as follows:

- Building detection recognizes the existence of buildings of interest and localizes them into regions.
- Building extraction delineates physical building boundaries in 2D or 3D, which may implicitly include the building detection problem.
- Building reconstruction detects building roof structures and delineates a complete description of building roof shapes in 3D object space as well as building boundaries.

Figure 17.1 illustrates the logical hierarchy for the acquisition of 3D building models. The building detection output can be used as an input for supporting the building

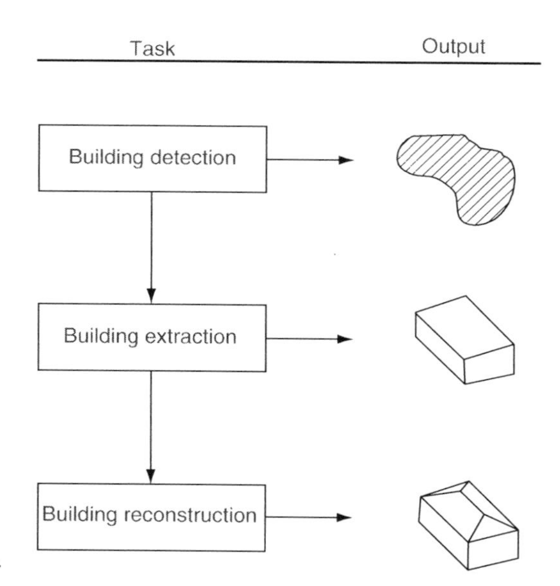

FIGURE 17.1
Illustration of hierarchical building reconstruction tasks.

extraction or the building reconstruction process. Similarly, 3D reconstruction of a building roof can start using a boundary delineated by the building extraction process. In some cases, however, these tasks can be achieved independently of one another; for example, it is possible to obtain building boundary fitting on regions with no indication of building location. Also, it is not strictly required for building extraction to be performed before the building reconstruction. Building boundaries can be obtained during or after the building roof shape is reconstructed. However, most building reconstruction systems follow the logical hierarchy of the three tasks mentioned earlier. The main reason for this is that the complexity of the building reconstruction problem can be reduced greatly as each task removes distracting features from the observed data in a coarse-to-fine manner. Thus, as the vision tasks proceed following a reconstruction hierarchy, each task can focus on its own target within a much more simplified data domain.

17.2.2 Building Object Representations

Generic knowledge about a geometrical shape of the building object is represented using a range of building models. There are two different models for the description of the building objects, namely, parametric and generic. The parametric building model uses prespecified model primitives, including rectangle, L-shape, I-shape, T-shape, gable roof, and hip roof (see Figure 17.2). The number of modeling cues, such as corner points, straight lines, and surface patches, used to form a model primitive is predetermined, and their topological relations (e.g., angles and relative distances) between the modeling cues are given from the specified model primitive. However, the geometry of the modeling cues is unknown. Once the modeling cues are extracted from the data sources using feature extraction algorithms, these are matched with the predetermined parametric model to the observed data. This is done in order to estimate the correct geometry of the model and thus, to reconstruct the building model. To model more complex buildings, different types of parametric models are integrated through a set operation such as intersection, union, and subtraction.

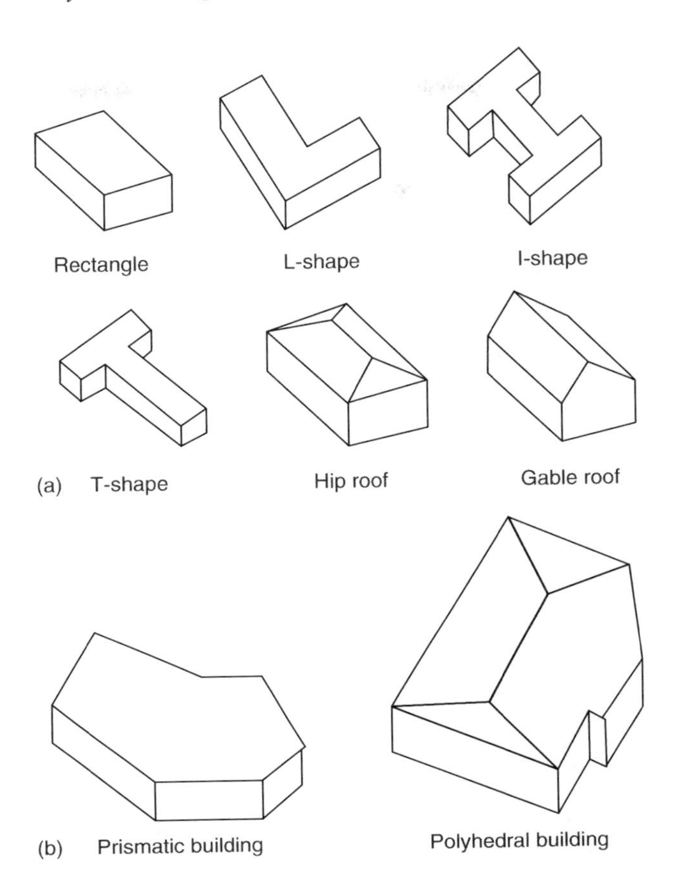

FIGURE 17.2
Building models.

The most benefit that can be derived from the parametric modeling approach is that the method provides specific object interpretation for internal structure and building classification, which is useful in some applications. The predetermined geometric and topological relations, often called constructive solid geometry, support constraints to group fragmentary features or infer additional cues, which are often missed by feature-extraction algorithms. However, since the number of building models used is prefixed, the parametric modeling approach is of limited use. Indeed, building shapes in a real-world setting usually show high irregularity and unpredictability. Thus it is a challenging issue to reconstruct these irregular building shapes by combining multiple parametric models. In addition, as the geometry of the parametric model is unknown, too many modeling cues can potentially be matched with the model. As a result, the number of building hypotheses generated is very large, and thus the selection of a correct building representation is always a bottleneck in this approach. For this reason, most parametric building models are used for the reconstruction of residential houses, whose shape variances can be represented by using few model instances.

In contrast to the parametric building model, the generic building model does not rely on geometric constraints driven from prespecified model primitives. Instead, the model specifies an internal structure of building object only using heuristic topological relations (e.g., proximity, orthogonality, symmetry, and parallelogram) where the geometric parameters are not prefixed in number. This approach is suitable for representing irregular building shapes with minimal geometric constraints. However, the major problem of generic models is that buildings can be reconstructed completely only if feature-extraction algorithms can detect all the modeling cues required. This problem often occurs in an urban environment since urban scenes normally contain rich object context with many occluding parts, which disturbs a perfect object interpretation. In this case, the missing cues cannot be recovered by the generic building model since generic models do not use a priori knowledge of building shapes, and thus complete building representation is hard to achieve. The generic building model has two subclasses: (1) the prismatic model and (2) the polyhedral model. The prismatic model represents a building shape with a set of edges and an associated height for the entire building. This model does not provide a roof structure. The polyhedral building model forms a roof structure by collecting polygonal surface patches. The internal relations of roof structure are obtained by intersecting the extracted polygonal surfaces. Figure 17.2 shows some examples of generic building models.

17.2.3 Previous Works

Most urban scenes usually contain very rich information, which includes a large amount of cues with geometric or chromatic cosimilarity to buildings, but which belongs to non-building objects. Due to the very dense and accurate 3D information provided by LiDAR, contrasts in heights or surface roughness between building rooftops and surrounding features can be greatly increased. Through this a much higher success rate of building detection can be achieved. For instance, Baltsavias et al. (1995) clustered a LiDAR DSM (Digital Surface Model) by height similarity in a compressed histogram space, and detected building regions by dividing the height histogram by a selected height threshold. Sagi Filin (2002) presented an algorithm to cluster irregular laser points into predefined classes according to attributes of surface textures. Through this, structured planar surfaces forming rooftops can be recognized. Rather than using discrete multiple-echoes, Ducic et al. (2006) recently suggested a decision-making tree to determine feature classes including buildings, relying on heuristic knowledge on feature attributes characterized by the analysis of digitized waveforms. Instead of directly detecting buildings, building regions can be detected by eliminating other features; this is suggested by a number of other researchers (Weidner and Forstner, 1995; Baillard and Maitre, 1999). This fore-to-background separation shares a common strategy of hierarchical object detection, in which the most dominant feature, i.e., terrain, is first eliminated from the scene, followed by differentiating vegetated features from buildings. In this context, automatic techniques to distinguish laser point impacts on the ground from other points have been intensively researched (Sithole and Vosselman, 2004). A critical issue to increase the success rate of terrain extraction is to develop a mechanism, which implements a criterion for differentiating terrain points from nonterrain points while being continuously adaptive to variances in terrain slopes. In earlier days, this criterion was determined in a deterministic way based on scene knowledge (Baltsavias et al., 1995). An adaptive algorithm was presented by selecting different classification criteria subject to underlying terrain slopes (Kraus and Pfeifer, 1998; Axelsson, 1999). An important study suggesting a global objective function for terrain extraction was conducted in the framework of Markov Random Field by Baillard and Maitre (1999). However, after an extensive comparison of different strategies, Sithole and Vosselman (2004) drew the conclusion that there is no universal terrain-filtering algorithm that can show a perfect detection rate in all instances of different landforms (i.e., with various slopes and irregularities). Hence, further manual classification, which removes features unnecessary for recognizing the terrain, is still required (Fowler, 2001).

Another important task in building detection is to distinguish between buildings and vegetation in point clouds. Hug and Wehr (1997) proposed a binary classification algorithm to separate trees from man-made objects using the laser reflectivity. Brunn and Weidner (1998) developed a classification algorithm to detect buildings and vegetation using differential geometric attributes based on Bayesian Network theory. Axelsson (1999) classified a regional feature into either vegetation or building classes by comparing an information length required for encoding each class using given LiDAR points. A final decision was made by selecting the minimum description length (MDL), which contains sufficient numbers of breakpoints for the vegetation class or, for the building class, shows an optimum balance between the numbers of breakpoints and surface smoothness. Building detection can be also achieved by integrating chromatic information, provided by optical imagery, with laser points. Complementary fusion principles for extracting building outlines were driven by integrating LiDAR data with multispectral IKONOS imagery (Kim and Muller, 2002). Rottensteiner et al. (2005) employed the Dempster–Shafer theory as a data fusion framework for building detection from airborne orthophotos with multispectral bands and LiDAR data. The building object is detected on a per-pixel classification by probabilistically investigating various properties (e.g., color, height variance, and surface roughness) of each pixel, which is later validated in a region-wise manner.

For the last few years, many enthusiastic researchers have made great efforts to develop techniques for automatically reconstructing building models from airborne LiDAR data. The model-driven approaches fit prespecified models to LiDAR data, thereby determining geometric parameters and topological relations across modeling cues. Wang and Schenk (2000) reconstructed a parametric model by analyzing triangulated facets constructed based on data-driven edges. Hu (2003) applied the well-known Hough transformation to obtain the four corner points of buildings for reconstructing a box-type model. A good example of a model-driven method for modeling polyhedral buildings was presented by Maas and Vosselman (1999), who were able to determine the parameters of a standard gable roof with small dorms through the analysis of invariant moments from LiDAR points. In order to cope with more complicated models, a number of different building primitives are hypothesized and tested within partitioned rectangles of existing ground planes, and a full description of the roof model is generated by integrating the verified parametric models (Brenner and Haala, 1998; Vosselman and Suveg, 2001).

The data-driven method reconstructs the boundary of prismatic or polyhedral building models by grouping extracted lines, corners, and planes based on minimal use of prior knowledge of the generic building model. For creating a prismatic model, Guo and Yasuoka (2002) combined monochromatic IKONOS imagery with LiDAR data for extracting buildings. After isolating building regions by subtracting terrain surface from LiDAR data, an active contour model was fitted to building outlines using the guidance of edges extracted from both LiDAR data and IKONOS imagery. Chen et al. (2004) detected step edges from LiDAR data, which are geometrically improved with the support of the linear features extracted from airborne imagery with a 10 cm ground sampling distance. These refined lines are grouped in a closed polygon using a splitting-and-merging algorithm. For reconstructing a polyhedral model, plane segments are obtained using the Hough transformation (Vosselman, 1999); RANSAC (Brenner, 2000); orthogonal projection of point clouds (Schwalbe et al., 2005); and region growing (Rottensteiner et al., 2005). Also, linear features are extracted by intersecting planes detected and the analysis of height discontinuity (Vosselman, 1999); approximating the boundary of planes (Alharthy and Bethel, 2004). The Gestalt laws (proximity, symmetry, parallelism, and regularity) are often used for bridging the gaps between the extracted features in order to reconstruct the model topology (Hu, 2003) or to snap adjacent planes (Alharthy and Bethel, 2004). In order to impose geometric regularity on reconstructed models, additional constraints are used where (1) a set of rules for intersecting adjacent planes are predetermined (Hofmann, 2004); (2) the roof structure is restricted to the dominant building orientation (Vosselman, 1999; Schwalbe et al., 2005); or (3) orientations of building outlines are derived from an existing ground plan (Brenner, 2000). A global optimality of shape regularity is achieved by the MDL principle (Weidner and Forstner, 1995).

Using an existing ground plan to subdivide a building region of interest into a set of rectangles can greatly reduce the complexity of the building reconstruction process (Brenner and Haala, 1998; Brenner, 2000; Vosselman and Suveg, 2001). The advantage of using the ground plans is to obtain a very accurate building extraction result, by which the remaining building reconstruction becomes much simpler. Brenner and Haala (1998) used a building ground plan to subdivide LiDAR DSM into a number of rectangles, then a 3D parametric building model including flat, hip, gable, and desk roof model was selected for each 2D rectangle after a number of trial-and-errors of fitting the models to the DSM. A wider range of building shapes may be difficult to model by relying only on the rectangular partitioning result. Brenner (2000) grouped planar roof patches using a rule-based filtering process developed based on the geometric information extracted from the building ground plans. Vosselman and Suveg (2001) refined rectangular partitioning by adding 3D lines detected by the Hough transformation from LiDAR DSM, where the final partitions can be obtained after merging similar roof patches. A parametric roof model is hypothesized and tested within partitioned rectangles, and a full description of the roof model is generated by collecting the verified models. The aforementioned research work has proven that the method based on an existing ground plan is a practical solution for reconstructing a large number of buildings (Brenner, 2000). However, as building complexity becomes higher, presumptions of coherent relations between the roof structures and the ground plan, which define the partitioning rules, may be invalid.

17.3 Methodology Overview

The shape of a building object with planar facets is decomposed into collections of more primitive elements (i.e., modeling cues) such as facets, edges, and vertices. A generic model of 3D buildings can be represented based on a combined topological and geometric description of the building object. Here, the topology of building models is obtained by recovering a set of relations that indicate how the faces, edges, and vertices are connected to each other, while geometry can be specified by estimating the appropriate geometric parameters (perimeter, intersecting angle, area, orientation, location, etc.) of modeling cues. The purpose of our research is to automatically model building objects by recovering both topological and geometrical relations across modeling cues, relying primarily on the information extracted from LiDAR data only, but with minimal use of a priori knowledge of targeted building shapes. This data-driven approach is more amenable to describing variants of building shapes than is the use of prespecified model primitives. However, a scientific challenge in the data-driven technique may lie in a reliable grouping of modeling cues that are completely missed or fragmented. This fragmentation can be due to a variety of reasons, including the noise inherited from laser-scanners, disadvantageous backscattering properties, and object complexity. In order to resolve this problem, we exploit the well-known data structure of the BSP-Tree as a unique framework for globally grouping incomplete features extracted from a given data space that yields successively finer approximations of the building object.

The BSP-Tree is a method for recursively subdividing n-dimensional space into convex sets with a homogeneous property by an $(n - 1)$-dimensional hyperplane. For the 2D case, a given polygon is divided into two subpolygons by intersecting it with a line found inside the polygon. The partitioning process continues until all the resulting polygons satisfy certain criteria that are differently specified depending on the purpose of an application targeted. A consequence of this hierarchical subdivision is formulated as a representation of the complex spatial scene by means of a binary tree data structure. The BSP-Tree is a widely used representation of solid geometry, and is extremely versatile due to its powerful classification structure. The most traditional application of the BSP-Tree is in 3D computer graphics where it is used to increase rendering efficiency for solid modeling, hidden surface removal, shadow generation, visibility orderings, and image representation (Fuchs et al., 1980; Gordon and Chen, 1991). The applications usually utilize the tree structure as a preprocessor to store a virtual environment comprising polygonal solid models, but are not usually used to create the tree for extracting 3D models directly from remotely sensed data.

The following sections describe the BSP-Tree as a special data structure consisting of convex hierarchical decompositions of 3D LiDAR points. With this tree structure, a generic building model is represented by the union of convex regions, which correspond to segmented planar roof-facets that are delineated by a rectilinear boundary with maximum response to real edges. In other words, the convex decompositions of the LiDAR space induced by the BSP-Tree method serve as a fusion framework for integrating area-features (i.e., planar clustering result) and edge-features (i.e., line extraction result) for representing the boundary of 3D building rooftops. Two building reconstruction systems will be introduced here. The first method concerns a low point density (i.e., less than one point per square meter), with significant irregularity in LiDAR data acquisition that hinders the extraction of a sufficient amount of modeling cues. To resolve this problem, additional cues to compensate for LiDAR data limitations are extracted by using high-resolution satellite imagery and integrated for creating a prismatic building model based on the BSP-Tree framework. Only using LiDAR data, the second method presents how the BSP-Tree can reconstruct a detailed model of polyhedral rooftops in a building cluster with a mixture of various building primitives as the point density of LiDAR data increases to a few points per square meter. Considering the increased modeling level of the details targeted, the method includes more sophisticated objective functions to obtain an optimality of the creation of the BSP-Tree for describing complicated rooftops. Both methods share a hierarchical building detection process. The main purposes of the building detection process are to classify the LiDAR data into building and nonbuilding points; and to provide coarse building models for individual building objects by bounding each isolated building region with a rectangle (i.e., initial building model) for the subsequent reconstruction processes. Different types of modeling cues and their extraction algorithms are implemented in order to deal with prismatic and polyhedral models respectively. As the coarse building shape (rectangle) is hierarchically subdivided by the extracted linear features, convex polygons with more fine scales are produced by the BSP-Tree. Finally, a geometric model of the building object is reconstructed when all partitioned polygons are verified as either building planes for the prismatic model or homogeneous roof-facets for the polyhedral model. This topology reconstruction is implemented by optimally controlling the generation of the BSP-Tree. The functional elements of the proposed building reconstruction system are schematically illustrated in Figure 17.3.

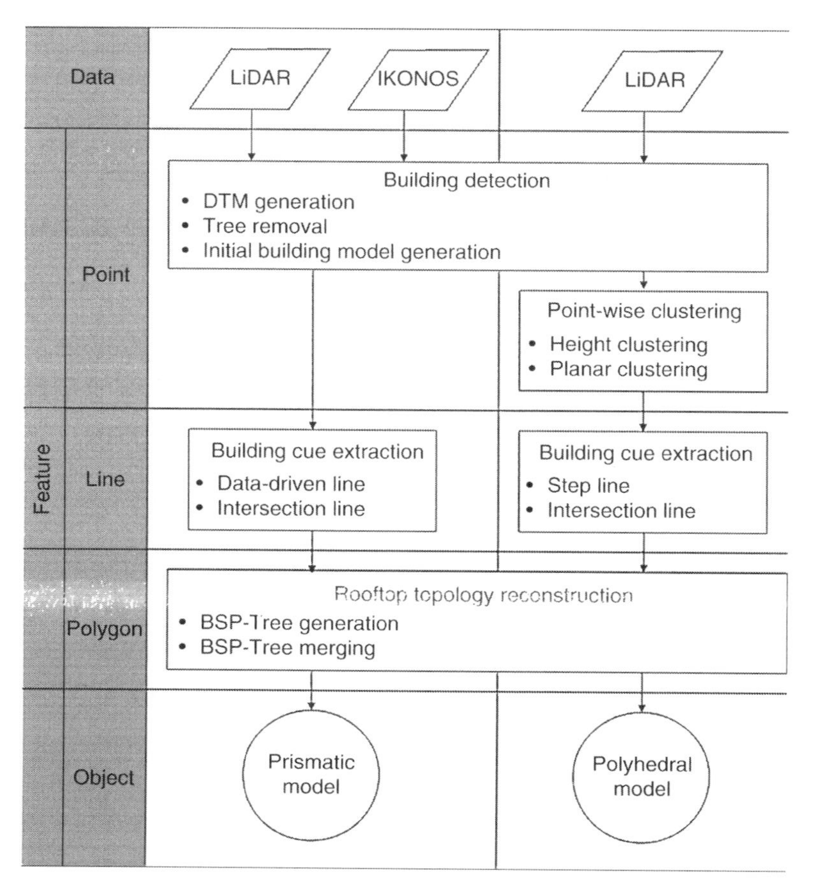

FIGURE 17.3
Schematic diagram of the proposed 3D building reconstruction system.

17.3.1 Prismatic Model Reconstruction

This section describes a new technique to automatically extract prismatic building models from airborne LiDAR data with low and irregular point density by integrating IKONOS Pan-sharpened imagery. The proposed method presents compensatory roles of active (laser scanning) and passive (optical) sensors in feature extraction and grouping for autonomous extraction of building outline vectors. The method consists of three steps: (1) building detection; (2) linear feature extraction; and (3) BSP-Tree construction for creating building models.

17.3.1.1 Building Detection

As previously discussed, the building detection process for discriminating building objects from other features greatly reduces urban scene complexity. The remaining reconstruction process can then focus on individual building regions. Rather than directly detecting building points (Filin, 2002), we followed a hierarchical strategy (Baillard and Maitre, 1999) where the targeted foreground objects (i.e., buildings) are detected by removing the background features including terrain and trees. The first step was to apply a model-based filtering algorithm, called recursive terrain fragmentation (RTF) filter, to airborne LiDAR data (Sohn and Dowman, 2008) in order to automatically identify laser points impacting only on the ground. This technique was developed, in particular, to allow a LiDAR filter to be self-adaptive to various landforms and slopes. This terrain filter employed a tetrahedral model to generally represent a terrain surface with a single slope (homogeneous terrain). A model-fitness between LiDAR points and the tetrahedral terrain model was measured in a framework of MDL. In a coarse-to-fine scheme, this model-fitness scores triggered to recursively fragment a LiDAR DEM convolved with heterogeneous terrain slopes into piecewise homogeneous subregions where the underlying terrain can be well characterized by the terrain model. A more detailed description of the aforementioned RTF filter can be found in Sohn and Dowman (2008). Once these homogeneous terrains were obtained, the separation of nonterrain points from the ground can be simply made by a height threshold, as terrain characteristics over subregions are uniformly regularized. Then, the building points were retained by removing points over the tree objects from nonterrain points. In order to eliminate the tree points, we compute the normalized difference vegetation indices (NDVI) by a combination of red and near-infrared channels of IKONOS imagery. Integrated with NDVI map, a tree point is identified if the point has an NDVI value that is larger than a predefined threshold. After removing tree features from the nonterrain ones, individual buildings comprising only building points are bounded by the rectangles that will be used as initial building models to the following building reconstruction procedures.

17.3.1.2 Extracting Linear Features

Straight lines are key features comprising most building outlines. Because of this reason, many building extraction systems concentrate on detecting straight lines as a low-level modeling cue. In this section, the straight lines comprising building boundaries are extracted by investigating abrupt changes of intensity gradients with support from LiDAR classification results. Also, we present a data fusion strategy to compensate missed or partly fragmented linear features by fusing both optical and laser sensors.

Data-Driven Line Cue Generation: As described earlier, we aim to develop a building extraction system, particularly using LiDAR data with very low and irregular density and IKONOS multispectral bands. Since the LiDAR data that is currently used in this experiment has disadvantageous characteristics for extracting lines and planes, we mainly rely on IKONOS data to extract lines, rather than using LiDAR data. First, straight lines are extracted by the Burns algorithm (Burns et al., 1986). Then these lines are filtered by a length criterion, by which only lines larger than a prespecified length threshold l_d remain for further processing (Figure 17.4a). Second, two rectangular boxes with a certain width l_w are generated along two orthogonal directions to the line vector, which is filtered in length. The determination of the boundary line can be given if nonbuilding and building points are simultaneously found in both boxes or if only building-label points are found in one of the boxes and no LiDAR point can be found in the other box (Figure 17.4b). As a final line filtering process, a geometric disturbance that has been corrupted by noise is regularized over boundary lines. A set of dominant line angles of boundary lines is analyzed from a gradient-weighted histogram, which is quantized in 255 discrete angular units. In order to separate a weak but significant peak from the other nearby dominant angles, a hierarchical histogram-clustering method is applied. Once the dominant angle, θ_d, is obtained, lines with angle discrepancies that are less than certain angle thresholds, θ_{th}, from θ_d are found. Then, their line geometries are modified as their angles are replaced with θ_d. These modified lines do not contribute to the succeeding dominant angle analysis

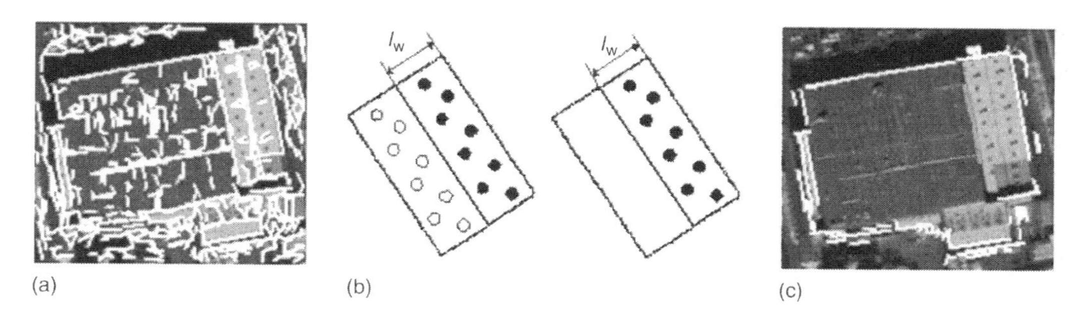

FIGURE 17.4
Extraction of data-driven cue extraction and verification from IKONOS image. (From Sohn, G. and Dowman, I., *ISPRS J. Photogram. Rem. Sens.*, 62, 43, 2007. With permission.)

and the next dominant angle is obtained. In this way, a set of dominant angles is obtained, by which geometric properties of boundary lines can be regularized (Figure 17.4c).

Model-Driven Line Cue Generation: Although a satisfactory result for extracting boundary lines could be obtained by the method presented in the previous section, the line cues produced in a data-driven way do not always provide sufficient cue density covering all parts of building edges. This is because significant boundary lines may be missed due to low contrast, shadow overcast, and occlusion effects, especially when 1 m satellite imagery of a complex scene is solely used. However, LiDAR data is relatively less affected by those disadvantageous factors than is the optical imagery. Thus, new lines are virtually extracted from LiDAR space in order to compensate for the lack of data-driven line density by employing specific building models. It is assumed that building outlines are comprised of parallel lines. Based on this, for each data-driven line, parallel lines and U-structured lines are inferred from LiDAR space. The process aims to acquire at most three model-driven lines starting from a data-driven line, but it may fail to generate any model-driven line. To detect model-driven lines, a small virtual box is generated from each data-driven line. Then, it grows over the building roof in a direction parallel to the selected data-driven line until the building points are maximally captured without including any nonbuilding point (Figure 17.5a). After detecting the parallel line, the virtual box continues to grow, but at this time toward two orthogonal directions parallel to the boundary line detected (Figure 17.5b). In this way, a set of model-driven lines can be newly produced from all data-driven lines (Figure 17.5c and d). As can be shown in Figure 17.5d, erroneous model-driven lines can be generated due to invalid assumption of symmetric constraints

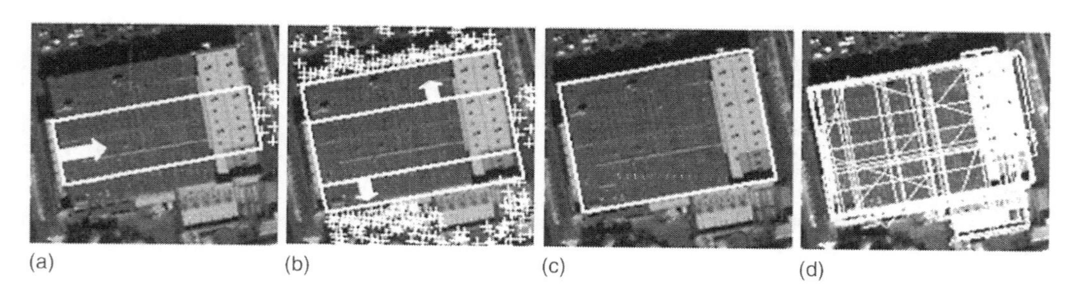

FIGURE 17.5
Results of model-driven line cue generation. (From Sohn, G. and Dowman, I., *ISPRS J. Photogram. Rem. Sens.*, 62, 43, 2007. With permission.)

on building shape and errors in data-driven line detection. A verification process to filter off such erroneous model-driven lines will be discussed later.

Edge-Constrained Data Interpolation: The spatial distribution of laser points is the major factor that affects final reconstruction results of the suggested methods. Thus, irregularity of laser point density will degrade the overall building description result, leading to wrong building shape identification. To prevent this problem, it is necessary to interpolate LiDAR data with irregular point density for the purpose of achieving higher and more regular point density. The interpolation process in the current experiment simply increases the amount of LiDAR data by adding new laser points where triangular sizes in the TIN constructed with LiDAR data are larger than a certain threshold. However, the construction of the TIN is constrained by both data-driven and model-driven lines so that the interpolation of laser points over the rooftops is limited to the inside of building outlines, while the terrain points increase outside buildings only.

17.3.1.3 BSP-Tree Algorithm for Building Description

The BSP-Tree is a binary data structure that is used to recursively partition a given region into a set of convex polygons with homogeneous attributes. In this section, we exploit the BSP-Tree as a special data structure consisting of convex hierarchical decompositions of 3D LiDAR points. With this tree structure, a prismatic building model is reconstructed through the union of convex polygons, which corresponds to segmented building regions that are delineated by a rectilinear boundary with maximum response to real edges. In other words, the convex decompositions of the LiDAR space induced by the BSP-Tree method serve as a fusion framework for integrating area-features (i.e., clustered building points) and edge-features (line extraction result) for the purpose of representing the boundary of 3D building rooftops. The BSP-Tree algorithm used for building reconstruction consists of the three components: (1) hierarchical binary partitioning, (2) hypothesis-test optimization, and (3) polygon merging. More detailed description of each process is given as follows:

Hierarchical Binary Partitioning: The rectangle bounding each building object is given as the initial building model P_0. All member points of P_0 are labeled as one of the binary labels, building and nonbuilding, based on the building detection result. Both data-driven and model-driven lines $\{l_i\}$ extracted in the previous section are formulated as the hyperlines $\{h_i\}$, each of which will be involved in the subdivision of the given building polygon. This is described by

$$h_i(\rho_i, \theta_i) = \{(x, y) \in R \mid x\cos\theta_i + y\sin\theta_i - \rho_i = 0\} \tag{17.1}$$

where (ρ_i, θ_i) means the distance of the origin from a line segment l_i, and the slope angle measured between the edge normal and the x-axis, respectively. A hyperline h_1 is then chosen to halve planar clustered points into the positive and negative regions, P_{1+} and P_{1-}. Each of the subpolygons is expressed by

$$\begin{aligned} P_{1+}(h_1; \rho_1, \theta_1) &= \{(x, y) \in R \mid x\cos\theta_1 + y\sin\theta_1 - \rho_1 > 0\} \\ P_{1-}(h_1; \rho_1, \theta_1) &= \{(x, y) \in R \mid x\cos\theta_1 + y\sin\theta_1 - \rho_1 < 0\} \end{aligned} \tag{17.2}$$

The normal vector of the hyperline h_1 is defined by $(\cos\theta_i, \sin\theta_i)$. The positive polygon P_{1+} corresponds to the one that lies in the direction of the normal vector of l_i; the negative polygon P_{1-} is located in the opposite direction. A BSP-Tree is now constructed as the root node holds the hyperline h_1, where all vertices comprising P_0 and the two subpolygons are represented by leaf nodes. Each leaf of the BSP-Tree contains binary attributes of a partitioning and terminating node, which are determined according to a polygon classification result over a given tree node. A convex region will be represented as the partitioning node in the BSP-Tree if the region contains the member points with a mixture of building and nonbuilding labels (open polygon). However, the terminating node will be given to the region if it has one of the following properties: (1) the member points in the polygon are attributed with building label only (closed polygon); (2) the polygon does not contain any member point (empty polygon); (3) the minimum length of the lateral sides or the area of the polygon is less than a certain threshold (garbage polygon). This node classification determines a triggering and terminating condition of the BSP-Tree over the node. That is, when a successive hyperline h_1 is selected, the line continues to partition a partitioning node of polygon P_{1+}. However, the partitioning process will be terminated to a terminating node of P_{1-}. This process continues until no partitioning node is generated by the BSP-Tree.

Hypothesis-Test Optimization: There are usually many erroneous hyperlines inherited in their formation. For instance, blunders in data-driven lines are caused when the quality of the building detection is low, and undesirable model-driven lines may be produced if the building object targeted does not have particular geometric symmetry. Those distracting hyperlines must be eliminated during the partitioning process. This optimality is achieved by the hypothesize-and-test scheme with a partition scoring function. All of the hyperlines are tested to partition a polygon, P_i, and the partitioning result generated by each hyperline is evaluated by a partition scoring function. An optimal hyperline, h^*, with the highest partitioning score, is finally selected to partition P_i by

$$h^* = \arg\max_{\forall\{h_i\}} \left\{\max\left\{H\left(P_{j+};h_i\right), H\left(P_{j-};h_i\right)\right\}\right\} \tag{17.3}$$

In Equation 17.3, $H(\ )$ is an objective function to evaluate a goodness of partitioning over P_i produced by an arbitrary hyperline candidate, h_i. For instance, if P_i is partitioned by h_i and yields the positive and negative polygons, P_{j+} and P_{j-}, the scoring function, H, evaluates the partitioning results respectively over two subdivided polygons, and determines a final score for h_i as the maximum value of the portioning results. H computes partitioning scores according to the ratio of label distribution of building to nonbuilding labels over a given polygon. It aims to assign a higher partitioning score when a "closed" polygon with a larger area is produced by h_i. This scoring function is described by

$$\begin{aligned} H(P_{j+};h_i) &= \frac{1}{2}\left\{\frac{N_{\text{bld}}(P_{j+};h_i)}{N_{\text{bld}}(P_i;h_i)} + \frac{N_{\text{non-bld}}(P_{j-};h_i)}{N_{\text{non-bld}}(P_i;h_i)}\right\} \\ H(P_{j-};h_i) &= \frac{1}{2}\left\{\frac{N_{\text{bld}}(P_{j-};h_i)}{N_{\text{bld}}(P_i;h_i)} + \frac{N_{\text{non-bld}}(P_{j+};h_i)}{N_{\text{non-bld}}(P_i;h_i)}\right\} \end{aligned} \tag{17.4}$$

where $N_{\text{non-bld}}$ and N_{bld} are functions to count the numbers of building labels and nonbuilding labels belonging to a corresponding polygon. Figure 17.6b shows a procedure for obtaining the best result of the initial polygon, P_0. A hyperline candidate, h_i, is selected from the hyperline list, by which P_0 is divided into the positive and negative

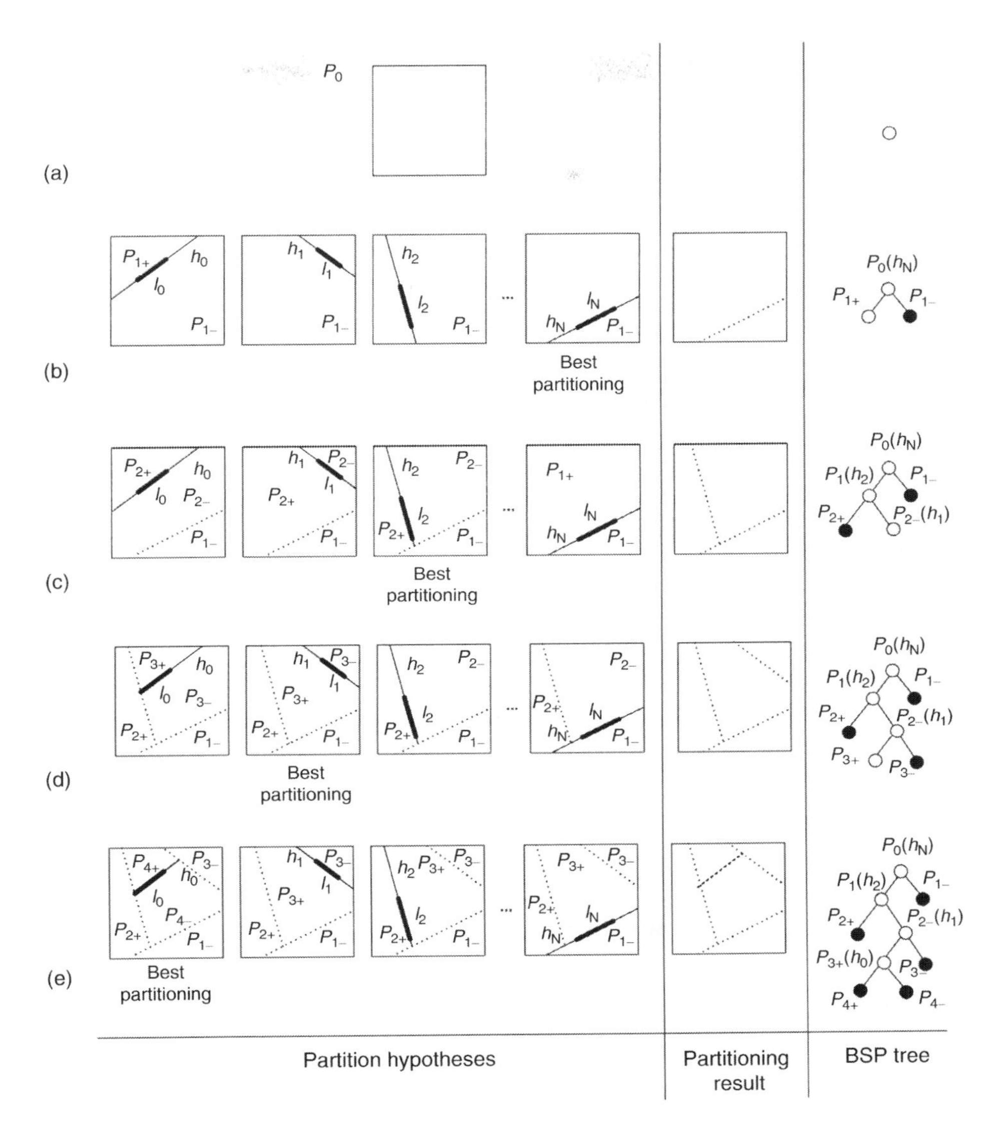

FIGURE 17.6
BSP-Tree construction. (From Sohn, G. and Dowman, I., *ISPRS J. Photogram. Rem. Sens.*, 62, 43, 2007. With permission.)

planes, P_{1+} and P_{1-}. Once the partitioning scores for h_i are computed by Equations 17.3 and 17.4, the remaining hyperlines are sequentially selected from the hyperline list and their partitioning scores are also measured by H. As can be seen in Figure 17.6b, a hyperline, h_N, with the maximum partitioning score, is finally selected to partition P_0. Then, all the vertices of P_0 and h_N, are stored as a root node of the BSP-Tree, which is expanded while new child nodes with vertices of P_{1+} and P_{1-} are added to the root node for further recursive partitioning. The same method used for the partitioning of P_0 is applied to P_{1+} and P_{1-} respectively, but to only an open polygon. In Figure 17.6b, P_{1-} is classified as the "closed" polygon and P_{1+} is classified as the open polygon. Thus, only the partitioning process over P_{1+} is triggered, and the partition of P_{1+} by a hyperline, h_2, is selected as the "best" result. The BSP-Tree of Figure 17.6b is expanded with new child polygons, P_{2+} and P_{2-}, and the selected hyperline, h_2 (Figure 17.6c). This process continues until no leaf node of the BSP-Tree can be partitioned by hyperlines (Figure 17.6d and e).

Polygon Merging: After constructing the BSP-Tree, a plane adjacency graph is created by collecting the final leaves of the BSP-Tree, where each node represents a segmented polygon feature and each arc represents the connectivity between neighboring polygons. Starting from the node with the largest area in the plane adjacency graph, a heuristic validation of the building polygon that belongs to the building structure is applied to its adjacent planes. All the closed polygons, which include only building points as the building polygon, are verified as the building polygon. In addition, an open polygon is accepted as a building polygon if it contains building labels with a significant ratio (ρ_{th}) although it includes nonbuilding labels. This heuristic grouping rule for an open polygon can be described as follows:

$$\forall P_i = \text{open polygon } P_i \mapsto \text{building polygon;}$$
$$\text{if} \quad \rho_{pt}(P_i) = \frac{N_{bld}(P_i)}{N_{mem}(P_i)} > \rho_{th} \tag{17.5}$$

where ρ_{pt} is a point ratio of building labels over the total number of member points of P_i; N_{bld} and N_{mem} are functions to count the numbers of building labels and nonbuilding labels belonging to P_i. Once all the "building" polygons are merged together, building boundaries are reconstructed. Figure 17.7 shows a sequence of the building reconstruction results obtained by the method presented based on the BSP-Tree.

17.3.1.4 Performance Assessment

Two datasets shown in Figure 17.8 are chosen for the research: a LiDAR DSM and a pan-sharpened multispectral (PSM) IKONOS image over the Greenwich area. The LiDAR DSM provided by Infoterra Co. covers a subsite of Greenwich industrial area with the size of 1.2 × 1.2 km² (~1,180,073.6 m²). It was acquired by the first pulse of Optech's 1020 airborne laser sensor and contains a total of 113,541 points, which approximately corresponds to a point density of 0.1 points/m². The PSM IKONOS image was also provided by Infoterra. The imagery combining the multispectral data with the panchromatic data was orthorectified by Space Imaging Co. and resampled with to a 1 m ground pixel. The area shows some characteristics of a typical urban environment; a number of industrial buildings

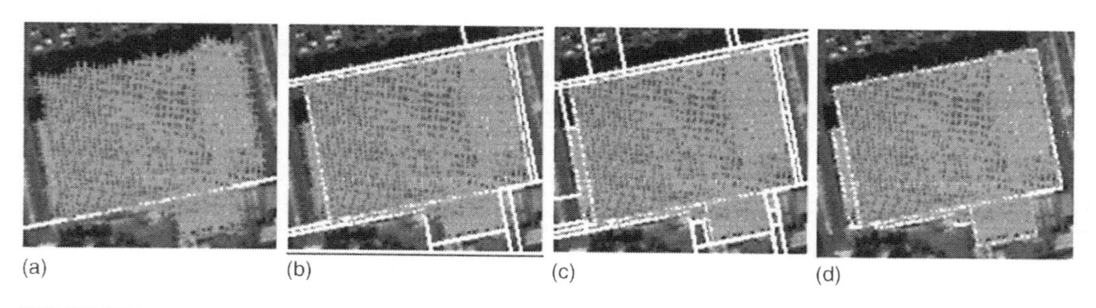

FIGURE 17.7
Illustration of polygon cue generation and grouping. (From Sohn, G. and Dowman, I., *ISPRS J. Photogram. Rem. Sens.*, 62, 43, 2007. With permission.)

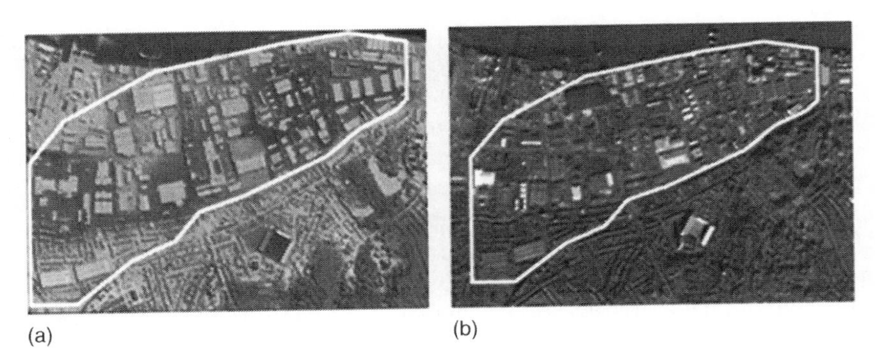

FIGURE 17.8
Greenwich test datasets. (From Sohn, G. and Dowman, I., *ISPRS J. Photogram. Rem. Sens.*, 62, 43, 2007. With permission.)

with different sizes spread over the area, from the lowest 7 m to the highest 39 m; a row of terraced houses can be found at one corner; and additionally a group of tall trees with various heights are located on the street. Figures also show a very irregular surface over some rooftops. Although the point density of the Greenwich data shows 0.1 points/m^2, this is much lower over some building rooftops and the ground near the buildings. Supposedly this is caused by certain factors such as poor weather conditions, high absorption by specific material properties, occlusion effects, laser scanning patterns, and flying height restrictions over London.

Figure 17.9a shows a building vector map (referred as UCL building map) delineating a total of 170 building outlines, which was automatically generated by applying the BSP-Tree

FIGURE 17.9
Building extraction results over the Greenwich dataset. (From Sohn, G. and Dowman, I., *ISPRS J. Photogram. Rem. Sens.*, 62, 43, 2007. With permission.)

algorithm over the Greenwich dataset. The positional accuracy of the UCL building map was assessed in comparison to the reference building polygon, namely MasterMap, provided by Ordnance Survey (see Figure 17.9b). First, we measured differences between two binary maps (UCL building maps and MasterMap) by a per-pixel manner. Based on the comparison result, each pixel is classified into four different error types; (1) TP (true positive) is a pixel that is classified as building by both datasets; (2) TN (true negative) is a pixel that is classified as nonbuilding by both datasets; (3) FP (false positive) is a pixel that is classified as "building" only by the UCL building map; and (4) FN (false negative) is classified as "building" only by MasterMap. Over a total of 2,886,499 pixels overlapped between the two binary maps, 9.8% (282,726 pixels) and 87.8% (2,535,114 pixels) are labeled as TP and TN, respectively, while 1.1% (31,142 pixels) and 1.3% (37,517 pixels) are labeled as FP and FN, respectively. Finally, based on the per-pixel classification result, we estimated the cartographical correctness of the UCL building map by the following objective metrics (Lee et al., 2003; Rottensteiner et al., 2005):

$$\begin{aligned}
&\text{Branching factor}: N(\text{FP})/N(\text{TP}) \\
&\text{Miss factor}: N(\text{FN})/N(\text{TP}) \\
&\text{Completeness (\%)}: 100\times N(\text{TP})/(N(\text{TP})+N(\text{FN})) \\
&\text{Correctness (\%)}: 100\times N(\text{TP})/(N(\text{TP})+N(\text{FP})) \\
&\text{Quality (\%)}: 100\times N(\text{TP})/(N(\text{TP})+N(\text{FN})+N(\text{FP}))
\end{aligned} \tag{17.6}$$

Here, $N(\ )$ counts the total number of pixels that are classified as one of the four error types. Each metric described in Equation 17.6 provides its own quantitative measure for evaluating the overall success of the building extraction algorithms. The branching factor and the miss factor are closely related to the boundary delineation performance, by measuring over-classified pixels (FP) and under-classified pixels (FN) produced by the automatic algorithm. The completeness and the correctness provide the measures of building detection performance, each of which is estimated by computing the ratio of correctly classified building pixels (TP) with respect to the total building pixels classified by MasterMap (TP + FN) and by the automatic process (TP + FP). The quality percentage combines aspects of both boundary delineation and building detection measures in order to summarize system performance.

Table 17.1 shows the evaluation measures on the UCL building map computed by Equation 17.6. In relation to other research, it can be concluded that the building detection performance achieved by the suggested technique is satisfactory. Lee et al. (2003) developed a method to extract outlines from IKONOS imagery with multispectral bands. Unsupervised classification results based on an ISODATA algorithm were used as evidences for detecting

TABLE 17.1

Performance Evaluation Results of the Proposed Building Extraction System

Performance Measure	Evaluation Result
Branching factor	0.11
Miss factor	0.13
Completeness	88.3%
Correctness	90.1%
Quality	80.5%

buildings, and a squaring algorithm regularized building outlines constrained with two dominant directions. They reported that the technique achieved a building detection rate of 64%, a branching factor of 0.39, and a quality percentage of 51%. Kim and Muller (2002) integrated LiDAR with multispectral bands of IKONOS for building extraction. Parish (2002) evaluated the performance of the system, which showed that the contribution of LiDAR data improved the building detection rate up to 86% compared to Lee et al.'s work, but ignorance of the geometric regulation in the building description process resulted in worse branching factor of 1.0, and thereby a rather poor quality percentage of 64%. Based on the theory of Dempster–Shafer, Rottensteiner et al. (2005) detected buildings from LiDAR data and an airborne orthophoto with multispectral bands, which were resampled at 1 m resolution. When the reference map was compared to detected buildings in a per-pixel evaluation, the system detected buildings with the completeness of 94% and the correctness of 85%. In a per-building evaluation, they reported that 95% of the buildings larger than 50 m^2 were detected, whereas about 89% of the detected buildings were correct.

Although the developed building extraction system showed a satisfactory result, it also produced a certain number of building extraction errors. These should be reduced in order to achieve a more accurate extraction of building objects. Most of the false negative pixels in the evaluation metrics (black pixels in Figure 17.9c) were generated by the under-detection of buildings. This problem is partly caused as the NDVI classification tends to over-remove building points, in particular over terraced houses with a long and narrow structure, and is partly due to the fact that LiDAR data is acquired with extremely low density over some buildings. False negatives can be reduced further if either LiDAR data are sufficiently acquired with higher density or other information, such as multiple-echoes, intensity, and surface roughness, is available for removing trees (Maas, 2001). Medium grey-colored pixels in Figure 17.9c represent the false positive errors, indicating the delineation quality of building outlines. Most of the boundary delineation errors stemmed from a deviation from the OS reference data of one or two pixels, where LiDAR measurements were acquired over buildings with high point density. However, as LiDAR points were acquired with less point density, more errors were produced at the boundaries. This is caused by the fact that the detection of data-driven lines and model-driven lines is more difficult over building rooftops with coarser point density than with denser point density. As a result, mislocation of both lines leads to the generation of delineation errors around building boundaries. Therefore, uniform acquisition of LiDAR data with higher point density will make the proposed system more controllable, and thus reduce building delineation errors. Also, some building extraction errors are due to the inherent faults in the OS building polygons. Although these errors were carefully removed from the UCL building map before performing the quality assessment, the inherent faults in the OS data can still be found. Thus some parts of buildings were missed in the OS MasterMap, although the UCL building map can successfully delineate building boundaries. The analysis of the reference errors suggests that the developed building extraction technique can also be used for applications detecting changes in an urban environment and supporting map compilation.

17.3.2 Polyhedral Building Reconstruction

The previous section has shown that the BSP-Tree algorithm is a useful framework for creating building outline vectors. The key elements comprising the framework are (1) linear features corresponding to building saliencies; (2) homogeneous attributes characterizing building objects; and (3) objective function to optimally control spatial partitioning. Due to the poor point density of LiDAR data used, the method was limited to the extraction of building outlines. In this section, we further exploit the BSP-Tree algorithm to reconstruct details of building rooftops with polyhedral models from LiDAR data with high point density. In particular, the proposed method targets 3D buildings with a mixture of multiple flat and sloped planes containing rooftop furniture. Considering this research goal, this section will discuss different strategies to collect linear and areal features comprising polyhedral buildings and an objective function to construct the BSP-Tree.

17.3.2.1 Extracting Initial Building Models

Similar to the hierarchical strategy described in Section 3.1, the reconstruction system starts with removing terrain features from the scene by applying the RTF filter to LiDAR points, which leads to the isolation of individual building objects. The only difference in the building detection process compared to the previous method is that the chromatic information derived from IKONOS multispectral bands is not available to remove tree points in the current experiment. Thus, the laser points impacting on tree objects are eliminated by using the geometric properties extracted from only LiDAR data. This includes the following procedures: (1) laser points showing a large height difference between the first and last returns were first removed; and (2) the connected component analysis was then applied to spatially clustered nonterrain points for investigating average height, boundary regularity, and surface smoothness. A nonpoint cluster is removed as the tree object if the aforementioned criteria are less than thresholds. After removing tree features from the nonterrain ones, individual buildings comprising only building points are bounded by the rectangles. These will be fed back as initial building models to the following building reconstruction procedures.

17.3.2.2 Extracting Planar Features

Our system aims to reconstruct 3D polyhedral building models. Thus, reliable clustering of planar attributes from LiDAR data is critical to achieving successful reconstruction of building models. Since we assume that the number of primitives and types that comprise the targeted building object are unknown, conventional planar clustering algorithms are disadvantageous for simultaneously targeting a large building object comprised of various building primitives. To resolve this problem, an individual building is first decomposed into a set of clusters according to height similarity, and then each height cluster is more finely segmented into a set of planar features.

Height Cluster Extraction: Since we aim to reconstruct 3D buildings comprising a mixture of multiple flat and sloped planes containing rooftop furniture, extracting modeling cues directly from an entire building may result in difficulties due to such a high degree of shape complexity. In order to reduce this complexity, LiDAR points collected for individual buildings are first decomposed into a set of clusters based on height similarity. A maximum height deviation for each point is measured from its neighboring points in a triangulated irregular network (TIN), which contributes to producing a height histogram of a certain bin size. After applying a simple moving box filter to the height histogram, a maximum height peak, δ_h, is automatically found from the histogram, within a height range between 1 and 3 m. Similarly to a conventional region growing algorithm, LiDAR points are grouped into one height cluster if a height discrepancy between a point and those connected to it in a TIN is less than δ_h. As a result, the data domain R of LiDAR points P located over a building rooftop are divided into a set of height clusters $\{R_i : i = 1, \ldots, N\}$ so that $R = \cup_{i=1}^{N} R_i$, $R_i \cap R_j = \phi$ if $i \neq j$, and P satisfies a homogeneity criterion on each R_i.

Planar Cluster Extraction: For reconstructing polyhedral building models, planar features play an important role in either extracting intersection lines or in optimally partitioning the data space, thereby reconstructing the rooftop topology across the extracted linear features. Thus, obtaining a reliable planar clustering result from LiDAR points is important. Following a method discussed by Vosselman et al. (2004), we implemented a hierarchical voting process in the parameter space for clustering planar features. A 3D plane passing through a point (x, y, z) in the data space can be represented by $x \cos \alpha + y \cos \beta + z \cos \gamma = \rho$. Here, (α, β, γ) is the plane normal vector angles measured between the plane and the x-, y-, and z-axis, respectively; satisfying $\cos^2 \alpha + \cos^2 \beta + \cos^2 \gamma = 1$ and ρ is the distance of the plane from the origin of the coordinate system. The voting process quantizes the plane parameter $(\alpha, \beta, \gamma, \rho)$ in a discrete parameter space with certain bin sizes. The position of a particular bin in the parameter space uniquely represents a plane in the data space. The plane parameter locally estimated from the data space votes for a corresponding bin in the parameter space. Thus, by searching maximum peaks in the parameter space, the dominant planes passing through laser points of interest can be found (Vosselman et al., 2004). The implemented plane clustering algorithm works on each height cluster independently. After assigning the plane parameters to the entire building, a plane adjacency graph is created for analyzing the connectivity of adjacent planes. The connected planes are merged if the plane parameters $(\alpha, \beta, \gamma, \rho)$ are almost the same and are recomputed using all of the points of the merged plane. This plane merging process continues until no similar adjacent planes are found.

17.3.2.3 Extracting Linear Features

An edge is a significant local change in LiDAR data, usually associated with a discontinuity in either height (step line) or principal curvature (intersection line). A multiple-story building only needs step lines to be modeled, while a gable-roof requires intersection lines to be modeled. For more complicated buildings with a mixture of various building primitives, both step and intersection lines must be extracted.

Step Line Extraction: A height jump line is produced where the height continuity changes abruptly over rooftop surfaces. The line can be found around the building boundary or the rooftops of multiple-story buildings, where two planes are adjacent to each other, but where there is a large height discontinuity. In the current experiment, we develop a straight line extractor, called a Compass Line Filter (CLF), for extracting height jump lines from irregularly distributed LiDAR points. The process starts by collecting step edges, which are points showing abrupt height discontinuities relative to their neighborhoods in the TIN. Since the proposed algorithm generates straight lines by approximating the step edges contour, obtaining a thinner contour yields a better approximation result. Thus, a thinning process is considered to a point P_c as the step edge if P_c satisfies two conditions: (1) its neighboring points P_n belongs to more than one height cluster and (2) P_c is the member of the highest plane. This relation can be described by a signed height difference function, $H(P_c, P_n) = P_c - P_n$, as follows:

$$\left| \max_{\forall P_N} H(P_c, P_N) \right| > \left| \min_{\forall P_N} H(P_c, P_N) \right| \tag{17.7}$$

Once all the step edges are extracted and grouped by each height cluster, the next step is to determine the local edge orientation for each step edge using a CLF. The local edge orientation provides directional information of the change of height discontinuity at a particular point. To this end, we developed a compass line operator shown in Figure 17.10b that has the whole set of eight filtering lines with different slopes $\{\theta_i : i = 1, \ldots, 8\}$, each of which, as illustrated in Figure 17.10c, is equally separated in steps of 22.5° (the first compass line corresponds to the horizontal line). Each line has two virtual boxes (line kernel) where length l is the same

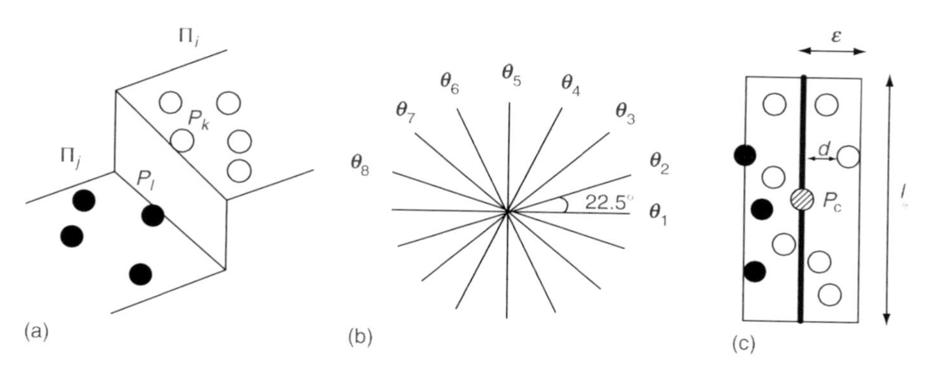

FIGURE 17.10
Compass line filter. (From Sohn, G., Huang, X., and Tao, V., *Photogrammetric Engineering and Remote Sensing*, in press, 2008. With permission.)

as the one of line and width corresponding to ε. Suppose that a height jump edge P_c belonging to a height cluster R_i is convolved with the whole set of eight line kernels. The local edge gradient is measured by the sum of distances d between the compass line θ_i and N kernel member points that are located within the line kernel K_i and belonging to R_i. A final local edge orientation θ^* is determined with the orientation of the kernel that yields the maximum local edge gradient. The values for the output orientation lie between one and eight, depending on which of the eight kernels had produced the maximum response. The computation of this local orientation for height jump edges is described by

$$G(\theta_i) = \sum_{k=1}^{N} \frac{2}{\left(1 + e^{\varepsilon d_k}\right)}, \quad \theta^* = \arg\max_{\forall \theta_i} \left(G(\theta_i)\right), \quad i = 1, \ldots, 8 \tag{17.8}$$

Once the local edge orientation is determined by the compass line operator, step edges with similar local edge orientation, belonging to the same height cluster, are grouped. The local edge orientation angle is explicitly assigned by a prespecified CLF. Finally the real orientations for θ_i are newly determined by the well-known Hough transformation.

Intersection Line Extraction: Once all of the planes are detected by the plane clustering process for each height cluster, the intersection lines are obtained by intersecting two adjacent planes. Thus, accuracy of the intersection line computation is subjective to the plane clustering result. The major problem of the plane clustering method presented in the previous section is that the process follows a winner-take-it-all approach, which does not consider a retrovalidation in the hierarchical voting scheme. To avoid this problem, a postvalidation process is conducted before extracting intersection lines. First, points located around the boundary between adjacent planes are collected, and then the orthogonal distances from each boundary point to the adjacent planes are measured. A true planar membership of the point is now determined by selecting the plane having the shortest distance measure of adjacent planes. This process continues until no boundary points change their planar membership. By reestimating the result of planar clustering, a more accurate intersection line computation can be obtained.

17.3.2.4 BSP-Tree Algorithm for Modeling Building Rooftops

Once the linear and planar features are extracted, a scheme of the BSP-Tree algorithm described in the Section 3.1.3 is applied for polygonal partitioning of LiDAR data.

The BSP-Tree framework is composed of three key components: (1) hierarchical binary partitioning, (2) hypothesis-test optimization, and (3) polygon merging. A rectangle bounding isolated building regions is regarded to be an initial building model, P_0. Both of the step and intersection lines extracted are integrated in order to make a list of hyperlines $\{h_i\}$. Entire member points within P_0 are attributed as one of plane labels. Note that terrain points classified by the building detection process are labeled as a single plane. Starting from a root node representing P_0, a BSP-Tree is hierarchically constructed until all leaf nodes cannot be partitioned any further. Compared to the previous method, a different partitioning criterion is used for controlling the generation of the BSP-Tree. Since the aim of the current system is to reconstruct polyhedral models, the criterion to govern the partitioning function is selected for obtaining maximal planar homogeneity over partitioned polygons, rather than for capturing maximum numbers of building points. Accordingly, the polygon classification to attribute tree nodes as either partitioning or terminating nodes is determined depending on the planar labeling property for a given polygon P_i. That is, the open polygon (partitioning node) is given to P_i if it contains different plane labels, while the closed polygon (terminating node) to P_i with single plane label. Based on the result of the polygon classification, it is determined to trigger or terminate the binary partitioning of P_i and, if partitioned, P_i is divided into two subpolygons, P_{i-} and P_{i+}, by intersecting P_i with a selected hyperline.

The partitioning result will be different when a different sequence of line segments is employed. Thus, it is necessary to have the hyperline selection criterion, which provides an optimal partitioning result over P_i. This optimality is achieved by the hypothesis-test scheme with a partition scoring function. The overall partitioning function is implemented similar to the previous method. However, rather than relying on a monotonic metric, the current objective function considers three different measurements to score a goodness of binary partitioning for each hyperline h_i. These three criteria include (1) plane homogeneity, (2) geometric regularity, and (3) edge correspondence.

Plane Homogeneity: This criterion controls hierarchical generation of the BSP-Tree by maximizing the numbers of points with similar planar properties for partitioned polygons. Suppose that an arbitrary polygon P_i contains points with N different plane labels $L_p = \{L_i : i = 1, \ldots, N\}$ by the plane clustering method presented in previous section. A representative plane label of P_i is determined as L_r, to which maximum numbers of plane labels of P_i are assigned. The plane labels L_p are then binarized into either foreground label L_{fore} or background label L_{back}. A member point of P_i is labeled as L_{fore} if its plane label corresponds to L_r, otherwise as L_{back}. As an arbitrary hyperline h_i partitions P_i into two subpolygons P_{i+} and P_{i-}, a higher score of plane homogeneity is given for h_i if maximum labeling homogeneity of both L_{fore} and L_{back} is obtained in each partitioned subpolygon. This normalized strength of the plane homogeneity SH is measured by

$$\begin{aligned} \mathrm{SH}(P_i;h_i) &= \max\left(\mathrm{SH}(P_{i+};h_i), \mathrm{SH}(P_{i-};h_i)\right) \\ \mathrm{SH}(P_{i+};h_i) &= \frac{1}{2}\left\{\frac{N_{\mathrm{fore}}(P_{i+};h_i)}{N_{\mathrm{fore}}(P_i;h_i)} + \frac{N_{\mathrm{back}}(P_{i-};h_i)}{N_{\mathrm{back}}(P_i;h_i)}\right\} \\ \mathrm{SH}(P_{i-};h_i) &= \frac{1}{2}\left\{\frac{N_{\mathrm{fore}}(P_{i-};h_i)}{N_{\mathrm{fore}}(P_i;h_i)} + \frac{N_{\mathrm{back}}(P_{i+};h_i)}{N_{\mathrm{back}}(P_i;h_i)}\right\} \end{aligned} \tag{17.9}$$

Edge Correspondence: A better partitioning result can be achieved when the boundary between two partitioned regions strongly corresponds to real edges. The strength of edge correspondence SE is measured by the ratio of lengths of a physical line l_i extracted from LiDAR data and the corresponding hyperline h_i. A higher score for the edge correspondence is assigned to h_i if a longer length of l_i is found in the polygon by

$$\text{SE}(P_i; h_i, l_i) = \frac{\text{Length}(l_i)}{\text{Length}(h_i)} \tag{17.10}$$

Geometric Regularity: Most building models have regular geometry (i.e., orthogonal, parallel, symmetric properly), rather than sharp corners. This heuristic preference toward the geometric regularity SG is measured by the minimum intersecting angle between P_i and h_i. A lower score is given for h_i when h_i intersects P_i with sharper intersecting angles; scores increase as the minimum intersecting angle increases. Note that the intersecting angles are measured only for one of two child polygons of P_i, which contains larger numbers of the foreground labels (Figure 17.11).

$$\begin{aligned} &\text{SG}(P_i; h_i) = \min(\text{Ang}(P_i; h_i)) \\ &\text{Ang}() = [0(0^\circ \le \vartheta \le 30^\circ),\ 0.5(30^\circ \le \vartheta \le 60^\circ),\ 1(60^\circ \le \vartheta \le 180^\circ)] \end{aligned} \tag{17.11}$$

All the hyperlines are tested to partition P_i, and the partitioning result generated by each hyperline is evaluated by a partition scoring function. A hyperline, h^*, with the highest partitioning score is finally selected to partition P_i by

$$h^* = \arg\max_{\forall\{h\}} (\alpha \times \text{SH}(P_i \mid h_i) + \beta \times \text{SL}(P_i \mid h_i, l_i) + \gamma \times \text{SG}(P_i \mid h_i)) \tag{17.12}$$

where $\alpha + \beta + \gamma = 1$ and usually (α, β, γ) is chosen as (0.5, 0.3, 0.2), respectively. After constructing the BSP-Tree, a plane adjacency graph is created by collecting the final leaves of the BSP-Tree, where each node represents a planar roof facet and each arc represents the connectivity between neighboring planes. Starting from the node with the largest area in the plane adjacency graph, a simple validation of normal vector compatibility is applied to its adjacent planes. The planes having similar normal vector angles

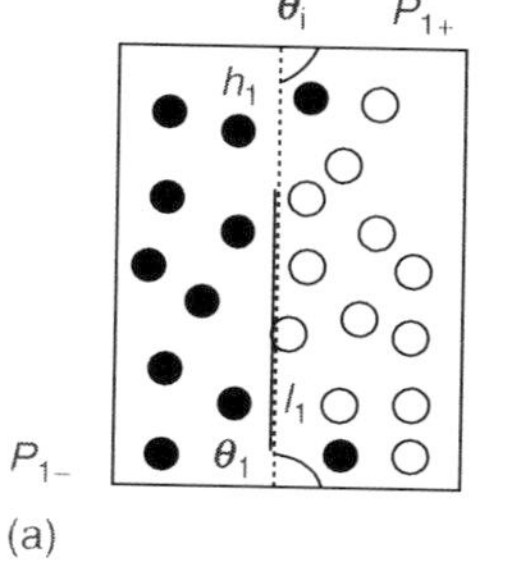

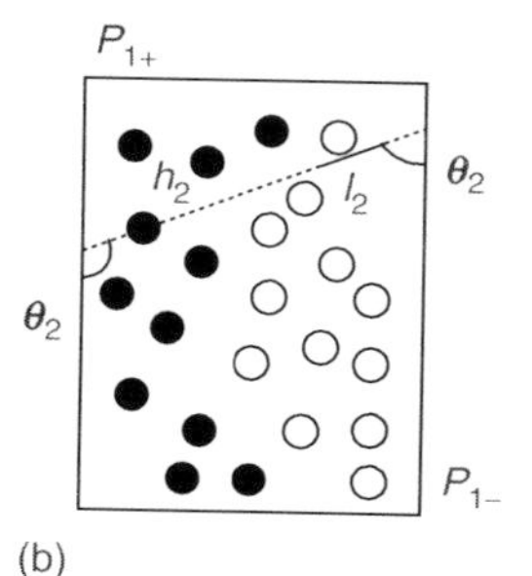

FIGURE 17.11
Partition scoring functions. (From Sohn, G., Huang, X., and Tao, V., *Photogrammetric Engineering and Remote Sensing*, in press, 2008. With permission.)

are merged and the planar parameters for merged planes are reestimated. This merging process continues until no plane can be accepted by the coplanar similarity test. Once all the polygons are merged together, building boundaries are reconstructed. Figure 17.12 shows a sequence of the building reconstruction results obtained by the method presented based on the BSP-Tree.

17.3.2.5 Experimental Results

In this section, we discuss and evaluate the performance of the building reconstruction technique that we propose here. Figure 17.13a shows an elevation model over downtown Toronto. The data was acquired by Optech Incorporated ALTM 2050 airborne laser scanner with a pulse repetition rate of 50 KHz at a flying height of 850 m above ground level. The Toronto test dataset covers approximately 1 km by 1 km where a total of 2,741,684 laser points were acquired with a mean density of 2.74 points/m^2 (~0.6 m point spacing). According to the sensor and flight specifications, 0.43 m horizontal and 0.15 m vertical accuracies are expected. The overall elevation is almost flat and the northern part of the data is slightly

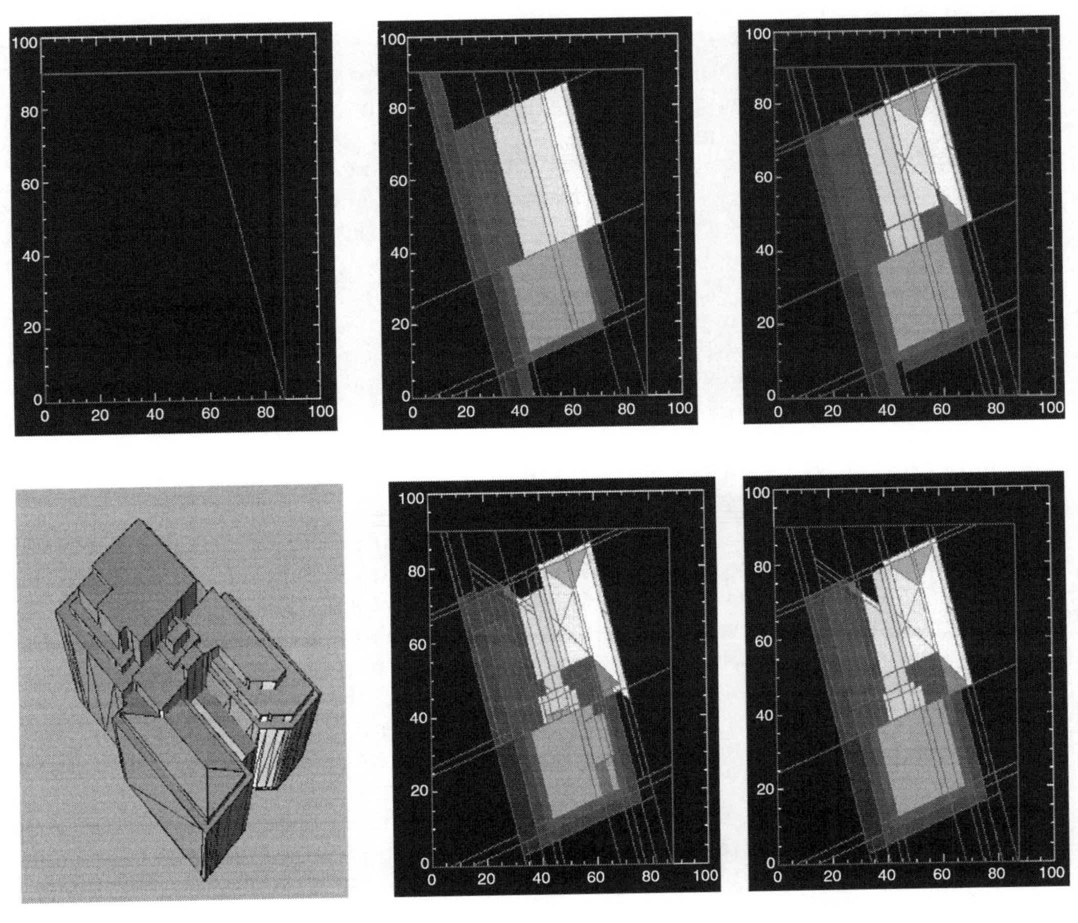

FIGURE 17.12
BSP building reconstruction results. (From Sohn, G., Huang, X., and Tao, V., *Photogrammetric Engineering and Remote Sensing*, in press, 2008. With permission.)

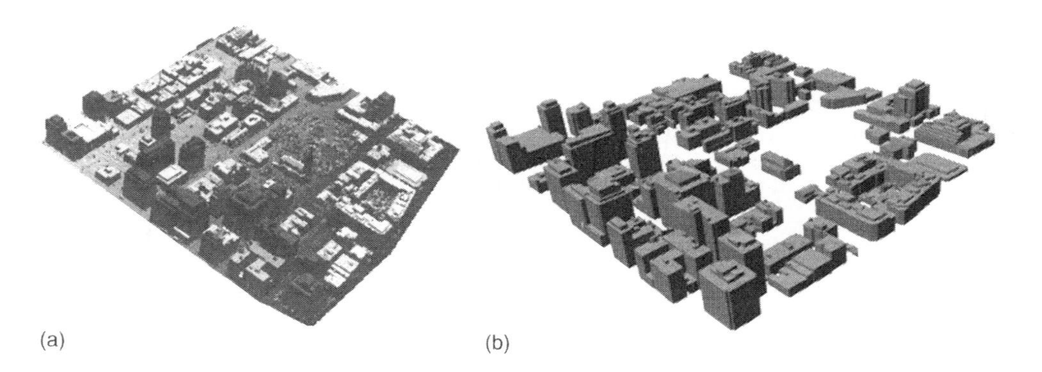

FIGURE 17.13
Toronto dataset and 3D building reconstruction results. (From Sohn, G., Huang, X., and Tao, V., *Photogrammetric Engineering and Remote Sensing*, in press, 2008. With permission.)

higher (≅100 m) than the southern area. The test scene was chosen to include many buildings with different orientations and roof structures; consequently, it is a good region for testing our reconstruction algorithm.

Figure 17.13b shows a final result of 3D building reconstruction from the Toronto dataset based on the proposed methods. As a prerequisite processing step for achieving the result shown in Figure 17.13b, a total of 53 buildings comprising 182,450 points were detected, each of which was bounded by an initial building model (i.e., rectangle). In Figure 17.13b, a total of 529 roof facets were reconstructed. Since the Toronto dataset includes buildings with a wide range of shape complexity, the number of roof facets reconstructed varies from one to 43 and an average of 10 planes per building, with a standard deviation of 12.3 planes. Out of a total 182,450 points, 94.48% (172,378 points) were recognized as planar points by the planar clustering algorithm (i.e., representing a particular planar roof-facet), while 5.52% (10,072 points) were not clustered as the planar points, but were eliminated during the planar clustering process. Also, 87.7% (160,078 points) of the planar points were actually utilized for the final reconstruction of the roof facets as the foreground points (those points are termed pure planar points), while 5.5% planar points (10,072 points), namely impure planar points, remained as the background points in under-partitioned roof facets.

The proposed algorithms were implemented and performed on a desktop PC with a Pentium-4 processor at a clockspeed of 2.40 GHz and with 512 MB RAM. We attempt to evaluate the efficiency of our method by measuring the total time consumed in reconstructing the entire test scene. The execution time of building reconstruction on this platform was a total of 461.1 s, which corresponds to 8.5 s per building. Most of the buildings (79% buildings) were reconstructed in less than 10 s, while the maximum execution time reached 85.3 s over a building, with the largest area being 8,270 m^2 and the maximum number of planes being 43. Note that this figure does not include all of the time taken for building detection (Figure 17.14a).

Since a ground truth of 3D building rooftops was not available over the test site, an absolute comparison of reconstruction accuracy between the ground truth and the reconstructed polygons was not possible. Thus, we examined the accuracy of the building reconstruction results in two relative measures by measuring the residuals of the pure planar points from a reconstructed plane (pure planar fitness) and the ratio of the number of the impure planar points against the total numbers of pure planar points in a reconstructed roof facet (planar impurity). The pure planar fitness will show a goodness-of-fit between observed data and reconstructed models, while the planar impurity can indicate

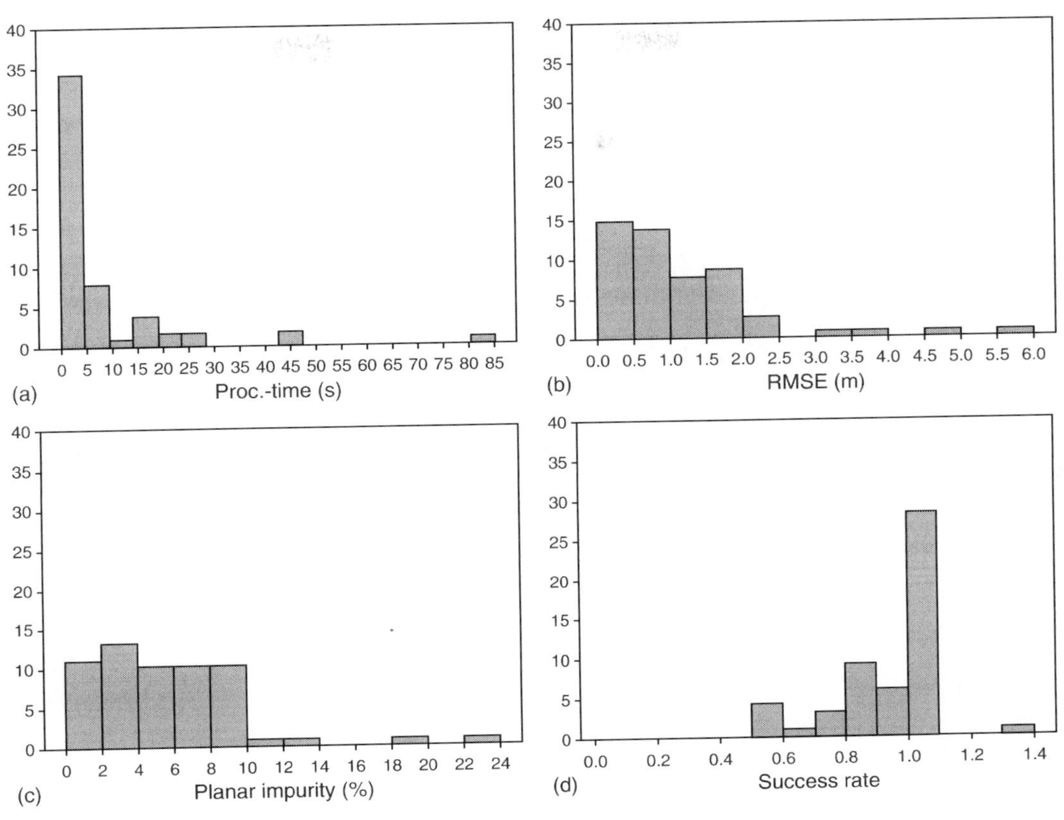

FIGURE 17.14
Building reconstruction performance. (From Sohn, G., Huang, X., and Tao, V., *Photogrammetric Engineering and Remote Sensing*, in press, 2008. With permission.)

a degree of planar homogeneity obtained by reconstructed building models. The root mean squared error of the pure planar fitness was measured as the mean $\mu = 0.38$ (m), and standard deviation $\sigma = 0.22$ m, while the average ratio of the planar impurity was measured as 5.1% with $\sigma = 4.4\%$ (Figure 17.14b and c). It was observed that most of the impure planar points were produced around roof boundary regions where adjacent roof facets were not correctly delineated by reconstructed models due to geometrically erroneous hyperlines or over roof facets, which were under-partitioned due to a lack of hyperlines extracted.

Finally we also evaluated the reliability of our building reconstruction system by comparing the number of automatically reconstructed roof facets to the system where buildings were manually counted. The success rate of the algorithm is defined by $R = 1 - (N_t - N_r)/N_t$. Here, N_r is the total number of planes reconstructed by our method, and N_t is the total number of planes that are visually found by a human operator. The success rate R will be larger than 1 if the algorithm over-reconstructs building roof facets, but is less than 1 if roof facets are under-reconstructed by the method. A perfect success rate of 1 will be achieved if no roof plane is missed by our method. Figure 17.14d shows that the success rate R can be achieved by our method as $\mu = 0.92$ ($\mu_{min} = 0.5$ and $\mu_{max} = 1.3$) with $\sigma = 0.2$. This suggests that when a building is comprised of 10 roof facets, our method, on average, fails to reconstruct approximately one plane. As seen in the figure, the algorithm tends to under-reconstruct building rooftops, rather than over-reconstruct the models.

It has been visually confirmed that the over-reconstruction of building models occurred in cases when either a small flat roof or superstructure was hard to distinguish in height from the neighboring large-area planes during the height clustering process; or step lines, usually having a short length of 1–3 m, could not be detected by the edge detector due to high irregular density of LiDAR data.

In Figure 17.14b and c, we visualized several examples of 3D buildings reconstructed by the proposed building reconstruction system. The buildings in this figure show very complicated shapes, where several building primitives are horizontally or vertically connected to each other. Also, it is evident that all of the buildings in the figures contain large amounts of data acquisition errors due to the small coverage. In fact, 3D structures of the buildings presented in these figures were not very easily interpreted by visual investigation, either from the optical imagery or LiDAR data. Although the shape complexity and data irregularity are high, Figure 17.14d showed that the overall quality of the reconstruction result based on the proposed technique is satisfactory, and that the reconstructed models effectively preserve the original geometry. Also the reconstruction results were achieved without requiring a priori knowledge of specific building models or of the topological relations between adjacent planes. Thus it can be concluded that the developed techniques would be useful for quickly providing an initial, yet fine rooftop model with very high shape complexity. However, it is also evident that the reconstructed models are not perfect. As seen in Figure 17.15a, details of the structures around the corners were lost and most of them were generalized too much as rectangular corners. Also, a superstructure in the middle of the building was wrongly reconstructed, although the original shape of this structure is not certain, even by visual investigation. Those problems may be caused by a significant lack of point density covering those particular regions, which results in missing linear features.

17.4 Conclusion

It is difficult task to extract sufficient cues to describe a complete object shape from remotely sensed data including airborne LiDAR data. This is because information acquired by the sensors is not normally disturbed by several factors such as the illumination condition, resolution effect, irregular point density, object complexity, and disadvantageous imaging perspective. This chapter has presented a new technique to recover a full delineation of 3D building shape from incomplete geometric features extracted from airborne LiDAR data. In the current research, this task has been achieved through the development of a unique data-driven framework that employs the BSP-Tree algorithm. The BSP-Tree method was used to recursively partition a given LiDAR data domain with linear features into a set of convex polygons. Each polygon is required to maximally satisfy a homogenous condition characterizing major parts of a building shape. A generic building model was reconstructed by merging portioned convex polygons and thus implicitly grouping fragmentary linear features without relying on prespecified parametric models or a number of heuristic grouping rules. The results showed that the method can represent either prismatic models or highly complicated polyhedral buildings, where multiple building primitives are horizontally or vertically connected with, and occluded from, each other. The prismatic building was reconstructed using LiDAR data with low and very irregular point density in a combination of high-resolution satellite imagery, while the polyhedral models were created only using LiDAR data with high point density. Starting from a coarse description, the developed techniques incrementally refine the model at

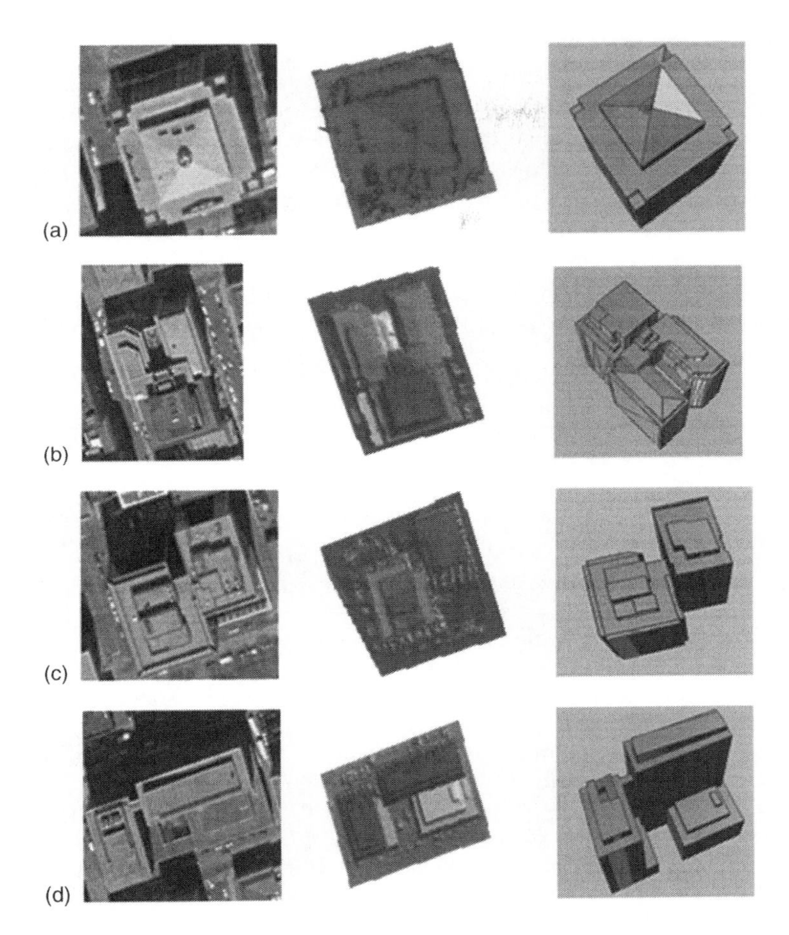

FIGURE 17.15
Building reconstruction results for multiple roof primitives. (From Sohn, G., Huang, X., and Tao, V., *Photogrammetric Engineering and Remote Sensing*, in press, 2008. With permission.)

different modeling scales by maximizing the coplanarity, edge correspondence, and geometric regularity. By this nature of global optimization, the system can simultaneously reconstruct a large number of connected roof facets, but does not require a local analysis of the primitive connectivity between adjacent facets that are independently generated. These characteristics are the most beneficial compared to existing model-driven techniques which may not provide good results under these circumstances. Another advantage of the BSP-Tree algorithm, which has not been studied yet in the current analysis, is the ability to combine a search structure with a representation scheme into one unique data structure. For instance, a building model can be represented in a different scale (i.e., by controlling the depth of BSP-Tree). This useful property can be further exploited for generalizing building shapes according to the level of detail. Due to the irregular nature of laser scanning and building occlusion problems, it is usually difficult to provide a complete set of linear features for representing building models from LiDAR data. The proposed BSP-Tree algorithm provides an alternative approach to the traditional feature-grouping approaches to overcome the difficult problem of predicting when linear features could be fragmented into an unpredictable order.

Acknowledgments

We would like to express our gratitude to Professor Ian Dowman at University College London, who advised the development of the BSP-Tree algorithm in 3D building reconstruction as part of Dr. Gunho Sohn's doctoral project at UCL. Also, we would like to acknowledge the support of Professor Peter Muller at UCL, Infoterra Ltd, United Kingdom, Ordnance Survey, United Kingdom, and Optech International, Canada, who kindly provided the Greenwich datasets and OS MaterMap and Toronto datasets respectively for this research. We gratefully acknowledge the support of the GEOIDE Phase III (Project #12) while writing this chapter.

References

Alharthy, A. and Bethel, J., 2004. Detailed building reconstruction from airborne laser data using a moving surface method. *International Archives of Photogrammetry and Remote Sensing*, 35(Part B3):85–92.

Ameri, B., 2000. Automatic recognition and 3D reconstruction of buildings from digital imagery. PhD thesis, Institute of Photogrammetry, University of Stuttgart, 110 pp.

Axelsson, P., 1999. Processing of laser scanner data - algorithms and applications. *ISPRS Journal of Photogrammetry and Remote Sensing*, 54(2–3):138–147.

Baltsavias, E., Mason, S., and Stallman, D., 1995. Use of DTMs/DSMs and orthoimages to support building extraction. In A. Grün, O. Kübler, and P. Agouris (Eds.), *Automatic Extraction of Man-Made Objects from Aerial and Space Images*, pp. 199–210, Birkhäuser-Verlag, Basel.

Baillard, C. and Maitre, H., 1999. 3D reconstruction of urban scenes from aerial stereo imagery: A focusing strategy. *Computer Vision and Image Understanding*, 76(3):244–258.

Brenner, C., 2000. Towards fully automatic generation of city models. *International Archives of Photogrammetry and Remote Sensing*, 33(Part B3):85–92.

Brenner, C. and Haala, N., 1998. Fast production of virtual reality city models. *International Archives of Photogrammetry and Remote Sensing*, 32(4):77–84.

Brunn, A. and Weidner, U., 1998. Hierarchical Bayesian nets for building extraction using dense digital surface models. *ISPRS Journal of Photogrammetry and Remote Sensing*, 53:296–307.

Burns, J.B., Hanson, A.R., and Riseman, E.M., 1986. Extracting straight lines. *IEEE Pattern Analysis and Machine Intelligence*, 8(4):425–455.

Chen, L., Teo, T., Shao, Y., Lai, Y., and Rau, J., 2004. Fusion of Lidar data and optical imagery for building modelling. *International Archives of Photogrammetry and Remote Sensing*, 35(Part B4):732–737.

Ducic, V., Hollaus, M., Ullrich, A., Wagner, W., and Melzer, T., 2006. 3D vegetation mapping and classification using full-waveform laser scanning. *In workshop on 3D Remote Sensing in Forestry*, February 14–15, 2006, pp. 211–217, Vienna, Austria.

Filin, S., 2002. Surface clustering from airborne laser scanning data. *International archives of Photogrammetry and Remote Sensing*, 34(Part 3):119–124.

Fowler, R., 2001. Topographic Lidar. In D.F. David (Ed.), *Digital Elevation Model Technologies and Applications: The DEM Users Manual*: pp. 207–236. American Society for Photogrammetry and Remote Sensing, Bethesda, MD.

Fuchs, H., Kedem, Z.M., and Naylor, B.F., 1980. On visible surface generation by a priori tree structures. *Computer Graphics*, 14(3):124–133.

Gordon, D. and Chen, S., 1991. Front-to-back display of BSP trees. *IEEE Computer Graphics and Applications*, 11:79–85.

Guo, T. and Yasuoka, Y., 2002. Snake-based approach for building extraction from high-resolution satellite images and height data in urban areas. Proceedings of the 23rd Asian Conference on Remote Sensing, November 25–29, Kathmandu. 7 pp (on CD-ROM).

Hofmann, A.D., 2004. Analysis of TIN-structure parameter spaces in airborne laser scanner data for 3-d building model generation. *International Archives of Photogrammetry and Remote Sensing*, 35(B3):302–307.

Hu, Y., 2003. Automated extraction of digital terrain models, roads and buildings using airborne lidar data, PhD dissertation, University of Calgary, Calgary, AB, 206 pp.

Hug, C. and Wehr, A., 1997. Detecting and identifying topographic objects in imaging laser altimetry data. *International Archives of Photogrammetry and Remote Sensing*, 32(Part 3–4):19–26.

Kim, J.R. and Muller, J.-P., 2002. 3D reconstruction from very high resolution satellite stereo and its application to object identification. *International Archives of Photogrammetry and Remote Sensing*, 34(Part 4): 7 pp (on CD-ROM).

Kraus, K. and Pfeifer, N., 1998. Determination of terrain models in wooded areas with airborne laser scanner data. *ISPRS Journal of Photogrammetry and Remote Sensing*, 53:193–203.

Lee, D.S., Shan, J., and Bethel, J.S., 2003. Class-guided building extraction from IKONOS imagery. *Photogrammetric Engineering and Remote Sensing*, 69(2):143–150.

Maas, H.-G., 2001. The suitability of airborne laser scanner data for automatic 3D object reconstruction. In A. Grün, E.P. Baltsavias and O. Henricsson (Eds.), *Automatic Extraction of Man-Made Objects from Aerial and Space Images (III)*: pp. 291–296. Balkema, Lisse, The Netherlands.

Maas, H.-G. and Vosselman, G., 1999. Two algorithms for extracting building models from raw laser altimetry data. *ISPRS Journal of Photogrammetry and Remote Sensing*, 54(2–3):153–163.

Parish, B., 2002. Quality assessment of automated object identification in IKONOS Imagery. M.Sc. Thesis, University of London, London.

Rottensteiner, F., Trinder, J., Clode, S., and Kubik, K., 2005. Using the Dempster-Shafer method for the fusion of LiDAR data and multi-spectral images for building detection. *Information Fusion*, 6(4):283–300.

Schwalbe, E., Maas, H.-G., and Seidel, F., 2005. 3D building model generation from airborne laser scanner data using 2D GIS data and orthogonal point cloud projections. *International Archives of Photogrammetry and Remote Sensing*, 36(Part 3/W19):209–214.

Sithole, G. and Vosselman, G., 2004. Experimental comparison of filter algorithms for bare-Earth extraction from airborne laser scanning point clouds. *ISPRS Journal of Photogrammetry and Remote Sensing*, 59 (1–2):85–101.

Sohn, G. and Dowman, I., 2007. Data fusion of high-resolution satellite imagery and LiDAR data for automatic building extraction. *ISPRS Journal of Photogrammetry and Remote Sensing*, 62:43–63.

Sohn, G. and Dowman, I., 2008. A model-based approach for reconstructing a terrain surface from airborne lidar data. *The Protogrammetric Record*, 23(122):170–193.

Sohn, G., Huang, X., and Tao, V., 2008. Using a binary space partitioning tree for reconstructing polyhedral building models from airborne LiDAR data. *Photogrammetric Engineering and Remote Sensing*, In press.

Vosselman, G., 1999. Building reconstruction using planar faces in very high density height data. *International Archives of Photogrammetry and Remote Sensing*, 32(Part 2):87–92.

Vosselman, G., Gorte, B.G.H., Sithole, G., and Rabbani, T., 2004. Recognising structure in laser scanner point clouds. *International Archives of Photogrammetry and Remote Sensing*, 36(8/W2):33–38.

Vosselman, G. and Suveg, I., 2001. Map based building reconstruction from laser data and images. In A. Grün, E.P. Baltsavias, and O. Henricsson (Eds.), *Automatic Extraction of Man-Made Objects from Aerial and Space Images (III)*: pp. 231–242. Balkema, Lisse, The Netherlands.

Wang, Z. and Schenk, T., 2000. Building extraction and reconstruction from lidar data. *International Archives of Photogrammetry and Remote Sensing*, 33(B3):958–964.

Weidner, U. and Forstner, W., 1995. Towards automatic building reconstruction from high resolution digital elevation models. *ISPRS Journal of Photogrammetry and Remote Sensing*, 50(4):38–49.

World Resources Institute, 1996. World Resources 1996–1997: The Urban Environment. Oxford University Press, New York/Oxford.

18

A Framework for Automated Construction of Building Models from Airborne LiDAR Measurements

Keqi Zhang, Jianhua Yan, and Shu-Ching Chen

CONTENTS

18.1 Introduction

Building footprint, height, volume, and three-dimensional (3D) shape information can be used to estimate energy demand, quality of life, urban populations, property tax, and surface roughness [1]. Three-dimensional building models are essential for 3D city or urban landscape models, urban flooding prediction, and assessment of urban heat island effects. Building models can be divided into two categories: simple and sophisticated. Geometric attributes for a simple building model consist of a footprint polygon and a height value.

The geometric attributes for a sophisticated building model include not only a footprint polygon but also planes or other types of surfaces for various parts of the roof as well as their projections (polygons) on the ground plane. Only one fixed building height exists for a simple model, while building heights of the sophisticated model are variable. The advantage of the simple model is that it requires fewer geometric attributes to delineate a building. The buildings are represented by various types of 3D boxes; therefore, the 3D rendering is fast. Many commercial GIS software packages such as ArcGIS (www.esri.com) can display simple building models effectively. The simple building model is sufficient for applications such as numerical modeling of urban flooding and heat island effect, estimating urban population and energy demand, and large-scale 3D visualization, all of which do not require the details of buildings. The key to extracting a simple building model from LiDAR measurements is to derive footprints. The building height value can be represented using statistical height values such as mean and median of LiDAR measurements within a footprint.

The disadvantage of the simple building model is the lack of detail and accuracy of building shapes. The sophisticated models overcome this limitation by offering more geometric information for 3D buildings. The sophisticated models are required by applications such as hurricane wind damage models for individual properties, property tax estimation, and detailed urban landscape modeling. The disadvantage of sophisticated models is that the 3D rendering is slow. Most existing commercial GIS software cannot display sophisticated models efficiently due to compound geometric structures.

High-resolution data needed for extracting 3D building models can be derived through airborne light and detection (LiDAR) mapping systems. However, airborne LiDAR systems generate voluminous and irregularly spaced 3D point measurements of objects, including ground, building, trees, and cars scanned by the laser beneath the aircraft. The sheer volumes of point data require dedicated algorithms for automated building reconstruction. This paper focuses on presenting a framework for the construction of simple and sophisticated building models from LiDAR measurements. The chapter is arranged as follows. Section 18.2 reviews previous work. Section 18.3 describes the algorithms that derive building models. Section 18.4 describes the sample LiDAR data set used in this study. Section 18.5 examines the results by applying the proposed framework to the sample data set and Section 18.6 includes conclusions.

18.2 Literature Review

18.2.1 Separation of Ground and Nonground Points

The critical step in building model construction is to identify building measurements from LiDAR data. Two ways are often utilized to identify building measurements from LiDAR data. One is to separate ground, buildings, trees, and other measurements from LiDAR data simultaneously using segmentation. Examples of this method can be found in Maas and Filin and Pfeifer [2,3].

The other way is to separate the ground from nonground LiDAR measurements first and then identify the building points from nonground measurements. Numerous algorithms have been developed to identify ground measurements from LiDAR data. For example, Vosselman proposed a slope-based filter to remove nonground measurements by comparing slopes between a LiDAR point and its neighbors [4]. Shan and Sampath applied a slope-based 1D bidirectional filter to LiDAR measurements along the cross track direction to label nonground points [5]. Zhang et al. used mathematical morphology to identify ground measurements [6]. The nonground measurements can be simply derived by removing identified ground data from a raw data set. To facilitate the data processing, digital surface models (DSMs) interpolated from raw LiDAR measurements and digital terrain model (DTM) interpolated from ground measurements are often produced. The image for nonground objects is derived by subtracting DTM from DSM. Examples using this method to derive nonground objects can be found in Refs. [7–10].

18.2.2 Building Measurement Identification

The next step is to extract building point measurements or pixels in the data set for nonground objects, which are dominated by trees and buildings. The distinct difference between buildings and trees is that the roof surfaces are approximately planar, while canopy surfaces are irregular. Several parameters based on this difference have been proposed for segmenting buildings and trees. For example, the first derivatives of heights are either a zero (flat roof) or constant (sloped roof) for a planar surface, and the second derivatives of a sloped planar surface are zero. The first and second derivatives of an irregular surface should be variable. Morgan and Tempfli applied Laplacian and Sobel operators to height surfaces to separate building and tree measurements [8]. The problem with this method is that the derivatives from LiDAR measurements for roofs are not constant because of measurement errors. Small features such as chimneys, water tanks, and pipes on a roof surface can also produce abnormal derivative values. In addition, the derivatives at the edge of a building have a large variation, making it difficult to separate buildings from adjacent trees.

Another technique to separate building and tree measurements involves using a least squares method to estimate the parameters for a plane that fits a LiDAR point and its neighbors within a local window [11–13]. It is expected that the deviations of roof points from their fitted planes will be small and the plane parameters will be similar and consistent, while deviations and plane parameters for tree points will be large and variable. Compared to derivatives, the plane parameters are less sensitive to individual outliers caused by chimneys and water tanks. The drawback of this method is that plane parameters are not robust at the boundary of the buildings because fewer points are available for parameter estimation.

Many airborne LiDAR systems are capable of deriving the first and last return measurements produced by multiple reflections of a laser pulse by the objects on the earth's surface. The height difference between the first and last return measurements can be used to separate building and tree measurements [14]. The height difference is usually large for tree measurements and close to zero for building measurements. A measurement is identified as a building measurement by comparing its height difference to a predefined threshold. However, this method does not work for areas covered by dense trees where laser pulses cannot penetrate. In addition, the elevation difference between the first and last returns less than 3–4 m is not reliable because of the influence of the laser pulse width, which is typically 10 ns [15].

Hough transform has also been used to separate building points from tree measurements or to identify them directly from a raw LiDAR data set [9,16,17]. Data in the physical space are transformed to and analyzed in the parameter space. The advantage of the Hough transform is its tolerance of gaps in the feature boundary. However, it is difficult to define the optimum cell size for voting in the parameter space, which is influenced by the error range of LiDAR measurements, the sampling density, and local height changes of a roof surface. Unfortunately, local height changes of individual surfaces are different; thus it is difficult to quantify these changes using a single value. Usually, the cell size is set empirically, and if the cell size is too large, several real-world planes could be merged into one during building identification. In contrast, if the cell size is too small, one real-world plane could be split into several smaller planes. In addition, the adjacency of point measurements is not considered by the Hough transform. Therefore, LiDAR measurements for separated but closely adjacent buildings with the same height could be mixed into the same category.

The LiDAR measurements for buildings and trees can be segmented based on one or more of the above-mentioned parameters. There are two ways to perform the segmentation task. One is the point- or pixel-level classification. Maas employed raw elevation, Laplace filtered height, and maximum slope from LiDAR measurements to perform a supervised maximum likelihood classification [2]. Filin separated LiDAR measurements for buildings, vegetation, ground, and other features using an unsupervised classification based on the position of a point, the parameters of the tangent plane to the point, and the relative height difference between the point and its neighbors [11]. Elberink and Maas used unsupervised *k*-means classification in their texture-based segmentation. The problem with a point-level classification is that the measurements for a building cannot be guaranteed to classify into the same category. Also, the selection of training data sets for a supervised classification can be very time consuming [18]. An alternative way is to find a building area using a region-growing algorithm based on a seed point [12,13]. This method considers the adjacency of LiDAR measurements and the robustness of region-growing processes [13].

18.2.3 Building Model Creation

After measurements for a building are identified, a raw footprint can be derived by connecting boundary points of LiDAR measurements for a building. The raw footprint polygon has to be generalized for building models because the raw footprint includes considerable noise due to irregularly spaced LiDAR points. Alharthy and Bethel employed the histogram of boundary points to generalize the footprint edges by assuming that the buildings have only two dominant directions that are perpendicular with each other [14]. Based on the same assumption, Sampath and Shan used the least squares model to regularize the footprint edges [19]. Recently, Zhang et al. have published a method to refine a footprint iteratively based on estimated dominant directions [13]. Footprints with oblique edges, which are not perpendicular to dominant directions, are allowed in this method as long as the total length of the oblique edges is less than the total length of the edges parallel or perpendicular to the dominant directions.

Simple building models can be derived by adding a uniform height value once a refined building footprint is derived. However, the process to derive sophisticated building models is more complicated. Schwalbe et al. categorized the methods to derive 3D building models into two categories: model-driven and data-driven [20]. In the model-driven method, building models are identified by fitting predefined models into the LiDAR measurements For example, Maas and Vosselman estimated parameters for primitive building models based on the invariant moment analysis [21]. Brenner extended this method to complex buildings by splitting a building into simple primitives first and then fitting individual primitives using point clouds [22,23]. However, building models for a study area are not always available in advance, which limits the application of the model-driven method.

In the data-driven method, building measurements are grouped first for different roof planes. Then, the 2D topology of each building, which is represented by a set of connected planar roof surfaces projected onto a horizontal plane, is derived. Rottensteiner and Jansa approximated pixels on edges with line segments first and then intersected these line segments to derive the vertices of the 2D topology [9]. Alharthy and Bethel derived the polygon for each roof plane by connecting the boundary points and simplifying the polygon edges using the Douglas–Peucker algorithm [14,24]. Polygons for each roof plane are then snapped into the 2D topology. Since neighboring roof polygons may overlap or be separated from each other, it is not easy to snap the neighboring polygons together.

Raw 3D building models can be directly created from the derived 2D topologies and identified roof planes. However, the quality of such building models is poor because a 2D topology is often noisy due to irregularly spaced LiDAR measurements. Uncertainty in estimated roof plane parameters due to errors in LiDAR measurements and segmentation also distort raw building models. A refinement of 2D topology and roof plane parameters is often needed to derive high quality building models. Many geometric constraints have been proposed to regularize and refine the 2D topology. Gruen and Wang proposed seven constraints formulated into weighted observation equations, and they enforced these constraints by using least squares adjustment (LSA) [25]. Before LSA is applied, points, lines, or planes related to each constraint should be grouped together manually, a step that limits the application of this method. Other methods realized the difficulty in enforcing many constraints simultaneously and only enforced important constraints, especially parallelism. The parallelism constraint assumes that a building has two dominant directions perpendicular to each other, and the building edges on the 2D topology should be parallel to either of the dominant directions [17]. However, this assumption is too strict and cannot be applied to buildings with edges oblique to both dominant directions.

The main objective of this chapter is to present a framework involving a series of algorithms for the extraction of simple and sophisticated building models from LiDAR measurements (Figure 18.1). The framework consists of three major steps. First, the nonground and ground measurements are separated. Second, building measurements are identified by region growing using a local plane-fitting technique. Third, simple building models are derived and adjusted based on estimated dominant directions, and sophisticated building models are derived and refined based on the 2D topology of roof facets.

18.3 Construction of Building Models

18.3.1 Separating Ground and Nonground Measurements

The first step in the proposed framework for constructing building models is to separate ground and nonground measurements. We selected the progressive morphological filter for this task because this filter identified the ground and nonground measurements well for the sample data set used in this chapter [6]. Other filters can also be used in this step if they produce a good classification. The progressive morphological filter classifies ground and nonground measurements by comparing the elevation difference between interpolated and filtered surfaces. By gradually increasing the window size, the filter separates the measurements for different-sized nonground objects from ground data.

A 2D array, whose elements represent points falling in cells of a mesh overlaying the data set, is employed to facilitate the filtering and building identification computations. The cell size (c_s) of the mesh is usually set to be less than the average spacing of the LiDAR points to reduce information loss. Each point measurement from the LiDAR data set is assigned to a cell in terms of its x and y coordinates. If more than one point falls in the same cell, the point with the lowest elevation is selected as the array element. If no point exists in a cell, the array element for the cell is assigned as its nearest neighbor.

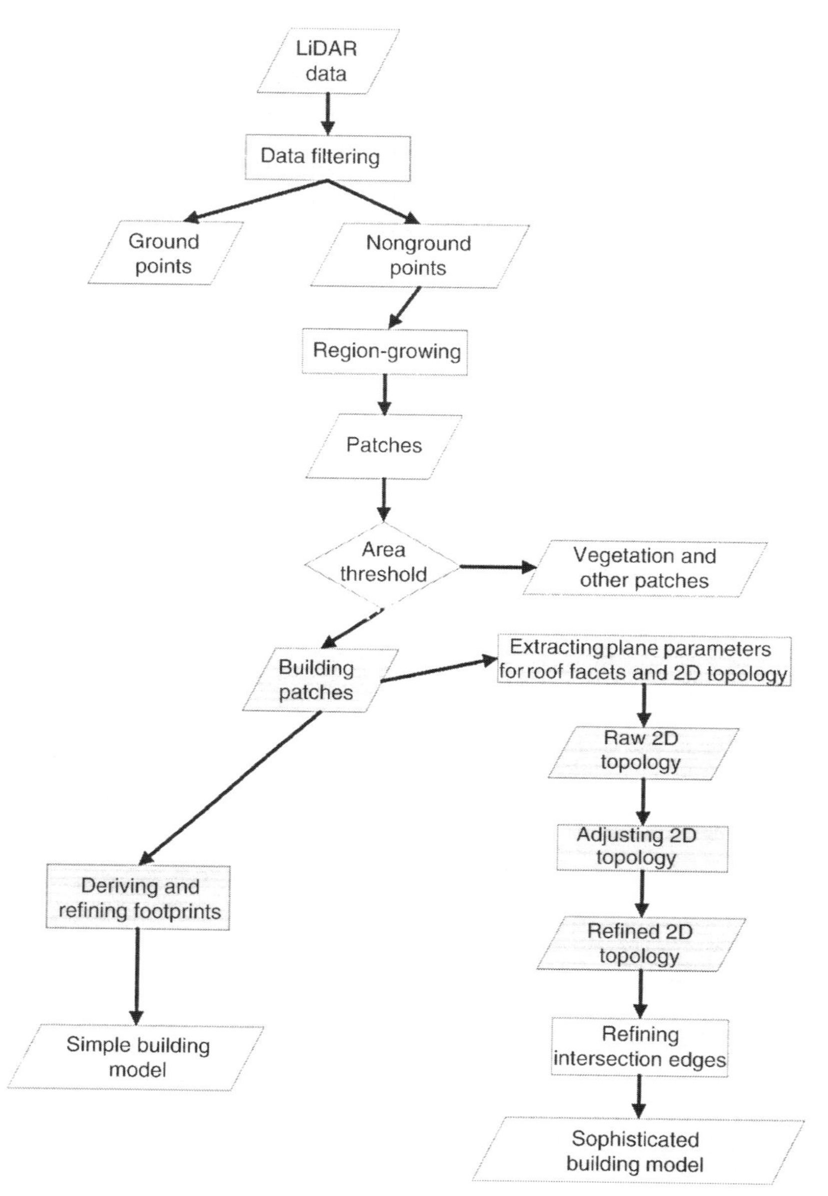

FIGURE 18.1
Framework for reconstruction of 3D building models from LiDAR measurements.

18.3.2 Segmenting Nonground Measurements and Identifying Building Patches

The second step of the framework is to separate building measurements from vegetation measurements using a region-growing algorithm based on a plane-fitting technique. The rationale behind this algorithm is to group nonground measurements, which are located on the same planes, into patches. Building patches are much larger in size because measurements for a roof facet are almost always located in the same plane, while vegetation patch sizes are small due to large changes in elevations for an irregular canopy. Region-growing segmentation of nonground measurements starts first with the selection of the seed point for a group. Then, the neighbors of the seed point are recursively added into the group by comparing the elevations of the neighboring points to the plane generated by fitting points in the group using the least squares method.

Both triangular irregular network (TIN) and grid data structures can be used to represent nonground measurements for segmentation. Only algorithms based on grid data structure are presented in detail in this chapter because the method based on TIN is similar to the method based on the grid. The only difference is that the region-growing algorithm based on a grid data structure recursively expands a building patch using the eight neighbors of a cell, while the region-growing algorithm based on a TIN structure expands a building patch using the three neighbors of a triangle. The region-growing algorithm for a grid data structure is presented as follows:

1. Create a set of areas for connected nonground measurements $A=\{S_1, S_2, \ldots, S_N\}$. We start an area creation by selecting any nonground measurement (typically it is the most left nonground measurement) as a seed nonground measurement. An area is created by connecting recursively nonground measurements among the eight neighbors of the seed measurement. The creation of an area stops when no additional nonground neighbors can be connected. The area creation process is repeated until every nonground measurement is included in S_i.
2. Partition S_i into PI_i, which represents a set of inside points, and PB_i, which represents a set of boundary points. If at least one of the eight neighbors of a point is a ground measurement, the point is defined as a boundary point. Otherwise, the point is an inside point.
3. Remove S_i from A if the area of S_i < predefined min_surface or the number of inside points = 0
4. For each S_i in A
   (a) Given an inside point $p_k(x_0, y_0, z_0)$, a Cartesian coordinate system (x, y, and z) is established using p_k as the origin. In this coordinate system, a best fitting plane for p_k and its eight neighbors is derived by using the least squares method. Assume that the plane is defined by

$$z = ax + by + c \tag{18.1}$$

   The parameters (a, b, c) can be derived by minimizing the sum of squares due to deviations (SSD), which is represented by ssd_k

$$\text{ssd}_k = \sum_{(p_k)\in K} (z_k - h_k)^2 \tag{18.2}$$

   where
   K is a set for p_k and its neighbors
   h_k and z_k are observed and plane-fitted surface elevations, respectively

   We derive a set $SSD_i = \{\text{ssd}_1, \text{ssd}_2, \ldots, \text{ssd}_{Ni}\}$, which consist of minimized ssds for all inside points in S_i.

   (b) While $SSD_i \neq \varnothing$ (empty), sort ssd in SSD_i in the ascending order, set $P_j=\varnothing$, and set $j=1$ for the first time.

(c) Select the point p_k with minimum ssd_k as a seed point for region-growing and add p_k to a set P_j.
(d) Label the neighbors of a seed point by examining whether they belong to the same set P_j through a plane-fitting technique. A plane is constructed based on the points in the category using a least squares fit. If the elevation from a neighbor point to this plane is less than a predefined threshold Δh_T, the neighbor is added to P_j. Δh_T is determined by the elevation error of the LiDAR survey and is usually 15–30 cm. The process is continued until no point can be added to P_j.
(e) Remove points in P_j from S_i and corresponding ssds from SSD_i, Go back to step (b).
(f) Continue the above process until $S_i = \emptyset$.

5. Let P represent a set of patch P_j derived from the above process, we have $P = \{P_1, P_2, \ldots, P_M\}$ and $P_j = \{p_{1j}, p_{2j}, \ldots, p_{Nj}\}$, where p_{kj} is a kth point in patch j. Remove P_j from P if the area of P_j is less than a predefined threshold min_surface, and P_j is not completely enclosed by other patches.
6. Merge patches in P if they are connected and derive a new set P_M. Through this process, adjacent roof surfaces from the same building, having different slopes, are merged into a large building patch. Remove P_{Mj} from P_M if the area of P_{Mj} is less than a predefined threshold min_building. The remaining patches in P_M are building patches.

The above algorithm indicates that the identification of building patches depends on the region-growing process, which, in turn, relies on the selection of seed points. Zhang et al. has demonstrated that the region-growing process is robust when starting with a seed point with a minimum SSD [13].

18.3.3 Deriving Simple Building Models

Raw footprints are derived by connecting the boundary points for building patches. The boundary of a raw footprint contains noise because of the irregularly spaced LiDAR point measurements (Figure 18.2b). A refinement of the footprint boundary is needed to reduce the noise. First, the Douglas–Peucker algorithm is employed to generalize line segments in the footprint [24]. This algorithm works well in reducing the noise, but it removes critical corner vertices of the raw footprint in some cases, leading to an orientation distortion of the segments in the footprint (Figure 18.2c). In order to recover the removed critical vertices and the distorted segments, the dominant directions of building footprints need to be estimated.

A method based on weighted line segment lengths is used in this study to estimate the dominant directions of a building footprint. Let x' and y' represent possible dominant directions in a 2D coordinate system x and y (Figure 18.3). The dominant directions x' and y' are related to the coordinate system x and y through a counterclockwise rotation by an angle φ ($0° \le \varphi < 90°$). Therefore, the key step to estimate the dominant directions is to find the rotation angle φ. Assuming that the counterclockwise intersection angle between a line segment and x axis is θ_i($0° \le \theta_i < 180°$), we define

$$\text{SL} = \sum_{i=1}^{N-1} g(L_i) f(\beta_i(\theta_i, \varphi)) \tag{18.3}$$

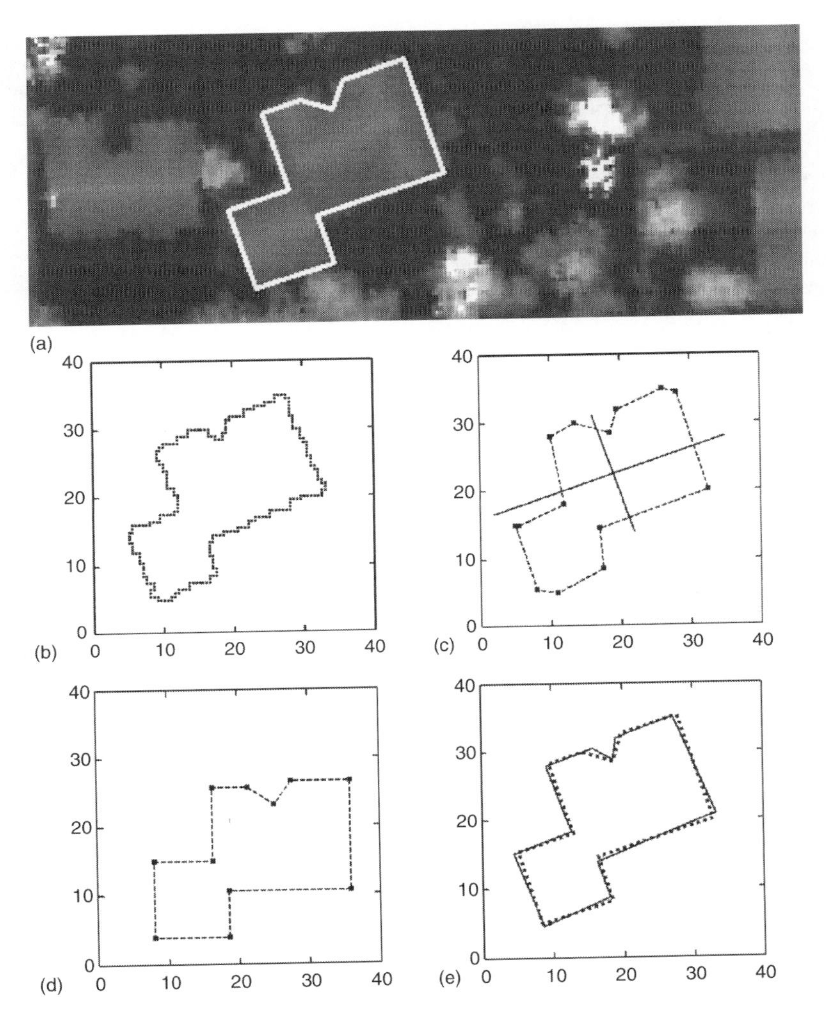

FIGURE 18.2
Example to illustrate the reconstruction process of a simple building model. The x and y coordinates are in meters. (a) The LiDAR image of a sample building with adjusted final footprint. (b) The raw footprint derived by connecting boundary points of identified building measurements through the region-growing algorithm. The raw footprint is noisy because of the interpolation of irregularly spaced LiDAR measurements. (c) The coarse footprint derived using the Douglas–Peucker algorithm. Two solid lines across the footprint are estimated dominant directions. (d) The adjusted footprint that recovered critical corner vertices removed by the Douglas–Peucker algorithm. The footprint was rotated clockwise according to the estimated dominant directions. (e) Comparison of the final footprint (solid) with corresponding known footprint (dash). (From Zhang, K., Yan, J., and Chen, S.C., *IEEE Trans. Geosci. Remote Sensing*, 44, 2523, 2006. With permission.)

where

- N is the total number of vertices of a building footprint
- L_i is the segment length
- β_i ($0° \le \beta_i < 45°$) is the minimum intersection angle between a segment and the nearest axis in the coordinate system x' and y', and is determined by θ_i and φ
- $g()$ is the weight function based on L_i
- $f()$ is the weight function based on θ_i and φ

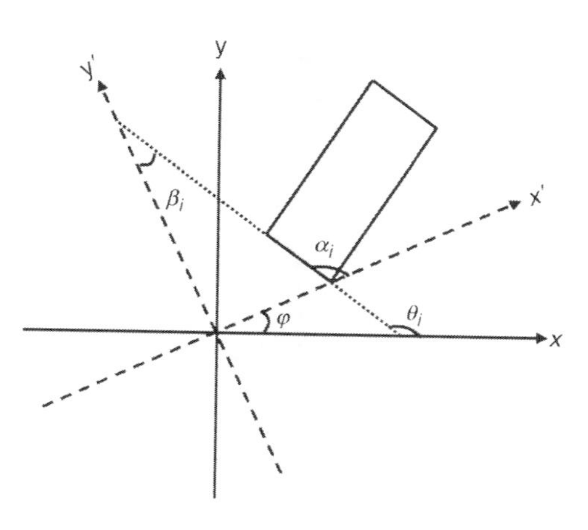

FIGURE 18.3
Relationship between angles β_i, θ_i, α_i, and φ. The coordinate system x' and y' is a counterclockwise rotation of the coordinate system x and y by an angle φ. θ_i is the counterclockwise intersection angle between a segment of a building footprint (solid square) and the axis x. α_i is the counterclockwise intersection angle between a segment and the axis x'. β_i is the minimum intersection angle between a segment and the axes x' and y'. (From Zhang, K., Yan, J., and Chen, S.C., *IEEE Trans. Geosci. Remote Sensing*, 44, 2523, 2006. With permission.)

The dominant building directions can be estimated by finding an optimum φ so that *SL* will reach a minimum.

A linear function is employed to represent $g()$

$$g(L_i) = \frac{L_i}{\sum_{i=1}^{N-1} L_i} \tag{18.4}$$

Finding the optimum φ depends heavily on $f()$, which can have many forms such as linear and exponential expressions. We would like to construct $f()$ such that the segments close to the dominant direction will have a small contribution to SL. Here, a linear function is used to represent $f()$:

$$f(\beta_i(\theta_i,\varphi)) = \frac{\beta_i}{45} \tag{18.5}$$

Obviously, the closer the segment is to the dominant direction, the smaller the $f()$. β_i is determined by

$$\beta_i = \begin{cases} \min(\alpha_i, 90-\alpha_i) & \text{if } \alpha_i \le 90^\circ \\ \min(180-\alpha_i, \alpha_i-90) & \text{if } \alpha_i > 90^\circ \end{cases} \tag{18.6}$$

where α_i ($0^\circ \le \alpha_i < 180^\circ$) is the counterclockwise intersection angle between a line segment and the axis x' and has the following relationship with θ_i and φ:

$$\alpha_i = \begin{cases} \theta_i - \varphi & \text{if } \theta_i \ge \varphi \\ 180 + \theta_i - \varphi & \text{if } \theta_i < \varphi \end{cases} \tag{18.7}$$

Numerically, the optimum φ is found by comparing SL values for angles between 0° and 90°. After φ is derived, the building footprint is rotated so that the x and y axes are aligned with the dominant directions of the buildings (Figure 18.2d).

It has been proven that the estimated dominant directions are the same as the directions of the parallel and perpendicular segments as long as the total length of the oblique lines is less than the total length of the parallel and perpendicular segments of a footprint [13]. After the dominant directions of a footprint are estimated, five operations including split, intersect, and three types of merge are employed to refine the building footprint. The selection of operations is mainly determined by the projections of a line segment in the two dominant directions [13]. Figure 18.2e displays an adjusted footprint using these five operations. Once a refined footprint is obtained, a simple building model can be created by adding a uniform height value for the building.

18.3.4 Deriving Sophisticated Building Models

18.3.4.1 2D Topology Extraction

In addition to the outline of the footprint, the inside boundaries between adjacent roof facets and equations describing these roof facets need to be derived for a sophisticated 3D building model. It is assumed that the roof facets follow plane equations in our 3D building model. The boundaries between roof facets are part of the 2D topology, which is represented by a set of connected polygons that are the projections of roof facets on a horizontal plane. The edges and vertices that form the topology are derived from grouped LiDAR measurements for individual facets of a roof. To facilitate this discussion, measurements for each roof facet for a building are assigned a unique positive integer label starting from the label "1" (Figure 18.4a and b). Nonbuilding measurements surrounding the building are labeled as "0". In order to obtain the boundaries of a roof facet, the cell size of the mesh covering the data set is reduced by half, and new points are inserted between the old cells.

The label of a newly inserted point p is determined by checking four pairs of neighbors of p as illustrated in Figure 18.5. If the label at the left bottom corner is the same as the label at the right upper corner, p is assigned the same label as these two neighbors (Figure 18.5a). If not, the other pairs of neighbors (Figure 18.5b through d) are checked following the same procedure. If all four pairs of neighbors of p have different labels, p is identified as a boundary point separating two or more neighboring roof facets and is assigned a label value "–1" (Figure 18.4c).

A boundary point that joins at least three different roof facets (Figure 18.4d) is classified as a vertex of the 2D topology, while a boundary point that separates only two neighboring roof facets is defined as an edge point. An edge is derived by connecting all edge points between two vertices. The vertices and edges from boundary points constitute a raw 2D topology. The raw 2D topology is rotated clockwise so that the dominant directions of the footprint align with the x and y axes. The raw 2D topology is simplified using the Douglas–Peucker algorithm to reduce the noise due to the interpolation of the irregularly spaced LiDAR points. Parallel and perpendicular edges in a simplified topology can be distorted in some cases because no geometric constraints are applied during the simplification (Figure 18.4e). Therefore, a refinement of the topology is needed to minimize this distortion.

18.3.4.2 Roof Facet Adjustment

Before performing a topological adjustment, plane parameters for a roof facet are estimated using the points within a segmented patch for the roof facet. Parallel and perpendicular

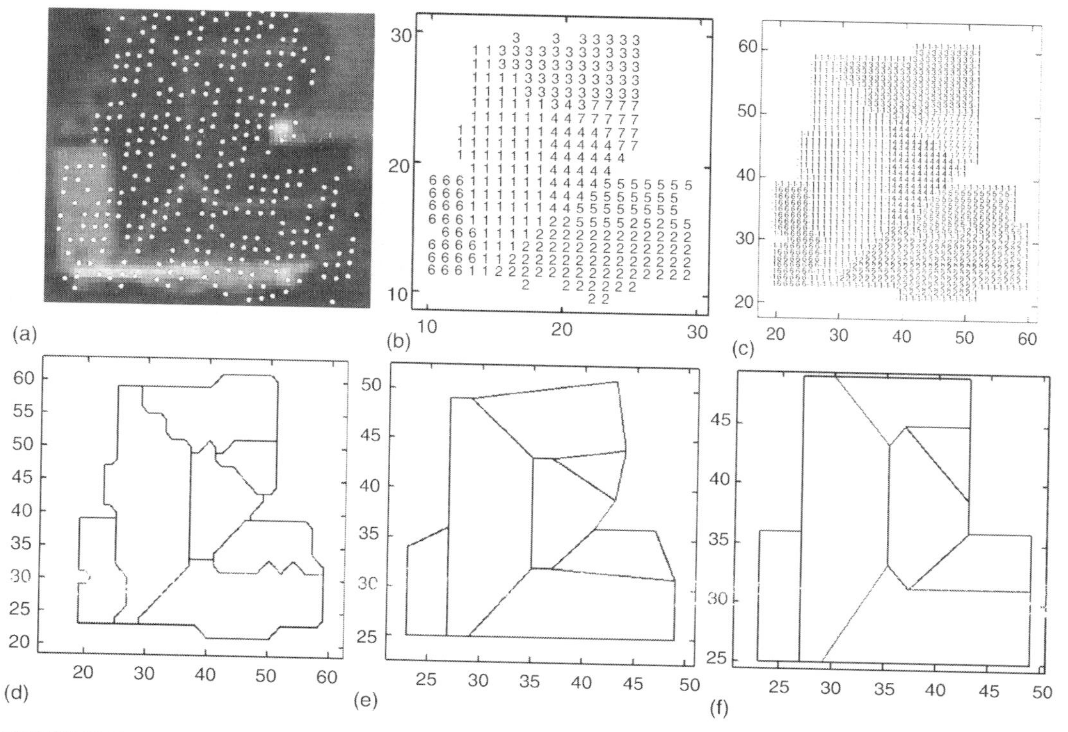

FIGURE 18.4
Example illustrating the reconstruction process of a sophisticated 3D building model. The x and y coordinates are in meters. (a) Raw LiDAR points overlaid on the aerial photograph, (b) polygons for segmented roof facets, which are labeled by different positive integers, (c) roof facet polygons with labeled (−1) boundary, (d) raw 2D topology of the building, (e) simplified 2D topology using the Douglas–Peucker algorithm, and (f) adjusted 2D topology by applying the snake algorithm.

properties of a plane are also enforced using the variation of elevation values. Each roof plane is first examined whether it is flat by analyzing the variation (Var) in the elevations of LiDAR measurements for the plane using the following equation:

$$\mathrm{Var} = \sum (z_k - \bar{z})^2, \quad \bar{z} = \frac{\sum z_k}{n} \tag{18.8}$$

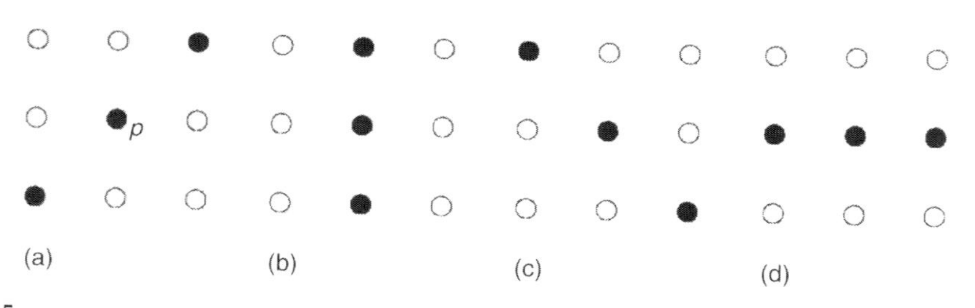

FIGURE 18.5
Patterns used to determine whether a point p is a boundary point. If four pairs of neighbors of p have different labels, p is a boundary point.

where

n is the number of measurements for a roof facet

$\bar{z}$ is the average elevations of the LiDAR measurements z_k

If Var is less than the threshold T_Var², the plane is classified as horizontal, and the plane equation is adjusted to be $z = \bar{z}$. If not, the roof facet is classified as nonflat. Then, we examine whether the plane for a nonflat roof facet is perpendicular to X–Z plane. Each 3D measurement (x, y, z) on the plane is projected onto the X–Z plane to form a 2D point (x, z). The line segment with equation $z = ax + b$ that fits the projected 2D points is derived using the least squares method. The square root of SSD (srssd) is used to determine the perpendicular property of a roof facet

$$\mathrm{SSD} = \sum (z'_k - z_k)^2, \quad \mathrm{srssd} = \sqrt{\mathrm{SSD}} \tag{18.9}$$

where z'_k and z_k are the estimated and observed elevations for point k, respectively. If srssd is less than the threshold T_Var, the plane is perpendicular to the X–Z plane and its plane equation is adjusted to be $z = ax + b$. If srssd is greater than T_Var, we then check whether the plane is perpendicular to the Y–Z plane using the same procedure. Each 3D measurement (x, y, z) is projected onto the Y–Z plane to form a 2D point (y, z). The linear equation $z = a'y + b'$ fitting the 2D points is estimated using the least squares method. If srssd from the observed and estimated z values is less than T_Var, the roof plane is perpendicular to Y–Z plane, and its plane equation is adjusted to be $z = a'y + b'$. Otherwise, the nonflat roof plane is neither perpendicular to X–Z nor Y–Z, and its plane equation is not adjusted.

18.3.4.3 Adjustment of 2D Topology

The snake algorithm is utilized to refine the 2D topology of a building. The snake algorithm was introduced by Kass et al. to locate features of interest in an image [26]. First, an initial contour enclosing the feature of interest is selected, and then the contour is pulled towards the target feature by minimizing energy functions that represent constraint forces. The total energy E_{total} of a contour with a parametric representation $v(s) = (x(s), y(s))$ consists of the internal energy E_{int} and the external energy E_{ext}, and can be written as

$$E_{\mathrm{total}} = \int_0^1 E_{\mathrm{int}}(v(s)) + E_{\mathrm{ext}}(v(s))\,\mathrm{d}s \tag{18.10}$$

The internal energy represents the forces that constrain the contour to be smooth and are formulated as

$$E_{\mathrm{int}}(v(s)) = (\alpha(s)\,|\,v_s(s)\,|^2 + \beta(s)\,|\,v_{ss}(s)\,|^2) \tag{18.11}$$

where the first-order energy term $v_s(s)$ moves the points closer to each other and the second-order energy term $v_{ss}(s)$ favors equidistant points. The external energy attracts the contour towards the feature of interest, such as the high intensity or high gradient areas. The minimization of energy can be implemented using finite difference, finite element, or dynamic programming methods [26–28]. Traditional snake algorithms only consider the outlines of components in an image. In order to adjust the 2D topology, both the deformation of the outline of a component and the inside structure of edges between polygons have to be considered.

A dynamic programming method for 1D snakes was extended to adjust 2D topology in our framework because the method guarantees a global minimum and is numerically stable [29]. The dynamic programming method assumes that each vertex on the contour is adjustable to a position that belongs to a set of positions around the vertex. The set of positions determines the range of vertex variation. Each position of the vertex corresponds to one state of the vertex. A set of the vertices connecting contours corresponding to the minimal energy represent the vertices for an optimum 2D topology. The key for deriving the optimum 2D topology is to define appropriate energy functions for the snake algorithm. Traditional inner energy functions for a smooth contour shown in Equation 18.11 are not suitable for building topology adjustment because there are intersection angles between the two adjacent edges and the transition between two edges is not smooth. In addition, the distances between the vertices of the topology are not equal in many cases.

The objective of building footprint adjustment is to enforce the requirement that the edges be parallel to the dominant directions in a 2D topology while keeping the adjusted footprint as close to its original location as possible. To meet this requirement, we are proposing new functions, direction energy E_{Dir}, to enforce the parallel constraint and, deviation energy E_{Dis}, to limit the deviation of the adjusted footprint from its original position. Given an edge $e' = (v', w')$ joining two vertices v' and w' on the footprint, and v' and w' are one of the possible states for vertices v and w, we define the direction energy E_{Dir} for the edge $e' = (v', w')$ as

$$E_{\text{Dir}}(e' = (v', w')) = \begin{cases} 0 & c(e) = 1, v'_x = w'_x \\ 0 & c(e) = 2, v'_y = w'_y \\ 0 & c(e) = 3, |v'_y - w'_y| > T\_\text{Projection} \quad \text{or} \\ & |v'_x - w'_x| > T\_\text{Projection} \\ |v' - w'| & \text{others} \end{cases} \tag{18.12}$$

where $c(e)$ with values of 1, 2, or 3 represents horizontal, vertical, and nonadjustable oblique edges, respectively. A zero value is assigned to the direction energy function E_{Dir} for edge e' in the 2D topology if e' is either parallel to the dominant directions or e' is a nonadjustable oblique line. A nonadjustable oblique edge is the line whose projection on x or y axis is larger than a threshold T_Projection. The purpose in introducing nonadjustable lines is to preserve oblique edges in the 2D topology. All remaining edges are classified as adjustable ones, and a penalty value proportional to the length of the edge is assigned to E_{Dir} for an adjustable edge e'. As a result of this energy function, the adjustable horizontal and vertical edges tend to deform and align with the dominant directions, and nonadjustable oblique edges tend to remain oblique.

The energy function E_{Dis} is defined as the sum of the distance values between points on the adjusted edge (v', w') and the original edge (v, w):

$$E_{\text{Dis}}(e' = (v', w')) = \sum_{p \in v'w'} D_p(v, v', w, w') \tag{18.13}$$

where p is a point located on the edge connecting points v' and w'. The smaller the distance energy of the adjusted edge e', the closer e' is to the original edge e. Only the distance values of points in an area close to the original 2D topology need to be calculated since each vertex on the 2D topology is only allowed to move within a small window $W_v \times W_v$. In order to compute E_{Dis}, a distance transform is applied to the 2D topology to derive a gray scale image whose pixel intensity indicates shortest distances to boundaries. The edges between roof facets are rasterized using the same grid mesh for 2D topology extraction described in the previous section. Distance values of edge cells are initialized as 0 and the distance values of their direct neighbors are assigned a value of 1. The distance values of direct neighbors of cells having values of 1 are assigned a value of 2 and so on. Figure 18.6 shows the distance value of the image after applying the distance transform to the 2D topology shown in Figure 18.4c. The distance value is calculated up to 3 cells. The total energy for each adjusted edge e' is determined as follows:

$$E(e' = (v', w')) = C_{Dir}E_{Dir}(e') + C_{Dis}E_{Dis}(e') \tag{18.14}$$

where C_{Dir} and C_{Dis} are weights for two energy terms. The snake algorithm transforms each adjustable vertex on the 2D topology and finds an optimum combination of vertices by minimizing the sum of the energy functions for all edges on the 2D topology. Figure 18.4f shows one example where a contour was adjusted using the snake algorithm based on defined energy functions.

A 2D snake algorithm based on the energy functions described above is required to refine a 2D topology for a building. Unfortunately, the general 2D snake algorithm is a nondeterministic polynomial (NP) time completeness problem. It is time consuming to adjust a complicated 2D topology using a brute force method. However, a subset of 2D snake problems

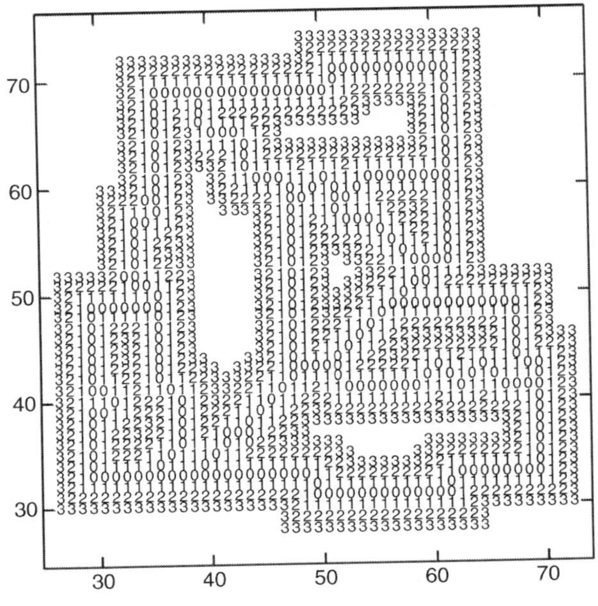

FIGURE 18.6
Distance values derived by applying distance transform to the 2D topology of a building. Label "0" indicates the edges of roof facets, and labels "1," "2," and "3" represent the distances from the edges. The x and y coordinates are in meters.

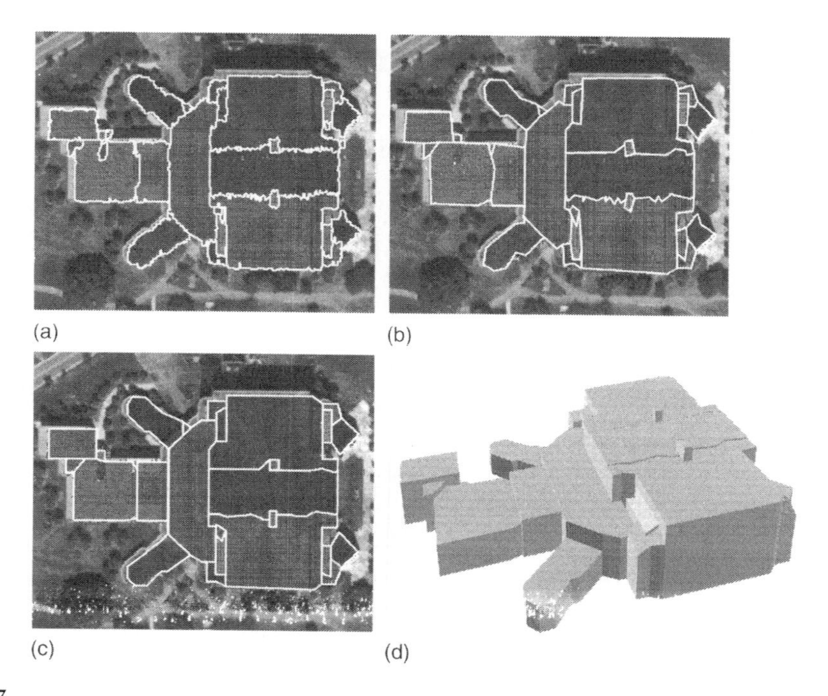

FIGURE 18.7
Derivation of a complex 3D building model from LiDAR measurements involves (a) raw 2D topology, (b) simplified 2D topology, (c) refined 2D topology through the 2D snake algorithm, and (d) reconstructed 3D building model. LiDAR measurements for various flat roof facets are displayed with different colors. The background of (a), (b), and (c) is a black-white area photograph. The details of the roof surface of this building can also be found in Figure 18.11.

can be solved in polynomial time using an algorithm based on the graph theory [29]. Our work demonstrates that more than 95% of building topologies can be adjusted by the graph-based algorithm. Figure 18.7 shows an example adjusting the 2D topology of a complicated building using the graph-based algorithm. The building roof consists of 46 vertices and 67 boundaries, and all the roof planes are flat. After applying the 2D snake algorithm based on the proposed energy functions, the 2D topology is adjusted and shown in Figure 18.7c.

18.3.4.4 Adjustment of Intersection Edges

Some buildings, such as residential houses, mainly consist of nonhorizontal roof facets that form many intersection edges. The edges between building facets can be classified into two categories: intersection and step edges. A step edge separates either two parallel planes or two intersection planes with a height discontinuity (Figure 18.8a and b). An intersection edge separates two neighboring roof planes with height continuity (Figure 18.8c). Obviously, all edges of the footprint outline are step edges.

The height values at an intersection edge from two adjacent roof facets may be different because the 2D topology adjustment algorithm does not enforce height continuity constraints. To overcome this inconsistency caused by nonflat roof facets, the edges from the snake-based adjustment within the outline of the building footprints are replaced with intersection segments of the neighboring planes. Before the replacement operation is performed, intersection edges have to be identified. We determine an intersection edge using the following equation:

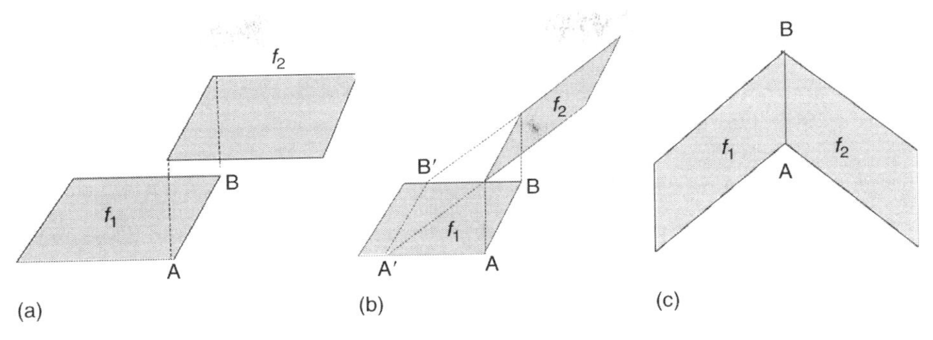

FIGURE 18.8
Types of edge (AB) between two roof facets. (a) Step edge separating two parallel planes, (b) step edge separating two intersection planes with height discontinuity, and (c) intersection edge separating two intersection planes with height continuity.

$$\mathrm{DH}(e(v,w)) = \sum_{p \in \overline{vw}} |h_1(p) - h_2(p)| / n \tag{18.15}$$

where
$\overline{vw}$ is the set of grid cells containing the edge e
n is the total number of grid cells in the set
p is a grid cell in the set
$h_1(p)$ and $h_2(p)$ are the elevation of p on its two neighboring planes, respectively

If $DH(e)$ is less than a predefined threshold T_Edge, edge e is classified as the intersection edge as shown in Figure 18.8c. If not, it is a step edge as shown in Figure 18.8b. The threshold T_Edge is determined by the error of LiDAR measurements. In our framework, T_Edge is set as $2\Delta h_T$. Δh_T is the elevation deviation threshold of a LiDAR point to a fitting plane and is used in the building patch segmentation algorithm in Section 18.3.2. Since the height deviation of a point from the fitting plane error is assumed to be Δh_T in the worst situation, the height difference between two neighboring roof planes with height continuity can reach $2\Delta h_T$.

18.4 Data Processing

The study area is located at and around the campus of Florida International University (FIU), covering $6\,\mathrm{km}^2$ of low relief topography. Surveyed features include residential houses, complex buildings, individual trees, forest stands, parking lots, open ground, ponds, roads, and canals. The data were collected on August 2003 with Optech ALTM 1233 systems operated by FIU. The Optech system recorded the coordinates (x, y, z) and the intensity of the point measurements corresponding to the first and last laser returns. The data set consists of five overlapping 340 m wide swaths of 13 cm diameter footprints spaced approximately 1 m apart. Three-dimensional building models were reconstructed for the FIU campus and an adjacent area to examine the effectiveness of the proposed framework. The thresholds used in our experiments for 3D building reconstruction are listed in Tables 18.1 and 18.2. A sensitivity analysis showed that small changes in most thresholds have little impact on the final results. It took 9, 2, 0.7, and 2 min for a personal

TABLE 18.1

Parameters for Extracting Simple Building Models

Parameters	Values
Cell size (c_s) for progressive morphological filter	0.5 m
Height difference to aggregate a point (Δh_T)	0.2 m
Minimal surface on the roof (Min_Surface)	5 m^2 (20 c_s^2)
Minimal building area (Min_Building)	60 m^2 (240 c_s^2)
Douglas distance (T_Douglas)	1.5 m (3 c_s)
Threshold for footprint classification (T_SL)	0.3
Threshold for split adjustment (T_Projection_Final)	2 m (4 c_s)
Threshold for deviation (T_Deviation)	2 m (4 c_s)
Threshold for triangle area ratio (T_Ratio)	0.1
Threshold for footprint evaluation (T_Footprint)	0.85

Source: From Zhang, K., Yan, J., and Chen, S.C., *IEEE Trans. Geosci. Remote Sensing*, 44, 2523, 2006. With permission.

computer with a 2.8-GHz processor and a 2-GB RAM to perform morphological filtering, building measurement identification, and simple and sophisticated 3D building reconstruction for the FIU campus dataset. A 2D array with about 7.2 million elements was employed to represent raw and interpolated points covering an area of 1.8 km^2. Aerial photographs and field investigation were used to help evaluate the reconstructed building models. The aerial photographs were collected in 1999 at a resolution of 0.3 m.

18.5 Results

18.5.1 Simple 3D Building Model

Both qualitative and quantitative methods were employed to measure the errors committed in extracting simple building models in this study. A qualitative method checked the

TABLE 18.2

Parameters for Reconstructing Sophisticated Building Models

Parameters	Values
Cell size (c_s) for snake algorithm	1 m
Height difference to aggregate a point (Δh_T)	0.15 m
Douglas distance (T_Douglas)	1.5 m (3 c_s)
Threshold for split adjustment (T_Projection)	2 m (4 c_s)
Threshold for plane adjusting (T_Var)	0.09 m
Weight for direction energy (C_{Dir})	1
Weight for deviation energy (C_{Dis})	5
Threshold for edge classification (T_Edge)	0.3 m
Window size for vertex adjusting (W_v)	5 m

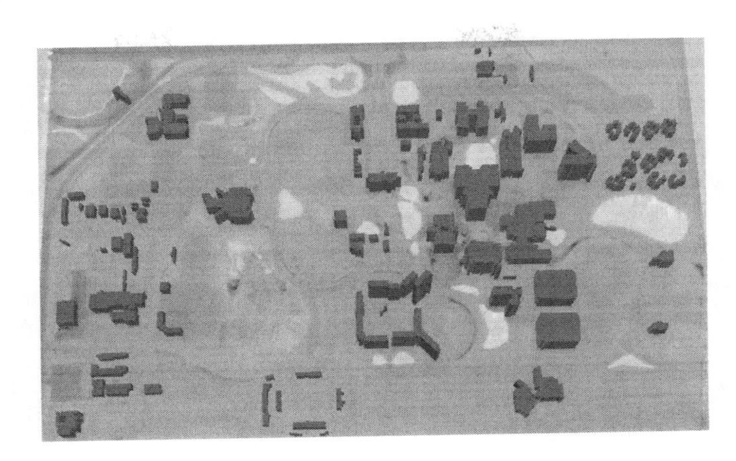

FIGURE 18.9
Simple 3D building models for FIU campus. Each building model was created using the final footprint and the average building height derived from LiDAR measurements. The DTM for building bases was derived by interpolating ground measurements identified by the progressive morphological filter. (From Zhang, K., Yan, J., and Chen, S.C., *IEEE Trans. Geosci. Remote Sensing*, 44, 2523, 2006. With permission.)

quality of the estimated dominant directions and derived footprints by visually comparing the extracted footprints with those in the maps and aerial photographs. The quantitative method examines the accuracy of extracted footprints using count-based and area-based metric methods [13]. The count-based metric method quantifies commission and omission errors in the number of footprints identified, while the area-based metric method measures the area differences between identified and known building footprints. Figure 18.9 shows a simple 3D building map for the FIU campus based on refined building footprints and heights. The refined footprints were derived by applying footprint adjustment algorithms from Section 18.3.3 to raw footprints that were obtained by connecting boundary points of an identified building patch. The heights of the buildings were derived by averaging the elevation differences between building measurements and the DTM interpolated from ground measurements.

Comparison of 62 adjusted footprints with footprints from the map provided by the FIU Planning and Facility Management Department showed that all buildings were identified correctly in terms of number of extracted buildings. However, 10% of the building footprints were mistakenly removed, and 2% of the footprints were incorrectly included into the final output in terms of the total area of extracted buildings. These building models have been used to construct a 3D synthetic visual environment to animate hurricane-induced freshwater flooding at the FIU campus. The simple building model extraction algorithm was also applied to residential areas adjacent to the FIU campus. The commission and omission errors of extracted building footprints are both around 6% [13].

The effectiveness of the footprint adjustment algorithm is well illustrated in Figures 18.10 and 18.11. Figure 18.10a shows that the raw building footprints have noise on their boundaries. Most of the noise was removed in the adjusted footprints as shown in Figure 18.10b, and the smoothness of the building outline was greatly improved. Figure 18.11 shows a complex building that consists of footprint segments parallel and oblique to the dominant directions. The portion of the building whose direction is different from the dominant directions was also adjusted appropriately.

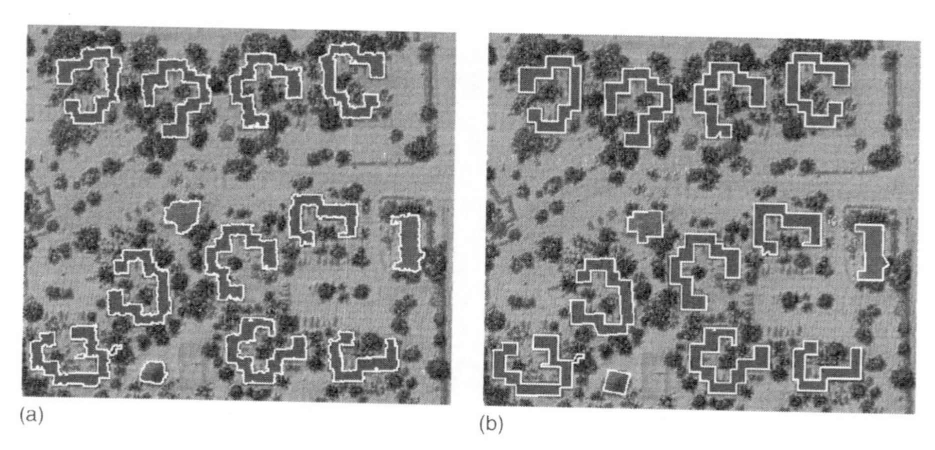

FIGURE 18.10
Comparison of raw (a) and adjusted (b) footprints for simple building models. The small zigzag noise in the raw footprints was removed in the adjusted footprints, making the adjusted footprints look more realistic. The background images were produced by interpolating point LiDAR measurements. (From Zhang, K., Yan, J., and Chen, S.C., *IEEE Trans. Geosci. Remote Sensing*, 44, 2523, 2006. With permission.)

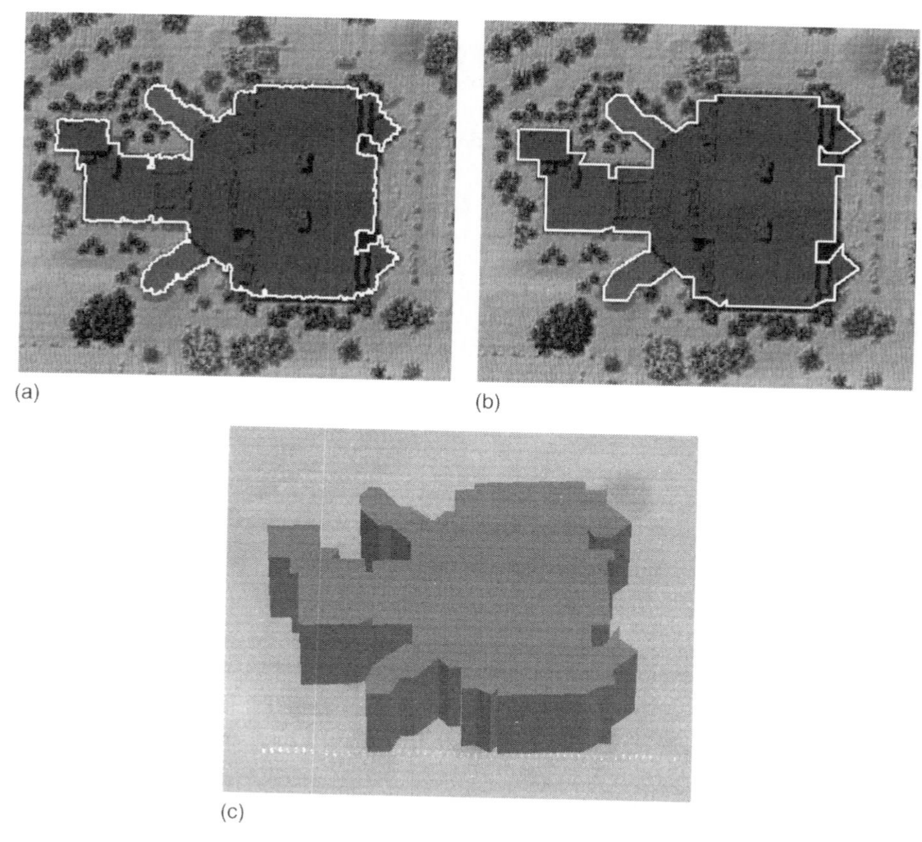

FIGURE 18.11
(a) Raw, (b) adjusted, and (c) 3D simple model for a complex building. Two dominant directions of the footprint are nearly horizontal and vertical. Note that oblique portions of the building that are not aligned with the dominant directions were well preserved in the simple building model.

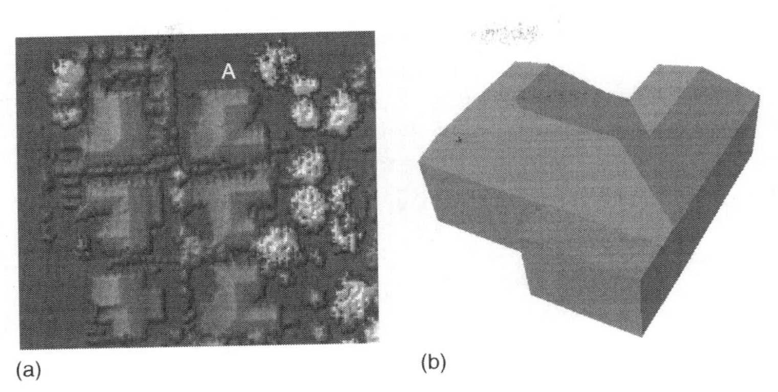

FIGURE 18.12
Building and tree images interpolated from LiDAR measurements for (a) a residential area and (b) a sophisticated 3D building model for a residential building. The sophisticated 3D building model was derived by applying the proposed building reconstruction framework to LiDAR measurements. The reconstructed model is for a building indicated by the letter A in (a).

18.5.2 Sophisticated 3D Building Model

Figure 18.7 shows an example of sophisticated 3D building models at the FIU campus. The noise in the edges of the building footprint was reduced greatly after the adjustment of the 2D topology was performed (Figure 18.7c). A 3D building model was created based on the adjusted roof facets, roof edges, and associated height data (Figure 18.7d). Compared to a simple building model (Figure 18.11), a sophisticated model includes much more details of roof facets and looks more realistic. Figure 18.12 shows a sample residential building with nonflat roof facets. Various nonflat roof facets were reconstructed well by the proposed algorithm.

Figure 18.13a and b show extracted 2D topologies and 3D models for commercial and residential buildings next to the FIU campus. It is very difficult to quantify the errors of the algorithms for reconstructing 3D building models since no ground-truth data with higher accuracy are available. Digitizing 3D building models manually from LiDAR measurements or DSM is impractical. Derivation of 3D building models from aerial photographs for the study areas is also impossible because no overlapped photographs are

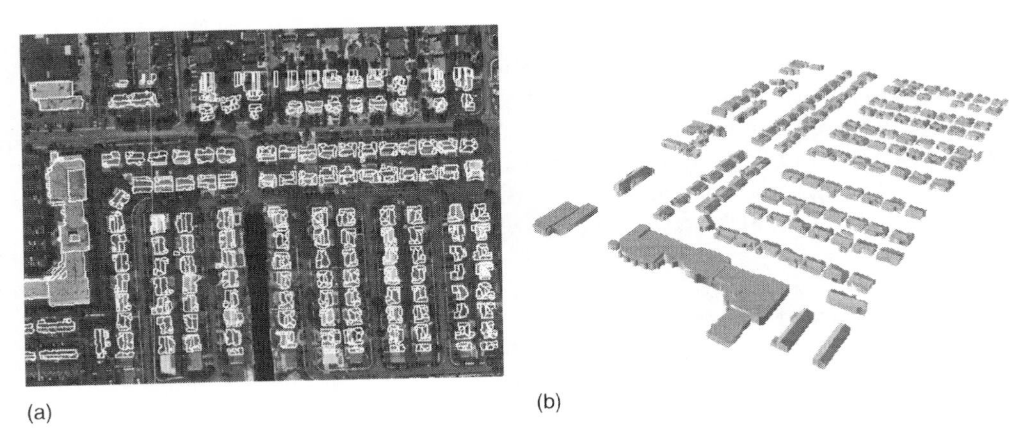

FIGURE 18.13
(a) A 2D topology and (b) reconstructed sophisticated 3D models for commercial and residential buildings next to the FIU campus.

available. We qualitatively examined the quality of extracted sophisticated building models by comparing the models with raw LiDAR measurements, LiDAR DSMs, and aerial photographs in ArcGIS. The results indicated that most extracted 2D topologies represent the boundaries of connected roof facets well. Most buildings were reconstructed properly after applying the building refinement algorithms.

18.6 Discussion and Conclusions

A framework including a series of algorithms has been developed to automatically reconstruct 3D building models from LiDAR measurements. The framework includes five major components: (1) the progressive morphological filter for separating the ground and nonground measurements, (2) a region-growing algorithm based on a local plane-fitting technique for segmenting building measurements, (3) an algorithm for estimating the dominant direction of a building, (4) a method for extracting 2D topology, and (5) a snake-based algorithm for adjusting the 2D topology. The entire process is highly automatic and requires little human input, which is very useful for processing voluminous LiDAR measurements.

The application of the framework to the FIU campus and adjacent residential areas shows that the region-growing algorithm identified building patches well, the snake algorithm adjusted most buildings properly, and simple and sophisticated 3D building models were reconstructed effectively. The quantitative accuracy analysis indicates that all buildings were extracted and about 12% errors in the total area of footprints were committed by the proposed algorithms in reconstructing simple building models, despite the fact that there are several complex buildings on the FIU campus.

Accurate segmentation of roof facets is critical for the extraction of 2D topology and reconstruction of sophisticated building models. Numerical experiments demonstrated that segmentation is sensitive to the errors in LiDAR measurements. The segmentation based on a single strip of LiDAR measurements is more robust than that of overlapped strips of LiDAR measurements because relatively large errors are introduced by the data from overlapped strips. Theoretically, the high density of LiDAR measurements will increase the accuracy of segmentation of building patches because more points are available for parameter estimation, but this is not the case for overlapped strip data. This poses a serious challenge for using multistrip data for reconstruction of sophisticated building models. Further improvement in segmentation is needed for reconstruction of better sophisticated building models. Changes in boundaries between roof facets within a building footprint have little effect on the extraction of the outline of the footprint. The derivation of the footprint outline is robust because of a distinctive difference in elevations between building and adjacent ground measurements. Therefore, the performance of the region-growing algorithm has much less effect on the reconstruction of simple building models than on the reconstruction of sophisticated building models.

Acknowledgments

The authors would like to thank the anonymous reviewers for valuable comments. This work was supported by the Florida Hurricane Alliance Research Program sponsored by the National Oceanic and Atmospheric Administration.

References

1. Jensen, J.R., 2000. *Remote Sensing of the Environment*. Prentice-Hall, Upper Saddle River, New Jersey, pp. 592
2. Maas, H.G., Fast determination of parametric house models from dense airborne laser scanner data, ISPRS Workshop on Mobile Mapping Technology, *International Archive of Photogrammetry and Remote Sensing*, XXXII, part 2W1, 5W1, IC5/3W, 1999.
3. Filin, S. and Pfeifer, N., Segmentation of airborne laser scanning data using a slope adaptive neighborhood, *ISPRS Journal of Photogrammetry and Remote Sensing*, 60, 71, 2006.
4. Vosselman, G., Slope based filtering of laser altimetry data, XIXth ISPRS Congress, Commission IV, *International Archive of Photogrammetry and Remote Sensing*, XXXIII, part B4, 2000, p. 958.
5. Shan, J. and Sampath, A., Urban DEM generation from raw LiDAR data: A labeling algorithm and its performance, *Photogrammetric Engineering and Remote Sensing*, 71, 217, 2005.
6. Zhang, K., Chen, S.C., Whitman, D., Shyu, M.L., Yan, J., and Zhang, C., A progressive morphological filter for removing non-ground measurements from airborne LiDAR data, *IEEE Transactions on Geoscience and Remote Sensing*, 41, 872, 2003.
7. Weidner, U. and Forstner, W., Towards automatic building reconstruction from high resolution digital elevation models, *ISPRS Journal of Photogrammetry and Remote Sensing*, 50, 38, 1995.
8. Morgan, M. and Tempfli, K., Automatic building extraction from airborne laser scanning data, XIXth ISPRS Congress Commission III, *International Archive of Photogrammetry and Remote Sensing*, XXXIII, part B3, 2000, p. 616.
9. Rottensteiner, F. and Jansa, J., Automatic extraction of buildings from LiDAR data and aerial images, ISPRS Commission IV Symposium, Geospatial Theory, Processing and Applications, *International Archive of Photogrammetry and Remote Sensing*, XXXIV, part 4, WG IV/7: Data Integration and Digital Mapping, 2002.
10. Elaksher, A.F. and Bethel, J.S., Reconstructing 3D Buildings from LiDAR Data, ISPRS Commission III Symposium, Photogrammetric and Computer Vision, *International Archive of Photogrammetry and Remote Sensing*, XXXIV, part 3A/B, 2002. A102.
11. Filin, S., Surface clustering from airborne laser scanning data, ISPRS Commission III Symposium, Photogrammetric and Computer Vision, *International Archive of Photogrammetry and Remote Sensing*, XXXIV part 3A/B, 2002. A119.
12. Morgan, M. and Habib, A., Interpolation of LiDAR data and automatic building extraction, *ACSM-ASPRS 2002 Annual Conference Proceedings*, 2002.
13. Zhang, K., Yan, J., and Chen, S.C., Automatic construction of building footprints from airborne LiDAR data, *IEEE Transactions on Geoscience and Remote Sensing*, 44, 2523, 2006.
14. Alharthy, A. and Bethel, J., Heuristic filtering and 3d feature extraction from LiDAR data, ISPRS Commission III Symposium, Photogrammetric Computer Vision, *International Archive of Photogrammetry and Remote Sensing*, XXXIV, part 3A/B, A-29, 2002.
15. Carter, W.E., Shrestha, R.L., and Slatton, K.C., Photon counting airborne laser swath mapping (PC-ALSM), *4th International Asia-Pacific Environmental Remote Sensing Symposium*, Vol. 5661, SPIE Proceedings: Remote Sensing Applications of the Global Positioning System, 2004, p. 78.
16. Overby, J., Bodum, L., Kjems, E., and Ilsøe, P.M., Automatic 3D building reconstruction from airborne laser scanning and cadastral data using hough transform, XXth ISPRS Congress, Geo-Imagery Bridging Continents, *International Archive of Photogrammetry and Remote Sensing*, XXXV, part B3, 2004.
17. Vosselman, G., Building reconstruction using planar faces in very high density height data, ISPRS WG III/2&3 Workshop, Automatic Extraction of GIS Objects from Digital Imagery, *International Archives of Photogrammetry and Remote Sensing*, XXXII, part 3–2W5, 1999, p. 87.
18. Elberink, S.O. and Maas, H.G., The use of anisotropic height texture measures for the segmentation of airborne laser scanner data, XIXth ISPRS Congress Commission III, *International Archive of Photogrammetry and Remote Sensing*, XXXIII, part B3/2, 2000, p. 678.
19. Sampath, A. and Shan, J., Building boundary tracing and regularization from airborne LiDAR point clouds, *Photogrammetric Engineering and Remote Sensing*, 73, 805, 2007.

20. Schwalbe, E., Maas, H.G., and Seidel, F., 3-D building model generation from airborne laser scanner data using 2-D data and orthogonal point cloud projections, ISPRS Workshop Laser Scanning, *International Archive of Photogrammetry and Remote Sensing*, XXXVI, part 3/W19, Session 10: Building Reconstruction, 2005.
21. Maas, H.G. and Vosselman, G., Two algorithms for extracting building models from raw laser altimetry data, *ISPRS Journal of Photogrammetry and Remote Sensing*, 54, 153, 1999.
22. Brenner, C., Towards fully automatic generation of city models, XIXth ISPRS Congress Commission III, *International Archive of Photogrammetry and Remote Sensing*, XXXIII, Part B3, 2000, p. 85.
23. Brenner, C., Modeling 3-D objects using weak CSG primitives, XXth ISPRS Congress, Geo-Imagery Bridging Continents, *International Archive of Photogrammetry and Remote Sensing*, XXXV, part B3, 2004.
24. Douglas, D.H. and Peucker, T.K., Algorithms for the reduction of the number of points required to represent a digitized line or its caricature, *The Canadian Cartographer*, 10, 112, 1973.
25. Gruen, A. and Wang, X., News from CyberCity-Modeler, *Third International Workshop on Automatic Extraction of Man-Made Objects from Aerial and Space Images*, 2001, p. 93.
26. Kass, M., Witkin, A., and Terzopoulos, D., Snakes: Active contour models, *International Journal of Computer Vision*, 1, 321, 1988.
27. Cohen, L.D. and Cohen, I., Finite-element methods for active contour models and balloons for 2-D and 3-D images, *IEEE Transactions on Pattern Analysis and Machine Learning*, 15, 1131, 1993.
28. Amini, A.A., Weymouth, T.E., and Jain, R.C., Using dynamic programming for solving variational problem in vision, *IEEE Transactions on Pattern Analysis and Machine Intelligence*, 12, 855, 1990.
29. Yan, J., Zhang, K., Zhang, C., Chen, S.C., and Narasimhan, G., A Graph Reduction Method for 2D Snake Problems, *IEEE Computer Society Conference on Computer Vision and Pattern Recognition*, 2007, p. 6.

19

Quality of Buildings Extracted from Airborne Laser Scanning Data: Results of an Empirical Investigation on 3D Building Reconstruction

Eberhard Gülch, Harri Kaartinen, and Juha Hyyppä

CONTENTS

19.1 Introduction

Since more than one decade, three-dimensional (3D) geographical information systems have been of increasing importance in urban areas in various applications such as urban planning, visualization, environmental studies and simulation (pollution, noise), tourism, facility management, telecommunication network planning, 3D cadastre, and vehicle or pedestrian navigation. In the late 1990s, OEEPE conducted a survey on 3D city models. The scope of that study was to find out the state of the art of generating and using 3D city data. The study was based on a questionnaire sent out to about 200 European institutions (Fuchs et al., 1998). Fifty-five responded to the questionnaire from 17 countries (curiously, 30% of the replies came from Germany, which was certainly influenced by the activities of three major mobile phone companies in 3D city modeling). The most important objects of interest according to the users were buildings (95%), traffic network (76%), and vegetation (71%). It was stated that there was a definite lack of economical techniques for producing 3D city data. The lack of knowledge of information sources, as well as high costs for building and vegetation acquisition hindered broader use at that time. The data sources used by the producers were aerial imagery (76%), map data (54%), classical survey data (46%), and aerial range data/laser scanning data (20%). At that time, range data were regarded by many producers as too expensive. In addition, the usual point density acquired at that time in airborne laser scanning (ALS) did not really allow for a broader application of object extraction in urban areas.

Semiautomatic and automatic methods for 3D city models are aimed at reducing the costs of providing these data with reasonable level of detail. Today, aerial photogrammetry is still one of the main techniques for obtaining 3D building information. Digital aerial photogrammetry supports accurate measurement of points and structures, which are usually defined by a human operator (Brenner, 2005). Despite significant research efforts in the past, the low degree of automation achieved has remained the major problem. Thus, the majority of development work has recently focused on semiautomatic systems, in which, for example, recognition and interpretation tasks are performed by the human operator, whereas modeling and precise measurement are supported by automation.

Due to the development of scanning systems and improvements in the accuracy of direct georeferencing, ALS became a feasible technology to provide range data in the early 1990s. At the end of that decade ALS was already considered as a mature technology (Baltsavias, 1999). ALS provides dense point clouds (more than 10 returns per square meter are possible today) with 3D coordinates. This makes range data segmentation relatively easy and feasible for the modeling of buildings (Brenner, 2005).

A new research field was opened with the integration of laser point clouds and photogrammetric processes with aerial photos. This also provided new technological solutions. By combining the expected good height assessment accuracy of the laser scanner and the expected good planimetric accuracy of aerial images, both high accuracy and higher automation can in theory be obtained. However, despite the progress that has been made with integrating laser scanning systems and digital images, automated processing of the resulting datasets is at a very early research stage (Brenner, 2005).

A short summary of the recent state of the art in building extraction methods can be obtained from Baltsavias (2004), Brenner (2001, 2005), Gülch (2004), and Haala (2004). More details on building extraction methods with emphasis on achievements or impact on the development of the last years can be obtained from Alharthy and Bethel (2004), Ameri and Fritsch (2000), Baillard and Dissard (2000), Baillard and Maître (1999), Braun et al. (1995), Brenner (2003), Brunn et al. (1998), Brunn and Weidner (1997, 1998), Centeno and Miqueles (2004), Chen et al. (2004), Cho et al. (2004), Cord and Declerq (2001), Cord et al. (2001), Dash et al. (2004), Dold and Brenner (2004), Elaksher and Bethel (2002), Elaksher et al. (2003), Fischer et al. (1998), Forlani et al. (2003, 2006), Fraser et al. (2002), Fuchs and Le-Men (2000), Fujii and Arikawa (2002), Förstner (1999), Grün (1997, 1998), Grün and Wang (1998a,b; 1999a,b), Grün et al. (2003), Haala et al. (1998), Haala and Brenner (1999), Haithcoat et al. (2001), Hofmann et al. (2001, 2003), Hofmann (2004), Jaynes et al. (2003), Jutzi and Stilla (2004), Jülge and Brenner (2004), Khoshelham (2004), Kim and Muller (1998), Lee and Choi (2004), Li et al. (2004), Maas and Vosselman (1999), Maas (2001), Madhavan et al. (2004), Mayer (1999), Morgan and Habib (2001, 2002), Neidhart and Sester (2003), Niederöst (2003), Noronha and Nevatia (2001), Oda et al. (2004), Oriot et al. (2004), Overby et al. (2003), Paparoditis et al. (1998), Peternell and Steiner (2004), Rottensteiner (2000, 2003), Rottensteiner and Briese (2002), Rottensteiner and Jansa (2002), Rottensteiner et al. (2003, 2004, 2005), Sahar and Krupnik (1999), Schwalbe (2004), Sequeira et al. (1999), Shufelt (1999), Sinning-Meister et al. (1996), Sohn (2004), Stilla and Jurkiewicz (1999), Süveg and Vosselman (2004), Söderman et al. (2004), Taillandier and Deriche (2004), Takase et al. (2004), Tan and Shibasaki (2002), Tsay (2001), Tseng and Wang (2003), Vosselman (2002), Vosselman and Dijkman (2001), Vosselman and Süveg (2001), Wang and Grün (2003), Weidner and Förstner (1995), Zhan et al. (2002), Zhang and Wang (2004), and Zhao and Shibasaki (2003). We currently observe a period of mutual exchange of experiences from laser scanner data analysis and multi-image matching to derive 3D point clouds, and hence we avoided sorting the example references into fixed categories.

All methods in the cited references are usually evaluated by the producer of the software only. Very rarely methods are compared to several other competing methods. Even less comparisons were done for comparing photogrammetric methods based on digital imagery and methods based on airborne laser scanning data.

Due to the rapid development of sensors and methods during the last years, it was proposed in 2003 and accepted that under the EuroSDR Commission III "Production Systems and Processes," headed by President Eberhard Gülch, a joint test would be undertaken in order to compare various methods. Juha Hyyppä was assigned project leader. The major part of the investigations was carried out by Harri Kaartinen. The Finnish Geodetic Institute (FGI) acted as the pilot center.

The objectives of the EuroSDR Building Extraction project were to evaluate the quality, accuracy, feasibility, and economical aspects of

1. Semiautomatic building extraction based on photogrammetric techniques, with the emphasis on commercial and operative systems.
2. Semiautomatic and automatic building extraction techniques based on high density laser scanner data, with the emphasis on commercial and research systems.
3. Semiautomatic and automatic building extraction techniques based on the integration of laser scanner data and aerial images, where mainly research systems were expected.

This book chapter summarizes the results obtained in the comparison. We first describe the test data and the software used. Then we present the analysis of the test. We will focus on the ALS aspect, but we will include the digital photogrammetric techniques based on image data and also the results of hybrid systems, as they seem to have a high potential for future developments.

We will also focus on buildings, as this is still the major object type in 3D city modeling. EuroSDR has also performed a test on tree extraction from airborne laser scanning data, headed by EuroSDR Commission 2 president Juha Hyyppä, which is, however, not part of the chapter here.

In a final section, we try to give recommendations for further development in the methods applied and the methods by which results are evaluated based on different available reference data.

19.2 Test Data Sets

The project consisted of three sites provided by the FGI, namely Espoonlahti, Hermanni, and Senaatti and one site, Amiens, provided by the Institut Geographique National (IGN). The test sites were selected according to various criteria, such as specific characteristics, namely, density, type, and complexity of buildings, as well as vegetation and undulating terrain aspects, and were also based on the availability of various input data sources and the possibility to generate ground truth data.

For each test site, the following data were available and could be downloaded from the FGI ftp-site:

- Aerial images
- Camera calibration and image orientation information
- Ground control point coordinates and jpg images of point locations (with the exception of Amiens)
- Airborne laser scanner data
- Cadastral map vectors of selected buildings (vector ground plans)
  - Espoonlahti: 11 buildings
  - Hermanni: 9 buildings
  - Senaatti: 6 buildings/blocks
  - Amiens: 7 buildings

The participants were requested to create the vectors of the 3D building models from the given areas in all the four test sites using the materials described above. The participants could use any methods and data combinations they wished. The 3D model should consist of permanent structures of the test area: buildings modeled in as much detail as possible (this mainly concerns roof structures) and terrain so that it would be possible to measure building heights using the model. The vector ground plans provided could be used in methods that require this information. They were also used by the pilot center to check the quality of the results.

19.2.1 FGI Test Sites Espoonlahti, Hermanni, and Senaatti

Each of the test sites had their own characteristics:

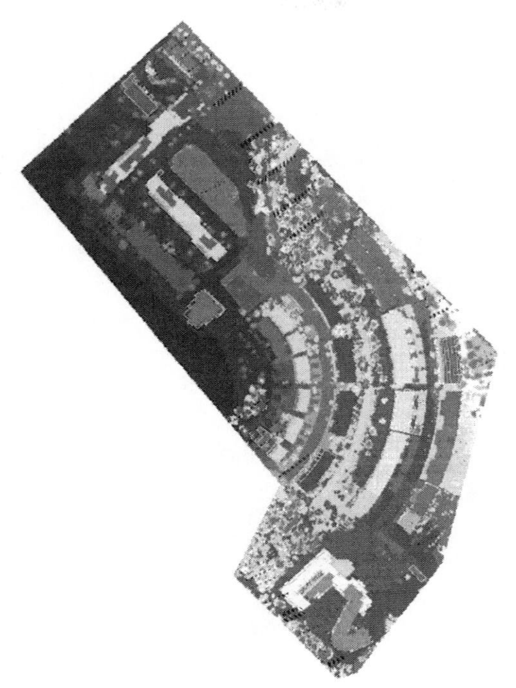

FIGURE 19.1
Airborne laser scanning data of the FGI test site Espoonlahti. (Courtesy of Finnish Geodetic Institute.)

Espoonlahti: With a significant variety of houses, partly row houses, undulating terrain, and a large number of trees, Espoonlahti is located in Espoo, about 15 km west of Helsinki with high-rise buildings and terraced houses. Laser scanner data and one aerial image are shown in Figures 19.1 and 19.2.

Hermanni: A large, simple block of flat or gabled houses with low vegetation, Hermanni is a residential area about 3 km from the main city center with four- to six-storey houses built mainly in the 1950s. Laser scanner data and one aerial image are shown in Figures 19.3 and 19.4.

Senaatti: A typical European city center with some complex, historic buildings and no vegetation, Senaatti includes the area around the Senate Square in Helsinki main city center, three- to six-storey houses and Lutheran Cathedral built mainly in the nineteenth century. Laser scanner data and an aerial image are shown in Figures 19.5 and 19.6.

All FGI test sites were flown with an aerial film camera in large image scales and digitized with a photogrammetric scanner yielding ground pixel sizes of 5.5–7.5 cm (cf. Table 19.1).

In addition to the construction type that differs from site to site, the test sites have also been flown with different laser scanners (TopEye, TopoSys-I, TopoSys-Falcon) and with different pulse densities (from 1.6 to about 20 pulses per m^2) as shown in Table 19.2. The last pulse data was not available for all the test sites, and therefore the first pulse data was considered as the major data.

Airborne images were in the TIFF format, with orientation parameters and ground point coordinates that could be used as control or check points. Laser scanner data

FIGURE 19.2
Digital aerial image data of the FGI test site Espoonlahti. (Courtesy of Finnish Geodetic Institute.)

was in the ASCII format (*XYZ* point data). Cadastral map vectors were in the DXF format.

In the preprocessing of Senaatti and Hermanni data, the laser scanner data was matched visually by the pilot center to map vectors in the *XY*-directions and height

TABLE 19.1
Features of Aerial Images of FGI Test Sites

	Espoonlahti	Hermanni	Senaatti
Photos	Stereo pair	Stereo pair	Stereo pair
Date	June 26, 2003	May 4, 2001	April 24, 2002
Camera	RC-30	RC-30	RC-30
Lens	15/4 UAG-S, no 13355	15/4 UAG-S, no 13260	15/4 UAG-S, no 13260
Calibration date	November 22, 2002	January 18, 2000	April 14, 2002
Flying height (m); scale	860; 1:5300	670; 1:4000	660; 1:4000
Pixel size (μm)	14	15	14
Pixel size on the ground (cm)	7.5	6	5.5

FIGURE 19.3
Airborne laser scanning data of the FGI test site Hermanni. (Courtesy of Terrasolid Ltd and Helsinki City Survey Division.)

was shifted using known height points. These shifts were done after the original laser scanner observations were transformed from WGS84-coordinates to a local rectangular coordinate system.

TABLE 19.2
Features of Laser Scanner Data of FGI Test Sites

	Espoonlahti	Hermanni	Senaatti
Acquisition	May 14, 2003	End of June 2002	June 14, 2000
Instrument	TopoSys Falcon	TopEye	TopoSys-1, pulse modulated
Flight altitude (m)	400	200	800
Pulse frequency (Hz)	83,000	7,000	83,000
Field of view (°)	±7.15	±20	±7.1
Measurement density	10–20/m^2	7–9 per m^2 on the average	1.6/m^2 on the average
Swath width (m)	100	Ab. 130	Ab. 200
Mode	First pulse	Two pulses	First pulse

FIGURE 19.4
Airborne laser scanning data of the FGI test site Hermanni. (Courtesy of Helsinki City Survey Division.)

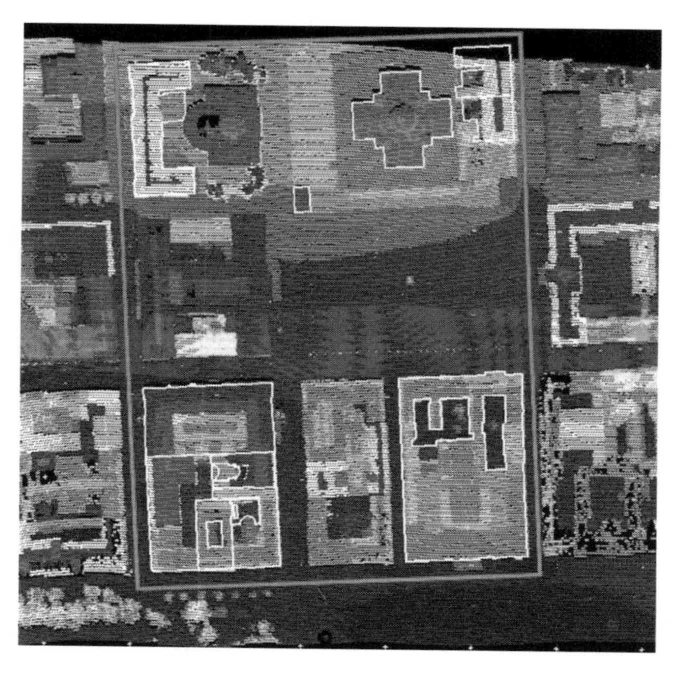

FIGURE 19.5
Airborne laser scanning data of the FGI test site Senaatti. (Courtesy of Blom Kartta Oy.)

FIGURE 19.6
Digital aerial image data of the FGI test site Senaatti. (Courtesy of Helsinki City Survey Division.)

19.2.2 IGN Test Site Amiens

This study site, with a high density of small buildings, is located in the city of Amiens in Northern France. Laser acquisition with a density of 4 points per sqm came from the TopoSys system (cf. Figure 19.7). Airborne images were part of a digital acquisition (June 23, 2001)

FIGURE 19.7
Airborne laser scanning data of the Amiens test site. (Courtesy of IGN, France.)

FIGURE 19.8
Digital aerial image data of the Amiens test site. (Courtesy of IGN, France.)

using IGN's France digital camera. The images were in the TIFF format and the ground pixel size was around 25 cm (cf. also Figure 19.8). The main characteristic of these acquisitions is the large overlap rate (around 80%) between the different images: each point of the terrain appears on a large number of images. Amiens consists of four strips with a total of eleven images. A digitized cadastral map, with a 2D description of buildings was provided in the DXF format.

19.3 Accuracy Evaluation

We decided to focus on two aspects of accuracy evaluation: (1) the analysis using reference points and (2) the analysis using reference raster ground plans. When using reference points we can derive information on the quality of location, building height and building length parameters, and roof inclination. With the comparison to ground plan reference we can derive information on shape similarity or dissimilarity.

19.3.1 Reference Data

On test sites Espoonlahti, Hermanni, and Senaatti, reference data was collected from November to December 2003 using a Trimble 5602 DR200 + tacheometer. Measured targets included corners of walls, roofs, chimneys, and equivalent constructions as well as ground points next to building corners. In this survey altogether, about 980 points were measured in Espoonlahti, 400 points in Hermanni, and 200 points in Senaatti. Known points were used to orientate the tacheometer to the test site's coordinate system.

On all three FGI test sites, repeated observations to the same targets from different station setups were made to control the uniformity and accuracy of reference measurements.

The differences in these repeated measurements were on average 4.7 cm in planimetry (max 8.3 cm) and 1.2 cm in height (max 3.5 cm) based on 19 control observations in total. It is a fact that the distance measurement directly to the surface of a building wall is affected by the angle between the wall and the measuring beam, especially at long distances (as the laser footprint expands). This effect can also be seen in those control measurements where measured distances were somewhat longer than those mainly used in this test, thus, giving a slightly more pessimistic value of the total accuracy expected.

The reference data for the Amiens test site included 32 roof points measured manually from aerial images. The given accuracy (standard deviation) of these points was 25 cm for X, Y, and Z. The reference data was measured and delivered by IGN, France.

After some minor updating and refinements on the basis of reference measurements in the field, the cadastral maps of Helsinki and Espoo City Survey Division were used to produce raster images of building ground plans with a pixel size of $10 \times 10\,cm^2$ for the test sites Hermanni and Espoonlahti.

19.3.2 Accuracy Evaluation Methods

Analysis Using Reference Points: Reference points were used to analyze the accuracy of the location (single point measurement), length (distance between two points), and roof inclination (slope between two points) of the modeled buildings. Single points were analyzed separately for planimetric and height errors. If a digital elevation model (DEM) was included, building heights were measured to the DEM using bilinear interpolation. If not, wall vectors were used. On some models the same ground height was used for whole buildings resulting in large errors in building height determination. These cases were marked at the building height analysis stage and omitted from the figures presenting the building height results.

Root mean squared error (RMSE, Equation 19.1) was calculated for building length, height, and roof inclination:

$$\text{RMSE} = \sqrt{\frac{\sum_{i=1}^{n} (e_{1i} - e_{2i})^2}{n}} \tag{19.1}$$

where

- e_{1i} is the result obtained with the described retrieved model
- e_{2i} is the corresponding reference measured value
- n is the number of samples

Additionally, minimum, maximum, medium, mean, standard deviation, and interquartile range (IQR, Equation 19.2) values were calculated. IQR values represent the range between the 25th and the 75th quartiles:

$$\text{IQR} = p_{75\text{th}} - p_{25\text{th}} \tag{19.2}$$

where

- $p_{75\text{th}}$ is the value at the 75th quartile
- $p_{25\text{th}}$ is the value at the 25th quartile

For example, if the IQR is 20 cm and the median value is 0, then 50% of the errors are within ±10 cm. Significantly deviating measurements (outliers) were detected using threshold levels: the lower bound at the 25th quartile minus 1.5*IQR and the upper bound at the 75th quartile plus 1.5*IQR. IQR is not as sensitive to large deviations (gross errors) as standard deviation/error. Additionally, the coefficient of determination R^2 was calculated to help separate cases between low variability of the reference data and high estimation accuracy and high variability of the reference data and low estimation accuracy. Descriptive statistics are presented before and after outliers have been removed.

Analysis Using Reference Raster Ground Plans: In the Espoonlahti and Hermanni test sites, reference raster ground plans were used to compute the total relative building area and total relative shape dissimilarity (Henricsson and Baltsavias, 1997). The total relative building area gives the difference between the modeled building area and the reference building area. The total relative shape dissimilarity is the sum of the area difference and the remaining overlap error, i.e., the sum of missing area and the extra area divided by the reference area. The vector models delivered by the participants were rasterized (pixel size 10 cm) and compared to reference raster ground plans. For those participants that modeled only buildings with given ground plans, the reference raster ground plans consisted of only of these buildings. The total relative building area and shape dissimilarity were computed in two ways. Firstly, all buildings in the test sites were included, giving the values for the whole test site. Secondly, only those buildings that existed on both reference and delivered model were included, giving the values for modeled buildings. The eaves of the roofs were not included in the reference raster ground plans, thus making the reference buildings somewhat smaller than the extracted buildings in the Hermanni test site (in Espoonlahti, there were no eaves, just straight walls from the ground to the roof).

19.4 Participants Results

Three-dimensional models were obtained from 10 participants coming from 11 organizations (Table 19.3). In this table, we give the used formats of the building vectors received, and we have indicated the cases where we also received DEM information by the participants.

19.5 Applied Methods and Software by Partners

The participants have applied quite different methods and Table 19.4 summarizes the data use, degree of automation, and time use of those applied methods. The participants and their specific method are sorted based on the data used. First we give the purely image-based methods, then some hybrid methods, and finally the methods relying on laser scanning data. Among these we present, at the end, the methods that require ground plan data. The level of automation is indicated as given by the participants and the ranges from low to high. We can already note here a tendency for high automation with the purely laser scanning methods using given ground plan information, whereas the methods relying on image data are mainly classified as low to medium. The time use was also given by the participants in one of the three categories: low, medium, and high. Absolute figures in hours or minutes were not received and partly no answers were given, or they were unclear. Surprisingly in the expected difficult sites Senaatti and Espoonlahti we can observe low time usage also, whereas in the easy to moderate case of Hermanni we can find medium to high time usage also. We have added the experience of the operator as a further feature, showing that some of the software can be handled without extensive training.

TABLE 19.3
Participant-Generated 3D-Models and Data Formats

	Test Site							
	Espoonlahti		Hermanni		Senaatti		Amiens	
Participant	**Building Vectors**	**DEM**	**Building Vectors**	**DEM**	**Building Vectors**	**DEM**	**Building Vectors**	**DEM**
CyberCity AG, Switzerland	DXF	ArcView	DXF	ArcView	DXF	ArcView		
Delft University of Technology, The Netherlands			DXF and VRML	VRML	DXF and VRML	VRML		
Hamburg University of Applied Sciences and Nebel + Partner, Germany	DXF		DXF		DXF		DXF	
Institut Geographique National, France (IGN)	DXF and VRML	DXF and VRML	DXF and VRML	DXF and VRML	DXF and VRML	DXF and VRML	DXF and VRML	DXF and VRML
Swedish Defence Research Agency (FOI)	DXF		DXF		DXF		DXF	
University of Aalborg, Denmark			VRML					
C+B Technik, Germany	DXF and ASCII		DXF and ASCII		DXF and ASCII		DXF and ASCII	
Institut Cartografic de Catalunya, Spain (ICC)	DGN		DGN		DGN		DGN	
Dresden University of Technology, Germany			VRML		VRML			
University of Applied Sciences, Stuttgart, Germany	DXF		DXF		DXF			

In the following, we describe the single methods that follow the above order. We base the descriptions on the description provided by the participants and the references in literature. Please note that some of the methods may be not described in a comparable level of detail, due to confidentiality reasons.

19.5.1 CyberCity AG

CyberCity used aerial images, camera calibration, and exterior orientation information. Work was done by experienced personnel. CyberCity methods are based on concepts reported by Grün and Wang (1998a,b; 1999a,b) and commercialization of the methods in

TABLE 19.4

Summary of Used Data, Level of Automation, and Time Use of Building Extraction

		Used Data				Time Use			
Participant	**Operator**	**Laser Data (%)**	**Aerial Images (%)**	**Ground Plan**	**Level of Automation**	**Espoonlahti**	**Hermanni**	**Senaatti**	**Amiens**
Cybercity	Exp.		100		Low	Medium	Medium	Medium	
Hamburg	Inexp.		100		Low	Medium		High	
Stuttgart	Inexp.		100		Low-high	Low	Low	n.a.	
IGN (Amiens)	Exp.		100		Medium				Low
IGN (Espl, Her, Sen)	Exp.	50	50		Medium	Low	Low	Low	
ICC laser + aerial	Inexp.	80	20		Low-medium	High	Medium	High	
Nebel + Partner	Inexp.	90	10		Medium	High	High	Medium	High
ICC laser	Inexp.	100			Low-medium	High	High	High	n.a.
FOI	Exp.	100			High	Low	Low	Low	Low
FOI outlines	Exp.	100		X	High	Low	Low	Low	Low
C + B Technik	Exp.	100		X	Medium-high	n.a.	n.a.	n.a.	n.a.
Delft	Exp.	100		X	Medium-high		n.a.	Medium	
Aalborg	Inexp.	100		X	High		n.a.		
Dresden	Exp.	100		X	High		Low	Low	

Note: n.a., not available.

CyberCity AG. The modeling is based on two steps. First, manual photogrammetric stereo measurements are carried out using an in-house developed digital photogrammetric workstation Visual Star. The point measurements are taken in a certain order and the points are attached certain codes using CyberCity's Point Cloud Coding System, which will help the more automatic process. Then, the resulting point clouds are imported into the main module CC-Modeler, which automatically triangulates the roof points and creates the 3D building models. The building footprints (i.e., derived from the roof outline) and walls are generated by projecting and intersecting the boundary roof points with the Digital Terrain Model (DTM), if that exists. Alternatively, cadastral building footprints can be projected back to the roof, which results in realistic roof overhangs and fulfills the consistency between 2D cadastral maps and 3D city models. With CC-Edit, geometric corrections can be carried out to get accurate and correct 3D city models with planar faces and without overlaps or gaps between neighboring buildings, etc. This is required to derive orthogonal and parallel lines of buildings from the manual stereo measurements.

19.5.2 Hamburg University of Applied Sciences

Hamburg University of Applied Sciences used a digital photogrammetric workstation DPW770 with SocetSet from BAE Systems/Leica Geosystems (BAE System, 2000) to extract buildings by manual stereo measurements using aerial images, camera calibration, and exterior orientation information. Photogrammetrically measured data was corrected and improved in an AutoCAD 2000 system. All measurements and modeling were performed by an inexperienced person. After importing the digital aerial images in the digital station using the given camera calibration and exterior orientation data, an automatic interior orientation of each image was performed. The quality of the exterior orientation of each stereo pair was checked for existing y-parallaxes and by comparing the measured and given control points. If the exterior orientation was accepted, each building was measured manually by the operator using the software module Feature Extraction of SOCET Set. The operator could select three different roof types (flat, peaked, and gabled) and the sequence of the photogrammetric measurements depends on the roof type. Feature Extraction (Auto Create Mode) automatically creates the walls of the buildings based on the value specified for the ground. Complex buildings were broken down into simpler components, which were later combined in AutoCAD. Finally, the measured data was transferred to AutoCAD 2000 for the correction of some measurements and for modeling of the complex buildings. This approach can be regarded as highly manual.

19.5.3 Stuttgart University of Applied Sciences

Stuttgart used inJECT1.9 (prerelease) software to semi-automatically extract buildings using aerial images, camera calibration, and exterior orientation information. For a description of the used automated tools see (Gülch and Müller, 2001). An inexperienced student performed the building extraction as a part of a diploma thesis, supervised by E. Gülch.

The general workflow was as follows:

- Preparation of orientation and image data and direct import to inJECT1.9 (prerelease)
- Measurement of buildings without stereo-viewing using parametric (Figure 19.9) or polyhedral (Figure 19.10) 3D building models with one common ground height for a building part or for a building composite. Partly, the building models were measured as CSG structure forming composite buildings. These were not merged inside inJECT, but were merged externally. The results presented here give the

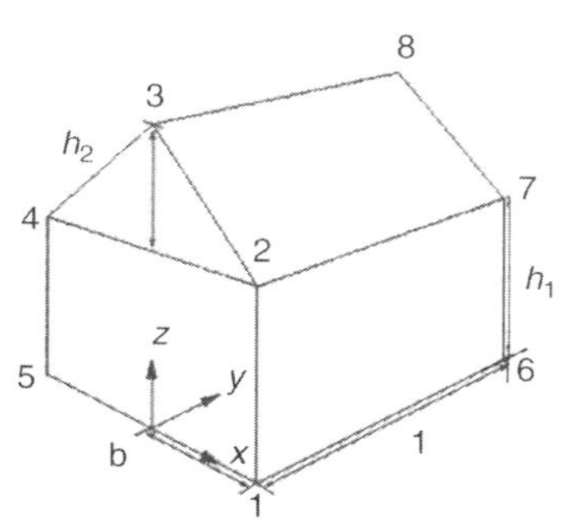

FIGURE 19.9
Parametric building models (here saddleback-roof building in the Hermanni test site). (Courtesy of University of Applied Sciences, Stuttgart.)

pure measurements inside inJECT. Snapping function, rectangular enforcement for polygonal measurements, and automatic enforcement of planarity were frequently used for the polyhedral buildings. Basically all ground height measurements were done by image matching. The rooftop height and some shape features of saddleback and hip-roof buildings were measured by area-based and feature-based image matching.

- Storage in GML3 and export to DXF and partly VRML (Figure 19.11).

In addition to the derived GML3 and DXF files, a visualization using VRML was automatically derived with automatically extracted texture (see example in Figure 19.11).

19.5.4 IGN

IGN used calibrated aerial images in the multiview context in the Amiens test site and calibrated aerial images and laser DSM in the Espoonlahti, Hermanni, and Senaatti test sites.

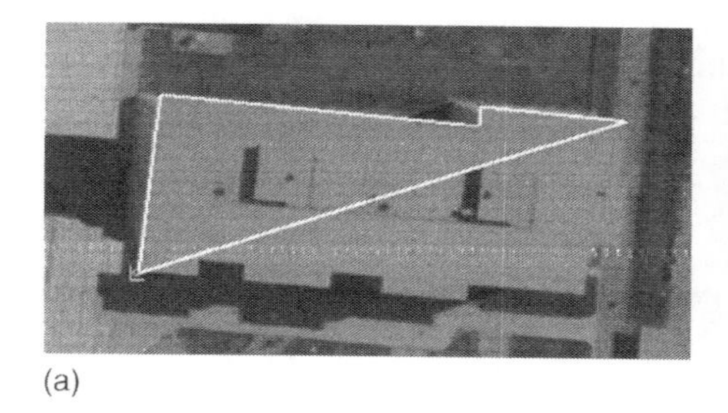

(a)

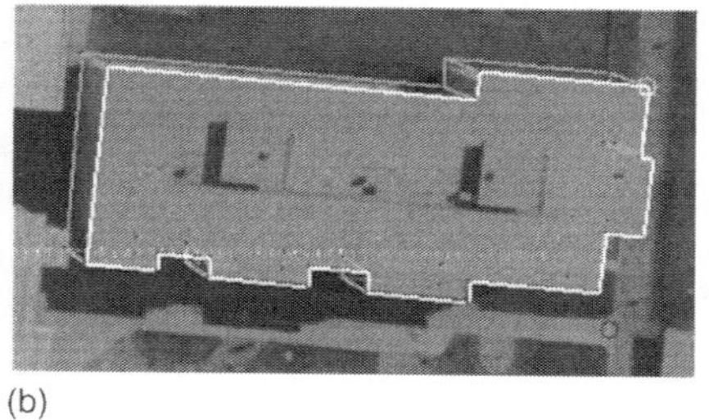

(b)

FIGURE 19.10
Polyhedral flat-roof building. (a) First five corners are measured, and (b) whole roof outline is measured and with one matched ground point the walls are derived. (Courtesy of University of Applied Sciences, Stuttgart.)

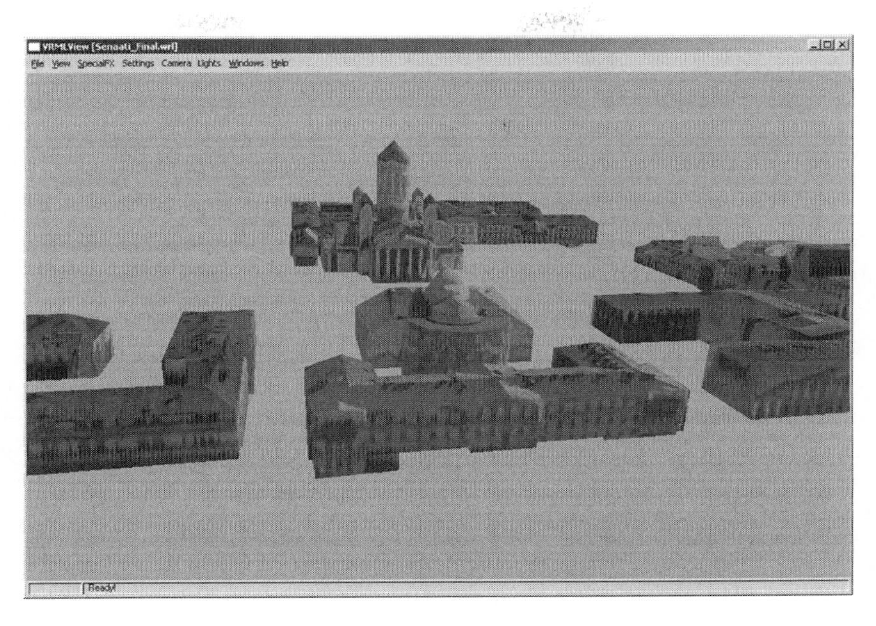

FIGURE 19.11
Building extraction in the FGI test site Senaatti with inJECT with parametric and polyhedral building models and textured 3D view in a VRML Browser. (Student exercise by A. Novacheva and S.H. Foo, supervisor E. Gülch.)

All work was carried out by experienced personnel: one person working with Amiens, Espoonlahti, and Hermanni and one with Senaatti. For each test site, IGN created a pseudo-cadastral map manually using aerial images.

The workflow was divided into preparation (fully automatic procedures such as DSM and true orthoprocessing), cadastral map edition and pruning, 3D reconstruction of buildings, and quality control. The materials and methods varied between test sites.

Amiens Test Site: First a DSM and the true ortho-image were processed by correlation using the multiview context. Then two methods of reconstruction were used. For prismatic models, the 2D shape was edited and the median height in the DSM was measured. For other models, the skeleton of the central ridge was edited manually in one image and the system automatically reconstructed the 3D shape. Finally, the quality was controlled with a difference image of 3D polygons and DSM.

Espoonlahti Test Site: Pseudo-cadastral maps were edited on a single image with height adjustment using the roll-button of the mouse. For each 2D polygon, a median height was measured on the laser DSM. Finally, the quality was controlled with a difference image of 3D polygons and DSM.

Hermanni Test Site: Pseudo-cadastral maps were edited on a single image with height adjustment using the roll-button of the mouse. Reconstruction was fully automatic using these 2D polygons and laser DSM.

Senaatti Test Site: On this test site, several modules were used. Sometimes pseudo-cadastral maps were edited on a single image with height adjustment using the roll-button of the mouse. After this, reconstruction was carried out using a model-driven approach (Flamanc et al., 2003). Moreover, manual tools were used for complex buildings. Finally, special tools such as dome edition mode were used to extract specific structures.

19.5.5 ICC

ICC used TerraScan, TerraPhoto, and TerraModeler software by Terrasolid (Terrasolid, 2003) to extract buildings using laser scan data with aerial images (marked with "ICC laser + aerial" in the results) and without aerial images (marked with "ICC laser" in the results). All work was carried out by one person. The major difference of the applied approach with the Nebel + Partner, who also used TerraScan for the test, was that in the ICC process, the orientation information of aerial images was applied. Thus, the building outlines could be derived using the image data.

19.5.6 Nebel + Partner GmbH

Nebel + Partner used TerraScan software by Terrasolid (Terrasolid, 2003) to extract buildings using laser scan data. All measurements and modeling was performed by an inexperienced person. Aerial images were not used for any measurements. Image crops were only used as superimposed images for better visual interpretation of the laser point clouds during the measurement of the roofs. Before manual measurements with the point clouds, each laser point set was automatically classified as buildings, ground elevation, and high, medium, and low vegetation by TerraScan. In the TerraScan software tool—Construct Building—an algorithm automatically finds roofs based on laser hits on the planar surfaces of the roofs, which results in vectorized planes of each roof. Roof boundaries can also be created or modified manually. In TerraScan, the operator can select three different building boundary types (rectangle, rectangular, and polygon). A rectangle boundary consists of exactly four edges, while a rectangular boundary has more than four edges; a polygon is a free-form polygon with angles also different from 90°.

19.5.7 FOI

FOI used their own in-house software and methods to extract buildings using laser data without a given ground plan (marked with "FOI" in the results) and with a given ground plan (marked "FOI outlines" in the results). All work was carried out by an experienced person.

Without a ground plan, the three preprocessing steps of the building extraction algorithm were as follows:

1. Rasterize the data to obtain digital surface models (DSMzmax and DSMzmin).
2. Estimate the terrain surface (DTM).
3. Classify the data above the ground surface (vegetation/buildings).

Each group of connected pixels classified as buildings were used as the ground plan for the building extraction algorithm.

When using the given ground plan, instead of classifying the data (step 3 above), the outlines were used to create a classification image with buildings. The classification image is used to specify the ground plans of the buildings. The outlines were not used when estimating the roof polygons along the contour of the building.

The raw laser scanner data was resampled into a regular grid ($0.25 \times 0.25\,m^2$). Two grids were saved, one containing the lowest height value in each cell (DSMzmin), and one containing the highest z-value (DSMzmax). The building reconstruction methods are based on DSMzmin in order to reduce noise from antennas, for example. The ground surface (DTM) is then estimated (Elmqvist, 2002). Having estimated the ground surface, the remaining pixels are further classified as buildings and vegetation. Each group of connected pixels classified as buildings used as ground plan for the building extraction algorithm. If the given ground plan is used for the building extraction algorithm, the classification of buildings and vegetation is not needed.

For each ground plan detected in the classification or obtained otherwise, elevation data is used to extract planar roof faces. The planar faces of the roof are detected using clustering of surface normals. The algorithm works on gridded but not interpolated data. For each pixel within the building ground plan, a window is formed. The surface normal is estimated by fitting a plane to the elevation data within the window. The plane parameters are estimated using a least squares adjustment.

A clustering of the parameters of the surface normals is then performed. In the clustering space, the cell with the largest number of points is used as initial parameters of a plane and an initial estimation of which pixels belong to the plane is made using these parameter values. Finally, an enhanced estimation of the plane parameters using a reweighted least squares adjustment is performed and a final estimation of the pixels that belong to the plane is obtained. These pixels form a roof segment. This clustering process is repeated on the remaining nonsegmented pixels within the ground plan. Only the roof segments having a certain minimum area are retained. After the clustering process, region growing is performed on the roof segments until all the pixels within the ground plan belong to a roof segment.

Having segmented the roof faces of a building, the relationship between the faces is defined. By following the outlines of the roof faces, the so-called "topological points" are first added. Topological points are defined as vertices where a roof face's neighbor changes. Next, each section of an outline between any two topological points is defined either as an intersection line, height jump section, or both. For sections defined as both an intersection line and jump section, a new topological point is added to split the section.

The jump sections are often noisy and difficult to model. Therefore, for each jump section, lines are estimated using the 2D Hough transform. The slopes of the estimated lines are then adjusted according to the orientation of the plane. The intersection points of the estimated lines are used as vertices along the jump sections. These vertices along the jump sections are used together with the topological points to define the roof polygons.

After having defined the roof polygons, a 3D model of the building can be created. The roof of the model is created using the defined roof polygons. The height value of the vertices is obtained from the z-value of the plane at the location of a vertex. Wall segments are inserted between any two vertices along the jump sections.

19.5.8 C+B Technik

C+B Technik used their own software to extract buildings using laser scan data and the given ground plans. In the first step, a triangle net is created from the laser scanner data, which builds the basic data structure for the modeling process. The basic computation method selects the laser scanner points within a building polygon. Triangles are combined to create surfaces and triangle sides are classified as edge lines. The resulting surfaces, edges, and corners are analyzed and edited to achieve the typical building objects, which are, for example, inner vertical walls, horizontal or tilted roof planes, and horizontal ridges. The surfaces are also adapted to the building polygon, so that, for example, an edge line and a house corner fit together. For each situation a complete automatic solution is not possible, so a complete check and possible editing of the result must be performed. The extraction of buildings from laser scanner data is split first into an automatic computation step and then into an interactive check and editing step. The results for each building have the following logical structure: the ground polygon, the outer walls, the inner walls, and the roof surfaces. The results can be output as ASCII- or DXF-data.

19.5.9 Delft University of Technology

Delft used their own software and methods to extract buildings using laser data and ground plans based on studies by Vosselman and Dijkman (2001) and Vosselman and Süveg (2001). All work was carried out by an experienced person. If building outlines were not available, they were manually drawn in a display of the laser points with colour-coded heights. In the Hermanni test site, only buildings with a given ground plan were modeled.

If the point cloud within a building polygon can be represented by a simple roof shape (flat, shed, gable, hip, gambrel, spherical, or cylindrical roof), the model of this roof is fitted to the points with a robust least squares estimation. Often, building models have to be decomposed interactively such that all the parts correspond to the above-mentioned shape primitives. If the point cloud is such that all roof planes can be detected automatically, an automatic reconstruction is attempted based on the intersection of the detected neighboring roof faces and the detection of height jump edges between roof faces.

If the two situations above do not apply, the building polygon is split into two or more parts until each part fulfills one of the two conditions mentioned above. Optionally, point clouds can be edited to remove outlier points that would disable an automatic roof reconstruction.

19.5.10 University of Aalborg

The modeling of buildings was carried out by three university students as part of their studies (Frederiksen et al., 2004). The applied method used laser scanner and 2D vector map data to extract and model the buildings. The applied method can be used to model rectangular buildings with gable roofs and rectangular buildings with gable roofs at different levels.

A 2D vector map is used to extract the data within the buildings and the 2D vector map is also used as building outline.

After the extraction of the data, the data is interpolated into a rectangular grid. For every point in the grid, a surface normal based on the point and its neighboring points are calculated (see Burrough and McDonnell, 1998, pp. 190–192). Now all points belonging to a roof plane have to be found. An assumption is made: the two longest parallel sides of the building each have one roof plane orientated towards the side of the building, which means a surface normal on the roof plane is perpendicular in the xy-plane to the side of the building. This means that every point is examined regarding its surface normal, and the surface normals perpendicular to the side of the building and with approximately the same orientation are grouped. The result is two groups of surface normals—each group representing a roof plane.

The points of each roof plane are now adjusted by the principles of least squares adjustment and a method called the Danish method (Juhl, 1980). The Danish method is a method used in adjustments and is used to automatically weight down points with large residuals compared to the determined roof plane from the least squares adjustment. This means that points that do not belong to the roof plane, e.g., points on chimneys are sorted out of the group automatically and will not influence the final determination of the parameters of the roof planes (Juhl, 1980).

Gable roofs at different levels will also show only two groups of points with approximately similar surface normals, because the method cannot distinguish between roof planes at different levels. To find the border between the buildings, a gradient filter is applied to the grid and large gradients are grouped by means of a connected components analysis.

Groups are adjusted by least squares adjustment to a line and thus a building with gable roofs at different levels is divided into rectangular buildings with two roof planes.

19.5.11 Dresden University of Technology

The Dresden method (Hofmann, 2004) only uses point clouds obtained by a presegmentation of the airborne laser scanner data. All work was carried out by an experienced person. It is a plane-based approach that presumes that buildings are characterized by planes. It utilizes a TIN-structure that is calculated into the point cloud. The method only uses point clouds of the laser scanner data that contain one building. In order to get such point clouds, polygons coarsely framing the building can be used to extract the points (e.g., in ArcGIS). The polygons can be created manually or map or ground plan information can be used.

The parameters of every TIN-mesh, which define its position in space uniquely, are mapped into a 3D triangle-mesh parameter space. As all triangles of a roof face have similar parameters, they form clusters in the parameter space.

Those clusters that represent roof faces are detected with a cluster analysis technique. By analyzing the clusters, significant roof planes are derived from the 3D triangle parameter space while taking common knowledge of roofs into account. However, no prior knowledge of the roof as, for example, the number of roof faces is required. The obtained roof planes are intersected in accordance to their position in space. The roof outlines are determined by analyzing the intersected roof faces and the ground plan is derived.

19.5.12 Discussion

From the descriptions above we can not only identify commercial systems but also in-house systems used by various organizations. We have only two commercial systems (CyberCity, inJECT) dedicated for some semiautomatic extraction from digital imagery. We have at least one in-house system (IGN) for it as well. Others are based on standard digital photogrammetric workstations. We have a majority of systems using airborne laser scanning data, but we can identify only few commercial systems like Terrascan and C+B, one in-house system (FOI), and several very advanced research approaches. Several participants based their results on inexperienced personnel, which seems to be an indication that those software packages can be handled without extensive training, unlike earlier stereo plotter measurements that required highly trained personnel. We can also observe the need for some approaches to have ground plan information available, which certainly has some effect on the overall applicability and costs which is restricted or influenced then by the availability and costs of such additional data. From the hybrid approaches, we can find various ways how the second data source (images, laser data) is actually used. There is so far no clear trend visible for a best strategy to involve both.

19.6 Results of the Empirical Test

We want to first present the results of the single approaches and then we want to compare laser scanning approaches with photogrammetric approaches based on image data. Here, we group also the different degrees of hybrid approaches using both input sources

We want to classify the results in each case in the following categories:

- Building location, height, and length
- Roof inclination

In a last major step, we focus on the effects of automation for the used methods.

We present the major results for different data sets and give comments on their analysis. For more details on the full evaluation, we refer to Kaartinen et al. (2005a,b) and Kaartinen and Hyyppä (2006).

19.7 Results of Single Approaches

19.7.1 Building Location, Height, and Length

Figure 19.12 shows the differences between various models with respect to elevation and planimetric errors for single point targets defined by IQR quality measure. The quality measures are given separately for north–south and east–west directions due to the different point spacing in along and across track directions of TopoSys systems. Building location accuracies for all modeled buildings in Espoonlahti are given for all and for selected buildings (same four buildings modeled by all participants).

Figure 19.13 gives information about building height defined by IQR and RMSE quality measures. Some methods do not use DEM information; in these cases no values are given. Figure 19.14 gives information about building length defined by IQR and RMSE measures.

CyberCity achieved a good quality in all three Finnish test sites. However, it should be remembered that measurements in CyberCity are done manually from stereoscopic images and automation is mainly used for connecting points to derive 3D models. Thus, the manual part of the process is verified in this analysis. The quality improved when image quality was better, but in general the accuracy variation within the test sites was low.

The Hamburg and Stuttgart models were affected more by the site and site-wise data characteristics. Since the Hermanni test site was relatively easy, the Stuttgart model used high automation in Hermanni resulting in decreased performance. With the most difficult test site, Espoonlahti, the Stuttgart model was almost as good with all buildings as the CyberCity model and even better with selected buildings (same buildings modeled by all partners). It has to be remembered that the Stuttgart process was done by a student and the CyberCity process was done by an expert. The Hamburg model showed lower performance than these two other photogrammetric techniques even though it was based on a more standard photogrammetric process but with an inexperienced operator.

Hybrid techniques using TerraScan (see ICC laser + aerial) showed a quality almost comparable to the best photogrammetric methods. The same method (i.e., TerraScan) used by Nebel and Partner showed typically lower accuracy, which can be explained by the lower amount of time spent and the lower amount of aerial image information used in the process. In ICC, the aerial images and image orientations were used to measure the outlines, and in Nebel and Partner, the images were used for visual interpretation purposes. It seems that good quality results can be obtained when building outlines are measured using image information and laser-derived planes.

In Hermanni, where all participants using laser data provided a model, the ICC model (based on TerraScan) showed the best accuracy. Also, the Delft and Aalborg models (the latter developed by undergraduate students within the project) resulted in reasonably good accuracy. Delft, Aalborg, and C+B Technik mainly used given ground plans to determine building location, so planimetric accuracy is directly affected by the ground plans that do not include roof eaves. This can also be seen in the results: the planimetric accuracy is almost the same for these three and also point deviations in Hermanni seem to have common points. It should be noted that the generation of the FOI model was fully

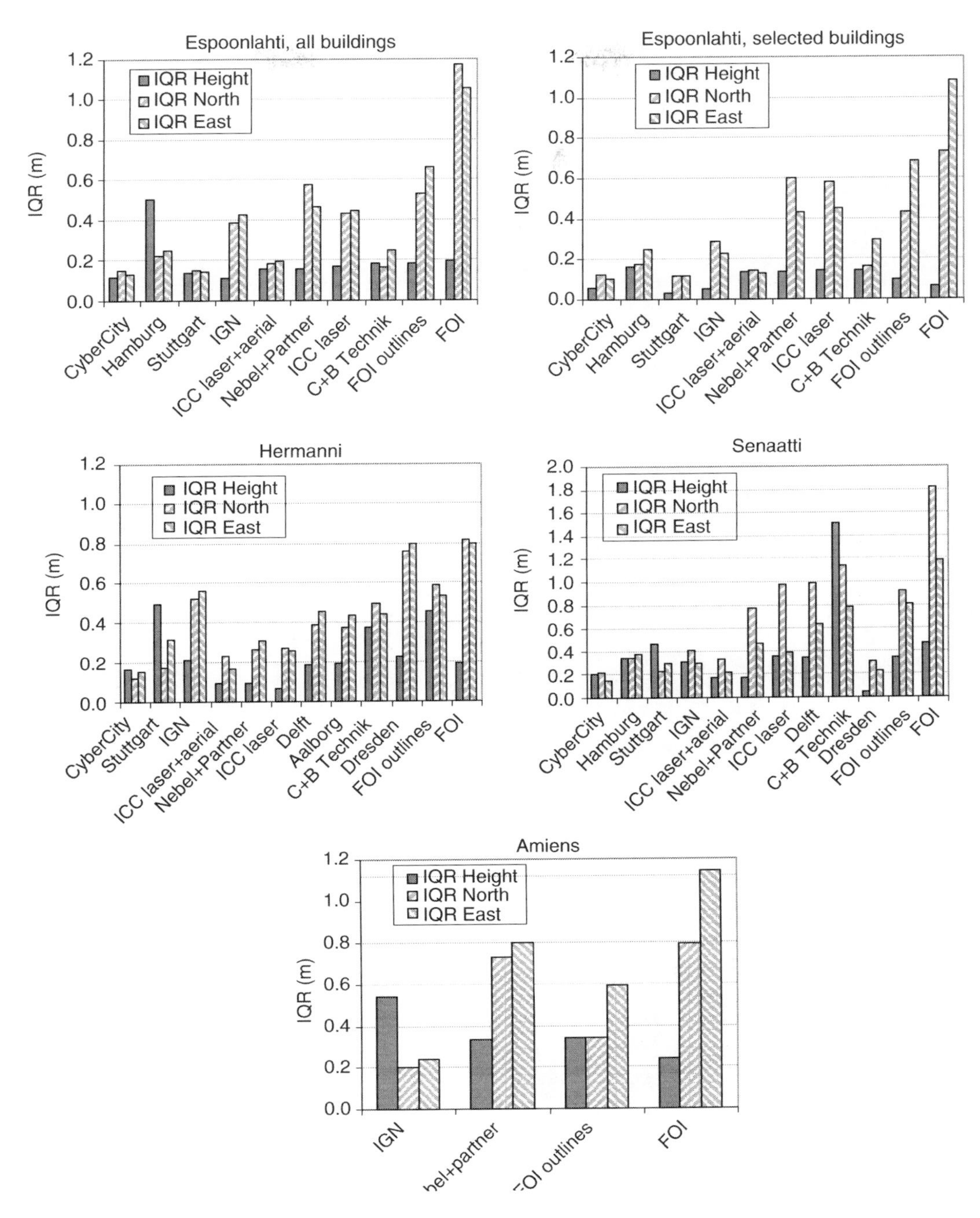

FIGURE 19.12
Location accuracy (IQR) of the models with respect to single points. The selected buildings (four) in Espoonlahti had been measured by all the participants.

automatic and their newest extraction algorithm was based on the use of the first and last pulse data, and the last pulse data was not provided in the test, since it was not available from all test sites. C+B did exceptionally well in Espoonlahti, which was the most difficult

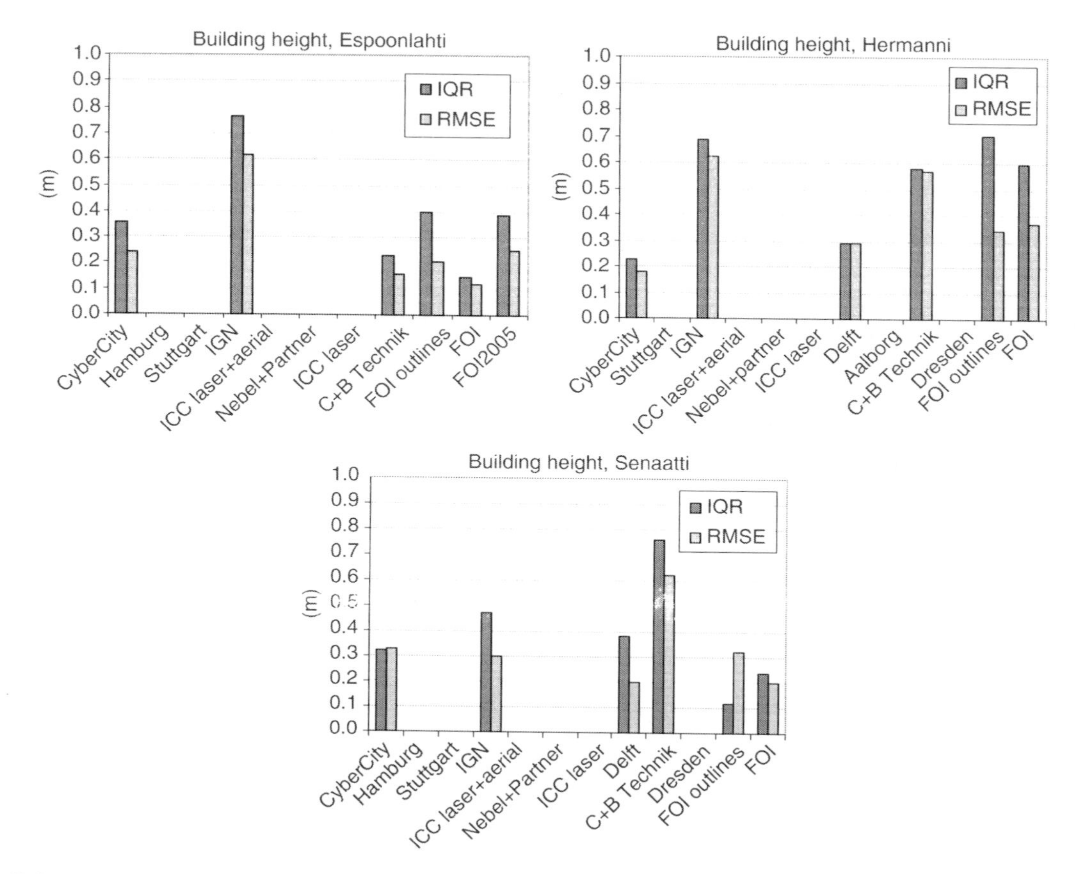

FIGURE 19.13
Building height accuracy. IQR: all observations, RMSE: outliers removed. For some approaches that did not use DEM information, no results are given.

area, but they provided only the buildings where outlines were given. One possible explanation for the variability of the C+B performance in various test sites is that since it is based on TINs, it works better than plane-based models (laser points defining a plane) with more complex and with small structures. Also, the Dresden model performed extremely well in Senaatti, but not in Hermanni. The number of points used in the analysis was extremely low for the Dresden model.

19.7.2 Roof Inclination

Figure 19.15 gives the roof inclination accuracy for each model, defined with IQR and RMSE quality measures.

In Hermanni, the best result was obtained with the Dresden model and basically all laser based methods resulted in less than 1° error (RMSE). The C+B approach had a larger error than the photogrammetric processes. Most probably the TIN-based principle of C+B does not give as good results as using all the points hitting one surface and defining a plane using all of them. In photogrammetry, the roof inclination is obtained from two measurements, and therefore the same accuracy as laser was clearly not achieved. When the building size was smaller and the pulse density was lower, this difference between photogrammetry and laser scanning was reduced or even disappeared (Figure 19.15).

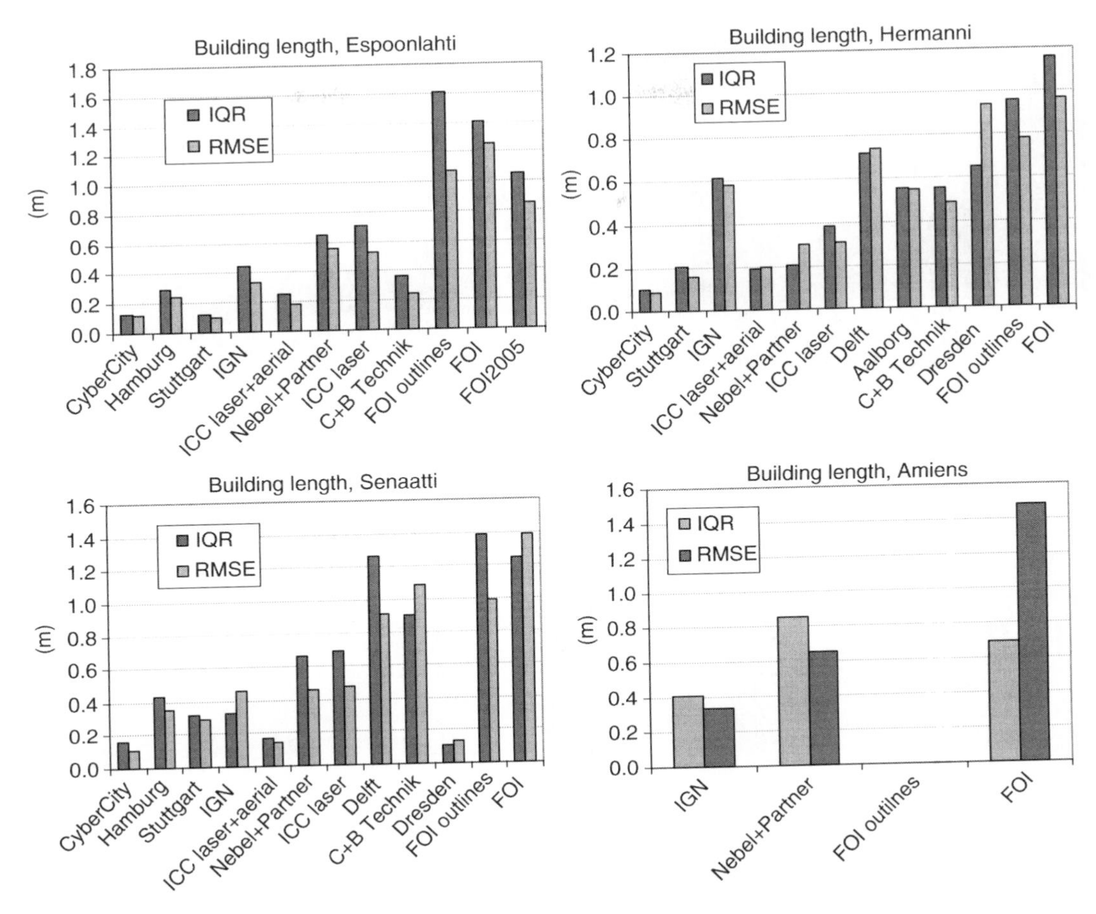

FIGURE 19.14
Building length accuracy. IQR: all observations, RMSE: outliers removed.

19.8 Comparison of Laser Scanning and Photogrammetry

The purpose here is to give an evaluation of the performance of the applied methods based on laser scanning data and those using aerial image data (photogrammetric methods) and all hybrid approaches. We group the results accordingly to derive conclusions for the two major competing approaches currently used.

19.8.1 Building Outlines, Height, and Length

In general, photogrammetric methods were more accurate in determining building outlines (mean of IQR North and IQR East), as may be seen in Figure 19.16. Taking all test sites into account, the IQR value of photogrammetric methods ranged from 14 to 36 cm (average 21 cm, median 22 cm and standard deviation (std.) 7.2 cm of IQR values). The corresponding values for aerial image assisted laser scanning ranged from 20 to 76 cm (mean 44 cm, median 46 cm, std. 18.5 cm). Laser scanning-based building outline errors ranged from 20 to 150 cm (mean 66 cm, median 60 cm, std. 33.2 cm). We also display the linear regression line where applicable in the following figures.

Point density, shadowing of trees, and the complexity of the structure were the major reasons for site-wise variation of the laser scanner based results. The lowest accuracy

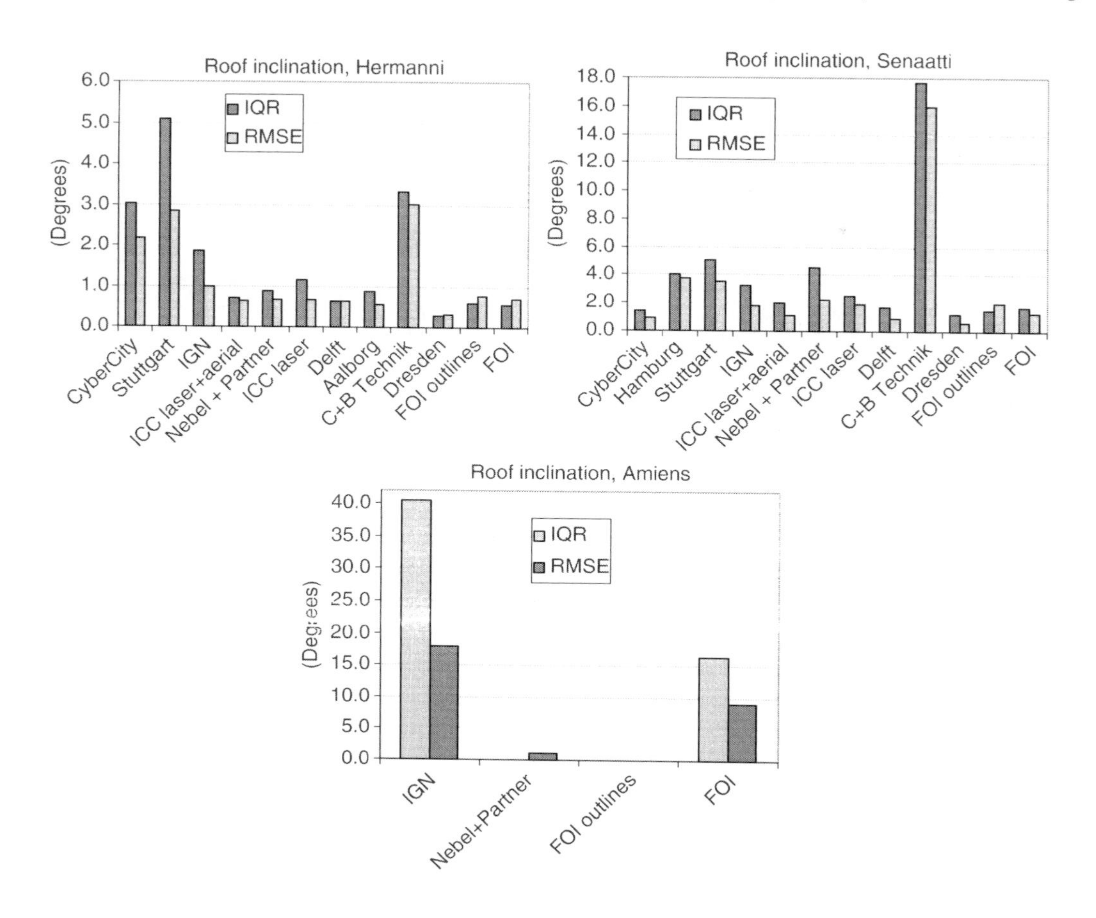

FIGURE 19.15
Roof inclination accuracy of the models. IQR: all observations, RMSE: outliers removed.

was obtained with the lowest pulse density (Senaatti). Also in Amiens, the complexity deteriorated the performance. It was almost impossible to reveal the transition from one house to another using DSM data in Amiens. The small number of trees, simple building structure, and relatively high pulse density resulted in the highest accuracy at the Hermanni test site.

The effect of the point density on the achieved average accuracy (planimetric, mean of IQR North and IQR East, and height errors) is depicted in Figure 19.17. The figure clearly shows how point density improvement can slowly reduce both height and planimetric errors. On average, we can state that a 10 times increase in pulse density will result in the reduction of errors by 50%.

In building length determination (Figure 19.18), laser-based methods were not as accurate as the photogrammetric methods, as could be expected from the above. The photogrammetrically derived lengths varied from 14 to 51 cm (RMSE, mean 26 cm, median 22 cm, std. 12.6 cm). Lengths obtained with aerial image assisted laser scanning varied from 19 to 108 cm (mean 59.4 cm, median 57 cm, std. 31.2 cm). The laser scanning based lengths varied from 13 to 292 cm (mean 93 cm, median 84.5 cm, std. 60.9 cm). With laser scanning, the complexity of the buildings was the major cause for site-wise variation rather than the point density.

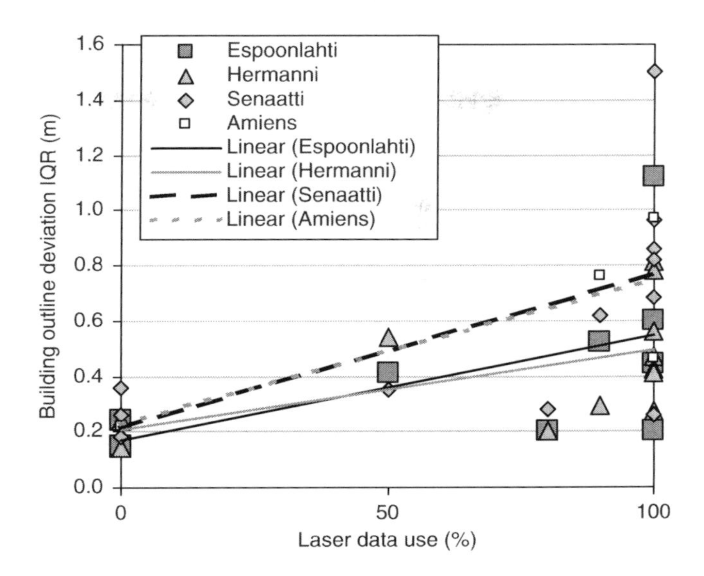

FIGURE 19.16
Building outline deviation, all targets. Laser data use of 0% refers to photogrammetric methods, 100% to fully laser scanning-based techniques, and intermediate values to hybrid techniques.

The IQR value for laser scanning height determination ranged from 4 to 153 cm (mean 32 cm, median 22 cm, std. 31.5 cm). One fully automatic method caused high errors, modifying the mean value. Laser scanning assisted by aerial images resulted in IQR values between 9 and 34 cm (mean 18 cm, median 16.5 cm, std. 8.5 cm). Photogrammetric height determination ranged from 14 to 54 cm (mean 33 cm, median 35 cm, std. 18 cm). The accuracy of height determination exactly followed the laser scanning point density. With the high-density data in Espoonlahti, all participants were able to provide average heights with a better accuracy than 20 cm IQR value.

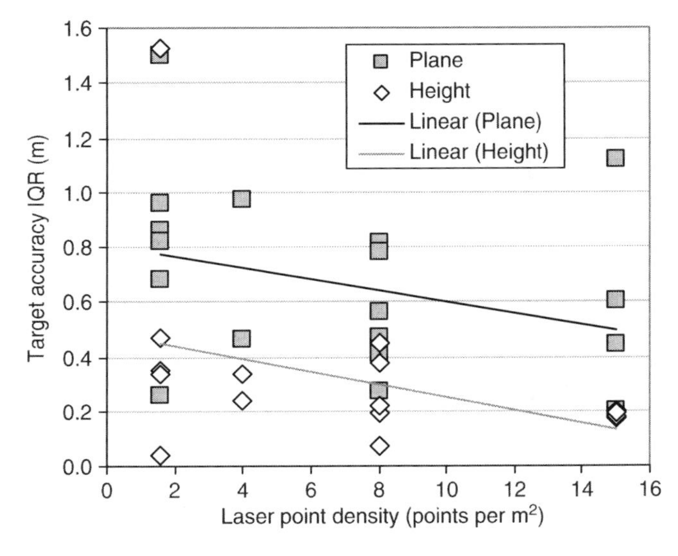

FIGURE 19.17
Average accuracy versus laser point density.

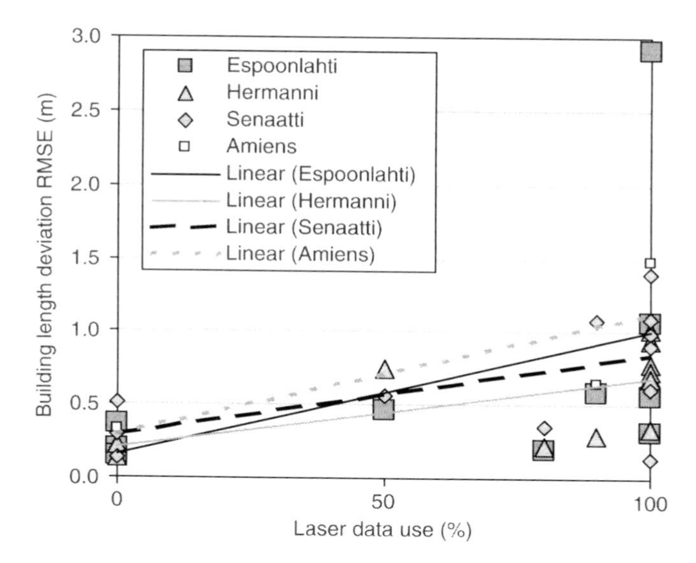

FIGURE 19.18
Building length deviation (all observations.)

The comparison of various laser point densities and methods for height determination and with photogrammetry is given in Figure 19.19. Photogrammetric methods are on the left side (laser data use 0%) and laser scanning methods on the right (laser data use 100%). In the cases of Hermanni, Amiens, and Espoonlahti, the regression line decreases, indicating that on average height determination is slightly more accurate using laser scanner than photogrammetric techniques. In Senaatti, the opposite conclusion can be drawn due to the low pulse density and to an unsuccessful fully automatic building extraction case.

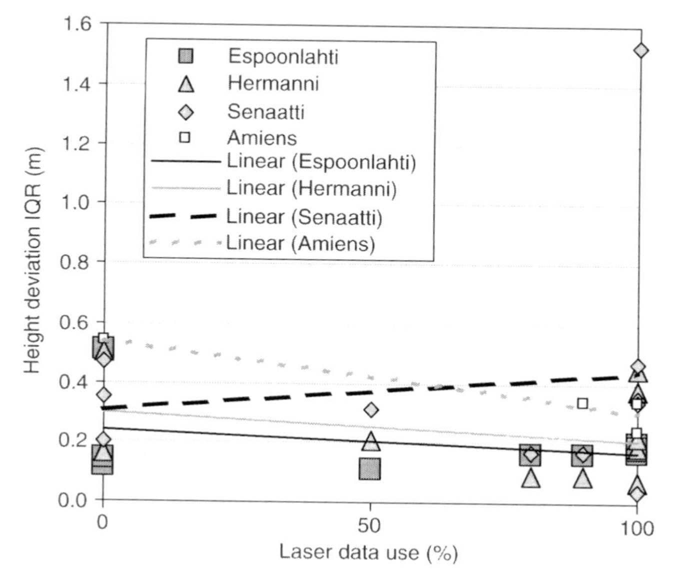

FIGURE 19.19
Target height deviation.

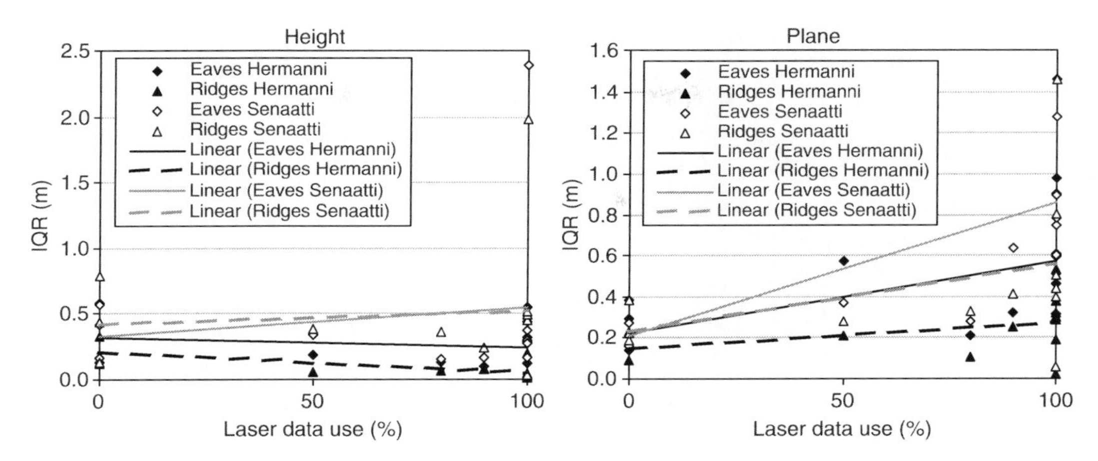

FIGURE 19.20
Target height and plane (mean of north/east) deviations for different targets.

In Figure 19.20, the target height and plane deviations are shown separately for eaves and ridges for Hermanni and Senaatti. As expected, for laser scanning methods, the accuracy is better for ridges, which are determined by the intersection of two planes. In theory, the height determination of ridges is extremely accurate if the number of hits defining the planes is high enough. The planimetric accuracy of ridges and eaves was more accurate with photogrammetry than with laser scanning.

19.8.2 Roof Inclination

Roof inclination determination was more accurate when using laser data than photogrammetry, but a large variation in quality exists due to methods and test sites (i.e., complex buildings). The RMSE (all observations included) using laser scanning for roof inclination varied from 0.3° to 15.9° (mean 2.7°, median 0.85°, std. 4.4°). The corresponding values for aerial image assisted laser scanning ranged from 0.6° and 2.3° (mean 1.3°, median 1.1°, std. 0.6°) and for photogrammetry ranged from 1.0° to 17.9° (mean 5.2°, median 3.2°, std. 6.3°). In Senaatti and Amiens, the roof inclinations are steep and roofs short, so even small errors in target height determination lead to large errors in inclination angle. The Hermanni test site is relatively easy for both methods. In Hermanni, the accuracy of roof inclination determination was about 2.5° for photogrammetric methods and about 1° (RMSE, all observations included) for laser-based methods.

19.8.3 Shape Similarity

There are buildings or parts of buildings missing on many models. The reasons for this are not always known, but at least the shadowing of big trees and the complexity of buildings can cause problems. Participants may also strive for different goals in building extraction; some want to model the buildings as detailed as possible and some aim to obtain only the basic form of buildings. For automatic methods, the vegetation causes problems as trees are also classified as buildings. In other methods, the extra buildings are mainly temporary constructions, such as site huts for construction sites, which are usually impossible to separate out from real buildings. Images showing the differences in 2D between reference raster and delivered models are given below.

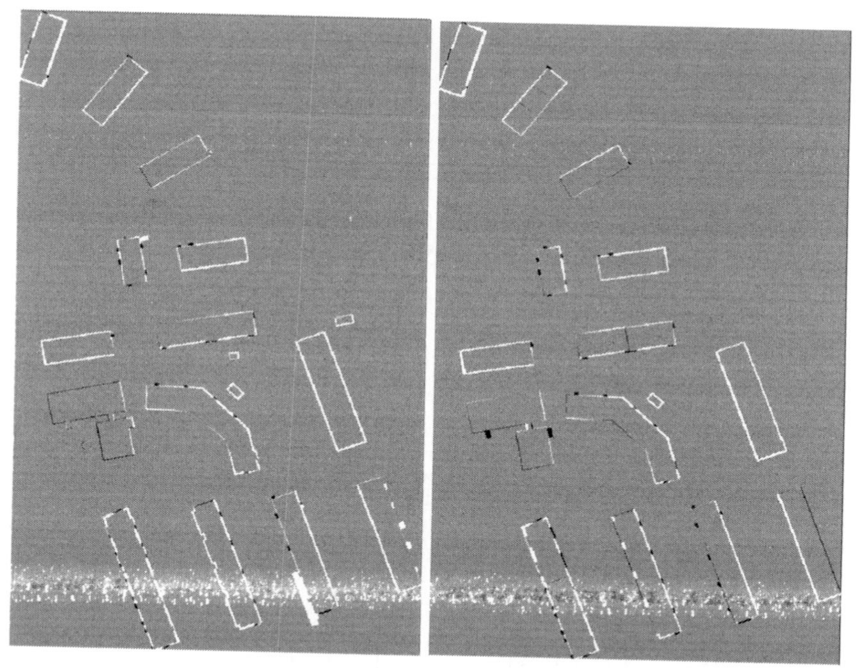

FIGURE 19.21
Image data only: CyberCity, Stuttgart (white, extra area; black, missing area).

Figure 19.21 shows the similarity in the FGI test site Hermanni for two photogrammetric approaches. Figure 19.22 shows the similarity in the FGI test site Hermanni for three laser scanning approaches. In the Hermanni site, the extracted buildings are large in

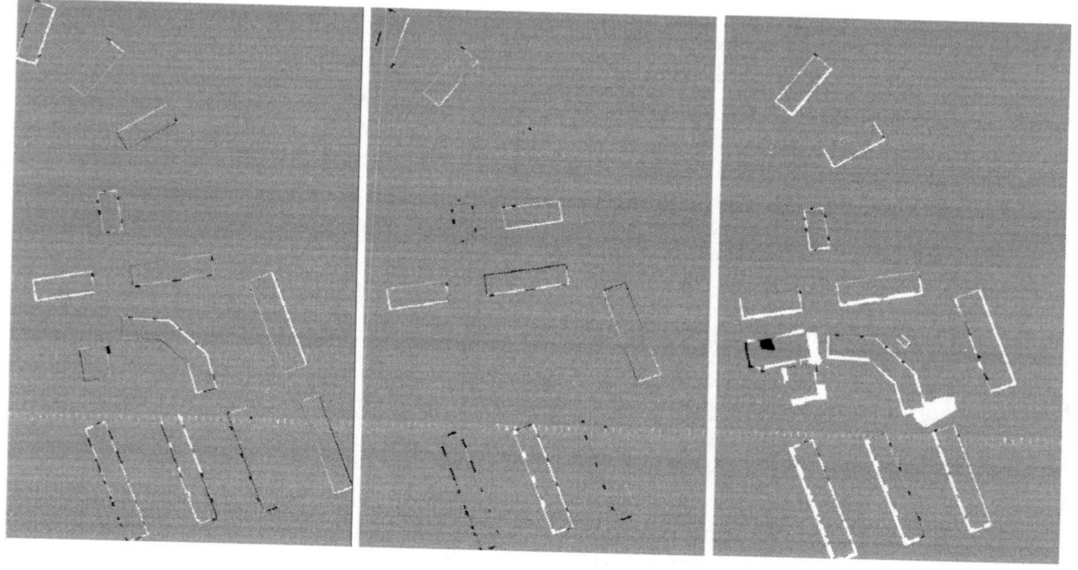

FIGURE 19.22
Laser data only: ICC laser, Aalborg, FOI (white, extra area; black, missing area).

size; thus, all the buildings have been modeled by partners in the examples and thus, the shape dissimilarity reveals quality difference of the models.

19.9 Effect of Automation

We want to now analyze the effects of automation in the applied methods. We therefore group methods according to their degree of automation and analyze overall planimetric and height accuracies.

The degree of automation varied significantly among the participants of this test. In general, the laser data allows higher automation in creating the models. Editing of the complex building models is needed to slow down the process. Although some laser-based processes are relatively automatic, the processes are still under development. In general, the planar target accuracy is affected by the degree of automation (and method, Figure 19.23). The accuracy of low automation methods is about 20–30 cm, while for high automation methods it is about 60–100 cm (IQR). The target height accuracy seems to be almost independent of the degree of automation (Figure 19.24). The degree of automation was estimated as a value from 0 to 10 based on the workflow charts and the procedure descriptions and comments provided by participants.

In general, the methods using aerial images and interactive processes are capable of producing more detail in building models, but only some providers modeled more detailed structures such as chimneys and ventilation equipment on roofs.

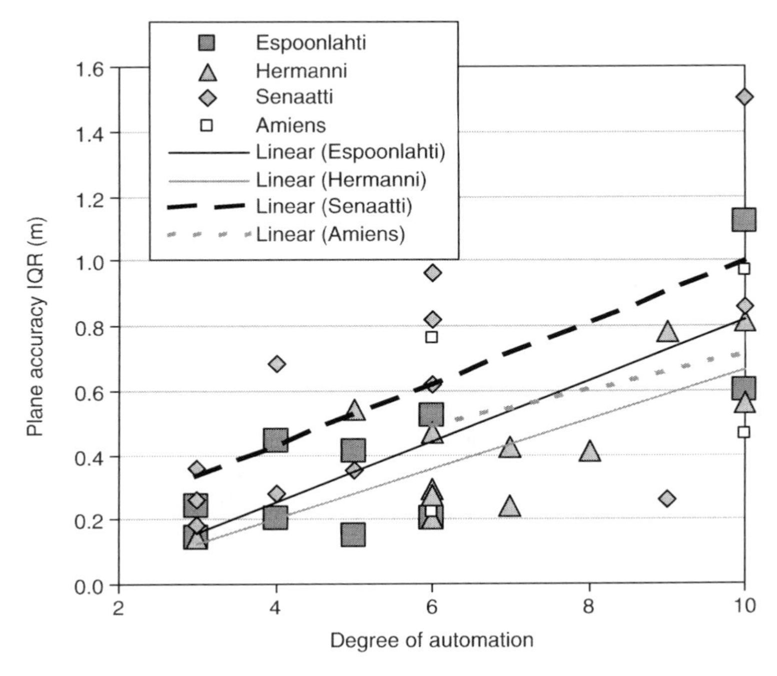

FIGURE 19.23
Obtained overall planimetric accuracy (mean of north/east) as a function of automation.

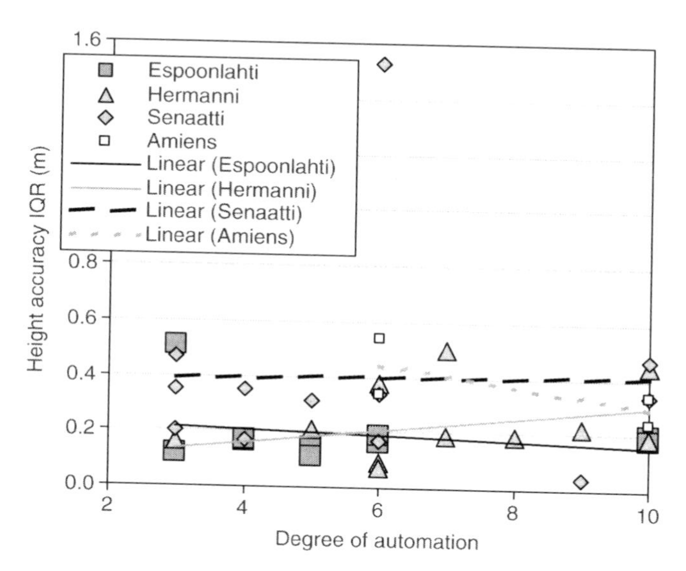

FIGURE 19.24
Obtained overall height accuracy as a function of automation.

19.10 Discussion and Conclusions

It can be concluded that, presently, photogrammetric techniques and hybrid techniques using both laser scanning and image data provide the highest accuracy and level of detail in 3D city reconstruction. Despite the high amount of research during the last decade, the level of automation is still relatively low.

Improvement in automation can be achieved most significantly by utilizing the synergy of laser and photogrammetry. The photogrammetric techniques are powerful for visual interpretation of the area, measurement of the building outlines, and of small details (e.g., chimneys), whereas laser scanning gives height, roof planes, and ridge information at its best. When the advantages of both methods are implemented well in a single system, high accuracy and a relatively high level of automation can be achieved. In this test, there was only one real hybrid system. However, several participants incorporated both major data sources in various ways. This only shows that there is no commonly agreed strategy available. We strongly recommend focusing on this aspect to derive commonly accepted and proofed methods for multisensor fusion in this field. Both ISPRS and EuroSDR working groups have started to deal with this subject.

The industry is following a similar idea, with the modern airborne laser scanners equipped with medium-format or even large-format digital aerial cameras with RGB and NIR (near-infrared) channels. This means that usually at least the swath width of the scanner is covered by image information. This must not necessarily allow for stereo, as single overlapping images would be sufficient to produce orthophotos, as the DTM is derived from simultaneously recorded laser scanning data. In summary, we can expect increasing R&D efforts in the multisensor fusion field in the near future not only from academia but also from companies.

The quality of the laser scanning data has a significant impact on the accuracy of the fully laser-based models. Since pulse density is increasing substantially (although the novel European Toposys system already had an 83 kHz system in 1997, the PRF of ALS produced by today's leading manufacturer Optech has almost followed Moore's Law. In 1995, the PRF was just a few kilohertz, while today it is 100 kHz) and there are new possibilities to use pulse information (multiple pulse, waveforms), it is expected that the accuracy of the laser-based techniques will follow this technical development. The rapid rate of implementation and development of laser scanning data can also be seen from the improvement of FOI's results during the test. On the other hand, when the study was initiated, the implementation level of the majority of laser-based techniques was low. It can be seen from Table 19.4 (time spent) that due to the good implementation of photogrammetric methods, the time spent on these models is not much higher than that by laser scanning.

We note that, based on the results presented and the development of the last 10 years,

- There is a need to integrate laser scanning and photogrammetric techniques in a more advanced way. The accuracy of single height measurements with the laser should be used in the photogrammetric process to increase the automation. There are currently some developments in this field.
- Quality improvement of digital camera data will increase the quality of photogrammetric techniques and it will also allow higher automation by using remote sensing methods in the image processing. This is supported by the RGB and NIR recording capabilities of medium- and large-format digital aerial cameras.
- Rapid technological development in laser scanning will allow higher accuracies to be obtained by techniques relying only on laser scanning.
- Laser scanning allows higher automation needed in the building extraction process. On the other hand side, some in-house developments on the derivation of Digital Surface Models from aerial imagery indicate a potential that is by far not exploited, and that could compete, at least in some aspects, with airborne laser scanning on building extraction.
- There is a huge need to implement laser-based techniques into practical working processes.
- New algorithmic innovations will emerge in all areas, due to digital camera data and laser scanning developments that will increase automation.

In the project, the first pulse data was used for the extraction. It is known already that the use of the last pulse and the combined use of the first and last pulse would lead to improvements.

In practice, the availability of the material (aerial image or laser point clouds) determines significantly the kind of techniques that can be used. The costs of laser scanning are significantly higher, although in many cities the material has already been collected. In many cities, building outline information exists as well as some kind of existing 3D models. Hence, there is not always a general recommendation or solution for all specific problems. The availability of the sources and their up-to-dateness definitely affect their potential usage.

In order to be able to separate methodological development from development due to improvement of data, it is recommended to use the same test sites, with old and new laser scanning and aerial data, for future verifications. The value of such empirical investigations is rather high and has been confirmed not only by the high number of participants but also by many fruitful discussions on the results of this investigation presented at various conferences and meetings.

In the future, an even bigger challenge is to find the answers to how existing models could be updated, how changes should be verified, and how existing information can be used optimally in a more automatic process.

Acknowledgments

All participants, namely, George Vosselman (International Institute for Geo-Information Science and Earth Observation), Alexandra Hofmann (Dresden University of Technology, presently Definiens), Urs Mäder (CyberCity AG), Åsa Persson, Ulf Söderman and Magnus Elmqvist (Swedish Defence Research Agency), Antonio Ruiz and Martina Dragoja (Institut Cartogràfic de Catalunya), Sylvain Airault, David Flamanc and Gregoire Maillet (MATIS IGN), Thomas Kersten (Hamburg University of Applied Sciences), Jennifer Carl and Robert Hau (Nebel + Partner), Emil Wild (C+B Technik GmbH) and Lise Lausten Frederiksen, Jane Holmgaard and Kristian Vester (University of Aalborg) and Eberhard Gülch (Stuttgart University of Applied Sciences) are gratefully acknowledged for their cooperation in the project. Hannu Hyyppä from Helsinki University of Technology and Leena Matikainen from FGI also contributed significantly. The support of Espoo and Helsinki City Survey Divisions, FM-Kartta Oy (Blom Kartta Oy) and Terrasolid Oy, in providing data for the project is gratefully acknowledged.

References

Alharthy, A. and Bethel, J., 2004. Detailed building reconstruction from airborne laser data using a moving surface method, in Proceedings of the 20th ISPRS Congress, July 2004, Istanbul, Turkey, *International Archives of Photogrammetry, Remote Sensing and Spatial Information Sciences*, XXXV(B3/III), 237–242.

Ameri, B. and Fritsch, D., 2000. Automatic 3D building reconstruction using plane roof structures, in *Proceedings of the ASPRS congress*, Washington. Available at www.itp.uni-stuttgart.de/publications/2000/pub2000.html.

BAE Systems, 2000. SOCET Set—User's Manual Version 4.3.

Baillard, C. and Maître, H., 1999. 3-D reconstruction of urban scenes from aerial stereo imagery: A focusing strategy, *Computer Vision and Image Understanding*, 76(3), 244–258.

Baillard, C. and Dissard, O., 2000. A stereo matching algorithm for urban digital elevation models, *Photogrammetric Engineering and Remote Sensing*, 66(9), 1119–1128.

Baltsavias, E.P., 1999. Airborne laser scanning: Existing systems and firms and other resources, *ISPRS Journal of Photogrammetry and Remote Sensing*, 54(1999), 164–198.

Baltsavias, E.P., 2004. Object extraction and revision by image analysis using existing geodata and knowledge: Current status and steps towards operational systems, *ISPRS Journal of Photogrammetry and Remote Sensing*, 58(3–4), 129–151.

Braun, C., Kolbe, T.H., Lang, F. et al., 1995. Models for photogrammetric building reconstruction, *Computers and Graphics*, 19(1), 109–118.

Brenner, C., 2001. City models—automation in research and practice, *Photogrammetric Week'01*, 149–158.

Brenner, C., 2003. Building reconstruction from laser scanning and images, in *Proceedings of ITC Workshop on Data Quality in Earth Observation Techniques*, Enschede, The Netherlands, November 2003, 8 pp.

Brenner, C., 2005. Building reconstruction from images and laser scanning, *International Journal of Applied Earth Observation and Geoinformation*, 6, 186–198.

Brunn, A., Gülch, E. et al., 1998. A hybrid concept for 3D building acquisition, *ISPRS Journal of Photogrammetry and Remote Sensing*, 53(2), 119–129.

Brunn, A. and Weidner, U., 1997. Extracting buildings from digital surface models, *International Archives of Photogrammetry, Remote Sensing and Spatial Information Sciences*, 32(3–4W2), 27–34.

Brunn, A. and Weidner, U., 1998. Hierarchical Bayesian nets for building extraction using dense digital surface models, *ISPRS Journal of Photogrammetry and Remote Sensing*, 53(5), 296–307.

Burrough, P. and McDonnell, R., 1998. *Principles of Geographical Information Systems*, Oxford University Press, ISBN: 0–19–823366–3.

Centeno, J.A.S. and Miqueles, J., 2004. Extraction of buildings in Brazilian urban environments using high resolution remote sensing imagery and laser scanner data, *International Archives of Photogrammetry, Remote Sensing and Spatial Information Sciences*, XXXV(B4), 4 pp.

Chen, L.-C., Teo, T.-A., Shao, Y.-C., Lai, Y.-C., and Rau, J.-Y. 2004. Fusion of lidar data and optical imagery for building modelling, *International Archives of Photogrammetry, Remote Sensing and Spatial Information Sciences*, XXXV(B4), 6 pp.

Cho, W. et al., 2004. Pseudo-grid based building extraction using airborne lidar data, *International Archives of Photogrammetry, Remote Sensing and Spatial Information Sciences*, XXXV(B3/III), 378–381.

Cord, M. and Declerq, D., 2001. Three-dimensional building detection and modeling using a statistical approach, *IEEE Transactions on Image Processing*, 10(5), 715–723.

Cord, M., Jordan, M., and Cocquerez, J.P., 2001. Accurate building structure recovery from high-resolution aerial imagery, *Computer Vision and Image Understanding*, 82(2), 138–173.

Dash, J., Steinle, E., Singh, R.P., and Bähr, H.P., 2004. Automatic building extraction from laser scanning data: An input tool for disaster management, *Advances in Space Research*, 33, 317–322.

Dold, C. and Brenner C., 2004. Automatic matching of terrestrial scan data as a basis for the generation of detailed 3D city models, *International Archives of Photogrammetry, Remote Sensing and Spatial Information Sciences*, XXXV(B3/III), 1091–1096.

Elaksher, A.F. and Bethel, J.S., 2002. Reconstructing 3D buildings from lidar data, *International Archives of Photogrammetry, Remote Sensing and Spatial Information Sciences*, 34(3A), 102–107.

Elaksher, A.F., Bethel, J.S., and Mikhail, E.M., 2003. Roof boundary extraction using multiple images, *Photogrammetric Record*, 18(101), 27–40.

Elmqvist, M., 2002. Ground surface estimation from airborne laser scanner data using active shape models, in Proceedings of the ISPRS Technical Commission III Symposium "Photogrammetric Computer Vision" PCV'02, September 2002, Graz, Austria, *International Archives of Photogrammetry, Remote Sensing and Spatial Information Sciences*, Vol. XXXIV, Part 3A.

Fischer, A., Kolbe, T.H., Lang, F., Cremers, A.B., Förstner, W., Plümer, L., and Steinhage, V., 1998. Extracting buildings from aerial images using hierarchical aggregation in 2D and 3D, *Computer Vision and Image Understanding*, 72(2), 185–203.

Flamanc, D., Maillet, G., and Jibrini, H., 2003. 3D city models: An operational approach using aerial images and cadastral maps, *International Archives of Photogrammetry, Remote Sensing and Spatial Information Sciences*, XXXIV, Part 3/W8.

Forlani, G., Nardinocchi, C., Scaioni, M., and Zingaretti, P., 2003. Building reconstruction and visualization from LIDAR data, *International Archives of Photogrammetry, Remote Sensing and Spatial Information Sciences*, XXXIV (5W12), 151–156.

Forlani, G., Nardinocchi, C., Scaioni, M., and Zingaretti, P., 2006. Complete classification of raw LIDAR data and 3D reconstruction of buildings, *Pattern Analysis and Applications*, 8(4), 357–374.

Fraser, C.S., Baltsavias, E., and Grün, A., 2002. Processing of Ikonos imagery for submetre 3D positioning and building extraction, *ISPRS Journal of Photogrammetry and Remote Sensing*, 56(3), 177–194.

Frederiksen, L., Holmgaard, J., and Vester, K., 2004. Automatiseret 3D-bygningsmodellering–fran laserscanning til bygningsmodel, University of Aalborg (in Danish).

Fuchs, C., Gülch, E., and Förstner, W., 1998. OEEPE Survey on 3D-City Models, OEEPE Publication No. 35:9–123. Bundesamt für Kartographie und Geodäsie. Frankfurt.

Fuchs, F. and Le-Men, H., 2000. Efficient subgraph isomorphism with 'a priori' knowledge: Application to 3D reconstruction of buildings for cartography, *Advances in Pattern Recognition*, LCNS, 1876, 427–436.

Fujii, K. and Arikawa, T., 2002. Urban object reconstruction using airborne laser elevation image and aerial image, *IEEE Transactions on Geoscience and Remote Sensing*, 40(10), 2234–2240.

Förstner, W., 1999. 3D-city models: Automatic and semiautomatic acquisition methods, Fritsch, D and Spiller, R (Eds.), *Photogrammetric Week'99*, CD-ROM, 291–303.

Grün, A., 1997. Automation in building extraction, *Photogrammetric Week'97*, 175–186.
Grün, A., 1998. TOBAGO—a semi-automated approach for the generation of 3-D building models, *ISPRS Journal of Photogrammetry and Remote Sensing*, 53(2), 108–118.
Grün, A. and Wang, X., 1998a. CC-Modeler: A topology generator for 3-D city models, *ISPRS Journal of Photogrammetry and Remote Sensing*, 53(5), 286–295.
Grün, A. and Wang, X., 1998b. Acquisition and Management of Urban 3D data using CyberCity Modeler. Available at www.cybercity.tv/pub/ccm_3D_data.pdf.
Grün, A. and Wang, X., 1999a. CyberCity Modeller: A tool for interactive 3-D city model generation, *Photogrammetric Week'99*, 317–327.
Grün, A. and X. Wang, 1999b. CyberCity Spatial Information System (CC-SIS): A new concept for the management of 3-D city models in a hybrid GIS, in *Proceedings of the 20th Asian Conference on Remote Sensing*, November 22–25, 1999, Hong Kong, China, 8 pp.
Grün, A., Li, Z., and Wang, X., 2003. Generation of 3D city models with linear array CCD-sensors. Available at http://carstad.gsfc.nasa.gov/topics/JBRESEARCH/STARLABO/Opt3D_citymod_new.pdf.
Gülch, E. and Müller, H., 2001. New applications of semi-automatic building extraction, in *Proceedings of the third International Workshop on* Automatic *Extraction of Man-Made Objects from Aerial and Space Images*, Ascona, Switzerland, Swets & Zeitlinger/Balkema, The Netherlands.
Gülch, E., 2004. Erfassung von 3D-Geodaten mit Digitaler Photogrammetrie, in Coors, V. and Zipf, A. (Eds.) *3D-Geoinformationssysteme—Grundlagen und Anwendungen*, Herbert Wichmann Verlag, Heidelberg, Germany, pp. 4–25.
Haala, N., 2004. Laserscanning zur dreidimensionalen Erfassung von Stadtgebieten, in Coors, V. and Zipf, A. (Eds.) *3D-Geoinformationssysteme—Grundlagen und Anwendungen*, Herbert Wichmann Verlag, Heidelberg, Germany, 26–38.
Haala, N., Brenner, C., and Anders, K.-H., 1998. 3D urban GIS from laser altimeter and 2D map data, *International Archives of Photogrammetry, Remote Sensing and Spatial Information Sciences*, 32(3/1), 339–346.
Haala, N. and Brenner, C., 1999. Extraction of buildings and trees in urban environments, *ISPRS Journal of Photogrammetry and Remote Sensing*, 54, 123–129.
Haithcoat, T.L., Song, W., and Hipple, J.D., 2001. Building footprint extraction and 3-D reconstruction from LIDAR data, *IEEE/ISPRS Joint Workshop on Remote Sensing and Data Fusion over Urban Areas*, 5 pp. CD-ROM.
Henricsson, O. and Baltsavias, E., 1997. 3-D building reconstruction with ARUBA: A qualitative and quantitative evaluation, in *Proceedings of International Workshop "Automatic Extraction of Man-Made Objects from Aerial and Space Images (II)*, Ascona, Switzerland. Available at http://e-collection.ethbib.ethz.ch/show?type=inkonf&nr=96.
Hofmann, A., Maas, H.-G., and Streilein, A., 2001. Knowledge-based building detection based on laser scanner data and topographic map information, in *Proceedings of the ISPRS Commission III, Symposium 2002, Photogrammetric Computer Vision, PCV02*, Graz Austria, 6 pp.
Hofmann, A., Maas, H.-G., and Streilein, A., 2003. Derivation of roof types by cluster analysis in parameter spaces of airborne laser scanner point clouds, in *Proceedings of the ISPRS Working Group III/3 Workshop, '3-D Reconstruction from Airborne Laserscanner and InSAR Data'*, Dresden, Germany, 2003. Available at http://www.isprs.org/commission3/wg3/workshop_laserscanning/.
Hofmann, A.D., 2004. Analysis of tin-structure parameter spaces in airborne laser scanner data for 3-D building model generation, *International Archives of Photogrammetry, Remote Sensing and Spatial Information Sciences*, XXXV(B3/III), 6 pp.
Jaynes, C., Riseman, E., and Hanson, A., 2003. Recognition and reconstruction of buildings from multiple aerial images, *Computer Vision and Image Understanding*, 90(1), 68–98.
Juhl, J., 1980. Det analytiske stråleudjævningsprogram sANA, Laboratoriet for fotogrammetri og landmåling, University of Aalborg (in Danish).
Jutzi, B. and Stilla, U., 2004. Extraction of features from objects in urban areas using spacetime analysis of recorded laser pulses, *International Archives of Photogrammetry, Remote Sensing and Spatial Information Sciences*, XXXV(B2), 6 pp.

Jülge, K. and Brenner, C., 2004. Object extraction from terrestrial laser scan data for a detailed building description, *International Archives of Photogrammetry, Remote Sensing and Spatial Information Sciences*, XXXV(B3/III), 1079–1084.

Kaartinen, H., Hyyppä, J., Gülch, E., Hyyppä, H., Matikainen, L., Vosselman, G., Hofmann, A.D., Mäder, U., Persson, Å., Söderman, U., Elmqvist, M., Ruiz, A., Dragoja, M., Flamanc, D., Maillet, G., Kersten, T., Carl, J., Hau, R., Wild, E., Frederiksen, L., Holmgaard, J., and Vester, K., 2005a. EuroSDR building extraction comparison, in *Proceedings of the ISPRS Hannover Workshop High-Resolution Earth Imaging for Geospatial Information*, May 17–20, 2005, CD-ROM, 6 pp.

Kaartinen, H., Hyyppä, J., Gülch, E., Vosselman, G., Hyyppä, H., Matikainen, L., Hofmann, A.D., Mäder, U., Persson, Å., Söderman, U., Elmqvist, M., Ruiz, A., Dragoja, M., Flamanc, D., Maillet, G., Kersten, T., Carl, J., Hau, R., Wild, E., Frederiksen, L., Holmgaard, J., and Vester, K., 2005b. Accuracy of 3D city models: EuroSDR comparison, *International Archives of Photogrammetry, Remote Sensing and Spatial Information Sciences*, XXXVI(Part 3/W19), 227–232, CD-ROM.

Kaartinen, H. and Hyyppä, J., 2006. EuroSDR-Project Commission 3 "Evaluation of Building Extraction", Final Report, *EuroSDR—European Spatial Data Research*, Official Publication No. 50:9–77.

Khoshelham, K., 2004. Building extraction from multiple data sources: A data fusion framework for reconstruction of generic models, *International Archives of Photogrammetry, Remote Sensing and Spatial Information Sciences*, XXXV(B3/III), 980–986.

Kim, T. and Muller, J.P., 1998. A technique for 3D building reconstruction, *Photogrammetric Engineering and Remote Sensing*, 64(9), 923–939.

Lee, I. and Choi, Y., 2004. Fusion of terrestrial laser scanner data and images for building reconstruction, *International Archives of Photogrammetry, Remote Sensing and Spatial Information Sciences*, XXXV(B5), 6 pp.

Li, B.J., Li, Q.Q., Shi, W.Z., and Wu, F.F., 2004. Feature extraction and modeling of urban building from vehicle-borne laser scanning data, *International Archives of Photogrammetry, Remote Sensing and Spatial Information Sciences*, XXXV(B5), 6 pp.

Maas, H.-G. and Vosselman, G., 1999. Two algorithms for extracting building models from raw laser altimetry data, *ISPRS Journal of Photogrammetry and Remote Sensing*, 54(2–3), 153–163.

Maas, H.-G., 2001. The suitability of airborne laser scanner data for automatic 3D object reconstruction, in Baltsavias, E. (Ed.), *Automatic Extraction of Man-Made Objects from Aerial and Space Images III*, Swets & Zeitlinger B.V., Lissie, The Netherlands, pp. 291–296.

Madhavan, B.B., Tachibana, K., Sasagawa, T., Okada, H., and Shimozuma, Y., 2004. Automatic extraction of shadow regions in high-resolution ADS40, *International Archives of Photogrammetry, Remote Sensing and Spatial Information Sciences*, XXXV(B3/III), 808–810.

Mayer, H., 1999. Automatic object extraction from aerial imagery—A survey focusing on buildings, *Computer Vision and Image Understanding*, 74(2), 138–149.

Morgan, M. and Habib, A., 2001. 3D TIN for automatic building extraction from airborne laser scanner data, in *Proceedings of the American Society of Photogrammetry and Remote Sensing* (ASPRS), St. Louis, MO.

Morgan, M. and Habib, A., 2002. Interpolation of lidar data and automatic building extraction, in *Proceedings of the ACSM-ASPRS Annual Conference*, Washington D.C. CD-ROM.

Neidhart, H. and Sester, M., 2003. Identifying building types and building clusters using 3D-laser scanning and GIS-data, *International Archives of Photogrammetry, Remote Sensing and Spatial Information Sciences*, XXXV(B4), 6 pp.

Niederöst, M., 2003. Detection and reconstruction of buildings for automated map updating, Diss., Technische Wissenschaften ETH Zurich. (Mitteilungen/Institut für Geodäsie und Photogrammetrie/ETH), 139 pp.

Noronha, S. and Nevatia, R., 2001. Detection and modeling of buildings from multiple aerial images, *IEEE Transactions on Pattern Analysis and Machine Intelligence*, 23(5), 501–518.

Oda, K., Takano, T., Doihara, T., and Shibasaki, R., 2004. Automatic building extraction and 3D city modeling from lidar data based on Hough transformation, *International Archives of Photogrammetry, Remote Sensing and Spatial Information Sciences*, 35(B3:III), 277–280.

Oriot, H. and Michel, A., 2004. Building extraction from stereoscopic aerial images, *Applied Optics*, 43(2), 218–226.

Overby, J., Bodum, L., Kjems, E., and Ilsøe, P.M., 2003. Automatic 3D building reconstruction from airborne laser scanning and cadastral data using Hough transform, *International Archives of Photogrammetry, Remote Sensing and Spatial Information Sciences*, XXXV(B3), 6 pp.

Paparoditis, N., Cord, M., Jordan, M., and Cocquerez, J.P., 1998. Building detection and reconstruction from mid- and high-resolution aerial imagery, *Computer Vision and Image Understanding*, 72(2), 122–142.

Peternell, M. and Steiner, T., 2004. Reconstruction of piecewise planar objects from point clouds, *Computer-Aided Design*, 36(4), 333–342.

Rottensteiner, F., 2000. Semi-automatic building reconstruction integrated in strict bundle block adjustment, *International Archives of Photogrammetry and Remote Sensing*, XXXIII(B2/2), 461–468.

Rottensteiner, F., 2003. Automatic generation of high-quality building models from lidar data, *IEEE Computer Graphics and Applications*, 23(6), 42–50.

Rottensteiner, F. and Briese, Ch., 2002. A new method for building extraction in urban areas from high-resolution lidar data, *International Archives of Photogrammetry, Remote Sensing and Spatial Information Sciences*, XXXIV(3A), 295–301.

Rottensteiner, F. and Jansa, J., 2002. Automatic extraction of buildings from lidar data and aerial images. Available at http://www.ipf.tuwien.ac.at/publications/fr_jj_ottawa_02.pdf.

Rottensteiner, F., Trinder, J., Clode, S., and Kubik, K., 2003. Detecting buildings and roof segments by combining LIDAR data and multispectral images. Available at http://sprg.massey.ac.nz/ivcnz/Proceedings/IVCNZ_12.pdf.

Rottensteiner, F., Trinder, J., Clode, S., and Kubik, K., 2004. Fusing airborne laser scanner data and aerial imagery for the automatic extraction of buildings in densely built-up areas, *International Archives of Photogrammetry, Remote Sensing and Spatial Information Sciences*, 35(B3:III), 512–517.

Rottensteiner, F., Trinder, J., Clode, S., and Kubik, K., 2005. Using the Dempster-Shafer method for the fusion of LIDAR data and multi-spectral images for building detection, *Information Fusion*, 6(4), 283–300.

Sahar, L. and Krupnik, A., 1999. Semiautomatic extraction of building outlines from large-scale aerial images, *Photogrammetric Engineering and Remote Sensing*, 65(4), 459–465.

Schwalbe, E., 2004. 3D building model generation from airborne laserscanner data by straight line detection in specific orthogonal projections, *International Archives of Photogrammetry, Remote Sensing and Spatial Information Sciences*, XXXV(B3/III), 249–254.

Sequeira, V., Ng, K., Wolfart, E., Goncalves, J.G.M., and Hogg, D., 1999. Automated reconstruction of 3D models from real environments, *ISPRS Journal of Photogrammetry and Remote Sensing*, 54(1), 1–22.

Shufelt, J.A., 1999. Performance evaluation and analysis of monocular building extraction from aerial imagery, *IEEE Transactions on Pattern Analysis and Machine Intelligence*, 21(4), 311–326.

Sinning-Meister, M., Grün, A., and Dan, H., 1996. 3D city models for CAAD-supported analysis and design for urban areas, *ISPRS Journal of Photogrammetry and Remote Sensing*, 51, 196–208.

Sohn, G., 2004. Extraction of buildings from high-resolution satellite data and lidar, *International Archives of Photogrammetry, Remote Sensing and Spatial Information Sciences*, XXXV(B3/III), 1036–1042.

Stilla, U. and Jurkiewicz, K., 1999. Reconstruction of building models from maps and laser altimeter data, *Integrated Spatial Databases*, LCNS, 1737, 34–46.

Süveg, I. and Vosselman, G., 2004. Reconstruction of 3D building models from aerial images and maps, *ISPRS Journal of Photogrammetry and Remote Sensing*, 58(3–4), 202–224.

Söderman, U., Ahlberg, S., Elmqvist, M., and Persson, Å., 2004. Three-dimensional environment models from airborne laser radar data, in *Proceedings of SPIE Defense and Security Symposium*, 5412, 12 pp. Available at http://www.sne.foi.se/documents/SNE_SPIE_DS_04.pdf.

Taillandier, F. and Deriche, R., 2004. Automatic building extraction from aerial images: A generic Bayesian framework, *International Archives of Photogrammetry, Remote Sensing and Spatial Information Sciences*, 35(B3:III), 343–348.

Takase, Y., Yano, K., Kawahara, N., et al., 2004. Reconstruction and visualization of "virtual time-space of Kyoto", A 4D-GIS of the city, *International Archives of Photogrammetry, Remote Sensing and Spatial Information Sciences*, XXXV(B5), 6 pp.

Tan, G. and Shibasaki, R., 2002. A research for the extraction of 3D urban building by using airborne laser scanner data, in *Proceedings of the 23rd Asian Conference on Remote Sensing 2002.*

TerraSolid, 2003. TerraScan User's Guide.

Tsay, J.-R., 2001. A concept and algorithm for 3D city surface modelling, in *Proceedings of the ISPRS Commission*, III/WGIII/2, 4 pp. Available at http://www.isprs.org/commission3/proceedings02/papers/paper092.pdf.

Tseng, Y.H. and Wang, S.D., 2003. Semiautomated building extraction based on CSG model-image fitting, *Photogrammetric Engineering and Remote Sensing*, 69(2), 171–180.

Vosselman, G., 2002. Fusion of laser scanning data, maps, and aerial photographs for building reconstruction. Available at http://www.geo.tudelft.nl/frs/papers/2002/vosselman.igarss02.pdf.

Vosselman, G. and Dijkman, S., 2001. 3D building model reconstruction from point clouds and ground plans, *International Archives of Photogrammetry, Remote Sensing and Spatial Information Sciences*, 34(3W4), 37–43.

Vosselman, G. and Süveg, I., 2001. Map based building reconstruction from laser data and images. *Automatic Extraction of Man-Made Objects from Aerial and Space Images (III)*, Ascona, Switzerland, June 11–15, Balkema, The Netherlands, pp. 231–239.

Wang, X. and Grün, A., 2003. A hybrid GIS for 3D city models, *International Archives of Photogrammetry, Remote Sensing and Spatial Information Sciences*, 33(4/3), 1165–1172.

Weidner, U. and Förstner, W., 1995. Towards automatic building extraction from high-resolution digital elevation models, *ISPRS Journal of Photogrammetry and Remote Sensing*, 50(4), 38–49.

Zhan, Q., Molenaar, M., and Tempfli, K., 2002. Building extraction from laser data by reasoning on image segments in elevation slices, *International Archives of Photogrammetry and Remote Sensing*, XXXIV(3,B), 305–308.

Zhang, Y. and Wang, R., 2004. Multi-resolution and multi-spectral image fusion for urban object extraction, *International Archives of Photogrammetry, Remote Sensing and Spatial Information Sciences*, XXXV(B3/III), 960–966.

Zhao, H.J. and Shibasaki, R., 2003. Reconstructing a textured CAD model of an urban environment using vehicle-born range scanners and line cameras, *Machine Vision and Applications*, 14(1), 35–41.

Index

D

G

H

M

S